JACARANDA MATHS QUEST
GENERAL MATHEMATICS 11

VCE UNITS 1 AND 2 | THIRD EDITION

T0342881

JACARANDA MATHS QUEST
GENERAL MATHEMATICS 11

VCE UNITS 1 AND 2 | THIRD EDITION

MICHAEL SHEEDY

MARK BARNES

BRANDON CHUAH

CAITLIN MAHONY

RAYMOND ROZEN

SUE MICHELL

CONTRIBUTING AUTHOR

James Smart

jacaranda
A Wiley Brand

Third edition published 2023 by
John Wiley & Sons Australia, Ltd
155 Cremorne Street, Cremorne, Vic 3121

First edition published 2016
Second edition published 2019

Typeset in 10.5/13 pt TimesLTStd

© John Wiley & Sons Australia, Ltd 2023

The moral rights of the authors have been asserted.

ISBN: 978-1-119-87623-6

Trademarks

Jacaranda, the JacPLUS logo, the learnON, assessON and studyON logos, Wiley and the Wiley logo, and any related trade dress are trademarks or registered trademarks of John Wiley & Sons Inc. and/or its affiliates in the United States, Australia and in other countries, and may not be used without written permission. All other trademarks are the property of their respective owners.

The covers of the *Jacaranda Maths Quest VCE Mathematics* series are the work of Victorian artist Lydia Bachimova.

Lydia is an experienced, innovative and creative artist with over 10 years of professional experience, including five years of animation work with Walt Disney Studio in Sydney. She has a passion for hand drawing, painting and graphic design.

Illustrated by diacriTech and Wiley Composition Services

Typeset in India by diacriTech

A catalogue record for this book is available from the National Library of Australia

Printed in Singapore
M121036_220822

Contents

About this resource

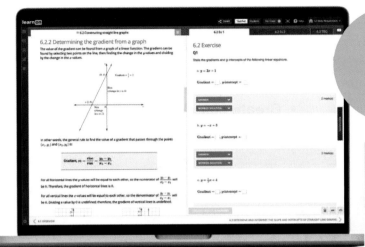

Everything you need for your students to succeed

JACARANDA MATHS QUEST
GENERAL
MATHEMATICS 11   VCE UNITS 1 AND 2 | THIRD EDITION

Developed by expert Victorian teachers for VCE students

Tried, tested and trusted. The NEW Jacaranda VCE Mathematics series continues to deliver curriculum-aligned material that caters to students of all abilities.

Completely aligned to the VCE Mathematics Study Design

Our expert author team of practising teachers ensures 100 per cent coverage of the new VCE Mathematics Study Design (2023–2027).

Everything you need for your students to succeed, including:

- **NEW!** Access targeted question sets including exam-style questions and all relevant past VCAA exam questions since 2013. Ensure assessment preparedness with practice Mathematical investigations.

- **NEW!** Be confident your students can get unstuck and progress, in class or at home. For every question online they receive immediate feedback and fully worked solutions.

- **NEW!** Teacher-led videos to unpack concepts, plus VCAA exam questions, exam-style questions and worked examples to fill learning gaps after COVID-19 disruptions.

Learn online with Australia's most

Everything you need for each of your lessons in one simple view

- Trusted, curriculum-aligned theory
- Engaging, rich multimedia
- All the teacher support resources you need
- Deep insights into progress
- Immediate feedback for students
- Create custom assignments in just a few clicks.

Practical teaching advice and ideas for each lesson provided in teachON

Each lesson linked to the Key Knowledge (and Key Skills) from the VCE Mathematics Study Design

Reading content and rich media including embedded videos and interactivities

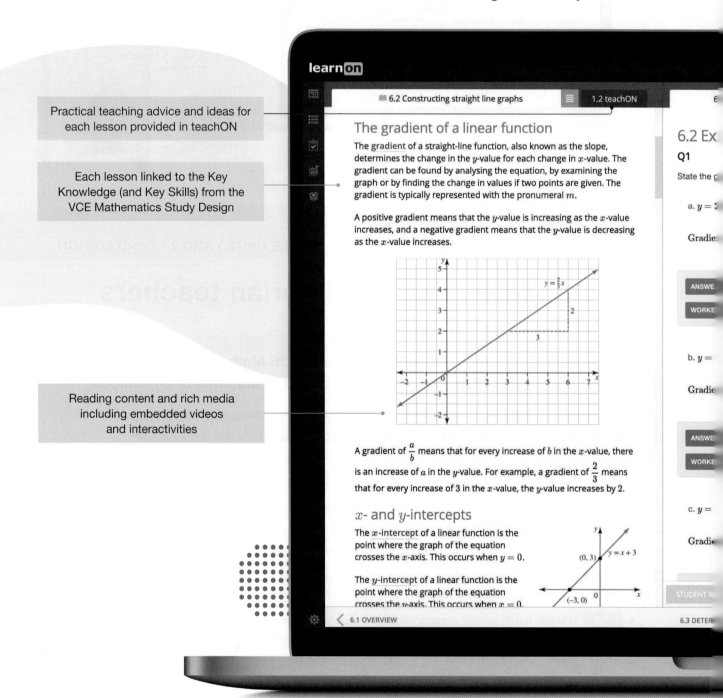

learnon

6.2 Constructing straight line graphs
1.2 teachON

The gradient of a linear function

The gradient of a straight-line function, also known as the slope, determines the change in the y-value for each change in x-value. The gradient can be found by analysing the equation, by examining the graph or by finding the change in values if two points are given. The gradient is typically represented with the pronumeral m.

A positive gradient means that the y-value is increasing as the x-value increases, and a negative gradient means that the y-value is decreasing as the x-value increases.

A gradient of $\frac{a}{b}$ means that for every increase of b in the x-value, there is an increase of a in the y-value. For example, a gradient of $\frac{2}{3}$ means that for every increase of 3 in the x-value, the y-value increases by 2.

x- and y-intercepts

The x-intercept of a linear function is the point where the graph of the equation crosses the x-axis. This occurs when $y = 0$.

The y-intercept of a linear function is the point where the graph of the equation crosses the y-axis. This occurs when $x = 0$.

6.1 OVERVIEW
6.3 DETER

6.2 Ex
Q1
State the g

a. $y = $
Gradie

ANSWE
WORKE

b. $y = $
Gradie

ANSWE
WORKE

c. $y = $
Gradi

STUDENT R
6.3 DETER

powerful learning tool, learnON

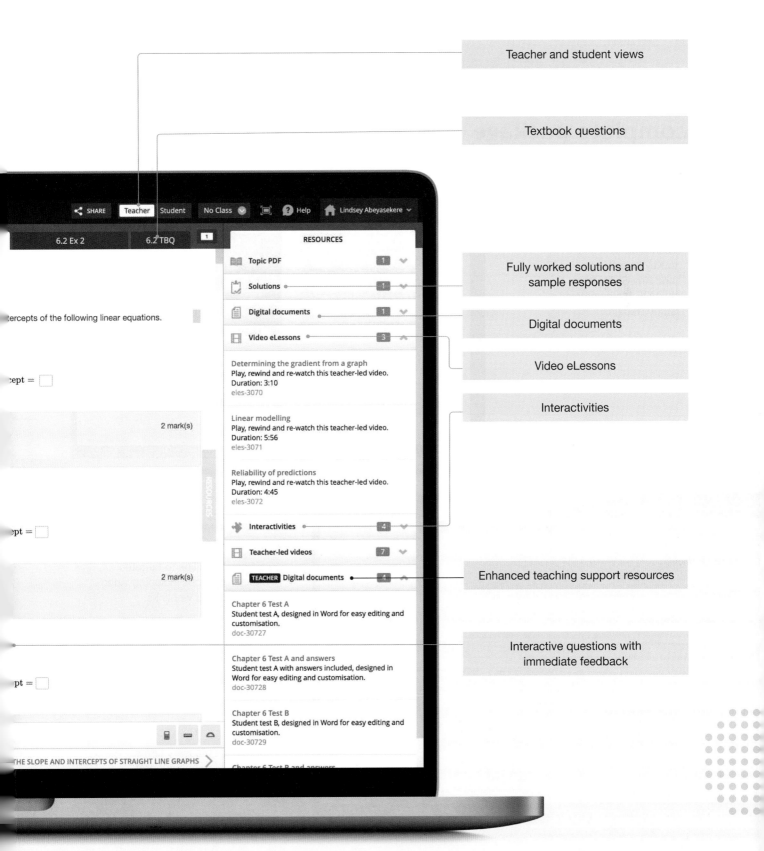

Teacher and student views

Textbook questions

Fully worked solutions and sample responses

Digital documents

Video eLessons

Interactivities

Enhanced teaching support resources

Interactive questions with immediate feedback

Get the most from your online resources

Online, these new editions are the complete package

Trusted Jacaranda theory, plus tools to support teaching and make learning more engaging, personalised and visible.

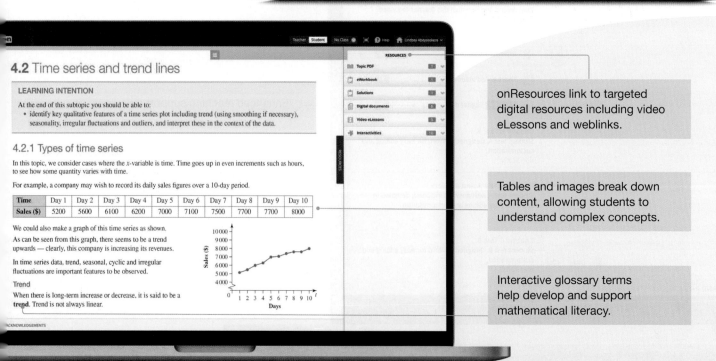

Each topic is linked to Key Knowledge (and Key Skills) from the VCE Mathematics Study Design.

onResources link to targeted digital resources including video eLessons and weblinks.

Tables and images break down content, allowing students to understand complex concepts.

Interactive glossary terms help develop and support mathematical literacy.

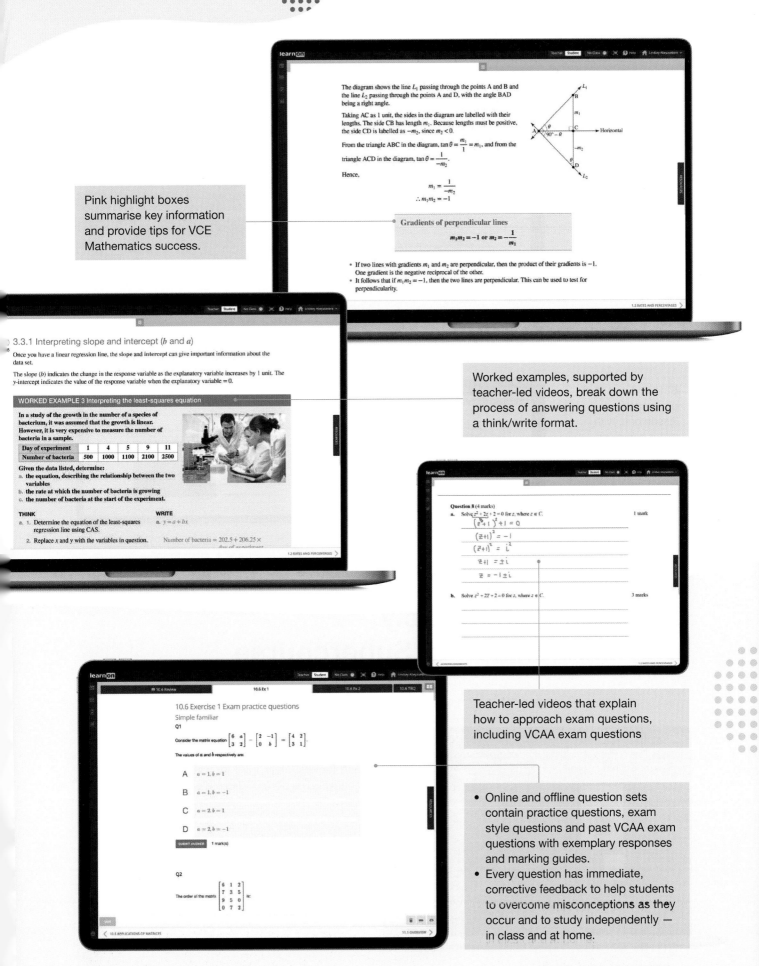

The diagram shows the line L_1 passing through the points A and B and the line L_2 passing through the points A and D, with the angle BAD being a right angle.

Taking AC as 1 unit, the sides in the diagram are labelled with their lengths. The side CB has length m_1. Because lengths must be positive, the side CD is labelled as $-m_2$, since $m_2 < 0$.

From the triangle ABC in the diagram, $\tan \theta = \dfrac{m_1}{1} = m_1$, and from the triangle ACD in the diagram, $\tan \theta = \dfrac{1}{-m_2}$.

Hence,

$$m_1 = \frac{1}{-m_2}$$

$$\therefore m_1 m_2 = -1$$

Gradients of perpendicular lines

$$m_1 m_2 = -1 \text{ or } m_2 = -\frac{1}{m_1}$$

- If two lines with gradients m_1 and m_2 are perpendicular, then the product of their gradients is -1. One gradient is the negative reciprocal of the other.
- It follows that if $m_1 m_2 = -1$, then the two lines are perpendicular. This can be used to test for perpendicularity.

Pink highlight boxes summarise key information and provide tips for VCE Mathematics success.

3.3.1 Interpreting slope and intercept (b and a)

Once you have a linear regression line, the slope and intercept can give important information about the data set.

The slope (b) indicates the change in the response variable as the explanatory variable increases by 1 unit. The y-intercept indicates the value of the response variable when the explanatory variable $= 0$.

WORKED EXAMPLE 3 Interpreting the least-squares equation

In a study of the growth in the number of a species of bacterium, it was assumed that the growth is linear. However, it is very expensive to measure the number of bacteria in a sample.

Day of experiment	1	4	5	9	11
Number of bacteria	500	1000	1100	2100	2500

Given the data listed, determine:
a. the equation, describing the relationship between the two variables
b. the rate at which the number of bacteria is growing
c. the number of bacteria at the start of the experiment.

THINK	WRITE
a. 1. Determine the equation of the least-squares regression line using CAS.	a. $y = a + bx$
2. Replace x and y with the variables in question.	Number of bacteria $= 202.5 + 206.25 \times$ day of experiment

Worked examples, supported by teacher-led videos, break down the process of answering questions using a think/write format.

Question 8 (4 marks)

a. Solve $z^2 + 2z + 2 = 0$ for z, where $z \in C$.   1 mark

$$\left(z + 1\right)^2 + 1 = 0$$

$$\left(z + 1\right)^2 = -1$$

$$\left(z + 1\right)^2 = i^2$$

$$z + 1 = \pm i$$

$$z = -1 \pm i$$

b. Solve $z^3 + 2z^2 + 2 = 0$ for z, where $z \in C$.   3 marks

Teacher-led videos that explain how to approach exam questions, including VCAA exam questions

10.6 Exercise 1 Exam practice questions

Simple familiar

Q1

Consider the matrix equation $\begin{bmatrix} 6 & a \\ 3 & 2 \end{bmatrix} - \begin{bmatrix} 2 & -1 \\ 0 & b \end{bmatrix} = \begin{bmatrix} 4 & 2 \\ 3 & 1 \end{bmatrix}$.

The values of a and b respectively are:

A $a = 1, b = 1$

B $a = 1, b = -1$

C $a = 2, b = 1$

D $a = 2, b = -1$

SUBMIT ANSWER   1 mark(s)

Q2

The order of the matrix $\begin{bmatrix} 6 & 1 & 2 \\ 7 & 3 & 5 \\ 9 & 5 & 0 \\ 0 & 7 & 2 \end{bmatrix}$ is:

- **Online and offline question sets contain practice questions, exam style questions and past VCAA exam questions with exemplary responses and marking guides.**
- **Every question has immediate, corrective feedback to help students to overcome misconceptions as they occur and to study independently — in class and at home.**

Topic reviews

Topic reviews include online summaries and topic level review exercises that cover multiple concepts. Topic level exam questions are structured just like the exams.

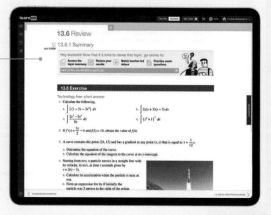

End-of-topic exam questions include relevant past VCE exam questions and are supported by teacher-led videos.

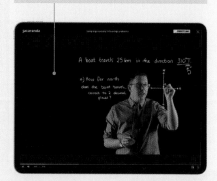

Get exam-ready!

Students can start preparing from lesson one, with exam questions embedded in every lesson — with relevant past VCAA exam questions since 2013.

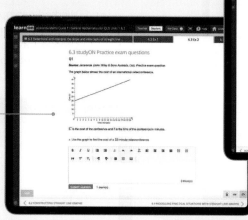

Practice, customisable Mathematical investigations available to build student competence and confidence.

Combine units flexibly with the Jacaranda Supercourse

Build the course you've always wanted with the Jacaranda Supercourse. You can combine all General Mathematics Units 1 to 4, so students can move backwards and forwards freely. Or Methods and General Units 1 & 2 for when students switch courses. The possibilities are endless!

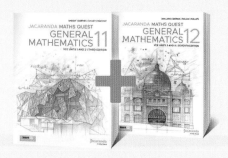

A wealth of teacher resources

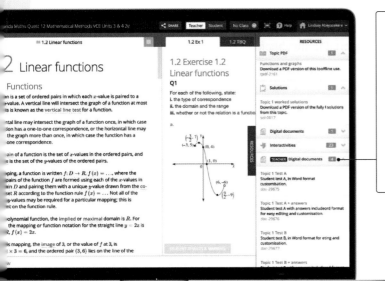

Enhanced teacher support resources, including:

- work programs and curriculum grids
- teaching advice and additional activities
- quarantined topic tests (with solutions)
- quarantined Mathematical investigations (with worked solutions and marking rubrics)

Customise and assign

A testmaker enables you to create custom tests from the complete bank of thousands of questions (including past VCAA exam questions).

Reports and results

Data analytics and instant reports provide data-driven insights into performance across the entire course.

Show students (and their parents or carers) their own assessment data in fine detail. You can filter their results to identify areas of strength and weakness.

Acknowledgements

The authors and publisher would like to thank the following copyright holders, organisations and individuals for their assistance and for permission to reproduce copyright material in this book.

Selected extracts from the VCE Mathematics Study Design (2023–2027) are copyright Victorian Curriculum and Assessment Authority (VCAA), reproduced by permission. VCE® is a registered trademark of the VCAA. The VCAA does not endorse this product and makes no warranties regarding the correctness and accuracy of its content. To the extent permitted by law, the VCAA excludes all liability for any loss or damage suffered or incurred as a result of accessing, using or relying on the content. Current VCE Study Designs and related content can be accessed directly at www.vcaa.vic.edu.au. Teachers are advised to check the VCAA Bulletin for updates.

- © Getty Images: **570** • © BlueRingMedia/Shutterstock: **566** • © Aleksey Stemmer/Shutterstock: **599** • © Alexlky/Shutterstock: **588** • © AlinaMD/Shutterstock: **118** • Andriy Malakhovsk/Shutterstock: **133** • arek_malang/Shutterstock: **570** • Artur Synenko/Shutterstock: **583** • Chris Hellyar/Shutterstock: **411** • fuyu liu/Shutterstock: **568** • Gorodenkoff/Shutterstock: **756** • Gypsytwitcher/Shutterstock: **601** • Iaroslav Neliubov/Shutterstock: **601** • Inked Pixels/Shutterstock: **586** • Joe Hendrickson/Shutterstock: **416** • koya979/Shutterstock: **590** • NigelSpiers/Shutterstock: **565** • Nuchylee/Shutterstock: **757** • Peter Turansky/Shutterstock: **780** • Route55/Shutterstock: **566** • saltodemata/Shutterstock: **567** • Shawn Hempel/Shutterstock: **588** • StockImageFactory/Shutterstock: **583** • Vlado Markov/Shutterstock: **589** • Wittybear/Shutterstock: **415** • Ye Liew/Shutterstock: **570** • © Anastasiia Malinich/Shutterstock: **448** • © Andy Dean Photography/Shutterstock: **395** • © Audio und werbung/Shutterstock: **387** • © Belyay/Shutterstock: **457** • © Digital Genetics/Shutterstock: **395** • © Hogan Imaging/Shutterstock: **462** • © Lukiyanova Natalia frenta/Shutterstock: **452** • © Maxx-Studio/Shutterstock: **394** • © Monkey Business Images/Shutterstock: **386, 465** • © Nataliia K/Shutterstock: **463** • © Rikard Stadler/Shutterstock: **431** • © SB Arts Media/Shutterstock: **473** • © Steve Mann/Shutterstock: **454** • © thitikan chuachan/Shutterstock: **380** • Cinematographer/Shutterstock: **601** • fritz16/Shutterstock: **589** • gorillaimages/Shutterstock: **449** • PK Studio/Shutterstock: **486** • Stefan Schurr/Shutterstock: **451** • Yershov Oleksandr/Shutterstock: **429** • ©/Getty Images: **132** • © 360b/Shutterstock: **148** • © AAIndia/Shutterstock: **336** • © adoc-photos/Getty Images: **424** • © Africa Studio/Shutterstock: **367** • © Aleksandar Mijatovic/Shutterstock: **663** • © Aleksandra Gigowska/Shutterstock: **182** • © Alena Ozerova/Shutterstock: **304** • © alessandro guerriero/Shutterstock: **191** • © Alex Cimbal/Shutterstock: **140** • © alexandra__artist/Shutterstock: **204** • © andersphoto/Shutterstock: **668** • © Andras Deak/Shutterstock: **43** • © Andrey Bayda/Shutterstock: **423** • © Andrey_Popov/Shutterstock: **179, 386, 427, 436** • © Andrey_Popov/Shutterstock: **113** • © Anggara dedy/Shutterstock: **779** • © Anna Kucherova/Shutterstock: **129** • © Anton Balazh/Shutterstock: **485** • © Anton_Ivanov/Shutterstock: **12** • © Antonio Guillem/Shutterstock: **402** • © Art3d/Shutterstock: **159** • © ArtFamily/Shutterstock: **201** • © ayzek/Shutterstock: **149** • © B Calkins/Shutterstock: **171** • © BearFotos/Shutterstock: **59, 309** • © Bettmann/Getty Images Australia: **262** • © Big Pants Production/Shutterstock: **786** • © Bignai/Shutterstock: **159** • © bikeriderlondon/Shutterstock: **150** • © Bloomicon/Shutterstock: **303** • © blue jean images RF/Getty Images: **68** • © Bojan Milinkov/Shutterstock: **28** • © BORTEL Pavel - Pavelmidi/Shutterstock: **629** • © Breadmaker/Shutterstock: **396** • © Brocreative/Shutterstock: **117** • © cassiede alain/Shutterstock: **679** • © Corepics VOF/Shutterstock: **244** • © Dani Vincek/Shutterstock: **249** • © Daniel M Ernst/Shutterstock: **403** • © Danny Xu/Shutterstock: **669** • © Daxiao Productions/Shutterstock: **111** • © DeiMosz/Shutterstock: **158** • © Dmitry Kalinovsky/Shutterstock: **406** • © Dmitry Lobanov/Shutterstock: **201** • © dotshock/Shutterstock: **310** • © Dragon Images/Shutterstock: **104** • © Dudarev Mikhail/Shutterstock: **427** • © Duplass/Shutterstock: **361** • © El Nariz/Shutterstock: **140** • © Elena Elisseeva/Shutterstock: **103** • © Elvira Kashapova/Shutterstock: **1** • © Empirephotostock/Shutterstock: **129** • © EpciStockMedia/Shutterstock: **117, 709** • © Eric Isselee/Shutterstock: **34** • © ericleefrancais/Shutterstock: **229** • © Ermolaev Alexander/Shutterstock: **186** • © ESB Professional/Shutterstock: **163** • © ESB Professional/Shutterstock: **153** • © Evgeniya Uvarova/Shutterstock: **133** • © FlashStudio/Shutterstock: **156** • © Fotofermer/Shutterstock: **229** • © Fotomay/Shutterstock: **324**

131 • © YAKOBCHUK VIACESLAV/Shutterstock: **7** • © YinYang/Getty Images: **184** • © Yuttana Jaowattana/Shutterstock: **138** • © Zerbor/Shutterstock: **344** • © ZUMA Press: Inc./Alamy Images, **51**

Every effort has been made to trace the ownership of copyright material. Information that will enable the publisher to rectify any error or omission in subsequent reprints will be welcome. In such cases, please contact the Permissions Section of John Wiley & Sons Australia, Ltd.

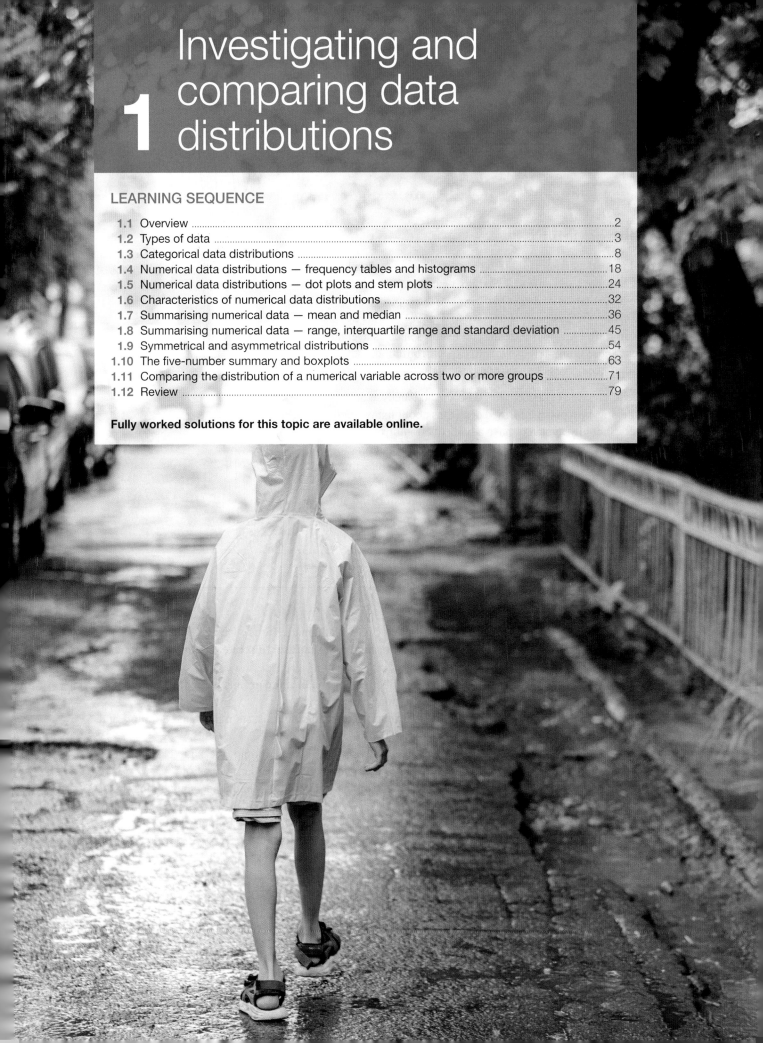

1 Investigating and comparing data distributions

LEARNING SEQUENCE

Fully worked solutions for this topic are available online.

1.1 Overview

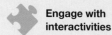

1.1.1 Introduction

Data is all around us; your height, the number of minutes it takes to you get to school and how many points your favourite team scored are all examples of data. On their own, these pieces of information don't tell us much, but if we look at collections of such data, we can start to recognise trends and features such as averages and ranges. These things tell us what to expect from future data; understanding data gives us an insight into the future.

Being able to process, organise, plot, describe and analyse data gives us the power to make informed decisions about many things in our lives. Should you plan to have a barbeque tomorrow? Is the tap water safe to drink? It would be possible to guess answers to these questions, but you might end up wet or sick. By collecting, analysing and interpreting data, these questions can be answered with a lot more certainty. The skills you learn in investigating and comparing data distributions will help you make more informed, better decisions in your life.

KEY CONCEPTS

This topic covers the following key concepts from the VCE Mathematics Study Design:
- types of data, including categorical (nominal or ordinal) or numerical (discrete or continuous, interval, ratio)
- display and description of categorical data distributions of one or more groups using frequency tables and bar charts, and the mode and its interpretation
- display and description of numerical data distributions using histograms, stem plots and dot plots and choosing between plots according to context and purpose
- summarising numerical data distributions, including use of and calculation of the sample summary statistics, median, range, and interquartile range (IQR) or mean and standard deviation
- the five-number summary and the boxplot as its graphical representation and display, including the use of the lower fence ($Q_1 - 1.5 \times IQR$) and upper fence ($Q_3 + 1.5 \times IQR$) to identify possible outliers
- consideration of a range of distributions (symmetrical, asymmetrical), their summary statistics and the percentage of data lying within several standard deviations of the mean
- use of back-to-back stem plots or parallel boxplots, as appropriate, to compare the distributions of a single numerical variable across two or more groups in terms of centre (median) and spread (IQR and range), and the interpretation of any differences observed in the context of the data.

Source: VCE Mathematics Study Design (2023–2027) extracts © VCAA; reproduced by permission.

1.2 Types of data

LEARNING INTENTION

At the end of this subtopic, you should be able to:
* identify types of data, including categorical (nominal or ordinal) or numerical (discrete or continuous, interval, ratio).

1.2.1 Data types — categorical and numerical

There are two major groups of data: **categorical data** and **numerical data**.

Categorical data is data that can be organised into categories such as colours, brands and opinions. For example, a survey asking you to rate your experience as 'bad' or 'average' or 'good' is collecting categorical data.

Numerical data contains numbers and can be counted or measured. For example, measuring the height of everyone in your class is collecting numerical data.

WORKED EXAMPLE 1 Identifying if data is categorical or numerical

Identify whether the following data is categorical or numerical.
a. The results of a survey asking favourite colours
b. The results of a survey asking number of pets

THINK	WRITE
a. 1. The results will be colours, which is an example of a category.	a. The data is categorical.
b. 1. The results will be numbers, which can be counted.	b. The data is numerical.

1.2.2 Categorical data

There are two major groups of categorical data: **ordinal data** and **nominal data**.

Data types

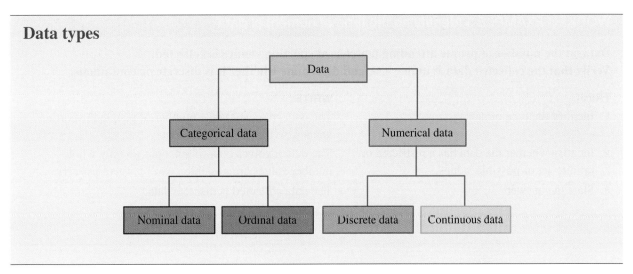

Nominal data has categories with no order or ranking. Nominal data is collected when taking observations (for example, type of bird) or asking for opinions (for example, favourite type of fruit).

Ordinal data has categories that have an order or ranking. Ordinal data is collected when people are asked to rate something (for example, if a product was 'bad', 'average' or 'good') or when ranked categories are used (for example, collecting data on whether the size was 'small', 'medium' or 'large').

WORKED EXAMPLE 2 Identifying the type of categorical data

Data on the different types of cars in a car yard is collected. Verify that the collected data is categorical, and determine whether it is ordinal or nominal.

THINK	WRITE
1. Identify the type of data.	The data collected is the brand or model of cars, so this is categorical data.
2. Identify whether the order of the data is relevant.	When assessing the types of different cars, the order is not relevant, so this is nominal data.
3. State the answer.	The data collected is nominal data.

1.2.3 Numerical data

There are two major groups of numerical data: **discrete data** and **continuous data**.

Discrete data is number data that is usually a whole number. Discrete data is collected when something has been counted (for example, number of pets). Discrete data has a restricted set of possible values.

Continuous data is number data that has been measured on a scale. Continuous data can be measured with low or high precision. For example, measuring the length of a table will give continuous data, but this data may have low precision (for example, 57 cm) or high precision (for example, 57.267 cm). Other examples of scales used for continuous data are weight, time and temperature. Continuous data has an infinite set of possible values.

WORKED EXAMPLE 3 Identifying the type of numerical data

Data on the number of people attending matches at sporting venues is collected. Verify that the collected data is numerical, and determine whether it is discrete or continuous.

THINK	WRITE
1. Identify the type of data.	The data collected is the number of people in sporting venues, so this is numerical data.
2. Identify whether the data has a restricted or infinite set of possible values.	The data involves counting people, so only whole number values are possible.
3. State the answer.	The data collected is discrete data.

1.2.4 Levels of measurement

There are four levels of measurement of variables. The four levels include nominal data, ordinal data, **interval data** and **ratio data**.

Interval data includes values found on a scale with no true zero. Temperature using the degrees Celsius scale is an example of interval data. This is because it has no true zero as temperatures below 0 °C exist; for example, −5 °C is a sensible value.

Ratio data includes values found on a scale with a true zero. Age using the scale of years is an example of ratio data. This is because it has a true zero representing when the person was born; negative age makes no sense.

Note that interval and ratio data are types of numerical data and can be discrete or continuous.

The four levels of measurement can be put in order. This order represents the increasing usefulness of the data for statistical analysis, with nominal data being the least useful and ratio data being the most useful.

The four levels are shown in this table.

	Nominal	Ordinal	Interval	Ratio
Categories	✓	✓	✓	✓
Order		✓	✓	✓
Scale			✓	✓
Scale starts at zero				✓
Examples	Colours, brands	Ratings: high, medium, low	Temperature in Celsius, year of birth	Length, weight

WORKED EXAMPLE 4 Determining the level of measurement

For the following sets of data, determine:
a. whether the data are nominal, ordinal, interval or ratio
b. which set of data is more useful for statistical analysis.
   i. The brand of shoes students at your school are wearing
   ii. The length of the right foot of students at your school

THINK	WRITE
a. i. The results will be brands, which are examples of categories.	a. i. The data are categorical.
ii. The results will be values, which are measured on a scale (length) that starts from zero.	ii. The data are ratio.
b. The levels of measurement from least to most useful for statistical analysis are nominal, ordinal, interval and ratio.	b. The data in ii is most useful for statistical analysis.

1. **WE1** Identify whether the following data is categorical or numerical.
   a. The results of a survey asking number of siblings
   b. The results of a survey asking favourite type of takeaway food
   c. The results from measuring the height of all Year 11 students

2. **WE2** Data on the different types of cereal on supermarket shelves is collected.
   Verify that the collected data is categorical, and determine whether it is ordinal or nominal.

3. **WE3** Data on the rating of hotels from 'one star' to 'five star' is collected.
   Verify that the collected data is categorical, and determine whether it is ordinal or nominal.

4. Identify whether the following numerical data sets are discrete or continuous.
   a. The amount of daily rainfall in Geelong
   b. The heights of players in the National Basketball League
   c. The number of children in families

5. Identify whether the following numerical data sets are nominal, ordinal, discrete or continuous.
   a. The times taken for the place getters in the Olympic 100 m sprinting final
   b. The number of gold medals won by countries competing at the Olympic Games
   c. The type of medals won by a country at the Olympic Games

6. Complete the following table by indicating the type of data.

Data	Type	
Example: The types of meat displayed in a butcher shop.	Categorical	Nominal
a. Wines rated as high, medium or low quality		
b. The number of downloads from a website		
c. Electricity usage over a three-month period		
d. The daily volume of petrol sold by a petrol station		

7. The birthplaces of 200 Australian citizens were recorded and are shown in the following **frequency table**.

Birthplace	Frequency
Australia	128
United Kingdom	14
India	10
China	9
Ireland	6
Other	33

Determine the type of data that is being collected.

8. **WE4** For the following sets of data, determine:

   a. whether the data are nominal, ordinal, interval or ratio
   b. which set of data is more useful for statistical analysis.

   i. The birth year of students at your school
   ii. The favourite type of ice cream of students at your school

9. You want to run a survey about sleep. Two possible questions you could ask are:
   Question 1: How did you sleep last night? Choose from well, medium, poorly.
   Question 2: How many hours did you sleep last night?

   a. Determine which type of data you will get from each question: nominal, ordinal, interval or ratio.
   b. Determine which question is better for statistical analysis. Explain why.

10. The winner of a weight loss reality TV show lost 70.1 kg on the show. Explain whether the weight lost is a discrete or continuous variable.

1.2 Exam questions

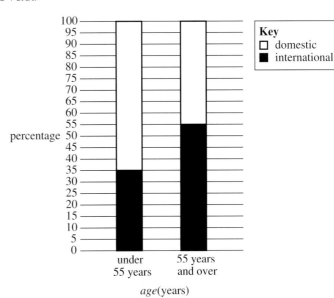

Question 1 (1 mark)

Source: VCE 2021, Further Mathematics Exam 1, Section A, Q1; © VCAA.

MC The percentaged segmented **bar chart** shows the *age* (under 55 years, 55 years and over) of visitors at a travel convention, segmented by *preferred travel destination* (domestic, international).

The variables *age* (under 55 years, 55 years and over) and *preferred travel destination* (domestic, international) are

A. both categorical variables.
B. both numerical variables.
C. a numerical variable and a categorical variable respectively.
D. a categorical variable and a numerical variable respectively.
E. a discrete variable and a continuous variable respectively.

Question 2 (1 mark)

Source: VCE 2019, Further Mathematics Exam 1, Section A, Q8; © VCAA.

MC Percy conducted a survey of people in his workplace. He constructed a two-way frequency table involving two variables.

One of the variables was *attitude towards shorter working days* (for, against). The other variable could have been

    A. *age* (in years).
    B. *sex* (male, female).
    C. *height* (to the nearest centimetre).
    D. *income* (to the nearest thousand dollars).
    E. *time* spent travelling to work (in minutes).

Question 3 (1 mark)

Source: VCE 2016, Further Mathematics Exam 1, Section A, Q2; © VCAA.

MC The blood pressure (low, normal, high) and the *age* (under 50 years, 50 years or over) of 110 adults were recorded. The results are displayed in the two-way frequency table below.

Blood pressure	Age	
	Under 50 years	**50 years or over**
Low	15	5
Normal	32	24
High	11	23
Total	58	52

The variables *blood pressure* (low, normal, high) and *age* (under 50 years, 50 years or over) are
    A. both nominal variables.
    B. both ordinal variables.
    C. a nominal variable and an ordinal variable respectively.
    D. an ordinal variable and a nominal variable respectively.
    E. a continuous variable and an ordinal variable respectively.

More exam questions are available online.

1.3 Categorical data distributions

LEARNING INTENTION

At the end of this subtopic, you should be able to:
- display and describe categorical data distributions of one or more groups using frequency tables and bar charts, and the mode and its interpretation.

1.3.1 Frequency tables

A **frequency table** is a way of organising raw data. A frequency table will usually have three columns:
- The first column contains the categories in the data set.
- The second column contains a tally.
- The third column contains the frequency value.

Frequency tables

In a frequency table a tally is used to aid in counting the raw data, and the frequency is found at the end by counting the tally's value. The total number of data can be found by adding the frequency values.

For example, here is the raw data collected from a survey asking students their favourite colour:

Red, Blue, Yellow, Red, Purple, Blue, Red, Yellow, Blue, Red

And here is the frequency table for that data:

Favourite colour	Tally	Frequency
Red	||||	4
Blue	|||	3
Yellow	||	2
Purple	|	1

WORKED EXAMPLE 5 Representing categorical data in a frequency table

Thirty students were asked to pick their favourite time of day between the following categories:
Morning (M), Early afternoon (A), Late afternoon (L), Evening (E)
The following data was collected:
A, E, L, E, M, L, E, A, E, M, E, L, E, A, L, M, E, E, L, M, E, A, E, M, L, L, E, E, A, E
Represent the data in a frequency table.

THINK

1. Create a frequency table to capture the data.

WRITE

Time of day	Tally	Frequency
Morning		
Early afternoon		
Late afternoon		
Evening		

2. Go through the data, filling in the tally column as you progress. Sum the tally columns to complete the frequency column.

Time of day	Tally	Frequency
Morning	ⅠⅠⅠⅠ	5
Early afternoon	ⅠⅠⅠⅠ	5
Late afternoon	ⅠⅠⅠⅠⅠⅠ	7
Evening	ⅠⅠⅠⅠ ⅠⅠⅠⅠⅠⅠⅠ	13

1.3.2 Bar charts

Bar charts (also known as bar graphs) are a way of displaying categorical data. Bar charts have a gap between bars. The categories are shown along the horizontal axis, which relates to the first column of a frequency table. The frequency is shown on the vertical axis, which displays the values from the frequency column of a frequency table.

Here is a bar chart based on the data that was collected regarding students' favourite colours:

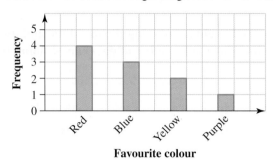

Bar charts display a lot of information. We need to know how to interpret and summarise this information. When describing a bar chart, the important features are: how the information was collected, the number of data points (found by adding the value of each column), the categories and the most/least common item(s).

Here is a description of the bar chart above:

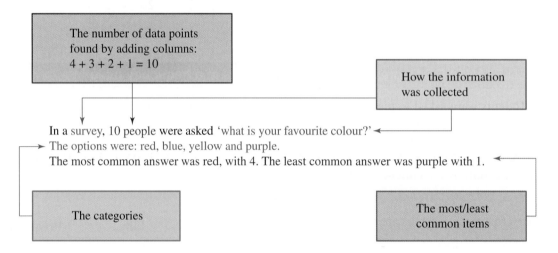

The number of data points found by adding columns:
$4 + 3 + 2 + 1 = 10$

How the information was collected

In a survey, 10 people were asked 'what is your favourite colour?'
The options were: red, blue, yellow and purple.
The most common answer was red, with 4. The least common answer was purple with 1.

The categories

The most/least common items

tlvd-3689

WORKED EXAMPLE 6 Displaying and describing data in a bar chart

Students from a particular school were surveyed to find out what sport they participated in on the weekend. The results are summarised in the frequency table shown.

Sport	Frequency
Tennis	40
Swimming	30
Cricket	60
Basketball	50
No sport	70

a. **Display the data in a bar chart.**
b. **Describe the information in the bar chart.**

THINK	**WRITE**
a. 1. Choose an appropriate scale for the bar chart. As the frequencies go up to 70 and all of the values are multiples of 10, we will mark our intervals in 10s. Display the different categories along the horizontal axis.	**a.** 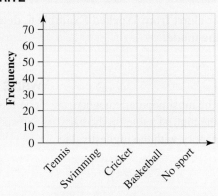
2. Draw bars to represent the frequency of each category, making sure there are spaces between the bars.	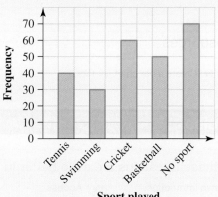
b. The important things to describe are: • how the information was collected • the number of data points, which is: $40 + 30 + 60 + 50 + 70 = 250$ • the categories • the most/least common items, which are: no sport with 70 and swimming with 30.	**b.** Students from a particular school were asked 'What sport do you participate in on the weekend?' There were 250 students surveyed. The options were: tennis, swimming, cricket, basketball and no sport. The most common answer was no sport with 70; the least common was swimming with 30.

1.3.3 The mode

The **mode** of categorical data is the most common category. In a frequency table, the mode is the category with the highest frequency value. In a bar chart, the mode is the category with the highest bar. Identifying the mode is a type of data analysis. The mode can also be called the modal category.

It is possible to have more than one mode.

Here is the bar chart from the data used previously. The mode of the bar chart is red because it has the highest bar.

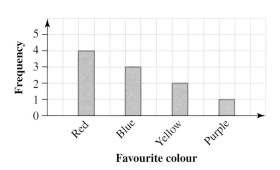

Using the bar chart from Worked Example 6, determine which category is the mode.

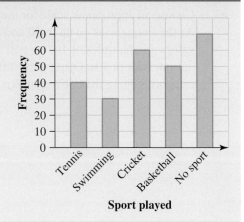

THINK

The mode is the most popular category, so will have the highest value on the bar chart. The category with the highest value is 'No sport'.

WRITE

The mode is 'No sport'.

 Resources

 Interactivity Create a bar chart (int-6493)

1.3 Exercise

Students, these questions are even better in jacPLUS

 Receive immediate feedback and access sample responses

 Access additional questions

 Track your results and progress

Find all this and MORE in jacPLUS ▶

1. **WE5** Thirty students were asked to pick their favourite type of music from the following categories: Pop (P), Rock (R), Classical (C), Folk (F), Electronic (E).
The following data was collected:

E, R, R, P, P, E, F, E, E, P, R, C, E, P, E, P, C, R, P, F, E, P, P, E, R, R, E, F, P, R

Represent the data in a frequency table.

2. **WE6** The preferred movie genre of 100 students is shown in the following frequency table.

Favourite movie genre	Frequency
Action	32
Comedy	19
Romance	13
Drama	15
Horror	7
Musical	4
Animation	10

a. Display the data in a bar chart.
b. Describe the information in the bar chart.

3. The favourite pizza type of 60 students is shown in the following bar chart.

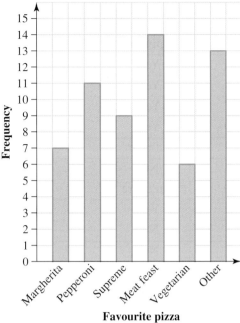

Favourite pizza

Display the data in a frequency table.

4. A group of students at a university were surveyed about their usual method of travel, with the results shown in the following table.

Student	Transport method	Student	Transport method
A	Bus	N	Car
B	Walk	O	Bus
C	Train	P	Car
D	Bus	Q	Bus
E	Car	R	Bicycle
F	Bus	S	Car
G	Walk	T	Train
H	Bicycle	U	Bus
I	Bus	V	Walk
J	Car	W	Car
K	Car	X	Train
L	Train	Y	Bus
M	Bicycle	Z	Bus

a. Determine the type of data being collected.
b. Organise the data into a frequency table.
c. Display the data as a bar chart.
d. Describe the information in the bar chart.

5. In a telephone survey people were asked the question, 'Do you agree that convicted criminals should be required to serve their full sentence and not receive early parole?' They were required to respond with either 'Yes', 'No' or 'Don't care' and the results are as follows.

Person	Opinion	Person	Opinion
A	Yes	N	Yes
B	Yes	O	No
C	Yes	P	No
D	Yes	Q	Yes
E	Don't care	R	Yes
F	No	S	Yes
G	Don't care	T	Yes
H	Yes	U	No
I	No	V	Yes
J	No	W	Yes
K	Yes	X	Don't care
L	No	Y	Yes
M	Yes	Z	Yes

a. Organise the data into an appropriate table.
b. Display the data as a bar chart.
c. Identify the data as either nominal or ordinal. Explain your answer.

6. **WE7** Twenty-five students were asked to pick their favourite type of animal to keep as a pet. The following data was collected.

Dog, Cat, Cat, Rabbit, Dog, Guinea pig, Dog, Cat, Cat, Rat, Rabbit, Ferret, Dog, Guinea pig, Cat, Rabbit, Rat, Dog, Dog, Rabbit, Cat, Cat, Guinea pig, Cat, Dog

a. Represent the data in a frequency table.
b. Draw a bar chart to represent the data.
c. Determine which category is the mode of the data.

7. The different types of coffee sold at a café in one hour are displayed in the following bar chart.

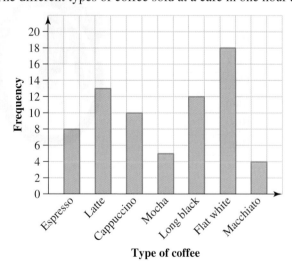

a. Determine the modal category of the coffees sold.
b. Calculate how many coffees were sold in that hour.

8. The result of an opinion survey are displayed in the bar chart.

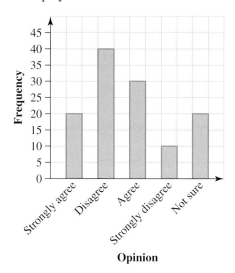

a. Determine the type of data being displayed.
b. Explain what is wrong with the current data display.
c. Redraw the bar chart displaying the data correctly.

9. Exam results for a group of students are shown in the following table.

Student	Result	Student	Result	Student	Result	Student	Result
1	A	6	C	11	B	16	C
2	B	7	C	12	C	17	A
3	D	8	C	13	C	18	C
4	E	9	E	14	C	19	D
5	A	10	D	15	D	20	E

a. Display the exam result data in a frequency table.
b. Display the data in a bar chart.
c. Determine the type of data collected.

10. The number of properties sold in the capital cities of Australia for a particular time period is shown in the following table.

City	Number of bedrooms			
	2	3	4	5
Adelaide	8	12	5	4
Brisbane	15	11	8	6
Canberra	8	12	9	2
Hobart	3	9	5	1
Melbourne	16	18	12	11
Sydney	23	19	15	9
Perth	7	9	12	3

Use the given information to create a bar chart that represents the number of bedrooms of properties sold in the capital cities during this time period.

11. The maximum daily temperatures (°C) in Adelaide during a 15-day period in February are listed in the following table.

Day	1	2	3	4	5	6	7	8	9	10	11	12	13	14	15
Temp (°C)	31	32	40	42	32	34	41	29	25	33	34	24	22	24	30

Temperatures greater than or equal to 39 °C are considered above average and those less than 25 °C are considered below average.

   a. Organise the data into three categories and display the results in a frequency table.
   b. Display the organised data in a bar chart.
   c. Determine the type of data displayed in your bar chart.

12. The frequency table shown displays the different categories of purchases in a shopping basket.

Category	Frequency
Fruit	6
Vegetables	8
Frozen goods	5
Packaged goods	11
Toiletries	3
Other	7

   a. Calculate how many items were purchased in total.
   b. Calculate the percentage of the total purchases that were fruit.

13. Data for the main area of education and study for a selected group of people aged 15 to 64 during a particular year in Australia is shown in the following table.

Number of people (thousands)					
Main area of education and study	15–19	20–24	25–34	35–44	45–64
Agriculture	10	9	14	5	5
Creative arts	36	51	20	10	9
Engineering	59	75	50	13	6
Health	44	76	64	32	32
Management and commerce	71	155	135	86	65

   a. Create separate bar charts for each age group to represent the data.
   b. State the modal age category for Health.

14. The bar chart shown represents the ages of attendees at a local sporting event.

   a. Represent the data in a frequency table.
   b. Determine the modal category.
   c. The age groups are changed to 'Under 20', '20–39', '40–59' and '60+'. Redraw the bar chart with these new categories.
   d. Determine whether this changes the modal category.

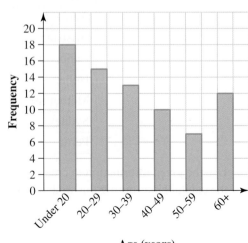

1.3 Exam questions

▶ **Question 1 (1 mark)**

Source: VCE 2012, Further Mathematics Exam 1, Section A, Q1; © VCAA.

MC The following bar chart shows the distribution of wind directions recorded at a weather station at 9.00 am on each of 214 days in 2011.

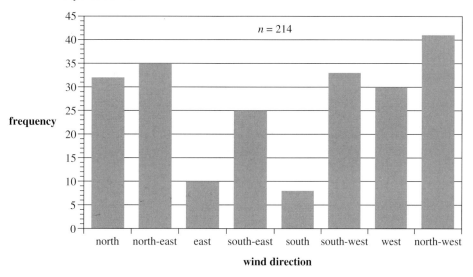

According to the bar chart, the most frequently observed wind direction was

A. south-east.　　　**B.** south.　　　**C.** south-west.　　　**D.** west.　　　**E.** north-west.

▶ **Question 2 (1 mark)**

Source: VCE 2012, Further Mathematics Exam 1, Section A, Q2; © VCAA.

MC The following bar chart shows the distribution of wind directions recorded at a weather station at 9.00 am on each of 214 days in 2011.

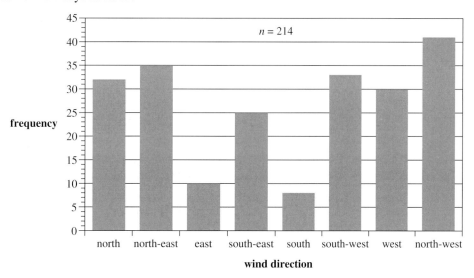

According to the bar chart, the percentage of the 214 days on which the wind direction was observed to be east or south-east is closest to

A. 10%　　　**B.** 16%　　　**C.** 25%　　　**D.** 33%　　　**E.** 35%

MC For the bar chart shown, determine which of the following
statements is false.
  A. There are five families with zero children.
  B. There are zero families with five children.
  C. There are more families with three children than there are
    families with four children.
  D. The modal number of children is two.
  E. There are ten families with four children.

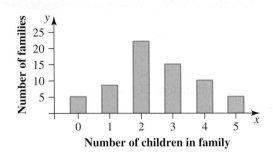

More exam questions are available online.

1.4 Numerical data distributions — frequency tables and histograms

> **LEARNING INTENTION**
>
> At the end of this subtopic you should be able to:
> • display and describe numerical data distributions using frequency tables and histograms.

1.4.1 Frequency tables — grouped data

A frequency table can be used to organise raw numerical data. This is usually done by grouping the data — the
mathematical name for these groups is intervals. To group data, first the size of the interval must be chosen. The
size of the interval can also be called interval width or class size. Generally, aim to have a class size that gives
between five and ten different intervals.

When writing intervals, first write the bottom value, then a dash, then the *less than* symbol ($<$), and then the top
value of the group. For example:

$$10 - {<}15$$

This interval starts at 10 and goes up to 15, not including 15; this interval has a class size of 5.

Note that it is possible to have a frequency table for ungrouped numerical data if there are a small number of
numerical values.

> **WORKED EXAMPLE 8 Representing numerical data in a frequency table**
>
> The following data represents the time (in seconds) it takes for each individual in a group of
> 20 students to run 100 m.
>
> 18.2, 20.1, 15.6, 13.5, 16.7, 15.9, 19.3, 22.5, 18.4, 15.9
> 12.4, 14.1, 17.7, 19.4, 21.0, 20.4, 18.2, 15.8, 16.1, 14.6
>
> **Group and display the data in a frequency table.**
>
> | THINK | WRITE |
> |---|---|
> | 1. Identify the smallest and largest values in the data set. This will help you choose your class size and decide what the first class should be. | Smallest value $= 12.4$
Largest value $= 22.5$
We will have class intervals of 2, starting with $12 - {<} 14$. |

2. Draw a frequency table to represent the data. Complete the tally column in your table and use this to fill in the frequency column.

Time (seconds)	Tally	Frequency
12– < 14	||	2
14– < 16	|||||	6
16– < 18	|||	3
18– < 20	||||	5
20– < 22	|||	3
22– < 24	|	1

1.4.2 Histograms

There are several ways to display numerical data. The first one we will look at is the **histogram**.

Histograms are similar to bar charts, but have two important differences: in histograms the columns touch each other to represent that there are no gaps in the data, and the categories along the horizontal axis are intervals (or groups) of values. For example:

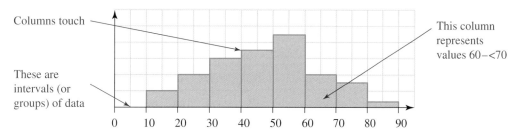

Columns touch

These are intervals (or groups) of data

This column represents values 60–<70

Histograms can be used to display continuous data that has been organised into groups (intervals) with the height of each column showing the frequency of that interval.

On a histogram, the mode is the tallest column. It is possible to display a histogram using a CAS calculator. See the worked example below.

tlvd-3690

WORKED EXAMPLE 9 Drawing a histogram

The following frequency table represents the heights of players in a basketball squad.

Height (cm)	175– < 180	180– < 185	185– < 190	190– < 195	195– < 200	200– < 205
Frequency	1	3	6	3	1	1

Draw a histogram to represent this data.

THINK	**WRITE**
1. Look at the data range and use the leading values from each interval in the table for the scale of the horizontal axis.	The height data in the table has intervals starting from 175 cm and increasing by 5 cm.

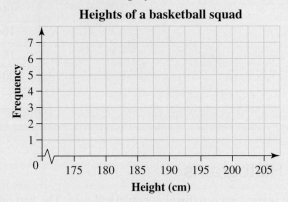

Heights of a basketball squad

THINK	**WRITE**
2. Draw rectangles for each interval to the height of the frequency indicated by the data in the table.	

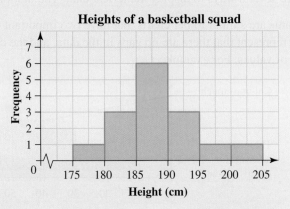

Heights of a basketball squad

TI | THINK	**DISPLAY/WRITE**	**CASIO | THINK**	**DISPLAY/WRITE**
1. In a Lists & Spreadsheet page, label the first column as *height* and the second column as *freq*. Enter the midpoint of each height interval in the first column and the frequencies in the second column.		1. On a Statistics screen, relabel list 1 as *height* and list 2 as *freq*. Enter the midpoint of each height interval in the first column and the frequencies in the second column.	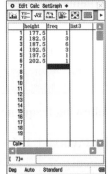
2. Press MENU, then select: 3: Data 8: Summary Plot Complete the fields as: X List: height Summary List: freq Display On: New Page then select OK.		2. Select: • SetGraph • Setting… Complete the fields as: Draw: On Type: HistogramXList: main\height Freq: main\freq then select Set.	

3. On the Summary Plot page, press MENU, then select:
   2: Plot Properties
   2: Histogram Properties
   2: Bin Settings
   1: Equal Bin Width
   Complete the fields as:
   Width: 5
   Alignment: 175
   then select OK.

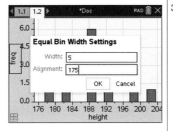

3. Press the y icon. Complete the fields as:
   HStart: 175
   HStep: 5
   then select OK.

4. The histogram is displayed on the screen.

4. The histogram is displayed on the screen.

1.4 Exercise

1. **WE8** The following data represents the time (in seconds) it takes for each individual in a group of 20 students to swim 50 m.

   48.5, 54.1, 63.0, 39.7, 51.3, 57.7, 68.4, 59.4, 37.5, 41.8, 72.3, 56.3, 45.4, 39.2, 60.3, 56.6, 48.1, 42.9, 53.3, 64.1

   Group and display the data in a frequency table.

2. The following data set indicates the time, in seconds, it takes for a tram to travel between two stops on 20 weekday mornings.

$$95,\ 112,\ \ 99,\ \ 91,\ 105,\ 110,\ 97,\ 122,\ 108,\ 101,$$
$$95,\ \ 89,\ 100,\ 115,\ 124,\ \ 98,\ 87,\ 111,\ 115,\ 106$$

a. Group and display the data in a frequency table with intervals of width 10 seconds.
b. Group and display the data in a frequency table with intervals of width 5 seconds.

3. **WE9** Draw histograms to represent the following data sets.

a. The cholesterol levels measured for a group of people

Cholesterol level (mmol/L)	1– < 2.5	2.5– < 4.0	4.0– < 5.5	5.5– < 7.0	7.0– < 8.5
Frequency	2	8	12	14	10

b. The distances travelled to school by a group of students

Distance travelled (km)	0– < 22	2– < 4	4– < 6	6– < 8	8– < 10
Frequency	18	26	14	8	2

4. Organise each of the following data sets into a frequency table using intervals of 5, commencing from the lowest value. Then draw a histogram to represent the data.

a. 5, 7, 14, 17, 13, 24, 22, 15, 12, 26, 17, 15, 14, 13, 15, 7, 8, 13, 17, 24, 22, 7, 13, 20, 12, 15, 23, 20, 17, 15, 17, 16, 20, 23, 15, 16, 18, 17, 14, 15

b. 34, 28, 45, 46, 13, 24, 11, 33, 41, 35, 16, 15, 35, 13, 14, 28, 27, 22, 36, 31, 11, 18, 24, 20, 12, 15, 41, 50, 27, 13, 14, 16, 20, 23, 31, 26, 25, 27, 34, 35

5. A group of 40 workers were surveyed on their average hours worked per week. The results were:

$$36,\ 40,\ 42,\ 40,\ 34,\ 33,\ 38,\ 36,\ 43,\ 39,\ 35,\ 36,\ 22,$$
$$39,\ 37, 37,\ 40,\ 38,\ 25,\ 27,\ 41,\ 34,\ 33,\ 28,\ 36,\ 25,$$
$$37,\ 39,\ 35, 36,\ 36,\ 37,\ 28,\ 42,\ 39,\ 40,\ 33,\ 35,\ 37,\ 38$$

a. Display the data in a frequency table using intervals of 5 commencing from 20 hours.
b. Use the frequency table to draw a histogram of the data.

6. **MC** This is a histogram about the ages of people who visited an ice-cream shop over a two-hour period. Determine which of the following is false.

A. A total of 46 people visited during the two-hour period
B. The most common age group was 40– < 50
C. No-one who visited was over 90 years old.
D. A total of 11 people under 30 visited during the two-hour period.
E. More than one person who visited was over 80 years old.

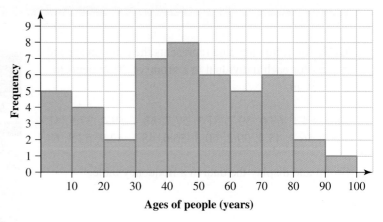

7. Students in Year 11 at Braybrae High School were asked how many hours they worked in the previous week. The results are shown in the histogram.

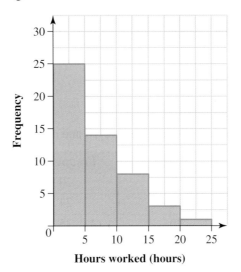

Hours worked (hours)

a. Determine the most common response.
b. Draw a frequency table representing this data.
c. Calculate the total number of students surveyed.

8. Explain which set of data the following histogram could represent. There may be more than one answer.

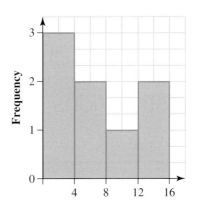

Data set 1: 0, 1, 2, 4, 6, 9, 13, 14
Data set 2: 1, 3, 4, 5, 7, 9, 10, 13, 14
Data set 3: 1, 2, 3, 5, 6, 11, 13, 15

1.4 Exam questions

Question 1 (1 mark)
A group of 28 students had their heights recorded. See the data set below.

| 123 | 125 | 126 | 124 | 111 | 135 | 147 | 125 | 128 | 123 | 126 | 128 | 124 | 124 |
| 120 | 122 | 125 | 123 | 110 | 130 | 140 | 127 | 128 | 123 | 120 | 149 | 125 | 127 |

Construct a frequency table using a suitable class interval.

Question 2 (1 mark)

A group of 28 students had their heights recorded. See the data set below.

| 123 | 125 | 126 | 124 | 111 | 135 | 147 | 125 | 128 | 123 | 126 | 128 | 124 | 124 |
| 120 | 122 | 125 | 123 | 110 | 130 | 140 | 127 | 128 | 123 | 120 | 149 | 125 | 127 |

Construct a histogram to represent these results.

Question 3 (3 marks)

The average number of hours worked each week was surveyed and a frequency table drawn up.

Hours	Frequency
0–	20
10–	24
20–	35
30–	46
40–	64
50–	18
60–	8
70–	5

a. State the number of workers surveyed. **(1 mark)**
b. Draw a histogram to show the information. **(1 mark)**
c. Determine the mode of the number of hours worked each week. **(1 mark)**

More exam questions are available online.

1.5 Numerical data distributions — dot plots and stem plots

LEARNING INTENTION

At the end of this subtopic you should be able to:
• display and describe numerical data distributions using stem plots, back-to-back stem plots and dot plots, and choose between plots according to context and purpose.

1.5.1 Stem plots

A **stem plot** is a way of displaying numerical data. Another name for stem plots is stem-and-leaf plots. In stem plots, raw data is organised by collecting all the numbers with the same stem and writing the leaves in order. Stem plots have a key. This is used to translate the stem and leaf into the number it represents. For example, Key: $1|4 = 14$ means that the stem is the tens value and the leaf is the unit (or ones) value.

If a stem value has a * it represents the second half of that stem. For example, in this stem plot the stem 1 contains values from 10 to 14 and the stem 1* contains values from 15 to 19. When drawing a stem plot, if there are 4 or less place values it is sensible to split stems into first and second half using a *.

The stem values are ordered with lowest value at the top. The leaf values are in order on each line, with lowest value closest to the stem.

Key: $1|4 = 14$

Stem	Leaf
1	4
1*	7 7 9
2	2 3 4 4
2*	6 8
3	0 1 3
3*	5 6 9
4	0

The following data set (of 31 values) shows the maximum daily temperature during the month of January in a particular area.

26, 22, 24, 26, 28, 28, 27, 42, 25, 25, 29, 31, 23, 33, 34, 27,
39, 44, 35, 34, 27, 30, 36, 30, 30, 28, 33, 23, 24, 34, 37

Draw a stem plot to represent the data.

THINK	WRITE
1. Identify the place values for the data. If there are 4 or less different place values, split each into two.	The temperature data has values in the 20s, 30s and 40s.

Stem	Leaf
2	
2*	
3	
3*	
4	

2. Write the units for each stem place value in numerical order, with the smallest values closest to the stem. Make sure to keep consecutive numbers level as they move away from the stem. Remember to add a key to your plot.

Key: $2 \mid 2 = 22°\,C$

Stem	Leaf
2	2 3 3 4 4
2*	5 5 6 6 7 7 7 8 8 8 9
3	0 0 0 1 3 3 4 4 4
3*	5 6 7 9
4	2 4

1.5.2 Back-to-back stem plots

A **back-to-back stem plot** is a way of displaying two sets of numerical data. Another name for back-to-back stem plots is back-to-back stem-and-leaf plots.

Back-to-back stem plots are similar to stem plots; the main difference is an extra leaf column on the left-hand side. The leaf values are again in order on each line, with lowest value closest to the stem.

Key: $1 \mid 4 = 14$

Leaf	Stem	Leaf
3 2	1	4
9 8 6 6	1*	7 7 9
4 2 1 0	2	2 3 4 4
7 6	2*	6 8
4 2 1	3	0 1 3
6 5	3*	5 6 9
	4	0

The following data sets show the heights of players on two basketball teams: Redbacks and Cobras.

Redbacks: 172, 173, 175, 176, 181, 185, 190, 193
Cobras: 169, 174, 174, 175, 178, 180, 189, 194

Draw a back-to-back stem plot to represent the data.

THINK	WRITE
1. Identify the place values for the data. If there are 4 or less different place values, split each into two.	The height values are in the 160s, 170s, 180s and 190s.

2. Draw a stem column in the middle and the leaf columns on either side. Write the units for each leaf place value in numerical order, with the smallest values closest to the stem. Make sure to keep consecutive numbers level as they move away from the stem.
Remember to add a key to your plot.

Redbacks		Cobras
Leaf	Stem	Leaf
	16*	
	17	
	17*	
	18	
	18*	
	19	

Key 17|1 = 171 cm

Redbacks		Cobras
Leaf	Stem	Leaf
	16*	9
3 2	17	4 4
6 5	17*	5 8
1	18	0
5	18*	9
3 0	19	4

1.5.3 Dot plots

Dot plots are another way of organising numerical data. Dot plots are similar to bar charts and histograms as they display the categories along a horizontal axis. However, dot plots do not have a vertical axis.

In dot plots, each piece of data is represented by one dot. Frequency can be read off a dot plot by counting the number of the dots.

For example, this dot plot shows that the category '5' has a frequency of 3.

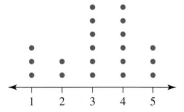

WORKED EXAMPLE 12 Drawing a dot plot

The frequency table shows the number of floors in apartment buildings in a particular area. Draw a dot plot to represent the data.

Number of floors	Frequency
2	2
3	5
4	3
5	0
6	4
7	2

THINK	WRITE
1. Draw a horizontal scale using the discrete data values shown.	The discrete data values are given by the number of floors. 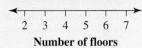
2. Place one dot directly above the number on the scale for each discrete data value present, making sure to keep corresponding dots at the same level.	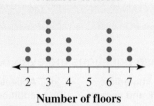

1.5.4 Choosing between plots

For a single set of data there are many choices of plots. The table shown summarises the different types of data and options for displays for each type.

		Bar chart	Histogram	Stem plot	Dot plot
Categorical	Nominal	✓			✓
	Ordinal	✓			✓
Numerical	Discrete		✓	✓	✓
	Continuous		✓	✓	

Note that bar charts can sometimes be used for numerical data, but this is not the preferred choice of plot. The grouped data can be classified as ordinal or continuous depending on the situation.

When deciding which plot to choose, the context and purpose must also be considered.

For example, a data set with high values may be difficult to display in a stem plot, so a histogram would be better. Another example is a data set with many discrete values; this would be difficult to show on a dot plot, so a stem plot or histogram would be better.

WORKED EXAMPLE 13 Choosing how to display data

For the following sets of data, choose an appropriate type of data display from bar chart, histogram, stem plot, back-to-back stem plot and dot plot. Note that there may be more than one correct answer.
a. The results of a survey about favourite type of sandwich in a class of 30 students
b. The sizes of 100 fish caught, grouped into 10 cm intervals

THINK	WRITE
a. 1. The data is categorical, ordinal. The data display choices are bar chart or dot plot.	a. Bar chart (dot plot is also correct).
b. 1. The data is numerical, continuous and it is grouped. The data display choices are stem plot or histogram.	b. Histogram (stem plot is also correct).

1.5 Exercise

1. **WE10** Draw a stem plot for each of the following data sets.

    a. The dollars spent per day on lunch by 15 people:

    22, 21, 22, 24, 19, 22, 24, 21, 22, 23, 25, 26, 22, 23, 22

    b. The number of hours spent per week playing computer games by a group of 20 students at a particular school:

    14, 21, 25, 7, 25, 20, 21, 14, 21, 20, 6, 23, 26, 23, 17, 13, 9, 24, 17, 24

2. **WE11** Draw a back-to-back stem plot for each of the following data sets.

    a. The number of passengers per day transported by a taxi driver:

    March: 33, 27, 44, 47, 23, 24, 22, 35, 42, 36, 17, 25, 34, 13, 15, 27, 28, 23, 37, 34

    April: 22, 27, 23, 20, 12, 15, 43, 30, 27, 15, 27, 36, 20, 23, 35, 36, 28, 17, 14

    b. The number of patients per day treated by two doctors.

    Dr. Hammond: 44, 38, 55, 56, 23, 34, 31, 43, 51, 45, 26, 25, 45, 23, 24, 38, 37, 32, 46, 41

    Dr. Valenski: 21, 28, 34, 30, 22, 25, 51, 60, 17, 23, 24, 26, 30, 33, 41, 26, 35, 17, 24, 25

3. The following set of data indicates the number of people who attend early morning fitness classes run by a business for its workers:

    14, 17, 13, 8, 16, 21, 25, 16, 19, 17, 21, 8, 13

    Display the data as a stem plot.

4. The total number of games played by the players from two basketball squads is shown in the following stem plots.

Key: 0 | 1 = 1 game played

Stem	Leaf
0	1
1	4 7
2	4 4 8
3	3 3 5 6
4	1 2 3
5	1 1
6	5
7	
8	
9	1

Key: 2 | 4 = 24 games played

Stem	Leaf
2	4
3	1 2 6
4	3 4 5
5	2
6	
7	
8	2 5 7
9	3

Draw a stem plot that combines the data for the two teams.

5. **WE12** Draw a dot plot to represent each of the following collections of data.

a. The number of wickets per game taken by a bowler in a cricket season

Number of wickets	Frequency
0	4
1	6
2	4
3	2
4	1
5	1

b. The number of hours per week spent checking emails by a group of workers at a particular company

Hours checking emails	Frequency
1	1
2	1
3	2
4	4
5	8
6	4

6. Draw dot plots to represent the following collections of data.

a. The scores per round of a golfer over a particular time period (40 values):

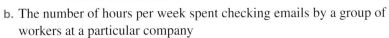

73, 77, 74, 77, 73, 74, 72, 75, 72, 76, 77, 75, 74, 73, 75, 77, 78, 73, 77, 74,
72, 77, 73, 70, 72, 75, 73, 70, 77, 75, 77, 76, 70, 73, 75, 76, 78, 77, 74, 75

b. The scores out of 10 in a multiple-choice test for a group of students (30 values):

6, 7, 4, 7, 3, 7, 7, 5, 7, 6,
7, 5, 1, 3, 5, 7, 8, 3, 7, 4,
9, 5, 4, 6, 7, 9, 10, 5, 7, 4

7. A group of 26 students received the following marks on a test:

$$6, 4, 3, 8, 6, 9, 5, 6, 9, 7, 7, 8, 5,$$
$$7, 4, 3, 8, 6, 5, 7, 9, 5, 6, 6, 7, 8$$

Display the data as a dot plot.

8. **WE13** For the following sets of data, choose an appropriate type of data display from bar chart, histogram, stem plot, back-to-back stem plot and dot plot. Note that there may be more than one correct answer.

   a. The weights of 200 apples at a fruit store, grouped into 10 g intervals
   b. The results of an online poll asking 'How was your day?' with the options: terrible, fine, great

9. Name the best type of data display for the results of a survey about number of pets in a class of 24 students.

10. Consider the set of data in the stem plot shown.

Key: $0|1 = 1$

Stem	Leaf
0	1
1	1 1 1 4 4 6 6 7 8
2	3 3 4 4 7 7 9

   a. Instead of grouping the data in 10s, the stems could be split in half to use groups of 5. The data values from 10 to 14 would be placed in a column labelled '1', and data values from 15 to 19 would be put in the column labelled '1*'.
   Use the data from the original stem plot to create the split stem plot.
   b. Comment on the effect of splitting the stem for the data in this question.

1.5 Exam questions

▶ **Question 1 (1 mark)**

Source: VCE 2020, Further Mathematics Exam 2, Section A, Q2; © VCAA.

The *neck size*, in centimetres, of 250 men was recorded and displayed in the dot plot below.

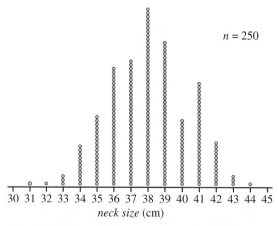

Data: RW Johnson, 'Fitting percentage of body fat to simple body measurements', *Journal of Statistics Education*, 4:1, 1996, <https://doi.org/10.1080/10691898.1996.11910505>

Write down the modal *neck size*, in centimetres, for these 250 men.

Question 2 (2 marks)

Source: VCE 2019, Further Mathematics Exam 2, Section A, Q1; © VCAA.

Table 1 shows the *day number* and the *minimum temperature*, in degrees Celsius, for 15 consecutive days in May 2017.

Table 1

Day number	Minimum temperature (°C)
1	12.7
2	11.8
3	10.7
4	9.0
5	6.0
6	7.0
7	4.1
8	4.8
9	9.2
10	6.7
11	7.5
12	8.0
13	8.6
14	9.8
15	7.7

Data: Australian Government, Bureau of Meteorology,

a. Which of the two variables in this data set is an ordinal variable? **(1 mark)**

b. The incomplete ordered stem plot below has been constructed using the data values for days 1 to 10.

key: $4|1 = 4.1$ $n = 15$

Minimum temperature (°C)

```
 4 | 1 8
 5 |
 6 | 0 7
 7 | 0
 8 |
 9 | 0 2
10 | 7
11 | 8
12 | 7
```

Complete the **stem plot above** by adding the data values for days 11 to 15. **(1 mark)**

(Answer on the stem plot above.)

Question 3 (1 mark)

Source: VCE 2016, Further Mathematics Exam 1, Section A, Q3; © VCAA

MC The stem plot below displays 30 temperatures recorded at a weather station.

The modal temperature is

A. 2.8 °C

B. 2.9 °C

C. 3.7 °C

D. 8.0 °C

E. 9.0 °C

Key: $2|2 = 2.2$°C

Temperature

```
2 | 2 2 4 4
2 | 5 7 8 8 8 8 8 8 9 9 9 9
3 | 1 2 3 3 4 4 4
3 | 5 6 7 7 7 7
4 | 1
```

More exam questions are available online.

1.6 Characteristics of numerical data distributions

1.6.1 Describing distributions

A number of key features should be included in any description of a distribution. These key features are: modality, outliers and shape.

The **modality** of a graph refers to the mode or most common value. A graph may have one mode, two modes (called bimodal) or many modes (called multimodal).

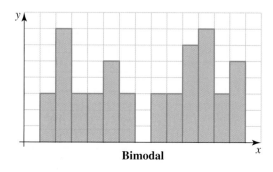

Bimodal

An **outlier** is a data point that has a very different value compared to the rest of the data. It is important to recognise outliers because sometimes these need to be removed from data before doing any analysis to avoid misleading results.

The shape of a distribution is generally either **symmetrical**, **positively skewed** or **negatively skewed**.

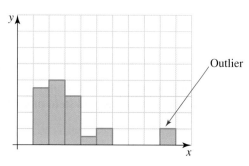

Outlier

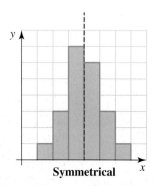

Symmetrical

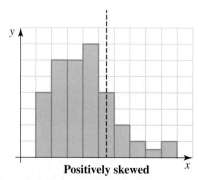

Positively skewed

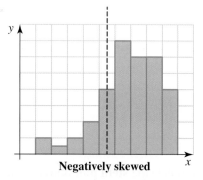

Negatively skewed

For a symmetrical graph, the data sits evenly (relatively) to the left and the right of the centre; for a positively skewed graph, more data sits to the left; and for a negatively skewed graph, more data sits to the right. We will study shape in more detail in later.

Characteristics of numerical data distributions

To describe distributions, the following characteristics should be used:
- modality — include the mode and type of modality
- outlier — if there are any outliers, include the value
- shape — symmetrical, positively skewed or negatively skewed.

There are other characteristics of numerical data distributions, such as mean, median, range, interquartile range and standard deviation. We will look at these in later subtopics.

WORKED EXAMPLE 14 Describing numerical distributions

Describe the distribution of the data shown in the following histogram.

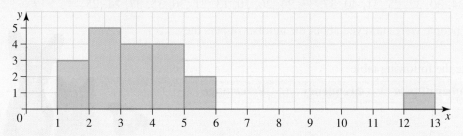

THINK	WRITE
1. Look for the mode and comment on its value.	The distribution has one mode with data values most frequently in the $2 \leq x < 3$ interval.
2. Identify the presence of any potential outliers.	There is one potential outlier in the interval between 12 and 13.
3. Describe the shape in terms of symmetry or skewness.	If we include the outlier, the data set can be described as positively skewed as it is clustered to the left. If we don't include the outlier, the distribution can be considered to be approximately symmetrical.

1.6 Exercise

1. **WE14** Describe the numerical distributions shown by the following histograms.

a.
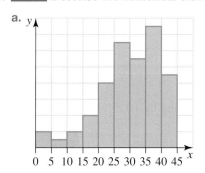

b.
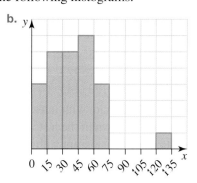

2. Describe the distribution of the following data sets after drawing histograms with intervals of 10, commencing with the smallest values.

   a. 105, 70, 140, 127, 132, 124, 122, 125, 123, 126, 107, 105, 104, 113, 125, 70, 88, 103, 107, 124, 122, 76, 103, 120, 112, 115, 123, 120, 117, 115, 107, 106, 120, 123, 115, 74, 128, 119

   b. 4, 18, 35, 26, 12, 25, 21, 34, 43, 37, 6, 25, 25, 23, 34, 38, 37, 22, 36, 31, 21, 28, 34, 30, 32, 25, 31, 40, 37, 33, 24, 26, 10, 13, 21, 36, 35, 37, 24, 25

3. A group of people were surveyed about the number of pets they owned.

   a. Complete the following table.

Number of pets	Frequency	Percentage
0–1		
2–3		30%
4–5	8	
More than 5	2	5%
Total		100%

   b. Describe the type of data collected in the survey.
   c. Calculate how many people were surveyed.
   d. Display the data in an appropriate graph.
   e. Describe the distribution.

4. The waiting time for patients to receive treatment in a hospital emergency department is shown in the following histogram.

   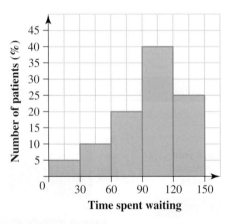

   a. According to the histogram, determine the percentage of patients who had to wait more than 90 minutes.
   b. Determine the percentage of patients who received treatment in less than one hour.
   c. Describe the distribution.

5. The systolic blood pressure readings for a group of 38 adults are listed below.
   118, 125, 130, 122, 123, 128, 135, 128, 117, 121, 123, 126, 129,
   142, 144, 148, 146, 122, 123, 118, 148, 126, 126, 144, 139, 147,
   144, 142, 118, 124, 122, 145, 144, 143, 124, 125, 140, 119

   a. Use the data to draw a histogram with intervals starting at 115 and increasing by 10.
   b. Describe the distribution shown in the histogram.
   c. Now use the data to draw a histogram with intervals starting at 115 and increasing by 3.
   d. Describe how the second histogram is different from the first.
   e. Describe what the second histogram might be demonstrating about this data.

6. The average price of a litre of petrol on 20 days during a three-month period is as shown.

$$\$ 1.67, \ \$ 1.77, \ \$ 1.61, \ \$ 1.78, \ \$ 1.73, \ \$ 1.56, \ \$ 1.66, \ \$ 1.63, \ \$ 1.82, \ \$ 1.72,$$
$$\$ 1.75, \ \$ 1.56, \ \$ 1.63, \ \$ 1.71, \ \$ 1.70, \ \$ 1.45, \ \$ 1.40, \ \$ 1.78, \ \$ 1.68, \ \$ 1.72$$

a. Use the data to draw a histogram with intervals starting at $1.40 and increasing by amounts of 5 cents.
b. Now use the data to draw a histogram with intervals starting at $1.40 and increasing by 10 cents.
c. Describe each display and comment on the differences.

7. The heights in centimetres of a sample of AFL footballers are shown below.

183, 182, 196, 175, 198, 186, 195, 184, 181,
193, 174, 181, 177, 194, 202, 196, 200, 176,
178, 188, 199, 204, 192, 193, 191, 183, 174,
187, 184, 176, 194, 195, 188, 180, 189, 191,
196, 189, 181, 181, 183, 185, 184, 185, 208

a. Use CAS to display the data as a histogram using:

i. intervals of 5 commencing with the smallest data value
ii. intervals of 10 commencing with the smallest data value
iii. intervals of 15 commencing with the smallest data value.

b. Describe each display and comment on the effect of changing the size of the intervals.

8. The average maximum temperature (in °C) in Victoria for 1993–2012 is shown in the following table.

Year	1993	1994	1995	1996	1997	1998	1999	2000	2001	2002
Temp.	22.3	22.6	21.8	22.1	22.4	22.7	22.1	22.7	23.1	23.1
Year	2003	2004	2005	2006	2007	2008	2009	2010	2011	2012
Temp.	22.7	23.4	23.4	23.1	22.7	22.1	22.9	22.6	22.6	22.7

Use CAS to display the data for the period 1993–2012 as a histogram using intervals of 0.4 °C commencing with the data value 20 °C. Describe the shape of the distribution.

1.6 Exam questions

Question 1 (1 mark)

Source: VCE 2019, Further Mathematics Exam 1, Section A, Q2; © VCAA.

MC The histogram below shows the distribution of the *population size* of 48 countries in 2018.

The shape of this histogram is best described as

A. positively skewed with no outliers.
B. positively skewed with outliers.
C. approximately symmetric.
D. negatively skewed with no outliers.
E. negatively skewed with outliers.

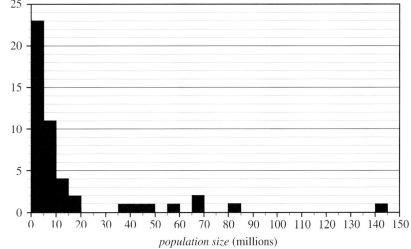

Data: Worldometers,

Question 2 (1 mark)

MC Describe the shape of the distribution of the data in the graph.
  A. Symmetrical
  B. Positively skewed
  C. Negatively skewed
  D. Bimodal
  E. Negatively skewed with an outlier

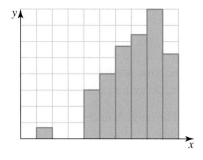

Question 3 (1 mark)

MC Describe the shape of the distribution of the data in the graph.
  A. Symmetrical
  B. Positively skewed
  C. Negatively skewed
  D. Bimodal
  E. Positively skewed with an outlier

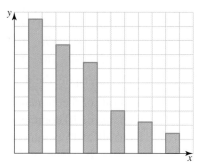

More exam questions are available online.

1.7 Summarising numerical data — mean and median

LEARNING INTENTION

At the end of this subtopic you should be able to:
  • summarise numerical data distributions by calculating mean and median.

1.7.1 The mean

One way to make raw numerical data more useful is to summarise the data. One tool we can use for summarising data is the **mean**.

The mean (also known as the average) is one way of determining the centre of the data. To calculate the mean, all the data values are added together and then divided by the number of data values.

Calculating the mean

$$\bar{x} = \frac{\text{sum of data values}}{\text{number of data values}}$$

$$= \frac{\sum x}{n}$$

$$\text{or } \mu = \left(\frac{\sum x}{n}\right)$$

The symbol for mean is $\bar{x}$, which is called x bar.

The symbol μ, which is called mu, can also represent mean.

The symbol $\sum$, which is called sigma, represents the sum, and the symbol n represents the number of data values.

tlvd-3691

WORKED EXAMPLE 15 Calculating the mean of the data

Calculate the mean of the following data set, correct to 2 decimal places.

6, 3, 4, 5, 7, 7, 4, 8, 5, 10, 6, 10, 9, 8, 3, 6, 5, 4

THINK	WRITE
1. Calculate the sum of the data values.	$6+3+4+5+7+7+4+8+5+10+6+10+9+$ $8+3+6+5+4=110$
2. Divide the sum by the number of data values.	$\bar{x} = \dfrac{\text{Sum of data values}}{\text{Number of data values}}$ $= \dfrac{110}{18}$ $= 6.111...$
3. Write the answer.	The mean of the data set is 6.11.

TI | THINK	DISPLAY/WRITE	CASIO | THINK	DISPLAY/WRITE
1. In a Lists & Spreadsheet page, label the first column as x. Enter the given data values in the first column.		1. On a Statistics screen, relabel list 1 as x. Enter the given data values in the first column.	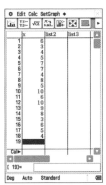
2. In a Calculator page, press MENU and select: 6: Statistics 1: Stat Calculations 1: One-Variable Statistics… When prompted, enter the number of lists as 1 and then select OK. Complete the fields as: X1 List: x Frequency List: 1 then select OK.	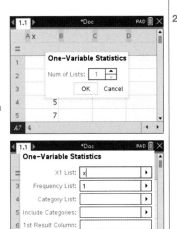	2. Select: • Calc • One-Variable Complete the fields as: XList: main\x Freq: 1 then select OK.	

3. The mean is displayed on the screen as $\bar{x}$.

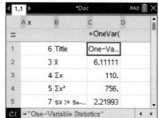

The mean is 6.11 (2 decimal places).

3. The mean is displayed on the screen as $\bar{x}$.

The mean is 6.11 (2 decimal places).

1.7.2 The median

Another tool we can use for summarising data is the **median**.

The median (also known as the middle value) is another way of determining the centre of the data. To calculate the median, all the data values are placed in order. If there is an odd number of data values (n is odd), then the median is the middle value. If there is an even number of data values (n is even), then the median is the average of the two middle values.

Calculating the median

To determine the position of the median, use this rule:

$$\text{Median} = \left(\frac{n+1}{2}\right) \text{th position}$$

WORKED EXAMPLE 16 Calculating the median of the data

Calculate the median of the following data sets.
a. 5, 3, 4, 5, 7, 7, 4, 8, 5, 10, 6, 10, 9, 8, 3, 6, 5, 4
b. 16, 3, 4, 5, 17, 27, 14, 18, 15, 10, 6, 10, 9, 8, 23, 26, 35

THINK	WRITE
a. 1. Put the data set in order from lowest to highest.	a. 3, 3, 4, 4, 4, 5, 5, 5, 5, 6, 6, 7, 7, 8, 8, 9, 10, 10
2. Identify the data value in the $\left(\frac{n+1}{2}\right)$ th position.	There are 18 data values, so the median will be in position $\left(\frac{18+1}{2}\right) = 9.5$, or halfway between position 9 and position 10.
	3, 3, 4, 4, 4, 5, 5, 5, $\boxed{5, 6,}$ 6, 7, 7, 8, 8, 9, 10, 10
	median = 5.5
3. Write the answer.	The median of the data set is 5.5.

b. 1. Put the data set in order from lowest to highest.

b. 3, 4, 5, 6, 8, 9, 10, 10, 14, 15, 16, 17, 18, 23, 26, 27, 35

2. Identify the data value in the $\left(\dfrac{n+1}{2}\right)$ th position.

There are 17 data values, so the median will be in position $\left(\dfrac{17+1}{2}\right) = 9$.

3, 4, 5, 6, 8, 9, 10, 10, 14, 15, 16, 17, 18, 23, 26, 27, 35

median = 14

3. Write the answer.

The median of the data set is 14.

TI	THINK	DISPLAY/WRITE	CASIO	THINK	DISPLAY/WRITE

a. 1. In a Lists & Spreadsheet page, label the first column as x. Enter the given data values in the first column.

a. 1. On a Statistics screen, relabel list 1 as x. Enter the given data values in the first column.

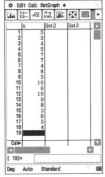

2. In a Calculator page, press MENU and select:
6: Statistics
1: Stat Calculations
1: One-Variable Statistics…
When prompted, enter the number of lists as 1.
Complete the fields as: X1 List: x
Frequency List: 1
then select OK.

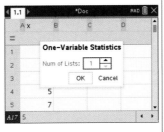

2. Select:
• Calc
• One-Variable
Complete the fields as:
XList: main\x
Freq: 1
then select OK.

3. The median is displayed on the screen as Median X.

The median is 5.5.

The median is displayed on the screen as Med.

The median is 5.5.

1.7.3 Choosing between the mean and the median

To decide whether to choose the mean or the median to describe a data set, two things need to be considered: outliers and shape.

Outliers

If the data has outliers, the median is the better choice to measure the centre of the data. This is because outliers have a strong influence on the mean but little to no effect on the median (unless there is more than one outlier). Here is an example that shows the difference:

Data A (no outlier): 3, 4, 5, 6, 7, 8, 9

Data B (with an outlier): 3, 4, 5, 6, 7, 8, 90

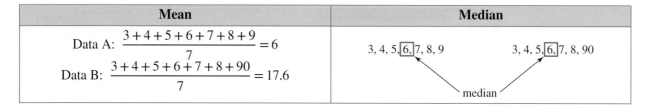

Mean	Median
Data A: $\dfrac{3+4+5+6+7+8+9}{7} = 6$ Data B: $\dfrac{3+4+5+6+7+8+90}{7} = 17.6$	3, 4, 5, 6, 7, 8, 9 3, 4, 5, 6, 7, 8, 90 median

In this example, the mean was impacted by the outlier (changed from 6 to 17.6) but the median did not change. This shows that the median is a better choice for data with an outlier.

Shape

The shape of data also impacts the choice between the mean and the median.

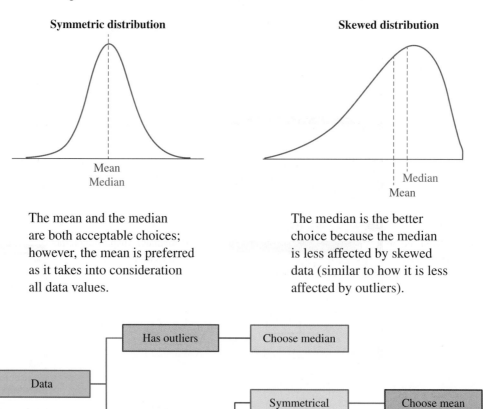

Symmetric distribution

Mean
Median

The mean and the median are both acceptable choices; however, the mean is preferred as it takes into consideration all data values.

Skewed distribution

Median
Mean

The median is the better choice because the median is less affected by skewed data (similar to how it is less affected by outliers).

Data → Has outliers → Choose median

Data → No outliers → Symmetrical → Choose mean

No outliers → Skewed → Choose median

The following histogram represents the IQ test results for a group of people.

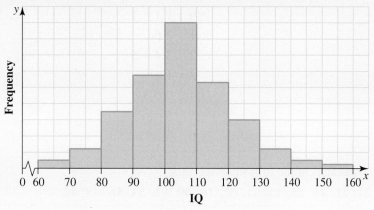

Determine which measure of centre is best to represent the data set.

THINK	WRITE
1. Look at the distribution of the data set.	The data set is approximately symmetrical and has no outliers.
2. If the data set is approximately symmetrical with no outliers, the mean is probably the best measure of centre. If there are outliers or the data is skewed, the median is probably the best measure of centre to use. State the answer.	The mean is the best measure of centre to represent this data set.

 Resources

 Interactivity Mean, median, mode and quartiles (int-6496)

1.7 Exercise

Students, these questions are even better in jacPLUS

 Receive immediate feedback and access sample responses

 Access additional questions

 Track your results and progress

Find all this and MORE in jacPLUS

1. **WE15** Calculate the mean of the following data set.

108, 135, 120, 132, 113, 138, 125, 138, 107, 131,
113, 136, 119, 152, 134, 158, 136, 132, 113, 128

2. a. Calculate the mean of the following data set correct to 2 decimal places.

25, 23, 24, 25, 27, 26, 23, 28, 24, 20, 25, 20, 29, 28, 23, 27, 24

  b. Replace the highest value in the data set from part a with the number 79, and then calculate the mean again, correct to 2 decimal places.

  c. Describe how changing the highest value in the data set affected the mean.

3. Calculate the mean of the following data set correct to 2 decimal places.

Key: 1|2 = 12

Stem	Leaf
0	1 1 5 7
1	2 6
2	3 4 4 5
3	1 3
4	0 0 3
5	5
6	5

4. **WE16** Calculate the medians of the following data sets.

a. 15, 3, 54, 53, 27, 72, 41, 85, 15, 11, 62, 16, 49, 81, 53, 56, 75, 42

b. 126, 301, 422, 567, 179, 267, 149, 198, 165, 170, 602, 180, 109, 85, 223, 206, 335

5. Answer the following questions.

a. Calculate the mean (correct to 2 decimal places) and median for the following data set.

Average annual rainfall in selected Australian cities	
City	**Rainfall (mm)**
Sydney	1276
Melbourne	654
Brisbane	1194
Adelaide	563
Perth	745
Hobart	576
Darwin	1847
Canberra	630
Alice Springs	326

b. Determine the most appropriate measure of centre to represent this data.

6. Answer the following questions.

a. Calculate the median of the following data set.

21, 22, 23, 24, 27, 26, 22, 27, 23, 21, 24, 20, 31, 25, 24, 28, 23

b. Replace the highest value in the data set from part **a** with the number 96 and then calculate the median again.

c. Determine how changing the highest value in the data set affected the median.

7. Calculate the median of the following data sets.

a.

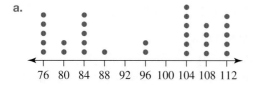

b. 1.02, 2.01, 3.21, 4.63, 1.49, 3.45, 1.17,
1.38, 1.47, 1.70, 5.02, 1.38, 1.91, 8.54

8. **WE17** The following stem plot represents the lifespan of different animals at an animal sanctuary. Determine which measure of centre is best to represent the data set.

Key: $1|2 = 12$

Stem	Leaf
0	3 5 9
1	2 4 6 8
2	0 1 4 5 5 7 9
3	0 2 6
4	
5	
6	0 3

9. The following data set represents the salaries (in $000s) of workers at a small business.

$$45, \quad 50, \quad 55, \quad 55, \quad 55, \quad 60, \quad 65, \quad 65, \quad 70, \quad 70, \quad 75, \quad 80, \quad 220$$

   a. Calculate the mean of the salaries correct to the nearest whole number.
   b. Calculate the median of the salaries.
   c. When it comes to negotiating salaries, the workers want to use the mean to represent the data and the management want to use the median. Explain why this might be the case.

10. On a particular weekend, properties sold at auction for the following 30 prices:

$4 700 000,	$3 160 000,	$2 725 000,	$2 616 000,	$2 560 000,	$241 000,
$265 000,	$266 000,	$310 000,	$320 000,	$3 010 000,	$2 580 000,
$2 450 000,	$2 300 000,	$2 275 000,	$286 000,	$325 000,	$330 000,
$435 500,	$456 000,	$1 350 000,	$1 020 000,	$900 000,	$735 000,
$733 000,	$305 000,	$330 000,	$347 000,	$357 000,	$408 000

   a. Calculate the mean and median for the data.
   b. Draw a histogram of the data using intervals commencing at the lowest value and increasing by amounts of $250 000.
   c. Mark the location of the mean and median on the histogram.
   d. Determine the more appropriate measure of centre to represent this data.

11. The waiting times in minutes and seconds for a group of passengers at a railway station are:

19:28, 17:35, 14:21, 16:22, 12:18, 11:09, 13:15, 16:21, 11:45, 12:26, 14:16, 17:12,
13:42, 14:51, 15:26, 15:13, 18:02, 11:22, 12:26, 13:10, 13:18, 13:41, 13:23, 14:06

Calculate the mean of the data correct to the nearest second.

12. The winning margins in the NRL over a particular period of time were as follows.

Winning margin	Frequency
2	4
4	12
6	8
8	5
10	4
12	4
16	1
20	1
34	1

   a. Calculate the mean and the median.
   b. Determine the most appropriate measure of centre for this data set. Explain why.

1.7 Exam questions

▶ **Question 1 (2 marks)**

Source: VCE 2020, Further Mathematics Exam 2, Section A, Q4; © VCAA.

The *age*, in years, *body density*, in kilograms per litre, and *weight*, in kilograms, of a sample of 12 men aged 23 to 25 years are shown in the table below.

Age (years)	Body density (kg/litre)	Weight (kg)
23	1.07	70.1
23	1.07	90.4
23	1.08	73.2
23	1.08	85.0
24	1.03	84.3
24	1.05	95.6
24	1.07	71.7
24	1.06	95.0
25	1.07	80.2
25	1.09	87.9
25	1.02	94.9
25	1.09	65.3

For these 12 men, determine

    a. their median *age*, in years   **(1 mark)**

    b. the mean of their body *density*, in kilograms per litre.   **(1 mark)**

▶ **Question 2 (1 mark)**

Source: VCE 2020, Further Mathematics Exam 2, Section A, Q1; © VCAA.

Body mass index (*BMI*), in kilograms per square metre, was recorded for a sample of 32 men and displayed in the ordered stem plot below.

```
Key: 21|6 = 21.6   n = 32
21 | 6  9  9
22 | 1  2  5  6
23 | 0  1  4  6  6  7  8
24 | 4  5  6  7  7  9
25 | 6  8
26 | 1  7  9
27 | 3  7
28 | 2
29 | 1  8
30 | 4
31 | 1
```

Determine the median *BMI* for this group of men.

Question 3 (2 marks)

Source: VCE 2015, Further Mathematics Exam 1, Section A, Q2; © VCAA.

MC For an ordered set of data containing an odd number of values, the middle value is always

  A. the mean.

  B. the median.

  C. the mode.

  D. the mean and the median.

  E. the mean, the median and the mode.

More exam questions are available online.

1.8 Summarising numerical data — range, interquartile range and standard deviation

LEARNING INTENTION

At the end of this subtopic you should be able to:
- summarise numerical data distributions by calculating range, interquartile range (IQR) and standard deviation.

1.8.1 Range

Range is another tool we can use to make raw numerical data more useful. Range is one way of measuring the spread of the data. We use measures of spread to tell us how far apart the values are from one another.

For example, these two sets of data have the same values of mean (50) and median (44), but they are very different data sets. Range helps us describe that difference.

Data set 1: 36, 43, 44, 59, 68

Data set 2: 1, 2, 44, 80, 123

Range

Range is determined by calculating the difference between the largest and smallest values in the data set.

$$\text{Range} = \text{largest value} - \text{smallest value}$$

WORKED EXAMPLE 18 Calculating the range of a set of data

For the following sets of data, calculate the range.

a. 36, 43, 44, 59, 68
b. 1, 2, 44, 80, 123

THINK

a. 1. The largest value is 68 and the smallest value is 36. Substitute these values into the range formula.

WRITE

a. Range = largest value − smallest value

     = 68 − 36

     = 32

▶

b. 1. The largest value is 123 and the smallest value is 1. Substitute these values into the range formula.

b. Range = largest value − smallest value
$$= 123 - 1$$
$$= 122$$

1.8.2 Interquartile range

Another measure of spread is **interquartile range**. Interquartile range tells us how spread out the data is relative to the centre.

To calculate interquartile range, first the data is split into quarters called **quartiles**. To do this, first the median is found and the data is split into lower and upper half. The median of the lower half is then calculated; this is called the **lower quartile** (Q_1). The median of the upper half is then calculated; this is called the **upper quartile** (Q_3). Note that the median is the Q_2 value, but we do not need it for this calculation.

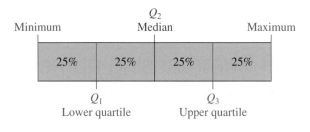

The interquartile range (IQR) is determined by calculating the difference between Q_3 and Q_1.

Interquartile range

$$IQR = Q_3 - Q_1$$

where Q_1 is the lower quartile and Q_3 is the upper quartile of the data set.

tlvd-3692

WORKED EXAMPLE 19 Calculating the interquartile range of a set of data

Calculate the interquartile range of the following set of data.

$$23, 34, 67, 17, 34, 56, 19, 22, 24, 56, 56, 34, 23, 78, 22, 16, 15, 35, 45$$

THINK	WRITE
1. Put the data in order.	15, 16, 17, 19, 22, 22, 23, 23, 24, 34, 34, 34, 35, 45, 56, 56, 56, 67, 78
2. Identify the median.	There are 19 data values, so the median will be in position $\left(\dfrac{19+1}{2}\right) = 10$

median
↓

15, 16, 17, 19, 22, 22, 23, 23, 24, ⟨34⟩
The median is 34.

3. Identify Q_1 by calculating the median of the lower half of the data.

There are 9 values in the lower half of the data, so Q_1 will be the 5th of these values

$$Q_1$$

15, 16, 17, 19, (22,) 22, 23, 23, 24

$Q_1 = 22$

4. Identify Q_3 by calculating the median of the upper half of the data.

There are 9 values in the upper half of the data, so Q_3 will be the 5th of these values.

$$Q_3$$

34, 34, 35, 45, (56,) 56, 56, 67, 78

$Q_3 = 56$

5. Calculate the interquartile range using $IQR = Q_3 - Q_1$.

$$IQR = Q_3 - Q_1$$
$$= 56 - 22$$
$$= 34$$

6. Write the answer in a sentence.

The interquartile range is 34.

TI	THINK	DISPLAY/WRITE

1. In a Lists & Spreadsheet page, label the first column as x. Enter the given data values in the first column.

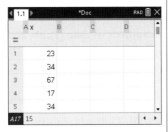

2. In a Calculator page, press MENU and select:
6: Statistics
1: Stat Calculations
1: One-Variable Statistics…
When prompted, enter the number of lists as 1.
Complete the fields as:
X1 List: x
Frequency List: 1
then select OK.

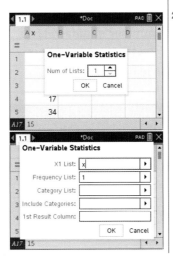

CASIO	THINK	DISPLAY/WRITE

1. On a Statistics screen, relabel list 1 as x. Enter the given data values in the first column.

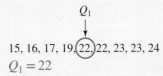

2. Select:
 • Calc
 • One-Variable
Complete the fields as:
X List: main\x
Freq:1
then select OK.

3. The first quartile is displayed as Q_1X and the third quartile is displayed as Q_3X.

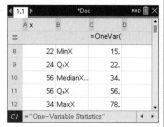

3. The first quartile is displayed as Q_1 and the third quartile is displayed as Q_3.

4. Calculate the interquartile range.

$$IQR = Q_3 - Q_1$$
$$= 56 - 22$$
$$= 34$$

Calculate the interquartile range.

$$IQR = Q_3 - Q_1$$
$$= 56 - 22$$
$$= 34$$

1.8.3 Standard deviation

Another measure of spread is **standard deviation**. Standard deviation gives a measure of spread from the mean. The value of the standard deviation indicates the distance from the centre that will include most data values. We will look at this in more detail in in the next subtopic.

Standard deviation is calculated by first summing the square of the difference between each data point and the mean, then dividing this number by one less than the number of data points and finally taking the square root. This calculation is usually done by using technology, but it is possible to do by hand using the following formula.

Sample standard deviation

$$s = \sqrt{\frac{\sum (x - \bar{x})^2}{n - 1}}$$

where $\bar{x}$ is the mean of the data values and n is the number of data values.

Note that we will only look at sample standard deviation in this course.

WORKED EXAMPLE 20 Calculating standard deviation of a data set

Calculate the standard deviation of the following set of data. Give your answer to 2 decimal places.
5, 9, 13, 25

THINK

1. Calculate the mean by substituting the values in the mean formula.

2. Calculate the sum of the difference between each point and the mean by using the formula $\sum (x - \bar{x})^2$.

WRITE

$$\bar{x} = \frac{\text{sum of data values}}{\text{number of data values}}$$
$$= \frac{5 + 9 + 13 + 25}{4}$$
$$= 13$$

$$\sum (x - \bar{x})^2 = (5 - 13)^2 + (9 - 13)^2 + (13 - 13)^2 + (25 - 13)^2$$
$$= 224$$

3. Substitute the values in the sample standard deviation formula to calculate the value of s.

$$s = \sqrt{\frac{\sum (x - \bar{x})^2}{n - 1}}$$
$$= \sqrt{\frac{224}{4 - 1}}$$
$$= 8.6401$$

4. Write the answer in a sentence.

The sample standard deviation is 8.64.

TI | THINK	DISPLAY/WRITE	CASIO | THINK	DISPLAY/WRITE
1. On a Lists & Spreadsheet page, enter the data and give it the title *data*.	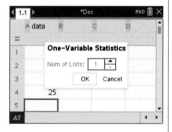	1. On the Statistic screen, label list1 as *score* and enter the values from the question. Press EXE after entering each value.	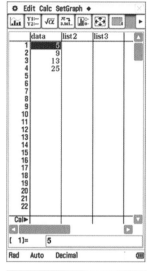
2. Press MENU, then select: 1: Stat calculations 4: Statistics 1: One-variable statistics For Num of Lists, leave the option as '1' and press ENTER.		2. To calculate the summary statistics, tap: • Calc • One-Variable	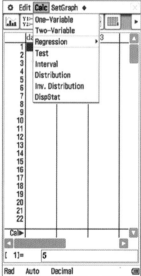

3. In X1 List select and choose the name of your list: *data*, then press ENTER.

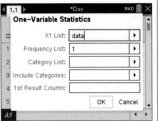

3. Set values as:
    - XLlist: main\score
    - Freq: 1

4. In column B, find the row labelled sx := s_n and read the value in column C. This is the standard deviation.

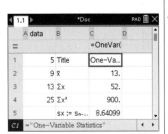

The standard deviation is 8.64 (rounded to 2 decimal places).

4. Tap OK.
    The standard deviation is shown as s_x.

The standard deviation is 8.64 (rounded to 2 decimal places).

 Resources

 Interactivities The median, the interquartile range, the range and the mode (int-6244)

The mean and the standard deviation (int-6246)

1.8 Exercise

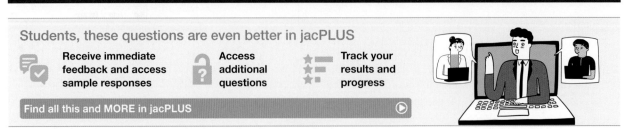

Students, these questions are even better in jacPLUS

Receive immediate feedback and access sample responses

Access additional questions

Track your results and progress

Find all this and MORE in jacPLUS

1. **WE18** Calculate the range of the following set of data.

421, 331, 127, 105, 309, 512, 129, 232, 124, 154, 246, 124, 313, 218, 112, 136, 155, 305, 415

2. **WE19** Calculate the interquartile range of the following set of data.

$$3.11, \ 3.16, \ 1.13, \ 1.56, \ 3.19, \ 4.43, \ 1.98, \ 4.89,$$
$$2.12, \ 4.78, \ 3.21, \ 8.88, \ 1.21, \ 5.67, \ 2.22, \ 3.34$$

3. The results for a multiple-choice test for 20 students in two different classes are as follows.

Class A: 7, 13, 14, 13, 14, 14, 12, 8, 18, 13, 14, 12, 16, 14, 12, 11, 13, 14, 13, 15

Class B: 18, 19, 12, 12, 11, 17, 9, 18, 17, 14, 13, 11, 17, 13, 17, 14, 14, 15, 13, 12

   a. Determine the spread of the marks for each class by using the range.
   b. Determine the spread of the marks for each class by using the interquartile range.

4. The competition ladder of the Australian and New Zealand netball championship is as follows.

Position	Team	Win	Loss	Goals for	Goals against
1	Adelaide Thunderbirds	12	1	688	620
2	Melbourne Vixens	9	4	692	589
3	Waikato BOP Magic	9	4	749	650
4	Queensland Firebirds	9	4	793	691
5	Central Pulse	8	5	736	706
6	Southern Steel	6	7	812	790
7	West Coast Fever	5	8	715	757
8	NSW Swifts	4	9	652	672
9	Canterbury Tactix	2	11	700	882
10	Northern Mystics	1	12	699	879

   a. Calculate the spread for the 'Goals for' column by using the range.
   b. Calculate the spread for the 'Goals for' column by using the interquartile range.
   c. Compare the spread of the 'Goals for' column with the spread of the 'Goals against' column.

5. **WE20** Calculate the standard deviation of the following set of data. Give your answer to 2 decimal places.

$$10, \ 11, \ 19, \ 21$$

6. A survey of the number of motor vehicles that pass a school between 8.30 am and 9.30 am on 10 days during a term are as follows:

$$72, \ 89, \ 94, \ 78, \ 83, \ 84, \ 88, \ 97, \ 82, \ 88$$

   a. Use CAS to calculate the standard deviation of the sample correct to 2 decimal places.
   b. Calculate the interquartile range of the sample.
   c. The lowest number is reduced by 10 and the highest value increased by 10.
      Recalculate the values of the standard deviation and interquartile range.
   d. State how each of the measures is affected by the change in the values.

7. Data collected on the number of daylight hours in Alice Springs is as shown.

10.3, 9.8, 9.6, 9.5, 8.5, 8.4, 9.1, 9.8, 10.0, 10.0, 10.1, 10.0, 10.1, 10.1, 10.6, 8.7, 8.8, 9.0, 8.0, 8.5, 10.6, 10.8, 10.5, 10.9, 8.5, 9.5, 9.3, 9.0, 9.4, 10.6, 8.3, 9.3, 9.0, 10.3, 8.4, 8.9

   a. Calculate the range of the data.
   b. Calculate the interquartile range of the data.
   c. Comment on the difference between the two measures and what this indicates.

8. The table shows the number of registered passenger vehicles in two particular years for the states and territories of Australia.

Number of passenger vehicles		
State	**Year 1**	**Year 2**
New South Wales	3 395 905	3 877 515
Victoria	2 997 856	3 446 548
Queensland	2 138 364	2 556 581
South Australia	915 059	1 016 590
Western Australia	1 205 266	1 476 743
Tasmania	271 365	305 913
Northern Territory	73 302	91 071
Australian Capital Territory	191 763	229 060

a. Calculate the interquartile range and standard deviation (correct to 1 decimal place) for both years.
b. Recalculate the interquartile range and standard deviation for both years after removing the three smallest values.
c. Comment on the effect of the removal of the three smallest values on the interquartile ranges and standard deviations.

9. The volume of wine ('000 litres) available for consumption in Australia for a random selection of months over a 10-year time period is shown in the following table.

38 595	41 301	44 212	39 362	38 914	38 273	39 456	38 823
41 123	42 981	44 567	41 675	41 365	42 845	43 987	41 583
39 347	42 673	44 835	39 773	38 586	38 833	39 756	39 095
42 946	46 382	44 892	41 038	41 402	42 587	43 689	41 209

a. Use CAS to calculate the mean and standard deviation of the data correct to 2 decimal places.
b. Calculate the median and interquartile range of the data.
c. Calculate the percentage, correct to 2 decimal places, of the actual data values from the sample that are within one standard deviation of the mean (i.e. between the number obtained by subtracting the standard deviation from the mean and the number obtained by adding the standard deviation to the mean).
d. Calculate the percentage of the actual data values from the sample that are between the first and third quartiles.
e. Comment on the differences between your answers for parts c and d.

10. A random sample of monthly consumer price indices in various cities of Australia is shown in the following table. Answer the following questions, giving answers correct to 2 decimal places where appropriate.

Sydney	Melbourne	Brisbane	Adelaide	Perth	Hobart	Darwin	Canberra
0.4	0.8	0.9	0.7	0.6	0.3	1.2	0.8
0.9	1.0	1.1	1.1	0.8	0.8	0.3	1.0
1.4	1.3	1.3	1.5	1.4	1.3	0.9	1.4
1.5	1.2	1.7	1.3	1.6	1.0	1.5	1.2
1.1	1.2	1.4	1.3	1.0	1.1	1.8	1.5
0.1	0.3	0.2	0.2	0.1	0.2	0.1	0.3
0.4	0.3	0.5	0.5	0.9	0.5	1.1	0.6
1.1	0.5	1.4	1.1	0.8	1.2	1.9	0.9
0.5	0.6	0.3	0.4	0.5	0.6	0.1	0.4
0.8	1.3	0.7	0.5	1.2	0.7	0.5	0.6

a. Use CAS to calculate the standard deviation and interquartile range of the entire data set.
b. Use CAS to calculate the standard deviation and interquartile range for each city.
c. State which city bears the closest similarity to the entire data set.
d. State which city bears the least similarity to the entire data set.

1.8 Exam questions

Question 1 (1 mark)

Source: VCE 2020, Further Mathematics Exam 2, Section A, Q3; © VCAA.

In a study of the association between **BMI** and *neck size*, 250 men were grouped by *neck size* (below average, average and above average) and their **BMI** recorded.

Five-number summaries describing the distribution of **BMI** for each group are displayed in the table below along with the group size.

The associated boxplots are shown below the table.

Neck size	Group size	BMI (kg/m^2)				
		Min.	Q_1	Median	Q_3	Max.
below average	50	18.1	20.6	21.6	23.2	26.8
average	124	19.8	23.4	24.6	26.0	33.9
above average	76	23.1	26.25	28.1	29.95	39.1

What is the interquartile range (IQR) of *BMI* for the men with an average neck size?

Question 2 (1 mark)

Source: VCE 2019, Further Mathematics Exam 1, Section A, Q5; © VCAA.

MC The stem plot shows the distribution of mathematics *test scores* for a class of 23 students.

For this class, the interquartile range (IQR) of *test scores* is
  A. 14.5
  B. 17.5
  C. 18
  D. 24
  E. 49

key: $4|2 = 42$ $n = 23$

```
4 | 0 1 4 4
5 | 2 7 9 9 9
6 | 5 6 8 8 9 9
7 | 0 0 5 6 7 8
8 | 5 9
```

Question 3 (1 mark)

MC Thirty Year 11 students were asked to measure the number of hours of homework they completed each week. The data are recorded below.

12	11	10	11	9	8	5	13	14	12	12	11	10	11	9
8	5	13	14	12	12	11	10	11	9	8	5	13	3	2

The standard deviation, correct to the nearest whole number, for this data set is
  A. 4   B. 10   C. 11   D. 12   E. 3

More exam questions are available online.

1.9 Symmetrical and asymmetrical distributions

1.9.1 Symmetrical and asymmetrical summary statistics

Symmetrical distributions

Symmetrical distributions are sets of data that, when displayed, show approximately the same shape and same amount of data to the left and right of the centre.

Here are some examples of approximately symmetrical data including a histogram, a stem plot, a dot plot and a histogram with a curved line.

Histogram	Stem plot
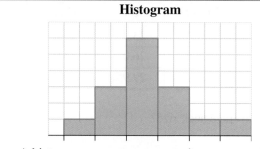	**Stem** **Leaf** 0* 8 8 1 3 4 4 1* 6 6 7 7 9 2 1 1 2* 5
A histogram can represent continuous numerical data.	**Stem plots** (or stem-and-leaf plots) can be used to display both discrete and continuous numerical data.
Dot plot	**Histogram with a curved line**
	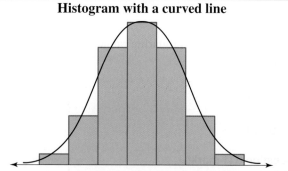
Discrete numerical and categorical data can also be displayed as a dot plot.	This example is called a **normal distribution** and is a classic example of a symmetrical distribution.

When summarising a symmetrical distribution, the best measure of centre is the mean and the best measure of spread is typically the standard deviation.

Asymmetrical distributions

Asymmetrical distributions are sets of data that, when displayed, show different shapes and different amounts of data above and below the centre. Some examples of asymmetrical data are as follows.

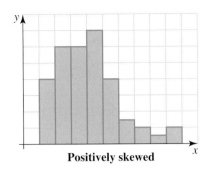

Positively skewed

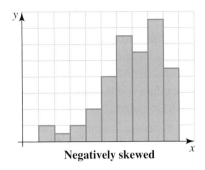

Negatively skewed

As discussed earlier, the distributions with higher frequencies on the left side of the graph are positively skewed, while those with higher frequencies on the right side are negatively skewed. Some examples of positively and negatively skewed stem and dot plots are as follows.

Positively skewed		Negatively skewed	
Stem	Leaf	Stem	Leaf
2	2 5 6	1	8
3	3 3 4 5 6 7 7 8 9	2	2
4	2 5 5 6 7 7 7 8 8 9	3	3 9
5	2 4 6 8 8	4	2 9
6	5 8 9	5	2 4 6 8 8
7	4 5	6	5 6 7 8 9
8	7	7	2 2 3 4 4 5 9
9	9	8	2 7

Positively skewed	Negatively skewed
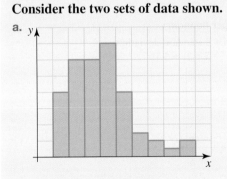	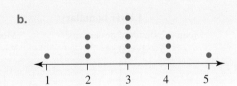

When summarising an asymmetrical distribution, the best measure of centre is the median and the best measure of spread is the interquartile range.

WORKED EXAMPLE 21 Determining type of distribution and summary statistics

Consider the two sets of data shown.

a.

b.

For the given sets of data:
 i. state whether the distribution is symmetrical or asymmetrical
 ii. state the best measure of centre (mean or median) and spread (standard deviation or interquartile range).

THINK	**WRITE**
a. i. 1. The data is asymmetrical because it is not evenly distributed above and below the centre.	**a. i.** Asymmetrical (positively skewed)
ii. 2. For asymmetrical distributions the best measure of centre is the median and the best measure of spread is the interquartile range.	**ii.** Median and interquartile range
b. i. 1. The data is symmetrical because it is evenly distributed above and below the centre.	**b. i.** Symmetrical
ii. 2. For symmetrical distributions the best measure of centre is the mean and the best measure of spread is the standard deviation.	**ii.** Mean and interquartile range

1.9.2 Using the standard deviation

The standard deviation (s) can be used to reveal how much data is located around the centre (mean, $\bar{x}$) of a symmetrical data distribution. For the following example, we will investigate this set of data with a mean of 3 and a standard deviation of 1.08.

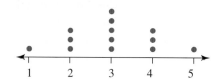

To examine the amount of data within one standard deviation, first calculate the lower boundary by subtracting the standard deviation from the mean, and then calculate the upper boundary by adding the standard deviation to the mean:

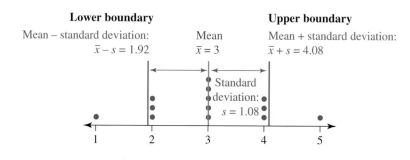

$$\text{lower boundary} = \bar{x} - s$$
$$\text{upper boundary} = \bar{x} + s$$

In this example, 11 dots are within the boundaries.

To examine the amount of data within multiple standard deviations, use the same process as for one standard deviation, but this time multiply s by the number of multiples required.

Boundaries for two standard deviations:

$$\text{lower boundary} = \bar{x} - 2 \times s$$
$$\text{upper boundary} = \bar{x} + 2 \times s$$

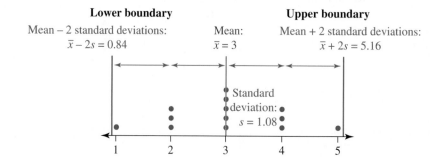

In this example, 13 dots are within the boundaries.

Boundaries for three standard deviations:

$$\text{lower boundary} = \bar{x} - 3 \times s$$
$$\text{upper boundary} = \bar{x} + 3 \times s$$

And so on.

Once the boundaries have been found, the amount of data within the boundaries can be given as a percentage using:

$$\text{percentage data} = \frac{\text{number of data within boundaries}}{\text{total number of data}} \times 100$$

WORKED EXAMPLE 22 Using the standard deviation

This is a dot plot with a mean of 4.67 and a standard deviation of 1.93.

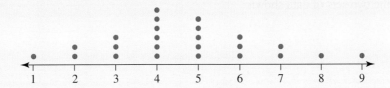

a. **Calculate the lower boundary for two standard deviations.**
b. **Calculate the upper boundary for two standard deviations.**
c. **Calculate the percentage of data within two standard deviations of the mean. Give your answer to 2 decimal places.**

THINK

a. 1. The mean is $\bar{x} = 4.67$ and the standard deviation is $s = 1.93$.
The lower boundary for two standard deviations is the mean minus two times the standard deviation.

b. 1. The mean is $\bar{x} = 4.67$ and the standard deviation is $s = 1.93$.
The upper boundary for two standard deviations is the mean plus two times the standard deviation.

c. 1. The number of data values within the boundaries is the number of dots between 0.81 and 8.53.

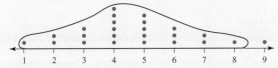

2. The total number of data values is 24. We can calculate the percentage.

WRITE

a. Lower boundary $= \bar{x} - 2 \times s$
$= 4.67 - 2 \times 1.93$
$= 0.81$

b. Upper boundary $= \bar{x} + 2 \times s$
$= 4.67 + 2 \times 1.93$
$= 8.53$

c. The number of data values within the boundaries is 23.

$$\text{Percentage data} = \frac{\text{number of data within boundaries}}{\text{total number of data}} \times 100$$

$$= \frac{23}{24} \times 100$$

$$= 95.83\%$$

| 3. Write the answer in a sentence, correct to 2 decimal places. | 95.83% of the data is within two standard deviations of the mean. |

1.9 Exercise

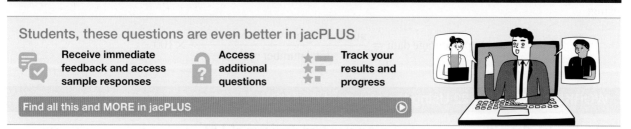

1. **WE21** Consider the two sets of data shown.

    a.
Stem	Leaf
7	5
8	9
9	2 9
10	1 2
11	0 2 3 4
12	0 1 3
13	2

    b.

    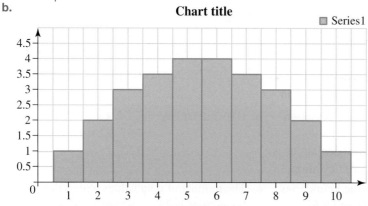

    Chart title

    For each of these sets of data, determine:

    i. whether the distribution is symmetrical or asymmetrical

    ii. the best measure of centre (mean or median) and spread (standard deviation or interquartile range).

2. A class survey of the number of red pens in each person's pencil case was conducted. These were the results:

    $$0, 0, 1, 3, 4, 2, 1, 1, 4, 3, 1, 0, 2, 2, 3, 2, 3, 4, 2, 8, 10$$

    a. Draw these results on a dot plot.
    b. State whether the distribution is symmetrical or asymmetrical.
    c. State the best measure of centre (mean or median) and spread (standard deviation or interquartile range) for this data.

3. **WE22** This is a dot plot with mean 12 and standard deviation 1.45.

   a. Calculate the lower boundary for two standard deviations from the mean.
   b. Calculate the upper boundary for two standard deviations from the mean.
   c. Calculate the percentage of data within two standard deviations of the mean. Give your answer to 2 decimal places.

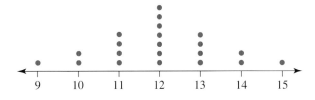

4. Students in a General Mathematics class discuss how many times they have bought lunch from the canteen this year. The results are shown in the dot plot. The mean is 6 and the standard deviation is 2.88.

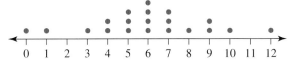

   a. Calculate the percentage of data within one standard deviation of the mean.
   b. Calculate the percentage of data within two standard deviations of the mean.
   c. Calculate the percentage of data within three standard deviations of the mean.
   d. Without doing any calculations, determine the percentage of data within four standard deviations of the mean. Explain how you got your answer.

5. The heights of acacia trees along a small section of the Werribee River are measured and recorded. The results are as follows.

Key: 12 | 2 = 122 cm

Stem	Leaf
12	2
13	0 9
14	7
15	5 8
16	2 6 7 9
17	1 1
18	9
19	5

This data has a mean of 160.07 cm and a standard deviation of 20.5.

   a. Calculate the percentage of data within one standard deviation of the mean. Give your answer to 2 decimal places.
   b. Calculate the percentage of data within two standard deviations of the mean.

6. An AI video device was used to collect data about the number of people in cars driving past a certain point on Calder Freeway in a five-minute period. The results are as follows.

Number of people in each car	1,	2,	2,	3,	4,	2,	2,	1,	3

   a. Draw this data as a dot plot.
   b. Calculate the mean. Give your answer to 2 decimal places.
   c. The standard deviation of this data is 0.97. Calculate the percentage of data within one standard deviation of the mean. Give your answer to 2 decimal places.

7. During a recent Call of Duty tournament, 50 viewers were asked to answer the question 'In a week, how many hours do you spend playing Call Of Duty on average?' The results of the poll are shown on this dot plot.

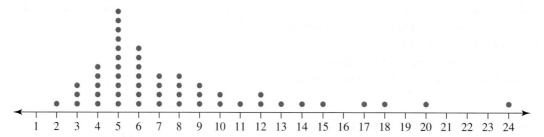

a. State whether the distribution is symmetrical or asymmetrical.
b. The mean of this data is 7.82 and the median is 6. Explain why the median is the preferred measure of centre.
c. The standard deviation of this data is 4.66.

    i. Calculate the lower boundary for two standard deviations from the mean.
    ii. Comment on what is unusual about your answer to i.
    iii. Comment on whether your answer to i makes sense in the context of this data. Explain your answer.

d. State the best measure of spread (standard deviation or interquartile range) for this data.

8. A normal distribution is a type of a symmetrical distribution. Normal distributions have the following properties.

Boundaries	Percentage of data
Within one standard deviation of the mean	68%
Within two standard deviations of the mean	95%
Within three standard deviations of the mean	99.7%

The following data has a mean of 3.5 and a standard deviation of 1.04.

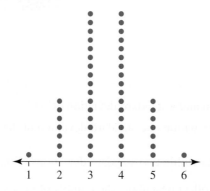

a. Calculate the percentage of data within one standard deviation of the mean.
b. Calculate the percentage of data within two standard deviations of the mean.
c. Calculate the percentage of data within three standard deviations of the mean.
d. Explain whether this data is approximately a normal distribution.

9. Billy is growing tomatoes. The following data shows the number of days between seed germination and fruit production for tomato plants in Billy's garden. The first set of data includes an outlier, and the second set of data has the outlier removed.

Data with outlier:

Key $5|4 = 54$ days

Stem	Leaf
5	4
6	3 4
7	0 3 4 7
8	2 9
9	3
10	
11	
12	
13	9

Data without outlier:

Stem	Leaf
5	4
6	3 4
7	0 3 4 7
8	2 9
9	3

a. Using the data **with the outlier**:

  i. calculate the mean
  ii. calculate the standard deviation
  iii. calculate the percentage of data within one standard deviation of the mean.

b. Using the data **with the outlier removed**:

  i. calculate the mean
  ii. calculate the standard deviation
  iii. calculate the percentage of data within one standard deviation of the mean.

c. Comment on the impact of removing the outlier on the mean and standard deviation.
d. Explain which mean you would use to describe the typical average time for a tomato plant to produce fruit in Billy's garden.

10. In a normal distribution, 68% of the data is within one standard deviation of the mean, and 95% of the data is within two standard deviations of the mean.

a. For a normal distribution with a mean of 90 and a standard deviation of 4:

  i. calculate the percentage of data between 86 and 94
  ii. calculate the percentage of data between 82 and 98.

b. For a normal distribution with a mean of 140 and a standard deviation of 20:

  i. calculate the percentage of data between 120 and 160
  ii. calculate the percentage of data between 100 and 180.

Question 1 (1 mark)

Source: VCE 2021, Further Mathematics Exam 1, Section A, Q9; © VCAA.

MC The heights of females living in a small country town are normally distributed:
- 16% of the females are more than 160 cm tall.
- 2.5% of the females are less than 115 cm tall.

The mean and the standard deviation of this female population, in centimetres, are closest to

   A. mean $= 135$ standard deviation $= 15$
   B. mean $= 135$ standard deviation $= 25$
   C. mean $= 145$ standard deviation $= 15$
   D. mean $= 145$ standard deviation $= 20$
   E. mean $= 150$ standard deviation $= 10$

Question 2 (1 mark)

Source: VCE 2019, Further Mathematics Exam 1, Section A, Q6; © VCAA.

MC For data that is normally distributed, the following is true:

Boundaries	Percentage of data
Within one standard deviation of the mean	68%
Within two standard deviations of the mean	95%
Within three standard deviations of the mean	99.7%

The time taken to travel between two regional cities is approximately normally distributed with a mean of 70 minutes and a standard deviation of 2 minutes.

The percentage of travel times that are between 66 minutes and 72 minutes is closest to

   A. 2.5%   **B.** 34%   **C.** 68%   **D.** 81.5%   **E.** 95%

Question 3 (1 mark)

Source: VCE 2015, Further Mathematics Exam 1, Section A, Q1; © VCAA.

MC The stem plot displays the average number of decayed teeth in 12-year-old children from 31 countries.

Based on this stem plot, the distribution of the average number of decayed teeth for these countries is best described as

   A. negatively skewed with a median of 15 decayed teeth and a range of 45
   B. positively skewed with a median of 15 decayed teeth and a range of 45
   C. approximately symmetric with a median of 1.5 decayed teeth and a range of 4.5
   D. negatively skewed with a median of 1.5 decayed teeth and a range of 4.5
   E. positively skewed with a median of 1.5 decayed teeth and a range of 4.5

key : $0|2 = 0.2$

```
0 | 2
0 | 5 6 7 7 8 9
1 | 0 0 0 0 1 4 4 4
1 | 5 6 7
2 | 3 3 4
2 | 7 7 8 9
3 | 0 4
3 | 5 6
4 | 1
4 | 7
```

More exam questions are available online.

1.10 The five-number summary and boxplots

LEARNING INTENTION

At the end of this subtopic you should be able to:
- calculate the five-number summary to create boxplots
- identify outliers by calculating lower and upper fences.

1.10.1 Boxplots — the five-number summary

A **boxplot** is another way of displaying numerical data. To create a boxplot, the first step is to determine the five-number summary, by calculating the following five values:

1. the lowest value (X_{min})
2. the lower quartile (Q_1)
3. the median (Q_2)
4. the upper quartile (Q_3)
5. the highest value (X_{max}).

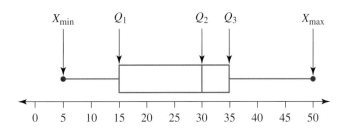

To draw a boxplot, the plot is placed either above or below a number line. A box is drawn between Q_1 and Q_3, with a vertical line at Q_2. Outside the box are lines called 'whiskers' that extend to the maximum and minimum values.

tlvd-3693

WORKED EXAMPLE 23 Drawing a boxplot for data contained in a stem plot

Draw a boxplot for the data contained in the following stem plot, which shows the number of coffees sold by a café each day over a 21-day period.

Key: 6 | 3 = 63 coffees

Stem	Leaf
6	3 5 8
7	0 2 4 5 7 9
8	1 1 3 6 8
9	0 1 5 6 7
10	1 4

THINK	WRITE
1. Determine the median of the data.	There are 21 values, so the median is in the $\left(\dfrac{21+1}{2}\right) = 11$th position.

median

63, 65, 68, 70, 72, 74, 75, 77, 79, 81, ⃝81

$Q_2 = 81$

2. Determine the value of the lower quartile.

$$\left(\frac{10+1}{2}\right) = 5.5$$

There are 10 values in the lower half of the data, so Q_1 will be between the 5th and 6th values.

$$Q_1$$

63, 65, 68, 70, 72, 74, 75, 77, 79, 81

$$Q_1 = \frac{72+74}{2}$$

$$= 73$$

3. Determine the value of the upper quartile.

$$\left(\frac{10+1}{2}\right) = 5.5$$

There are 10 values in the upper half of the data, so Q_3 will be between the 5th and 6th values.

$$Q_3$$

83, 86, 88, 90, 91, 95, 96, 97, 101, 104

$$Q_3 = \frac{91+95}{2}$$

$$= 93$$

4. Write the five-number summary.

$$X_{\min} = 63$$
$$Q_1 = 73$$
$$Q_2 = 81$$
$$Q_3 = 93$$
$$X_{\max} = 104$$

5. Rule a suitable scale for your boxplot, which covers the full range of values. Draw the central box first (from Q_1 to Q_3, with a line at Q_2) and then draw in the whiskers from the edge of the box to the minimum and maximum values.

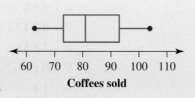

Coffees sold

TI	THINK	DISPLAY/WRITE	CASIO	THINK	DISPLAY/WRITE

1. In a Lists & Spreadsheet page, label the first column as *coffees*. Enter the given data values in the first column.

1. On a Statistics screen, relabel list 1 as *coffees*. Enter the given data values in the first column.

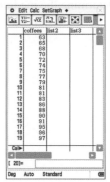

2. In a Data & Statistics page, click on the label of the horizontal axis and select *coffees*.

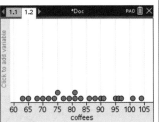

2. Select:
   - SetGraph
   - Setting …
   Complete the fields as:
   Draw: On
   Type: MedBox
   X List: main\coffees
   Freq: 1
   Tick the Show Outliers box, then select Set.

3. Press MENU, then select:
   1: Plot Type
   2: Box Plot

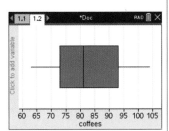

Click the y icon.

1.10.2 Identifying possible outliers — upper and lower fences

When using boxplots there is a precise way to identify if a piece of data is an outlier. To do this, we use a **lower fence** and an **upper fence**. Values between the lower and upper fences are included in the boxplot, and values outside the lower and upper fences are outliers. These outliers are shown next to the boxplot as a cross ($\times$) on the number line.

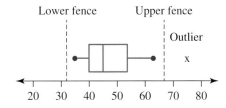

To calculate the values of the upper and lower fences, first calculate the interquartile range (IQR) (see subtopic 1.8). Then use these formulas:

Calculating lower and upper fence values

$$\text{Lower fence} = Q_1 - 1.5 \times \text{IQR}$$

$$\text{Upper fence} = Q_3 + 1.5 \times \text{IQR}$$

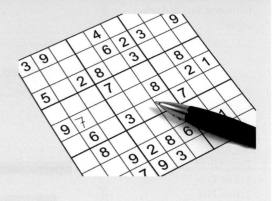

WORKED EXAMPLE 24 Calculating the values of the lower and upper fences

The following stem plot represents the time taken (in minutes) for 25 students to finish a maths puzzle.

Key: 1 | 5 = 1.5 minutes

Stem	Leaf
1	5
2	8
3	4 6 6 9
4	0 0 2 3 5 5 7 8 8
5	0 2 4 4 5 8 9
6	0 4
7	
8	5

a. **Calculate the values of the lower and upper fences.**
b. **Identify any outliers in the data set.**
c. **Draw a boxplot to represent the data.**

THINK

a. 1. Determine the median of the data.

2. Determine the value of the lower quartile.

3. Determine the value of the upper quartile.

WRITE

a. There are 25 values, so the median is in the

$$\left(\frac{25+1}{2}\right) = 13\text{th position.}$$

median

1.5, 2.8, 3.4, 3.6, 3.6, 3.9, 4.0, 4.0, 4.2, 4.3, 4.5, 4.5, ④.7

$Q_2 = 4.7$

$$\left(\frac{12+1}{2}\right) = 6.5$$

There are 12 values in the lower half of the data, so Q_1 will be between the 6th and 7th values.

Q_1

1.5, 2.8, 3.4, 3.6, 3.6, 3.9, 4.0, 4.0, 4.2, 4.3, 4.5, 4.5

$$Q_1 = \frac{3.9 + 4.0}{2}$$

$$= 3.95$$

$$\left(\frac{12+1}{2}\right) = 6.5$$

There are 12 values in the upper half of the data, so Q_3 will be between the 6th and 7th values.

Q_3

4.8, 4.8, 5.0, 5.2, 5.4, 5.4, 5.5, 5.8, 5.9, 6.0, 6.4, 8.5

$$Q_3 = \frac{5.4 + 5.5}{2}$$

$$= 5.45$$

4. Calculate the IQR.

$$IQR = Q_3 - Q_1$$
$$= 5.45 - 3.95$$
$$= 1.5$$

5. Calculate the values of the lower and upper fences.

$$\text{Lower fence} = Q_1 - 1.5 \times IQR$$
$$= 3.95 - 1.5 \times 1.5$$
$$= 3.95 - 2.25$$
$$= 1.7$$

$$\text{Upper fence} = Q_3 + 1.5 \times IQR$$
$$= 5.45 + 1.5 \times 1.5$$
$$= 5.45 + 2.25$$
$$= 7.7$$

b. 1. Identify whether any values lie below the lower fence or above the upper fence.

2. State the answer.

b. Values below the lower fence (1.7): 1.5
Values above the upper fence (7.7): 8.5

There are two outliers: 1.5 and 8.5.

c. 1. Write the five-number summary, giving the minimum and maximum values as those that lie within the lower and upper fences.

c. $X_{min} = 2.8$, $Q_1 = 3.95$, $Q_2 = 4.7$, $Q_3 = 5.45$ and $X_{max} = 6.4$

2. Rule a suitable scale for your boxplot to cover the full range of values. Draw the central box first (from Q_1 to Q_3, with a line at Q_2) and then draw in the whiskers from the edge of the box to the minimum and maximum values. Mark the outliers with an 'x'.

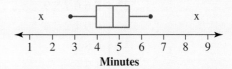

 Resources

 Interactivity Boxplots (int-6245)

1.10 Exercise

Students, these questions are even better in jacPLUS

 Receive immediate feedback and access sample responses

 Access additional questions

Track your results and progress

Find all this and MORE in jacPLUS

1. The boxplot shows the temperatures in Melbourne over a 23-day period.

   a. State the median temperature.
   b. State the range of the temperatures.
   c. State the interquartile range of temperatures.

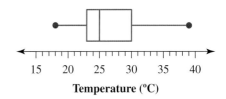

2. **WE23** Draw a boxplot for the data contained in the following stem plot, which shows the number of sandwiches sold by a café per day over a 21-day period.

Key: 2 | 9 = 29 sandwiches

Stem	Leaf
2	9
3	1 3 6 8 9
4	2 4 5 5 6 7 7 8
5	0 0 3 5 8
6	1 2

3. The following stem plot shows the ages of 25 people when they had their first child.

Key: 1* | 7 = 17 years old

Stem	Leaf
1*	7 8 8
2	0 2 3 3 4
2*	5 6 6 7 8 9
3	0 0 1 2 2 4
3*	6 8 9
4	1 3

a. Prepare a five-number summary of the data.
b. Draw a boxplot of the data.
c. Comment on the distribution of the data.

4. **MC** The five-number summary for a data set is 45, 56, 70, 83, 92. Select which of the following statements is definitely false.

A. There are no outliers in the data set.
B. Half of the scores are between 56 and 70.
C. The range is 47.
D. The value of the lower fence is 15.5.
E. The data has no noticeable skew.

5. Match the items in the list on the left to the list on the right.

1. Range	a. Median
2. Q_2	b. $Q_1 - 1.5 \times IQR$
3. Interquartile range	c. $Q_3 - Q_1$
4. Lower fence	d. $Q_3 + 1.5 \times IQR$
5. Upper fence	e. $X_{max} - X_{min}$

6. Determine whether the following statements are true or false.

a. You can always determine the median from a boxplot.
b. A stem plot contains every piece of data from a data set.
c. Boxplots show the complete distribution of scores within a data set.

7. **WE24** The stem plot represents the time taken (in minutes) for 25 students to finish a logic problem.
   a. Calculate the values of the lower and upper fences.
   b. Identify any outliers in the data set.
   c. Draw a boxplot to represent the data.

Key: 4 |4 = 4.4 minutes

Stem	Leaf
4	4
5	
6	2 6 9
7	0 4 7 7 8
8	0 3 3 5 6 8 9
9	1 2 4 6 7
10	2 4 4
11	5

8. The boxplot represents the scores made by an Australian football team over a season.
   a. State the highest number of points the team scored in the season.
   b. State the lowest number of points the team scored in the season.
   c. State the range of points scored.
   d. State the interquartile range of points scored.

9. Here is a five-number summary with two values missing:
   $X_{min} = ?$, $Q_1 = 9$, $Q_2 = 11$, $Q_3 = ?$, $X_{max} = 17$
   The range of this data set is 10, and the interquartile range is 4. Calculate the missing values.

10. Commuters were surveyed about the number of minutes it takes them to drive to work. The five-number summary for a set of data is:
    $X_{min} = 9$, $Q_1 = 13$, $Q_2 = 20$, $Q_3 = 29$, $X_{max} = 35$
    a. Calculate the range of this data.
    b. Calculate the interquartile range of this data.
    c. State the median of this data.
    d. Calculate the upper and lower fences of this data.
    e. Use your answer to d to explain what would make a commute time an outlier in this data set.

1.10 Exam questions

Question 1 (1 mark)
Source: VCE 2020, Further Mathematics Exam 1, Section A, Q3; © VCAA.

MC The times between successive nerve impulses (time), in milliseconds, were recorded.

The table shows the mean and the five-number summary calculated using 800 recorded data values.

	Time (milliseconds)
Mean	220
Minimum value	10
First quartile (Q_1)	70
Median	150
Third quartile (Q_3)	300
Maximum value	1380

Data: adapted from P Fatt and B Katz, 'Spontaneous subthreshold activity at motor nerve endings', *The Journal of Physiology*, 117, 1952, pp. 109–128

The shape of the distribution of these 800 times is best described as
  A. approximately symmetric.
  B. positively skewed.
  C. positively skewed with one or more outliers.
  D. negatively skewed.
  E. negatively skewed with one or more outliers.

Question 2 (2 marks)

Source: VCE 2020, Further Mathematics Exam 2, Section A, Q2c; © VCAA.

The *neck size*, in centimetres, of 250 men was recorded and displayed in the dot plot below.

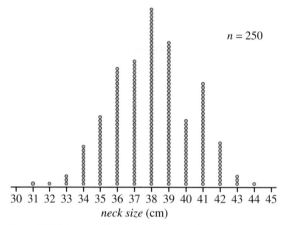

$n = 250$

neck size (cm)

Data: RW Johnson, 'Fitting percentage of body fat to
simple body measurements', *Journal of Statistics
Education*, 4:1, 1996,
<https://doi.org/10.1080/10691898.1996.11910505>

The five-number summary for this sample of neck sizes, in centimetres, is given below.

Minimum	First quartile (Q_1)	Median	Third quartile (Q_3)	Maximum
31	36	38	39	44

Use the five-number summary to construct a boxplot, showing any outliers if appropriate, on the grid below.

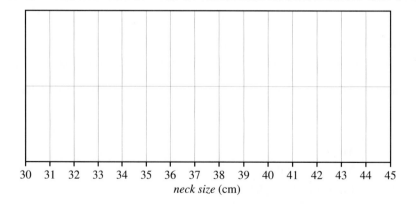

neck size (cm)

Question 3 (1 mark)
Source: VCE 2017, Further Mathematics Exam 1, Section A, Q2; © VCAA.

MC The boxplot below shows the distribution of the forearm *circumference*, in centimetres, of 252 people.

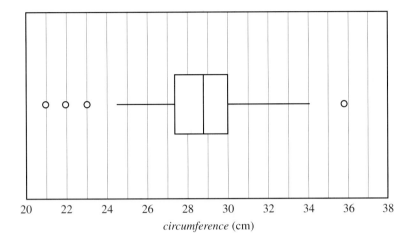

circumference (cm)

The five-number summary for the forearm *circumference* of these 252 people is closest to

A. 21, 27.4, 28.7, 30, 34   **B.** 21, 27.4, 28.7, 30, 35.9   **C.** 24.5, 27.4, 28.7, 30, 34

D. 24.5, 27.4, 28.7, 30, 35.9   **E.** 24.5, 27.4, 28.7, 30, 36

More exam questions are available online.

1.11 Comparing the distribution of a numerical variable across two or more groups

---

LEARNING INTENTION

At the end of this subtopic you should be able to:
- use back-to-back stem plots or parallel boxplots, as appropriate, to compare the distributions of a single numerical variable across two or more groups in terms of centre (median) and spread (IQR and range), and interpret any differences observed in the context of the data.

1.11.1 Comparing distributions

As we have already seen, back-to back stem plots are a way of displaying two sets of the same data side by side. To compare two sets of data, we compare the centre and spread of each set of data.

Parallel boxplots are useful tools to compare two sets of data. Parallel boxplots are two or more boxplots displayed with the same scale.
For example, this is a parallel boxplot that displays data about the ages of squash players in two different clubs.

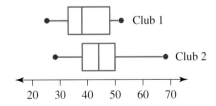

To compare the sets of data, we compare median, interquartile range (IQR) and range. When comparing median values, include context about what the data is measuring, for example which data set took longer on average, or had the greater length on average, or was hotter on average, and so on.

When comparing interquartile range and range values, include whether the values are smaller, larger or similar, and what this says in relation to the data. Higher interquartile range or range indicates higher variability.

Comparing the data about squash players above, we could write:

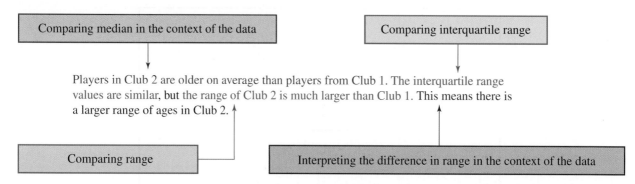

| Comparing median in the context of the data |

Players in Club 2 are older on average than players from Club 1. The interquartile range values are similar, but the range of Club 2 is much larger than Club 1. This means there is a larger range of ages in Club 2.

| Comparing interquartile range |

| Comparing range |

| Interpreting the difference in range in the context of the data |

tlvd-3695

WORKED EXAMPLE 25 Comparing a back-to-back stem plot and a parallel boxplot

The following back-to-back stem plot shows the size (in kg) of two different breeds of dog.
a. **Draw parallel boxplots of the two sets of data.**
b. **Compare and contrast the two sets of data.**

Key: $2 \mid 6 = 26 \, \text{kg}$

Breed X		Breed Y
Leaf	Stem	Leaf
9 8 7 7 6 4 4	1	
8 7 5 4 3 3 1 0	2	6 9
	3	3 5 5 7 8
	4	0 2 4 5 6 9
	5	1 3

THINK

a. 1. Calculate the five-number summary for the first data set (Breed X).

WRITE

a. $X_{\min} = 14$

$X_{\max} = 28$

There are 15 pieces of data, so the median is the $\left(\dfrac{15 + 1}{2}\right) = $ 8th piece of data.

$Q_2 = 20$

There are 7 pieces of data in the lower half, so Q_1 is the 4th value.

$Q_1 = 17$

There are 7 pieces of data in the upper half, so Q_3 is the 4th value.

$Q_3 = 24$

Five-number summary: 14, 17, 20, 24, 28

2. Calculate the five-number summary for the second data set (Breed Y).

$X_{min} = 26$

$X_{max} = 53$

There are 15 pieces of data, so the median is the $\left(\dfrac{15+1}{2}\right) = 8$th piece of data.

$Q_2 = 40$

There are 7 pieces of data in the lower half, so Q_1 is the 4th value.

$Q_1 = 35$

There are 7 pieces of data in the upper half, so Q_3 is the 4th value.

$Q_3 = 46$

Five-number summary: 26, 35, 40, 46, 53

3. Use the five-number summaries to plot the parallel boxplots. Use a suitable scale that will cover the full range of values for both data sets.

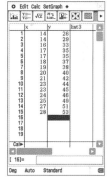

b. Compare and contrast the data sets, looking at where the key points of each data set lie. Comment on any noticeable differences in the centre and spread of the scores, as well as the shape of the distributions.

b. On the whole, Breed X is considerably lighter than Breed Y, with only a small overlap in the data sets. Breed X has a smaller interquartile range than Breed Y, although both spreads are balanced with no noticeable skew.

TI	THINK	DISPLAY/WRITE

1. In a Lists & Spreadsheet page, label the first column as x and the second column as y. Enter the values for breed X in the first column and the values for breed Y in the second column.

2. In a Data & Statistics page, click on the label of the horizontal axis and select x.

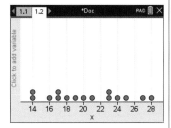

CASIO	THINK	DISPLAY/WRITE

1. On a Statistics screen, relabel list 1 as x and list 2 as y. Enter the values for breed X in the first column and the values for breed Y in the second column.

2. Select:
 • SetGraph
 • Setting …
 Select tab 1, then complete the fields as:
 Draw: On
 Type: MedBox
 XList: main\x
 Freq: 1
 Tick the Show Outliers box.

3. Press MENU, then select:
   1: Plot Type
   2: Box Plot

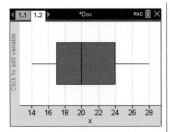

3. Select tab 2, then complete the fields as:
   Draw: On
   Type: MedBox
   XList: main\y
   Freq: 1
   Tick the Show Outliers box, then select Set.

4. Press MENU, then select:
   2: Plot Properties
   5: Add X Variable
   then select y.

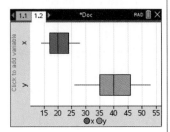

4. Click the y icon.

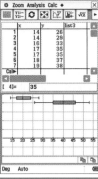

 Resources

 Interactivites Back-to-back stem plots (int-6252)
Parallel boxplots (int-6248)

1.11 Exercise

Students, these questions are even better in jacPLUS

 Receive immediate feedback and access sample responses

 Access additional questions

Track your results and progress

Find all this and MORE in jacPLUS

1. **WE25** The back-to-back stem plot shows the amount of sales (in $000s) for two different high street stores.
   a. Draw a parallel boxplot of the two sets of data.
   b. Compare and contrast the two sets of data.

Key: 3 | 4 = $3400

Store 1 Leaf	Stem	Store 2 Leaf
	3	4 7 9
7 4	4	2 4 6 8
6 2 1	5	1 2 5 5 9
8 8 5 5	6	3 5 7
5 3 2	7	
6 1	8	
0	9	

2. The parallel boxplot shows the difference in grades (out of 100) between students at two schools.

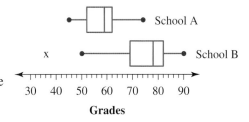

   a. Identify which school had the highest overall grade.
   b. Identify which school had the lowest overall grade.
   c. Calculate the difference between the interquartile ranges of the grades of the two schools.

3. The parallel boxplot shows the performance of two leading brands of battery in a test of longevity.

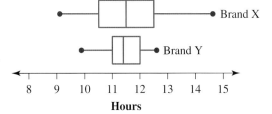

   a. State the brand that had the better median performance.
   b. State the brand that gave the most consistent performance.
   c. State the brand that had the worst performing battery.
   d. State the brand that had the best performing battery.

4. The prices of main meals at two restaurants that appear in the Good Food Guide are shown in the back-to-back stem plot.

Key: 1 | 8 = $18

Restaurant A Leaf	Stem	Restaurant B Leaf
	1	8 9
9 9 8 5 5 4	2	2 5 5 7
8 6 5 5 2	3	0 0 2 5 5 8
2 0 0	4	0 3 6
	5	
9	6	

   a. Identify any outliers in either set of data.
   b. Prepare five-number summaries for the price of the meals at each restaurant.
   c. Draw a parallel boxplot to compare the two data sets.
   d. Compare and contrast the cost of the main meals at each restaurant.

5. The table displays the number of votes that two political parties received in 15 different constituencies in local elections.

Party A	425	630	813	370	515	662	838	769
Party B	632	924	514	335	748	290	801	956

Party A	541	484	745	833	497	746	651
Party B	677	255	430	789	545	971	318

   a. Prepare five-number summaries for both parties' votes.
   b. Display the data sets on a parallel boxplot.
   c. Comment on the distributions of both data sets.

6. The back-to-back stem plot shows the share prices (in $) of two companies from 18 random months out of a 10-year period.

   a. Display the data in two frequency tables with intervals of 5.
   b. Display the data on a parallel boxplot.
   c. Comment on the distributions of both data sets.

Key: $1^* \mid 7 = \$17$

Company A Leaf	Stem	Company B Leaf
4 2	1	
9 7 5	1*	7 9
4 4 1	2	0 3 4
9 8 6 6	2*	7 8 8
3 3 2 0	3	1 2 4
8 6	3*	6 9
	4	0 1 2
	4*	5 6

7. The parallel boxplot details the weight of strawberries harvested in kg at two different farms for the month of March over a 15-year period.

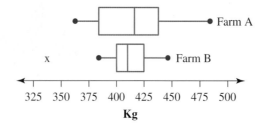

Decide whether the following statements are true or false.

   a. Farm A produced a larger harvest of strawberries in March than Farm B more often than not.
   b. The strawberry harvest at Farm B in March is much more reliable than the strawberry harvest at farm A.
   c. Farm A had the highest producing month for strawberries on record.
   d. Farm A had the lowest producing month for strawberries on record.

8. The parallel boxplot shows the weekly sales figures of three different mobile phones across a period of six months.

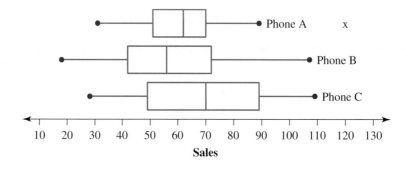

   a. State the phone that had the highest weekly sales overall.
   b. State the phone that had the most consistent sales.
   c. State the phone that had the largest range in sales.
   d. State the phone that had the largest interquartile range in sales.
   e. State the phone that had the highest median sales figure.

9. The following data sets show the daily sales figures for three new drinks across a 21-day period.

Drink 1: 35, 51, 47, 56, 53, 64, 44, 39, 50, 47, 62, 66, 58, 41, 39, 55, 52, 59, 47, 42, 60

Drink 2: 48, 53, 66, 51, 37, 44, 70, 59, 41, 68, 73, 62, 56, 40, 65, 77, 74, 63, 54, 49, 61

Drink 3: 57, 49, 51, 49, 52, 60, 46, 48, 53, 56, 52, 49, 47, 54, 61, 50, 33, 48, 54, 57, 50

a. Prepare a five-number summary for each drink, excluding any outliers.
b. Plot a parallel boxplot to compare the sales of the three drinks.
c. Compare and contrast the sales of the three drinks.

10. The following back-to-back stem plot displays the rental price (in $) of one-bedroom apartments in two different suburbs.

Key: 25 | 0 = $250

Suburb A Leaf	Stem	Suburb B Leaf
	25	0
	26	5 9
5 5	27	0 0 5
9 9 5 0	28	5 9 9
5 5 0	29	0
5 5 0 0 0	30	0 0 0
5 5 0 0	31	0 5 5
	32	9 9
	33	
	34	0 0
0	35	

a. Prepare a five-number summary for each suburb, excluding any outliers.
b. Plot a parallel boxplot to compare the data sets.
c. Compare and contrast the rental price in the two suburbs.
d. The rental prices in a third suburb, Suburb C, were also analysed, with the data having a five-number summary of 280, 310, 325, 340, 375. Add the third data set to your parallel boxplot.
e. Compare the rent in the third suburb with the other two suburbs.

1.11 Exam questions

Question 1 (1 mark)
Source: VCE 2021, Further Mathematics Exam 1, Section A, Q6; © VCAA.

MC The relationship between *resting pulse rate*, in beats per minute, and *age group* (15–20 years, 21–30 years, 31–50 years, over 50 years) is best displayed using

A. a histogram.
B. a scatterplot.
C. parallel boxplots.
D. a time series plot.
E. a back-to-back stem plot.

Question 2 (4 marks)

Source: VCE 2019, Further Mathematics Exam 2, Section A, Q2; © VCAA.

The parallel boxplots below show the *maximum daily temperature* and *minimum daily temperature*, in degrees Celsius, for 30 days in November 2017.

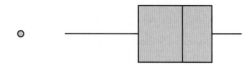

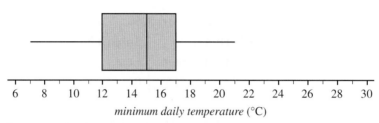

minimum daily temperature (°C)

Data: Australian Government, Bureau of Meteorology,

a. Use the information in the boxplots to complete the following sentences.
For November 2017

   i. the interquartile range for the *minimum daily temperature* was ⬚ °C   **(1 mark)**
   ii. the median value for *maximum daily temperature* was ⬚ °C higher than the median value for *minimum daily temperature*   **(1 mark)**
   iii. the number of days on which the *maximum daily temperature* was less than the median value for *minimum daily temperature* was ⬚.   **(1 mark)**

b. The *temperature difference* between *the minimum daily temperature* and the *maximum daily temperature* in November 2017 at this location is approximately normally distributed with a mean of 9.4 °C and a standard deviation of 3.2 °C.
Determine the number of days in November 2017 for which this *temperature difference* is expected to be greater than 9.4 °C.   **(1 mark)**

Question 3 (1 mark)

Source: VCE 2016, Further Mathematics Exam 1, Section A, Q8; © VCAA.

MC Parallel boxplots would be an appropriate graphical tool to investigate the association between the monthly median rainfall, in millimetres, and the
   A. monthly median wind speed, in kilometres per hour.
   B. monthly median temperature, in degrees Celsius.
   C. month of the year (January, February, March, etc.).
   D. monthly sunshine time, in hours.
   E. annual rainfall, in millimetres.

More exam questions are available online.

1.12 Review

1.12.1 Summary

doc-37611

1.12 Exercise

1. **MC** The interquartile range of the data distribution shown in the stem plot is:

Key: $2 \mid 6 = 26$

Stem	Leaf
0	2
1	1 5
2	6 6 7 8
3	8 8 9
4	3 4
5	2

A. 41   **B.** 50   **C.** 28   **D.** 20.5   **E.** 26

2. **MC** For the parallel boxplots shown, select which statement is correct.

A. Group A has a smaller IQR than Group B.
B. Group B has a greater range than Group A.
C. Group A has a higher median than Group B.
D. Group A has a smaller highest value than Group B.
E. 25% of Group B is greater than Group A's Q_1 and less than its median.

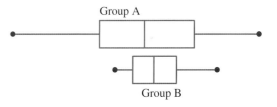

3. **MC** Data gathered on the number of home runs in a baseball season would be classified as:

A. discrete.   **B.** nominal.   **C.** continuous.   **D.** ordinal.   **E.** categorical.

4. **MC** For the data set 789, 211, 167, 321, 432, 222, 234, 456, 456, 234, the five-figure summary in order from the smallest value to the largest is:

A. 167, 222, 321, 456, 789   **B.** 167, 222, 277.5, 456, 789
C. 167, 234, 432, 456, 789   **D.** 167, 234, 432, 456, 789
E. 167, 222, 287.5, 444, 789

5. **MC** The boxplot shown would best be described as:

A. positively skewed.
B. symmetrical with an outlier.
C. positively skewed with an outlier.
D. negatively skewed with an outlier.
E. negatively skewed.

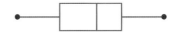

6. **MC** For the sample data set 2, 3, 5, 2, 3, 6, 3, 8, 9, 2, 8, 9, 2, 6, 7, the mean and standard deviation respectively would be closest to:

   A. 5 and 6   **B.** 5 and 2.6   **C.** 2.6 and 5   **D.** 5 and 2.7   **E.** 2.7 and 5

7. **MC** For the stem plot shown, the median and range respectively are:

   A. 73 and 41
   B. 73.5 and 41
   C. 71 and 39
   D. 71 and 41
   E. 73 and 39

   Key: 5 | 1 = 51

Stem	Leaf
5	1 2
6	2 3 4
7	3 4 4 5
8	6 6
9	2

8. **MC** For the parallel boxplots shown, select which statement is **not** correct.

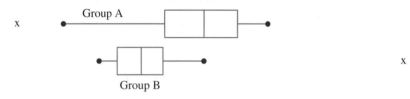

   A. 75% of Group A is larger than 75% of group B.
   B. 25% of Group A has a larger spread than 75% of group B.
   C. The median of Group A is equal to the highest value of Group B.
   D. Group A is negatively skewed and Group B is positively skewed.
   E. The middle 50% of both groups are approximately symmetrical.

9. **MC** For the data set 21, 56, 110, 15, 111, 45, 250, 124, 78, 24, the number of outliers and the value of $1.5 \times \text{IQR}$ will respectively be:

   A. 0 and 87   **B.** 1 and 87   **C.** 1 and 111   **D.** 1 and 130.5   **E.** 0 and 130.5

Short answer

10. The price of a barrel of oil in US dollars over a particular 18-month time period is shown in the following table.

Month	Price (US$)
Jan	102.96
Feb	97.63
Mar	108.76
Apr	105.25
May	106.17
Jun	83.17
Jul	83.72
Aug	88.99
Sep	95.34

Month	Price (US$)
Oct	92.44
Nov	87.05
Dec	88.69
Jan	93.14
Feb	97.46
Mar	90.71
Apr	97.1
May	90.74
Jun	93.41

   a. Calculate the mean and median for this data set. Give your answers correct to 1 decimal place.
   b. Calculate the interquartile range and the standard deviation for this data set. Give your answers correct to 2 decimal places.
   c. Explain which would be the best measures of centre and spread for this data set.

11. Use the data on the incidence of communicable diseases in Australia to answer the following questions.

Incidence of communicable diseases in Australia over two consecutive years

Disease	Year 1	Year 2
Hepatitis C	11 089	7286
Typhoid fever	116	96
Legionellosis	302	298
Meningococcal disease	259	230
Tuberculosis	1324	1327
Influenza (laboratory confirmed)	59 090	13 419
Measles	104	70
Mumps	165	95
Chickenpox	1753	1743
Shingles	2716	2978
Dengue virus infection	1406	1201
Malaria	508	399
Ross River virus infection	4796	5147

a. Calculate the mean (correct to 1 decimal place) and median number of cases of communicable diseases of the sample for each year.

b. Comment on the differences between the mean and median values calculated in part a.

12. The number of passengers arriving from overseas during a particular time period at various airports in Australia is shown in the following table.

Airport	Number of passengers
Adelaide	5743
Brisbane	480 625
Cairns	5110
Coolangatta	7 655
Darwin	5318
Melbourne	594 286
Perth	318 493

a. Calculate the mean and standard deviation for the sample. Give your answers correct to 1 decimal place.

b. Compare the mean and standard deviation to the median and interquartile range for the sample.

13. The following table shows data on the top 10 tourist destinations in Europe.

Country	Nights in country (× 1000)
Spain	200 552
Italy	158 527
France	98 700
United Kingdom	80 454
Austria	72 225
Germany	54 097
Greece	46 677
Portugal	25 025
Netherlands	25 014
Czech Republic	17 747

a. Display the data as a boxplot.
b. Describe the distribution of the data using the five-figure summary and identify any outliers.

14. The biggest winning margins in AFL grand finals up to the year 2019 are shown in the following table.

Biggest winning margins

Winning margin	Year	Winning team	Winning score	Losing team	Losing score
119	2007	Geelong	163	Port Adelaide	44
96	1988	Hawthorn	152	Melbourne	56
83	1983	Hawthorn	140	Essendon	57
81	1980	Richmond	159	Collingwood	78
80	1994	West Coast	143	Geelong	63
78	1985	Essendon	170	Hawthorn	92
73	1949	Essendon	125	Carlton	52
73	1956	Melbourne	121	Collingwood	48
63	1946	Essendon	150	Melbourne	87
61	1995	Carlton	141	Geelong	80
61	1957	Melbourne	116	Essendon	55
60	2000	Essendon	135	Melbourne	75

a. i. Display the winning margin data as a boxplot.
  ii. Display the winning score data as a boxplot.
  iii. Display the losing score data as a boxplot.
b. Describe each boxplot from part a.

15. The following table shows data on military expenditure by region (US $billion) for the years 1999–2012.

Year	Africa	The Americas	Asia and Oceania	Europe	Middle East
1999	22	441	198	348	69
2000	19	458	202	360	80
2001	20	466	213	362	85
2002	21	515	224	374	80
2003	21	571	234	380	84
2004	23	620	247	383	89
2005	24	652	260	387	98
2006	25	665	275	397	105
2007	26	685	296	408	110
2008	30	737	312	419	106
2009	32	793	348	428	109
2010	34	817	355	419	115
2011	38	808	369	411	117
2012	38	768	382	419	128

a. Display the data for the Americas, Asia and Oceania, and Europe as parallel boxplots.
b. Use the boxplots to compare military spending between the three regions.

Extended response

16. Use the data shown to answer the following questions.

Women who gave birth and Indigenous status by states and territories

Status	NSW	Vic	Qld	WA	SA	Tas	ACT	NT	Aust
Indigenous	2904	838	3332	1738	607	284	107	1474	11 284
Non-Indigenous	91 958	70 328	57 665	29 022	18 994	5996	5601	2369	281 933

a. Display the data in an appropriate display.
b. Calculate the mean births per state/territory of Australia for both Indigenous and non-Indigenous groups. Give your answers correct to 1 decimal place.
c. Calculate the median births per state/territory of Australia for both Indigenous and non-Indigenous groups.
d. Calculate the standard deviation (correct to 1 decimal place) and IQR for the data on births per state/territory of Australia for both Indigenous and non-Indigenous groups.
e. Comment on the measures of centre and spread you have calculated for this data.

17. Use the data on Tokyo's average maximum temperatures to answer the questions.

Tokyo average maximum temperature, 1980–2012

Year	Temp. (°C)	Year	Temp. (°C)	Year	Temp. (°C)	Year	Temp. (°C)
1980	19.3	1985	19.4	2003	19.6	2008	20.1
1981	19.0	1986	18.8	2004	21.2	2009	20.3
1982	19.6	1987	20.0	2005	20.4	2010	20.6
1983	19.6	1988	19.0	2006	19.9	2011	20.2
1984	18.8	1989	19.9	2007	20.6	2012	20.0

a. Calculate the mean and standard deviation of the temperature data for the two 10-year periods of 1980–89 and 2003–12. Give your answers correct to 2 decimal places.

b. State what the means and standard deviations calculated indicate about the two 10-year periods.

c. Calculate the mean and standard deviation of the total 20 years of the sample data. Give your answers correct to 2 decimal places.

d. Comment on how these measurements compare to the calculations you made in part **a**.

18. The following table shows the AFL grand final statistics for a sample of players who have kicked a total of 5 or more goals from the clubs Carlton and Collingwood.

Player	Team	Kicks	Marks	Handballs	Disposals	Goals	Behinds
Alex Jesaulenko	Carlton	23	11	9	32	11	0
John Nicholls	Carlton	29	3	1	30	13	1
Wayne Johnston	Carlton	78	19	17	95	5	7
Robert Walls	Carlton	19	9	5	24	11	1
Craig Bradley	Carlton	61	11	37	98	6	2
Mark MacLure	Carlton	34	16	14	48	5	4
Stephen Kernahan	Carlton	44	26	8	52	17	5
Ken Sheldon	Carlton	36	5	12	48	5	2
Syd Jackson	Carlton	13	3	1	14	5	1
Rodney Ashman	Carlton	25	4	10	35	5	2
Greg Williams	Carlton	30	6	29	59	6	4
Alan Didak	Collingwood	46	17	24	70	6	2
Peter Moore	Collingwood	42	22	13	55	11	7
Ricky Barham	Collingwood	42	15	16	58	5	5
Travis Cloke	Collingwood	26	16	9	35	5	4
Ross Dunne	Collingwood	17	6	6	23	5	2
Craig Davis	Collingwood	27	8	8	35	6	3

a. Use the data for goals to compare the two clubs using parallel boxplots.

b. Comment on what the parallel boxplots indicate about the data for goals.

c. Compare the data for kicks and handballs using parallel boxplots.

d. Comment on what the parallel boxplots indicate about the data for kicks and handballs.

1.12 Exam questions

Question 1 (1 mark)

Source: VCE 2021, Further Mathematics Exam 1, Section A, Q5; © VCAA.

MC The stem plot below shows the *height*, in centimetres, of 20 players in a junior football team.

key : 14|2 = 142 cm $n = 20$

```
14 | 2 2 4 7 8 8 9
15 | 0 0 1 2 5 5 6 8
16 | 0 1 1 2
17 | 9
```

A player with a height of 179 cm is considered and outlier because 179 cm is greater than

   A. 162 cm   **B.** 169 cm   **C.** 172.5 cm   **D.** 173 cm   **E.** 175.5 cm

Question 2 (1 mark)

Source: VCE 2020, Further Mathematics Exam 1, Section A, Q2; © VCAA.

MC The times between successive nerve impulses (time), in milliseconds, were recorded.

The table shows the mean and the five-number summary calculated using 800 recorded data values.

	Time (milliseconds)
Mean	220
Minimum value	10
First quartile (Q_1)	70
Median	150
Third quartile (Q_3)	300
Maximum value	1380

Data: adapted from P Fatt and B Katz, 'Spontaneous subthreshold activity at motor nerve endings', *The Journal of Physiology*, 117, 1952, pp. 109–128

Of these 800 times, the number of times that are longer than 300 milliseconds is closest to

A. 20 **B.** 25 **C.** 75 **D.** 200 **E.** 400

Question 3 (2 marks)

Source: VCE 2019, Further Mathematics Exam 2, Section A, Q3; © VCAA.

The five-number summary for the distribution of *minimum daily temperature* for the months of February, May and July in 2017 is shown in Table 2.

The associated boxplots are shown below the table.

Table 2. Five-number summary for *minimum daily temperature*

Month	**Minimum**	Q_1	**Median**	Q_3	**Maximum**
February	5.9	9.5	10.9	13.9	22.2
May	3.3	6.0	7.5	9.8	12.7
July	1.6	3.7	5.0	5.9	7.7

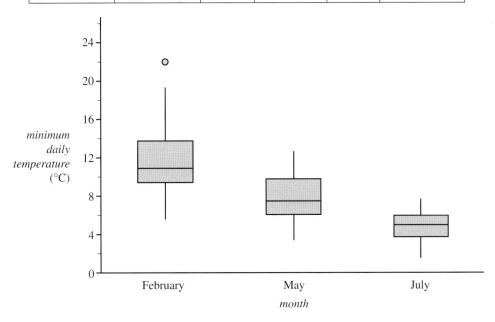

Data: Australian Government, Bureau of Meteorology,

Explain why the information given above supports the contention that *minimum daily temperature* is associated with the *month*. Refer to the values of an appropriate statistic in your response.

Source: VCE 2017, Further Mathematics Exam 2, Section A, Q2; © VCAA.

The back-to-back stem plot below displays the *wingspan*, in millimetres, of 32 moths and their *place of capture* (forest or grassland).

```
          Key : 1|8 = 18      wingspan (mm)
        forest (n = 13)      grassland (n = 19)
                    6 | 1 | 8
          2 1 1 0 0 0 0 | 2 | 2 2 4 4
                    7 | 2 | 5 5 9
                  4 0 | 3 | 0 0 1 2 3 4
                    5 | 3 | 6 8
                      | 4 | 0 3
                      | 4 | 5
                    2 | 5 |
```

a. Which variable, *wingspan* or *place of capture, is* a categorical variable? **(1 mark)**

b. Write down the modal wingspan, in millimetres, of the moths captured in the forest. **(1 mark)**

c. Use the information in the back-to-back stem plot to complete the table below. **(2 marks)**

	Wingspan (mm)				
Place of capture	minimum	Q_1	median (M)	Q_3	maximum
forest		20	21	32	52
grassland	18	24	30		45

d. Show that the moth captured in the forest that had a *wingspan* of 52 mm is an outlier. **(2 marks)**

e. The back-to-back stem plot suggests that w*ingspan is* associated with *place of capture.* **(2 marks)**
Explain why, quoting the values of an appropriate statistic.

Source: VCE 2016, Further Mathematics Exam 2, Section A, Q2; © VCAA.

The weather station also records daily maximum temperatures.

a. The five-number summary for the distribution of maximum temperatures for the month of February is displayed in the table below.

	Temperature (°C)
Minimum	16
Q_1	21
Median	25
Q_3	31
Maximum	38

There are no outliers in this distribution.

i. Use the five-number summary above to construct a boxplot on the grid below.　**(1 mark)**

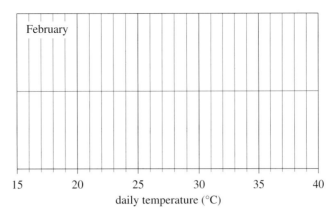

February

daily temperature (°C)

ii. What percentage of days had a maximum temperature of 21°C, or greater, in this particular February?

(1 mark)

b. The boxplots below display the distribution of maximum daily *temperature* for the months of May and July.

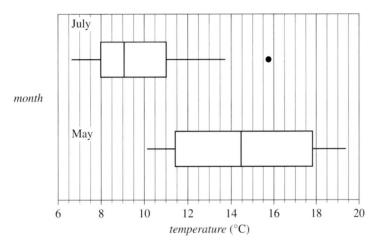

month

July

May

temperature (°C)

i. Describe the shapes of the distributions of daily *temperature* (including outliers) for July and for May.

(1 mark)

July _____

May _____

ii. Determine the value of the upper fence for the July boxplot.　**(1 mark)**

iii. Using the information from the boxplots, explain why the maximum daily *temperature* is associated with the *month* of the year. Quote the values of appropriate statistics in your response.　**(1 mark)**

More exam questions are available online.

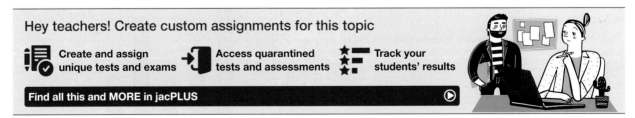

Hey teachers! Create custom assignments for this topic

Create and assign unique tests and exams

Access quarantined tests and assessments

Track your students' results

Find all this and MORE in jacPLUS

Answers

Topic 1 Investigating and comparing data distributions

1.2 Types of data

1.2 Exercise

1. a. Numerical
   b. Categorical
   c. Numerical
2. Nominal
3. Ordinal
4. a. Continuous
   b. Continuous
   c. Discrete
5. a. Continuous
   b. Discrete
   c. Ordinal

6.

Data	Type	
a. Wines rated as high, medium or low quality	Categorical	Ordinal
b. The number of downloads from a website	Numerical	Discrete
c. The electricity usage over a three-month period	Numerical	Continuous
d. A volume of petrol sold by a petrol station per day	Numerical	Continuous

7. Nominal categorical
8. a. i. Interval   ii. Nominal
   b. Data set i is more useful.
9. a. Q1 ordinal, Q2 ratio
   b. Question 2 because ratio is better than ordinal.
10. The weight is a continuous data. The weight is measured on a scale. Continuous data has an infinite set of possible values.

1.2 Exam questions

Note: Mark allocations are available with the fully worked solutions online.
1. A
2. B
3. B

1.3 Categorical data distributions

1.3 Exercise

1.

Favourite type of music	Frequency
Pop	9
Rock	7
Classical	2
Folk	3
Electronic	9

2. a.

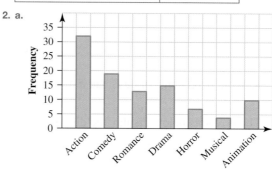

Favourite movie genre

b. 100 people surveyed; most common 'action'; least common 'musical'

3.

Favourite pizza	Frequency
Margherita	7
Pepperoni	11
Supreme	9
Meat feast	14
Vegetarian	6
Other	13

4. a. Nominal categorical

b.

Transport method	Frequency
Bus	9
Walk	3
Train	4
Car	7
Bicycle	3

c.

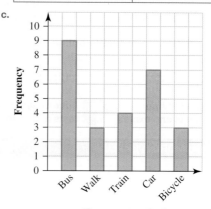

Transport method

d. 26 people surveyed; most common 'bus'; least common 'walk' and 'bicycle'

5. a.

Response	Frequency
Yes	16
Don't care	3
No	7

b.

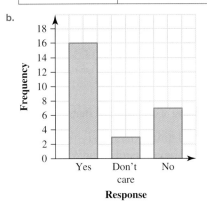

c. Ordinal, as it makes sense to arrange the data in order from 'Yes' to 'No', with 'Don't care' between them.

6. a.

Favourite animal	Frequency
Dog	7
Cat	8
Rabbit	4
Guinea pig	3
Rat	2
Ferret	1

b.

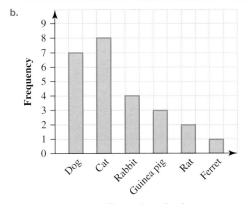

c. Cat

7. a. Flat white   b. 70

8. a. Ordinal categorical

b. The data should be in order from 'Strongly agree' through to 'Strongly disagree'.

c.

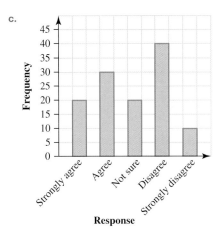

9. a.

Result	Frequency
A	3
B	2
C	8
D	4
E	3

b.

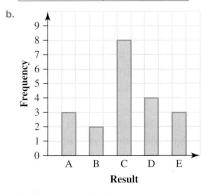

c. Ordinal categorical

10.

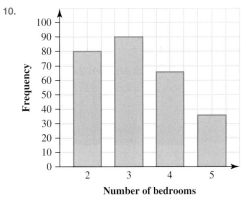

11. a.

Temperature	Frequency
Above average	3
Average	9
Below average	3

b.

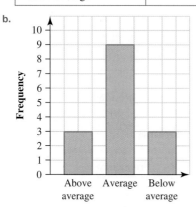

Temperature

c. Ordinal categorical

12. a. 40 **b.** 15%

13. a. 15–19

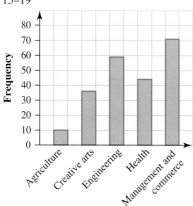

Main area of education and study

20–24

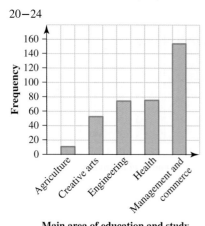

Main area of education and study

25–34

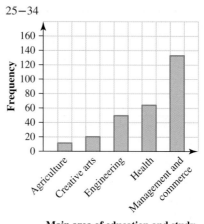

Main area of education and study

35–44

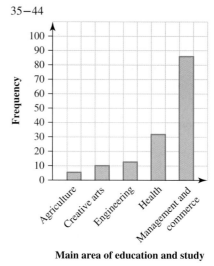

Main area of education and study

45–64

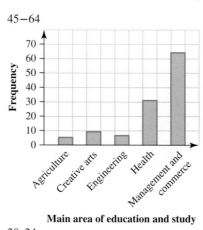

Main area of education and study

b. 20–24

14. a.

Age group	Frequency
Under 20s	18
20–29	15
30–39	13
40–49	10
50–59	7
60+	12

b. Under 20

c.

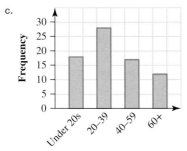

Age group

d. Yes, the modal category is now 20–39.

1.3 Exam questions

Note: Mark allocations are available with the fully worked solutions online.

1. E
2. B
3. B

1.4 Numerical data distributions — frequency tables and histograms

1.4 Exercise

1.

Time (seconds)	Frequency
30 – < 40	3
40 – < 50	5
50 – < 60	7
60 – < 70	4
70 – < 80	1

2. a.

Time (seconds)	Frequency
80 – < 90	2
90 – < 100	6
100 – < 110	5
110 – < 120	5
120 – < 130	2

b.

Time (seconds)	Frequency
85 – < 90	2
90 – < 95	1
95 – < 100	5
100 – < 105	2
105 – < 110	3
110 – < 115	3
115 – < 120	2
120 – < 125	2

3. a.

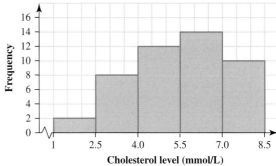

b.

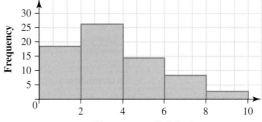

4. a.

Class interval	Frequency
5 – < 10	5
10 – < 15	9
15 – < 20	16
20 – < 25	9
25 – < 30	1

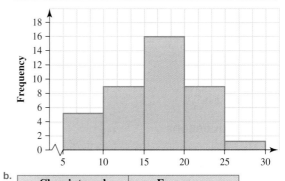

b.

Class interval	Frequency
11 – < 16	10
16 – < 21	5
21 – < 26	5
26 – < 31	6
31 – < 36	8
36 – < 41	1
41 – < 46	3
46 – < 51	2

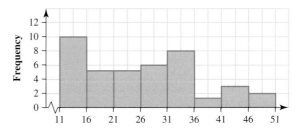

5. a.

Average hours	Frequency
$20 - <25$	1
$25 - <30$	5
$30 - <35$	5
$35 - <40$	21
$40 - <45$	8

b.

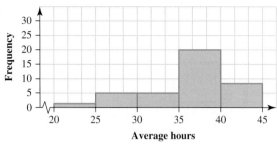

c. The distribution has one mode with data values that are most frequent in the $35 - <40$ interval. There are no obvious outliers, and the distribution has a negative skew.

6. C

7. a. $0 - <5$

b.

Number of hours worked	Frequency
$0 - <5$	25
$5 - <10$	14
$10 - <15$	8
$15 - <20$	3
$20 - <25$	1

c. 51

8. Data sets 1 and 3

1.4 Exam questions

Note: Mark allocations are available with the fully worked solutions online.

1.

Class interval (5 cm)	Frequency
110–114	2
115–119	0
120–124	10
125–129	11
130–134	1
135–139	1
140–144	1
145–149	2

2.

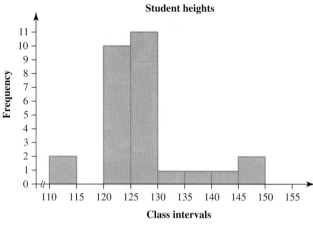

3. a. 220 [1 mark]

b. [1 mark]

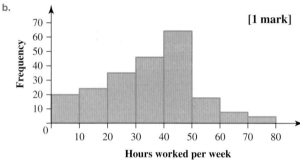

c. 45 hours per week [1 mark]

1.5 Numerical data distributions — dot plots and stem plots

1.5 Exercise

1. a. Key: $1*|9 = \$19$

Stem	Leaf
1*	9
2	1 1 2 2 2 2 2 2 3 3 4 4
2*	5 6

b. Key: $0*|6 = 6$ hours

Stem	Leaf
0*	6 7 9
1	3 4 4
1*	7 7
2	0 0 1 1 1 3 3 4 4
2*	5 5 6

2. a. Key: $1|2 = 12$ passengers

March leaf	Stem	April Leaf
3	1	2 4
7 5	1*	5 5 7
4 3 3 2	2	0 0 2 3 3
8 7 7 5	2*	7 7 7 8
4 4 3	3	0
7 6 5	3*	5 6 6
4 2	4	3
7	4*	

b.

Key: 1|7 = 17 patients

Dr. Hammond Leaf	Stem	Dr. Valenski Leaf
	1*	7 7
4 3 3	2	1 2 3 4 4
6 5	2*	5 5 6 6 8
4 2 1	3	0 0 3 4
8 8 7	3*	5
3 1	4	1
6 5 5 4	4*	
1	5	1
6 5	5*	
	6	0

3. Key: 0*|8 = 8 people

Stem	Leaf
0*	8 8
1	3 4 4
1*	6 6 7 7 9
2	1 1
2*	5

4.

Key: 0|1 = 1 game played

Squad 1 leaf	Stem	Squad 2 Leaf
1	0	
4	1	
7	1*	
4 4	2	4
8	2*	
3 3	3	1 2
6 5	3*	6
3 2 1	4	3 4
	4*	5
1 1	5	2
5	6*	
	7	
	8	2
	8*	5 7
1	9	3

5. a.

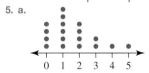

Number of wickets

b.

Hours checking email

6. a.

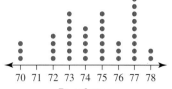

Round score

b.

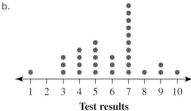

Test results

7.

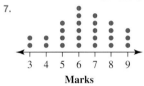

Marks

8. a. Histogram

b. Bar chart or dot plot

9. Histogram or dot plot

10. a. Key: 0|1 = 1

Stem	Leaf
0	1
0*	
1	1 1 1 4 4
1*	6 6 7 8
2	3 3 4 4
2*	7 7 9

b. Splitting the stem for this data gives a clearer picture of the spread and shape of the distribution of the data set.

1.5 Exam questions

Note: Mark allocations are available with the fully worked solutions online.

1. 38 cm

2. a. Day number

b. key : 4|1 = 4.1 $n - 15$

minimum temperature (°C)

4	1 8
5	
6	0 7
7	0 5 7
8	0 6
9	0 2 8
10	7
11	8
12	7

3. A

1.6 Characteristics of numerical data distributions

1.6 Exercise

1. a. The distribution has one mode with data values that are most frequent in the 35−<40 interval. There are no obvious outliers, and there is a negative skew to the distribution.

   b. The distribution has one mode with data values that are most frequent in the 45−<60 interval. There are two potential outliers in the 120−<135 interval, and the distribution is either symmetrical (excluding the outliers) or has a slight positive skew (including the outliers).

2. a.

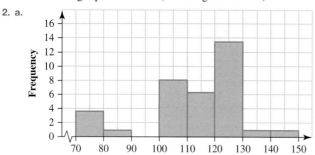

   The distribution has one mode with data values that are most frequent in the 120− <130 interval. There are potential outliers in the 70− <80 interval, and there is a negative skew to the distribution.

   b.

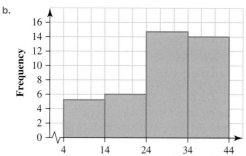

   The distribution has one mode with data values that are most frequent in the 24−<34 interval. There are no obvious outliers, and there is a negative skew to the distribution.

3. a.

Number of pets	Frequency	Percentage
0–1	18	45%
2–3	12	30%
4–5	8	20%
More than 5	2	5%
Total	40	100%

   b. Discrete numerical

   c. 40

d.

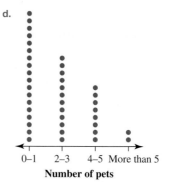

Number of pets

   e. The distribution has one mode with data values that are most frequent in the 0–1 interval. There are no obvious outliers, and the distribution has a positive skew.

4. a. 65%

   b. 15%

   c. The distribution has one mode with data values that are most frequent in the 90−<120 interval. There are no obvious outliers, and the distribution has a negative skew.

5. a.

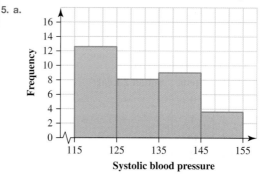

Systolic blood pressure

   b. The distribution has one mode with data values that are most frequent in the 115−<125 interval. There are no obvious outliers, and the distribution has a positive skew.

   c.

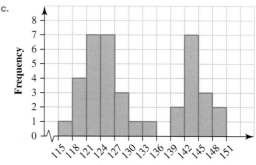

Systolic blood pressure

   d. The second histogram is split into two distinct groups with three modes. The lower group is symmetrical around the interval 121−<128. The upper group has a slight positive skew.

e. The second histogram might demonstrate that there are two distinct groups present in the data, for example a younger age group and an older age group.

6. a. See the image at the bottom of the page.*

b.
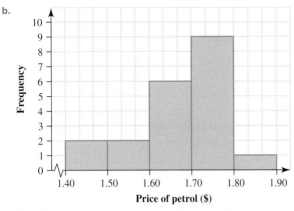
Price of petrol ($)

c. Both histograms have one mode with a negative skew to the distribution. The first histogram gives the impression that there may be outliers at the start of the data set, but this is not as evident in the second histogram.

7. a. i.
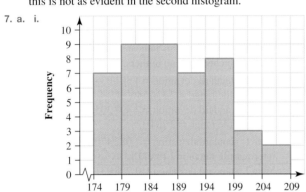
Height (cm)

ii.
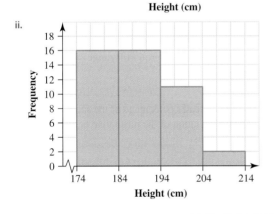
Height (cm)

iii.
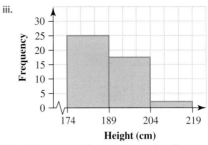
Height (cm)

b. The histogram with more intervals allows you to see the shape and details of the distribution better, although the overall shape is fairly similar in all three histograms.
   i. This distribution has two modes (181.5 and 186.5), with a positive skew and no obvious outliers.
   ii. This distribution also has two modes (179 and 189), with a positive skew and no obvious outliers.
   iii. This distribution has one mode (181.5), with a positive skew and no obvious outliers.

8.
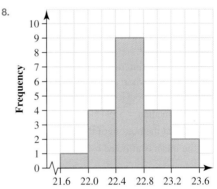
Temperature (°C)

The distribution for the interval 1993–2012 is approximately symmetrical, with a slight negative skew.

1.6 Exam questions

Note: Mark allocations are available with the fully worked solutions online.
1. B
2. E
3. B

*6. a.
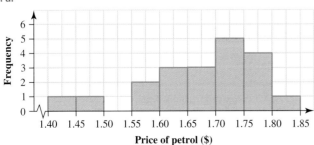
Price of petrol ($)

1.7 Summarising numerical data — mean and median

1.7 Exercise

1. 128.4
2. a. 24.76
   b. 27.71
   c. The mean increased significantly.
3. 26.18
4. a. 51   b. 198
5. a. Mean = 867.89 mm, median = 654 mm
   b. The median, as it is not affected by the extreme values present in the data set.
6. a. 24
   b. 24
   c. The median is unchanged.
7. a. 100   b. 1.805
8. The median, as the data set has two clear outliers.
9. a. $74 231
   b. $65 000
   c. It would be in the workers' interest to use a higher figure when negotiating salaries, whereas it would be in the management's interest to use a lower figure.
10. a. Mean = $1 269 850, median = $594 500
    b. and c. See the image at the bottom of the page.*
    d. The median, as the mean is affected by a few very high values.
11. Mean = 14 : 23
12. a. Mean = 7.55, Median = 6
    b. The median would be the preferred choice due to the extreme value of 34.

1.7 Exam questions

Note: Mark allocations are available with the fully worked solutions online.
1. a. 24 years   b. 1.065 kg/L
2. 24.55 kg/m^2
3. B

1.8 Summarising numerical data — range, interquartile range and standard deviation

1.8 Exercise

1. 407
2. 2.555
3. a. Class A = 11, Class B = 10
   b. Class A = 2, Class B = 5
4. a. 160
   b. 57
   c. Goals against: range = 293, interquartile range = 140
      The 'Goals against' column is significantly more spread out than the 'Goals for' column.
5. 5.56
6. a. 7.37
   b. 7
   c. Standard deviation = 11.49, interquartile range = 7
   d. The standard deviation increased by 4.12, while the interquartile range was unchanged.
7. a. 2.9
   b. 1.25
   c. The range is less than double the value of the interquartile range. This indicates that the data is quite tightly bunched with no outliers.
8. a. Year 1: standard deviation = 1 301 033.5, interquartile range = 2 336 546
      Year 2: standard deviation = 1 497 303.5, interquartile range = 2 734 078
   b. Year 1: standard deviation = 1 082 470.9, interquartile range = 2 136 718
      Year 2: standard deviation = 1 228 931, interquartile range = 2 415 365
   c. Both values are reduced, but there is a bigger impact on the interquartile range than the standard deviation.
9. a. Mean = 41 440.78, standard deviation = 2248.92
   b. Median = 41 333, interquartile range = 3609
   c. 59.38%
   d. 50%
   e. There is a greater percentage of the sample within one standard deviation of the mean than between the first and third quartiles.

*10. b, c

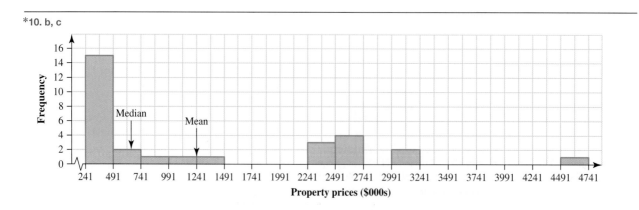

10. a. Standard deviation $= 0.46$, interquartile range $= 0.7$

b. See the table at the bottom of the page.*

c. Sydney

d. Darwin

1.8 Exam questions

Note: Mark allocations are available with the fully worked solutions online.

1. 2.6

2. C

3. E

1.9 Symmetrical and asymmetrical distributions

1.9 Exercise

1. a. i. Asymmetrical

ii. Median and interquartile range

b. i. Symmetrical

ii. Mean and standard deviation

2. a.

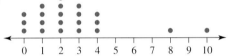

b. Asymmetrical

c. Median and interquartile range

3. a. 9.1 **b.** 14.9 **c.** 90.48%

4. a. 65%

b. 90%

c. 100%

d. 100% because 3 standard deviations contained 100%

5. a. 64.29% **b.** 100%

6. a.

b. 2.22

c. 66.67%

7. a. Asymmetrical

b. Median

c. i. -1.5

ii. It is negative.

iii. Can't play for negative hours

d. Interquartile range

8. a. 68% **b.** 96% **c.** 100% **d.** Yes

9. a. i. 79.8 **ii.** 22.7 **iii.** 81.82%

b. i. 73.9 **ii.** 12.0 **iii.** 70%

c. The mean and standard deviation decrease.

d. 73.9 — mean without outlier

10. a. i. 68% **ii.** 95%

b. i. 68% **ii.** 95%

1.9 Exam questions

Note: Mark allocations are available with the fully worked solutions online.

1. C

2. D

3. E

1.10 The five-number summary and boxplots

1.10 Exercise

1. a. 25 °C **b.** 21 °C **c.** 7 °C

2.

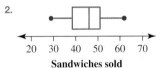

3. a. 17, 23, 28, 33, 43

b.

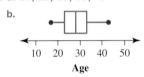

c. The data is fairly symmetrical with no obvious outliers.

4. B

5. 1. matches e.

2. matches a.

3. matches c.

4. matches b.

5. matches d.

6. a. True **b.** True **c.** False

7. a. Lower fence $= 4.625$, upper fence $= 12.425$

b. 4.4 is an outlier.

c.

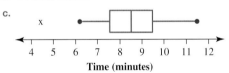

8. a. 104 **b.** 43 **c.** 61 **d.** 17

9. $X_{min} = 7, Q_3 = 13$

10. a. 26

b. 16

c. 20

d. $-11, 53$

e. A commute longer than 53 minutes.

***10. b.**

	Sydney	**Melbourne**	**Brisbane**	**Adelaide**	**Perth**	**Hobart**	**Darwin**	**Canberra**
Std dev.	0.46	0.40	0.51	0.45	0.44	0.38	0.67	0.41
IQR	0.7	0.7	0.9	0.8	0.6	0.6	1.2	0.6

1.10 Exam questions

Note: Mark allocations are available with the fully worked solutions online.

1. C

2. Please see the answer in the Worked solution.

3. B

1.11 Comparing the distribution of a numerical variable across two or more groups

1.11 Exercise

1. a.

b. On the whole, store 2 has less sales than store 1; however, the sales of store 2 are much more consistent than store 1's sales.

The sales of store 1 have a negative skew, while the sales of store 2 are symmetrical. There are no obvious outliers in either data set.

2. a. School B

b. School B

c. 3 (School B has a bigger interquartile range.)

3. a. Brand X   **b.** Brand Y   **c.** Brand X   **d.** Brand X

4. a. $69 in Restaurant A is an outlier.

b. Restaurant A: 24, 28, 33.5, 38, 42
Restaurant B: 18, 25, 30, 38, 46

c.

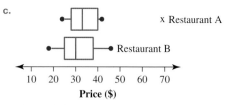

d. The meals in Restaurant A are more consistently priced, but are also generally higher priced. The distribution of prices at Restaurant A has a positive skew, while the distribution of prices at Restaurant B is nearly symmetrical.

5. a. Party A: 370, 497, 651, 769, 838
Party B: 255, 335, 632, 801, 971

b.

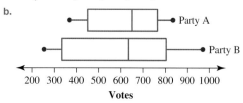

c. The spread of votes for Party B is far larger than it is for Party A. Party A polled more consistently and had a higher median number of votes. Party A had a nearly symmetrical distribution of votes, while Party B's votes had a slight negative skew.

6. a. Company A

Share price ($)	Frequency
10−< 15	2
15−< 20	3
20−< 25	3
25−< 30	4
30−< 35	4
35−< 40	2

Company B

Share price ($)	Frequency
15−< 20	2
20−< 25	3
25−< 30	3
30−< 35	3
35−< 40	2
40−< 45	3
45−< 50	2

b.

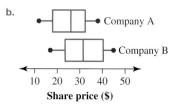

c. On the whole, the share price of Company B is greater than the share price of Company A. However, the share price of Company A is more consistent than the share price of Company B. The share price of Company A has a negative skew, while the share price of Company B has a nearly symmetrical distribution.

7. a. True   **b.** True   **c.** True   **d.** False

8. a. Phone A   **b.** Phone A   **c.** Phone B
d. Phone C   **e.** Phone C

9. a. Drink 1: 35, 43, 51, 58.5, 66
Drink 2: 37, 48.5, 59, 67, 77
Drink 3: 46, 49, 51.5, 55, 61

b.

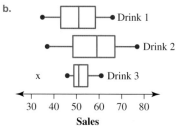

c. The sales of Drink 3 are by far the most consistent, although overall Drink 2 has the highest sales. Drink 2's sales are also the most inconsistent of all the drinks. There is one outlier in the data sets (33 in Drink 3).

10. a. Suburb A: 275, 289, 300, 307.5, 315
Suburb B: 250, 272.5, 295, 315, 340

b.

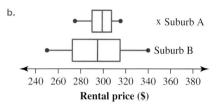

Rental price ($)

c. The rental prices in Suburb A are far more consistent than the rental prices in Suburb B. There is one outlier in the data sets ($350 in Suburb A). Although Suburb A has a higher median rental price, you could not say that it was definitely more expensive than Suburb B.

d.

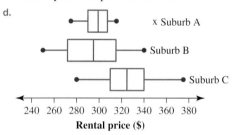

Rental price ($)

e. Suburb C has a higher average rental price than either Suburb A or B. The spread of the prices in Suburb C is more similar to those in Suburb B than Suburb A.

1.11 Exam questions

Note: Mark allocations are available with the fully worked solutions online.

1. C

2. a. i. 5 °C   ii. 10 °C   iii. One

   b. 15 days

3. E

1.12 Review

1.12 Exercise

Multiple choice

1. D
2. E
3. A
4. B
5. B
6. D

7. B

8. C

9. B

Short answer

10. a. Mean = 94.6, median = 93.3

   b. IQR = 8.64, standard deviation = 7.49

   c. For this sample, the mean and the standard deviation would be preferred as there are no clear outliers and no apparent skew.

11. a. Year 1: mean = 6432.9, median = 1324
    Year 2: mean = 2637.6, median = 1201

   b. The mean values are significantly different but the medians are very similar. This would seem to indicate the presence of extreme values in the data.

12. a. Mean = 202 461.4, standard deviation = 257 819.6

   b. Median = 7655, IQR = 475 307
    The median and interquartile range are not influenced by the presence of the much smaller values for Darwin and Adelaide.

13. a.

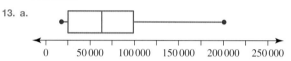

   b. Five-figure summary:
    17 747, 25 025, 63 161, 98 700, 200 552
    The data shows a positive skew with the upper 25% having a much greater spread than the lower 25%. The middle 50% is approximately symmetrical. There are no outliers.

14. a. i.

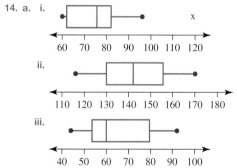

   b. i. Negative skew with an upper outlier

    ii. Symmetrical with no outliers

    iii. Positive skew with no outliers

15. a. See the image at the bottom of the page*

---

*15. a.

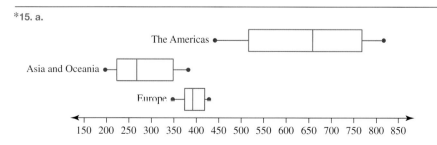

b. Military spending is greater but more variable in the Americas as indicated by the larger range and IQR. Europe has the least variable spending. Asia and Oceania spend the least but are more variable than Europe.

Extended response

16. a. See the image at the bottom of the page.*

    b. Indigenous mean = 1410.5,
       Non-indigenous mean = 35 241.6

    c. Indigenous median = 1156,
       Non-indigenous median = 24 008

    d. Indigenous: standard deviation = 1193.8,
       IQR = 1875.5
       Non-indigenous: standard deviation = 33 949.03,
       IQR = 58 198

    e. The median and IQR are probably more appropriate due to the presence of potential extreme values in the data.

17. a. 1980–89: mean = 19.34, standard deviation = 0.44
       2003–12: mean = 20.29, standard deviation = 0.45

    b. The mean temperature is about one degree higher in the period 2003–12, but the standard deviations indicate that the data have similar spreads.

    c. Total data: mean = 19.82, standard deviation = 0.65

    d. The mean of the total data is halfway between the two separate time periods. The standard deviation indicates a much greater variation from the mean for the total data.

18. a.

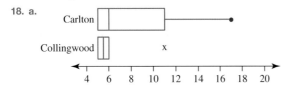

b. Goals scored in grand finals for this sample of players is greater but more variable among the Carlton players, as indicated by the larger range and IQR. Collingwood players are concentrated at 5 or 6 with the exception of the one upper outlier of 11.

c. See the image at the bottom of the page.*

d. Kicks for this sample of players are greater but more variable than the handballs, as indicated by the larger range and IQR. Both are positively skewed with one upper outlier.

1.12 Exam questions

Note: Mark allocations are available with the fully worked solutions online.

1. E

2. D

3. The *month* is the explanatory variable and the *minimum daily temperature* is the response variable. The median values decrease with the month, which is expected as the year moves from summer into winter months.

4. a. Place of capture

    b. 20 mm

    c.

Place of capture	Wingspan (mm)				
	minimum	Q_1	median (M)	Q_3	maximum
Forst	**16**	20	21	32	52
Grassland	18	24	30	**36**	45

    d. Please see the worked solution in the online resources.

    e. Please see the worked solution in the online resources.

---

*16. a.

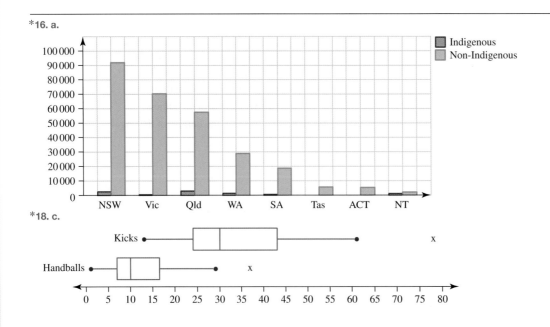

*18. c.

5. a. i.

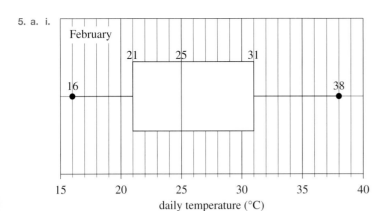

ii. 75%

b. **i.** July: Positively skewed with an outlier at the high end
May: Approximately symmetric with no outliers

ii. 15.5

iii. The median maximum daily temperature for May is approximately 14.4 °C and for July it is approximately 9.1 °C. These are very different temperatures, which we would expect as July is in winter and May is in autumn. Therefore, the maximum daily temperature is associated with the month of the year.

2 Linear relations and equations

Fully worked solutions for this topic are available online.

2.1 Overview

2.1.1 Introduction

Linear equations have been around for over 4000 years. A simple 2 × 2 linear equation system with two unknowns was first solved by the Babylonians. Around 200 BC, the Chinese demonstrated the ability to solve a 3 × 3 system of equations. However, it wasn't until the 17th century that progress was made in linear algebra by the inventor of calculus, Leibniz. This was followed by work by Cramer and later Gauss.

Linear equations themselves were invented in 1843 by Irish mathematician Sir William Rowan Hamilton. He was reputedly a genius: at the young age of 13, he spoke 13 languages, and at 22 he was a professor at the University of Dublin. His work has been applied in many fields, given that there are many situations in which there is a direct relationship between two variables. Classic examples are of water being added to a tank at a constant rate or a taxi trip being charged at a constant rate per kilometre.

A linear equation to model the cost of a taxi trip can be used to compare one taxi company to another. The break-even point refers to the point at which the cost is the same for each taxi company. This is the point at which the two linear graphs intersect and it can be found using a graphical technique, or using substitution or elimination techniques of simultaneous equations.

KEY CONCEPTS

This topic covers the following key concepts from the VCE Mathematics Study Design:

- the linear function $y = a + bx$, its graph, and interpretation of the parameters a and b in terms of initial value and constant rate of change respectively
- formulation and analysis of linear models from worded descriptions or relevant data (including simultaneous linear equations in two variables) and their application to solve practical problems including domain of interpretation
- piecewise linear (line segment, step) graphs and their application to modelling practical situations, including tax scales and charges and payment.

Note: Concepts shown in grey are covered in other topics.

Source: VCE Mathematics Study Design (2023–2027) extracts © VCAA; reproduced by permission.

2.2 Linear relations and solving linear equations

2.2.1 Identifying linear relations

A **linear relation** is a relationship between two **variables** that when plotted give a straight line. When a linear relation is expressed as an equation, it is called a *linear equation*. Linear equations come in many different forms and can be easily identified because the powers on the pronumerals are always 1. A linear equation defines or describes relationships between two or more variables.

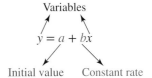

Identifying a linear equation

Linear	**Non-linear**
• **Equations contain one, or at most, two variables.**	• **Equations involve pronumerals with powers other than 1 (e.g. roots, squares, cubes, exponents).**
• **The powers on the pronumerals are always 1.**	• **Equations include the scenario when two pronumerals are multiplied together.**
• **The pronumerals cannot be multiplied together.**	
Examples of linear equations are:	**Examples of non-linear equations are:**
$6x = 24, \ y = 11 + 4x, \ \dfrac{3(t+1)}{2} = 12$	$y = 5x^2, \ y = 4^{2x}, \ xy = 7, \ \sqrt{y} = 9$

Many real-life situations can be described by linear relations, such as water being added to a tank at a constant rate, money being saved when the same amount is deposited at regular time intervals or converting between degrees Celsius and Fahrenheit.

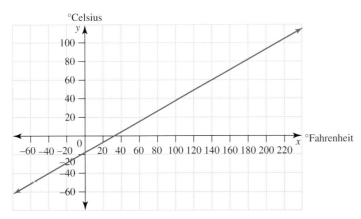

tlvd-3696

Identify which of the following equations are linear.

a. $y = 1 + 4x$

b. $b = c^2 - 5c + 6$

c. $y = \sqrt{x}$

d. $m^2 = 6(n - 10)$

e. $d = \dfrac{3t + 8}{7}$

f. $y = 5^x$

THINK	WRITE
a. 1. Identify the variables.	**a.** y and x
2. Write the power of each variable.	y has a power of 1. x has a power of 1.
3. Determine whether the equation is linear.	Since both variables have a power of 1, this is a linear equation.
b. 1. Identify the variables.	**b.** b and c
2. Write the power of each variable.	b has a power of 1. c has a power of 2.
3. Determine whether the equation is linear.	c has a power of 2, so this is not a linear equation.
c. 1. Identify the variables.	**c.** y and x
2. Write the power of each variable. *Note:* A square root is a power of $\dfrac{1}{2}$.	y has a power of 1. x has a power of $\dfrac{1}{2}$.
3. Determine whether the equation is linear.	x has a power of $\dfrac{1}{2}$, so this is not a linear equation.
d. 1. Identify the variables.	**d.** m and n
2. Write the power of each variable.	m has a power of 2. n has a power of 1.
3. Determine whether the equation is linear.	m has a power of 2, so this is not a linear equation.
e. 1. Identify the variables.	**e.** d and t
2. Write the power of each variable.	d has a power of 1. t has a power of 1.
3. Determine whether the equation is linear.	Since both variables have a power of 1, this is a linear equation.
f. 1. Identify the two variables.	**f.** y and x
2. Write the power of each variable.	y has a power of 1. x is the power.
3. Determine whether the equation is linear.	Since x is the power, this is not a linear equation.

2.2.2 Transposing linear equations

If we are given a **linear equation** between two variables, we are able to **transpose** this relationship. That is, we can change the equation so that the variable on the right-hand side of the equation becomes the stand-alone variable on the left-hand side of the equation (the subject of the equation). To transpose the linear equation $y = a + bx$ to make x the subject of the equation means the subject of the equation is the variable that is being calculated. It is the letter on its own on one side of the equals sign.

WORKED EXAMPLE 2 Transposing linear equations

Transpose the linear equation $y = 7 + 4x$ to make x the subject of the equation.

THINK	WRITE
1. Isolate the variable on the right-hand side of the equation (by subtracting 7 from both sides).	$y - 7 = 7 + 4x - 7$ $y - 7 = 4x$
2. Divide both sides of the equation by the coefficient of the variable on the right-hand side (in this case 4).	$\dfrac{y-7}{4} = \dfrac{4x}{4}$ $\dfrac{y-7}{4} = x$
3. Transpose the relation by interchanging the left-hand side and the right-hand side.	$x = \dfrac{y-7}{4}$

TI | THINK	DISPLAY/WRITE	CASIO | THINK	DISPLAY/WRITE
1. On a Calculator page, press MENU and select: 3: Algebra 1: Solve Complete the entry line as: solve $(y = 7 + 4x, x)$ then press ENTER.	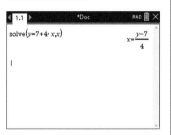	1. On the Main screen, complete the entry line as: solve $(y = 7 + 4x, x)$ then press EXE.	
2. The answer appears on the screen.	$x = \dfrac{y-7}{4}$	2. The answer appears on the screen.	$x = \dfrac{y}{4} - \dfrac{7}{4}$

2.2.3 Solving linear equations with one variable

To solve linear equations with one variable, all operations performed on the variable need to be identified in order, and then the opposite operations need to be performed in reverse order.

In practical problems, solving linear equations can answer everyday questions such as the time required to have a certain amount in the bank, the time taken to travel a certain distance or the number of participants needed to raise a certain amount of money for charity.

Opposite operations

A list of mathematical operations and their corresponding opposite operation are listed below. This information will be useful when solving linear equations.

Operation	+	−	×	÷
Opposite operation	−	+	÷	×

Solve the following linear equations to determine the unknowns.

a. $5x = 12$ 　　　　b. $8t + 11 = 20$ 　　　　c. $12 = 4(n-3)$ 　　　　d. $\dfrac{4x-2}{3} = 5$

THINK

a. 1. Identify the operations performed on the unknown.

2. Write the opposite operation.

3. Perform the opposite operation on both sides of the equation.

4. Write the answer in its simplest form.

b. 1. Identify the operations performed in order on the unknown.

2. Write the opposite operations.

3. Perform the opposite operations in reverse order on both sides of the equation, one operation at a time.

4. Write the answer in its simplest form.

c. 1. Identify the operations performed in order on the unknown. (Remember operations in brackets are performed first.)

2. Write the opposite operations.

3. Perform the opposite operations on both sides of the equation in reverse order, one operation at a time.

4. Write the answer in its simplest form.

d. 1. Identify the operations performed in order on the unknown.

WRITE

a. $5x = 5 \times x$

So the operation is $\times 5$.

The opposite operation is $\div 5$.

Step 1 ($\div 5$):

$5x = 12$

$\dfrac{5x}{5} = \dfrac{12}{5}$

$x = \dfrac{12}{5}$

$x = \dfrac{12}{5}$

b. $8t + 11$

The operations are $\times 8, +11$.

$\div 8, -11$

Step 1 (-11):

$8t + 11 = 20$

$8t + 11 - 11 = 20 - 11$

$8t = 9$

Step 2 ($\div 8$):

$\dfrac{8t}{8} = \dfrac{9}{8}$

$t = \dfrac{9}{8}$

$t = \dfrac{9}{8}$

c. $4(n-3)$

The operations are $-3, \times 4$.

$+3, \div 4$

Step 1 ($\div 4$):

$12 = 4(n-3)$

$\dfrac{12}{4} = \dfrac{4(n-3)}{4}$

$3 = n - 3$

Step 2 ($+3$):

$3 + 3 = n - 3 + 3$

$6 = n$

$n = 6$

d. $\dfrac{4x-2}{3}$

The operations are $\times 4, -2, \div 3$.

2. Write the opposite operations.

$\div 4, +2, \times 3$

3. Perform the opposite operations on both sides of the equation in reverse order, one operation at a time.

Step 1 ($\times 3$):
$$\frac{4x-2}{3} = 5$$
$$3 \times \frac{4x-2}{3} = 5 \times 3$$
$$4x - 2 = 15$$

Step 2 ($+2$):
$$4x - 2 + 2 = 15 + 2$$
$$4x = 17$$

Step 3 ($\div 4$):
$$\frac{4x}{4} = \frac{17}{4}$$
$$x = \frac{17}{4}$$

4. Write the answer in its simplest form.

$$x = \frac{17}{4}$$

TI	THINK	DISPLAY/WRITE	CASIO	THINK	DISPLAY/WRITE
d. 1. On a Calculator page, press MENU and select: 3: Algebra 1: Solve Complete the entry line as: $\text{solve}\left(\frac{-2+4x}{3} = 5, x\right)$ then press ENTER.		d. 1. On the Main screen, complete the entry line as: $\text{solve}\left(\frac{-2+4x}{3} = 5, x\right)$ then press EXE.	
2. The answer appears on the screen.	$x = \dfrac{17}{4}$	2. The answer appears on the screen.	$x = \dfrac{17}{4}$

2.2.4 Substituting into linear equations

If we are given a linear equation between two variables and have the value of one of the variables, we can **substitute** this into the equation to determine the other value.

WORKED EXAMPLE 4 Substituting numbers into a linear equation

Substitute $x = 3$ into the linear equation $y = 5 + 2x$ to determine the value of y.

THINK	WRITE
1. Substitute the variable (x) with the given value.	$y = 5 + 2(3)$
2. Evaluate the right-hand side of the equation.	$y = 5 + 6$ $y = 11$

TI	THINK	DISPLAY/WRITE	CASIO	THINK	DISPLAY/WRITE
1. On a Calculator page, complete the entry line as: $y = 5 + 2x \mid x = 3$ then press ENTER. *Note:* The 'given' or 'with' symbol $(\mid)$ is found by pressing CTRL, then the $=$ button, then selecting $\mid$.		1. On the Main screen, complete the entry line as: $y = 5 + 2x \mid x = 3$ then press EXE. *Note:* The 'given' or 'with' symbol $(\mid)$ is found in the 'Math 3' menu on the Keyboard menu.	
2. The answer appears on the screen.	$y = 11$	2. The answer appears on the screen.	$y = 11$

2.2.5 Literal linear equations

A **literal equation** is an equation that includes several pronumerals or variables. Literal equations often represent real-life situations. The equation $y = a + bx$ is an example of a literal linear equation where the parameters a and b represent the initial value and constant rate of change respectively.

To solve literal linear equations, you need to isolate the variable you are trying to solve for.

WORKED EXAMPLE 5 Solving literal linear equations

Solve the literal linear equation $y = a + bx$ for x.

THINK	WRITE
1. Isolate the terms containing the variable you want to solve on one side of the equation.	$y - a = bx$
2. Divide by the coefficient of the variable you want to solve for.	$\dfrac{y - a}{b} = x$
3. Transpose the equation.	$x = \dfrac{y - a}{b}$

TI	THINK	DISPLAY/WRITE	CASIO	THINK	DISPLAY/WRITE
1. On a Calculator page, press MENU and select: 3: Algebra 1: Solve Complete the entry line as: solve $(y = a + bx, x)$ then press ENTER. *Note:* Be sure to enter the multiplication operation between the variables m and x.		1. On the Main screen, complete the entry line as: solve $(y = a + b \times x, x)$ then press EXE. *Note:* Be sure to enter the multiplication operation between the variables m and x.	
2. The answer appears on the screen.	$x = \dfrac{y - a}{b}$	2. The answer appears on the screen.	$x = \dfrac{y}{b} - \dfrac{a}{b}$

  Resources

Interactivities Transposing linear equations (int-6449)

Solving linear equations (int-6450)

2.2 Exercise

1. **WE1** Identify which of the following equations are linear.

   a. $y^2 = 7x + 1$
   b. $t = 7x^3 - 6x$
   c. $y = 3(x + 2)$
   d. $m = 2^{x+1}$
   e. $4x + 5y - 9 = 0$

2. Bethany was asked to identify which equations from a list were linear. The following table shows her responses.

Equation	Bethany's response
$y = 1 + 4x$	Yes
$y^2 = 5x - 2$	Yes
$y + 6x = 7$	Yes
$y = x^2 - 5x$	No
$t = 6d^2 - 9$	No
$m^3 = n + 8$	Yes

   a. Insert another column into the table and add your responses identifying which of the equations are linear.
   b. Provide advice to Bethany to help her to correctly identify linear equations.

3. Identify which of the following are linear equations.

   a. $y = 5 + 2t$
   b. $x^2 = 2y + 5$
   c. $m = 3(n + 5)$
   d. $d = 25 + 80t$
   e. $\sqrt{y} = x + 5$
   f. $s = \dfrac{100}{t}$

4. Samson was asked to identify which of the following were linear equations. His responses are shown in the table.

Equation	Samson's response
$y = 6 + 5x$	Yes
$y^2 = 6x - 1$	Yes
$y = x^2 + 4$	No
$y^3 = 7(x + 3)$	Yes
$y = 6 + \dfrac{1}{2}x$	Yes
$\sqrt{y} = 4x + 2$	Yes
$y^2 + 5x^3 + 9 = 0$	No
$10y - 11x = 12$	Yes

   a. Based on Samson's responses, determine whether he would state that $6y^2 + 7x = 9$ is linear. Justify your answer.
   b. Provide advice to Samson to ensure that he can correctly identify linear equations.

5. **WE2** Transpose the linear equation $y = -3 + 6x$ to make x the subject of the equation.

6. Transpose the following linear equations to make x the subject.

   a. $y = 5 + 2x$ b. $3y = 8 + 6x$ c. $p = -6 + 5x$ d. $6y = 1 + 3x$

7. **WE3** Solve the following linear equations to determine the unknowns.

   a. $2(x + 1) = 8$ b. $n - 12 = -2$ c. $4d - 7 = 11$ d. $\dfrac{x+1}{2} = 9$

8. a. Write the operations in order that have been performed on the unknowns in the following linear equations.

   i. $10 = 4a + 3$ ii. $3(x + 2) = 12$ iii. $\dfrac{s+1}{2} = 7$ iv. $16 = 2(3c - 9)$

   b. Calculate the exact values of the unknowns in part **a** by solving the equations. Show all of the steps.

9. Calculate the exact values of the unknowns in the following linear equations.

   a. $14 = 5 - x$ b. $\dfrac{4(3y-1)}{5} = -2$ c. $\dfrac{2(3-x)}{3} = 5$

10. Solve the following literal linear equations for the pronumerals given in brackets.

    a. $v = u + at$ (a) b. $xy - k = m$ (x) c. $\dfrac{x}{p} - r = s$ (x)

11. **WE4** Substitute $x = 5$ into the linear equation $y = 5 - 6x$ to determine the value of y.

12. Substitute $x = -3$ into the linear equation $y = 3 + 3x$ to determine the value of y.

13. The equation $w = 120 + 10t$ represents the amount of water in a tank, w (litres), at any time, t (minutes). Calculate the time, in minutes, that it takes for the tank to have the following amounts of water.

    a. 450 litres b. 1200 litres

14. **WE5** Solve the literal linear equation $-q + px = r$ for x.

15. Solve the literal linear equation $C = \pi d$ for d.

16. Yorx was asked to solve the linear equation $5w - 13 = 12$. His solution is shown.

    Step 1: $\times 5, -13$

    Step 2: Opposite operations $\div 5, +13$

    Step 3: $5w - 13 = 12$
    $$\dfrac{5w - 13}{5} = \dfrac{12}{5}$$
    $$w - 13 = 2.4$$

    Step 4: $w - 13 + 13 = 2.4 + 13$
    $$w = 15.4$$

    a. Show that Yorx's answer is incorrect by calculating the value of w.
    b. Determine what advice you would give to Yorx so that he can solve linear equations correctly.

17. The literal linear equation $F = 1.8(K - 273) + 32$ converts the temperature in Kelvin (K) to Fahrenheit (F). Solve the equation for K to give the formula for converting the temperature in Fahrenheit to Kelvin.

18. Consider the linear equation $y = \dfrac{3x+1}{4}$. Calculate the value of x for each of the following y-values.

    a. 2 b. -3 c. $\dfrac{1}{2}$ d. 10

19. The distance travelled, d (kilometres), at any time t, (hours), can be found using the equation $d = 95t$. Calculate the time, in hours, that it takes to travel the following distances. Give your answers correct to the nearest minute.

    a. 190 km   b. 250 km   c. 65 km   d. 356.5 km   e. 50 000 m

20. The amount, A, in dollars in a bank account at the end of any month, m, can found using the equation $A = 400 + 150m$.

    a. Calculate how many months (to the nearest month) it would take to have the following amounts of money in the bank account.

       i. $1750   ii. $3200

    b. Calculate how long it would take to have $10 000 in the bank account. Give your answer correct to the nearest month.

21. The temperature, C, in degrees Celsius can be found using the equation $C = \dfrac{5(F - 32)}{9}$, where F is the temperature in degrees Fahrenheit. Nora needs to set her oven at 190 °C, but her oven's temperature is measured in Fahrenheit.

    a. Write the operations performed on the variable F.
    b. Write the order in which the operations need to be performed to calculate the value of F.
    c. Determine the temperature in Fahrenheit that Nora should set her oven to.

22. The equation that determines the surface area of a cylinder with a radius of 3.5 cm is $A = 3.5\pi(3.5 + h)$. Determine the height in cm of cylinders with radii of 3.5 cm and the following surface areas. Give your answers correct to 2 decimal places.

    a. 200 cm²   b. 240 cm²   c. 270 cm²

2.2 Exam questions

Question 1 (1 mark)
MC The interest, I, owed on a loan of $1000 borrowed for 1 year depends on the rate of interest $r\%$ p.a., according to the formula $I = 10r$. For an interest rate of 5%, the interest owed will be
   A. $0.50   **B.** $5   **C.** $50   **D.** $500   **E.** $5000

Question 2 (1 mark)
Solve the equation $2(x + 1) = 5(x - 2)$.

Question 3 (1 mark)
MC Determine which solution satisfies all the following equations.

$$3x - \frac{1}{2} = \frac{1}{4}$$
$$4x + 5 = 6$$
$$4 - \frac{8x}{3} = \frac{10}{3}$$

A. $x = \dfrac{1}{4}$ **B.** $x = 4$ **C.** $x = \dfrac{1}{12}$ **D.** $x = -\dfrac{1}{4}$ **E.** $x = -4$

More exam questions are available online.

2.3 Developing linear equations

LEARNING INTENTION

At the end of this subtopic you should be able to:
- develop linear equations from worded problems with one or more than one unknown
- generate tables of values from equations.

2.3.1 Developing linear equations from word descriptions

To write a worded statement as a linear equation, we must first identify the unknown and choose a **pronumeral** to represent it. We can then use the information given in the statement to write a linear equation in terms of the pronumeral.

The linear equation can then be solved as before, and we can use the result to answer the original question.

tlvd-3698

WORKED EXAMPLE 6 Setting up and solving a linear equation with one unknown

Soft drink multipacks containing 24 × 375 mL cans are sold at SupaSave for $34.00. Form and solve a linear equation to determine the price of 1 can of soft drink correct to the nearest cent.

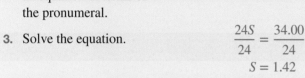

THINK	WRITE
1. Identify the unknown and choose a pronumeral to represent it.	S = price of a can of soft drink
2. Use the given information to write an equation in terms of the pronumeral.	$24S = 34.00$
3. Solve the equation.	$\dfrac{24S}{24} = \dfrac{34.00}{24}$ $S = 1.42$
4. Interpret the solution in terms of the original problem.	The price of 1 can of soft drink is $1.42.

2.3.2 Tables of values

Tables of values can be generated from formulas by entering given values of one variable into the formula. Tables of values can be used to solve problems and to draw graphs representing situations (as covered in more detail in Topic 5).

WORKED EXAMPLE 7 Constructing a table of values

The amount of water filling a tank is found by the rule $W = 20 + 100t$, where W is the amount of water in the tank in litres and t is the time in hours.

a. Generate a table of values that shows the amount of water, W, in the tank every hour for the first 8 hours (i.e. $t = 0, 1, 2, 3, ..., 8$).

b. Using your table, determine how long it will take, in hours, for there to be over 700 litres in the tank.

THINK	**WRITE**

a. 1. Enter the required values of t into the formula to calculate the values of W.

a.

$t = 0$:
$W = 20 + 100\,(0)$
$\quad = 20$

$t = 1$:
$W = 20 + 100\,(1)$
$\quad = 120$

$t = 2$:
$W = 20 + 100\,(2)$
$\quad = 220$

$t = 3$:
$W = 20 + 100\,(3)$
$\quad = 320$

$t = 4$:
$W = 20 + 100\,(4)$
$\quad = 420$

$t = 5$:
$W = 20 + 100\,(5)$
$\quad = 520$

$t = 6$:
$W = 20 + 100\,(6)$
$\quad = 620$

$t = 7$:
$W = 20 + 100\,(7)$
$\quad = 720$

$t = 8$:
$W = 20 + 100\,(8)$
$\quad = 820$

2. Enter the calculated values into a table of values.

t	0	1	2	3	4	5	6	7	8
W	20	120	220	320	420	520	620	720	820

b. 1. Use your table of values to locate the required column.

b.

t	0	1	2	3	4	5	6	⑦	8
W	20	120	220	320	420	520	620	⑦⑳	820

2. Read the corresponding values from your table and write the answer.

$t = 7$

It will take 7 hours for there to be over 700 litres of water in the tank.

TI | THINK

a. 1. In a Lists & Spreadsheet page, label the first column t and the second column w. Enter the values 0 to 8 in the first column. In the formula cell underneath the label w, enter the rule for w starting with an $=$ sign, then press ENTER.

DISPLAY/WRITE

1.1 ▶		*Doc	RAD 🔲 ✕
	A t	B w	C	D
=		$100 \cdot t + 20$		
1	0			
2	1			
3	2			
4	3			
5	4			
B	$w := 100 \cdot t + 20$			◀ ▶

CASIO | THINK

a. 1. On a Spreadsheet screen, type t into cell A1 and w into cell B1. Enter the values 0 to 8 in cells A2 to A10.
In cell B2, complete the entry line as:
$= 100 \times A2 + 20$
then press EXE.

DISPLAY/WRITE

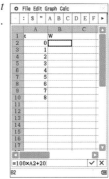

2. When prompted, select Variable Reference for t and select OK.

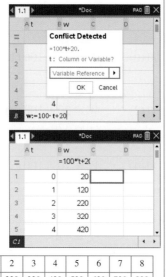

2. Click on cell B2 and select:
   • Edit
   • Copy
   Highlight cells B3 to B10, then select:
   • Edit
   • Past

3. The table of values can be read from the screen.

t	0	1	2	3	4	5	6	7	8
W	20	120	220	320	420	520	620	720	820

3. The table of values can be read from the screen.

t	0	1	2	3	4	5	6	7	8
W	20	120	220	320	420	520	620	720	820

2.3 Exercise

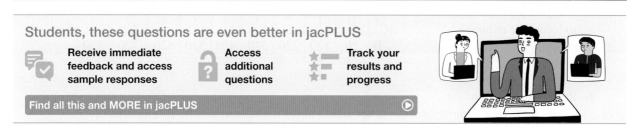

Students, these questions are even better in jacPLUS

Receive immediate feedback and access sample responses

Access additional questions

Track your results and progress

Find all this and MORE in jacPLUS

1. **WE6** Art pencils at the local art supply store sell in packets of 8 for $17.92. Form and solve a linear equation to determine the price of 1 art pencil.

2. Natasha is trying to determine which type of cupcake is the best value for money. The three options Natasha is considering are:
   • 4 red velvet cupcakes for $9.36
   • 3 chocolate delight cupcakes for $7.41
   • 5 caramel surprise cupcakes for $11.80.

Form and solve linear equations for each type of cupcake to determine which has the cheapest price per cupcake.

3. Three is added to a number and the result is then divided by four, giving an answer of nine. Determine the number.

4. One pair of sides of a parallelogram are each 3 times the length of the sides in the other pair. Calculate the side lengths if the perimeter of the parallelogram is 84 cm.

5. One week Jordan bought a bag of his favourite fruit-and-nut mix at the local market. The next week he saw that the bag was on sale for 20% off the previously marked price. Jordan purchased two more bags at the reduced price. Jordan spent $20.54 in total for the three bags. Calculate the original price of a bag of fruit-and-nut mix.

6. Six times the sum of four plus a number is equal to 126. Determine the number.

7. Libby enjoys riding along Beach Road on a Sunday morning. She rides at a constant speed of 0.4 kilometres per minute.

   a. Generate a table of values that shows how far Libby has travelled for each of the first 10 minutes of her journey.
   b. One Sunday Libby meets a friend 3 kilometres into her journey. Determine the minutes between which Libby meets her friend.

8. **WE7** Tommy is saving for a remote-controlled car that is priced at $49. He has $20 in his piggy bank. Tommy saves $3 of his pocket money every week and puts it in his piggy bank.
   The amount of money in dollars, M, in his piggy bank after w weeks can be found using the rule $M = 20 + 3w$.

   a. Generate a table of values that shows the amount of money, M, in Tommy's piggy bank every week for 12 weeks (i.e. $w = 0, 1, 2, 3, ..., 12$).
   b. Using your table, determine how many weeks it will take for Tommy to have saved enough money to purchase the remote-controlled car.

9. Sabrina is a landscape gardener and has been commissioned to work on a rectangular piece of garden. The length of the garden is 6 metres longer than the width, and the perimeter of the garden is 64 m. Calculate the dimensions of the garden.

10. Jett is starting up a small business selling handmade surfboard covers online. The start-up cost is $250. He calculates that each cover will cost $14.50 to make.
    The rule that finds the cost, C, to make n covers is $C = 250 + 14.50n$.

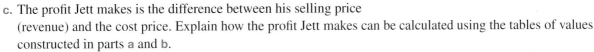

    a. Using CAS or otherwise, generate a table of values to determine the cost of producing 10 to 20 surfboard covers.
    b. If Jett sells the covers for $32.95, construct a table of values to determine the revenue for selling 10 to 20 surfboard covers.
    c. The profit Jett makes is the difference between his selling price (revenue) and the cost price. Explain how the profit Jett makes can be calculated using the tables of values constructed in parts a and b.
    d. Using your explanation in part c and your table of values, determine the profits made by Jett if he sells 10 to 20 surfboard covers.

2.3 Exam questions

Question 1 (1 mark)
MC Think of a number, add 3, then multiply the result by 6. The answer is 24. Determine which equation can be solved to find the initial number.

   A. $6x + 3 = 24$
   B. $6(x - 3) = 24$
   C. $6x - 3 = 24$
   D. $6(x + 3) = 24$
   E. $3x + 6 = 24$

Question 2 (1 mark)
MC In Year 11 at Orana Secondary College, there are 20% more girls than boys. There are 44 students altogether in Year 11. If b represents the number of boys in the year group, determine which equation can be used to find the number of boys.

   A. $b + 0.2b = 44$
   B. $b + 20b = 44$
   C. $b + 1.2b = 44$
   D. $b = 44 - 20b$
   E. $b = 44 + 1.2b$

`MC` The table shows the perimeters of paddocks whose lengths are twice their widths.

Length (m)	Width (m)	Perimeter (m)
100	50	300
300	150	900
500	250	1500
700	350	2100
900	450	2700
1100	550	3300
1300	650	3900
1500	750	4500
1700	850	5100
1900	950	5700
2100	1050	6300

Determine the perimeter of the paddock with length 1.1 km.

    A. 400 m   **B.** 550 m   **C.** 310 m   **D.** 2200 m   **E.** 3300 m

More exam questions are available online.

2.4 Simultaneous linear equations

LEARNING INTENTION

At the end of this subtopic you should be able to:
- determine the point of intersection of two linear graphs
- solve simultaneous equations using the substitution method
- solve simultaneous equations using the elimination method
- develop linear equations from worded problems with more than one unknown.

2.4.1 Solving simultaneous equations graphically

Simultaneous equations are sets of equations that can be solved at the same time. They often represent practical problems that have two or more unknowns. For example, you can use simultaneous equations to calculate the costs of individual apples and oranges when different amounts of each are bought.

Point of intersection

Solving simultaneous equations gives the set of values that is common to all of the equations. If these equations are presented graphically, then the set of values common to all equations is the point of intersection.

To solve a set of simultaneous equations graphically, the equations must be sketched on the same set of axes to determine the point of intersection. If the equations do not intersect, then there is no solution for the simultaneous equations.

tlvd-3699

WORKED EXAMPLE 8 Determining the point of intersection of two linear graphs

Consider the following simultaneous equations.

$$y = 4 + 2x \text{ and } y = 3 + 3x$$

Using CAS or otherwise, sketch both graphs on the same set of axes and solve the equations.

THINK

1. Use CAS or another method to sketch the graphs $y = 4 + 2x$ and $y = 3 + 3x$.

2. Locate the point where the graphs intersect (or cross over).

3. Using the graph, identify the x- and y-values of the point of intersection.

WRITE

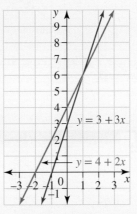

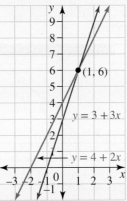

The solution is $(1, 6)$, or $x = 1$ and $y = 6$.

TI | THINK

1. On a Graphs page, complete the entry line for function 1 as:
$f1(x) = 4 + 2 \cdot x$ then press ENTER.

DISPLAY/WRITE

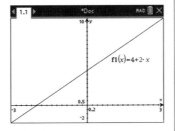

CASIO | THINK

1. On a Graph & Table screen, complete the entry line for function 1 as:
$y1 = 4 + 2 \cdot x$
then click the tick box.

DISPLAY/WRITE

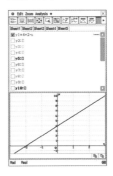

2. Complete the entry line for function 2 as:
$f2(x) = 3 + 3 \cdot x$
then press ENTER.
Note: press the e button to bring up the entry line.

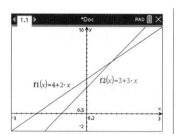

2. Complete the entry line for function 2 as:
$y2 = 3 + 3 \cdot x$
then click the tick box.

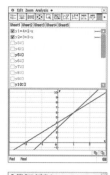

3. Press MENU and select:
6: Analyze Graph
4: Intersection
Move to the left of the point of intersection and press the CLICK button.

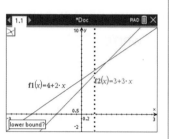

3. Select:
- – Analysis
- – G-Solve
- – Intersection
The coordinates of the point of intersection appear on the screen.

4. Move to the right of the point of intersection and press the CLICK button.

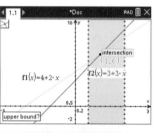

4. Write the answer.   The solution is (1, 6), or $x = 1$ and $y = 6$.

5. The coordinates of the point of intersection appear on the screen.

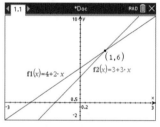

6. Write the answer.   The solution is (1, 6), or $x = 1$ and $y = 6$.

2.4.2 Solving simultaneous equations using substitution

Simultaneous equations can also be solved algebraically. One algebraic method is known as substitution. This method requires one of the equations to be substituted into the other by replacing one of the variables. The second equation is then solved and the value of one of the variables is determined.

The substitution method is often used when one or both of the equations are transposed so that one of the variables has been made the subject, for example, $c = 15 + 12b$ and $2c + 3b = -3$, or $y = 6 + 4x$ and $y = 2 + 6x$.

WORKED EXAMPLE 9 Solving simultaneous equations using the substitution method

Solve the following pairs of simultaneous equations using substitution.

a. $c = -15 + 12b$ and $2c + 3b = -3$

b. $y = 6 + 4x$ and $y = 2 + 6x$

c. $3x + 2y = -1$ and $y = -8 + x$

THINK	WRITE
a. 1. Identify which variable will be substituted into the other equation.	a. $c = -15 + 12b$
2. Substitute $c = -15 + 12b$ into the other equation.	$2c + 3b = -3$ $2(12b - 15) + 3b = -3$
3. Expand and simplify the left-hand side, and solve the equation for the unknown variable.	$-30 + 24b + 3b = -3$ $27b - 30 = -3$ $27b = -3 + 30$ $27b = 27$ $b = 1$
4. Substitute the value of $b = 1$ into one of the equations.	$c = -15 + 12b$ $= -15 + 12(1)$ $= -3$
5. Write the answer.	The solution is $b = 1$ and $c = -3$.
b. 1. Both equations are in the form $y = a + bx$, so set them equal each other.	b. $6 + 4x = 2 + 6x$
2. Move all of the pronumerals to one side.	$6 + 4x - 4x = 2 + 6x - 4x$ $6 = 2 + 6x - 4x$ $6 = 2 + 2x$
3. Solve for the unknown.	$6 - 2 = 2 - 2 + 2x$ $4 = 2x$ $\dfrac{4}{2} = \dfrac{2x}{2}$ $2 = x$
4. Substitute the value found into either of the original equations.	$y = 6 + 4x$ $= 6 + 4 \times 2$ $= 6 + 8$ $= 14$
5. Write the answer.	The solution is $x = 2$ and $y = 14$.
c. 1. One equation is in the form $y = a + bx$, so substitute this equation into the other.	c. $3x + 2y = -1$ $3x + 2(-8 + x) = -1$
2. Expand and simplify the equation.	$3x - 16 + 2x = -1$ $5x - 16 = -1$
3. Solve for the unknown.	$5x - 16 + 16 = -1 + 16$ $5x = 15$ $x = 3$

| 4. Substitute the value found into either of the original equations. | $y = -8 + x$
$= -8 + 3$
$= -5$ |
| 5. Write the answer. | The solution is $x = 3$ and $y = -5$. |

2.4.3 Solving simultaneous equations using elimination

Solving simultaneous equations using **elimination** requires the equations to be added or subtracted so that one of the pronumerals is eliminated or removed. Simultaneous equations that have both pronumerals on the same side are often solved using elimination.

For example, $3x + y = 5$ and $4x - y = 2$ both have x and y on the same side of the equation, so they can be solved with this method.

tlvd-3701

WORKED EXAMPLE 10 Solving simultaneous equations using elimination

Solve the following pairs of simultaneous equations using elimination.
a. $3x + y = 5$ and $4x - y = 2$
b. $2a + b = 7$ and $a + b = 5$
c. $3c + 4d = 5$ and $2c + 3d = 4$

THINK	WRITE
a. 1. Write the simultaneous equations with one on top of the other.	a. $3x + y = 5 \quad [1]$ $4x - y = 2 \quad [2]$
2. Select one pronumeral to be eliminated.	Select y.
3. Check the coefficients of the pronumeral being eliminated.	The coefficients of y are 1 and -1.
4. If the coefficients are the same number but with different signs, add the equations together.	$[1] + [2]$: $3x + 4x + y - y = 5 + 2$ $\qquad\qquad 7x = 7$
5. Solve the equation for the unknown pronumeral.	$7x = 7$ $\dfrac{7x}{7} = \dfrac{7}{7}$ $x = 1$
6. Substitute the pronumeral back into one of the equations.	$3x + y = 5$ $3(1) + y = 5$
7. Solve the equation to determine the value of the other pronumeral.	$3 + y = 5$ $3 - 3 + y = 5 - 3$ $y = 2$
8. Write the answer.	The solution is $x = 1$ and $y = 2$.
b. 1. Write the simultaneous equations with one on top of the other.	b. $2a + b = 7 \quad [1]$ $a + b = 5 \quad [2]$
2. Select one pronumeral to be eliminated.	Select b.

3. Check the coefficients of the pronumeral being eliminated.

The coefficients of b are both 1.

4. If the coefficients are the same number with the same sign, subtract one equation from the other.

$[1] - [2]$:
$2a - a + b - b = 7 - 5$

5. Solve the equation for the unknown pronumeral.

$a = 2$

6. Substitute the pronumeral back into one of the equations.

$a + b = 5$
$2 + b = 5$

7. Solve the equation to determine the value of the other pronumeral.

$b = 5 - 2$
$b = 3$

8. Write the answer.

The solution is $a = 2$ and $b = 3$.

c. 1. Write the simultaneous equations with one on top of the other.

c. $3c + 4d = 5$ $[1]$
$2c + 3d = 4$ $[2]$

2. Select one pronumeral to be eliminated.

Select c.

3. Check the coefficients of the pronumeral being eliminated.

The coefficients of c are 3 and 2.

4. If the coefficients are different numbers, then multiply them both by another number, so they both have the same value.

$3 \times 2 = 6$
$2 \times 3 = 6$

5. Multiply the equations (all terms in each equation) by the numbers selected in step 4.

$[1] \times 2$:
$6c + 8d = 10$
$[2] \times 3$:
$6c + 9d = 12$

6. Check the sign of each coefficient for the selected pronumeral.

$6c + 8d = 10$ $[3]$
$6c + 9d = 12$ $[4]$
Both coefficients of c are 6.

7. If the signs are the same, subtract one equation from the other and simplify.

$[3] - [4]$:
$6c - 6c + 8d - 9d = 10 - 12$
$-d = -2$

8. Solve the equation for the unknown.

$d = 2$

9. Substitute the pronumeral back into one of the equations.

$2c + 3d = 4$
$2c + 3(2) = 4$

10. Solve the equation to determine the value of the other pronumeral.

$2c + 6 = 4$
$2c + 6 - 6 = 4 - 6$
$2c = -2$
$\dfrac{2c}{2} = \dfrac{-2}{2}$
$c = -1$

11. Write the answer.

The solution is $c = -1$ and $d = 2$.

TI	THINK	DISPLAY/WRITE	CASIO	THINK	DISPLAY/WRITE

c. 1. On a Calculator page, press MENU and select:
3: Algebra
1: Solve
Complete the entry line as:
solve $(3 \cdot c + 4 \cdot d = 5$
and $2 \cdot c + 3 \cdot d = 4, \{c, d\})$
then press ENTER.

solve($3 \cdot c + 4 \cdot d = 5$ and $2 \cdot c + 3 \cdot d = 4, \{c, d\}$)
$c = -1$ and $d = 2$

c. 1. On the Main screen, complete the entry line as:
solve
$(\{3c + 4d = 5, 2c + 3d = 4\},$
$\{c, d\})$ then press EXE.

2. The answer appears on the screen.

$c = -1$ and $d = 2$.

2. The answer appears on the screen.

$c = -1$ and $d = 2$.

2.4.4 Word problems with more than one unknown

In some instances a word problem might contain more than one unknown. If we are able to express both unknowns in terms of the same pronumeral, we can create a linear equation as before and solve it to determine the value of both unknowns.

tlvd-3702

WORKED EXAMPLE 11 Setting up and solving a linear equation with two unknowns

Nyibol is counting the number of insects and spiders she can find in her back garden. All insects have 6 legs and all spiders have 8 legs. In total, Nyibol finds 43 bugs with a total of 290 legs.
By forming a linear equation, determine exactly how many insects and spiders Nyibol found.

THINK	WRITE
1. Identify one of the unknowns and choose a pronumeral to represent it.	Let $s =$ the number of spiders.
2. Define the other unknown in terms of this pronumeral.	Let $43 - s =$ the number of insects.
3. Write expressions for the total numbers of spiders' legs and insects' legs.	Total number of spiders' legs $= 8s$ Total number of insects' legs $= 6(43 - s)$ $= 258 - 6s$
4. Create an equation for the total number of legs of both types of creature.	$8s + (258 - 6s) = 290$
5. Solve the equation.	$8s + 258 - 6s = 290$ $8s - 6s = 290 - 258$ $2s = 32$ $s = 16$
6. Substitute this value back into the second equation to determine the other unknown.	The number of insects $= 43 - 16$ $= 27$
7. Write the answer in a sentence.	Nyibol found 27 insects and 16 spiders.

Resources

Interactivities Solving simultaneous equations graphically (int-6452)

Solving simultaneous equations using substitution (int-6453)

2.4 Exercise

Students, these questions are even better in jacPLUS

Receive immediate feedback and access sample responses

Access additional questions

Track your results and progress

Find all this and MORE in jacPLUS

1. **WE8** Consider the following simultaneous equations.

$$y = 1 + 5x \text{ and } y = -5 + 2x$$

Using CAS or otherwise, sketch both graphs on the same set of axes and solve the equations.

2. Using CAS or otherwise, sketch and solve the following three simultaneous equations.

$$y = 7 + 3x, \ y = 8 + 2x \text{ and } y = 12 - 2x$$

3. A pair of simultaneous equations can be solved graphically as shown in the diagram. From the diagram, determine the solution for this pair of simultaneous equations.

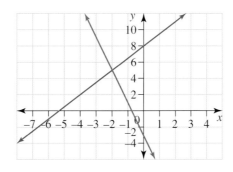

4. Using CAS or otherwise, solve the following sets of simultaneous equations graphically.

   a. $y = 1 + 4x$ and $y = -1 + 3x$

   b. $y = -5 + x$ and $y = 3 - 3x$

   c. $y = 3(x - 1)$ and $y = 2(2x + 1)$

   d. $y = -1 + \dfrac{x}{2}$ and $y = \dfrac{x}{4 + 2}$

   e. $y = 4 + 3x, \ y = 3 + 2x$ and $y = -x$

5. **WE9** Solve the following pairs of simultaneous equations using substitution.

   a. $y = 1 + 2x$ and $2y - x = -1$
   b. $m = 5 + 2n$ and $m = -1 + 4n$
   c. $2x - y = 5$ and $y = 1 + x$

6. Determine the solutions to the following pairs of simultaneous equations using substitution.

   a. $2(x + 1) + y = 5$ and $y = -6 + x$

   b. $\dfrac{x + 5}{2} + 2y = 11$ and $y = -2 + 6x$

7. **MC** Select which of the following pairs of simultaneous equations would best be solved using the substitution method.

   A. $4y - 5x = 7$ and $3x + 2y = 1$
   B. $3c + 8d = 19$ and $2c - d = 6$
   C. $12x + 6y = 15$ and $9x - y = 13$
   D. $3(t + 3) - 4s = 14$ and $2(s - 4) + t = -11$
   E. $n = 9m + 12$ and $3m + 2n = 7$

8. Use the substitution method to solve the following pairs of simultaneous equations.

   a. $y = 5 + 2x$ and $y = -2 + 3x$
   b. $y = -2 + 5x$ and $y = 2 + 7x$
   c. $y = 2(3x + 1)$ and $y = 4(2x - 3)$
   d. $y = -9 + 5x$ and $3x - 5y = 1$
   e. $3(2x + 1) + y = -19$ and $y = -1 + x$
   f. $\dfrac{3x + 5}{2} + 2y = 2$ and $y = -2 + x$

9. A student chose to solve the following pair of simultaneous equations using the elimination method.

   $$3x + y = 8 \text{ and } 2x - y = 7$$

   a. Explain why this student's method would be the most appropriate method for this pair of simultaneous equations.
   b. Show how these equations would be solved using this method.

10. **WE10** Solve the following pairs of simultaneous equations using elimination.

    a. $3x + y = 5$ and $4x - y = 2$
    b. $2a + b = 7$ and $a + b = 5$
    c. $3c + 4d = 5$ and $2c + 3d = 4$

11. Consider the following pair of simultaneous equations.

    $$ax - 3y = -16 \text{ and } 3x + y = -2.$$

    If $y = 4$, determine the values of a and x.

12. Using the elimination method, solve the following pairs of simultaneous equations.

    a. $4x + y = 6$ and $x - y = 4$
    b. $x + y = 7$ and $x - 2y = -5$
    c. $2x - y = -5$ and $x - 3y = -10$
    d. $4x + 3y = 29$ and $2x + y = 13$
    e. $5x - 7y = -33$ and $4x + 3y = 8$
    f. $\dfrac{x}{2} + y = 7$ and $3x + \dfrac{y}{2} = 20$

13. **MC** The first step when solving the following pair of simultaneous equations using the elimination method is:

    $$2x + y = 3 \quad [1]$$
    $$3x - y = 2 \quad [2]$$

    A. add equations [1] and [2] together.
    B. multiply both equations by 2.
    C. subtract equation [1] from equation [2].
    D. multiply equation [1] by 2 and multiply equation [2] by 3.
    E. subtract equation [2] from equation [1].

14. **WE11** Fredo is buying a large bunch of flowers for his mother in advance of Mother's Day. He picks out a bunch of roses and lilies, with each rose costing $6.20 and each lily costing $4.70. In total he picks out 19 flowers and pays $98.30.
    Form a linear equation to determine exactly how many roses and lilies Fredo bought.

15. Miriam has a sweet tooth, and her favourite sweets are strawberry twists and chocolate ripples. The local sweet shop sells both as part of their pick and mix selection, so Miriam fills a bag with them.
    Each strawberry twist weighs 5 g and each chocolate ripple weighs 9 g.
    Miriam's bag has 28 sweets, weighing a total of 188 g.

    Determine the number of each type of sweet that Miriam bought by forming and solving a linear equation.

16. Michelle and Lydia live 325 km apart. On a Sunday they decide to drive to each other's respective towns. They pass each other after 2.5 hours. If Michelle drives an average of 10 km/h faster than Lydia, calculate the speed at which they are both travelling.

17. A shopper Yuri is doing their weekly grocery shop and is buying both carrots and potatoes. They calculate that the average weight of a carrot is 60 g and the average weight of a potato is 125 g.

Furthermore, they calculate that the average weight of the carrots and potatoes that they purchase is 86 g. If the shopping weighed 1.29 kg in total, determine how many of each they purchased.

18. Brendon and Marcia were each asked to solve the following pair of simultaneous equations.

$$3x + 4y = 17 \quad [1]$$
$$4x - 2y = 19 \quad [2]$$

Marcia decided to use the elimination method. Her solution steps are shown.

Step 1: [1] × 4:
12x + 16y = 68   [3]
[2] × 3:
12x − 6y = 57   [4]

Step 2: [3] + [4]:
10y = 125

Step 3: y = 12.5

Step 4: Substitute y = 12.5 into [1]:
3x + 4(12.5) = 17

Step 5: Solve for x:
3x = 17 − 50
3x = −33
3x = 17 − 50   3x = −33   x = −11

Step 6: The solution is x = −11 and y = 12.5.

a. Marcia made an error in step 2. Explain where she made her error, and hence correct her mistake.
b. Using the correction you made in part a, determine the correct solution to this pair of simultaneous equations.
c. Brendon decided to eliminate y instead of x. Using Brendon's method of eliminating y first, show all the appropriate steps involved to reach a solution.

19. In a ball game, a player can kick the ball into the net to score a goal or place the ball over the line to score a behind. The scores in a game between the Rockets and the Comets were:

Rockets: 6 goals 12 behinds, total score 54
Comets: 7 goals 5 behinds, total score 45

The two simultaneous equations that can represent this information are shown.

Rockets: 6x + 12y = 54
Comets: 7x + 5y = 45

a. By solving the two simultaneous equations, determine the number of points that are awarded for a goal and a behind.
b. Using the results from part a, determine the scores for the game between the Jetts, who scored 4 goals and 10 behinds, and the Meteorites, who scored 6 goals and 9 behinds.

20. Milan and Yashab both work part time at an ice-cream shop. The simultaneous equations shown represent the number of hours Milan (x) and Yashab (y) work each week.

Equation 1: Total number of hours worked by Yashab and Milan: $x + y = 15$

Equation 2: Number of hours worked by Yashab in terms of Milan's hours: $y = 2x$

a. Explain why substitution would be the best method to use to solve these equations.

b. Use substitution to determine the number of hours worked by Milan and Yashab each week.

c. To ensure that he has time to do his Mathematics homework, Milan changes the number of hours he works each week. He now works $\dfrac{1}{3}$ of the number of hours worked by Yashab. An equation that can be used to represent this information is $x = \dfrac{y}{3}$.

Determine the number of hours worked by Milan, given that the total number of hours that Milan and Yashab work does not change.

2.4 Exam questions

Question 1 (1 mark)

MC The graphs of $y = 3 + 2x$ and $y = -2 - 3x$ are shown. Their point of intersection is

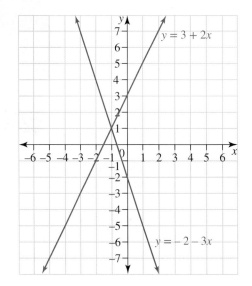

A. $(1, -1)$ **B.** $(0, 3)$ **C.** $(0, -2)$ **D.** $(0, 0)$ **E.** $(-1, 1)$

Question 2 (1 mark)

MC Using the substitution method, the solution to the simultaneous equations $x + 2y = 9$ and $y = -1 + 5x$ is

A. $(1, 4)$ **B.** $(9, 5)$ **C.** $(4, 1)$ **D.** $(-1, 5)$ **E.** $(2, 9)$

Question 3 (1 mark)

MC Using the elimination method, the solution to the simultaneous equations $2x + y = 8$ and $3x - y = 17$ is

A. $(-5, 2)$ **B.** $(2, 5)$ **C.** $(5, -2)$ **D.** $(5, 2)$ **E.** $(-5, -2)$

More exam questions are available online.

2.5 Problem solving with simultaneous equations

LEARNING INTENTION

At the end of this subtopic you should be able to:
- construct simultaneous equations from worded problems
- use simultaneous equations to determine the break-even point.

2.5.1 Setting up simultaneous equations

The solutions to a set of simultaneous equations satisfy all equations that were used. Simultaneous equations can be used to solve problems involving two or more variables or unknowns, such as the cost of 1 kilogram of apples and bananas, or the number of adults and children attending a show.

WORKED EXAMPLE 12 Constructing simultaneous equations from worded problems

At a fruit shop, 2 kg of apples and 3 kg of bananas cost \$13.16, and 3 kg of apples and 2 kg of bananas cost \$13.74. Represent this information in the form of a pair of simultaneous equations.

THINK	WRITE
1. Identify the two variables.	The cost of 1 kg of apples and the cost of 1 kg of bananas
2. Select two pronumerals to represent these variables. Define the variables.	a = cost of 1 kg of apples b = cost of 1 kg of bananas
3. Identify the key information and rewrite it using the pronumerals selected.	2 kg of apples can be written as $2a$. 3 kg of bananas can be written as $3b$.
4. Construct two equations using the information.	$2a + 3b = 13.16$ $3a + 2b = 13.74$

2.5.2 Break-even points

A **break-even point** is a point where the costs equal the selling price. It is also the point where there is zero profit.

For example, if the equation $C = 45 + 3t$ represents the production cost to sell t shirts, and the equation $R = 14t$ represents the revenue from selling the shirts for \$14 each, then the break-even point is the number of shirts that need to be sold to cover all costs.

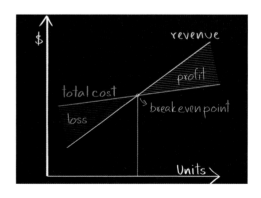

Determining the break-even point

To determine the break-even point, the equations for cost and revenue are solved simultaneously. That is, solve the equation $R = C$.

tlvd-3702

WORKED EXAMPLE 13 Using simultaneous equations to determine the break-even point

Santo sells shirts for $25. The revenue, R, for selling n shirts is represented by the equation $R = 25n$. The cost to make n shirts is represented by the equation $C = 2200 + 3n$.

a. Solve the equations simultaneously to determine the break-even point.

b. Determine the profit or loss, in dollars, for the following shirt orders.

   i. **75 shirts**   ii. **220 shirts**

THINK	WRITE
a. 1. Write the two equations.	a. $C = 2200 + 3n$ $R = 25n$
2. Equate the equations ($R = C$).	$2200 + 3n = 25n$
3. Solve for the unknown.	$2200 + 3n - 3n = 25n - 3n$ $2200 = 22n$ $\dfrac{2200}{22} = n$ $n = 100$
4. Substitute back into either equation to determine the values of C and R.	$R = 25n$ $= 25 \times 100$ $= 2500$
5. Write the answer in the context of the problem.	The break-even point is (100, 2500). Therefore, 100 shirts need to be sold to cover the production cost, which is $2500.
b. i. 1. Write the two equations.	b. i. $C = 2200 + 3n$ $R = 25n$
2. Substitute the given value into both equations.	$n = 75$ $C = 2200 + 3 \times 75$ $= 2425$ $R = 25 \times 75$ $= 1875$
3. Determine the profit/loss.	Profit/loss $= R - C$ $= 1875 - 2425$ $= -550$
4 Write the answer in the context of the problem.	Since the answer is negative, it means that Santo lost $550 (i.e. selling 75 shirts did not cover the cost to produce the shirts).

ii. 1. Write the two equations.	**ii.** $C = 2200 + 3n$	
	$R = 25n$	
	$n = 220$	
2. Substitute the given value into both equations.	$C = 2200 + 3 \times 220$	
	$\qquad = 2860$	
	$R = 25 \times 220$	
	$\qquad = 5500$	
3. Determine the profit/loss.	$\text{Profit/loss} = R - C$	
	$\qquad = 5500 - 2860$	
	$\qquad = 2640$	
4 Write the answer in the context of the problem.	Since the answer is positive, it means that Santo made \$2640 profit from selling 220 shirts.	

 Resources

 Interactivity Break-even points (int-6454)

2.5 Exercise

Students, these questions are even better in jacPLUS

 Receive immediate feedback and access sample responses

 Access additional questions

 Track your results and progress

Find all this and MORE in jacPLUS

1. **WE12** Mary bought 4 donuts and 3 cupcakes for \$10.55, and Sharon bought 2 donuts and 4 cupcakes for \$9.90. Let d represent the cost of a donut and c represent the cost of a cupcake. Represent this information in the form of a pair of simultaneous equations.

2. A pair of simultaneous equations representing the number of adults and children attending a zoo is shown below.

    Equation 1: $a + c = 350$
    Equation 2: $25a + 15c = 6650$

    a. By solving the pair of simultaneous equations, determine the total number of adults and children attending the zoo.
    b. In the context of this problem, identify what equation 2 represents.

3. **WE13** Yolanda sells handmade bracelets at a market for \$12.50. The revenue, R, for selling n bracelets is represented by the equation $R = 12.50n$.
   The cost to make n bracelets is represented by the equation $C = 80 + 4.50n$.

    a. Solve the equations simultaneously to determine the break-even point.
    b. Determine the profit or loss, in dollars, if Yolanda sells:

    i. 8 bracelets
    ii. 13 bracelets.

4. The entry fee for a charity fun run event is $18. It costs event organisers $2550 for the hire of the tent and $3 per entry for administration. Any profit will be donated to local charities. An equation to represent the revenue for the entry fee is $R = an$, where R is the total amount collected in entry fees, in dollars, and n is the number of entries.

   a. Write an equation for the value of a.

   The equation that represents the cost for the event is $C = 2550 + bn$.

   b. Write an equation for the value of b.
   c. By solving the equations simultaneously, determine the number of entries needed to break even.
   d. A total of 310 entries are received for this charity event. Show that the organisers will be able to donate $2100 to local charities.
   e. Determine the number of entries needed to donate $5010 to local charities.

5. A school group travelled to the city by bus and returned by train. The two equations show the adult, a, and student, s, ticket prices to travel on the bus and train.

$$\text{Bus: } 3.5a + 1.5s = 42.50$$
$$\text{Train: } 4.75a + 2.25s = 61.75$$

   a. Write the cost of a student bus ticket, s, and an adult bus ticket, a.
   b. Determine the most suitable method to solve these two simultaneous equations.
   c. Using your method from part b, solve the simultaneous equations and hence determine the number of adults and the number of students in the school group.

6. The following pair of simultaneous equations represents the number of adult and concession tickets sold and the respective ticket prices for the premier screening of the blockbuster *Aliens Attack*.

$$\text{Equation 1: } a + c = 544$$
$$\text{Equation 2: } 19.50a + 14.50c = 9013$$

   a. Identify the costs, in dollars, of an adult ticket, a, and a concession ticket, c.
   b. In the context of this problem, determine what equation 1 represents.
   c. By solving the simultaneous equations, determine how many adult and concession tickets were sold for the premier screening.

7. Charlotte has a babysitting service and charges $12.50 per hour. After Charlotte calculated her set-up and travel costs, she constructed the cost equation $C = 45 + 2.50h$, where C represents the cost in dollars per job and h represents the hours Charlotte babysits for.

   a. Write an equation that represents the revenue, R, earned by Charlotte in terms of number of hours, h.
   b. By solving the equations simultaneously, determine the number of hours Charlotte needs to babysit to cover her costs (that is, the break-even point).
   c. In one week, Charlotte had four babysitting jobs as shown in the table.

Babysitting job	1	2	3	4
Number of hours (h)	5.0	3.5	4.0	7.0

   i. Determine whether Charlotte made a profit or loss for each individual babysitting job.
   ii. Calculate whether Charlotte made a profit for the week.

d. Charlotte made a $50 profit on one job. Determine the total number of hours she babysat for.

8. Trudi and Mia work part time at the local supermarket after school. The following table shows the number of hours worked for both Trudi and Mia and the total wages, in dollars, paid over two weeks.

Week	Trudi's hours worked	Mia's hours worked	Total wages
Week 1	15	12	$400.50
Week 2	9	13	$328.75

   a. Construct two equations to represent the number of hours worked by Trudi and Mia and the total wages paid for each week. Write your equations using the pronumerals t for Trudi and m for Mia.
   b. In the context of this problem, determine what t and m represent.
   c. By solving the pair of simultaneous equations, determine the values of t and m.

9. Saksham uses carrots and apples to make his special homemade fruit juice. One week he buys 5 kg of carrots and 4 kg of apples for $31.55. The next week he buys 4 kg of carrots and 3 kg of apples for $24.65.

   a. Set up two simultaneous equations to represent the cost of carrots, x, in dollars per kilogram, and the cost of apples, y, in dollars per kilogram.
   b. By solving the simultaneous equations, determine how much Saksham spends on 1 kg each of carrots and apples.
   c. Determine the amount Saksham spends the following week when he buys 2 kg of carrots and 1.5 kg of apples. Give your answer correct to the nearest 5 cents.

10. The table shows the number of 100 g serves of strawberries and grapes and the total kilojoule intake.

Fruit	100 g serves	
Strawberries, s	3	4
Grapes, g	2	3
Total kilojoules	**1000**	**1430**

   a. Construct two equations to represent the number of serves of strawberries, s, and grapes, g, and the total kilojoules using the pronumerals shown.
   b. By solving the pair of simultaneous equations constructed in part **a**, determine the number of kilojoules (kJ) for a 100 g serve of strawberries.

11. Two budget car hire companies offer the following deals for hiring a medium size family car.

Car company	Deal
FreeWheels	$75 plus $1.10 per kilometre travelled
GetThere	$90 plus $0.90 per kilometre travelled

   a. Construct two equations to represent the deals for each car hire company. Write your equations in terms of cost, C, and kilometres travelled, k.
   b. By solving the two equations simultaneously, determine the value of k at which the cost of hiring a car will be the same for both companies.
   o. Rex and Jan hire a car for the weekend. They expect to travel 250 km over the weekend. Determine which car hire company should they use and explain why. Justify your answer using calculations.

12. The following table shows the number of boxes of three types of cereal bought each week for a school camp, as well as the total cost for each week.

Cereal	Week 1	Week 2	Week 3
Corn Pops, c	2	1	3
Rice Crunch, r	3	2	4
Muesli, m	1	2	1
Total cost, $	**27.45**	**24.25**	**36.35**

Wen is the cook at the camp. She decides to work out the cost of each box of cereal using simultaneous equations. She incorrectly sets up the following equations:

$$2c + c + 3c = 27.45$$
$$3r + 2r + 4r = 24.25$$
$$m + 2m + m = 36.35$$

a. Explain why these simultaneous equations will not determine the cost of each box of cereal.
b. Write the correct simultaneous equations.
c. Using CAS or otherwise, solve the three simultaneous equations and hence write the total cost for cereal for week 4's order of 3 boxes of Corn Pops, 2 boxes of Rice Crunch and 2 boxes of muesli.

13. Sally and Nem decide to sell cups of lemonade to the neighbourhood children. The cost to make the lemonade using their own lemons can be represented using the equation $C = 2 + 0.25n$, where C is the cost in dollars and n is the number of cups of lemonade sold.

a. If they sell cups of lemonade for 50 cents, write an equation to represent the selling price, S, for n cups of lemonade.
b. By solving two simultaneous equations, determine the number of cups of lemonade Sally and Nem need to sell to break even (i.e. cover their costs).
c. Sally and Nem increase their selling price. If they make a $7 profit for selling 20 cups of lemonade, calculate the new selling price.

14. The CotX T-Shirt Company produces T-shirts at a cost of $7.50 each after an initial set-up cost of $810.
a. Determine the cost to produce 100 T-shirts.
b. Using CAS or otherwise, complete the following table that shows the cost of producing T-shirts.

Number of T-shirts, n	0	20	30	40	50	60	80	100	120	140
Cost, C										

c. Write an equation that represents the cost, C, to produce n T-shirts.
d. CotX sells each T-shirt for $25.50. Write an equation that represents the amount of sales, S, in dollars for selling n T-shirts.
e. By solving two simultaneous equations, determine the number of T-shirts that must be sold for CotX to break even.
f. If CotX needs to make a profit of at least $5000, determine the minimum number of T-shirts they need to sell.

15. There are three types of fruit for sale at the market: starfruit, s, mango, m, and papaya, p. The following table shows the number of each fruit bought and the total cost in dollars.

Starfruit, s	Mango, m	Papaya, p	Total cost, $
5	3	4	19.40
4	2	5	17.50
3	5	6	24.60

  a. Using the pronumerals s, m and p, represent this information with three equations.

  b. Using CAS or otherwise, calculate the cost of one starfruit, one mango and one papaya.

  c. Using your answer from part b, determine the cost of two starfruit, four mangoes and four papayas.

16. The Comet Cinema offers four types of tickets to the movies: adult, concession, senior and member. The table below shows the number and types of tickets bought to see four different movies and the total amount of tickets sales in dollars.

Movie	Adult, a	Concession, c	Seniors, s	Members, m	Total sales, $
Wizard Boy	24	52	12	15	1071.00
Champions	35	8	45	27	1105.50
Pixies on Ice	20	55	9	6	961.50
Horror Nite	35	15	7	13	777.00

  a. Represent this information in four simultaneous equations, using the pronumerals given in the table.

  b. Using CAS or otherwise, determine the cost, in dollars, for each of the four different movie tickets.

  c. The blockbuster movie *Love Hurts* took the following tickets sales: 77 adults, 30 concessions, 15 seniors and 45 members. Using your values from part b:

    i. write the expression that represents this information

    ii. determine the total ticket sales in dollars and cents.

2.5 Exam questions

Question 1 (1 mark)

MC The value of x in the following rectangle is

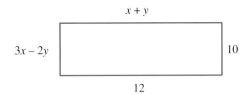

  A. 6.8

  B. 5.2

  C. 6

  D. 4

  E. 7

Question 2 (1 mark)

MC x and y are two consecutive even numbers. The relationship between x and y is

A. $x + y = 2$

B. $y = 1 + x$

C. $x + y = 1$

D. $y = 2 + x$

E. $y = 2x$

Question 3 (3 marks)

River Car Sales sells sedans and wagons. The profit made on selling six sedans and four wagons is $25 600. The profit made on selling three sedans and one wagon is $10 000.

    a. Write simultaneous equations showing the relationship between the profit on one sedan, s, and the profit on one wagon, w.   **(1 mark)**

    b. Solve the equations for s and w.   **(1 mark)**

    c. Hence, calculate the profit on selling 10 sedans and 8 wagons.   **(1 mark)**

More exam questions are available online.

2.6 Review

2.6.1 Summary

doc-37612

2.6 Exercise

Multiple choice

1. **MC** Select which of the following is the correctly transposed version of $y = -6 + 3x$ that makes x the subject of the equation.

   A. $x = \dfrac{y-6}{3}$
   B. $x = \dfrac{y+6}{3}$
   C. $x = 3y - 6$
   D. $x = 3y + 6$
   E. $x = 3y + 2$

2. **MC** The value of x in the linear equation $3(2x+5) = 12$ is:

   A. -1.5
   B. -1
   C. -0.5
   D. 2
   E. 4.5

3. **MC** The literal linear equation $v = u + at$ is an equation for motion, given an initial velocity, a rate of acceleration and a period of time. The correction solution to this equation for a is:

   A. $a = \dfrac{v-u}{t}$
   B. $a = \dfrac{u-v}{t}$
   C. $a = v - tu$
   D. $a = u - tv$
   E. $a = \dfrac{v}{t} - u$

The following information relates to questions 4 and 5.

Juliana and Alyssa live in two towns 232 km apart. One weekday they both have to drive to each other's town for a business meeting, leaving at the same time. Alyssa drives an average of 12 km/h faster than Juliana, and they pass each other after 2 hours.

4. **MC** If Juliana drives at an average speed of j km/h, determine Alyssa's average driving speed.

   A. $(232 - j)$ km/h
   B. $(232 + j)$ km/h
   C. $(j - 12)$ km/h

   D. $(j + 12)$ km/h
   E. $\dfrac{232}{j}$ km/h

5. **MC** Select Juliana's average driving speed from the following options.

   A. 50 km/h
   B. 52 km/h
   C. 60 km/h
   D. 64 km/h
   E. 72 km/h

6. **MC** The solution to the following pair of simultaneous equations is:

$$x + 5y = 7$$
$$y = 5 - 2x$$

   A. $x = -8, y = 3$
   B. $x = 2.25, y = 0.5$
   C. $x = 2, y = 1$
   D. $x = 1, y = 2$
   E. $x = 3, y = -8$

7. **MC** Two adults and four children went to the circus. They paid a total of $55.00 for their tickets. One adult and three children paid $35.00 to enter the same circus. Select the set of simultaneous equations that represents this situation, where a is the cost for an adult and c is the cost of a child's entry.

   A. $a + c = 55$
      $a + c = 35$
   B. $2a + c = 55$
      $a + 3c = 35$
   C. $2a + 3c = 55$
      $a + 4c = 35$
   D. $2a + 4c = 35$
      $a + 3c = 55$
   E. $2a + 4c = 55$
      $a + 3c = 35$

8. Solve each of the following equations for the unknown.

   a. $7(y+4)=35$

   b. $\dfrac{5-2m}{3}=-2$

   c. $\dfrac{3s}{4}+6=10$

   d. $\dfrac{-2(3t+1)}{5}+3=9$

9. Solve the following simultaneous equations by using either substitution or elimination.

   a. $y=-5+3x$ and $3x+2y=17$

   b. $2x+y=7$ and $3x-y=3$

   c. $3x-2y=-16$ and $y=2(x+9)$

   d. $4x+3y=17$ and $3x+2y=13$

10. Determine the break-even points for the following cost and revenue equations. Where appropriate, give your answers correct to 2 decimal places.

    a. $C=150+2x$ and $R=7.5x$

    b. $C=25+13.5x$ and $R=19.7x$

    c. $C=3500+22.5x$ and $R=35x$

11. Using CAS or otherwise, solve the following equations to determine the unknowns. Express your answer in exact form.

    a. $\dfrac{2-5x}{8}=\dfrac{3}{5}$

    b. $\dfrac{6(3y-2)}{11}=\dfrac{5}{9}$

    c. $\left(\dfrac{4x}{5}-\dfrac{3}{7}\right)+8=2$

    d. $\dfrac{7x+6}{9}+\dfrac{3x}{10}=\dfrac{4}{5}$

12. Using CAS or otherwise, solve the following groups of simultaneous equations. Write your answers correct to 2 decimal places.

    a. $y=6+5x$ and $3x+2y=7$

    b. $4(x+6)=y-6$ and $2(y+3)=x-9$

    c. $6x+5y=8.95$, $y=-1.36+3x$ and $2x+3y=4.17$

13. Petra is doing a survey of how many humans and pets are in her extended family. She gives the following information to her friend Juliana.

    - There are 33 humans and pets (combined).
    - The combined number of legs (pets and owners) is 94.
    - Each human has 2 legs and each pet has 4 legs.

    a. If h is the number of humans, express the number of pets in terms of h.

    b. Write expressions for the total number of human legs and the total number of pet legs in terms of h.

    c. Determine how many humans and pets there are in Petra's extended family.

Extended response

14. The height of a plant can be found using the equation $h=\dfrac{2(3t+15)}{3}$, where h is the height in cm and t is time in weeks.

    a. Using CAS or otherwise, determine the time the plant takes to grow to the following heights. Give your answers correct to the nearest week.

    i. 20 cm

    ii. 30 cm

    iii. 35 cm

    iv. 50 cm

b. Determine how high the plant is initially.

When the plant reaches 60 cm it is given additional plant food. The plant's growth each week for the next 4 weeks is found using the equation $g = 2 + t$, where g is the growth each week in cm and t is the time in weeks since additional plant food was given.

c. Determine the height of the plant in cm for the next 4 weeks.

15. Consider the following groups of graphs.

   i. $y_1 = -4 + 5x$ and $y_2 = 8 + 6x$
  ii. $y_1 = -5 - 3x$ and $y_2 = 1 + 3x$
 iii. $y_1 = 6 + 2x$ and $y_2 = -4 + 2x$
 iv. $y_1 = 3 - x$, $y_2 = 5 + x$ and $y_3 = 6 + 2x$

  a. Where possible, determine the point of intersection for each set of graphs using any method.
  b. Determine whether there are solutions for all of these sets of graphs. If not, identify which set of graphs has no solution, and explain why this is.

16. Hank is cooking a Sunday dinner of roast lamb and roast beef for 20 guests. He has a 2.5-kg leg of lamb and a 4.2-kg cut of beef. The recommended cooking time for the lamb is 62.5 minutes; the recommended cooking time for the beef is 105 minutes. Hank's cookbook recommends that the meat be left to rest for 15 minutes before carving.
The cooking time is the same per kilogram for both cuts of meats and increases at a constant rate per kilogram.

  a. Calculate the cooking time in minutes, t, per kilogram of meat, k.
  b. Construct an equation that finds the cooking time, in minutes, per kilogram of meat (lamb and beef) including the resting time.
  c. Using a spreadsheet and your equation from part **b**, complete the following table for different-sized cuts of meat. Write your answers correct to the nearest whole minute.

Weight (g)	Cooking time (minutes)	Weight (g)	Cooking time (minutes)
500		2250	
750		2500	
1000		2750	
1250		3000	
1500		3250	
1750		3500	
2000		3750	

  d. Marcia uses the equation from part **b** to help her with the cooking time of her Christmas turkey, which weighs 5.5 kg. Using the equation you found in part **b**, determine the cooking time in hours and minutes for the turkey.
  e. Marcia finds that the cooking time is incorrect. Explain why the equation did not help her to accurately determine the cooking time.
  f. Marcia finds a cookbook that suggests that the cooking time for a turkey is $\dfrac{3}{4}$ of an hour per kilogram. If the resting time for roast turkey is 30 minutes, construct an equation that finds the cooking time per kilogram of turkey.
  g. Using the equation you found in part **f**, determine the recommended cooking time, in hours and minutes, for Marcia's turkey.

17. Suzanne is starting a business selling homemade cupcakes. It will cost her $250 to buy all of the equipment, and each cupcake will cost $2.25 to make. Suzanne models her costs, C, in dollars, to make n cupcakes using the following equation:

$$C = b + 2.25n$$

a. Identify the value of b.

b. Determine how much it will cost Suzanne to make 50 cupcakes.

c. Suzanne receives an order to supply cupcakes for an afternoon tea. If it costs Suzanne $373.75 to make the order, determine how many cupcakes she has to make.

d. Suzanne sells each cupcake for $6.50. Write an equation to represent the revenue Suzanne earns, R, from selling n cupcakes.

e. One week, Suzanne sells 150 cupcakes. Determine the total profit, in dollars, that Suzanne makes for that week.

f. By solving a pair of simultaneous equations, calculate the total number of cupcakes Suzanne needs to sell to break even. Give your answer correct to the nearest whole number.

g. The graph shows Suzanne's cost, C, and revenue, R, for making and selling n cupcakes.

On the graph, add labels for:

   i. the line that represents the cost, C

   ii. the line that represents the revenue, R

   iii. the break-even point.

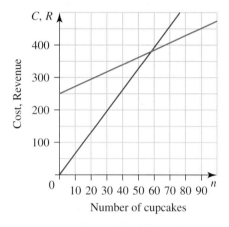

18. There are two types of tickets available for the GoodFeel Festival: full price and concession. On the first day, 6473 people attended the festival. The total ticket revenue, in dollars, was $109 559.80. A full-price ticket cost $19.80 and a concession ticket cost $14.60.

a. Construct two equations that represent the number of people attending and the ticket sales in terms of full price tickets, f, and concession tickets, c.

b. By solving your equations from part a, determine the number of each type of ticket sold on the first day.

It costs the organisers $43 500 plus $5.90 for each ticket sold to run the festival.

c. Construct an equation to represent the cost to the organisers, C, for the festival in terms of the number of tickets sold, x.

d. Construct an equation that shows the relationship between variables x, f and c.

e. Construct an equation to represent the revenue from ticket sales, R, in terms of f and c.

f. Using the ticket sales for the first day of the festival, determine if the organisers made a profit.

Question 1 (1 mark)

 Using the formula $F = \dfrac{9}{5}C + 32$ (where F represents degrees Fahrenheit and C represents degrees Celsius).

The number of degrees F when C is 25 is

    A. 257   **B.** 45   **C.** 57   **D.** 77   **E.** 0

Question 2 (1 mark)

MC The solution to the linear equation $115 - 5p = 45$ is

    A. -32 **B.** 23   **C.** 14   **D.** -23 **E.** -14

Question 3 (1 mark)

MC A number is multiplied by 7 and then divided by 3. The result is 2 less than the number multiplied by 4. The equation that best represents this is

    A. $\dfrac{7x}{3} = 2 - 4x$ **B.** $\dfrac{7x}{3} = 4x - 2$ **C.** $\dfrac{3x}{7} = 4x - 2$

    D. $\dfrac{3x}{7} = 4x + 2$ **E.** $\dfrac{7x}{3} = 4x + 2$

Question 4 (1 mark)

MC In a particular isosceles triangle, the two sides of equal length are each 3 cm longer than the third side. If the perimeter of the triangle is 24 cm, the length of the third side labelled x is

    A. $9\,\text{cm}$

    B. $\dfrac{24}{7}\,\text{cm}$

    C. $6\,\text{cm}$

    D. $8\,\text{cm}$

    E. $12\,\text{cm}$

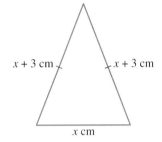

Question 5 (5 marks)

Different combinations are available when purchasing tickets to enter a theme park. The total cost for 2 adults and 3 children is $24.75. The total cost of 2 adults and 1 child is $16.50.

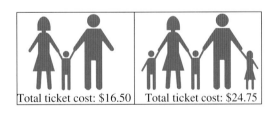

Total ticket cost: $16.50   Total ticket cost: $24.75

Use simultaneous equations to calculate the cost of an individual child ticket and an adult ticket. Give answers correct to the nearest 5 cents.

More exam questions are available online.

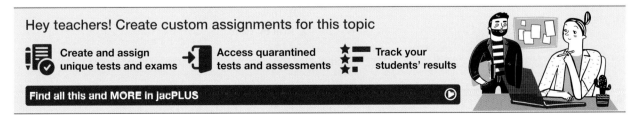

Hey teachers! Create custom assignments for this topic

Create and assign unique tests and exams   Access quarantined tests and assessments   Track your students' results

Find all this and MORE in jacPLUS

Answers

Topic 2 Linear relations and equations

2.2 Linear relations and solving linear equations

2.2 Exercise

1. a. Non-linear　　　　b. Non-linear
   c. Linear　　　　　　 d. Non-linear
   e. Linear

2. a.

Equation	Bethany's response	Correct response
$y = 1 + 4x$	Yes	Yes
$y^2 = 5x - 2$	Yes	No
$y + 6x = 7$	Yes	Yes
$y = x^2 - 5x$	No	No
$t = 6d^2 - 9$	No	No
$m^3 = n + 8$	Yes	No

   b. Bethany should look at both variables (pronumerals or letters). Both variables need to have a highest power of 1.

3. a. Linear　　　　　　 b. Non-linear
   c. Linear　　　　　　 d. Linear
   e. Non-linear　　　　 f. Non-linear

4. a. Yes, as the power of x is 1. (Note that this equation is not linear.)
   b. The power of both variables in a linear relation must be 1.

5. $x = \dfrac{y + 3}{6}$

6. a. $x = \dfrac{y - 5}{2}$　　　　b. $x = \dfrac{3y - 8}{6}$

   c. $x = \dfrac{p + 6}{5}$　　　　d. $x = 2y - \dfrac{1}{3}$

7. a. $x = 3$　　　　　　　b. $n = 10$
   c. $d = 4.5$　　　　　　d. $x = 17$

8. a. i. $\times 4, \, + 3$　　　　ii. $+ 2, \times 3$
      iii. $+ 1, \, \div 2$　　　　iv. $\times 3, \, - 9, \times 2$

   b. i. $a = \dfrac{7}{4}$　　　　ii. $x = 2$

      iii. $s = 13$　　　　iv. $c = \dfrac{17}{3}$

9. a. $x = -9$　　b. $y = -0.5$　　c. $x = -4.5$

10. a. $a = \dfrac{v - u}{t}$　　b. $x = \dfrac{m + k}{y}$　　c. $x = p(r + s)$

11. $y = -25$

12. $y = -6$

13. a. 33 minutes　　　　　　b. 108 minutes

14. $x = \dfrac{r + q}{p}$

15. $d = \dfrac{C}{\pi}$

16. a. $w = 5$
    b. Operations need to be performed in reverse order.

17. $K = \dfrac{F - 32}{1.8} + 273$

18. a. $x = \dfrac{7}{3}$　　　　　　b. $x = \dfrac{-13}{3}$

    c. $x = \dfrac{1}{3}$　　　　　　d. $x = 13$

19. a. 2 hours　　　　　　b. 2 hours 38 minutes
    c. 41 minutes　　　　　d. 3 hours 45 minutes
    e. 32 minutes

20. a. i. 9 months
       ii. 19 months ($= 18.67$ months)
    b. 5 years, 4 months

21. a. $- 32, \times 5, \div 9$
    b. $\times 9, \div 5, + 32$
    c. 374 °F

22. a. 14.69 cm　　b. 18.33 cm　　c. 21.06 cm

2.2 Exam questions

Note: Mark allocations are available with the fully worked solutions online.

1. C
2. $x = 4$
3. A

2.3 Developing linear equations

2.3 Exercise

1. $2.24
2. The red velvet cupcakes are the cheapest per cupcake.
3. 33
4. 10.5 cm and 31.5 cm
5. $7.90
6. 17
7. a.

Minute	1	2	3	4	5	6	7	8	9	10
Distance (km)	0.4	0.8	1.2	1.6	2	2.4	2.8	3.2	3.6	4

   b. Between the 7th and 8th minute
8. a. See the table at the foot of the page.*
   b. 10 weeks
9. 13 m by 19 m

*8. a.

Week	0	1	2	3	4	5	6	7	8	9	10	11	12
Money ($)	20	23	26	29	32	35	38	41	44	47	50	53	56

10. a. See the table at the foot of the page.*

   b. See the table at the foot of the page.**

   c. Subtract the values in the first table from the values in the second table.

   d. See the table at the foot of the page.***

2.3 Exam questions

Note: Mark allocations are available with the fully worked solutions online.

1. D

2. C

3. E

2.4 Simultaneous linear equations

2.4 Exercise

1.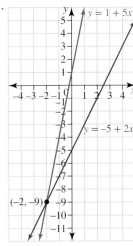

2.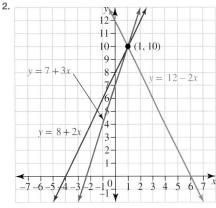

3. $x = -2$, $y = 5$

4. a. $x = -2$, $y = -7$ b. $x = 2$, $y = -3$
   c. $x = -5$, $y = -18$ d. No solution
   e. $x = -1$, $y = 1$

5. a. $x = -1$, $y = -1$
   b. $m = 11$, $n = 3$
   c. $x = 6$, $y = 7$

6. a. $x = 3$, $y = -3$ b. $x = 1$, $y = 4$

7. E

8. a. $x = 7$, $y = 19$ b. $x = -2$, $y = -12$
   c. $x = 7$, $y = 44$ d. $x = 2$, $y = 1$
   e. $x = -3$, $y = -4$ f. $x = 1$, $y = -1$

9. a. Both unknowns are on the same side.
   b. Add the two equations and solve for x, then substitute x into one of the equations to solve for y. $x = 3$, $y = -1$

10. a. $x = 1$, $y = 2$ b. $a = 2$, $b = 3$
    c. $c = -1$, $d = 2$

11. $a = 2$ and $x = -2$

12. a. $x = 2$, $y = -2$ b. $x = 3$, $y = 4$
    c. $x = -1$, $y = 3$ d. $x = 5$, $y = 3$
    e. $x = -1$, $y = 4$ f. $x = 6$, $y = 4$

*10. a.

Number of boards	10	11	12	13	14	15	16	17	18	19	20
Cost ($)	395	409.50	424	438.50	453	467.50	482	496.50	511	525.50	540

**10. b.

Number of boards	10	11	12	13	14	15	16	17	18	19	20
Revenue ($)	329.50	362.45	395.40	428.35	461.30	494.25	527.20	560.15	593.10	626.05	659

***10. d.

Number of boards	10	11	12	13	14	15	16	17	18	19	20
Profit ($)	−65.50	−47.05	−28.60	−10.15	8.30	26.75	45.20	63.65	82.10	100.55	119

13. A

14. 6 roses and 13 lilies

15. 16 strawberry twists and 12 chocolate ripples

16. Michelle: 70 km/h, Lydia: 60 km/h

17. 9 carrots and 6 potatoes

18. a. Marcia added the equations together instead of subtracting (and did not perform the addition correctly). The correct result for step 2 is $[1] - [2]$: $22y = 11$.

    b. $x = 5$, $y = \dfrac{1}{2}$

    c. $x = 5$, $y = \dfrac{1}{2}$

19. a. Goal $= 5$ points, behind $= 2$ points
    b. Jetts 40 points, Meteorites 48 points

20. a. The equation has unknowns on each side of the equals sign.

    b. Milan works 5 hours and Yashab works 10 hours.

    c. 3 hours 45 minutes (3.75 hours)

2.4 Exam questions

Note: Mark allocations are available with the fully worked solutions online.

1. E

2. A

3. C

2.5 Problem solving with simultaneous equations

2.5 Exercise

1. $4d + 3c = 10.55$ and $2d + 4c = 9.90$

2. a. 140 adults and 210 children

   b. The cost of an adult's ticket is $25, the cost of a children's ticket costs $15, and the total ticket sales is $6650.

3. a. (10, 125) Yolanda needs to sell 10 bracelets to cover her costs.

   b. i. $16 loss   ii. $24 profit

4. a. $a = 18$
   b. $b = 3$
   c. 170 entries
   d. $R = \$5580$, $C = \$3480$, $P = \$2100$
   e. 504 entries

5. a. $s = \$1.50$, $a = \$3.50$
   b. Elimination method
   c. 4 adults and 19 students

6. a. $a = \$19.50$, $c = \$14.50$

   b. The total number of tickets sold (both adult and concession)

   c. 225 adult tickets and 319 concession tickets

7. a. $R = 12.50h$

   b. 4.5 hours

   c. i. Charlotte made a profit for jobs 1 and 4, and a loss for jobs 2 and 3.

      ii. Yes, she made $15 profit.
      $(25 + 5 - (10 + 5)) = 30 - 15 = \$15$

   d. 9.5 hours

8. a. $15t + 12m = 400.50$ and $9t + 13m = 328.75$

   b. t represents the hourly rate earned by Trudi and m represents the hourly rate earned by Mia.

   c. $t = \$14.50$, $m = \$15.25$

9. a. $5x + 4y = 31.55$ and $4x + 3y = 24.65$
   b. $x = \$3.95$, $y = \$2.95$
   c. $12.35

10. a. $3s + 2g = 1000$ and $4s + 3g = 1430$
    b. 140 kJ

11. a. $C = 75 + 1.10k$ and $C = 90 + 0.90k$

    b. $k = 75$ km

    c. $C_{\text{FreeWheels}} = \$350$, $C_{\text{GetThere}} = \$315$. They should use GetThere.

12. a. The cost is for the three different types of cereal, but the equations only include one type of cereal.

    b. $2c + 3r + m = 27.45$
    $c + 2r + 2m = 24.25$
    $3c + 4r + m = 36.35$

    c. $34.15

13. a. $S = 0.5n$
    b. 8 cups of lemonade
    c. 70 cents

14. a. $1560
    b. See the table at the foot of the page.*
    c. $C = 810 + 7.5n$
    d. $S = 25.50n$
    e. 45 T-shirts
    f. 323 T-shirts

15. a. $5s + 3m + 4p = 19.4$
    $4s + 2m + 5p = 17.5$
    $3s + 5m + 6p = 24.6$
    b. $s = \$1.25$, $m = \$2.25$, $p = \$1.60$
    c. $17.90

16. a. $24a + 52c + 12s + 15m = 1071$
    $35a + 8c + 45s + 27m = 1105.5$
    $20a + 55c + 9s + 6m = 961.5$
    $35a + 15c + 7s + 13m = 777$

    b. Adult ticket $= \$13.50$, concession $= \$10.50$, seniors $= \$8.00$, members $= \$7.00$

    c. i. $77 \times 13.50 + 30 \times 10.50 + 15 \times 8.00 + 45 \times 7.00$
       ii. $1789.50

*14. b.

n	0	20	30	40	50	60	80	100	120	140
C	810	960	1035	1110	1185	1260	1410	1560	1710	1860

2.5 Exam questions

Note: Mark allocations are available with the fully worked solutions online.

1. A
2. D
3. a. $6s + 4w = \$25\,600$ [1]
     $3s + w = \$10\,000$ [2]
   b. $s = \$2400$, $w = \$2800$
   c. $\$46\,400$

2.6 Review

2.6 Exercise

Multiple choice

1. B   2. C   3. A   4. D   5. B
6. C   7. E

Short answer

8. a. $y = 1$ b. $m = 5.5$
   c. $s = 5\dfrac{1}{3}$ d. $t = -5\dfrac{1}{3}$
9. a. $x = 3$, $y = 4$
   b. $x = 2$, $y = 3$
   c. $x = -20$, $y = -22$
   d. $x = 5$, $y = -1$
10. a. $x = 27.27$
   b. $x = 4.03$
   c. $x = 280$
11. a. $x = \dfrac{-14}{25}$ b. $y = \dfrac{163}{162}$
   c. $x = \dfrac{-195}{28}$ d. $x = \dfrac{12}{97}$
12. a. $x = -0.38$, $y = 4.08$
   b. $x = -10.71$, $y = -12.86$
   c. $x = 0.75$, $y = 0.89$
13. a. $33 - h$
   b. Total number of human legs $= 2h$; total number of pet legs $= 4(33 - h)$
   c. 19 humans and 14 pets

Extended response

14. a.   i. 5 weeks   ii. 10 weeks
     iii. 13 weeks   iv. 20 weeks
   b. 10 cm
   c. 63 cm, 67 cm, 72 cm, 78 cm
15. a.   i. $(-12, -64)$
     ii. $(-1, -2)$

     iii. No solution
     iv. $(-1, 4)$
   b. No, the graphs in part iii are parallel (they have the same gradient).
16. a. 25 minutes per kilogram
   b. $t = 15 + 25k$
   c. See the table at the foot of the page.*
   d. 2 hours, 33 minutes
   e. This equation will probably vary for different types of meat.
   f. $t = 30 + 45k$
   g. 4 hours, 38 minutes
17. a. 250
   b. $\$362.50$
   c. 55
   d. $R = 6.5n$
   e. $\$387.50$
   f. 59
   g.

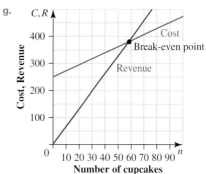

18. a. $f + c = 6473$, $19.8f + 14.6c = 109\,559.8$
   b. 2895 full price tickets and 3578 concession tickets
   c. $C = 43\,500 + 5.9x$
   d. $x = f + c$
   e. $R = 19.8f + 14.6c$
   f. Yes, they made a profit of $\$27\,869.10$.

2.6 Exam questions

Note: Mark allocations are available with the fully worked solutions online.

1. D
2. C
3. B
4. C
5. Child ticket = $\$4.15$; adult ticket = $\$6.20$ (to the nearest 5 cents)

*16. c.

Weight (g)	Cooking time (minutes)	Weight (g)	Cooking time (minutes)
500	27.5	2250	71.25
750	33.75	2500	77.5
1000	40	2750	83.75
1250	46.25	3000	90
1500	52.5	3250	96.25
1750	58.75	3500	102.5
2000	65	3750	108.75

3 Financial mathematics

LEARNING SEQUENCE

Fully worked solutions for this topic are available online.

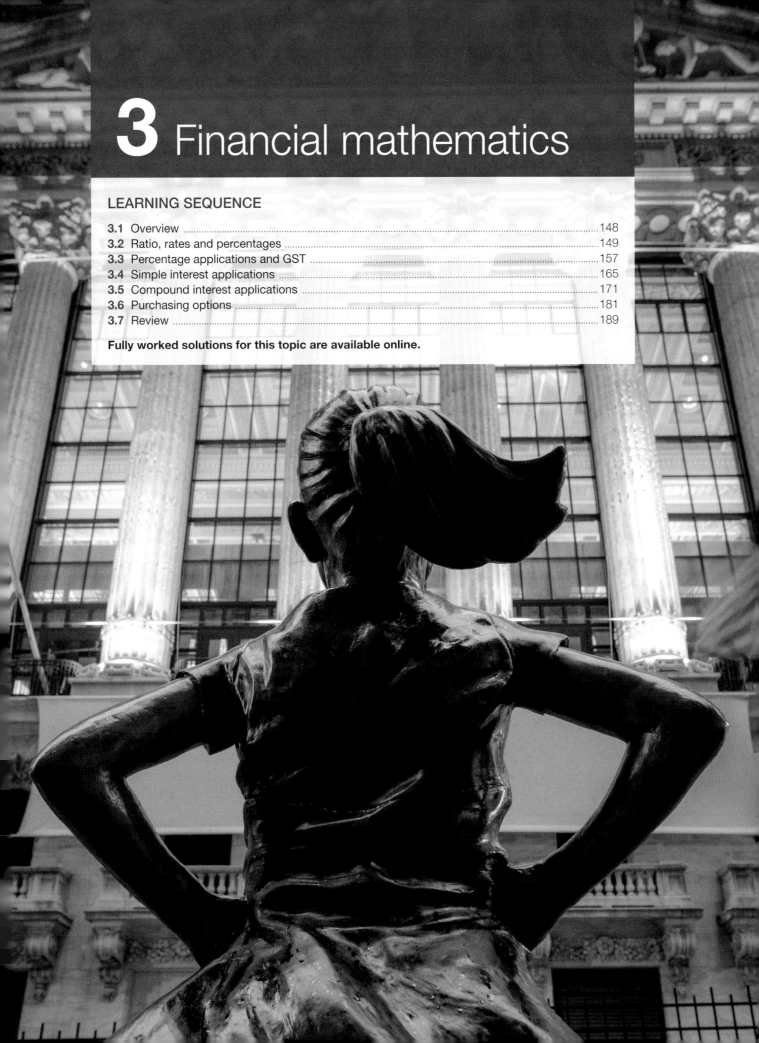

3.1 Overview

3.1.1 Introduction

Bank interest today is very different to what it originally was thousands of years ago. The basic premise, however, remains the same. The early loans and interest, around 10 000 to 5000 BC, were used in agriculture. Loans were made in seeds, grains, animals and tools to farmers. Since one seed could generate a plant with over 100 new seeds after the harvest, this allowed farmers to pay back their loans with interest. When animals were loaned, interest would be claimed based on the scenario. For example, a stud could be loaned for breeding purposes, and the lender would be repaid with an agreed number of newborn animals. Else, if animals were to be used for meat, the lender might expect the borrower to raise the animal to a certain weight and return this animal to them, and then reimburse the borrower for any additional weight gained on the animal.

Compound interest was first implemented in ancient Babylon. Compounding was the adding of accumulated interest back to the principal, so that interest was earned on interest. This is very different to simple interest, where the principal remains separate from the interest.

The *Wall Street Journal* in 1976 published an opinion article that attributed to Einstein the statement that compound interest was man's greatest invention. There is no proof that the renowned academic ever said this, but it's worth thinking about the importance of the statement nevertheless.

KEY CONCEPTS

This topic covers the following key concepts from the VCE Mathematics Study Design:
- percentage increase and decrease, mark-ups and discounts, and calculating GST in various financial contexts
- determining the impact of inflation on costs and the spending power of money over time
- the unitary method and its use in making comparisons and solving practical problems involving percentages and finance
- comparison of purchase options including cash, credit and debit cards, personal loans, 'buy now and pay later' schemes.

Source: VCE Mathematics Study Design (2023–2027) extracts © VCAA; reproduced by permission.

3.2 Ratio, rates and percentages

LEARNING INTENTION

At the end of this subtopic you should be able to:
- calculate percentage
- use ratios to compare two or more quantities
- calculate rates
- apply the unitary method.

3.2.1 Percentages

Percentages are fractions of 100. The percentage of a given value is calculated by multiplying it by the percentage expressed as a fraction or decimal.

You can write a value as a percentage of another value by expressing it as a fraction and multiplying by 100.

tlvd-3703

WORKED EXAMPLE 1 Calculating percentages

A teacher finds that 12% of students in their class obtain an A$^+$ for a test. To get an A$^+$, students need to score at least 28 marks. If there were 25 students in the class and the test was out of 32 marks:
a. calculate the minimum percentage needed to obtain an A$^+$
b. determine how many students received an A$^+$.

THINK	WRITE
a. 1. Write the minimum number of marks needed as a fraction of the total number of marks.	a. $\dfrac{28}{32}$
2. Multiply the fraction by 100 and simplify where possible.	$\dfrac{28}{32} \times 100 = \dfrac{^7 \cancel{28}}{^8 \cancel{32}} \times \dfrac{100}{1}$ $= \dfrac{7}{\cancel{2}} \times \dfrac{^{25}\cancel{100}}{1}$ $= \dfrac{175}{2}$ $= 87.5\%$
3. Write the answer as a sentence.	Students had to obtain a minimum of 87.5% to receive an A$^+$.

b. 1. Write the percentage as a fraction.

b. $12\% = \dfrac{12}{100}$

2. Multiply the fraction by the total number in the class and simplify.

$$\dfrac{12}{100} \times 25 = \dfrac{12}{^{4}\cancel{100}} \times \dfrac{^{1}\cancel{25}}{1}$$

$$= \dfrac{^{3}\cancel{12}}{^{1}\cancel{4}} \times \dfrac{1}{1}$$

$$= 3$$

3. Write the answer as a sentence.

Three students obtained an A^{+}.

3.2.2 Ratios

Ratios are used to compare quantities that are measured in the same unit. For example, if the ratio of bicycles to cars on a particular road during rush hour is $1:4$, there are 4 times as many cars on the road as bicycles.

Ratios

Ratios compare two or more quantities, and are in their simplest form when all parts are expressed using whole numbers and the highest common factor (HCF) of all the numbers is 1. A simplified ratio is equivalent to the original ratio.

WORKED EXAMPLE 2 Simplifying ratios

Simplify the following ratios by first determining the highest common factor.
a. $14:6$
b. $1.5:2:3.5$

THINK

a. 1. Consider factors of each quantity. Both values are divisible by 2.

2. Divide both values by 2.

b. 1. To work with whole numbers, multiply all quantities by 10. (We multiply by 10 as there is 1 decimal place value. If there were 2 decimal places, then we would multiply by 100.)

2. Express the whole number quantities as a ratio.

3. Identify the highest common factor (HCF) for the three parts (in this case 5). Simplify by dividing each part by the HCF.

WRITE

a. $\begin{array}{cc} 14 & : & 6 \\ \div 2 & & \div 2 \\ \downarrow & & \downarrow \\ = 7 & : & 3 \end{array}$

b. $\begin{array}{ccccc} 1.5 & : & 2 & : & 3.5 \\ \times 10 & & \times 10 & & \times 10 \\ \downarrow & & \downarrow & & \downarrow \\ 15 & : & 20 & : & 35 \end{array}$

$\begin{array}{ccccc} 15 & : & 20 & : & 35 \\ \div 5 & & \div 5 & & \div 5 \\ \downarrow & & \downarrow & & \downarrow \\ = 3 & : & 4 & : & 7 \end{array}$

3.2.3 Ratios of a given quantity

We can use ratios to determine required proportions of a given quantity. This can be useful when splitting a total between different shares.

WORKED EXAMPLE 3 Dividing an amount into a ratio

Carlos, Maggie and Gary purchased a winning lottery ticket; however, they did not each contribute to the ticket in equal amounts. Carlos paid $6, Maggie $10 and Gary $4. They agree to divide the $845 winnings according to their contributions.
a. Express the purchase contributions as a ratio in simplest form.
b. Determine the winnings each person is entitled to.

THINK	WRITE
a. Write the purchase contributions as a ratio, remembering to simplify by identifying the highest common factor.	a. $6 : 10 : 4$ HCF $= 2$ Ratio $= 3 : 5 : 2$
b. Add the numbers 3, 5 and 2 to calculate the total number of parts. Multiply the winning amount by the fraction representing each person's contribution.	b. Carlos $= 845 \times \dfrac{3}{10}$ $= \$253.50$ Maggie $= 845 \times \dfrac{5}{10}$ $= \$422.50$ Gary $= 845 \times \dfrac{2}{10}$ $= \$169.00$

3.2.4 Rates

A rate is a measure of change between two variables of different units.

Common examples of rates include speed in kilometres per hour (km/h) or metres per second (m/s), costs and charges in dollars per hour ($/h), and electricity usage in kilowatts per hour (kW/h).

Rates

Rates are usually expressed in terms of how much the first quantity changes with one unit of change in the second quantity.

tlvd-3704

WORKED EXAMPLE 4 Using rates to calculate speed

Calculate the rate (in km/h) at which you are moving if you are on a bus that travels 11.5 km in 12 minutes.

THINK	WRITE
1. Identify the two measurements: distance and time. As speed is commonly expressed in km/h, convert the time quantity unit from minutes to hours.	The quantities are 11.5 km and 12 minutes. $\dfrac{12}{60} = \dfrac{1}{5}$ or 0.2 hours

▶

2. Write the rate as a fraction and express in terms of one unit of the second value.	$$\frac{11.5\,\text{km}}{0.2\,\text{hrs}} = \frac{11.5}{0.2} \times \frac{5}{5}$$ $$= \frac{57.5}{1}$$ $$= 57.5$$
3. Write the answer as a sentence.	You are travelling at 57.5 km/h.

3.2.5 Unit cost calculations

In order to make accurate comparisons between the costs of differently priced and sized items, we need to identify how much a single unit of the item would be. This is known as the **unit cost**. For example, in supermarkets, similar cleaning products may be packaged in different sizes, making it difficult to tell which option is cheaper.

The unitary method

Unit cost calculations are an application of the unitary method, which is the same mathematical process we follow when simplifying a rate. If x items cost $\$y$, divide the cost by x to calculate the price of one item.

Unitary method

x items = $\$y$

1 item = $\$\dfrac{y}{x}$

tlvd-3705

WORKED EXAMPLE 5 Using the unitary method to calculate unit cost

Calculate the cost per 100 grams of pet food if a 1.25-kg box costs $7.50.

THINK	WRITE
1. Identify the cost and the weight. As the final answer is to be referenced in grams, convert the weight from kilograms to grams.	Cost: $7.50 Weight: 1.250 kg = 1250 g
2. Calculate the unit cost for 1 gram by dividing the cost by the weight.	$\dfrac{7.50}{1250} = 0.006$
3. Calculate the cost for 100 g by multiplying the unit cost by 100.	$0.006 \times 100 = 0.60$ Therefore, the cost per 100 grams is $0.60.

Resources

 interactivities Percentages (int-6458)
Speed (int-6457)

3.2 Exercise

1. **WE1** A teacher finds that 15% of students in one class obtain a B$^+$ for a test. To get a B$^+$, students needed to score at least 62 marks. If there were 20 students in the class and the test was out of 80 marks:

   a. calculate the minimum percentage needed to obtain a B$^+$
   b. determine how many students received a B$^+$.

2. A salesman is paid according to how much he sells in a week. He receives 3.5% of the total sales up to \$10 000 and 6.5% for amounts over \$10 000.

   a. Calculate his monthly pay if his total sales in four consecutive weeks are \$8900, \$11 300, \$13 450 and \$14 200.
   b. Calculate the percentage of his total pay for this time period represented by each week. Give your answers correct to 2 decimal places.

3. A real-estate agent is paid 4.25% of the sale price of any property she sells. Calculate how much she gets paid for selling properties costing:

   a. \$250 000
   b. \$310 500
   c. \$454 755
   d. \$879 256.

   Where necessary, give your answers correct to the nearest cent.

4. A student's test results in Mathematics are shown in the table.

	Test 1	Test 2	Test 3	Test 4	Test 5	Test 6	Test 7	Test 8
Mark	$\frac{16}{20}$	$\frac{14}{21}$	$\frac{26}{34}$	$\frac{36}{45}$	$\frac{14.5}{20}$	$\frac{13}{39}$	$\frac{42}{60}$	$\frac{26}{35}$
Percentage								

   a. Complete the table by calculating the percentage for each test, giving values correct to 2 decimal places where necessary.
   b. Calculate the student's overall result from all eight tests as a percentage, correct to 2 decimal places.

5. A person has to pay the following bills out of their weekly income of $1100.

Food	$280
Electricity	$105
Telephone	$50
Petrol	$85
Rent	$320

Giving answers correct to 2 decimal places where necessary:

a. express each bill as a percentage of the total bills
b. express each bill as a percentage of the weekly income.

6. **WE2** Simplify the following ratios.

a. $81:27:12$ b. $4.8:9.6$

7. A recipe for Mars bar slice requires 195 grams of chopped Mars bar pieces and 0.2 kilograms of milk chocolate. Express the weight of the Mars bar pieces to the weight of the milk chocolate as a ratio in simplest form.

8. Simplify the following ratios by first converting to the same unit where necessary.

a. $36:84$ b. $49:77:105$ c. $3.225\,\text{kg}:1875\,\text{g}$ d. $2.4\,\text{kg}:960\,\text{g}:1.2\,\text{kg}$

9. **WE3** In a bouquet of flowers, the ratio of red, yellow and orange flowers was $5:8:3$. If there were 48 flowers in the bouquet, determine how many of each colour were included.

10. Mark, Henry, Dale and Ben all chip in to buy a race car. The cost of the car was $18 000 and they contributed in the ratio of $3:1:4:2$.

a. Calculate the amount each person contributed.
b. They also purchased a trailer to tow the car. Mark put in $750, Henry $200, Dale $345 and Ben $615. Express these amounts as a ratio in simplest form.

11. The Murphys are driving from Melbourne to Adelaide for a holiday. They plan to have two stops before arriving in Adelaide. First they will drive from Melbourne to Ballarat, then to Horsham, and finally to Adelaide. The total driving time, excluding stops, is estimated to be 7 hours and 53 minutes.

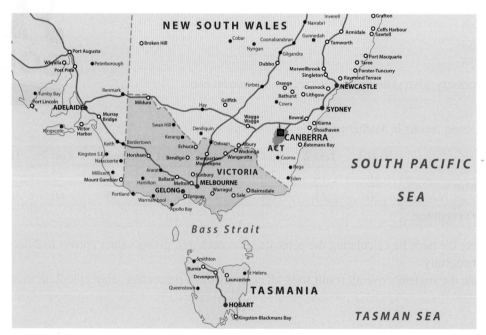

If the distance between the locations is in the ratio 44 : 67 : 136, determine the driving time between each location correct to the nearest minute.

12. **WE4** Calculate the rate (in km/h) at which you are moving if you are in a passenger aircraft that travels 1770 km in 100 minutes.

For questions 13–15, *give answers correct to* 2 *decimal places where appropriate.*

13. Calculate the rates in the units stated for:

   a. a yacht that travels 1.375 km in 165 minutes expressed in km/h

   b. a tank that loses 1320 mL of water in $2\frac{1}{3}$ hours expressed in mL/min

   c. a 3.6-metre-long carpet that costs $67.14 expressed in $/m

   d. a basketball player who has scored a total of 833 points in 68 games expressed in points/game.

14. **WE5** Calculate the cost in dollars per 100 grams for:

   a. a 650-g box of cereal costing $6.25   b. a 350-g packet of biscuits costing $3.25

   c. a 425-g jar of hazelnut spread costing $3.98   d. a 550-g container of yoghurt costing $3.69.

15. In a game of cricket, batsman A scored 48 runs from 66 deliveries, while batsman B scored 34 runs from 42 deliveries.

   a. State which batsman is scoring at the fastest rate (runs per delivery).

   b. Determine the combined scoring rate of the two batsmen in runs per 100 deliveries correct to 2 decimal places.

16. Change the following rates to the units indicated. Where necessary, give answers correct to 2 decimal places.

   a. 1.5 metres per second to kilometres per hour

   b. 60 kilometres per hour to metres per second

   c. 65 cents per gram to dollars per kilogram

   d. $5.65 per kilogram to cents per gram

17. Calculate the amount paid per hour for the following incomes, giving all answers correct to the nearest cent.

   a. $75 000 per annum for a 38-hour week   b. $90 000 per annum for a 40-hour week

   c. $64 000 per annum for a 35-hour week   d. $48 000 per annum for a 30-hour week

18. A particular car part is shipped in containers that hold 2054 items. Give answers to the following questions, correct to the nearest cent.

   a. If each container costs the receiver $8000, calculate the cost of each item.

   b. If the car parts are sold for a profit of 15%, determine how much is charged for each.

   c. The shipping company also has smaller containers that cost the receiver $7000 but only hold 1770 items. If the smaller containers are the only ones available, determine how much must the car part seller charge to make the same percentage profit.

19. A butcher has the following pre-packed meat specials.

BBQ lamb chops in packs of 12 for $15.50
Steak in packs of 5 for $13.80
Chicken drumsticks in packs of 11 for $11.33

a. Calculate the price per individual piece of meat for each of the specials, correct to the nearest cent.

b. The weights of two packages of meat are shown in the table.

Meat	Package 1	Package 2
BBQ lamb chops	2535 grams	2602 grams
Steak	1045 grams	1068 grams
Chicken drumsticks	1441 grams	1453 grams

Calculate the price per kilogram for each package correct to the nearest cent.

20. The ladder for the top four teams in the A-League is shown in the table.

Team		Win	Loss	Draw	Goals for	Goals against
1.	Western Sydney Wanderers	18	6	3	41	21
2.	Central Coast Mariners	16	5	6	48	22
3.	Melbourne Victory	13	9	5	48	45
4.	Adelaide United	12	10	5	38	37

Use CAS to:

a. express the win, loss and draw columns as a percentage of the total games played, correct to 2 decimal places

b. express the goals for as a percentage of the goals against, correct to 2 decimal places.

Question 1 (1 mark)

MC Angelique, Jerome and Summer were mentioned as beneficiaries in a will. The total inheritance was $50 000. This money was to be divided in the ratio of 2 : 3 : 1. The amount of money received by Summer is

A. $500 **B.** $5000 **C.** $16 666 67 **D.** $25 000 **E.** $8333.34

Question 2 (1 mark)

MC Simon divides his pay into three parts in the ratio 3 : 2 : 1, with 3 parts for rent, 2 parts for living and the rest for saving. He earns $900 per week. Select from the following the amount that he pays in rent.

A. $300 **B.** $450 **C.** $600 **D.** $200 **E.** $250

Question 3 (5 marks)

A group of Year 11 students was shopping together to gather provisions for their Outward Bound experience. They were trying to determine which sized jar of pasta sauce to purchase. The sizes and costs are:

150 g for $2.09; 250 g for $2.89; 500 g for $5.99.

   a. State which of the three jars of sauce was the most economical. Explain your reasoning. **(4 marks)**

   b. State whether the most economical purchase was the one you expected. **(1 mark)**

More exam questions are available online.

3.3 Percentage applications and GST

LEARNING INTENTION

At the end of this subtopic you should be able to:
- calculate percentage change
- calculate the final selling value
- determine mark-ups and discounts
- calculate the goods and services tax (GST).

3.3.1 Calculating percentage change

Percentages can be used to give an indication of the amount of change that has taken place, which makes them very useful for comparison purposes. Percentages are frequently used in comments in the media.

For example, a company might report that its profits have fallen by 6% over the previous year.

The percentage change is found by taking the actual amount of change that has occurred and expressing it as a percentage of the starting value.

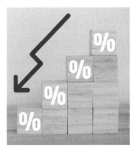

The rule for calculating percentage change

$$\text{Percentage change} = \frac{\text{finishing value} - \text{starting value}}{\text{starting value}} \times 100$$

---

tlvd-3706

WORKED EXAMPLE 6 Calculating the percentage change

The price of petrol was \$1.40 per litre but has now risen to \$1.65 per litre. Calculate the percentage change in the price of petrol, correct to 2 decimal places.

THINK	WRITE
1. Identify the amount of change.	$1.65 - 1.40 = 0.25$ The price of petrol has increased by 25 cents per litre.
2. Express the change as a fraction of the starting point, and simplify the fraction if possible.	$\dfrac{0.25}{1.40} = \dfrac{25}{140}$ $= \dfrac{5}{28}$
3. Convert the fraction to a percentage by multiplying by 100.	$\dfrac{5}{28} \times 100 = \dfrac{5}{7} \times \dfrac{25}{1}$ $= \dfrac{125}{7}$ ≈ 17.86
4. Write the answer as a sentence.	The price of petrol has increased by approximately 17.86%.

3.3.2 Calculating the final selling value

In the business world, percentages are often used to determine the final selling value of an item. For example, during a sale period a store might decide to advertise '25% off everything' rather than specify actual prices in a brochure.

At other times, when decisions are being made about the financial returns needed in order for a business to remain viable, the total production cost plus a percentage might be used. In either case, the required selling price can be obtained by multiplying by an appropriate percentage.

Consider the situation of the price of an item being reduced by 18% when it would normally sell for \$500. The reduced selling price can be found by evaluating the amount of the reduction and then subtracting it from the original value, as shown in the following calculations:

Reduction of 18%: $\dfrac{18}{100} \times 500 = \$90$

Reduced selling price: $\$500 - \$90 = \$410$

The selling price can also be obtained with a one-step calculation of $\dfrac{82}{100} \times 500 = \$410$.

In other words, reducing the price by 18% is the same as multiplying by 82% or $(100 - 18)\%$.

> ### Increasing or decreasing by a percentage
>
> **To reduce something by $x\%$, multiply by $(100 - x)\%$.**
>
> **To increase something by $x\%$, multiply by $(100 + x)\%$.**

Increase $160 by 15%.

THINK	WRITE
1. Add the percentage increase to 100.	$100 + 15 = 115$
2. Express the result as a percentage (by dividing by 100) and multiply by the value to be increased.	$\dfrac{115}{100} \times \$160 = \dfrac{115}{100} \times \dfrac{160}{1}$ $= \dfrac{23}{2} \times \dfrac{16}{1}$ $= 23 \times 8$ $= \$184$
3. Write the answer as a sentence.	Increasing $160 by 15% gives $184.

When a large number of values are being considered in a problem involving percentages, spreadsheets or other technologies can be useful to help carry out most of the associated calculations.

For example, a spreadsheet can be set up so that entering the original price of an item will automatically calculate several different percentage increases for comparison.

	A	B	C	D	E
1	Original		Increase by:		
2	price	+5%	+8%	+12%	+15%
3	$100.00	$105.00	$108.00	$112.00	$115.00
4	$150.00	$157.50	$162.00	$168.00	$172.50
5	$200.00	$210.00	$216.00	$224.00	$230.00
6	$300.00	$315.00	$324.00	$336.00	$345.00
7	$450.00	$472.50	$486.00	$504.00	$517.50

3.3.3 Mark-ups and discounts

When deciding on how much to charge customers, businesses have to take into account all of the costs they incur in providing their services. If their costs increase, they must mark up their own charges in order to remain viable.

For example, businesses that rely on the delivery of materials by road transport are susceptible to fluctuations in fuel prices, and they will take these into account when pricing their services.

If fuel prices increase, they will need to increase their charges; conversely, if fuel prices decrease, they might consider introducing discounts.

WORKED EXAMPLE 8 Calculating percentage mark-up

A transport company adjusts its charges as the price of petrol changes. Determine by what percentage, correct to 2 decimal places, the company's fuel costs change if the price per litre of petrol increases from $1.36 to $1.42.

THINK	WRITE
1. Calculate the amount of change.	$\$1.42 - \$1.36 = \$0.06$
2. Express the change as a fraction of the starting point.	$\dfrac{0.06}{1.36}$

3. Simplify the fraction where possible and then multiply by 100 to calculate the percentage change.

$$\frac{0.06}{1.36} = \frac{6}{136}$$
$$= \frac{3}{68}$$

$$\frac{3}{68} \times 100 = \frac{3}{17} \times \frac{25}{1}$$
$$= \frac{75}{17}$$
$$\approx 4.41$$

4. Write the answer as a sentence.

The company's fuel costs increase by 4.41%.

3.3.4 Goods and services tax (GST)

In Australia we have a 10% tax that is charged on most purchases, known as the **goods and services tax** (or **GST**). Some essential items, such as medicine, education and certain types of food, are exempt from GST, but for all other goods GST is added to the cost of items bought or services paid for. If a price is quoted as being 'inclusive of GST', the amount of GST paid can be evaluated by dividing the price by 11.

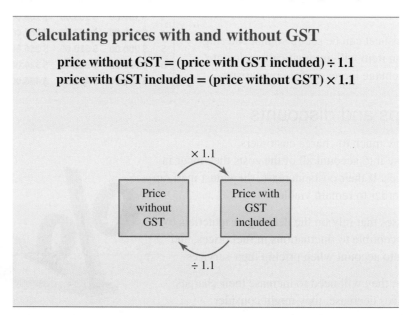

Calculating prices with and without GST

price without GST = (price with GST included) ÷ 1.1
price with GST included = (price without GST) × 1.1

WORKED EXAMPLE 9 Calculating the amount of GST

Calculate the amount of GST included in an item purchased for a total of $280.50.

THINK	WRITE
1. Determine whether the price already includes GST.	Yes, GST is included.
2. If GST is not included, calculate 10% of the value. If GST is included, divide the value by 11.	GST is included, so divide $280.50 by 11. $280.50 \div 11 = 25.5$
3. Write the answer as a sentence.	The amount of GST is $25.50.

 Resources

3.3 Exercise

Unless directed otherwise, give all answers to the following questions correct to 2 decimal places or the nearest cent, where appropriate.

1. **WE6** If the price of bananas was $2.65 per kg, calculate the percentage change (increase or decrease) if the price is now:
   a. $3.25 per kg
   b. $4.15 per kg
   c. $1.95 per kg
   d. $2.28 per kg.

2. Calculate the percentage change in the following situations.

   a. A discount voucher of 4 cents per litre was used on petrol advertised at $1.48 per litre.
   b. A trade-in of $5200 was applied to a car originally selling for $28 500.
   c. A shop owner purchases confectionary from the manufacturer for $6.50 per kilogram and sells it for 75 cents per 50 grams.
   d. A piece of silverware has a price tag of $168 at a market, but the seller is bartered down and sells it for $147.

3. The graph shows the change in the price of gold (in US dollars per ounce) from 27 July to 27 August in 2012.

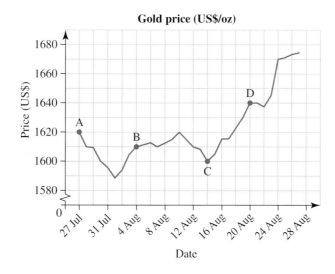

Gold price (US$/oz)

   a. Calculate the percentage change from:
      i. the point marked A to the point marked B
      ii. the point marked C to the point marked D.
   b. Calculate the percentage change from the point marked A to the point marked D.

4. A car yard offers three different vehicles for sale. The first car was originally priced at $18 750 and is now on sale for $14 991.

   The second car was originally priced at $12 250 and is now priced $9999, and the third car was originally priced at $23 990 and is now priced $19 888.

   Determine which car has had the largest percentage reduction.

5. **WE7** Increase:

   a. $35 by 8%
   b. $96 by 12.5%
   c. $142.85 by 22.15%
   d. $42 184 by 0.285%.

6. Decrease:

   a. $54 by 16%
   b. $7.65 by 3.2%
   c. $102.15 by 32.15%
   d. $12 043 by 0.0455%.

7. **WE8** A coffee shop adjusts its charges as the price of electricity changes. Determine by what percentage its power cost changes if the price of electricity increases from:

   a. 88 cents to 94 cents per kWh
   b. 92 cents to $1.06 per kWh.

8. An electrical goods department store charges $50 plus n cents per km for delivery of its products, where $n =$ the number of cents over $1.20 of the price per litre of petrol. Calculate the percentage increase in the total delivery charge for a distance of 25 km when the petrol price changes from $1.45 to $1.52 per litre.

9. **WE9** Calculate the amount of GST included in an item purchased for a total of:

   a. $34.98   b. $586.85   c. $56 367.85   d. $2.31.

10. Two companies are competing for the same job. Company A quotes a total of $5575 inclusive of GST. Company B quotes $5800 plus GST, but offers a 10% reduction on the total price for payment in cash. State which one is the cheaper offer, and by how much.

11. A plumber quotes their clients the cost of any parts required plus $74.50 per hour for their labour, and then adds on the required GST.

    a. Determine how much the plumber quotes for a job that requires $250 in parts (excluding GST) and should take 4 hours to complete.
    b. If the job ends up being faster than the plumber first thought and they end up charging the client for only 3 hours of labour, calculate the percentage discount on the original quote.

12. The price of a box of chocolates was originally $19.95. After it received an award for chocolate of the year, the price was increased by 12.25%. Twelve months later, the price was reduced by 15.5%.

    a. Calculate the final price of a box of this chocolate.
    b. Calculate the percentage change of the final price from the original price.

13. a. An advertisement for bedroom furniture states that you save $55 off the recommended retail price when you buy it for $385. Calculate the percentage by which the price has been reduced.

b. If another store was advertising the same furniture for 5% less than the sale price of the first store, determine by what percentage this has been reduced from the recommended retail price.

14. A bracelet is sold for $127.50. If this represents a 15% reduction from the RRP, calculate the original price.

15. Over a period of time prices in a store increased by 15%, then decreased by 10%, and finally increased by a further 5%. Calculate the overall percentage change over this time period, correct to the nearest whole percentage.

16. A power company claims that if you install solar panels for $1800, you will make this money back in savings on your electricity bill in 2 years. If you usually pay $250 per month, determine what percentage your bill will be reduced by if the company's claims are correct.

17. A house originally purchased for $320 000 is sold to a new buyer at a later date for $377 600.

a. Calculate the percentage change in the value of the house over this time period.

b. The new buyer pays a deposit of 15% and borrows the rest from a bank. Each year they are required to pay the bank 5% of the total amount borrowed. If they purchased the house as an investment, determine how much they should charge in rent per month in order to fully cover their bank payments.

18. Julie is shopping for groceries and buys the following items.
Bread — $3.30*
Fruit juice — $5.50*
Meat pies — $5.80
Ice cream — $6.90
Breakfast cereal — $5.00*
Biscuits — $2.90
All prices are listed before GST has been added on.

a. The items marked with an asterisk (*) are exempt from GST. Calculate the total amount of GST Julie has to pay for the shopping.

b. Calculate the additional amount Julie would have to pay if all of the items were eligible for GST.

c. Julie has a voucher that gives her a 10% discount from this shop. Use your answer from part a to calculate how much Julie pays for the groceries.

19. A carpet company offers a trade discount of 12.5% to a builder for supplying the floor coverings on a new housing estate.

a. If the builder spends $32 250, determine how much the carpet cost before the discount was applied. Round your answer to the nearest 5 cents.

b. If the builder charges his customers a total of $35 000, calculate the percentage discount the customers received compared to buying direct from the carpet company.

20. The Australian government is considering raising the GST from 10% to 12.5% in order to raise funds and cut the budget deficit.

The following shopping bill lists all items exclusive of GST.

Note: GST must be paid on all of the items in this bill.

1 litre of soft drink — $2.80

Large bag of pretzels — $5.30

Frozen lasagne — $6.15

Bottle of shampoo — $7.60

Box of chocolate — $8.35

2 tins of dog food — $3.50

Calculate the amount by which this shopping bill would increase if the rise in GST did go through.

21. The following table shows the number of participants in selected non-organised physical activities in Australia over a ten-year period.

Activity	Year 1	Year 2	Year 3	Year 4	Year 5	Year 6	Year 7	Year 8	Year 9	Year 10
Walking	4283	4625	5787	6099	5875	5724	5309	6417	6110	6181
Aerobics	1104	1273	1340	1551	1623	1959	1876	2788	2855	3126
Swimming	2170	2042	2066	2295	2070	1955	1738	2158	2219	2153
Cycling	1361	1342	1400	1591	1576	1571	1532	1850	1809	1985
Running	989	1067	1094	1242	1143	1125	1171	1554	1771	1748
Bushwalking	737	787	824	731	837	693	862	984	803	772
Golf	695	733	690	680	654	631	488	752	703	744
Tennis	927	818	884	819	792	752	602	791	714	736
Weight training	313	230	274	304	233	355	257	478	402	421
Fishing	335	337	387	349	312	335	252	356	367	383

a. Calculate the percentage change in the number of participants swimming from year 1 to year 10.

b. Calculate the percentage change in the number of participants walking from year 1 to year 10.

c. Calculate the overall percentage change in the number of participants swimming and walking combined during the time period.

22. The following table shows the changes in an individual's salary over several years.

Year	Annual salary	Percentage change
2013	$34 000	
2014	$35 750	
2015	$38 545	
2016	$42 280	
2017	$46 000	

Use CAS or a spreadsheet to answer these questions.

a. Evaluate the percentage change of each salary from the previous year.

b. State in which year the individual received the biggest percentage increase in salary.

3.3 Exam questions

⊳ **Question 1 (1 mark)**

MC The price of petrol was $1.38 per litre, but has now risen to $1.59 per litre. The percentage change in the price of petrol is

    A. 21%
    B. $\dfrac{21}{138}$%
    C. $\dfrac{7}{46}$%

    D. 350%
    E. $15\dfrac{5}{23}$%

⊳ **Question 2 (1 mark)**

MC During a supermarket sale campaign, the price of bread went 'down, down, down'. If the original price of the bread was $1.95 and the new price is $1.05, the percentage decrease in price is

    A. 90%
    B. $46\dfrac{2}{13}$%
    C. $\dfrac{6}{13}$%

    D. 600%
    E. $\dfrac{90}{195}$%

⊳ **Question 3 (1 mark)**

MC In one week, Max earns $127 working at a drive-through coffee shop. If he saved $34, the percentage of his weekly earnings that he spent, correct to the nearest percent, is

    A. 127%
    B. 34%
    C. 73%
    D. 93%
    E. 161%

More exam questions are available online.

3.4 Simple interest applications

LEARNING INTENTION

At the end of this subtopic you should be able to:
- calculate the amount of simple interest earned on an investment
- calculate the principal, rate or time
- evaluate the amount of simple interest investments and loans.

3.4.1 The simple interest formula

When you invest money and receive a return on your investment, the amount of money you receive on top of your original investment is known as the interest. Similarly, when you take out a loan, the additional amount that you pay back on top of the loan value is known as the interest. In simplest terms, interest can be considered the cost of borrowing money from a person or bank.

Interest is usually calculated as a percentage of the amount that is borrowed or invested, which is known as the **principal**. **Simple interest** involves a calculation based on the original amount borrowed or invested; as a result, the amount of simple interest for a particular loan is constant. For this reason, simple interest is often called 'flat rate' interest.

The simple interest formula

The formula to calculate simple interest is

$$I = \frac{Prn}{100}$$

where

I is the amount of interest earned

P is the principal (initial amount invested or borrowed)

r is the interest rate

n is the time period.

It is important to remember that the rate and the time must be compatible. For example, if the rate is per annum (yearly, abbreviated 'p.a.'), the time must also be in years.

The value of a simple interest investment can be evaluated by adding the total interest earned to the value of the principal.

tlvd-3707

WORKED EXAMPLE 10 Calculating the amount of simple interest

Calculate the amount of simple interest earned on an investment of $4450 that returns 6.5% per annum for 3 years.

THINK	WRITE
1. Identify the components of the simple interest formula.	$P = \$4450$ $r = 6.5\%$ $n = 3$
2. Substitute the values into the formula and evaluate the amount of interest.	$I = \dfrac{Prn}{100}$ $= \dfrac{4450 \times 6.5 \times 3}{100}$ $= 867.75$
3. Write the answer in a sentence.	The amount of simple interest earned is $867.75.

TI | THINK	DISPLAY/WRITE	CASIO | THINK	DISPLAY/WRITE
1. On the Calculator page, complete the entry line as:$i = \dfrac{prn}{100}$|$p = 4450$ and $r = 6.5$ and $n = 3$ Then press ENTER. *Note:* Be sure to enter the multiplication operation between the variables p, r and n.	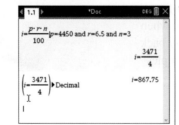	1. On the Financial screen, select: Calc(1) Simple Interest Complete the fields as: Days: 1095 I%: 6.5 PV: −4450 Then click the SI icon. *Note:* Be sure to include a negative sign with the $4550 for the present value to indicate that this money has been paid, not received.	

2. The answer can be read from the screen.	The amount of simple interest is $867.75.	2. The answer can be read from the screen. *Note:* The value for the simple interest is positive because money is being received.	The amount of simple interest is $867.75.

3.4.2 Calculating the principal, rate or time

The simple interest formula can be transposed (rearranged) to calculate other missing values in problems.

For example, we might want to know how long it will take for a simple interest investment of $1500 to grow to $2000 if we are being offered a rate of 7.5% per annum, or the interest rate needed for an investment to grow from $4000 to $6000 in 3 years.

The following formulas are derived from transposing the simple interest formula.

Alternative forms of the simple interest formula

To calculate the time: $n = \dfrac{100I}{Pr}$

To calculate the interest rate: $r = \dfrac{100I}{Pn}$

To calculate the principal: $P = \dfrac{100I}{rn}$

WORKED EXAMPLE 11 Determining the duration of an investment

Calculate how long it will take an investment of $2500 to earn $1100 with a simple interest rate of 5.5% p.a.

THINK	WRITE
1. Identify the components of the simple interest formula.	$P = 2500$ $I = 1100$ $r = 5.5$
2. Substitute the values into the formula and evaluate for n.	$n = \dfrac{100I}{Pr}$ $= \dfrac{100 \times 1100}{2500 \times 5.5}$ $= \dfrac{110\,000}{13\,750}$ $= 8$
3. Write the answer in a sentence.	It will take 8 years for the investment to earn $1100.

TI	THINK	DISPLAY/WRITE	CASIO	THINK	DISPLAY/WRITE
1. On the Calculator page, press MENU, then select: 3: Algebra 1: Solve Then complete the entry line as: $\text{solve}\left(i = \dfrac{p \times r \times n}{100}, n\right)$ $\|p = 2500$ and $i = 1100$ and $r = 5.5$. Press ENTER. *Note:* Be sure to enter the multiplication operation between the variables p, r and n.	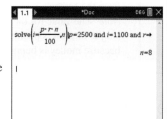	1. On the Main screen, complete the entry line as: $\text{solve}\left(i = \dfrac{p \times r \times n}{100}, n\right)$ $\|p = 2500$ and $i = 1100$ and $r = 5.5$. Then press EXE. *Note:* Be sure to enter the multiplication operation between the variables p, r and n.	 
2. The answer can be read from the screen.	It will take 8 years.	2. The answer can be read from the screen.	It will take 8 years.

3.4.3 Simple interest loans

For a simple interest loan, the total interest to be paid is usually calculated when the loan is taken out, and repayments are calculated from the total amount to be paid back (i.e. the principal plus the interest). For example, if a loan is for $3000 and the total interest after 2 years is $1800, the total to be paid back will be $4800. Monthly repayments on this loan would therefore be $4800 ÷ 24 = $200.

Total value of a simple interest investment or loan

The amount of a simple interest investment can be found by adding the simple interest to the principal. This can be expressed as $A = P + I$, where A represents the total amount of the investment.

WORKED EXAMPLE 12 Determining the repayment for a simple interest loan

Calculate the monthly repayments for a $14 000 loan that is charged simple interest at a rate of 8.45% p.a. for 4 years.

THINK	WRITE
1. Calculate the amount of interest charged.	$I = \dfrac{Prn}{100}$ $= \dfrac{14\,000 \times 8.45 \times 4}{100}$ $= 4732$
2. Add the interest to the principal to evaluate the total amount to be paid back.	$A = P + I$ $= 14\,000 + 4732$ $= 18\,732$
3. Divide by the number of months.	$\dfrac{18\,732}{48} = 390.25$
4. Write the answer as a sentence.	The monthly repayments will be $390.25.

Resources

3.4 Exercise

Unless otherwise directed, give all answers to the following questions correct to 2 decimal places or the nearest cent, where appropriate.

1. **WE10** Calculate the amount of simple interest earned on an investment of:

   a. $2575, returning 8.25% per annum for 4 years

   b. $12 250, returning 5.15% per annum for $6\frac{1}{2}$ years

   c. $43 500, returning 12.325% per annum for 8 years and 3 months

   d. $103 995, returning 2.015% per annum for 105 months.

2. Calculate the value of a simple interest investment of:

   a. $500, after returning 3.55% per annum for 3 years

   b. $2054, after returning 4.22% per annum for $7\frac{3}{4}$ years

   c. $3500, after returning 11.025% per annum for 9 years and 3 months

   d. $10 201, after returning 1.008% per annum for 63 months.

3. **WE11** Calculate how long it will take an investment of:

   a. $675 to earn $216 with a simple interest rate of 3.2% p.a.

   b. $1000 to earn $850 with a simple interest rate of 4.25% p.a.

   c. $5000 to earn $2100 with a simple interest rate of 5.25% p.a.

   d. $2500 to earn $775 with a simple interest rate of 7.75% p.a.

4. a. If $2000 earns $590 in 5 years, calculate the simple interest rate.

   b. If $1800 earns $648 in 3 years, calculate the simple interest rate.

   c. If $408 is earned in 6 years with a simple interest rate of 4.25%, determine how much was invested.

   d. If $3750 is earned in 12 years with a simple interest rate of 3.125%, determine how much was invested.

5. **WE12** Calculate the monthly repayments for:

   a. a $8000 loan that is charged simple interest at a rate of 12.25% p.a. for 3 years

   b. a $23 000 loan that is charged simple interest at a rate of 15.35% p.a. for 6 years

   c. a $21 050 loan that is charged simple interest at a rate of 11.734% p.a. for 6.25 years

   d. a $33 224 loan that is charged simple interest at a rate of 23.105% p.a. for 54 months.

6. Calculate the monthly repayments for:
   a. a $6225 loan that is charged simple interest at a rate of 7.025% p.a. for 130 weeks
   b. a $13 328 loan that is charged simple interest at a rate of 9.135% p.a. for 1095 days.

7. A savings account with a minimum monthly balance of $800 earns $3.60 interest in a month. Calculate the annual rate of simple interest.

8. $25 000 is invested for 5 years in an account that pays 6.36% p.a. simple interest.
   a. Calculate the amount of interest earned each year.
   b. Calculate the value of the investment after 5 years.
   c. If the money was reinvested for a further 2 years, determine the simple interest rate that would result in the investment amounting to $35 000 by the end of that time.

9. A borrower has to pay 7.8% p.a. simple interest on a 6-year loan. If the total interest paid is $3744:
   a. calculate the amount that was borrowed
   b. determine the repayments if they have to be made fortnightly.

10. $19 245 is invested in a fund that pays a simple interest rate of 7.8% p.a. for 42 months.
    a. Calculate the simple interest earned on this investment.
    b. The investor considers an alternative investment with a bank that offers a simple interest rate of 0.625% per month for the first 2.5 years and 0.665% per month after that. Determine which investment is better.

11. A bank offers a simple interest loan of $35 000 with monthly repayments of $545.
    a. Calculate the rate of simple interest if the loan is paid in full in 15 years.
    b. After 5 years of payments, the bank offers to reduce the total time of the loan to 12 years if the monthly payments are increased to $650. Calculate the interest to be paid over the life of the loan under this arrangement.
    c. Calculate the average rate of simple interest over the 12 years under the new arrangement.

12. $100 is invested in an account that earns $28 of simple interest in 8 months.
    a. Evaluate the annual rate of simple interest.
    b. Calculate the amount of interest that would have been earned in the 8 months if the annual interest rate was increased by 0.75%.

3.4 Exam questions

Question 1 (1 mark)
MC Ragnar invests $1500 at 15% p.a. simple interest for 18 months. At the end of the investment, the total amount that Ragnar will have in the investment account is
   A. $337.50　　　　　　　B. $4050　　　　　　　　C. $5550
   D. $1837.50　　　　　　E. $33 750

Question 2 (1 mark)
MC Alice's bank account pays 0.01% interest per month on the minimum monthly balance. She never has less than $275 in the bank. The least interest she earns is
   A. $27.50　　　B. $2.75　　　C. 27.5 cents　　　D. 2.75 cents　　　E. 0.275 cents

Question 3 (1 mark)
MC Michelle borrowed $500 for 2 years at a simple interest rate of 8.5%. The amount she had to pay back was
   A. $670　　　B. $585　　　C. $542.50　　　D. $85　　　E. $42.50

More exam questions are available online.

3.5 Compound interest applications

3.5.1 The compound interest formula

Simple interest rates calculate interest on the starting value. However, it is more common for interest to be calculated on the changing value throughout the time period of a loan or investment. This is known as compounding.

In compounding, the interest is added to the balance, and then the next interest calculation is made on the new value.

For example, consider an investment of \$5000 that earns 5% p.a. compounding annually. At the end of the first year,

the interest amounts to $\dfrac{5}{100} \times 5000 = \$250$, so the total investment will become \$5250. At the end of the second

year, the interest now amounts to $\dfrac{5}{100} \times 5250 = \$262.50$. As time progresses, the amount of interest becomes

larger at each calculation. In contrast, a simple interest rate calculation on this balance would be a constant, unchanging amount of \$250 each year.

It is easier to calculate **compound interest** by using the following formula.

The compound interest formula for annual compounding

$$A = P\left(1 + \frac{r}{100}\right)^n$$

where

A is the final value of the investment or loan

P is the principal (initial amount invested or borrowed)

r is the interest rate per annum (% p.a.)

n is the number of years.

As with the simple interest formula, we need to ensure that the rate of interest and the number of compounding periods are compatible. If we want to calculate the amount of compound interest, we need to subtract the principal from the final amount at the end of the compounding periods.

> ## The amount of compound interest
>
> $$I = A - P$$
>
> where
>
> A is the final value of the investment or loan
>
> P is the principal (initial amount invested or borrowed)
>
> I is the amount of compound interest.

Almost all questions involving compound interest can be solved using the financial solver on a CAS calculator.

> ## The financial solver
>
> - Using the TI-Nspire CX II CAS, on the calculator page, select:
>   - Menu
>   - 8: Finance
>   - 1: Finance Solver
> - Using the CASIO ClassPad II, on the financial page, select:
>   - Calc (1)
>   - Compound Interest
> - All financial solvers will require all but one of the following fields to be filled in:
>   - N: the total number of periods
>   - I%: annual interest rate
>   - PV: The principal value of the loan or investment
>   - PMT: The value of any regular payments being made
>   - FV: The final value of the loan or investment
>   - PpY or P/Y: Payments per year
>   - CpY or C/Y: Compounding periods per year
> - For General Maths, PpY and CpY are always the same value.
> - Money given to the bank has a negative value in the financial solver.
> - Money the bank gives to the customer has a positive value in the financial solver.
>
> | Loan | Investments (savings account) |
> | --- | --- |
> | PV is positive. | PV is negative. |
> | PMT is negative. | PMT is negative. |
> | FV is negative or 0. | FV is positive. |

tlvd-3708

WORKED EXAMPLE 13 Calculating the amount of compound interest earned

Use the compound interest formula to calculate the amount of interest on an investment of $2500 at 3.5% p.a. compounded annually for 4 years, correct to the nearest cent.

THINK	WRITE
1. Identify the components of the compound interest formula.	$P = 2500$ $r = 3.5$ $n = 4$

2. Substitute the values into the formula and evaluate the amount of the investment.

$$A = P\left(1 + \frac{r}{100}\right)^n$$

$$= 2500\left(1 + \frac{3.5}{100}\right)^4$$

$$= 2868.81 \text{ (to 2 decimal places)}$$

3. Subtract the principal from the final amount of the investment to calculate the interest.

$$I = A - P$$

$$= 2868.81 - 2500$$

$$= 368.81$$

4. Write the answer.

The amount of compound interest is $368.81.

TI	THINK	DISPLAY/WRITE	CASIO	THINK	DISPLAY/WRITE

1. On the Calculator page, press MENU, then select:
8: Finance
1: Finance Solver
Complete the fields as:
N: 4
I(%): 3.5
PV: −2500
Pmt: 0
PpY: 1
CpY: 1
Then move the cursor to the FV field and press ENTER.

2. The value of the investment after 4 years can be read from the screen:
FV = $2868.8075 …

3. The compound interest accumulated in 4 years can be found by subtracting the initial value of the investment from the investment value after 4 years.

Interest
= $2868.81 − $2500
= $368.81
After 4 years, the amount of compound interest accumulated is $368.81.

1. On the Financial screen, select:
Calc(1)
Compound Interest
Complete the fields as:
N: 4
I%: 3.5
PV: −2500
PMT: 0
P/Y: 1
C/Y: 1

2. The value of the investment after 4 years from the screen:
FV = $2868.8075…

3. The compound interest accumulated in 4 years can be found by subtracting the initial value of the investment from the investment value after 4 years.

Interest
= $2868.81 − $2500
= $368.81
After 4 years, the amount of compound interest accumulated is $368.81.

3.5.2 Calculating the interest rate or principal

As with the simple interest formula, the compound interest formula can be transposed if we need to calculate the interest rate or principal required to answer a particular problem. Transposing the compound interest formula gives the following formulas.

Rules for the compound interest rate and principal

To calculate the interest rate:

$$r = 100\left[\left(\frac{A}{P}\right)^{\frac{1}{n}} - 1\right]$$

To calculate the principal:

$$P = \frac{A}{\left(1 + \dfrac{r}{100}\right)^{n}}$$

WORKED EXAMPLE 14 Calculating the principal of an investment

Use the compound interest formula to calculate the principal required, correct to the nearest cent, to have a final amount of \$10 000 after compounding at a rate of 4.5% p.a. for 6 years.

THINK	WRITE
1. Identify the components of the compound interest formula.	$A = \$10\ 000$ $r = 4.5$ $n = 6$
2. Substitute the values into the formula to evaluate the principal.	$P = \dfrac{A}{\left(1 + \frac{r}{100}\right)^{n}}$ $= \dfrac{10\ 000}{\left(1 + \frac{4.5}{100}\right)^{6}}$ $= 7678.96$ (to 2 decimal places)
3. Write the final answer.	The principal required is \$7678.96.

TI | THINK	DISPLAY/WRITE	CASIO | THINK	DISPLAY/WRITE
1. On the Calculator page, press MENU, then select: 8: Finance 1: Finance Solver Complete the fields as: N: 6 I(%): 4.5 Pmt: 0 FV: 10 000 PpY: 1 CpY: 1 Then move the cursor to the PV field and press ENTER.	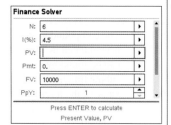	1. On the Financial screen, select: Calc(1) Compound Interest Complete the fields as: N: 6 I%: 4.5 PMT: 0 FV: 10 000 P/Y: 1 C/Y: 1 Then click the PV icon.	

2. The principal can be read from the screen: PV = $7678.96.
Note: The negative sign implies that the principal was paid, but does not need to be included in the answer.

2. The principal can be read from the screen: PV = $7678.96.
Note: The negative sign implies that the principal was paid, but does not need to be included in the answer.

Note: It is also possible to transpose the compound interest formula to calculate the number of compounding periods (*n*), but this requires logarithms and is outside the scope of this course.

3.5.3 Non-annual compounding

Interest rates are usually expressed per annum (yearly), but compounding often takes place at more regular intervals, such as quarterly, monthly or weekly. When this happens, adjustments need to be made when applying the formula to ensure that the rate is expressed in the same period of time.

The compound interest formula with non-annual compounding

$$A = P\left(1 + \frac{r}{C \times 100}\right)^{n \times c}$$

where

A is the final value of the investment or loan

P is the principal (initial amount invested or borrowed)

r is the interest rate per annum (% p.a.)

n is the number of years

c is the number of compounding periods per year.

- **Biannually** $c = 2$

- **Quarterly** $c = 4$

- **Monthly** $c = 12$

- **Fortnightly** $c = 26$

- **Weekly** $c = 52$

- **Daily** $c = 365$

The ratio of $\dfrac{r}{c}$ will give the interest rate per period and the product $t \times c$ will give the total number of periods. These values can be put into a simplified version of the rule:

$$A = P\left(1 + \frac{r}{100}\right)^{n}, \text{ where } r \text{ is the rate per period and } n \text{ is the number of periods.}$$

tlvd-3709

WORKED EXAMPLE 15 Calculating interest when compounding multiple times a year

Use the compound interest formula to calculate the amount of interest accumulated on $1735 at 7.2% p.a. for 4 years if the compounding occurs monthly. Give your answer correct to the nearest cent.

THINK	WRITE
1. Identify the components of the compound interest formula.	$P = \$1735$ $r = 7.2$ $n = 48$ (monthly periods)
2. Substitute the values into the formula and evaluate the amount.	$A = P\left(1 + \dfrac{r}{1200}\right)^n$ $= 1735\left(1 + \dfrac{7.2}{1200}\right)^{48}$ $= 2312.08$ (to 2 decimal places)
3. Subtract the principal from the amount of the investment.	$I = A - P$ $= 2312.08 - 1735$ $= 577.08$
4. Write the answer.	The amount of interest accumulated is $577.08.

TI | THINK	DISPLAY/WRITE	CASIO | THINK	DISPLAY/WRITE
1. On the Calculator page, press MENU, then select: 8: Finance 1: Finance Solver Complete the fields as: N: 48 I(%): 7.2 PV: −1735 Pmt: 0 PpY: 12 CpY: 12 Then move the cursor to the FV field and press ENTER.		1. On the Financial screen, select: Calc(1) Compound Interest Complete the fields as: N: 48 I%: 7.2 PV: −1735 PMT: 0 P/Y: 12 C/Y: 12 Then click the FV icon.	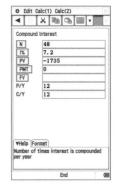
2. The value of the investment after 4 years can be read from the screen: FV = $2312.08		2. The value of the investment after 4 years can be read from the screen: FV = $2312.08	
3. The compound interest accumulated in 4 years can be found by subtracting the initial value of the term deposit from the investment value after 4 years.	Interest = $2312.08 − $1735 = $577.08 After 4 years, the amount of compound interest accumulated is $577.08.	3. The compound interest accumulated in 4 years can be found by subtracting the initial value of the term deposit from the investment value after 4 years.	Interest = $2312.08 − $1735 = $577.08 After 4 years, the amount of compound interest accumulated is $577.08.

3.5.4 Inflation

Inflation is a term used to describe a general increase in prices over time that effectively decreases the purchasing power of a currency. Inflation can be measured by the inflation rate, which is an annual percentage change of the **Consumer Price Index (CPI)**.

Inflation needs to be taken into account when analysing profits and losses over a period of time. It can be analysed by using the compound interest formula.

The increase in price of goods due to inflation can be calculated using the compounding interest formula, where the rate of inflation takes the place of the compound interest rate. Inflation is calculated once per year, so it is equivalent to compounding annually.

Spending power

As inflation increases, the **spending power** of a set amount of money will decrease. For example, if the cost of a loaf of bread was $4.00 and rose with the rate of inflation, in 5 years it might cost $4.50. As inflation gradually decreases the spending of the dollar, people's salaries often increase in line with inflation. This counterbalances the decreasing spending power of money.

WORKED EXAMPLE 16 Calculating inflation on an asset

An investment property is purchased for $300 000 and is sold 3 years later for $320 000. If the average annual inflation is 2.5% p.a., determine whether this has been a profitable investment.

THINK	WRITE
1. Recall that inflation is an application of compound interest and identify the components of the formula.	$P = 300\,000$ $r = 2.5$ $n = 3$
2. Substitute the values into the formula and evaluate the amount.	$A = P\left(1 + \dfrac{r}{100}\right)^n$ $= 300\,000\left(1 + \dfrac{2.5}{100}\right)^3$ $= 323\,067.19$ (to 2 decimal places)
3. Compare the inflated amount to the selling price.	Inflated amount: $323 067.19 Selling price: $320 000
4. Write the answer.	This has not been a profitable investment, as the selling price is less than the inflated purchase price.

TI | THINK	DISPLAY/WRITE	CASIO | THINK	DISPLAY/WRITE
1. On the Calculator page, press MENU, then select: 8: Finance 1: Finance Solver Complete the fields as: N: 3 I(%): 2.5 PV: $-300\,000$ Pmt: 0 PpY: 1 CpY: 1 Then move the cursor to the FV field and press ENTER.	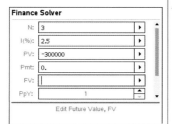	1. On the Financial screen, select: Calc(1) Compound Interest Complete the fields as: N: 3 I%: 2.5 PV: $-300\,000$ PMT: 0 P/Y: 1 C/Y: 1 Then click the FV icon.	

2. The value of the investment property after 3 years (the inflated price) can be read from the screen:

FV = $323 067.19

2. The value of the investment property after 3 years (the inflated price) can be read from the screen:

FV = $323 067.19

3. Write the answer.

This has not been a profitable investment, as the selling price ($320 000) is less than the inflated price ($323 067.19).

3. Write the answer.

This has not been a profitable investment, as the selling price ($320 000) is less than the inflated price ($323 067.19).

 Resources

 Interactivities Simple and compound interest (int-6265)

Non-annual compounding (int-6462)

3.5 Exercise

Students, these questions are even better in jacPLUS

 Receive immediate feedback and access sample responses

 Access additional questions

 Track your results and progress

Find all this and MORE in jacPLUS ▶

Unless otherwise directed, where appropriate give all answers to the following questions correct to 2 decimal places or the nearest cent.

1. **WE13** Use the compound interest formula to calculate the amount of compound interest on an investment of:

   a. $358 invested at 1.22% p.a. for 6 years
   b. $1276 invested at 2.41% p.a. for 4 years
   c. $4362 invested at 4.204% p.a. for 3 years
   d. $275 950 invested at 6.18% p.a. for 16 years.

2. Use the compound interest formula to calculate the future amount of:

   a. $4655 at 4.55% p.a. for 3 years
   b. $12 344 at 6.35% p.a. for 6 years
   c. $3465 at 2.015% p.a. for 8 years
   d. $365 000 at 7.65% p.a. for 20 years.

3. **WE14** Use the compound interest formula to calculate the principal required to yield a final amount of:

   a. $15 000 after compounding at a rate of 5.25% p.a. for 8 years
   b. $22 500 after compounding at a rate of 7.15% p.a. for 10 years
   c. $1000 after compounding at a rate of 1.25% p.a. for 2 years
   d. $80 000 after compounding at a rate of 6.18% p.a. for 15 years.

4. Use the compound interest formula to calculate the compound interest rate p.a. that would be required to grow:
   a. $500 to $1000 in 2 years
   b. $850 to $2500 in 3 years
   c. $1600 to $2900 in 4 years
   d. $3490 to $9000 in 3 years.

5. **WE15** Use the compound interest formula to calculate the amount of interest accumulated on:
   a. $2876 at 3.12% p.a. for 2 years, if the compounding occurs monthly
   b. $23 560 at 6.17% p.a. for 3 years, if the compounding occurs monthly
   c. $85.50 at 2.108% p.a. for 2 years, if the compounding occurs monthly
   d. $12 345 at 5.218% p.a. for 6 years, if the compounding occurs monthly.

6. Use the compound interest formula to calculate the final amount for:
   a. $675 at 2.42% p.a. for 2 years compounding weekly
   b. $4235 at 6.43% p.a. for 3 years compounding quarterly
   c. $85 276 at 8.14% p.a. for 4 years compounding fortnightly
   d. $53 412 at 4.329% p.a. for 1 year compounding daily.

7. An $8000 investment earns 7.8% p.a. compound interest over 3 years. Calculate the amount of interest earned if the amount is compounded:
   a. annually   b. monthly   c. weekly   d. daily.

8. **WE16** An investment property is purchased for $325 000 and is sold 5 years later for $370 000. If the average annual inflation is 2.73% p.a., determine whether this has been a profitable investment.

9. A business is purchased for $180 000 and is sold 2 years later for $200 000. If the annual average inflation is 1.8% p.a., determine whether a real profit has been made.

10. a. Calculate the interest accrued on a $2600 investment attracting a compound interest rate of 9.65%, compounded annually. Show your results in the following table.

Year	1	2	3	4	5	6	7	8
Interest accrued ($)								

   b. Show your results in a graph.

11. A parking fine that was originally $65 requires the payment of an additional late fee of $35. If the fine was paid 14 days late and interest had been compounding daily, calculate the annual rate of interest being charged.

12. A person has $1000 and wants to have enough to purchase something worth $1450.

   a. If they invest the $1000 in a bank account paying compound interest monthly and the investment becomes $1450 within 3 years, calculate the interest rate that the account is paying.
   b. If the price of the item increased in line with an average annual inflation rate of 2%, determine how much the person would have needed to invest to have enough to purchase it at the end of the same time period, using the same compound rate of interest as in part a.

13. The costs of manufacturing a smart watch decrease by 10% each year.

   a. If the watch initially retails at $200 and the makers decrease the price in line with the manufacturing costs, determine how much it will cost at the end of the first 3 years.
   b. Inflation is at a steady rate of 3% over each of these years, and the price of the watch also rises with the rate of inflation. Recalculate the cost of the watch for each of the 3 years according to inflation. (*Note:* Apply the manufacturing cost decrease before the inflation price increase.)

14. In 2006 Matthew earned approximately $45 000 after tax and deductions. In 2016 he earned approximately $61 000 after tax and deductions. If inflation over the 10-year period from 2006 to 2016 averaged 3%, determine whether Matthew's earning was comparatively more in 2006 or 2016.

15. Francisco is a purchaser of fine art, and his two favourite pieces are a sculpture he purchased in 1998 for $12 000 and a series of prints he purchased in 2007 for $17 000.

   a. If inflation averaged 3.3% for the period between 1998 and 2007, determine which item cost more in real terms.
   b. The value of the sculpture has appreciated at a rate of 7.5% since 1998, and the value of the prints has appreciated at a rate of 6.8% since 2007. Determine how much both were worth in 2015. Round your answers correct to the nearest dollar.

16. Use the compound interest formula to complete the following table. Assume that all interest is compounded annually.

Principal ($)	Final amount ($)	Interest earned	Interest rate (p.a.)	Number of years
11 000	12 012.28			2
14 000			3.25	3
22 050	25 561.99	3511.99		5
		2700.00	2.5	1

17. a. Using CAS, tabulate and graph an investment of $200 compounding at rate of 6.1% p.a. over 25 years.
    b. Evaluate, giving your answers to the nearest year, how long it will take the investment to:
       i. double
       ii. triple
       iii. quadruple.

18. Using CAS, compare compounding annually with compounding quarterly for $1000 at a rate of 12% p.a. over 5 years.

    a. Show the information in a graph or a table.

    b. Determine the effect of compounding at regular intervals during the year while keeping the annual rate the same.

3.5 Exam questions

Question 1 (1 mark)

Source: VCE 2020, Further Mathematics Exam 1, Section A, Q27; © VCAA.

MC Gen invests $10 000 at an interest rate of 5.5% per annum, compounding annually.

Determine after how many years her investment will be more than double its original value.

    A. 12   **B.** 13   **C.** 14   **D.** 15   **E.** 16

Question 2 (1 mark)

MC When considering the compound interest formula, select which of the following statements is false.

    A. The formula calculates only the interest earned.

    B. P represents principal, r represents rate and n represents the number of compounding periods.

    C. The A in the formula represents the final amount in the investment account at the end of the investment.

    D. When using the formula, the order of operations must be followed.

    E. When considering compounding periods you must calculate the number of times that interest is calculated and added to the account over the life of the loan.

Question 3 (1 mark)

MC If $5000 is invested at 8% p.a. compound interest for 3 years, compounding bi-annually, the interest earned is

    A. $6326.60   **B.** $5624.32   **C.** $1326.60

    D. $624.32   **E.** $5826.60

More exam questions are available online.

3.6 Purchasing options

LEARNING INTENTION

At the end of this subtopic you should be able to:
- compare different types of purchase options including cash, credit and debit cards, personal loans, 'buy now and pay later' schemes.

3.6.1 Cash purchases

Buying goods with cash is the most straightforward type of purchase you can make. The buyer owns the goods outright and no further payments are necessary. Some retailers or services offer a discount if you pay with cash.

WORKED EXAMPLE 17 Calculating cash discounts

A plumber offers a 5% discount if his customers pay with cash. Calculate how much a customer would be charged if they paid in cash and the fee before the discount was $139.

THINK	WRITE
1. Determine the percentage of the fee that the customer will pay after the discount is taken into account.	$100\% - 5\% = 95\%$
2. Multiply the fee before the discount by the percentage the customer will pay. Turn the percentage into a fraction.	$139 \times 95\% = 139 \times \dfrac{95}{100}$
3. Evaluate the amount to be paid.	$= 132.05$
4. Write the answer.	The customer will be charged $132.05.

3.6.2 Credit and debit cards

Credit cards

A **credit card** is an agreement between a financial institution (usually a bank) and an individual to loan an amount of money up to a pre-approved limit. Credit cards can be used to pay for transactions until the amount of debt on the credit card reaches the agreed limit of the credit card.

If a customer pays off the debt on their credit card within a set period of time after purchases are made, known as an interest-free period, they will pay no interest on the debt. Otherwise they will pay a high interest rate on the debt (usually 20–30% p.a.), with the interest calculated monthly. Customers are obliged to pay at least a minimum monthly amount of the debt (for example 3% of the balance).

It is worth noting that you still pay interest on the balance owing if you only pay off the minimum monthly amount. Not paying the minimum monthly amount will result in you incurring extra charges!

Credit cards often charge an annual fee, but customers can also earn rewards by using them, such as frequent flyer points for major airlines or discounts at certain retailers.

Debit cards

Debit cards are usually linked to bank accounts, although they can also be preloaded with set amounts of money. When a customer uses a debit card, the money is debited directly from their bank account or from the preloaded amount.

If a customer tries to make a transaction with a debit card that exceeds the balance in their bank account, then either their account will become overdrawn (which typically incurs a fee from the banking facility), or the transaction will be declined.

WORKED EXAMPLE 18 Calculating credit card interest

Heather has a credit card that charges an interest rate of 19.79% p.a. She tries to ensure that she always pays off the full amount at the end of the interest-free period, but an expensive few months over the Christmas holidays leaves the outstanding balance on her card at $635, $427 and $155 for three consecutive months. Calculate the total amount of interest Heather has to pay over the three-month period. Give your answer correct to the nearest cent.

THINK	WRITE
1. Use the simple interest formula to determine the amount of interest charged each month.	1st month: $I = \dfrac{Prn}{100}$ $= \dfrac{635 \times 19.79 \times \frac{1}{12}}{100}$ ≈ 10.47 2nd month: $I = \dfrac{Prn}{100}$ $= \dfrac{427 \times 19.79 \times \frac{1}{12}}{100}$ ≈ 7.04 3rd month: $I = \dfrac{Prn}{100}$ $= \dfrac{155 \times 19.79 \times \frac{1}{12}}{100}$ ≈ 2.56
2. Calculate the sum of the interest for the three months.	$10.47 + 7.04 + 2.56 = 20.07$
3. Write the answer.	Heather has to pay $20.07 in interest over the three-month period.

3.6.3 Personal loans

A **personal loan** is a loan made by a lending institution to an individual. A personal loan will usually have a fixed interest rate attached to it, with the interest paid by the customer calculated on a reduced balance. This means that the interest for each period will be calculated on the amount still owing, rather than the original amount of the loan.

WORKED EXAMPLE 19 Calculating the balance of a reducing balance loan

Francis takes out a loan of $3000 to help pay for a business management course. The loan has a fixed interest rate of 7.75% p.a. and Francis agrees to pay back $275 a month. Assuming that the interest is calculated before Francis's payments, calculate the outstanding balance on the loan after Francis's third payment. Give your answer correct to the nearest cent.

THINK	WRITE
1. Calculate the interest payable for the first month of the loan.	$I = \dfrac{Prn}{100}$ $= \dfrac{3000 \times 7.75 \times \frac{1}{12}}{100}$ ≈ 19.38
2. Calculate the total value of the loan before Francis's first payment.	$3000 + \$19.38 = \$3019.38$
3. Calculate the total value of the loan after Francis's first payment.	$\$3019.38 - \$275 = \$2744.38$
4. Calculate the interest payable for the second month of the loan.	$I = \dfrac{Prn}{100}$ $= \dfrac{2744.38 \times 7.75 \times \frac{1}{12}}{100}$ ≈ 17.72
5. Calculate the total value of the loan before Francis's second payment.	$\$2744.38 + \$17.72 = \$2762.10$
6. Calculate the total value of the loan after Francis's second payment.	$\$2762.10 - \$275 = \$2487.10$
7. Calculate the interest payable for the third month of the loan.	$I = \dfrac{Prn}{100}$ $= \dfrac{2487.1 \times 7.75 \times \frac{1}{12}}{100}$ ≈ 16.06
8. Calculate the total value of the loan before Francis's third payment.	$\$2487.10 + \$16.06 = \$2503.16$
9. Calculate the total value of the loan after Francis's third payment.	$\$2503.16 - \$275 = \$2228.16$
10. Write the answer.	The outstanding balance of the loan after Francis's third payment is $2228.16.

TI | THINK	DISPLAY/WRITE	CASIO | THINK	DISPLAY/WRITE
1. On the Calculator page, press MENU, then select: 8: Finance 1: Finance Solver Complete the fields as: N: 3 I(%): 7.75 PV: 3000 Pmt: −275 PpY: 12 CpY: 12 Then move the cursor to the FV field and press ENTER.	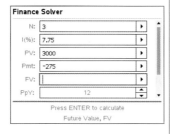 **Finance Solver** N: 3 I(%): 7.75 PV: 3000 Pmt: −275 FV: PpY: 12 Press ENTER to calculate Future Value, FV	1. On the Financial screen, select: Calc(1) Compound Interest Complete the fields as: N: 3 I%: 7.75 PV: 3000 PMT: −275 P/Y: 12 C/Y: 12 Then click the FV icon.	
2. The outstanding balance can be read from the screen.	 **Finance Solver** N: 3 I(%): 7.75 PV: 3000 Pmt: −275 FV: −2228.1616034885 PpY: 12 Finance Solver info stored into tvm.n, tvm.i, tvm.pv, tvm.pmt, ... The outstanding balance is $2228.16.	2. The outstanding balance can be read from the screen.	 The outstanding balance is $2228.16.

Time payments (hire purchase)

A **time payment**, or hire purchase, can be used when a customer wants to make a large purchase but doesn't have the means to pay up front. Time payments usually work by paying a small amount up front, and then paying weekly or monthly instalments.

3.6 Exercise

Students, these questions are even better in jacPLUS

- **Receive immediate feedback and access sample responses**
- **Access additional questions**
- **Track your results and progress**

Find all this and MORE in jacPLUS

Unless otherwise directed, where appropriate give all answers to the following questions correct to 2 decimal places or the nearest cent.

1. **WE17** An electrician offers a discount of 7.5% if his customers pay with cash. Calculate how much a customer would be charged if they paid in cash and the charge before the discount was applied was:
   a. $200
   b. $312
   c. $126.

2. **MC** George runs a pet-care service where he looks after cats and dogs on weekend afternoons. He charges a fee of $20 per pet plus $9 per hour. He also gives his customers a 6% discount if they pay in cash.

Charlene asks George to look after her two cats between 1 pm and 5 pm on a Saturday afternoon. Calculate how much she would have to pay if she paid in cash.

A. $33.85   **B.** $52.65   **C.** $71.45

D. $72.95   **E.** $73.85

3. **WE18** Barney is struggling to keep control of his finances and starts to use his credit card to pay for purchases. At the end of three consecutive months his outstanding credit card balance is $311.55, $494.44 and $639.70 respectively. If the interest rate on Barney's credit card is 22.75% p.a., calculate how much interest he is charged for the three-month period.

4. Dawn uses her credit card while on an overseas trip and returns with an outstanding balance of $2365.24 on it. Dawn can only afford to pay the minimum monthly balance of $70.96 before the interest-free period expires.

   a. Dawn's credit card charges an interest rate of 24.28% p.a. Calculate how much Dawn will be charged in interest for the next month.
   b. If Dawn spent $500 less on her overseas trip, determine by how much the interest she would be charged on her credit card can be reduced. (*Note:* Assume that Dawn still pays the minimum monthly balance of $70.96.)

5. **WE19** Petra takes out a loan of $5500 to help pay for a business management course. The loan has a fixed interest rate of 6.85% p.a. and Petra agrees to pay back $425 a month. Assuming that the interest is calculated before Petra's payments, calculate their outstanding balance on the loan after the third payment.

6. Calculate the total amount of interest paid after the sixth payment on a $2500 personal loan if the rate is 5.5% p.a. and $450 is paid off the loan each month. (Assume that the interest is calculated before the monthly payments.)

7. Drew has a leak in his water system and gets quotes from 5 different plumbers to try to find the best price for the job. From previous experience he believes it will take a plumber 90 minutes to fix his system. Help Drew decide which plumber to go with by calculating approximately how much each will charge.
   - Plumber A: A call-out fee of $100 plus an hourly charge of $80, with a 5% discount for payment in cash
   - Plumber B: A flat fee of $200 with no discount
   - Plumber C: An hourly fee of $130, with a 10% discount for payment in cash
   - Plumber D: A call-out fee of $70 plus an hourly fee of $90, with an 8% discount for payment in cash
   - Plumber E: An hourly fee of $120 with no discount

8. Items in an online store advertised for more than $100 can be purchased for a 12.5% deposit, with the balance payable 9 months later at a rate of 7.5% p.a. compounding monthly.

Determine how much the following items cost the purchaser under this arrangement.

   a. A sewing machine advertised at $150
   b. A portable air conditioner advertised at $550
   c. A treadmill advertised at $285
   d. A BBQ advertised at $675

9. Divya's credit card has a low interest rate of 13.55% p.a. but no interest-free period on purchases. Calculate the total interest she has to pay after making the following purchases.
   - New sound system — $499 — paid back after 7 days
   - 3 Blu-Ray films — $39 — paid back after 12 days
   - Food shopping — $56 — paid back after 2 days
   - Coffee machine — $85 — paid back after 18 days

10. An electrical goods store allows purchasers to buy any item priced at $1000 or more for a 10% deposit, with the balance payable 6 months later at a simple interest rate of 7.64% p.a. Calculate the final cost of each of the following items under this arrangement.
    a. An entertainment system priced at $1265
    b. An oven priced at $1450
    c. A refrigerator priced at $2018
    d. A washing machine priced at $3124

11. Elise gets a new credit card that has an annual fee of $100 and earns 1 frequent flyer point per $1 spent. In her first year using the card she spends $27 500 and has to pay $163 in interest on the card. Elise exchanged the frequent flyer points for a gift card to her favourite store, which values each point as being worth 0.8 cents. Determine whether using the credit card over the year was a profitable investment.

12. Michelle uses all of the $12 000 in her savings account to buy a new car worth $25 000 on a time payment scheme. The purchase also requires 24 monthly payments of $750.

    a. Calculate how much Michelle pays in total for the car.

    Michelle gets a credit card to help with her cash flow during this 24-month period, and over this time her credit card balance averages $215 per month. The credit card has an interest rate of 23.75% p.a.

    b. Calculate how much in interest Michelle pays on her credit card over this period.
    c. In another 18 months Michelle could have saved the additional $13 000 she needed to buy the car outright. Determine how much she would have saved by choosing to save this money first.

3.6 Exam questions

Question 1 (1 mark)
Source: VCE 2016, Further Mathematics Exam 1, Section A, Q23; © VCAA.

MC Sarah invests $5000 in a savings account that pays interest at the rate of 3.9% per annum compounding quarterly. At the end of each quarter, immediately after the interest has been paid, she adds $200 to her investment.

After two years, the value of her investment will be closest to
   A. $5805   **B.** $6600   **C.** $7004   **D.** $7059   **E.** $9285

Question 2 (1 mark)
Source: VCE 2015, Further Mathematics Exam 1, Section B, Module 4, Q1; © VCAA.

MC Fong's gas bill is $368.40. If he pays this bill on time, it will be reduced by 5%.

In this case, the bill would be reduced by
   A. $1.84   **B.** $5.00   **C.** $18.42   **D.** $184.20   **E.** $349.98

Question 3 (12 marks)

A Year 11 student wishes to invest some money in order to help him save for a holiday more quickly.

He is looking at two different investment options.

Option 1: Simple interest at 5% per annum for 4 years

Option 2: Compound interest at 3.5% per annum for 2 years
  a. Calculate the interest earned if he invests $3500 in the account offering option 1.   **(1 mark)**
  b. Calculate the interest earned if he invests $3500 in the account offering option 2.   **(2 marks)**

He decides that he really doesn't want to lock his money away for 4 years, but he definitely wants more interest than option 2 offers. He does some negotiating with the bank and they agree to change option 2 slightly so that it compounds biannually.
  c. Calculate the interest earned if he selects the new and improved option 2.   **(3 marks)**

The interest is better! But it is not quite enough. The bank agrees to extend the lending period on option 2.
  d. Determine for how many compounding periods he needs to invest the money to have a total of $5000 (investment and interest).   **(5 marks)**

Before he decides, he needs to calculate how long this investment will be for.
  e. Determine for how long he needs to invest his money in this investment for him to achieve his goals, and state whether you think he will select this option.   **(1 mark)**

More exam questions are available online.

3.7 Review

3.7.1 Summary

3.7 Exercise

Multiple choice

1. **MC** If a salesman earns 3.7% from the sale of each car, for a $14 990 sale he would be paid:

   A. $5546.30   B. $4051.35   C. $554.63   D. $550.12   E. $405.14

2. **MC** When fully simplified, $1\frac{4}{5} : 4\frac{1}{5}$ becomes:

   A. $18:42$ B. $1:4$ C. $1:3$ D. $8:2$ E. $3:7$

3. **MC** The unit cost (per gram) of a120-gram tube of toothpaste sold for $3.70 is:

   A. $32.43   B. $0.03   C. $0.44   D. $0.05   E. $0.31

4. **MC** If the price of petrol increased from 118.4 cents to 130.9 cents, the percentage change is:

   A. 10.6%   B. 9.5%   C. 90%   D. 1.1%   E. 12.5%

5. **MC** A basketball ring is sold for $28.50. If this represents a 24% reduction from the RRP, the original price was:

   A. $90.25   B. $118.75   C. $52.50   D. $37.50   E. $26.67

6. **MC** When the simple interest formula is transposed to calculate r, the correct formula is:

   A. $r = 100IPn$ B. $r = \dfrac{Pn}{100I}$ C. $r = \dfrac{100}{PnI}$ D. $r = \dfrac{100I}{Pn}$ E. $r = \dfrac{100IP}{n}$

7. **MC** A tradesman offers a 6.8% discount for customers who pay in cash. Determine how much would a customer pay if they paid their bill of $244 in cash.

   A. $16.59   B. $218.48   C. $227.41   D. $261.59   E. $237.20

8. **MC** Meredith walks dogs at the weekend. She charges $14.00 per dog plus $6.00 an hour. She offers her clients a 5% discount for paying in cash. Determine how much she would charge for someone paying cash to walk 3 dogs for 2 hours.

   A. $51.30   B. $131.10   C. $54   D. $2.70   E. $48

9. **MC** Ruhan borrows $12 000 in order to purchase a car that he needs for a new job. He agrees to pay the loan back with interest, in total, after 30 months. The interest rate is calculated using 4.5% p.a. simple interest. The amount Ruhan has to pay back is closest to:

   A. $12 000   B. $16 200   C. $1350   D. $28 200   E. $13 350

10. **MC** Sotiris invests $29 000 into a high interest saving account earning 6.2% p.a. compound interest, compounded monthly. He is planning to buy a house soon and needs $45 000 for the deposit. Determine how long he will have to wait until he has enough in the investment account to afford the deposit.

    A. 7 months   B. 8 months   C. 85 months   D. 86 months   E. 87 months

Short answer

11. While camping, the Blake family use powdered milk. They mix the powder with water in the ratio of 1:24. Determine how much of each ingredient they need to make up:

    a. 600 mL   b. 1.2 L.

12. For each of the following, calculate the unit price for the quantity shown in brackets.

    a. 750 g of Weetbix for $4.99 (per 100 g)
    b. $16.80 for 900 g of jelly beans (per 100 g)
    c. $4.50 for 1.5 L of milk (per 100 mL)
    d. $126.95 for 15 L of paint (per L)

13. Complete the following percentage changes.

    a. Increase $65.85 by 12.6%.   b. Decrease $14.56 by 23.4%.
    c. Increase $150.50 by 2.83%.   d. Decrease $453.25 by 0.65%.

14. Determine the GST that's included in or needs to be added to the price for these amounts.

    a. $45.50 with GST included   b. $109.00 plus GST
    c. $448.75 with GST included   d. $13.25 plus GST

15. Determine the unknown variable for each of the following scenarios.

    a. Calculate the amount of simple interest earned on an investment of $4500 that returns 6.87% per annum for 5.5 years.
    b. Determine how long it will take an investment of $1260 to earn $360 with a simple interest rate of 4.08%.
    c. Calculate the simple interest rate on an investment that earns $645 in 3 years when the initial principal was $5300.
    d. Calculate the monthly repayments for a $6250 loan that is charged simple interest at a rate of 9.32% per annum for 7.25 years.

16. Use the compound interest formula to calculate:

    a. the amount of interest on an investment of $3655 at 6.54% per annum for 2.5 years
    b. the future amount of $478 invested at 2.27% per annum for 10 years
    c. the compound interest rate per annum required to grow $1640 to $3550 in 3.25 years
    d. the principal required to yield a final amount of $22 000 after compounding at a rate of 11.2% per annum for 15 years.

17. Sophie bought an investment property for $250 000, and 4 years later she sold it for $275 000. If the average annual inflation was 2.82% per annum, determine whether this was a profitable investment for Sophie. Give your answer correct to 2 decimal places.

Extended response

18. The monthly repayment for a $250 000 property loan is $1230.

    a. Calculate the monthly repayment as a percentage of the loan.
    b. Calculate the overall yearly repayment as a percentage of the loan.
    c. If the repayments have to be made every month for 25 years, determine how much extra has to be paid back compared to the amount that was borrowed.
    d. Express the extra amount to be paid back over the 25 years as a percentage of the amount borrowed.

19. John is comparing different brands of lollies at the local supermarket. A packet of Brand A lollies costs $7.25 and weighs 250 g. A packet of Brand B lollies weighs 1.2 kg and costs $22.50.

    a. Determine which brand is the best value for money. Provide mathematical evidence to support your answer.
    b. If the company producing the more expensive brand was to reconsider their price, determine the price for their lollies that would match the unit price of the cheaper brand.

20. For a main course at a local restaurant, guests can select from a chicken, fish or vegetarian dish. On Friday night the kitchen served 72 chicken plates, 56 fish plates and 48 vegetarian plates.

a. Express the number of dishes served as a ratio in the simplest form.
b. On a Saturday night the restaurant can cater for 250 people. If the restaurant was full, determine how many people would be expected to order a non-vegetarian dish.
c. The Elmir family of five and the Cann family of three dine together. The total bill for the table was $268.

   i. Calculate the cost of dinner per head.
   ii. If the bill is split according to family size, determine the proportion of the bill that the Elmir family will pay.

21. Four years ago a business was for sale at $130 000. Amanda and Callan had the money to purchase the business but missed out at the auction. Four years later the business is again for sale, but now at $185 000.

a. Determine the percentage increase in the price over the 4 years.
b. Amanda and Callan will now need to borrow the increase in the price amount. Determine how much interest they will have to pay on a loan compounded annually over 5 years with a rate of 12.75%.

22. Manny has received three quotes for the painting of her house:
   • Painter 1: $290 per day (3 days needed) plus
     $20 per litre of paint (120 litres needed), with no further discounts
   • Painter 2: A flat fee of $4000 with a 10% discount
     for cash
   • Painter 3: An hourly fee of $80 (7 hours a day for
     3 days) plus $25 per litre of paint (120 litres needed) with a 5% discount
     for cash
     a. Calculate the cost of each painter.
     b. Based on price, state which painter Manny
        should select.

To pay for the painter, Manny withdraws $4000 from her bank account. The account had an opening balance of $6000 and earns a simple interest rate of 12.4%. The interest is calculated on the minimum monthly balance.

Date	Details	Amount
1st	Withdrawal	$4000
10th	Deposit	$151
15th	Withdrawal	$220
22nd	Deposit	$1500
29th	Withdrawal	$50
30th	Deposit	$250

c. Use the account information to calculate the amount of interest earned on the account this month.
d. Determine the percentage change of the account balance from the start to the end of the month.

3.7 Exam questions

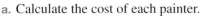

Question 1 (1 mark)

Source: VCE 2015, Further Mathematics Exam 1, Section B, Module 4, Q2; © VCAA.

MC An investment property was purchased for $600 000. Over a 10-year period, its value increased to $850 000.

The increase in value, as a percentage of the purchase price, is closest to
   A. 4.2%   **B.** 25.0%   **C.** 29.4%   **D.** 41.7%   **E.** 70.6%

Question 2 (3 marks)

Source: VCE 2015, Further Mathematics Exam 2, Module 4, Q1; © VCAA.

Jane and Michael have started a business that provides music at parties.

The business charges customers $88 per hour.

The $88 per hour includes a 10% goods and services tax (GST).

   a. Calculate the amount of GST included in the $88 hourly rate. **(1 mark)**

   b. Jane and Michael's first booking was a party where they provided music for four hours. Calculate the total amount they were paid for this booking. **(1 mark)**

   c. After six months of regular work, Jane and Michael decided to increase the hourly rate they charge by 12.5%.

   Calculate the new hourly rate (including GST). **(1 mark)**

Question 3 (1 mark)

MC A newspaper reported that the price of petrol had increased by 35% in the last five years. If the current price is $1.59, the original price of petrol mentioned in the article was

   A. $0.56　　　**B.** $2.14　　　**C.** $1.18　　　**D.** $2.94　　　**E.** $1.35

Question 4 (1 mark)

MC A television was advertised as being marked down by $33\frac{1}{3}$%. If the original price of the television is $2999, the discounted price will be

   A. $2666.33　　　**B.** $999.67　　　**C.** $2000　　　**D.** $1999.33　　　**E.** $1500

Question 5 (7 marks)

On occasion items can be purchased by paying a deposit and then paying monthly instalments.

Leonard was planning to purchase a surround sound entertainment system worth $5495 and was investigating different ways in which this purchase could be made. If he paid a 20% deposit, he would be able to pay the system off over the next year.

   a. Calculate a 20% deposit on this system. **(1 mark)**

   b. Calculate the balance owing. **(1 mark)**

   c. Leonard is able to make monthly payments for 12 months in order to pay the remainder of the cost. Calculate each monthly instalment, correct to the nearest dollar. **(1 mark)**

   d. The shop assistant offers Leonard a discount to try to entice him to buy the system now. The shop assistant offers a 32% discount on the total cost.

   i. Calculate the amount of the discount. **(1 mark)**

   ii. Calculate the cost of the entertainment system with the discount applied. **(1 mark)**

   e. Leonard decides to go with the payment plan, but he decides to pay slightly more each month. He decides to pay $425 each month.

   i. Determine how quickly Leonard will pay off the balance remaining. **(1 mark)**

   ii. Calculate the final payment. **(1 mark)**

More exam questions are available online.

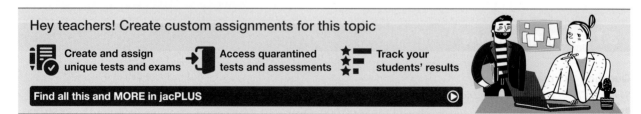

Hey teachers! Create custom assignments for this topic

Create and assign unique tests and exams　　Access quarantined tests and assessments　　Track your students' results

Find all this and MORE in jacPLUS

Answers

Topic 3 Financial mathematics

3.2 Ratio, rates and percentages

3.2 Exercise

1. a. 77.5% b. 3

2. a. $1943.25
   b. Week 1: 16.03%; week 2: 22.36%; week 3: 29.55%; week 4: 32.06%

3. a. $10 625 b. $13 196.25
   c. $19 327.09 d. $37 368.38

4. a. See the table at the bottom of the page.*
   b. 68.43%

5. a. Food: 33.33%; electricity: 12.5%; telephone: 5.95%; petrol: 10.12%; rent: 38.10%
   b. Food: 25.45%; electricity: 9.55%; telephone: 4.55%; petrol: 7.73%; rent: 29.09%

6. a. $27:9:4$ b. $1:2$

7. $39:40$

8. a. $3:7$ b. $7:11:15$
   c. $43:25$ d. $10:4:5$

9. 15 red, 24 yellow, 9 orange

10. a. Mark: $5400; Henry: $1800; Dale: $7200; Ben: $3600
    b. $150:40:69:123$

11. Melbourne to Ballarat: 1 hour, 24 minutes; Ballarat to Horsham: 2 hours, 8 minutes; Horsham to Adelaide: 4 hours, 20 minutes

12. 1062 km/h

13. a. 0.5 km/h b. 9.43 mL/ min
    c. $18.65/m d. 12.25 points/game

14. a. $0.96 b. $0.93 c. $0.94 d. $0.67

15. a. Batsman B
    b. 75.93 runs/ 100 deliveries

16. a. 5.4 km/h b. 16.67 m/s
    c. $650/kg d. 0.57 cents/gram

17. a. $37.96 b. $43.27 c. $35.16 d. $30.77

18. a. $3.89 b. $4.47 c. $4.54

19. a. BBQ lamb chops: $1.29
    Steak: $2.76
    Chicken drumsticks: $1.03

b. Pack A: $16.18/kg
   Pack B: $15.86/kg

20. a. See the table at the bottom of the page.*
    b.

Team	Goal percentage
1. Western Sydney Wanderers	195.24%
2. Central Coast Mariners	218.18%
3. Melbourne Victory	106.67%
4. Adelaide United	102.70%

3.2 Exam questions

Note: Mark allocations are available with the fully worked solutions online.

1. E
2. B
3. a. The medium-sized jar is the cheapest per 100 grams of sauce. The students should purchase this one.
   b. It should be expected that the largest jar would be the cheapest item; however, this is not always the case.

3.3 Percentage applications and GST

3.3 Exercise

1. a. 22.64% increase b. 56.60% increase
   c. 26.42% decrease d. 13.96% decrease

2. a. 2.70% decrease b. 18.25% decrease
   c. 130.77% increase d. 12.5% decrease

3. a. i. 0.62% ii. 2.5%
   b. 1.23%

4. The first car has the largest percentage reduction at 20.05%.

5. a. $37.80 b. $108
   c. $174.49 d. $42 304.22

6. a. $45.36 b. $7.41
   c. $69.31 d. $12 037.52

7. a. 6.82% b. 15.22%

8. 3.11%

9. a. $3.18 b. $53.35
   c. $5124.35 d. $0.21

10. Company A by $167

11. a. $602.80 b. 13.59%

12. a. $18.92 b. 5.16% decrease

13. a. 12.5% b. 16.88%

14. $150

*4. a.

	Test 1	Test 2	Test 3	Test 4	Test 5	Test 6	Test 7	Test 8
Percentage	80%	66.67%	76.47%	80%	72.5%	33.33%	70%	74.29%

*20. a.

	Team	Win	Loss	Draw
1.	Western Sydney Wanderers	66.67%	22.22%	11.11%
2.	Central Coast Mariners	59.26%	18.52%	22.22%
3.	Melbourne Victory	48.15%	33.33%	18.52%
4.	Adelaide United	44.44%	37.04%	18.52%

15. An overall increase of 9%

16. 30%

17. a. 18% **b.** $1337.33

18. a. $1.56 **b.** $1.38 **c.** $27.86

19. a. $36 857.15 **b.** 5.04%

20. $0.84

21. a. 0.79% decrease **b.** 44.31% increase
c. 29.15% increase

22. a.

Year	Annual salary	Percentage change
2013	$34 000	
2014	$35 750	5.15%
2015	$38 545	7.82%
2016	$42 280	9.69%
2017	$46 000	8.80%

b. 2016

3.3 Exam questions

Note: Mark allocations are available with the fully worked solutions online.

1. E

2. B

3. C

3.4 Simple interest applications

3.4 Exercise

1. a. $849.75 **b.** $4100.69
c. $44 231.34 **d.** $18 335.62

2. a. $553.25 **b.** $2725.76
c. $7069.34 **d.** $10 740.84

3. a. 10 years **b.** 20 years
c. 8 years **d.** 4 years

4. a. 5.9% **b.** 12% **c.** $1600 **d.** $10 000

5. a. $303.89 **b.** $613.65
c. $486.50 **d.** $1254.96

6. a. $243.94 **b.** $471.68

7. 5.4%

8. a. $1590 **b.** $32 950 **c.** 3.11%

9. a. $8000 **b.** $75.28

10. a. $5253.89
b. The first investment is the best (7.8% p.a.).

11. a. 12.02% **b.** $52 300 **c.** 12.45%

12. a. 42% p.a. **b.** $28.50

3.4 Exam questions

Note: Mark allocations are available with the fully worked solutions online.

1. D

2. D

3. B

3.5 Compound interest applications

3.5 Exercise

1. a. $385.02 **b.** $1403.52
c. $4935.59 **d.** $720 300.86

2. a. $664.76 **b.** $5515.98
c. $599.58 **d.** $1 229 312.85

3. a. $9961.26 **b.** $11 278.74
c. $975.46 **d.** $32 542.37

4. a. 41.42% **b.** 43.28% **c.** 16.03% **d.** 37.13%

5. a. $184.93 **b.** $4777.22
c. $3.68 **d.** $4526.95

6. a. $708.47 **b.** $5128.17
c. $118 035.38 **d.** $55 774.84

7. a. $2021.81 **b.** $2101.50
c. $2107.38 **d.** $2108.90

8. The inflated value is $371 851.73, so it was not profitable.

9. The inflated value is $186 538.32, so it is profitable.

10. a. See the table at the bottom of the page.*
b.

11. 1140.57% p.a.

12. a. 12.45% p.a. **b.** $1061.19

13. a. Year 1: $180, Year 2: $162, Year 3: $145.80
b. Year 1: $185.40, Year 2: $171.86, Year 3: $159.32

14. 2016

---

*10. a.

Year	1	2	3	4	5	6	7	8
Interest accrued ($)	250.90	275.11	301.66	330.77	362.69	397.69	436.07	478.15

*16.

Principal ($)	Final amount ($)	Interest earned ($)	Interest rate (p.a.)	Number of years
11 000	12 012.28	1012.28	4.5	2
14 000	15 408.84	1409.84	3.25	3
22 050	25 561.99	3511.99	3	5
108 000	110 070.00	2700.00	2.5	1

15. a. The series of prints

b. Sculpture: $41 032, prints: $28 775

16. See the table at the bottom of the page.*

17. a. See the table at the bottom of the page.*

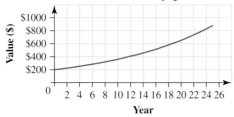

b. i. 12 years   **ii.** 19 years   **iii.** 24 years

18. a.

Compounding annually	
Year	Amount
1	$1000.00
2	$1120.00
3	$1254.40
4	$1404.93
5	$1573.52

Compounding quarterly	
Year	Amount
1	$1000.00
2	$1030.00
3	$1060.90
4	$1092.73
5	$1125.51
6	$1159.27
7	$1194.05
8	$1229.87
9	$1266.77
10	$1304.77
11	$1343.92
12	$1384.23
13	$1425.76
14	$1468.53
15	$1512.59
16	$1557.97
17	$1604.71
18	$1652.85
19	$1702.43
20	$1753.51

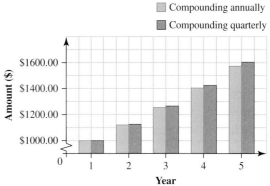

Note: The graph shows the amounts at the beginning of each year.

b. Compounding at regular intervals during the year accumulates more interest than compounding only once a year.

3.5 Exam questions

Note: Mark allocations are available with the fully worked solutions online.

1. B

2. A

3. C

3.6 Purchasing options

3.6 Exercise

1. a. $185   **b.** $288.60   **c.** $116.55

2. C

3. $27.41

4. a. $46.42   **b.** $10.12

5. $4312.44

6. $38.42

7. Plumber C

8. a. $157.57 **b.** $577.76 **c.** $299.38 **d.** $709.07

9. $2.08

10. a. $1308.49   **b.** $1499.85
    c. $2087.38   **d.** $3231.40

11. No, Elise loses $43.

12. a. $30 000   **c.** $5102.13
    b. $102.13

---

*17. a.

Year	0	1	2	3	4	5	6	7	8	9	10	11	12
Value ($)	200.00	212.20	225.14	238.88	253.45	268.91	285.31	302.72	321.18	340.78	361.56	383.62	407.02

Year	13	14	15	16	17	18	19	20	21	22	23	24	25
Value ($)	431.85	458.19	486.14	515.79	547.26	580.64	616.06	653.64	693.51	735.81	780.70	828.32	878.85

3.6 Exam questions

Note: Mark allocations are available with the fully worked solutions online.

1. D
2. C
3. a. $700

   b. $249.29

   c. $251.51

   d. 11

   e. He will not select this option.

3.7 Review

3.7 Exercise

Multiple choice

1. C
2. E
3. B
4. A
5. D
6. D
7. C
8. A
9. E
10. D

Short answer

11. a. 24 g of powder and 576 mL of water

    b. 48 g of powder and 1152 mL of water

12. a. $0.67   b. $1.87   c. $0.30   d. $8.46
13. a. $74.15   b. $11.15   c. $154.76   d. $450.30
14. a. $4.14   b. $10.90   c. $40.80   d. $1.33
15. a. $1700.33   b. 7 years

    c. 4.06%   d. $120.38

16. a. $627.22   b. $598.29

    c. 26.82%   d. $4475.60

17. In real terms she made a loss of $4415.44 when inflation is taken into account.

Extended response

18. a. 0.492%   b. 5.904%

    c. $119 000   d. 47.6%

19. a. Brand B ($1.88 compared to $2.90 per 100 g)

    b. $4.70

20. a. 9 : 7 : 6

    b. 182

    c.i. $33.50   ii. 62.5%

21. a. 42.31%   b. $45 217.93
22. a. Painter 1: $3270; Painter 2: $3600; Painter 3: $4446

    b. Painter 1

    c. $19.95

    d. 39.48%

3.7 Exam questions

Note: Mark allocations are available with the fully worked solutions online.

1. D
2. a. $8   b. $352   c. $99
3. C
4. D
5. a. $1099

   b. $4396

   c. $366

   d. i. $1758.40   ii. $3736.60

   e. i. 11 months   ii. $146

4 Matrices

Fully worked solutions for this topic are available online.

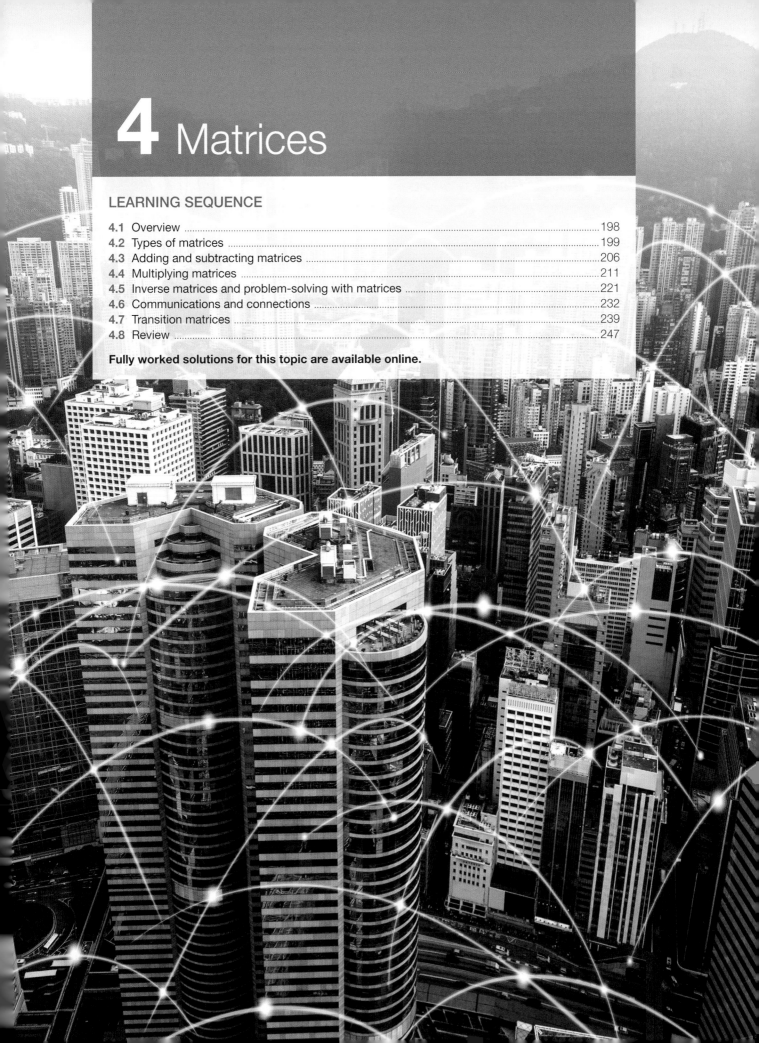

4.1 Overview

4.1.1 Introduction

A matrix (plural *matrices*) is a rectangular group of numbers arranged in rows and columns. Matrices can be used to evaluate electronic circuits, solve systems of equations, analyse networks, encrypt data and predict probabilities of events happening in the future. We will explore some of these applications while learning the basic theory of matrices.

One famous use of matrices is in search engines. Google started off by utilising an algorithm called Page Rank (named by Larry Page, Google's co-founder and CEO) that used matrices to calculate the importance and relevance of results based on the weblinks to and from each webpage.

In the beginning, there were many different search engines, but Google has risen to be the most widely used, in part due to its algorithm. Although what Google does these days may be a little different to what it did at the beginning, its foundation will always be a search engine that utilised the power of matrices.

KEY CONCEPTS

This topic covers the following key concepts from the VCE Mathematics Study Design:
- use of matrices to store and display information that can be presented in a rectangular array of rows and columns such as databases and links in social networks and road networks
- types of matrices (row, column, square, zero and identity) and the order of a matrix
- matrix addition, subtraction, multiplication by a scalar, and matrix multiplication including determining the power of a square matrix, using technology as applicable
- use of matrices, including matrix products and powers of matrices, to model and solve problems, for example costing or pricing problems, and squaring a matrix to determine the number of ways pairs of people in a network can communicate with each other via a third person
- inverse matrices and their applications including solving a system of simultaneous linear equations
- introduction to transition matrices (assuming the next state only relies on the current state), working with iterations of simple models linked to, for example, population growth or decay, including informal consideration of long-run trends and steady state.

Source: VCE Mathematics Study Design (2023–2027) extracts © VCAA; reproduced by permission.

4.2 Types of matrices

4.2.1 Defining matrices

Matrices come in different *dimensions* (sizes). We define the dimensions of a matrix by the number of rows and columns. Rows run horizontally and columns run vertically. This is known as the **order** of the matrix and is written as rows $\times$ columns.

In the following matrix A, there are 2 rows and 3 columns, which means it has an order of 2×3 (two by three).

$$
\begin{array}{c}
\text{Column 1 } 2 \; 3 \\
A = \begin{array}{c} \text{Row 1} \\ \text{Row 2} \end{array}\begin{bmatrix} 5 & 8 & 2 \\ 1 & 3 & 0 \end{bmatrix}
\end{array} \quad \rightarrow \quad \textbf{rows} \times \textbf{columns} \quad \rightarrow \quad 2 \times 3 \text{ matrix}
$$

Matrices are represented by capital letters. In this example, A represents a matrix.

Instead of only naming matrices by the number of rows and columns, we also use names for specific types of matrix *dimensions*.

Types of matrices		
Type of matrix	**Description**	**Example**
Column matrix	A matrix that has a single column is known as a **column matrix**.	$B = \begin{bmatrix} 1 \\ 4 \\ 2 \end{bmatrix}$ $C = \begin{bmatrix} 2 \\ 8 \\ 5 \\ 9 \end{bmatrix}$ B and C are both column matrices.
Row matrix	A matrix that has a single row is known as a **row matrix**.	$D = \begin{bmatrix} 2 & 3 \end{bmatrix}$ $E = \begin{bmatrix} 9 & 4 & 3 \end{bmatrix}$ D and E are both row matrices.
Square matrix	A matrix that has the same number of columns as rows is known as a **square matrix**.	$F = \begin{bmatrix} 8 \end{bmatrix}$ $G = \begin{bmatrix} 5 & 9 \\ -4 & 1 \end{bmatrix}$ $H = \begin{bmatrix} 7 & 6 & 4 \\ -3 & 1 & 7 \\ 4 & 9 & 8 \end{bmatrix}$ F, G and H are all square matrices.
Identity matrix	A square matrix in which all of the elements on the diagonal line from the top left to bottom right are 1s and all of the other elements are 0s is known as the **identity matrix**.	$A = \begin{bmatrix} 1 & 0 \\ 0 & 1 \end{bmatrix}$ and $B = \begin{bmatrix} 1 & 0 & 0 \\ 0 & 1 & 0 \\ 0 & 0 & 1 \end{bmatrix}$ A and B are both identity matrices.
Zero matrix	A square matrix that consists entirely of 0 elements is known as the **zero matrix**.	$A = \begin{bmatrix} 0 & 0 \\ 0 & 0 \end{bmatrix}$ A is a zero matrix.

tlvd-3796

At High Vale College, 150 students are studying General Mathematics and 85 students are studying Mathematical Methods. Construct a column matrix to represent the number of students studying General Mathematics and Mathematical Methods, and state the order of the matrix.

THINK	WRITE
1. Read the question and highlight the key information.	150 students study General Mathematics. 85 students study Mathematical Methods.
2. Display this information in a column matrix.	$\begin{bmatrix} 150 \\ 85 \end{bmatrix}$
3. How many rows and columns are there in this matrix?	The order of the matrix is 2×1.

4.2.2 Elements of matrices

The numbers in a matrix are called **elements**. They are represented by a *lowercase* letter with the row and column in *subscript*. For matrix B, b_{ij} is the element in row i, column j.

$$\text{row 1} \begin{array}{c} \text{column 3} \\ B = \begin{bmatrix} 1 & 2 & ③ \\ 4 & 5 & 6 \\ 7 & 8 & 9 \end{bmatrix} \end{array}$$

b_{13} is an element of matrix B, in row 1 and column 3.

The respective elements in matrix B are:

$$B = \begin{bmatrix} 1 & 2 & 3 \\ 4 & 5 & 6 \\ 7 & 8 & 9 \end{bmatrix} \rightarrow \begin{bmatrix} b_{11} & b_{12} & b_{13} \\ b_{21} & b_{22} & b_{23} \\ b_{31} & b_{32} & b_{33} \end{bmatrix}$$

For matrix C shown, determine the following elements.

$$C = \begin{bmatrix} 3 & 8 & 2 & 4 \\ -5 & 1 & -7 & 0 \\ 4 & 9 & 8 & -4 \end{bmatrix}$$

a. c_{12} b. c_{31} c. c_{13}

THINK	WRITE
a. c_{12} is the element of matrix C that is in row 1 and column 2.	a. $c_{12} = 8$
b. c_{31} is the element of matrix C that is in row 3 and column 1.	b. $c_{31} = 4$
c. c_{13} is the element of matrix C that is in row 1 and column 3.	c. $c_{13} = 2$

4.2.3 Describing matrices

A matrix is usually displayed in square brackets with no borders between the rows and columns.

The following table shows the number of participants attending three different dance classes (hip-hop, salsa and bachata) over the two days of a weekend. The matrix below displays the information presented in the table.

Number of participants attending the dance classes:

	Saturday	Sunday
Hip-hop	9	13
Salsa	12	8
Bachata	16	14

The number of participants attending the dance classes can be represented in a matrix:

$$\begin{bmatrix} 9 & 13 \\ 12 & 8 \\ 16 & 14 \end{bmatrix}$$

WORKED EXAMPLE 3 Creating a matrix from a table

The table shows the number of adults and children who attended three different events over the school holidays. Construct a matrix to represent this information.

	Circus	Zoo	Show
Adults	140	58	85
Children	200	125	150

THINK

1. A matrix is like a table that stores information. Determine what information needs to be displayed.

2. Write down how many adults and children attend each of the three events.

3. Write this information in a matrix. Remember to use square brackets.

WRITE

The information to be displayed is the number of adults and children attending the three events: circus, zoo and show.

	Circus	Zoo	Show
Adults	140	58	85
Children	200	125	150

$$\begin{bmatrix} 140 & 58 & 85 \\ 200 & 125 & 150 \end{bmatrix}$$

4.2.4 Interpreting matrices

At the Moomba festival, attendees (parents and children) visit three stages (A, B and C). The information is presented in the following matrix, E.

$$E = \begin{array}{c} \\ \text{Stage A} \\ \text{Stage B} \\ \text{Stage C} \end{array} \overset{\text{Parents Children}}{\begin{bmatrix} 654 & 620 \\ 543 & 600 \\ 320 & 300 \end{bmatrix}}$$

Columns	Rows
• Column 1 gives the number of parents at each stage. • Column 2 gives the number of children at each stage.	• Row 1 gives the number of attendees at stage A. • Row 2 gives the number of attendees at stage B. • Row 3 gives the number of attendees at stage C.

From the matrix above, we can also get the following information:
- There are a total of $320 + 300 = 620$ attendees at stage C.
- At Stage A there are 620 children.
- The total number of parents is $654 + 543 + 320 = 1517$.

WORKED EXAMPLE 4 Interpreting information from a matrix

A friend runs a cupcake business from home and sells cupcakes throughout the week. Their sales figures over two days were recorded in the following matrix.

$$S = \begin{array}{c} \\ \end{array} \overset{\text{Red velvet | Chocolate | Vanilla}}{\begin{bmatrix} 7 & 12 & 5 \\ 4 & 8 & 13 \end{bmatrix}} \begin{array}{l} \text{Saturday} \\ \text{Sunday} \end{array}$$

a. State the order of matrix S.
b. State what information the element s_{12} shows.
c. State which element gives the number of red velvet cupcakes sold on Sunday.
d. Determine the total number of vanilla cupcakes sold over the weekend.
e. Determine the total number of cupcakes sold on Sunday.

THINK	WRITE
a. The order is rows × columns.	a. The order of matrix S is 2×3.
b. The element s_{12} is in the first row (Saturday) and second column (chocolate). So, it represents the amount of chocolate cupcakes sold on Saturday.	b. There were 12 chocolate cupcakes sold on Saturday.
c. Sunday is row 2 and red velvet is column 1, so the element that gives the number of red velvet cupcakes sold on Sunday must be in row 2, column 1.	c. The number of red velvet cupcakes sold on Sunday is given by s_{21}.
d. To find the total number of vanilla cupcakes, add the numbers in column 3.	d. $5 + 13 = 18$
e. To find the total number of cupcakes sold on Sunday, add the numbers in row 2.	e. $4 + 8 + 13 = 25$

4.2 Exercise

1. **WE1** An energy-saving store stocks shower water savers and energy-saving light globes. In one month they sold 45 shower water savers and 30 energy-saving light globes. Construct a column matrix to represent the number of shower water savers and energy-saving light globes sold during this month, and state the order of the matrix.

2. State the order of the following matrices.

   a. $\begin{bmatrix} 7 & 3 \\ 8 & 5 \end{bmatrix}$

   b. $\begin{bmatrix} 3 \\ 1 \end{bmatrix}$

   c. $[2 \ -5 \ 8 \ 7]$

   d. $\begin{bmatrix} 7 & 4 & 9 \\ -5 & 2 & 4 \end{bmatrix}$

   e. $\begin{bmatrix} 9 & 2 & 3 \\ 4 & 5 & 6 \\ 8 & 1 & 9 \end{bmatrix}$

   f. $\begin{bmatrix} 4 & 1 \\ 6 & 4 \\ 8 & 3 \\ 7 & 7 \end{bmatrix}$

3. **WE2** For matrix Y, determine the following elements.

   $$Y = \begin{bmatrix} 8 & 7 & 5 & 4 \\ 1 & 3 & 9 & 8 \\ 1 & 6 & 2 & 4 \\ 2 & 3 & 0 & 7 \end{bmatrix}$$

   a. y_{13}
   b. y_{24}
   c. y_{31}
   d. y_{43}
   e. y_{41}
   f. y_{12}

4. **WE3** Cheap Auto sells three types of vehicles: cars, vans and motorbikes. They have two outlets at Valley Heights and Hill Vale. The number of vehicles in stock at each of the two outlets is shown in the table.

	Cars	Vans	Motorbikes
Valley Heights	18	12	8
Hill Vale	13	10	11

   Construct a matrix to represent this information.

5. Newton and Isaacs played a match of tennis. Newton won the match in five sets with a final score of 6–2, 4–6, 7–6, 3–6, 6–4. Construct a matrix to represent this information.

6. Determine the following elements in the matrices shown.

   $A = \begin{bmatrix} 8 & 1 \\ 6 & 9 \end{bmatrix}$ $B = \begin{bmatrix} 1 & 5 \\ 1 & 6 \end{bmatrix}$ $C = \begin{bmatrix} 6 & 3 \\ 7 & 5 \end{bmatrix}$

   a. b_{12}
   b. c_{11}
   c. a_{21}
   d. a_{11}
   e. c_{12}
   f. b_{22}

7. Consider the matrix $E = \begin{bmatrix} \dfrac{2}{3} & 0 & \dfrac{1}{4} \\ -1 & -\dfrac{1}{2} & -3 \end{bmatrix}$.

   a. Explain why the element e_{24} does not exist.
   b. State which element has a value of -3.

c. Nadia was asked to write down the value of element e_{12} and wrote -1. Explain Nadia's mistake and state the correct value of element e_{12}.

8. **WE4** Tickets were sold for a series of concerts by popular band Pink and Black. The tickets were separated into Section A and Section B tickets, sold at different prices. The concerts would be playing over three days: Thursday, Friday and Saturday. The ticket sales are given in matrix K.

$$K = \begin{array}{c} \\ \text{Thursday} \\ \text{Friday} \\ \text{Saturday} \end{array} \begin{array}{cc} A & B \\ \begin{bmatrix} 300 & 600 \\ 350 & 800 \\ 400 & 700 \end{bmatrix} \end{array}$$

a. State the order of matrix K.
b. State the information shown by the element k_{22}.
c. Determine which element gives the number of Section A tickets sold on Saturday.
d. Calculate the total number of Section B tickets sold over Thursday, Friday and Saturday.

9. Happy Greens Golf Club held a three-day competition from Friday to Sunday. Participants were grouped into three different categories: experienced, beginner and club member. The table shows the total entries for each type of participant on each of the days of the competition.

Category	Friday	Saturday	Sunday
Experienced	19	23	30
Beginner	12	17	18
Club member	25	33	36

a. State the number of participants in the competition on Friday.
b. Calculate the total number of entries for the three-day competition.
c. Construct a row matrix to represent the number of beginners participating in the competition on each of the three days.

10. The elements of matrix H are shown.
$h_{12} = 3$ $h_{11} = 4$ $h_{21} = -1$
$h_{31} = -4$ $h_{32} = 6$ $h_{22} = 7$

a. State the order of matrix H.   b. Construct matrix H.

11. The land area and population of each Australian state and territory were recorded and summarised in the table.

State/territory	Land area (km²)	Population (millions)
Australian Capital Territory	2 358	0.4
Queensland	1 727 200	4.2
New South Wales	801 428	6.8
Northern Territory	1 346 200	0.2
South Australia	984 000	1.6
Western Australia	2 529 875	2.1
Tasmania	68 330	0.5
Victoria	227 600	5.2

a. Construct an 8×1 matrix that displays the population, in millions, of each state and territory in the order shown in the table.
b. Construct a row matrix that represents the land area of each of the states in ascending order.

c. Town planners place the information on land area, in km^2, and population, in millions, for New South Wales, Victoria and Queensland respectively in a matrix.

    i. State the order of this matrix.   ii. Construct this matrix.

12. The estimated number of Aboriginal and Torres Strait Islander persons living in each state and territory in Australia in 2006 is shown in the table.

State and territory	Number of Aboriginal and Torres Strait Islander persons	% of population that are Aboriginal or Torres Strait Islander persons
New South Wales	148 178	2.2
Victoria	30 839	0.6
Queensland	146 429	3.6
South Australia	26 044	1.7
Western Australia	77 928	3.8
Tasmania	16 900	3.4
Northern Territory	66 582	31.6
Australian Capital Territory	4 043	1.2

a. Construct an 8×2 matrix to represent this information.
b. Determine the total number of Aboriginal and Torres Strait Islander persons living in the following states and territories in 2006.

    i. Northern Territory
    ii. Tasmania
    iii. Queensland, New South Wales and Victoria (combined)

c. Determine the total number of Aboriginal and Torres Strait Islander persons who were estimated to be living in Australia in 2006.

4.2 Exam questions

Question 1 (1 mark)

MC Select the order of the following matrix.

$$\begin{bmatrix} 1 & 2 \\ 5 & 4 \\ 3 & 7 \end{bmatrix}$$

A. 6   **B.** 2×3 **C.** 3×2 **D.** 9×13 **E.** 22

Question 2 (2 marks)

Determine the sum of the elements a_{21} and a_{32} in matrix $\begin{bmatrix} 5 & 2 \\ 1 & 0 \\ 3 & -2 \end{bmatrix}$.

Question 3 (1 mark)

MC Select the row matrix from the following matrices.

A. $\begin{bmatrix} 10 & -4 \\ 0 & 0 \end{bmatrix}$ **B.** $\begin{bmatrix} 7 & 0 & 0 \\ 0 & 12 & 0 \\ 0 & 0 & -10 \end{bmatrix}$ **C.** $\begin{bmatrix} 6 \\ -9 \\ 3 \end{bmatrix}$ **D.** $\begin{bmatrix} -4 & 6 & 5 & 2 \end{bmatrix}$ **E.** $[24]$

More exam questions are available online.

4.3 Adding and subtracting matrices

4.3.1 Adding matrices

To add matrices, you need to add the corresponding elements of each matrix together (that is, the numbers in the same position).

Matrices must be of the same order to perform addition (they must have the same number of rows and columns).

tlvd-3797

WORKED EXAMPLE 5 Adding matrices

If $A = \begin{bmatrix} 4 & 2 \\ 3 & -2 \end{bmatrix}$ and $B = \begin{bmatrix} 1 & 0 \\ 5 & 3 \end{bmatrix}$, determine the matrix sum $A + B$.

THINK	WRITE
1. Write down the two matrices in a sum.	$\begin{bmatrix} 4 & 2 \\ 3 & -2 \end{bmatrix} + \begin{bmatrix} 1 & 0 \\ 5 & 3 \end{bmatrix}$
2. Identify the elements in the same position. For example, 4 and 1 are both in the first row and first column. Add the elements in the same position.	$= \begin{bmatrix} 4+1 & 2+0 \\ 3+5 & -2+3 \end{bmatrix}$
3. Work out the sums and write the answer.	$\begin{bmatrix} 5 & 2 \\ 8 & 1 \end{bmatrix}$

TI | THINK	DISPLAY/WRITE	CASIO | THINK	DISPLAY/WRITE
1. On a Calculator page, complete the entry line as: $\begin{bmatrix} 4 & 2 \\ 3 & -2 \end{bmatrix} + \begin{bmatrix} 1 & 0 \\ 5 & 3 \end{bmatrix}$ then press ENTER. *Note:* The matrix templates can be found by pressing the matrix button 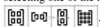 and then selecting one of the matrices .	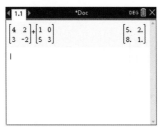	1. On the Main screen, complete the entry line as: $\begin{bmatrix} 4 & 2 \\ 3 & -2 \end{bmatrix} + \begin{bmatrix} 1 & 0 \\ 5 & 3 \end{bmatrix}$ then press EXE. *Note:* The matrix templates can be found in the Math 2 tab in the Keyboard menu.	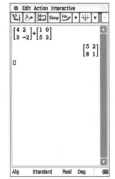
2. The answer appears on the screen.	$A + B = \begin{bmatrix} 5 & 2 \\ 8 & 1 \end{bmatrix}$	2. The answer appears on the screen.	$A + B = \begin{bmatrix} 5 & 2 \\ 8 & 1 \end{bmatrix}$

4.3.2 Subtracting matrices

To subtract matrices, you need to subtract the corresponding elements in the same order as presented in the question.

Matrices should be of the same order to perform subtraction (they should have the same number of rows and columns).

WORKED EXAMPLE 6 Subtracting matrices

If $A = \begin{bmatrix} 6 & 0 \\ 2 & -2 \end{bmatrix}$ and $B = \begin{bmatrix} 4 & 2 \\ 1 & 3 \end{bmatrix}$, determine the matrix subtraction $A - B$.

THINK	WRITE
1. Write the two matrices.	$\begin{bmatrix} 6 & 0 \\ 2 & -2 \end{bmatrix} - \begin{bmatrix} 4 & 2 \\ 1 & 3 \end{bmatrix}$
2. Subtract the elements in the same position.	$\begin{bmatrix} 6-4 & 0-2 \\ 2-1 & -2-3 \end{bmatrix}$
3. Work out the subtractions and write the answer.	$\begin{bmatrix} 2 & -2 \\ 1 & -5 \end{bmatrix}$

4.3 Exercise

Students, these questions are even better in jacPLUS

Receive immediate feedback and access sample responses

Access additional questions

Track your results and progress

Find all this and MORE in jacPLUS

1. a. **WE5** If $A = \begin{bmatrix} 2 & -3 \\ -1 & -8 \end{bmatrix}$ and $B = \begin{bmatrix} -1 & 9 \\ 0 & 11 \end{bmatrix}$, determine the matrix sum $A + B$.

   b. If $A = \begin{bmatrix} 0.5 \\ 0.1 \\ 1.2 \end{bmatrix}$, $B = \begin{bmatrix} -0.5 \\ 2.2 \\ 0.9 \end{bmatrix}$ and $C = \begin{bmatrix} -0.1 \\ -0.8 \\ 2.1 \end{bmatrix}$, determine the matrix sum $A + B + C$.

2. Consider the matrices $C = \begin{bmatrix} 1 & -3 \\ 7 & 5 \\ b & 8 \end{bmatrix}$ and $D = \begin{bmatrix} 0 & a \\ -5 & -4 \\ 2 & -9 \end{bmatrix}$.

   If $C + D = \begin{bmatrix} 1 & 1 \\ 2 & 1 \\ -4 & -1 \end{bmatrix}$, determine the values of a and b.

3. **WE6** If $A = \begin{bmatrix} 1 & 1 \\ -3 & -1 \end{bmatrix}$ and $B = \begin{bmatrix} 2 & 3 \\ -2 & 4 \end{bmatrix}$, determine the matrix subtraction $A - B$.

4. If $A = \begin{bmatrix} 5 \\ 4 \\ -2 \end{bmatrix}$, $B = \begin{bmatrix} -1 \\ 0 \\ 4 \end{bmatrix}$ and $C = \begin{bmatrix} -4 \\ 3 \\ 2 \end{bmatrix}$, calculate the following.

   a. $A + C$ b. $B + C$ c. $A - B$ d. $A + B - C$

5. Consider the following.

$$B - A = \begin{bmatrix} 4 \\ 0 \\ 2 \end{bmatrix}, \; A + B = \begin{bmatrix} 4 \\ 2 \\ 8 \end{bmatrix} \text{ and } A = \begin{bmatrix} 0 \\ 1 \\ 3 \end{bmatrix}$$

a. Explain why matrix B must have an order of 3×1.
b. Determine matrix B.

6. State whether the following are defined. If they are defined, evaluate them.

$$W = \begin{bmatrix} -2 & 5 \\ 3 & 8 \end{bmatrix}, \quad X = \begin{bmatrix} 3 \\ 6 \\ 8 \end{bmatrix}, \quad Y = \begin{bmatrix} 5 & -6 \\ 1 & 9 \end{bmatrix}, \quad Z = \begin{bmatrix} -9 & 2 \end{bmatrix}$$

a. $W + Y$ b. $X - Y$ c. $Z - W$ d. $Y + Z$

7. Evaluate the following.

a. $\begin{bmatrix} 0.5 & 0.25 & 1.2 \end{bmatrix} - \begin{bmatrix} 0.75 & 1.2 & 0.9 \end{bmatrix}$

b. $\begin{bmatrix} 1 & 0 \\ 3 & 1 \end{bmatrix} + \begin{bmatrix} 2 & -1 \\ 6 & 0 \end{bmatrix}$

c. $\begin{bmatrix} 12 & 17 & 10 \\ 35 & 20 & 25 \\ 28 & 32 & 29 \end{bmatrix} - \begin{bmatrix} 13 & 12 & 9 \\ 31 & 22 & 22 \\ 25 & 35 & 31 \end{bmatrix}$

d. $\begin{bmatrix} 11 & 6 & 9 \\ 7 & 12 & -1 \end{bmatrix} + \begin{bmatrix} 2 & 8 & 8 \\ 6 & 7 & 6 \end{bmatrix} - \begin{bmatrix} -2 & -1 & 10 \\ 4 & 9 & -3 \end{bmatrix}$

8. If $\begin{bmatrix} 3 & 0 \\ 5 & a \end{bmatrix} + \begin{bmatrix} 2 & 2 \\ -b & 1 \end{bmatrix} = \begin{bmatrix} c & 2 \\ 3 & -4 \end{bmatrix}$, determine the values of a, b and c.

9. If $\begin{bmatrix} 12 & 10 \\ 25 & 13 \\ 20 & a \end{bmatrix} - \begin{bmatrix} 9 & 11 \\ 26 & c \\ b & 9 \end{bmatrix} = \begin{bmatrix} 3 & -1 \\ -1 & 8 \\ 21 & -3 \end{bmatrix}$, determine the values of a, b and c.

10. Using the order of each of the following matrices, identify which of the matrices can be added to and/or subtracted from each other and explain why.

$$A = \begin{bmatrix} 1 & -5 \end{bmatrix}, \quad B = \begin{bmatrix} 2 \\ -8 \end{bmatrix}, \quad C = \begin{bmatrix} -1 \\ -9 \end{bmatrix}, \quad D = \begin{bmatrix} -4 \end{bmatrix}, \quad E = \begin{bmatrix} -3 & 6 \end{bmatrix}$$

11. Hard Eggs sells both free-range and barn-laid eggs in three different egg sizes (small, medium and large) to two shops, Appleton and Barntown. The number of cartons ordered for the Appleton shop is shown in the table below.

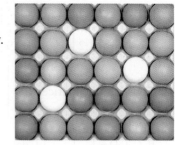

Eggs	Small	Medium	Large
Free-range	2	3	5
Barn-laid	4	6	3

a. Construct a 2×3 matrix to represent the egg order for the Appleton shop.

The total orders for both shops are shown in the table below.

Eggs	Small	Medium	Large
Free-range	3	4	8
Barn-laid	6	8	5

b. i. Set up a matrix sum to determine the order for the Barntown shop.
  ii. Use the matrix sum from part b. i. to determine the order for the Barntown shop. Show the order in a table.

12. A student was asked to complete the matrix sum $\begin{bmatrix} 8 & 126 & 59 \\ 17 & 102 & -13 \end{bmatrix} + \begin{bmatrix} 22 & 18 & 38 \\ 16 & 27 & 45 \end{bmatrix}$.

They gave $\begin{bmatrix} 271 \\ 194 \end{bmatrix}$ as the answer.

   a. By referring to the order of matrices, explain why their answer must be incorrect.
   b. By explaining how to add matrices, write simple steps for them to follow so that they are able to add and subtract any matrices. Use the terms 'order of matrices' and 'elements' in your explanation.

13. Frederick, Harold, Mia and Petra are machinists who work for Stitch in Time. The table below shows the hours worked by each of the four employees and the number of garments completed each week for the last three weeks.

Employee	Week 1		Week 2		Week 3	
	Hours worked	Number of garments	Hours worked	Number of garments	Hours worked	Number of garments
Frederick	35	150	32	145	38	166
Harold	41	165	36	152	35	155
Mia	38	155	35	135	35	156
Petra	25	80	30	95	32	110

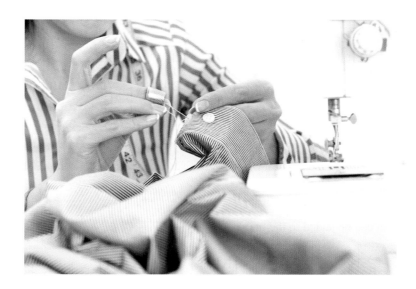

   a. Construct a 4×1 matrix to represent the number of garments each employee made in week 1.
   b. i. Create a matrix sum that would determine the total number of garments each employee made over the three weeks.
      ii. Using your matrix sum from part b. i., determine the total number of garments each employee made over the three weeks.
   c. Nula is the manager of Stitch in Time. She uses the following matrix sum to determine the total number of hours worked by each of the four employees over the three weeks.

$$\begin{bmatrix} 35 \\ 38 \\ 25 \end{bmatrix} + \begin{bmatrix} 36 \\ 30 \end{bmatrix} + \begin{bmatrix} 38 \\ 35 \\ 35 \end{bmatrix} = \begin{bmatrix} \\ \\ \end{bmatrix}$$

Complete the matrix sum by filling in the missing values.

14. There are three types of fish in a pond: speckles, googly eyes and fantails. At the beginning of the month there were 12 speckles, 9 googly eyes and 8 fantails in the pond. By the end of the month there were 9 speckles, 6 googly eyes and 8 fantails in the pond.

    a. Construct a matrix sum to represent this information.
    b. After six months, there were 12 speckles, 4 googly eyes and 10 fantails in the pond. Starting from the end of the first month, construct another matrix sum to represent this information.

15. Consider the following matrix sum: $A - C + B = D$. Matrix D has an order of 3×2.

    a. State the order of matrices A, B and C. Justify your answer.

    A has elements $a_{11} = x$, $a_{21} = 20$, $a_{31} = 3c_{31}$, $a_{12} = 7$, $a_{22} = y$ and $a_{32} = -8$.
    B has elements $b_{11} = x$, $b_{21} = 2x$, $b_{31} = 3x$, $b_{12} = y$, $b_{22} = 5$ and $b_{32} = 6$
    C has elements $c_{11} = 12$, $c_{21} = \frac{1}{2}a_{21}$, $c_{31} = 5$, $c_{12} = 9$, $c_{22} = 2y$ and $c_{32} = 2x$.

    b. Define the elements of D in terms of x and y.

    c. If $D = \begin{bmatrix} -8 & 1 \\ 14 & 2 \\ 16 & -8 \end{bmatrix}$, show that $x = 2$ and $y = 3$.

16. Using CAS, evaluate the following matrix calculation.

$$\begin{bmatrix} \frac{1}{2} & \frac{3}{4} & \frac{5}{6} \\ \frac{3}{5} & \frac{2}{7} & \frac{1}{3} \\ \frac{2}{3} & \frac{1}{4} & \frac{2}{9} \end{bmatrix} + \begin{bmatrix} \frac{1}{4} & \frac{3}{8} & \frac{1}{3} \\ \frac{1}{10} & \frac{3}{14} & \frac{4}{9} \\ \frac{1}{6} & \frac{1}{2} & \frac{2}{3} \end{bmatrix} - \begin{bmatrix} \frac{1}{8} & \frac{1}{2} & \frac{7}{6} \\ \frac{2}{15} & \frac{8}{21} & \frac{3}{4} \\ \frac{2}{9} & \frac{5}{8} & \frac{10}{9} \end{bmatrix}$$

4.3 Exam questions

Question 1 (1 mark)

MC Select the matrix that can be added to the matrix $\begin{bmatrix} 3 \\ -4 \end{bmatrix}$.

A. $\begin{bmatrix} 1 & 5 & 3 \\ 2 & -4 & 6 \end{bmatrix}$
B. $\begin{bmatrix} 1 & 2 \\ 5 & -4 \\ 3 & 6 \end{bmatrix}$
C. $\begin{bmatrix} 1 & 5 \\ 2 & -4 \end{bmatrix}$
D. $\begin{bmatrix} 1 \\ 2 \end{bmatrix}$
E. [12]

Question 2 (1 mark)

MC If $\begin{bmatrix} a & b \\ c & d \end{bmatrix} = \begin{bmatrix} -2 & 4 \\ 3 & 1 \end{bmatrix} + \begin{bmatrix} -2 & 1 \\ 5 & 2 \end{bmatrix}$, determine the value of a.

A. $a = 0$
B. $a = -4$
C. $a = 4$
D. $a = 2$
E. $\begin{bmatrix} -4 & 5 \\ 8 & 3 \end{bmatrix}$

Question 3 (2 marks)

If $A = \begin{bmatrix} 2 & 3 \\ -4 & 5 \end{bmatrix}$, $B = \begin{bmatrix} 1 & 3 \\ -3 & 5 \end{bmatrix}$, $C = \begin{bmatrix} -4 & 2 \\ 4 & 5 \end{bmatrix}$ and $O = \begin{bmatrix} 0 & 0 \\ 0 & 0 \end{bmatrix}$, calculate $A + B - C + O$.

More exam questions are available online.

4.4 Multiplying matrices

LEARNING INTENTION

At the end of this subtopic you should be able to:
- perform scalar matrix multiplication
- determine the product matrices
- perform multiplication by the identity matrix
- calculate the power of square matrices.

4.4.1 Scalar multiplication

Matrices can be multiplied by single numbers, called scalar quantities or scalars. For example, the number 2 is a **scalar quantity**.

When we perform scalar matrix multiplication, we multiply all elements inside the matrix by the scalar quantity. Scalar quantities can be any real number, including negative and positive numbers, fractions and decimals.

$$A = \begin{bmatrix} 3 & 2 \\ 5 & 1 \\ 0 & 7 \end{bmatrix}$$

$$2A = 2 \times \begin{bmatrix} 3 & 2 \\ 5 & 1 \\ 0 & 7 \end{bmatrix} = \begin{bmatrix} 2 \times 3 & 2 \times 2 \\ 2 \times 5 & 2 \times 1 \\ 2 \times 0 & 2 \times 7 \end{bmatrix}$$

tlvd-3799

WORKED EXAMPLE 7 Evaluating scalar matrix multiplication

Consider the matrix $A = \begin{bmatrix} 120 & 90 \\ 80 & 60 \end{bmatrix}$.

Evaluate the following.

a. $\dfrac{1}{4}A$

b. $0.1A$

THINK	WRITE
a. 1. Identify the scalar. In this case it is $\dfrac{1}{4}$, which means that each element in A is multiplied by $\dfrac{1}{4}$ (or divided by 4).	a. $\dfrac{1}{4}\begin{bmatrix} 120 & 90 \\ 80 & 60 \end{bmatrix}$
2. Multiply each element in A by the scalar.	$\begin{bmatrix} \dfrac{1}{4} \times 120 & \dfrac{1}{4} \times 90 \\ \dfrac{1}{4} \times 80 & \dfrac{1}{4} \times 60 \end{bmatrix}$
3. Simplify each multiplication by finding common factors and write the answer.	$\begin{bmatrix} \cancel{120}^{30} \times \dfrac{1}{\cancel{4}^1} & \cancel{90}^{45} \times \dfrac{1}{\cancel{4}^2} \\ \cancel{80}^{20} \times \dfrac{1}{\cancel{4}^1} & \cancel{60}^{15} \times \dfrac{1}{\cancel{4}^1} \end{bmatrix}$ $= \begin{bmatrix} 30 & \dfrac{45}{2} \\ 20 & 15 \end{bmatrix}$

b. 1. Identify the scalar. In this case it is 0.1, which means that each element in A is multiplied by 0.1 (or divided by 10).

b. $0.1 \begin{bmatrix} 120 & 90 \\ 80 & 60 \end{bmatrix}$

2. Multiply each element in A by the scalar.

$\begin{bmatrix} 0.1 \times 120 & 0.1 \times 90 \\ 0.1 \times 80 & 0.1 \times 60 \end{bmatrix}$

3. Determine the values for each element and write the answer.

$\begin{bmatrix} 12 & 9 \\ 8 & 6 \end{bmatrix}$

TI	THINK	DISPLAY/WRITE	CASIO	THINK	DISPLAY/WRITE

a. 1. On a Calculator page, complete the entry line as:
$\frac{1}{4} \begin{bmatrix} 120 & 90 \\ 80 & 60 \end{bmatrix}$
then press ENTER.
Note: The matrix templates can be found by pressing the matrix button 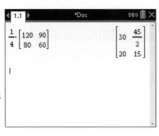 and then selecting one of the matrices.

a. 1. On the Main screen, complete the entry line as:
$\frac{1}{4} \begin{bmatrix} 120 & 90 \\ 80 & 60 \end{bmatrix}$
then press EXE.
Note: The matrix templates can be found in the Math 2 tab in the Keyboard menu.

2. The answer appears on the screen.

$\frac{1}{4} A = \begin{bmatrix} 30 & \frac{45}{2} \\ 20 & 15 \end{bmatrix}$

2. The answer appears on the screen.

$\frac{1}{4} A = \begin{bmatrix} 30 & \frac{45}{2} \\ 20 & 15 \end{bmatrix}$

4.4.2 The product matrix and its order

When two matrices are multiplied together, the result is called the product matrix.

Rules for multiplying matrices

For matrices to be able to be multiplied together, the *number of columns* in the first matrix must equal the *number of rows* in the second matrix.

For example, consider matrices A and B.	$A = \begin{bmatrix} 7 & 5 \\ 1 & 3 \end{bmatrix}$ $\quad B = \begin{bmatrix} 8 \\ 6 \end{bmatrix}$
Consider the order of each matrix to determine whether the product matrix is defined. Because the number of columns in matrix A and the number of rows in matrix B are the same, the product matrix is defined. The order of the resultant matrix is the remaining two numbers (number of rows in A and the number of columns in B).	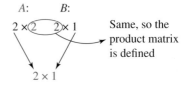

4.4.3 Matrix multiplication

To multiply matrices together, use the following steps.

Step 1: Confirm that the product matrix exists (that is, the number of columns in the first matrix equals the number of rows in the second matrix).

Step 2: Multiply the elements of each row of the first matrix by the elements of each column of the second matrix.

Step 3: Sum the products in each element of the product matrix.

Consider matrices C and D.	$C = \begin{bmatrix} 3 & 4 \\ 2 & 5 \end{bmatrix}$ and $D = \begin{bmatrix} 1 & 7 \\ 2 & 6 \end{bmatrix}$
Check the orders to see if the product matrix is defined. The remaining numbers make up the order of the product matrix.	Orders: $2 \times \boxed{2}$ and $\boxed{2} \times 2$ Same, so the product matrix is defined 2×2 ⟶ Order of product matrix
Multiply the elements in the first row of matrix C with the elements in the first column of matrix D, then sum their products into the corresponding element of the product matrix.	$C = \begin{bmatrix} 3 & 4 \\ 2 & 5 \end{bmatrix}$ and $D = \begin{bmatrix} 1 & 7 \\ 2 & 6 \end{bmatrix}$ $\begin{bmatrix} 3 & 4 \\ 2 & 5 \end{bmatrix} \times \begin{bmatrix} 1 & 7 \\ 2 & 6 \end{bmatrix} = \begin{bmatrix} 3 \times 1 + 4 \times 2 & - \\ - & - \end{bmatrix}$
Repeat as above with the first row of matrix C and the second column of matrix D.	$\begin{bmatrix} 3 & 4 \\ 2 & 5 \end{bmatrix} \times \begin{bmatrix} 1 & 7 \\ 2 & 6 \end{bmatrix} = \begin{bmatrix} 11 & 3 \times 7 + 4 \times 6 \\ - & - \end{bmatrix}$
Repeat as above with the second row of matrix C and the first column of matrix D.	$\begin{bmatrix} 3 & 4 \\ 2 & 5 \end{bmatrix} \times \begin{bmatrix} 1 & 7 \\ 2 & 6 \end{bmatrix} = \begin{bmatrix} 11 & 45 \\ 2 \times 1 + 5 \times 2 & - \end{bmatrix}$
Repeat as above with the second row of matrix C and the second column of matrix D.	$\begin{bmatrix} 3 & 4 \\ 2 & 5 \end{bmatrix} \times \begin{bmatrix} 1 & 7 \\ 2 & 6 \end{bmatrix} = \begin{bmatrix} 11 & 45 \\ 12 & 2 \times 7 + 5 \times 6 \end{bmatrix}$
We now have the product matrix.	$CD = \begin{bmatrix} 3 & 4 \\ 2 & 5 \end{bmatrix} \times \begin{bmatrix} 1 & 7 \\ 2 & 6 \end{bmatrix}$ $= \begin{bmatrix} 11 & 45 \\ 12 & 44 \end{bmatrix}$

Unlike the multiplication of real numbers, the multiplication of matrices is not commutative. This means that, in most cases, $AB \neq BA$. So, it's very important to keep in mind the order of matrix multiplication.

Try to test this out using matrices C and D from earlier. If you try to perform the matrix multiplication $D \times C$, is the resultant product matrix the same? You can check this by hand or using your CAS.

If $A = \begin{bmatrix} 3 & 5 \end{bmatrix}$ and $B = \begin{bmatrix} 2 \\ 6 \end{bmatrix}$, determine the product matrix AB.

THINK	WRITE
1. Set up the product matrix.	$\begin{bmatrix} 3 & 5 \end{bmatrix} \times \begin{bmatrix} 2 \\ 6 \end{bmatrix}$
2. Determine the order of the product matrix AB by writing the order of each matrix A and B.	$A \times B$ ① $\times 2$ and $2 \times$ ① AB has an order of 1×1.
3. Multiply each element in the first row by the corresponding element in the first column, then calculate the sum of the results.	3×2 and $5 \times 6 = 36$
4. Write the answer as a matrix.	$[36]$

tlvd-3711

Determine the product matrix MN if $M = \begin{bmatrix} 3 & 6 \\ 5 & 2 \end{bmatrix}$ and $N = \begin{bmatrix} 1 & 8 \\ 5 & 4 \end{bmatrix}$.

THINK	WRITE
1. Set up the product matrix.	$\begin{bmatrix} 3 & 6 \\ 5 & 2 \end{bmatrix} \times \begin{bmatrix} 1 & 8 \\ 5 & 4 \end{bmatrix}$
2. Determine the order of the product matrix MN by writing the order of each matrix M and N.	$M \times N$ ② $\times 2$ and $2 \times$ ② MN has an order of 2×2.
3. To determine the element mn_{11}, multiply the corresponding elements in the first row and first column and calculate the sum of the results.	$\begin{bmatrix} 3 & 6 \\ 5 & 2 \end{bmatrix} \times \begin{bmatrix} 1 & 8 \\ 5 & 4 \end{bmatrix}$ $3 \times 1 + 6 \times 5 = 33$
4. To determine the element mn_{12}, multiply the corresponding elements in the first row and second column and calculate the sum of the results.	$\begin{bmatrix} 3 & 6 \\ 5 & 2 \end{bmatrix} \times \begin{bmatrix} 1 & 8 \\ 5 & 4 \end{bmatrix}$ $3 \times 8 + 6 \times 4 = 48$
5. To determine the element mn_{21}, multiply the corresponding elements in the second row and first column and calculate the sum of the results.	$\begin{bmatrix} 3 & 6 \\ 5 & 2 \end{bmatrix} \times \begin{bmatrix} 1 & 8 \\ 5 & 4 \end{bmatrix}$ $5 \times 1 + 2 \times 5 = 15$
6. To determine the element mn_{22}, multiply the corresponding elements in the second row and second column and calculate the sum of the results.	$\begin{bmatrix} 3 & 6 \\ 5 & 2 \end{bmatrix} \times \begin{bmatrix} 1 & 8 \\ 5 & 4 \end{bmatrix}$ $5 \times 8 + 2 \times 4 = 48$
7. Construct the matrix MN by writing in each of the elements.	$\begin{bmatrix} 33 & 48 \\ 15 & 48 \end{bmatrix}$

TI	THINK	DISPLAY/WRITE	CASIO	THINK	DISPLAY/WRITE

TI | THINK   **DISPLAY/WRITE**   **CASIO | THINK**   **DISPLAY/WRITE**

1. On a Calculator page, complete the entry line as: $\begin{bmatrix} 3 & 6 \\ 5 & 2 \end{bmatrix} \times \begin{bmatrix} 1 & 8 \\ 5 & 4 \end{bmatrix}$ then press ENTER.
Note: The matrix templates can be found by pressing the matrix button .

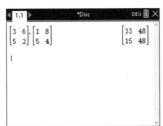

1. On the Main screen, complete the entry line as: $\begin{bmatrix} 3 & 6 \\ 5 & 2 \end{bmatrix} \times \begin{bmatrix} 1 & 8 \\ 5 & 4 \end{bmatrix}$ then press EXE.
Note: The matrix templates can be found in the Math 2 tab in the Keyboard menu.

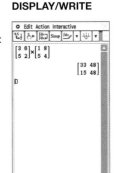

2. The answer appears on the screen. $MN = \begin{bmatrix} 33 & 48 \\ 15 & 48 \end{bmatrix}$

2. The answer appears on the screen. $MN = \begin{bmatrix} 33 & 48 \\ 15 & 48 \end{bmatrix}$

4.4.4 Multiplying by the identity matrix

When a matrix is multiplied by an identity matrix, the result is the same as if a real number is multiplied by 1. The matrix will remain the same.

So, for a matrix A, the following will apply:

$$AI = IA = A$$

For example, if the matrix $A = \begin{bmatrix} 2 & 4 & 6 \\ 3 & 5 & 7 \end{bmatrix}$ is multiplied by an identity matrix on it's left (that is, IA), it will be multiplied by a 2×2 identity matrix (because A has 2 rows). If A is multiplied by an identity matrix on it's right (that is AI), then it will be multiplied by a 3×3 identity matrix (because A has 3 columns).

$$\begin{array}{ccccc} A & I & I & A & A \end{array}$$

$$\begin{bmatrix} 2 & 4 & 6 \\ 3 & 5 & 7 \end{bmatrix} \begin{bmatrix} 1 & 0 & 0 \\ 0 & 1 & 0 \\ 0 & 0 & 1 \end{bmatrix} = \begin{bmatrix} 1 & 0 \\ 0 & 1 \end{bmatrix} \begin{bmatrix} 2 & 4 & 6 \\ 3 & 5 & 7 \end{bmatrix} = \begin{bmatrix} 2 & 4 & 6 \\ 3 & 5 & 7 \end{bmatrix}$$

$$\begin{bmatrix} 2\times1+4\times0+6\times0 & 2\times0+4\times1+6\times0 & 2\times0+4\times0+6\times1 \\ 3\times1+5\times0+7\times0 & 3\times0+5\times1+7\times0 & 3\times0+5\times0+7\times1 \end{bmatrix}$$

$$= \begin{bmatrix} 1\times2+0\times3 & 1\times4+0\times5 & 1\times6+0\times7 \\ 0\times2+1\times3 & 0\times4+1\times5 & 0\times6+1\times7 \end{bmatrix} = \begin{bmatrix} 2 & 4 & 6 \\ 3 & 5 & 7 \end{bmatrix}$$

Multiplication with the identity matrix is an exception to the rule that most matrix multiplication isn't commutative.

Multiplying by the identity matrix

For a matrix A, the following will apply:

$$AI = IA = A$$

where I is an identity matrix with 1s in the top left to bottom right diagonal and 0s for all other elements.

For example, $[1]$, $\begin{bmatrix} 1 & 0 \\ 0 & 1 \end{bmatrix}$ and $\begin{bmatrix} 1 & 0 & 0 \\ 0 & 1 & 0 \\ 0 & 0 & 1 \end{bmatrix}$ are all identity matrices.

Evaluate the following matrix multiplication.

$$\begin{bmatrix} 6 & 2 \\ 8 & 3 \end{bmatrix} \times \begin{bmatrix} 1 & 0 \\ 0 & 1 \end{bmatrix}$$

THINK	WRITE
1. Multiply the elements of the first row of the first matrix with the elements of the first column of the second matrix. Sum the products.	$\begin{bmatrix} 6 & 2 \\ 8 & 3 \end{bmatrix} \times \begin{bmatrix} 1 & 0 \\ 0 & 1 \end{bmatrix} = \begin{bmatrix} 6 \times 1 + 2 \times 0 & - \\ - & - \end{bmatrix}$
2. Repeat step 1 with the first row of the first matrix and the second column of the second matrix	$\begin{bmatrix} 6 & 2 \\ 8 & 3 \end{bmatrix} \times \begin{bmatrix} 1 & 0 \\ 0 & 1 \end{bmatrix} = \begin{bmatrix} 6 & 6 \times 0 + 2 \times 1 \\ - & - \end{bmatrix}$
3. Repeat step 1 with the second row of the first matrix and the first column of the second matrix.	$\begin{bmatrix} 6 & 2 \\ 8 & 3 \end{bmatrix} \times \begin{bmatrix} 1 & 0 \\ 0 & 1 \end{bmatrix} = \begin{bmatrix} 6 & 2 \\ 8 \times 1 + 3 \times 0 & - \end{bmatrix}$
4. Repeat step 1 with the second row of the first matrix and the second column of the second matrix.	$\begin{bmatrix} 6 & 2 \\ 8 & 3 \end{bmatrix} \times \begin{bmatrix} 1 & 0 \\ 0 & 1 \end{bmatrix} = \begin{bmatrix} 6 & 2 \\ 8 & 8 \times 0 + 3 \times 1 \end{bmatrix}$ $= \begin{bmatrix} 6 & 2 \\ 8 & 3 \end{bmatrix}$

4.4.5 Powers of square matrices

When a square matrix is multiplied by itself, the order of the resultant matrix is equal to the order of the original square matrix. Because of this fact, whole-number powers of square matrices always exist.

You can use CAS to quickly determine powers of square matrices.

If $A = \begin{bmatrix} 3 & 5 \\ 5 & 1 \end{bmatrix}$, calculate the value of A^3.

THINK	WRITE
1. Write the matrix multiplication in full.	$A^3 = AAA$ $= \begin{bmatrix} 3 & 5 \\ 5 & 1 \end{bmatrix} \begin{bmatrix} 3 & 5 \\ 5 & 1 \end{bmatrix} \begin{bmatrix} 3 & 5 \\ 5 & 1 \end{bmatrix}$
2. Calculate the first matrix multiplication (AA).	$AA = \begin{bmatrix} 3 & 5 \\ 5 & 1 \end{bmatrix} \begin{bmatrix} 3 & 5 \\ 5 & 1 \end{bmatrix}$ $= \begin{bmatrix} 3 \times 3 + 5 \times 5 & 3 \times 5 + 5 \times 1 \\ 5 \times 3 + 1 \times 5 & 5 \times 5 + 1 \times 1 \end{bmatrix}$ $= \begin{bmatrix} 34 & 20 \\ 20 & 26 \end{bmatrix}$

3. Rewrite the full matrix multiplication, substituting the answer found in the previous part.
Calculate the second matrix multiplication (AAA).

$$A^3 = AAA$$
$$= \begin{bmatrix} 34 & 20 \\ 20 & 26 \end{bmatrix} \begin{bmatrix} 3 & 5 \\ 5 & 1 \end{bmatrix}$$
$$= \begin{bmatrix} 34 \times 3 + 20 \times 5 & 34 \times 5 + 20 \times 1 \\ 20 \times 3 + 26 \times 5 & 20 \times 5 + 26 \times 1 \end{bmatrix}$$
$$= \begin{bmatrix} 202 & 190 \\ 190 & 126 \end{bmatrix}$$

4. Write the answer.

$$A^3 = \begin{bmatrix} 202 & 190 \\ 190 & 126 \end{bmatrix}$$

TI	THINK	DISPLAY/WRITE	CASIO	THINK	DISPLAY/WRITE

TI | THINK

1. On a Calculator page, complete the entry line as:
$$\begin{bmatrix} 3 & 5 \\ 5 & 1 \end{bmatrix}^3$$
then press ENTER.
Note: The matrix templates can be found by pressing the matrix button and then select one of these matrix .

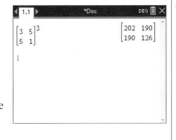

2. The answer appears on the screen.

$$A^3 = \begin{bmatrix} 202 & 190 \\ 190 & 126 \end{bmatrix}$$

CASIO | THINK

1. On the Main screen, complete the entry line as:
$$\begin{bmatrix} 3 & 5 \\ 5 & 1 \end{bmatrix}^3$$
then press EXE.
Note: The matrix templates can be found in the Math 2 tab in the Keyboard menu.

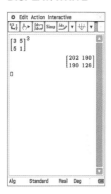

2. The answer appears on the screen.

$$A^3 = \begin{bmatrix} 202 & 190 \\ 190 & 126 \end{bmatrix}$$

4.4 Exercise

Students, these questions are even better in jacPLUS

Receive immediate feedback and access sample responses

Access additional questions

Track your results and progress

Find all this and MORE in jacPLUS

1. **WE7** Consider the matrix $C = \begin{bmatrix} 2 & 3 & 7 \\ 1 & 4 & 6 \end{bmatrix}$. Evaluate the following.

a. $4C$

b. $\dfrac{1}{5}C$

c. $0.3C$

2. Matrix D was multiplied by the scalar quantity x.

If $3D = \begin{bmatrix} 15 & 0 \\ 21 & 12 \\ 33 & 9 \end{bmatrix}$ and $xD = \begin{bmatrix} 12.5 & 0 \\ 17.5 & 10 \\ 27.5 & 7.5 \end{bmatrix}$, find the value of x.

3. **MC** Consider the matrix $M = \begin{bmatrix} 12 & 9 & 15 \\ 36 & 6 & 21 \end{bmatrix}$. Determine which of the following is equal to the matrix M.

A. $0.1 \begin{bmatrix} 1.2 & 0.9 & 1.5 \\ 3.6 & 0.6 & 2.1 \end{bmatrix}$

B. $3 \begin{bmatrix} 3 & 3 & 5 \\ 9 & 2 & 7 \end{bmatrix}$

C. $3 \begin{bmatrix} 4 & 3 & 5 \\ 12 & 2 & 7 \end{bmatrix}$

D. $3 \begin{bmatrix} 36 & 27 & 45 \\ 108 & 18 & 63 \end{bmatrix}$

E. $10 \begin{bmatrix} 120 & 90 & 15 \\ 36 & 6 & 21 \end{bmatrix}$

4. **a.** If $X = \begin{bmatrix} 3 & 5 \end{bmatrix}$ and $Y = \begin{bmatrix} 4 \\ 2 \end{bmatrix}$, show that the product matrix XY exists and state the order of XY.

   b. Determine which of the following matrices can be multiplied together and state the order of any product matrices that exist.

$$D = \begin{bmatrix} 7 & 4 \\ 3 & 5 \\ 1 & 2 \end{bmatrix}, C = \begin{bmatrix} 5 & 7 \\ 8 & 9 \end{bmatrix}, E = \begin{bmatrix} 4 & 1 & 2 \\ 6 & 2 & 6 \end{bmatrix}$$

5. The product matrix ST has an order of 3×4. If matrix S has 2 columns, write down the order of matrices S and T.

6. Determine which of the following matrices can be multiplied together. Justify your answers by finding the order of the product matrices.

$$D = \begin{bmatrix} 3 \\ 7 \\ 8 \end{bmatrix}, E = \begin{bmatrix} 5 & 8 \\ 7 & 1 \\ 9 & 3 \end{bmatrix}, F = \begin{bmatrix} 12 & 7 & 3 \\ 15 & 8 & 4 \end{bmatrix}, G = \begin{bmatrix} 13 & 15 \end{bmatrix}$$

7. **WE8** If $M = \begin{bmatrix} 4 \\ 3 \end{bmatrix}$ and $N = \begin{bmatrix} 7 & 12 \end{bmatrix}$, determine the product matrix MN.

8. Matrix $S = \begin{bmatrix} 1 & 4 & 3 \end{bmatrix}$, matrix $T = \begin{bmatrix} 2 \\ 3 \\ t \end{bmatrix}$ and the product matrix $ST = \begin{bmatrix} 5 \end{bmatrix}$. Find the value of t.

9. **WE9** For matrices $P = \begin{bmatrix} 3 & 7 \\ 8 & 4 \end{bmatrix}$ and $Q = \begin{bmatrix} 2 & 1 \\ 5 & 6 \end{bmatrix}$, determine the product matrix PQ.

10. For a school show, three different types of tickets can be purchased: adult, senior and child. The cost of each type of ticket is $12.50, $8.50 and $6.00 respectively.
The number of people attending the concert is shown in the following table.

Ticket type	Number of people
Adult	65
Senior	40
Child	85

   a. Construct a column matrix to represent the cost of the three different tickets in the order of adult, senior and child. If the number of people attending the concert is written as a row matrix, a matrix multiplication can be performed to determine the total amount in ticket sales for the concert.

   b. By determining the order of each matrix and then the product matrix, explain why this is the case.

   c. By completing the matrix multiplication from part **b**, determine the total amount (in dollars) in ticket sales for the concert.

11. Calculate the product matrices when the following pairs of matrices are multiplied together.

a. $\begin{bmatrix} 6 & 9 \end{bmatrix}$ and $\begin{bmatrix} 5 \\ 4 \end{bmatrix}$

b. $\begin{bmatrix} 5 \\ 4 \end{bmatrix}$ and $\begin{bmatrix} 6 & 9 \end{bmatrix}$

c. $\begin{bmatrix} 7 \\ 2 \\ 9 \end{bmatrix}$ and $\begin{bmatrix} 10 & 15 \end{bmatrix}$

d. $\begin{bmatrix} 6 & 5 \\ 8 & 3 \end{bmatrix}$ and $\begin{bmatrix} 2 \\ 9 \end{bmatrix}$

e. $\begin{bmatrix} 3 & 5 \\ 1 & 2 \end{bmatrix}$ and $\begin{bmatrix} 6 & 3 \\ 4 & 2 \end{bmatrix}$

12. **WE10** Evaluate the following matrix multiplications.

a. $\begin{bmatrix} 4 & 6 \\ 2 & 3 \end{bmatrix} \begin{bmatrix} 1 & 0 \\ 0 & 1 \end{bmatrix}$

b. $\begin{bmatrix} 1 & 0 \\ 0 & 1 \end{bmatrix} \begin{bmatrix} 4 & 6 \\ 2 & 3 \end{bmatrix}$

13. **WE11** If $P = \begin{bmatrix} 8 & 2 \\ 4 & 7 \end{bmatrix}$, calculate the value of P^2.

14. If $T = \begin{bmatrix} 3 & 5 \\ 0 & 6 \end{bmatrix}$, calculate the value of T^3.

15. The 3×3 identity matrix, $I_3 = \begin{bmatrix} 1 & 0 & 0 \\ 0 & 1 & 0 \\ 0 & 0 & 1 \end{bmatrix}$.

a. Calculate the value of I_3^2.

b. Calculate the value of I_3^3.

c. Calculate the value of I_3^4.

d. Comment on your answers to parts a–c.

16. The table below shows the percentage of students who are expected to be awarded grades A–E on their final examinations for Mathematics and Physics.

Grade	A	B	C	D	E
Percentage of students	5	18	45	25	7

The number of students studying Mathematics and Physics is 250 and 185 respectively.

a. Construct a column matrix, S, to represent the number of students studying Mathematics and Physics.

b. Construct a 1×5 matrix, A, to represent the percentage of students expected to receive each grade, expressing each element in decimal form.

c. In the context of this problem, state what product matrix SA represents.

d. Determine the product matrix SA. Write your answers correct to the nearest whole number.

e. In the context of this problem, state what element SA_{12} represents.

17. A product matrix, $N = MPR$, has order 3×4. Matrix M has m rows and n columns, matrix P has order $1 \times q$, and matrix R has order $2 \times s$. Determine the values of m, n, s and q.

18. A student was asked to perform the following matrix multiplication to determine the product matrix GH.

$$GH = \begin{bmatrix} 6 & 5 \\ 3 & 8 \\ 5 & 9 \end{bmatrix} \begin{bmatrix} 10 \\ 13 \end{bmatrix}$$

Their answer was $\begin{bmatrix} 60 & 65 \\ 30 & 104 \\ 50 & 117 \end{bmatrix}$.

a. By stating the order of product matrix GH, explain why the student's answer is obviously incorrect.

b. Determine the product matrix GH.

c. Explain the student's method of multiplying matrices and why this method is incorrect.

d. Provide simple steps to help the student multiply matrices.

19. Dodgy Bros sell vans, utes and sedans. The average selling price for each type of vehicle is shown in the table below.

Type of vehicle	Monthly sales ($)
Vans	$4000
Utes	$12 500
Sedans	$8500

The table below shows the total number of vans, utes and sedans sold at Dodgy Bros in one month.

Type of vehicle	Number of sales
Vans	5
Utes	8
Sedans	4

Stan is the owner of Dodgy Bros and wants to determine the total amount of monthly sales.

a. Explain how matrices could be used to help Stan determine the total amount, in dollars, of monthly sales.
b. Perform a matrix multiplication that finds the total amount of monthly sales.
c. Brian is Stan's brother and the accountant for Dodgy Bros. In calculating the total amount of monthly sales, he performs the following matrix multiplication.

$$\begin{bmatrix} 5 \\ 8 \\ 4 \end{bmatrix} \begin{bmatrix} 4000 & 12\,500 & 8500 \end{bmatrix}$$

Explain why this matrix multiplication is not valid for this problem.

20. The number of adults, children and seniors attending the zoo over Friday, Saturday and Sunday is shown in the table.

Day	Adults	Children	Seniors
Friday	125	245	89
Saturday	350	456	128
Sunday	421	523	102

Entry prices for adults, children and seniors are $35, $25 and $20 respectively.

a. Using CAS, perform a matrix multiplication that will find the entry fee collected for each of the three days.
b. Write the calculation that finds the entry fee collected for Saturday.
c. Determine if it is possible to perform a matrix multiplication that would find the total for each type of entry fee (adults, children and seniors) over the three days. Explain your answer.

⊳ **Question 1 (1 mark)**

MC If $A = \begin{bmatrix} -1 & 2 \\ -3 & 2 \end{bmatrix}$ and $B = \begin{bmatrix} -1 & 2 \\ 1 & -2 \end{bmatrix}$, determine $2A - 3B$.

A. $\begin{bmatrix} -2 & 4 \\ -6 & 4 \end{bmatrix}$ **B.** $\begin{bmatrix} -3 & 6 \\ 3 & -6 \end{bmatrix}$ **C.** $\begin{bmatrix} -5 & 10 \\ -3 & -2 \end{bmatrix}$ **D.** $\begin{bmatrix} 1 & -2 \\ -9 & 10 \end{bmatrix}$ **E.** $\begin{bmatrix} -5 & -2 \\ -9 & 2 \end{bmatrix}$

⊳ **Question 2 (1 mark)**

MC If $A = \begin{bmatrix} 3 & 4 \\ -5 & 0 \end{bmatrix}$ and $B = \begin{bmatrix} 4 & -1 & 5 \\ 0 & 5 & 0 \end{bmatrix}$ then $AB = A \times B$ equals

A. $\begin{bmatrix} 12 & 5 \\ -20 & 5 \end{bmatrix}$ **B.** $\begin{bmatrix} 12 & 17 & 15 \\ -20 & 5 & -25 \end{bmatrix}$ **C.** $\begin{bmatrix} 32 \\ -40 \end{bmatrix}$

D. $\begin{bmatrix} 16 & 5 & 15 \\ 20 & 5 & -25 \end{bmatrix}$ **E.** $[72]$

⊳ **Question 3 (3 marks)**

Given $X = \begin{bmatrix} 2 & -1 \\ 0 & 4 \end{bmatrix}$, determine the matrix $X^2 + 2X$.

More exam questions are available online.

4.5 Inverse matrices and problem-solving with matrices

LEARNING INTENTION

At the end of this subtopic you should be able to:
- calculate the determinant of a matrix and use it to determine whether the matrix has an inverse
- determine the inverse of a matrix
- solve a system of simultaneous linear equations.

4.5.1 Inverses and the determinant

In the real number system, a number multiplied by its reciprocal (inverse) results in 1. For example, $3 \times \dfrac{1}{3} = 1$. In this case, $\dfrac{1}{3}$ is the reciprocal (multiplicative inverse) of 3.

Multiplicative inverse

If the product matrix is the identity matrix, then one of the matrices is the multiplicative inverse of the other.

If we multiply matrix A by its inverse, the result is the identity matrix.

$$\begin{bmatrix} 2 & 5 \\ 1 & 3 \end{bmatrix} \begin{bmatrix} 3 & -5 \\ -1 & 2 \end{bmatrix} = \begin{bmatrix} 2 \times 3 + 5 \times -1 & 2 \times -5 + 5 \times 2 \\ 1 \times 3 + 3 \times -1 & 1 \times -5 + 3 \times 2 \end{bmatrix}$$

$$= \begin{bmatrix} 1 & 0 \\ 0 & 1 \end{bmatrix} \qquad \text{(the } 2 \times 2 \text{ identity matrix)}$$

Hence, we can say for any matrix:

$$AA^{-1} = I = A^{-1}A$$

Inverse matrices only exist for square matrices and can be easily found for matrices of order 2×2. Inverses can also be found for larger matrices, but they are much more complicated, so we will rely on using technology to determine them.

The determinant

To help us get the inverse, we need to calculate the **determinant**. The determinant of matrix A is usually written as det A.

The determinant tells us if there is an inverse and, if there is, it helps us calculate it. If the determinant is zero, it means there is no inverse.

If $A = \begin{bmatrix} a & b \\ c & d \end{bmatrix}$, the determinant of matrix A, also written as $\det(A)$ or $|A|$, can be found using:

$$\det(A) = ad - bc$$

WORKED EXAMPLE 12 Calculating the determinant of a matrix

Calculate the determinant of each of the following matrices and use it to determine whether the matrix has an inverse.

a. $N = \begin{bmatrix} 6 & 4 \\ 9 & 8 \end{bmatrix}$

b. $P = \begin{bmatrix} 4 & 3 \\ 8 & 6 \end{bmatrix}$

THINK	WRITE
a. 1. Identify the values of a, b, c and d.	a. $a = 6$ $b = 4$ $c = 9$ $d = 8$
2. Substitute the values for a, b, c and d to calculate the determinant of matrix N.	$\det(N) = ad - bc$ $= 6 \times 8 - 4 \times 9$ $= 12$
3. Since the determinant is not zero, matrix N does have an inverse.	The matrix N has an inverse.
b. 1. Identify the values of a, a, c and d.	b. $a = 4$ $b = 3$ $c = 8$ $d = 6$
2. Substitute the values for a, b, c and d to calculate the determinant of matrix P.	$\det(P) = ad - bc$ $= 4 \times 6 - 3 \times 8$ $= 0$
3. Since the determinant is zero, we can say that matrix P does not have an inverse.	The matrix P does not have an inverse.

TI	THINK	DISPLAY/WRITE	CASIO	THINK	DISPLAY/WRITE

a. On a calculator page complete the entry line as:
$$\det\left(\begin{bmatrix} 6 & 4 \\ 9 & 8 \end{bmatrix}\right)$$
Then press ENTER.

The determinant of matrix N is 12. Since the determinant is non-zero, there is an inverse.

a. On the Main screen, go to: Action > Matrix > Calculation > det Complete the entry line as:
$$\det\left(\begin{bmatrix} 6 & 4 \\ 9 & 8 \end{bmatrix}\right)$$
Then press EXE.

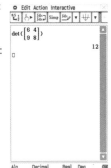

The determinant of matrix N is 12. Since the determinant is non-zero, there is an inverse.

b. On a calculator page complete the entry line as:
$$\det\left(\begin{bmatrix} 4 & 3 \\ 8 & 6 \end{bmatrix}\right)$$
Then press ENTER.

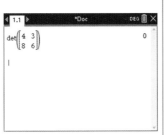

The determinant of matrix P is 0. Since the determinant is zero, there is no inverse.

b. On the Main screen, go to: Action > Matrix > Calculation > det Complete the entry line as: Then press EXE.

The determinant of matrix P is 0. Since the determinant is zero, there is no inverse.

4.5.2 Determining the inverse matrix

Inverse matrices only exist for square matrices and can be easily found for matrices of order 2×2. Inverses can also be found for larger square matrices; however, the processes to determine these are more complicated, so technology is often used to determine larger inverses.

> ### Determining the inverse matrix
>
> To determine the inverse for a matrix $A = \begin{bmatrix} a & c \\ b & d \end{bmatrix}$:
>
> **Step 1:** Swap the elements a_{11} and a_{22}: $\begin{bmatrix} d & c \\ b & a \end{bmatrix}$
>
> **Step 2:** Multiply elements a_{12} and a_{21} by -1: $\begin{bmatrix} d & -c \\ -b & a \end{bmatrix}$
>
> **Step 3:** Multiply by $\dfrac{1}{ad - bc}$:
>
> $$A^{-1} = \frac{1}{ad - bc}\begin{bmatrix} d & -c \\ -b & a \end{bmatrix}$$

If $A = \begin{bmatrix} 7 & 2 \\ 4 & 1 \end{bmatrix}$, determine A^{-1}.

THINK	WRITE
1. Swap elements a_{11} and a_{22}.	$\begin{bmatrix} 7 & 2 \\ 4 & 1 \end{bmatrix} \rightarrow \begin{bmatrix} 1 & 2 \\ 4 & 7 \end{bmatrix}$
2. Multiply elements a_{12} and a_{21} by -1.	$\begin{bmatrix} 1 & 2 \\ 4 & 7 \end{bmatrix} \rightarrow \begin{bmatrix} 1 & -2 \\ -4 & 7 \end{bmatrix}$
3. Find the determinant ($\det(A) = ad - bc$).	$a = 7,\ b = 2,\ c = 4,\ d = 1$ $ad - bc = 7 \times 1 - 2 \times 4$ $\qquad\quad = -1$
4. Multiply the matrix from step 2 by $\dfrac{1}{\det(A)}$.	$\dfrac{1}{-1} \begin{bmatrix} 1 & -2 \\ -4 & 7 \end{bmatrix} = -1 \begin{bmatrix} 1 & -2 \\ -4 & 7 \end{bmatrix}$ $= \begin{bmatrix} -1 & 2 \\ 4 & -7 \end{bmatrix}$

TI | THINK	DISPLAY/WRITE	CASIO | THINK	DISPLAY/WRITE
1. On a Calculator page, complete the entry line as: $\begin{bmatrix} 7 & 2 \\ 4 & 1 \end{bmatrix}^{-1}$ then press ENTER. *Note:* The matrix templates can be found by pressing the matrix button 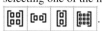 and then selecting one of the matrices 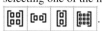.	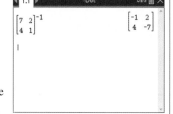	1. On the Main screen, complete the entry line as: $\begin{bmatrix} 7 & 2 \\ 4 & 1 \end{bmatrix}^{-1}$ then press EXE. *Note:* The matrix templates can be found in the Math 2 tab in the Keyboard menu.	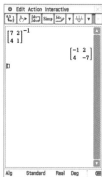
2. The answer appears on the screen.	$A^{-1} = \begin{bmatrix} -1 & 2 \\ 4 & -7 \end{bmatrix}$	2. The answer appears on the screen.	$A^{-1} = \begin{bmatrix} -1 & 2 \\ 4 & -7 \end{bmatrix}$

4.5.3 Solving simultaneous equations with matrices

Matrices are useful because they can solve large systems of simultaneous equations. In meteorology, values are input into systems that solve numerous equations simultaneously to predict what the weather will be in the future.

tlvd-3712

If $\begin{bmatrix} 4 & 2 \\ 5 & 3 \end{bmatrix} \begin{bmatrix} x \\ y \end{bmatrix} = \begin{bmatrix} 8 \\ 11 \end{bmatrix}$, determine the values of x and y.

THINK	WRITE
1. The matrix equation is in the form $AX = B$. Identify matrices A, B and X.	$A = \begin{bmatrix} 4 & 2 \\ 5 & 3 \end{bmatrix}$ $B = \begin{bmatrix} 8 \\ 11 \end{bmatrix}$ $X = \begin{bmatrix} x \\ y \end{bmatrix}$

2. To determine X we need to multiply both sides of the equation by the inverse A^{-1}.

$$A^{-1}AX = A^{-1}B$$

3. Determine the inverse A^{-1}.

$$A^{-1} = \frac{1}{12-10}\begin{bmatrix} 3 & -2 \\ -5 & 4 \end{bmatrix}$$

$$= \frac{1}{2}\begin{bmatrix} 3 & -2 \\ -5 & 4 \end{bmatrix}$$

$$= \begin{bmatrix} 1.5 & -1 \\ -2.5 & 2 \end{bmatrix}$$

4. Calculate $A^{-1}B$.

$$A^{-1}B = \begin{bmatrix} 1.5 & -1 \\ -2.5 & 2 \end{bmatrix}\begin{bmatrix} 8 \\ 11 \end{bmatrix}$$

$$= \begin{bmatrix} 1.5 \times 8 + -1 \times 11 \\ -2.5 \times 8 + 2 \times 11 \end{bmatrix}$$

$$= \begin{bmatrix} 1 \\ 2 \end{bmatrix}$$

5. Solve for x and y.

$$X = A^{-1}B$$

$$\begin{bmatrix} x \\ y \end{bmatrix} = \begin{bmatrix} 1 \\ 2 \end{bmatrix}$$

$$x = 1 \text{ and } y = 2$$

TI	THINK	DISPLAY/WRITE	CASIO	THINK	DISPLAY/WRITE
1. On a Calculator page, press MENU, then select: 3: Algebra 1: Solve Complete the entry line as: solve $\left(\begin{bmatrix} 4 & 2 \\ 5 & 3 \end{bmatrix} \times \begin{bmatrix} x \\ y \end{bmatrix}\right.$ $\left. = \begin{bmatrix} 8 \\ 11 \end{bmatrix}, \{x, y\}\right)$ then press ENTER. *Note:* The matrix templates can be found by pressing the matrix button 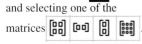 and selecting one of the matrices.	$x = 1$ and $y = 2$	1. On the Main screen, complete the entry line as: solve $\left(\begin{bmatrix} 4 & 2 \\ 5 & 3 \end{bmatrix} \times \begin{bmatrix} x \\ y \end{bmatrix} = \begin{bmatrix} 8 \\ 11 \end{bmatrix}, \{x, y\}\right)$ then press EXE. *Note:* The matrix templates can be found in the Math 2 tab in the Keyboard menu.	
2. The answer appears on the screen.	$x = 1$ and $y = 2$	2. The answer appears on the screen.	$x = 1$ and $y = 2$

4.5.4 Using inverse matrices to solve a system of simultaneous equations

If you have a pair of simultaneous equations, they can be set up as a matrix equation and solved using inverse matrices.

Take the pair of simultaneous equations $ax + by = c$ and $dx + ey = f$.

These can be set up as the matrix equation

$$\begin{bmatrix} a & b \\ d & e \end{bmatrix}\begin{bmatrix} x \\ y \end{bmatrix} = \begin{bmatrix} c \\ f \end{bmatrix}.$$

If we let $A = \begin{bmatrix} a & b \\ d & e \end{bmatrix}$, $X = \begin{bmatrix} x \\ y \end{bmatrix}$ and $B = \begin{bmatrix} c \\ f \end{bmatrix}$, this equation is of the form $AX = B$, which can be solved as $X = A^{-1}B$ (as determined previously).

WORKED EXAMPLE 15 Using inverse matrices to solve simultaneous equations

Solve the following pair of simultaneous equations by using inverse matrices.
$$2x + 3y = 6$$
$$4x - 6y = -4$$

THINK	WRITE
1. Set up the simultaneous equations as a matrix equation.	$\begin{bmatrix} 2 & 3 \\ 4 & -6 \end{bmatrix} \begin{bmatrix} x \\ y \end{bmatrix} = \begin{bmatrix} 6 \\ -4 \end{bmatrix}$
2. Determine the inverse of the matrix A, A^{-1}.	$A = \begin{bmatrix} 2 & 3 \\ 4 & -6 \end{bmatrix}$ $A^{-1} = \dfrac{1}{-12-12} \begin{bmatrix} -6 & -3 \\ -4 & 2 \end{bmatrix}$ $= \dfrac{1}{-24} \begin{bmatrix} -6 & -3 \\ -4 & 2 \end{bmatrix}$ $= \begin{bmatrix} \dfrac{1}{4} & \dfrac{1}{8} \\ \dfrac{1}{6} & -\dfrac{1}{12} \end{bmatrix}$
3. Calculate $A^{-1}B$.	$A^{-1}B = \begin{bmatrix} \dfrac{1}{4} & \dfrac{1}{8} \\ \dfrac{1}{6} & -\dfrac{1}{12} \end{bmatrix} \begin{bmatrix} 6 \\ -4 \end{bmatrix}$ $= \begin{bmatrix} \dfrac{1}{4} \times 6 + \dfrac{1}{8} \times -4 \\ \dfrac{1}{6} \times 6 + -\dfrac{1}{12} \times -4 \end{bmatrix}$ $= \begin{bmatrix} 1 \\ \dfrac{4}{3} \end{bmatrix}$
4. Write the answer.	$x = 1, \ y = \dfrac{4}{3}$

On an excursion, a group of students and teachers travelled to the city by train and returned by bus. On the train, the cost of a student ticket was $3 and the cost of a teacher ticket was $4.50, with the total cost for the train tickets being $148.50. On the bus, the cost of a student ticket was $2.75 and the cost of a teacher ticket was $3.95, with the total cost for the bus tickets being $135.60.

By solving a matrix equation, determine how many students and teachers attended the excursion.

THINK	WRITE
1. Identify the two unknowns in the problem. Assign a pronumeral to represent each unknown.	Number of students $= s$ Number of teachers $= t$
2. Construct a matrix to represent the unknowns.	$\begin{bmatrix} s \\ t \end{bmatrix}$
3. Highlight the key information — that is, how much the two different types of tickets were for students and teachers.	Student train ticket $= \$3$ Teacher train ticket $= \$4.50$ Student bus ticket $= \$2.75$ Teacher bus ticket $= \$3.95$
4. Construct a matrix to represent the information. Note that each row represents the two different types of travel.	$\begin{bmatrix} 3 & 4.50 \\ 2.75 & 3.95 \end{bmatrix} \begin{matrix} \text{Train} \\ \text{Bus} \end{matrix}$
5. Construct a matrix to represent the total cost in the same row order as in step 4.	$\begin{bmatrix} 148.50 \\ 135.60 \end{bmatrix}$
6. Set up a matrix equation in the form $AX = B$, remembering that X will represent the unknowns (the values that need to be found).	$\begin{bmatrix} 3 & 4.50 \\ 2.75 & 3.95 \end{bmatrix} \begin{bmatrix} s \\ t \end{bmatrix} = \begin{bmatrix} 148.50 \\ 135.60 \end{bmatrix}$
7. Solve the matrix equation by determining A^{-1} and multiplying it by B.	$A^{-1} = \dfrac{1}{3 \times 3.95 - 2.75 \times 4.50} \begin{bmatrix} 3.95 & -4.50 \\ -2.75 & 3 \end{bmatrix}$ $= \dfrac{1}{-0.525} \begin{bmatrix} 3.95 & -4.50 \\ -2.75 & 3 \end{bmatrix}$ $\begin{bmatrix} s \\ t \end{bmatrix} = A^{-1}B$ $= \dfrac{1}{-0.525} \begin{bmatrix} 3.95 & -4.50 \\ -2.75 & 3 \end{bmatrix} \begin{bmatrix} 148.50 \\ 135.60 \end{bmatrix}$ $= \dfrac{1}{-0.525} \begin{bmatrix} -23.625 \\ -1.575 \end{bmatrix}$ $= \begin{bmatrix} 45 \\ 3 \end{bmatrix}$
8. Write the answer as a sentence.	There were 45 students and 3 teachers on the excursion.

TI | THINK	DISPLAY/WRITE	CASIO | THINK	DISPLAY/WRITE
On a calculator page, press MENU, then select: 3: Algebra 1: Solve Complete the entry line as: $$\text{solve}\left(\begin{bmatrix} 3 & 4.5 \\ 2.75 & 3.95 \end{bmatrix} \times \begin{bmatrix} s \\ t \end{bmatrix} = \begin{bmatrix} 148.5 \\ 135.6 \end{bmatrix}, \{s, t\}\right)$$ Then press ENTER.	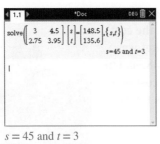 $s = 45$ and $t = 3$ There were 45 students and 3 teachers on the excursion.	On the Main screen, complete the entry line as: $$\text{solve}\left(\begin{bmatrix} 3 & 4.5 \\ 2.75 & 3.95 \end{bmatrix} \times \begin{bmatrix} s \\ t \end{bmatrix}\right.$$ $$\left.= \begin{bmatrix} 148.5 \\ 135.6 \end{bmatrix}, \{s, t\}\right)$$ Then press EXE.	 $s = 45$ and $t = 3$ There were 45 students and 3 teachers on the excursion.

4.5 Exercise

Students, these questions are even better in jacPLUS

- Receive immediate feedback and access sample responses
- Access additional questions
- Track your results and progress

Find all this and MORE in jacPLUS

1. By calculating the product matrix AB, determine whether the following matrices are multiplicative inverses of each other.

$$A = \begin{bmatrix} 4 & 5 \\ 2 & 3 \end{bmatrix}, B = \begin{bmatrix} 1.5 & -2.5 \\ -1 & 2 \end{bmatrix}$$

2. Calculate the determinant of each of the following matrices and use it to determine whether the matrix has an inverse.

   a. $\begin{bmatrix} 3 & 5 \\ 8 & 2 \end{bmatrix}$
   b. $\begin{bmatrix} -2 & -6 \\ 3 & 5 \end{bmatrix}$
   c. $\begin{bmatrix} 3 & 6 \\ 4 & 8 \end{bmatrix}$
   d. $\begin{bmatrix} -9 & 2 \\ 4 & 3 \end{bmatrix}$

3. Determine the inverses of the following matrices.

   a. $\begin{bmatrix} 5 & 2 \\ 2 & 1 \end{bmatrix}$
   b. $\begin{bmatrix} 7 & 4 \\ 3 & 2 \end{bmatrix}$
   c. $\begin{bmatrix} 2 & 1 \\ -3 & -2 \end{bmatrix}$

4. Consider the matrix $B = \begin{bmatrix} 3 & 2 \\ 9 & 6 \end{bmatrix}$. By calculating the value of the determinant, explain why B^{-1} does not exist.

5. a. Calculate the determinants of the following matrices to determine which of them have inverses.

   $$A = \begin{bmatrix} 2 & 1 \\ 5 & 3 \end{bmatrix}, B = \begin{bmatrix} 3 & 2 \\ 9 & 6 \end{bmatrix}, C = \begin{bmatrix} -3 & 6 \\ 4 & -8 \end{bmatrix}, D = \begin{bmatrix} 0 & 1 \\ 3 & -5 \end{bmatrix}$$

   b. For those matrices that have inverses, determine the inverse matrices.

6. If $\begin{bmatrix} 3 & 1 \\ 2 & 4 \end{bmatrix} \begin{bmatrix} x \\ y \end{bmatrix} = \begin{bmatrix} 3 \\ -8 \end{bmatrix}$, find the values of x and y.

7. A matrix equation is represented by $XA = B$, where $B = \begin{bmatrix} 1 & 2 \end{bmatrix}$ and $A^{-1} = \begin{bmatrix} 1 & -1 \\ -1.5 & 2 \end{bmatrix}$.

   a. State the order of matrix X and hence find matrix X.
   b. Determine matrix A.

8. Consider the matrix equation $\begin{bmatrix} 4 & 1 \\ 3 & 1 \end{bmatrix} \begin{bmatrix} x \\ y \end{bmatrix} = \begin{bmatrix} 9 \\ 7 \end{bmatrix}$.

   a. Explain how this matrix equation can be solved using the inverse matrix.
   b. State the inverse matrix used to solve this matrix equation.
   c. Calculate the values of x and y, clearly showing your working.

9. A student bought two donuts and three cupcakes for $14. The next week they bought three donuts and two cupcakes for $12.25. This information is shown in the following matrices, where d and c represent the cost of a donut and cupcake respectively.

$$\begin{bmatrix} 2 & 3 \\ 3 & 2 \end{bmatrix} \begin{bmatrix} d \\ c \end{bmatrix} = \begin{bmatrix} 14.00 \\ 12.25 \end{bmatrix}$$

   By solving the matrix equation, determine how much the student would pay for four donuts and three cupcakes.

10. **WE15** Solve the following pair of simultaneous equations by using inverse matrices.

$$x + 2y = 4$$
$$3x - 5y = 1$$

11. Show that there is no solution to the following pair of simultaneous equations by attempting to solve them using inverse matrices.

$$3x + 5y = 4$$
$$4.5x + 7.5y = 5$$

12. **WE16** For Ben's 8th birthday party, he and some friends went ice skating and ten-pin bowling. The price for ice skating was $4.50 per child and $6.50 per adult, with the total cost for the ice skating being $51. For the ten-pin bowling, the children were charged $3.25 each and the adults were charged $4.95 each, with the total cost for the bowling being $37.60.
By solving a matrix equation, determine how many children (including Ben) attended the party.

13. At the cinema, one group of friends bought 5 drinks and 4 bags of popcorn, spending $14. Another group bought 4 drinks and 3 bags of popcorn, spending $10.80. By solving a matrix equation, determine the price of 2 drinks and 2 bags of popcorn.

14. Jeremy has an interest in making jewellery and makes bracelets and necklaces that he sells to friends. Jeremy charges the same amount for each bracelet and necklace, regardless of the quantity sold.
Johanna buys 3 bracelets and 2 necklaces from Jeremy for $31.80.
Mystique buys 5 bracelets and 3 necklaces from Jeremy for $49.80.

   a. Construct a pair of simultaneous equations and use an inverse matrix to help determine the prices that Jeremy charges for each bracelet and each necklace.
   b. Determine how much Jeremy would charge for 7 bracelets and 4 necklaces.

15. A student was asked to solve the following matrix equation.

$$\begin{bmatrix} x & y \end{bmatrix} \begin{bmatrix} 2 & 0 \\ 1 & 3 \end{bmatrix} = \begin{bmatrix} 5 & 9 \end{bmatrix}$$

Their first step was to evaluate $\begin{bmatrix} 2 & 0 \\ 1 & 3 \end{bmatrix}^{-1}$. They wrote $6 \begin{bmatrix} 2 & 0 \\ 1 & 3 \end{bmatrix}$, which was incorrect.

a. Explain one of the errors the student made in determining the inverse matrix.

b. Hence, determine the correct inverse matrix $\begin{bmatrix} 2 & 0 \\ 1 & 3 \end{bmatrix}^{-1}$.

Their next step was to perform the following matrix multiplication.

$$6 \begin{bmatrix} 2 & 0 \\ 1 & 3 \end{bmatrix} \begin{bmatrix} 5 & 9 \end{bmatrix}$$

c. By determining the order of both matrices, explain why this multiplication is not possible.

d. Write the steps the student should have used to calculate this matrix multiplication.

e. Using your steps from part d, determine the values of x and y.

16. Whole Foods distribute two different types of apples, Sundowners and Pink Ladies, to two supermarkets, Foodsale and Betafoods. Foodsale orders 5 boxes of Sundowners and 7 boxes of Pink Ladies, with their order totalling $156.80. Betafoods pays $155.40 for 6 boxes each of Sundowners and Pink Ladies.

The following matrix equation represents part of this information, where s and p represent the price for a box of Sundowners and Pink Ladies respectively.

$$\begin{bmatrix} 5 & \\ 6 & 6 \end{bmatrix} \begin{bmatrix} s \\ p \end{bmatrix} = \begin{bmatrix} 156.80 \\ \end{bmatrix}$$

a. Complete the matrix equation.

b. By solving the matrix equation using CAS, determine the cost of a box of Sundowner apples.

There are 5 kg of apples in each box. Betafoods sells Sundowner apples for $3.49 per kilogram and Pink Ladies for $4.50 per kilogram.

c. Construct a row matrix, K, to represent the number of kilograms of Sundowners and Pink Ladies in Betafood's order.

A matrix representing the selling price, S, of each type of apple is constructed. A matrix multiplication is then performed that determines the total selling price (in dollars) for both types of apples.

d. Write the order of matrix S.

e. By performing the matrix multiplication, determine the total amount of revenue (in dollars) if all the apples are sold at the price stated.

f. Determine the profit, in dollars, made by Betafoods if all apples are sold at the stated selling prices.

17. Four hundred tickets were sold for the opening of the movie *The Robbit* at the Dendy Cinema. Two types of tickets were sold: adult and concession. Adult tickets were $15.00 and concession tickets were $9.50. The total revenue from the ticket sales was $5422.50.

   a. Identify the two unknowns and construct a pair of simultaneous equations to represent this information.
   b. Set up a matrix equation representing this information.
   c. Using CAS, determine the number of adult tickets sold.

18. Using CAS, determine the inverses of the following matrices.

   a. $\begin{bmatrix} 1 & -1 & 2 \\ 3 & 4 & 5 \\ 2 & 0 & 1 \end{bmatrix}$

   b. $\begin{bmatrix} 2 & 1 & -1 & -2 \\ -1 & 2 & 0 & 2 \\ 0 & 3 & 5 & 3 \\ 1 & 1 & 4 & 1 \end{bmatrix}$

19. Using CAS, solve the following matrix equation to determine the values of a, b, c and d.

$$\begin{bmatrix} 1 & 2 & 3 & 1 \\ 2 & -1 & -1 & 3 \\ 3 & 1 & 1 & -2 \\ -1 & 2 & -4 & -1 \end{bmatrix} \begin{bmatrix} a \\ b \\ c \\ d \end{bmatrix} = \begin{bmatrix} 7 \\ -4 \\ 12 \\ -14 \end{bmatrix}$$

4.5 Exam questions

Question 1 (1 mark)

MC The determinant of $A = \begin{bmatrix} 1 & -3 \\ 2 & 0 \end{bmatrix}$ is

A. -7 B. -6 C. 5 D. 6 E. -5

Question 2 (1 mark)

MC Solve this matrix equation for X.

$$\begin{bmatrix} 2 & 2 \\ 2 & -5 \end{bmatrix} X = \begin{bmatrix} 8 & 6 \\ -13 & -1 \end{bmatrix}$$

A. $\begin{bmatrix} -1 & -2 \\ -3 & -1 \end{bmatrix}$

B. $\begin{bmatrix} 1 & 2 \\ 3 & 1 \end{bmatrix}$

C. $\dfrac{1}{-14} \begin{bmatrix} -14 & -28 \\ -42 & -14 \end{bmatrix}$

D. $\dfrac{1}{-14} \begin{bmatrix} -14 & -28 \\ -42 & -14 \end{bmatrix}$

E. $\dfrac{1}{-14} \begin{bmatrix} -52 & -4 \\ 68 & 24 \end{bmatrix}$

Question 3 (1 mark)

MC To solve the simultaneous equations $3x - y = 0$ and $3y - 6x = 7$ using the matrix equation $AX = B$, the coefficient matrix A is

A. $\begin{bmatrix} 3 & -1 \\ 3 & -6 \end{bmatrix}$

B. $\begin{bmatrix} 3 & 1 \\ -6 & 3 \end{bmatrix}$

C. $\begin{bmatrix} 3 & 1 \\ 3 & 6 \end{bmatrix}$

D. $\begin{bmatrix} 0 \\ 7 \end{bmatrix}$

E. $\begin{bmatrix} 3 & -1 \\ -6 & 3 \end{bmatrix}$

More exam questions are available online.

4.6 Communications and connections

LEARNING INTENTION

At the end of this subtopic you should be able to:
- construct an adjacency matrix
- determine two-step connections.

4.6.1 Adjacency matrices

When matrices are used to analyse a graph, the graph must first be converted into a square matrix. This matrix is called the **adjacency matrix**, and it is used to represent the information in a graph.

The following are the general steps that need to be taken to create an adjacency matrix:

Step 1: Start with an $n \times n$ matrix where n is the number of objects/places.

Step 2: Input the number of connections between objects in the respective element in the adjacency matrix.

Take, for example, the following graph.

$\begin{array}{c} \\ A \\ B \\ C \end{array}\begin{array}{ccc} A & B & C \\ \left[\begin{array}{ccc} - & - & - \\ - & - & - \\ - & - & - \end{array}\right. \end{array}$	First start with a 3×3 matrix, because there are 3 points in the network.
$\begin{array}{c} \\ A \\ B \\ C \end{array}\begin{array}{ccc} A & B & C \\ \left[\begin{array}{ccc} 0 & 2 & 1 \\ - & - & - \\ - & - & - \end{array}\right. \end{array}$	Input the number of connections between objects. A connects to A 0 times. A connects to B 2 times. A connects to C 1 time.
$\begin{array}{c} \\ A \\ B \\ C \end{array}\begin{array}{ccc} A & B & C \\ \left[\begin{array}{ccc} 0 & 2 & 1 \\ 2 & 0 & 0 \\ - & - & - \end{array}\right. \end{array}$	B connects to A 2 times. B connects to B 0 times. B connects to C 0 times.

	A	B	C	
A	0	2	1	C connects to A 1 time.
B	2	0	0	C connects to B 0 times.
C	1	0	0	C connects to C 0 times.
				The adjacency matrix is now complete.

Note: Adjacency matrices for two-way communications diagrams are diagonally symmetrical. In *General Mathematics Units 3 & 4* we will encounter matrices that aren't diagonally symmetrical, due to having one-way communications.

WORKED EXAMPLE 17 Constructing an adjacency matrix

The diagram shows the number of roads connecting between four towns, A, B, C and D.
Construct an adjacency matrix to represent this information.

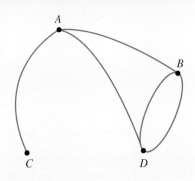

THINK

1. Since there are four connecting towns, a 4×4 adjacency matrix needs to be constructed. Label the rows and columns with the relevant towns A, B, C and D.

2. There is one road connecting town A to town B, so enter 1 in the cell from A to B.

3. There is also only one road between town A and towns C and D; therefore, enter 1 in the appropriate matrix positions. There are no loops at town A (i.e. no road connecting A to A); therefore, enter 0 in this position.

4. Repeat this process for towns B, C and D. Note that there are two roads connecting towns B and D, and that town C only connects to town A

WRITE

1.
$$\begin{array}{c} \\ A \\ B \\ C \\ D \end{array} \begin{array}{c} \begin{array}{cccc} A & B & C & D \end{array} \\ \begin{bmatrix} - & - & - & - \\ - & - & - & - \\ - & - & - & - \\ - & - & - & - \end{bmatrix} \end{array}$$

2.
$$\begin{array}{c} \\ A \\ B \\ C \\ D \end{array} \begin{array}{c} \begin{array}{cccc} A & B & C & D \end{array} \\ \begin{bmatrix} - & ① & - & - \\ - & - & - & - \\ - & - & - & - \\ - & - & - & - \end{bmatrix} \end{array}$$

3.
$$\begin{array}{c} \\ A \\ B \\ C \\ D \end{array} \begin{array}{c} \begin{array}{cccc} A & B & C & D \end{array} \\ \begin{bmatrix} 0 & 1 & 1 & 1 \\ - & - & - & - \\ - & - & - & - \\ - & - & - & - \end{bmatrix} \end{array}$$

4.
$$\begin{array}{c} \\ A \\ B \\ C \\ D \end{array} \begin{array}{c} \begin{array}{cccc} A & B & C & D \end{array} \\ \begin{bmatrix} 0 & 1 & 1 & 1 \\ 1 & 0 & 0 & 2 \\ 1 & 0 & 0 & 0 \\ 1 & 2 & 0 & 0 \end{bmatrix} \end{array}$$

4.6.2 Two-step communications

When you square the adjacency matrix (multiply it by itself), you get the number of two-step communications. This tells us how many times we can communicate with someone through someone else.

Take the following example of a communication diagram and adjacency matrix (L) showing connections between four friends: Edward, Jacob, Bella and Rebecca.

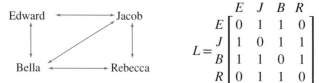

$$L = \begin{array}{c} \\ E \\ J \\ B \\ R \end{array} \begin{array}{cccc} E & J & B & R \\ \left[\begin{array}{cccc} 0 & 1 & 1 & 0 \\ 1 & 0 & 1 & 1 \\ 1 & 1 & 0 & 1 \\ 0 & 1 & 1 & 0 \end{array}\right] \end{array}$$

If we square the adjacency matrix, we will find the number of two-step communications between the four friends.

$$L = \begin{array}{c} \\ E \\ J \\ B \\ R \end{array} \begin{array}{cccc} E & J & B & R \\ \left[\begin{array}{cccc} 0 & 1 & 1 & 0 \\ 1 & 0 & 1 & 1 \\ 1 & 1 & 0 & 1 \\ 0 & 1 & 1 & 0 \end{array}\right] \end{array} \qquad L^2 = \begin{array}{c} \\ E \\ J \\ B \\ R \end{array} \begin{array}{cccc} E & J & B & R \\ \left[\begin{array}{cccc} 2 & 1 & 1 & 2 \\ 1 & 3 & 2 & 1 \\ 1 & 2 & 3 & 1 \\ 2 & 1 & 1 & 2 \end{array}\right] \end{array}$$

We can see from element l_{32} that there is a single one-step communication between Bella and Jacob.

From L^2, in the same position, we can see that there are two two-step communications between Bella and Jacob. They could communicate via Edward or Rebecca.

Bella → Edward → Jacob

Bella → Rebecca → Jacob

There are two ways Bella can give Jacob a message through someone else. If Bella and Jacob were unable to see or contact each other directly, they could still pass each other a message through Edward or Rebecca.

WORKED EXAMPLE 18 Determining two-step connections

The following adjacency matrix shows four locations and the number of connecting routes they have. The locations are Tokyo (T), France (F), Dubai (D) and Australia (A).

$$\begin{array}{c} \\ T \\ F \\ A \\ D \end{array} \begin{array}{cccc} T & F & A & D \\ \left[\begin{array}{cccc} 0 & 1 & 1 & 1 \\ 1 & 0 & 1 & 1 \\ 1 & 1 & 0 & 2 \\ 1 & 1 & 2 & 0 \end{array}\right] \end{array}$$

Using CAS, determine in how many ways a tourist can travel from Tokyo to Australia via one of the other two destinations.

THINK	WRITE
1. Use CAS to calculate the square of the adjacency matrix and evaluate the number of two-step connections.	$\begin{bmatrix} 0 & 1 & 1 & 1 \\ 1 & 0 & 1 & 1 \\ 1 & 1 & 0 & 2 \\ 1 & 1 & 2 & 0 \end{bmatrix}^2 = \begin{bmatrix} 3 & 2 & 3 & 3 \\ 2 & 3 & 3 & 3 \\ 3 & 3 & 6 & 2 \\ 3 & 3 & 2 & 6 \end{bmatrix}$

2. Interpret the information in the matrix and answer the question by locating the required value.

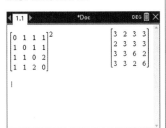

$$\begin{bmatrix} 3 & 2 & 3 & 3 \\ 2 & 3 & 3 & 3 \\ 3 & 3 & 6 & 2 \\ 3 & 3 & 2 & 6 \end{bmatrix}$$

There are three ways in which a tourist can travel from Australia to Tokyo while going through another destination.

TI	THINK	DISPLAY/WRITE

1. On a Calculator page, complete the entry line as:
$$\begin{bmatrix} 0 & 1 & 1 & 1 \\ 1 & 0 & 1 & 1 \\ 1 & 1 & 0 & 2 \\ 1 & 1 & 2 & 0 \end{bmatrix}^2$$
then press ENTER.
Note: The matrix templates can be found by pressing the matrix button and then selecting one of the matrices

  .

CASIO	THINK	DISPLAY/WRITE

1. On the Main screen, complete the entry line as:
$$\begin{bmatrix} 0 & 1 & 1 & 1 \\ 1 & 0 & 1 & 1 \\ 1 & 1 & 0 & 2 \\ 1 & 1 & 2 & 0 \end{bmatrix}^2$$
then press EXE.
Note: The matrix templates can be found in the Math 2 tab in the Keyboard menu. Clicking on the matrix icons multiple times will increase the order of the matrix.

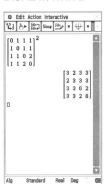

4.6 Exercise

1. **WE17** Construct adjacency matrices for the following graphs.

a.   b.   c.   d.

2. Draw networks for the following adjacency matrices.

a.
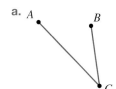
$$\begin{array}{c} \\ A \\ B \\ C \end{array} \begin{array}{ccc} A & B & C \\ \begin{bmatrix} 0 & 1 & 2 \\ 1 & 0 & 0 \\ 2 & 0 & 0 \end{bmatrix} \end{array}$$

b.
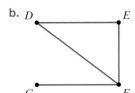
$$\begin{array}{c} \\ D \\ E \\ F \end{array} \begin{array}{ccc} D & E & F \\ \begin{bmatrix} 0 & 1 & 1 \\ 1 & 0 & 1 \\ 1 & 1 & 1 \end{bmatrix} \end{array}$$

c.
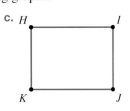
$$\begin{array}{c} \\ G \\ H \\ I \\ J \end{array} \begin{array}{cccc} G & H & I & J \\ \begin{bmatrix} 0 & 1 & 2 & 0 \\ 1 & 1 & 1 & 0 \\ 2 & 1 & 0 & 1 \\ 0 & 0 & 1 & 0 \end{bmatrix} \end{array}$$

d.
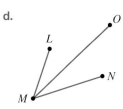
$$\begin{array}{c} \\ K \\ L \\ M \\ N \end{array} \begin{array}{cccc} K & L & M & N \\ \begin{bmatrix} 0 & 1 & 1 & 1 \\ 1 & 1 & 1 & 2 \\ 1 & 1 & 0 & 1 \\ 1 & 2 & 1 & 0 \end{bmatrix} \end{array}$$

3. Construct an adjacency matrix and determine the number of two-step communications between Karina and Winter. Write what the connections are.

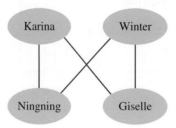

4. Five friends, Amira, Brian, Catherine, Dev and Emma, just had a disagreement that split the group up. The following network shows the communications in the group.

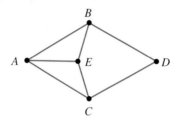

Determine in how many ways Amira can get a message to Dev via one other person (i.e. by asking someone to pass a message to Dev).

5. The diagram shows the network cable between five main computers (A, B, C, D and E) in an office building.

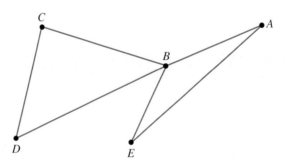

Construct an adjacency matrix to represent this information.

6. Peta, Seth, Tran, Ned and Wen chat on a social media site. The number of communications made between these friends in the last 24 hours is shown in the adjacency matrix below.

$$
\begin{array}{c@{}c}
 & \begin{array}{ccccc} P & S & T & N & W \end{array} \\
\begin{array}{c} P \\ S \\ T \\ N \\ W \end{array} &
\left[\begin{array}{ccccc}
0 & 1 & 3 & 1 & 0 \\
1 & 0 & 0 & 0 & 4 \\
3 & 0 & 0 & 2 & 1 \\
1 & 0 & 2 & 0 & 0 \\
0 & 4 & 1 & 0 & 0
\end{array}\right]
\end{array}
$$

a. Determine the number of times Peta and Tran communicated over the last 24 hours.
b. Determine whether Seth communicated with Ned at any time during the last 24 hours.
c. In the context of this problem, explain the existence of the zeros along the diagonal.
d. Using the adjacency matrix, construct a diagram that shows the number of communications between the five friends.

7. Airlink flies charter flights in the Cape Lancaster region. The direct flights between Williamton, Cowal, Hugh River, Kokialah and Archer are shown in the diagram.

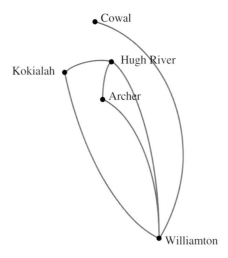

a. Using the diagram, construct an adjacency matrix that shows the number of direct flights between the five towns.
b. Determine the number of ways a person can travel between Williamton and Kokialah via another town.
c. Determine if it is possible to fly between Cowal and Archer and stop over at two other towns. Justify your answer.

8. **WE18** The following adjacency matrix shows the number of roads between three country towns: Glenorchy (G), St Arnaud (S) and Campbells Bridge (C).

$$\begin{array}{c c c c} & G & S & C \\ \begin{matrix} G \\ S \\ C \end{matrix} & \begin{bmatrix} 0 & 1 & 1 \\ 1 & 0 & 2 \\ 1 & 2 & 0 \end{bmatrix} \end{array}$$

Using CAS, determine the number of ways a person can travel from Glenorchy to St Arnaud via Campbells Bridge.

9. The direct Cape Air flights between five cities, Boston (B), Hyannis (H), Martha's Vineyard (M), Nantucket (N) and Providence (P), are shown in the following adjacency matrix.

$$\begin{array}{c c c c c c} & B & H & M & N & P \\ \begin{matrix} B \\ H \\ M \\ N \\ P \end{matrix} & \begin{bmatrix} 0 & 1 & 1 & 1 & 0 \\ 1 & 0 & 1 & 1 & 0 \\ 1 & 1 & 0 & 1 & 1 \\ 1 & 1 & 1 & 0 & 1 \\ 0 & 0 & 1 & 1 & 0 \end{bmatrix} \end{array}$$

a. Construct a diagram to represent the direct flights between the five cities.
b. Construct a matrix that determines the number of ways a person can fly between two cities via another city.
c. Explain how you would determine the number of ways a person can fly between two cities via two other cities.
d. Determine if it is possible to fly from Boston and stop at every other city. Explain how you would answer this question.

10. The adjacency matrix below shows the number of text messages sent between three friends, Stacey (*S*), Ruth (*R*) and Toiya (*T*), immediately after school one day.

$$\begin{array}{c} & \begin{array}{ccc} S & R & T \end{array} \\ \begin{array}{c} S \\ R \\ T \end{array} & \left[\begin{array}{ccc} 0 & 3 & 2 \\ 3 & 0 & 1 \\ 2 & 1 & 0 \end{array} \right] \end{array}$$

a. State the number of text messages sent between Stacey and Ruth.
b. Determine the total number of text messages sent between all three friends.

4.6 Exam questions

Question 1 (1 mark)

Source: VCE 2019, Further Mathematics Exam 1, Section B, Module 1, Q7; © VCAA.

MC The communication matrix below shows the direct paths by which messages can be sent between two people in a group of six people, *U* to *Z*.

$$\begin{array}{c} & \begin{array}{cccccc} U & V & W & X & Y & Z \end{array} \\ \begin{array}{c} U \\ V \\ W \\ X \\ Y \\ Z \end{array} & \left[\begin{array}{cccccc} 0 & 1 & 1 & 0 & 1 & 1 \\ 1 & 0 & 1 & 0 & 1 & 0 \\ 1 & 1 & 0 & 1 & 0 & 1 \\ 0 & 1 & 0 & 0 & 1 & 1 \\ 0 & 0 & 1 & 1 & 0 & 1 \\ 1 & 1 & 0 & 1 & 1 & 0 \end{array} \right] \end{array}$$

A '1' in the matrix shows that the person named in that row can send a message directly to the person named in that column.

For example, the '1' in row 4, column 2 shows that *X* can send a message directly to *V*.

In how many ways can *Y* get a message to *W* by sending it directly to one other person?
A. 0   **B.** 1   **C.** 2   **D.** 3   **E.** 4

Question 2 (1 mark)

MC Select the network from the following that can be described by the adjacency matrix.

$$\begin{array}{c} & \begin{array}{ccc} A & B & C \end{array} \\ \begin{array}{c} A \\ B \\ C \end{array} & \left[\begin{array}{ccc} 0 & 1 & 1 \\ 1 & 0 & 1 \\ 1 & 1 & 0 \end{array} \right] \end{array}$$

A. 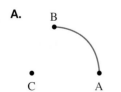   **B.** 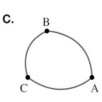   **C.** 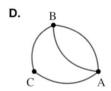   **D.** 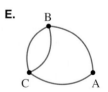   **E.**

Question 3 (1 mark)

Fill in the matrix below, which describes the numbers of roads out of each of towns A, B, C and D, as represented by the network.

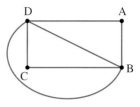

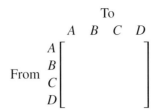

More exam questions are available online.

4.7 Transition matrices

LEARNING INTENTION

At the end of this subtopic you should be able to:
- identify the properties of transition matrices (assuming the Markov condition)
- draw a transition table and transition matrix
- draw a transition diagram
- determine states and the steady-state matrix.

4.7.1 Introduction to transition matrices and transition tables

A **transition matrix** is a square matrix that describes the way that transitions are made between two *states*.

This can be illustrated via the example of two countries, country A and country B, where it's predicted:

70% of the population from country A will migrate to country B the next year, and 30% will stay.

20% of the population from country B will migrate to country A, and 80% will stay.

We can lay this information out in a **transition table**.

		From	
		A	**B**
To	**A**	0.3	0.2
	B	0.7	0.8

Note: The columns in the transition table always add up to 1.

The following transition matrix can then be created from the transition table.

$$\begin{bmatrix} 0.3 & 0.2 \\ 0.7 & 0.8 \end{bmatrix}$$

tlvd-3713

WORKED EXAMPLE 19 Drawing a transition table and transition matrix

Students have the choice to study Maths or English in their study periods.
70% of students who are studying Maths in one study period will also study Maths in the next study period.
40% of students who are studying English in one study period will also study English in their next study period.

Draw a transition table to represent the scenario and then determine the transition matrix.

THINK

WRITE

1. First, we'll draw up a table with 'From' on the vertical axis and 'To' on the horizontal axis.

		From	
		English	**Maths**
To	**English**		
	Maths		

2. 70% of students studying Maths will study Maths in the next period.
So, we put 0.7 (70% in decimal form) into the cell intersection of 'From Maths' and 'To Maths'.

		From	
		English	**Maths**
To	**English**		
	Maths		0.7

3. If 70% of students studying Maths continue with it in the next period, what happens to the rest? Because there are two options here, we can conclude that the remainder of students, 30%, go to study English in the next study period.

		From	
		English	**Maths**
To	**English**		0.3
	Maths		0.7

4. 40% of students studying English will study English in their next study period, so we put 0.4 in the corresponding spot in the table.

		From	
		English	**Maths**
To	**English**	0.4	0.3
	Maths		0.7

5. Using what we know, we can say that the remainder of students studying English, 60%, will go on to study Maths in the next study period.

		From	
		English	**Maths**
To	**English**	0.4	0.3
	Maths	0.6	0.7

6. Extract the numbers from the transition table to get the transition matrix.

$$\begin{bmatrix} 0.4 & 0.3 \\ 0.6 & 0.7 \end{bmatrix}$$

Note: When we calculate with transition matrices, we assume something called the **Markov condition**. The Markov condition basically means that any decisions made in the past *will not* affect decisions to be made in the present or future.

If we didn't assume the Markov condition, we'd have to calculate with a different transition matrix for every iteration.

4.7.2 Transition diagrams

In addition to transition tables and matrices, we can draw diagrams to visually see the transitions that are made between the objects.

We can draw a diagram to represent the above migration situation between countries A and B.

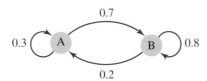

In the transition diagram, we use arrows to represent the transitions to and from each country.

WORKED EXAMPLE 20 Drawing a transition diagram

Consider the situation from Worked example 19, where 70% of students studying Maths in one period will go on to do so in the next, and 40% of students studying English in one period will also do so in the next.

Draw a transition diagram to represent the situation.

THINK	WRITE
1. 70% of students who study Maths will study Maths in their next study period, so we put 0.7 in the arrow that goes from Maths and leads back into Maths.	
2. If 70% of students studying Maths continue studying Maths, that means that the remaining 30% will go on to study English. We put in a 0.3 for the arrow leading from Maths to English.	
3. Next, we look at the second statement. 40% of students studying English will study English in the next period. We put an arrow with 0.4 leading from English back to English.	
4. Lastly, we think about the remainder of the students who studied English. If 40% stayed, then 60% left to study Maths.	

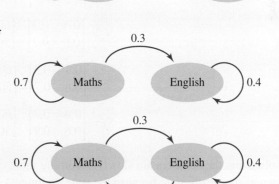

4.7.3 Determining states and the steady-state matrix

When working with transition diagrams, we have to be aware of the different states of the diagram. The **initial state** is the state of things at the beginning. To find the next state, we multiply the transition matrix and the **initial state matrix** together. This is because the initial state (what we start with) is being *changed* by the transition matrix.

Determining the nth state

To determine the nth state, apply the following formula:

$$S_n = T^n S_0$$

where:
S_n is the nth term
T^n is the transition matrix raised to the power of n
S_0 is the initial state matrix.

Following on from Worked example 20, if we started with 500 students studying Maths and 600 students studying English, our initial state matrix would be:

$$S_0 = \begin{matrix} \text{English} \\ \text{Maths} \end{matrix} \begin{bmatrix} 600 \\ 500 \end{bmatrix}$$

Note: The positions of the number of Maths students and the number of English students should be consistent with the transition table and matrix.

Steady-state matrix

We find that, if we go through multiple states, the student numbers will reach an **equilibrium** or **steady state**. This is where the number of students studying English or Maths remains the same, as the number of students transitioning *away* is the same as the number of students transitioning *into* each subject.

Let's see what happens to the student numbers as we progress through multiple states of transition:

$$S_1 = T S_0$$

$$S_1 = \begin{bmatrix} 0.4 & 0.3 \\ 0.6 & 0.7 \end{bmatrix} \begin{bmatrix} 600 \\ 500 \end{bmatrix} = \begin{bmatrix} 390 \\ 710 \end{bmatrix}$$

$$S_2 = \begin{bmatrix} 0.4 & 0.3 \\ 0.6 & 0.7 \end{bmatrix}^2 \begin{bmatrix} 600 \\ 500 \end{bmatrix} = \begin{bmatrix} 369 \\ 731 \end{bmatrix}$$

$$S_3 = \begin{bmatrix} 0.4 & 0.3 \\ 0.6 & 0.7 \end{bmatrix}^3 \begin{bmatrix} 600 \\ 500 \end{bmatrix} = \begin{bmatrix} 366.9 \\ 733.1 \end{bmatrix}$$

$$S_4 = \begin{bmatrix} 0.4 & 0.3 \\ 0.6 & 0.7 \end{bmatrix}^4 \begin{bmatrix} 600 \\ 500 \end{bmatrix} = \begin{bmatrix} 366.69 \\ 733.31 \end{bmatrix}$$

$$S_5 = \begin{bmatrix} 0.4 & 0.3 \\ 0.6 & 0.7 \end{bmatrix}^5 \begin{bmatrix} 600 \\ 500 \end{bmatrix} = \begin{bmatrix} 366.669 \\ 733.331 \end{bmatrix}$$

You may notice that the numbers are slowly becoming more consistent as they reach the equilibrium point. This is also known as the steady state because the state no longer changes.

Estimating the steady state

The steady-state matrix, S, is given by:

$$S = T^n S_0$$

as 1 approaches infinity.

Note: **This means that you can get a fairly accurate steady-state solution by evaluating this formula with a large value of 1. A value of $n = 50$ will generally get a good approximation of the steady state.**

WORKED EXAMPLE 21 Calculating a future state using the transition matrix

Scientists are observing the weather patterns in a desert.
They conclude that when it rains one day, there is an
80% chance that the weather will be hot the next day.
They also observe that there is a 5% chance of having
rain on any hot day.
If today is rainy, calculate the chance that it will rain
again in two days' time.

THINK

WRITE

1. First, we need to set up the transition matrix.

		From	
		Hot	**Rain**
To	**Hot**	0.95	0.8
	Rain	0.05	0.2

$$T = \begin{bmatrix} 0.95 & 0.8 \\ 0.05 & 0.2 \end{bmatrix}$$

2. Identify the elements of the initial state matrix.

$$S_0 = \begin{bmatrix} 0 \\ 1 \end{bmatrix}$$

3. Calculate the second state.

$$S_2 = T^2 S_0$$

$$S_2 = \begin{bmatrix} 0.95 & 0.8 \\ 0.05 & 0.2 \end{bmatrix}^2 \begin{bmatrix} 0 \\ 1 \end{bmatrix}$$

$$= \begin{bmatrix} 0.92 \\ 0.08 \end{bmatrix}$$

4. Interpret and write the answer.

There is an 8% chance of rain in two days' time.

WORKED EXAMPLE 22 Determining the steady state

At the end of every month, a logistics company moves 30% of their trucks from depot G to depot H,
and 60% of their trucks from depot H to depot G. In the first week, the company started with 120
trucks in depot G and 90 trucks in depot H.
In the long term, determine how many trucks will be located at each depot.

THINK

WRITE

1. Create a transition table.
   Remember that columns must add to 1 .

		From	
		H	**G**
To	**H**	0.4	0.3
	G	0.6	0.7

▶

2. Generate the transition matrix.

$$\begin{bmatrix} 0.4 & 0.3 \\ 0.6 & 0.7 \end{bmatrix}$$

3. Apply the equation for the steady-state solution, using a large value for n (50).

$$S = T^{50}S_0$$

$$= \begin{bmatrix} 0.4 & 0.3 \\ 0.6 & 0.7 \end{bmatrix}^{50} \begin{bmatrix} 90 \\ 120 \end{bmatrix}$$

$$= \begin{bmatrix} 70 \\ 140 \end{bmatrix}$$

4. Write the answer.

In the long term, there will be 70 trucks at depot H and 140 trucks at depot G.

4.7 Exercise

Students, these questions are even better in jacPLUS

Receive immediate feedback and access sample responses

Access additional questions

Track your results and progress

Find all this and MORE in jacPLUS

1. **WE19** Create a transition table and transition matrix to describe the following:
   - The probability of rain after a rainy day is 30%.
   - The probability of rain after a sunny day is 20%.

2. Create a transition matrix and describe in words the following transition table showing the movement of trains between two train depots, East and West.

		From	
		East	**West**
To	**East**	0.25	0.4
	West	0.75	0.6

3. **WE20** Each year, 20% of the residents of country A migrate to country B, 40% of the residents of country B stay in country B and the rest opt to migrate to country A. Draw a transition diagram to represent the migration between the two countries.

4. For the following diagram showing the movement of people between Melbourne and Canberra each year, describe the transitions being made.

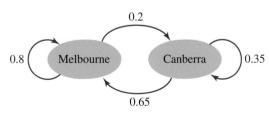

5. Create a transition matrix for the following transition diagram about transitions between Cooktown and Dieterville.

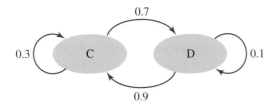

6. **WE21** Town J experiences a loss of 25% of its residents every year to Town K, while 20% of Town K residents move to Town J each year. If Town J started with 2000 residents and Town K started with 5000 residents, determine the number of residents there would be in each town after two years.

7. For the given transition matrix and initial state, calculate the fifth state.

$$T = \begin{bmatrix} 0.6 & 0.55 \\ 0.4 & 0.45 \end{bmatrix} \quad S_0 = \begin{bmatrix} 400 \\ 300 \end{bmatrix}$$

8. In a new experimental school, students are given a choice of which class they want to be in each day between class X and class Y. 85% of students from class X stay in class X. 55% of students in class Y move to class X the next day. If class X starts with 16 students and class Y starts with 28, determine the steady state of both classes.

9. A flock of birds migrates from Australia to Antarctica every year. At the same time, a group of the same species migrates from Antarctica to Australia each year. Use the table and initial state matrix to determine the steady state of the bird populations.

		From	
		Australia	**Antarctica**
To	**Australia**	0.35	0.2
	Antarctica	0.65	0.8

$$S_0 = \begin{bmatrix} 800 \\ 300 \end{bmatrix}$$

10. Two supermarkets, Coldes and Sheepies, are competing heavily with one another. A survey was conducted that showed the following:
70% of customers who shopped at Coldes returned to shop there the following week.
65% of customers who shopped at Sheepies returned to shop there the following week.
Assume that customers will shop at either Coldes or Sheepies.

   a. Construct the transition matrix for this situation.
   b. If Coldes started off with 1000 customers and Sheepies started off with 3000, determine the number of customers at Sheepies in the third week.
   c. Using the number of customers found in part b, determine the number of customers shopping at each supermarket at the steady state.

Question 1 (1 mark)

Source: VCE 2019, Further Mathematics Exam 1, Section B, Module 1, Q8; © VCAA.

MC An airline parks all of its planes at Sydney airport or Melbourne airport overnight.

The transition diagram below shows the change in the location of the planes from night to night.

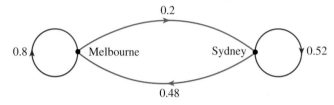

There are always m planes parked at Melbourne airport. There are always s planes parked at Sydney airport.

Of the planes parked at Melbourne airport on Tuesday night, 12 had been parked at Sydney airport on Monday night. How many planes does the airline have?

   A. 25   **B.** 37   **C.** 62   **D.** 65   **E.** 85

Question 2 (1 mark)

Source: VCE 2019, Further Mathematics Exam 2, Section B, Module 1, Q2a; © VCAA.

The theme park has four locations, Air World (A), Food World (F), Ground World (G) and Water World (W).

The number of visitors at each of the four locations is counted every hour.

By 10 am on Saturday the park had reached its capacity of 2000 visitors and could take no more visitors. The park stayed at capacity until the end of the day.

The state matrix, S_0, below, shows the number of visitors at each location at 10 am on Saturday.
$$S_0 = \begin{bmatrix} 600 \\ 600 \\ 400 \\ 400 \end{bmatrix} \begin{matrix} A \\ F \\ G \\ W \end{matrix}$$

What percentage of the park's visitors were at Water World (W) at 10 am on Saturday?

Question 3 (1 mark)

Source: VCE 2019, Further Mathematics Exam 2, Section B, Module 1, Q2b; © VCAA.

Let S_n be the state matrix that shows the number of visitors expected at each location n hours after 10 am on Saturday.

The number of visitors expected at each location n hours after 10 am on Saturday can be determined by the matrix recurrence relation below.

$$S_0 = \begin{bmatrix} 600 \\ 600 \\ 400 \\ 400 \end{bmatrix}, \quad S_{n+1} = T \times S_n \qquad \text{where} \quad T = \begin{bmatrix} 0.1 & 0.2 & 0.1 & 0.2 \\ 0.3 & 0.4 & 0.6 & 0.3 \\ 0.1 & 0.2 & 0.2 & 0.1 \\ 0.5 & 0.2 & 0.1 & 0.4 \end{bmatrix} \begin{matrix} A \\ F \\ G \\ W \end{matrix}$$

$$\begin{matrix} & this\ hour \\ & \begin{matrix} A & F & G & W \end{matrix} \end{matrix}$$

next hour

Complete the state matrix, S_1, below to show the number of visitors expected at each location at 11 am on Saturday.

$$S_1 = \begin{bmatrix} — \\ — \\ 300 \\ — \end{bmatrix} \begin{matrix} A \\ F \\ G \\ W \end{matrix}$$

More exam questions are available online.

4.8 Review

4.8.1 Summary

4.8 Exercise

Multiple choice

1. **MC** Consider the matrix equation $\begin{bmatrix} 6 & a \\ 3 & 2 \end{bmatrix} - \begin{bmatrix} 2 & -1 \\ 0 & b \end{bmatrix} = \begin{bmatrix} 4 & 2 \\ 3 & 1 \end{bmatrix}$.

   The values of a and b respectively are:

   A. $a = 1$, $b = 1$ **B.** $a = 1$, $b = -1$ **C.** $a = 2$, $b = 1$

   D. $a = 2$, $b = -1$ **E.** $a = 3$, $b = 1$

2. **MC** The order of the matrix $\begin{bmatrix} 6 & 1 & 2 \\ 7 & 3 & 5 \\ 9 & 5 & 0 \\ 0 & 7 & 2 \end{bmatrix}$ is:

   A. 3×4 **B.** 4×2 **C.** 4×3 **D.** 8   **E.** 12

3. **MC** Select the determinant of $\begin{bmatrix} 5 & 4 \\ 2 & 7 \end{bmatrix}$.

   A. 25   **B.** 26   **C.** 27   **D.** 30   **E.** 31

4. **MC** Select the number of ways that person A can communicate with person D via one other person.

$$\begin{array}{c} \\ A \\ B \\ C \\ D \end{array} \begin{array}{c} \begin{array}{cccc} A & B & C & D \end{array} \\ \begin{bmatrix} 0 & 1 & 1 & 1 \\ 1 & 0 & 0 & 1 \\ 1 & 0 & 0 & 1 \\ 1 & 1 & 1 & 0 \end{bmatrix} \end{array}$$

   A. 0   **B.** 1   **C.** 2   **D.** 3   **E.** 4

5. **MC** Sarah purchased 2 apples and 3 bananas from the local market for \$3.80. Later that week Sarah went back to the market and purchased 4 more apples and 4 more bananas for \$6.20.

   A matrix equation that could be set up to determine the price of a single apple and banana is:

   A. $\begin{bmatrix} 2 & 3 \\ 4 & 4 \end{bmatrix} \begin{bmatrix} x \\ y \end{bmatrix} = \begin{bmatrix} 3.8 \\ 6.2 \end{bmatrix}$ **B.** $\begin{bmatrix} x \\ y \end{bmatrix} \begin{bmatrix} 2 & 3 \\ 4 & 4 \end{bmatrix} = \begin{bmatrix} 3.8 \\ 6.2 \end{bmatrix}$ **C.** $\begin{bmatrix} 2 & 4 \\ 3 & 4 \end{bmatrix} \begin{bmatrix} x \\ y \end{bmatrix} = \begin{bmatrix} 3.8 \\ 6.2 \end{bmatrix}$

   D. $\begin{bmatrix} 2 & 3 \\ 4 & 4 \end{bmatrix} \begin{bmatrix} 3.8 \\ 6.2 \end{bmatrix} = \begin{bmatrix} x \\ y \end{bmatrix}$ **E.** $\begin{bmatrix} 3.8 \\ 6.2 \end{bmatrix} \begin{bmatrix} 2 & 3 \\ 4 & 4 \end{bmatrix} = \begin{bmatrix} x \\ y \end{bmatrix}$

6. MC Select the matrix that represents the product $\begin{bmatrix} 5 & 1 & 2 \\ 3 & 7 & 8 \end{bmatrix} \begin{bmatrix} 12 \\ 10 \\ 9 \end{bmatrix}$.

A. $\begin{bmatrix} 88 \\ 178 \end{bmatrix}$

B. $\begin{bmatrix} 60 & 10 & 18 \\ 36 & 63 & 72 \end{bmatrix}$

C. $\begin{bmatrix} 88 & 178 \end{bmatrix}$

D. $\begin{bmatrix} 48 \\ 49 \end{bmatrix}$

E. It does not exist.

7. MC The 2010 population age structure (in percentages) of selected countries is shown in the following table.

Country	Age groups by percentage of population		
	0–14 years	**15–64 years**	**Over 65 years**
Australia	18.9	67.6	13.5
China	16.4	69.5	14.1
Indonesia	27.0	67.4	5.6

A 3×1 matrix that could be used to represent the percentage of population in Indonesia across the three age groups is:

A. $\begin{bmatrix} 18.9 \\ 16.4 \\ 27.0 \end{bmatrix}$

B. $\begin{bmatrix} 27.0 & 67.4 & 5.6 \end{bmatrix}$

C. $\begin{bmatrix} 18.9 & 67.6 & 13.5 \\ 16.4 & 69.5 & 14.1 \\ 27.0 & 67.4 & 5.6 \end{bmatrix}$

D. $\begin{bmatrix} 16.4 \\ 69.5 \\ 14.1 \end{bmatrix}$

E. $\begin{bmatrix} 27.0 \\ 67.4 \\ 5.6 \end{bmatrix}$

8. MC Matrix A has an order of 3×2. Matrix B has an order of 1×3. Matrix C has an order of 2×1. Select which one of the following matrix multiplications is not possible.

A. AC **B.** BA **C.** BC **D.** CB **E.** ACB

9. MC If matrix $A = \begin{bmatrix} 16 & 15 \\ 10 & 9 \end{bmatrix}$, then A^{-1} is:

A. $\begin{bmatrix} 1 & 0 \\ 0 & 1 \end{bmatrix}$

B. $\begin{bmatrix} -16 & 10 \\ 15 & -9 \end{bmatrix}$

C. $\begin{bmatrix} 9 & -15 \\ -10 & 10 \end{bmatrix}$

D. $\begin{bmatrix} -54 & 90 \\ 60 & -60 \end{bmatrix}$

E. $\begin{bmatrix} \dfrac{-3}{2} & \dfrac{5}{2} \\ \dfrac{5}{3} & \dfrac{-8}{3} \end{bmatrix}$

10. MC Select the matrix that is the next state of the matrices shown.

$$T = \begin{bmatrix} 0.4 & 0.2 \\ 0.6 & 0.8 \end{bmatrix} \quad S_0 = \begin{bmatrix} 200 \\ 400 \end{bmatrix}$$

A. $\begin{bmatrix} 400 \\ 200 \end{bmatrix}$

B. $\begin{bmatrix} 160 \\ 440 \end{bmatrix}$

C. $\begin{bmatrix} 120 \\ 480 \end{bmatrix}$

D. $\begin{bmatrix} 440 \\ 160 \end{bmatrix}$

E. $\begin{bmatrix} 480 \\ 120 \end{bmatrix}$

Short answer

11. Given matrices $A = \begin{bmatrix} 2 & 3 \\ 1 & 4 \end{bmatrix}$, $B = \begin{bmatrix} 1 \\ 0 \end{bmatrix}$ and $C = \begin{bmatrix} 3 \\ 2 \end{bmatrix}$, evaluate the following.

a. $C + B$ **b.** $B - 2C$ **c.** $AB + C$ **d.** $1.5A$

12. Matrix D has an order of 3×2, matrix E has an order of $1 \times p$ and matrix F has an order of 2×2.

a. Calculate the value of p for which the product matrix ED exists.
b. If the product matrix H exists and $H = EDF$, state the order of H.

13. The table below shows the three different ticket prices, in dollars, and the number of tickets sold for a school concert.

Ticket type	Ticket price	Number of tickets sold
Adult	$12.50	140
Child/student	$6.00	225
Teacher	$10.00	90

a. Construct a column matrix to represent the ticket prices for adults, children/students and teachers respectively.
b. Perform a matrix multiplication to determine the total amount of ticket sales in dollars.

14. a. For each of the following pairs of matrices, state the order of the matrices, and hence state the order of the product matrix.

i. $\begin{bmatrix} 3 & 5 \end{bmatrix} \begin{bmatrix} 2 \\ 6 \end{bmatrix}$ ii. $\begin{bmatrix} 4 & 6 \end{bmatrix} \begin{bmatrix} 2 & 3 \\ 4 & 7 \end{bmatrix}$ iii. $\begin{bmatrix} -1 & 9 \\ 10 & 5 \end{bmatrix} \begin{bmatrix} 3 & -2 \\ 5 & 11 \end{bmatrix}$ iv. $\begin{bmatrix} 2 \\ 7 \\ 8 \end{bmatrix} \begin{bmatrix} 5 & 3 & 4 \end{bmatrix}$

b. Calculate the product matrices of the matrix multiplications given in part a.

15. a. Calculate the determinant of each of the following matrices.

i. $A = \begin{bmatrix} 6 & 3 \\ 5 & 3 \end{bmatrix}$ ii. $B = \begin{bmatrix} -2 & 2 \\ -1 & 3 \end{bmatrix}$

b. Calculate the inverse of each matrix given in part a.

c. Matrix $M = \begin{bmatrix} 1 \\ 6 \\ 9 \end{bmatrix}$. Explain why M^{-1} does not exist.

16. Solve the following simultaneous equations using matrix methods.

a. $2x + 3y = 1$ and $4x - y = 9$ b. $3x - 2y = 12$ and $x + 2y = 19$

Extended response

17. The energy content and amounts of fat and protein contained in each slice of bread and cheese and one teaspoon of margarine is shown in the table below.

Food	Energy content (kilojoules)	Fat (grams)	Protein (grams)
Bread	410	0.95	3.7
Cheese	292	5.5	1.6
Margarine	120	3.3	0.5

Pedro made toasted cheese sandwiches for himself and his friends for lunch. The total amounts of fat and protein (in grams) for each of the three foods — bread, cheese and margarine — in the prepared lunch were recorded in the following matrix.

$$\begin{bmatrix} 7.6 & 29.6 \\ 44.0 & x \\ 13.2 & 2 \end{bmatrix}$$

a. Calculate how many bread slices Pedro used.
b. If each sandwich used two pieces of bread, determine how many cheese sandwiches Pedro made.
c. Show that each sandwich had two slices of cheese.
d. Hence, determine the exact value of x.
e. Construct a 1×3 matrix to represent the number of slices of bread and cheese and teaspoons of margarine for each sandwich.

18. Tootin' Travel Agents sell three different types of train travel packages on the Midnight Express: Platinum, Gold and Red. The price for each travel package is shown in the table.

Class	Price
Platinum	$3890
Gold	$2178
Red	$868

a. Construct a column matrix, C, to represent the price of each of the three travel packages. State the order of C.

b. In the last month, Tootin' Travel Agents sold the following train travel packages:
   - Platinum: 62
   - Gold: 125
   - Red: 270.
   Construct a row matrix, P, to represent the number of train travel packages sold over the last month.

c. To determine the total amount in dollars for train travel packages in the month, product matrix PC is found. State the order of product matrix PC.

Travellers who book a year in advance receive a 5% discount. To calculate the discounted price, matrix C is multiplied by a scalar product, d.

d. Write down the value of d.

e. Using your value for d, construct a new matrix, E, that represents the discounted travel prices. Write your answer correct to the nearest cent.

19. TruSport owns two stores, LeisureLand and SportLand. The number of tennis racquets, baseball bats and soccer balls sold in the last week at the two stores is shown in the table below.

Store	Tennis rackets	Baseball bats	Soccer balls
LeisureLand	10	8	9
SportLand	9	12	11

The selling price of each item is shown in the table below.

	Tennis racket	Baseball bat	Soccer ball
Selling price	$45.95	$25.50	$18.60

a. Construct a 3×2 matrix to represent the number of tennis rackets, baseball bats and soccer balls sold at each of the two stores.

b. Construct a row matrix to represent the selling prices of the items.

c. i. Set up a matrix multiplication that finds the total amount, in dollars, that each store made in the last week.

   ii. Hence, find the total amount, in dollars, that each store made in the last week.

20. Two major newspapers, *The Old Times* and *Herald Moon*, compete against each other. It was found that 60% of the *Herald Moon* readers would go on to buy *The Old Times* the following week. 80% of *The Old Times* readers would continue to buy *The Old Times* the next week.

a. Create a transition table to represent the situation.
b. Create a transition matrix for the situation.
c. Determine the number of people who will be buying the *Herald Moon* after 3 weeks if initially 5000 people buy the *Herald Moon* and 1000 buy *The Old Times*.
d. Determine the number of people who will purchase each newspaper at the steady state.

4.8 Exam questions

Question 1 (1 mark)
Source: VCE 2020, Further Mathematics Exam 1, Section B, Module 1, Q6; © VCAA.

MC The element in row i and column j of matrix M is m_{ij}.

M is a 3×3 matrix. It is constructed using the rule $m_{ij} = 3i + 2j$.

M is

A. $\begin{bmatrix} 5 & 7 & 9 \\ 7 & 9 & 11 \\ 11 & 13 & 15 \end{bmatrix}$
B. $\begin{bmatrix} 5 & 7 & 9 \\ 8 & 10 & 12 \\ 11 & 13 & 15 \end{bmatrix}$
C. $\begin{bmatrix} 5 & 7 & 10 \\ 8 & 10 & 13 \\ 11 & 13 & 16 \end{bmatrix}$

D. $\begin{bmatrix} 5 & 8 & 11 \\ 7 & 10 & 13 \\ 9 & 12 & 15 \end{bmatrix}$
E. $\begin{bmatrix} 5 & 8 & 11 \\ 8 & 11 & 14 \\ 11 & 14 & 17 \end{bmatrix}$

Question 2 (1 mark)
Source: VCE 2020, Further Mathematics Exam 1, Section B, Module 1, Q5; © VCAA.

MC The diagram below shows the direct communication links that exist between Sam (S), Tai (T), Umi (U) and Vera (V). For example, the arrow from Umi to Vera indicates that Umi can communicate directly with Vera.

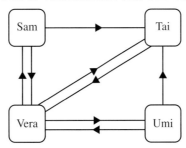

A communication matrix can be used to convey the same information.

In this matrix:
- a '1' indicates that a direct communication link exists between a sender and a receiver
- a '0' indicates that a direct communication link does not exist between a sender and a receiver.

The communication matrix could be

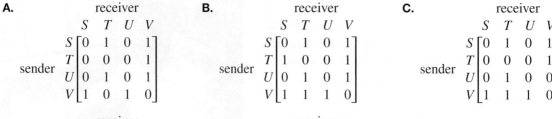

A.

receiver

$$\begin{array}{c} & S \ \ T \ \ U \ \ V \\ \text{sender} \begin{array}{c} S \\ T \\ U \\ V \end{array}\left[\begin{array}{cccc} 0 & 1 & 0 & 1 \\ 0 & 0 & 0 & 1 \\ 0 & 1 & 0 & 1 \\ 1 & 0 & 1 & 0 \end{array}\right]\end{array}$$

B.

receiver

$$\begin{array}{c} & S \ \ T \ \ U \ \ V \\ \text{sender} \begin{array}{c} S \\ T \\ U \\ V \end{array}\left[\begin{array}{cccc} 0 & 1 & 0 & 1 \\ 1 & 0 & 0 & 1 \\ 0 & 1 & 0 & 1 \\ 1 & 1 & 1 & 0 \end{array}\right]\end{array}$$

C.

receiver

$$\begin{array}{c} & S \ \ T \ \ U \ \ V \\ \text{sender} \begin{array}{c} S \\ T \\ U \\ V \end{array}\left[\begin{array}{cccc} 0 & 1 & 0 & 1 \\ 0 & 0 & 0 & 1 \\ 0 & 1 & 0 & 0 \\ 1 & 1 & 1 & 0 \end{array}\right]\end{array}$$

D.

receiver

$$\begin{array}{c} & S \ \ T \ \ U \ \ V \\ \text{sender} \begin{array}{c} S \\ T \\ U \\ V \end{array}\left[\begin{array}{cccc} 0 & 1 & 0 & 1 \\ 0 & 0 & 0 & 1 \\ 0 & 1 & 0 & 1 \\ 1 & 1 & 1 & 0 \end{array}\right]\end{array}$$

E.

receiver

$$\begin{array}{c} & S \ \ T \ \ U \ \ V \\ \text{sender} \begin{array}{c} S \\ T \\ U \\ V \end{array}\left[\begin{array}{cccc} 0 & 1 & 0 & 2 \\ 0 & 0 & 0 & 2 \\ 0 & 1 & 0 & 2 \\ 2 & 2 & 2 & 0 \end{array}\right]\end{array}$$

Question 3 (1 mark)

Source: VCE 2019, Further Mathematics Exam 1, Section B, Module 1, Q1; © VCAA.

MC Consider the following four matrix expressions.

$$\begin{bmatrix} 8 \\ 12 \end{bmatrix} + \begin{bmatrix} 4 \\ 2 \end{bmatrix} \qquad\qquad \begin{bmatrix} 8 \\ 12 \end{bmatrix} + \begin{bmatrix} 4 & 0 \\ 0 & 2 \end{bmatrix}$$

$$\begin{bmatrix} 8 & 0 \\ 12 & 0 \end{bmatrix} + \begin{bmatrix} 4 \\ 2 \end{bmatrix} \qquad\qquad \begin{bmatrix} 8 & 0 \\ 12 & 0 \end{bmatrix} + \begin{bmatrix} 4 & 0 \\ 0 & 2 \end{bmatrix}$$

How many of these four matrix expressions are defined?

    A. 0   **B.** 1   **C.** 2   **D.** 3   **E.** 4

Question 4 (3 marks)

Source: VCE 2019, Further Mathematics Exam 2, Section B, Module 1, Q1; © VCAA.

The car park at a theme park has three areas, A, B and C.

The number of empty (E) and full (F) parking spaces in each of the three areas at 1 pm on Friday are shown in matrix Q below.

$$Q = \begin{array}{c} \\ \\ \\ \\ \end{array}\begin{array}{cc} E & F \\ \end{array}\left[\begin{array}{cc} 70 & 50 \\ 30 & 20 \\ 40 & 40 \end{array}\right]\begin{array}{c} A \\ B \\ C \end{array}$$

    a. What is the order of matrix Q?   **(1 mark)**
    b. Write down a calculation to show that 110 parking spaces are full at 1 pm.   **(1 mark)**
    c. Drivers must pay a parking fee for each hour of parking.
Matrix P, below, shows the hourly fee, in dollars, for a car parked in each of the three areas.

$$\begin{array}{c} & area \\ & A \quad\ B \quad\ C \\ P = & \left[\begin{array}{ccc} 1.30 & 3.50 & 1.80 \end{array}\right]\end{array}$$

The total parking fee, in dollars, collected from these 110 parked cars if they were parked for one hour is calculated as follows.

$$P \times L = [207.00]$$

where matrix L is a 3×1 matrix.
Write down matrix L.   **(1 mark)**

Question 5 (1 mark)

Source: VCE 2018, Further Mathematics Exam 1, Section B, Module 1, Q2; © VCAA.

MC The matrix product $\begin{bmatrix} 4 & 2 & 0 \end{bmatrix} \times \begin{bmatrix} 4 \\ 12 \\ 8 \end{bmatrix}$ is equal to

A. $[144]$

B. $\begin{bmatrix} 16 \\ 24 \\ 0 \end{bmatrix}$

C. $4 \times \begin{bmatrix} 1 & 2 & 0 \end{bmatrix} \times \begin{bmatrix} 1 \\ 12 \\ 8 \end{bmatrix}$

D. $2 \times \begin{bmatrix} 2 & 1 & 0 \end{bmatrix} \times \begin{bmatrix} 2 \\ 6 \\ 4 \end{bmatrix}$

E. $4 \times \begin{bmatrix} 2 & 1 & 0 \end{bmatrix} \times \begin{bmatrix} 2 \\ 6 \\ 4 \end{bmatrix}$

More exam questions are available online.

Answers

Topic 4 Matrices

4.2 Types of matrices

4.2 Exercise

1. $\begin{bmatrix} 45 \\ 30 \end{bmatrix}$, order 2×1

2. a. 2×2 b. 2×1 c. 1×4
   d. 2×3 e. 3×3 f. 4×2

3. a. $y_{13} = 5$ b. $y_{24} = 8$ c. $y_{31} = 1$
   d. $y_{43} = 0$ e. $y_{41} = 2$ f. $y_{12} = 7$

4. $\begin{bmatrix} 18 & 12 & 8 \\ 13 & 10 & 11 \end{bmatrix}$

5. $\begin{bmatrix} 6 & 4 & 7 & 3 & 6 \\ 2 & 6 & 6 & 6 & 4 \end{bmatrix}$

6. a. $b_{12} = 5$
   b. $c_{11} = 6$
   c. $a_{21} = 6$
   d. $a_{11} = 8$
   e. $c_{12} = 3$
   f. $b_{22} = 6$

7. a. There is no 4th column.
   b. e_{23}
   c. Nadia thought that e_{12} was read as 1st column, 2nd row. The correct value is 0.

8. a. 3×2
   b. Element k_{22} shows that 800 Section B tickets were sold on Friday.
   c. k_{31}
   d. 2100 Section B tickets were sold in total over Thursday, Friday and Saturday.

9. a. 56 participants b. 213 entries   c. $\begin{bmatrix} 12 & 17 & 18 \end{bmatrix}$

10. a. 3×2 b. $H = \begin{bmatrix} 4 & 3 \\ -1 & 7 \\ -4 & 6 \end{bmatrix}$

11. a. $\begin{bmatrix} 0.4 \\ 4.2 \\ 6.8 \\ 0.2 \\ 1.6 \\ 2.1 \\ 0.5 \\ 5.2 \end{bmatrix}$

   b. See matrix at the bottom of the page.*
   c. i. 3×2
   ii. $\begin{bmatrix} 801\ 428 & 6.8 \\ 227\ 600 & 5.2 \\ 1\ 727\ 200 & 4.2 \end{bmatrix}$

12. a. $\begin{bmatrix} 148\ 178 & 2.2 \\ 30\ 839 & 0.6 \\ 146\ 429 & 3.6 \\ 26\ 044 & 1.7 \\ 77\ 928 & 3.8 \\ 16\ 900 & 3.4 \\ 66\ 582 & 31.6 \\ 4\ 043 & 1.2 \end{bmatrix}$

   b. i. 66 582
   ii. 16 900
   iii. 325 446
   c. 516 943

4.2 Exam questions

Note: Mark allocations are available with the fully worked solutions online.

1. C
2. -1
3. D

4.3 Adding and subtracting matrices

4.3 Exercise

1. a. $\begin{bmatrix} 1 & 6 \\ -1 & 3 \end{bmatrix}$ b. $\begin{bmatrix} -0.1 \\ 1.5 \\ 4.2 \end{bmatrix}$

2. $a = 4,\ b = -6$

3. $\begin{bmatrix} -1 & -2 \\ -1 & -5 \end{bmatrix}$

4. a. $\begin{bmatrix} 1 \\ 7 \\ 0 \end{bmatrix}$ b. $\begin{bmatrix} -5 \\ 3 \\ 6 \end{bmatrix}$ c. $\begin{bmatrix} 6 \\ 4 \\ -6 \end{bmatrix}$ d. $\begin{bmatrix} 8 \\ 1 \\ 0 \end{bmatrix}$

5. a. Both matrices must be of the same order for it to be possible to add and subtract them.
   b. $B = \begin{bmatrix} 4 \\ 1 \\ 5 \end{bmatrix}$

6. a. $\begin{bmatrix} 3 & -1 \\ 4 & 17 \end{bmatrix}$ b. Not defined
   c. Not defined   d. Not defined

7. a. $\begin{bmatrix} -0.25 & -0.95 & 0.3 \end{bmatrix}$
   b. $\begin{bmatrix} 3 & -1 \\ 9 & 1 \end{bmatrix}$
   c. $\begin{bmatrix} -1 & 5 & 1 \\ 4 & -2 & 3 \\ 3 & -3 & -2 \end{bmatrix}$
   d. $\begin{bmatrix} 15 & 15 & 7 \\ 9 & 10 & 8 \end{bmatrix}$

8. $a = -5,\ b = 2,\ c = 5$
9. $a = 6,\ b = -1,\ c = 5$
10. A and E have the same order, 1×2.
    B and C have the same order, 2×1.

---

*11. b. $\begin{bmatrix} 2358 & 68\ 330 & 227\ 600 & 801\ 428 & 984\ 000 & 1\ 346\ 200 & 1\ 727\ 200 & 2\ 529\ 875 \end{bmatrix}$

11. a. $\begin{bmatrix} 2 & 3 & 5 \\ 4 & 6 & 3 \end{bmatrix}$

   b. i. $\begin{bmatrix} 3 & 4 & 8 \\ 6 & 8 & 5 \end{bmatrix} - \begin{bmatrix} 2 & 3 & 5 \\ 4 & 6 & 3 \end{bmatrix}$

   ii. $\begin{bmatrix} 1 & 1 & 3 \\ 2 & 2 & 2 \end{bmatrix}$

Eggs	Small	Medium	Large
Free-range	1	1	3
Barn-laid	2	2	2

12. a. Both matrices are of the order 2×3; therefore, the answer matrix must also be of the order 2×3. The student's answer matrix is of the order 2×1, which is incorrect.

   b. A possible response is:
   Step 1: Check that all matrices are of the same order.
   Step 2: Add or subtract the corresponding elements.

13. a. $\begin{bmatrix} 150 \\ 165 \\ 155 \\ 80 \end{bmatrix}$

   b. i. $\begin{bmatrix} 150 \\ 165 \\ 155 \\ 80 \end{bmatrix} + \begin{bmatrix} 145 \\ 152 \\ 135 \\ 95 \end{bmatrix} + \begin{bmatrix} 166 \\ 155 \\ 156 \\ 110 \end{bmatrix}$

   ii. $\begin{bmatrix} 461 \\ 472 \\ 446 \\ 285 \end{bmatrix}$

   c. $\begin{bmatrix} 35 \\ 41 \\ 38 \\ 25 \end{bmatrix} + \begin{bmatrix} 32 \\ 36 \\ 35 \\ 30 \end{bmatrix} + \begin{bmatrix} 38 \\ 35 \\ 35 \\ 32 \end{bmatrix} = \begin{bmatrix} 105 \\ 112 \\ 108 \\ 87 \end{bmatrix}$

14. a. $\begin{bmatrix} 12 \\ 9 \\ 8 \end{bmatrix} + \begin{bmatrix} -3 \\ -3 \\ 0 \end{bmatrix} = \begin{bmatrix} 9 \\ 6 \\ 8 \end{bmatrix}$ or $\begin{bmatrix} 12 \\ 9 \\ 8 \end{bmatrix} - \begin{bmatrix} 9 \\ 6 \\ 8 \end{bmatrix} = \begin{bmatrix} 3 \\ 3 \\ 0 \end{bmatrix}$

   b. $\begin{bmatrix} 9 \\ 6 \\ 8 \end{bmatrix} + \begin{bmatrix} 3 \\ -2 \\ 2 \end{bmatrix} = \begin{bmatrix} 12 \\ 4 \\ 10 \end{bmatrix}$ or $\begin{bmatrix} 12 \\ 4 \\ 10 \end{bmatrix} - \begin{bmatrix} 9 \\ 6 \\ 8 \end{bmatrix} = \begin{bmatrix} 3 \\ -2 \\ 2 \end{bmatrix}$

15. a. To add or subtract matrices, all matrices must be of the same order. Since the resultant matrix D is of the order 3×2, all other matrices must also be of the order 3×2.

   b. $\begin{bmatrix} 2x - 12 & y - 2 \\ 2x + 10 & 5 - y \\ 3x + 10 & -2 - 2x \end{bmatrix}$

   c. When the equations $2x - 12 = -8$ and $y - 2 = 1$ are solved, the answer is $x = 2$, $y = 3$.

16. $\begin{bmatrix} \dfrac{5}{8} & \dfrac{5}{8} & 0 \\ \dfrac{17}{30} & \dfrac{5}{42} & \dfrac{1}{36} \\ \dfrac{11}{18} & \dfrac{1}{8} & \dfrac{-2}{9} \end{bmatrix}$

4.3 Exam questions

Note: Mark allocations are available with the fully worked solutions online.

1. D
2. B
3. $\begin{bmatrix} 7 & 4 \\ -11 & 5 \end{bmatrix}$

4.4 Multiplying matrices

4.4 Exercise

1. a. $\begin{bmatrix} 8 & 12 & 28 \\ 4 & 16 & 24 \end{bmatrix}$

   b. $\begin{bmatrix} \dfrac{2}{5} & \dfrac{3}{5} & \dfrac{7}{5} \\ \dfrac{1}{5} & \dfrac{4}{5} & \dfrac{6}{5} \end{bmatrix}$

   c. $\begin{bmatrix} 0.6 & 0.9 & 2.1 \\ 0.3 & 1.2 & 1.8 \end{bmatrix}$

2. $x = 2.5$

3. C

4. a. X has order 1×2. Y has order 2×1. Number of columns = number of rows, so XY exists and is of order 1×1.

   b. $DE : 3 \times 3$, $DC : 3 \times 2$, $ED : 2 \times 2$, $CE : 2 \times 3$

5. $S : 3 \times 2$, $T : 2 \times 4$

6. $DG : 3 \times 2$, $FD : 2 \times 1$, $FE : 2 \times 2$, $EF : 3 \times 3$, $GF : 1 \times 3$

7. $MN = \begin{bmatrix} 28 & 48 \\ 21 & 36 \end{bmatrix}$

8. $t = -3$

9. $PQ = \begin{bmatrix} 41 & 45 \\ 36 & 32 \end{bmatrix}$

10. a. $\begin{bmatrix} 12.50 \\ 8.50 \\ 6.00 \end{bmatrix}$

   b. Total number of tickets requires an order of 1×1, and the order of the ticket price is 3×1. The number of people must be of order 1×3 to result in a product matrix of order 1×1. Therefore, the answer must be a row matrix.

   c. $1662.50

11. a. $[66]$ b. $\begin{bmatrix} 30 & 45 \\ 24 & 36 \end{bmatrix}$ c. $\begin{bmatrix} 70 & 105 \\ 20 & 30 \\ 90 & 135 \end{bmatrix}$

   d. $\begin{bmatrix} 57 \\ 43 \end{bmatrix}$ e. $\begin{bmatrix} 38 & 19 \\ 14 & 7 \end{bmatrix}$

12. a. $\begin{bmatrix} 4 & 6 \\ 2 & 3 \end{bmatrix}$

   b. $\begin{bmatrix} 4 & 6 \\ 2 & 3 \end{bmatrix}$

13. $\begin{bmatrix} 72 & 30 \\ 60 & 57 \end{bmatrix}$

14. $\begin{bmatrix} 27 & 315 \\ 0 & 216 \end{bmatrix}$

15. a. $I_3{}^2 = \begin{bmatrix} 1 & 0 & 0 \\ 0 & 1 & 0 \\ 0 & 0 & 1 \end{bmatrix}$

b. $I_3{}^3 = \begin{bmatrix} 1 & 0 & 0 \\ 0 & 1 & 0 \\ 0 & 0 & 1 \end{bmatrix}$

c. $I_3{}^4 = \begin{bmatrix} 1 & 0 & 0 \\ 0 & 1 & 0 \\ 0 & 0 & 1 \end{bmatrix}$

d. Whatever power you raise I_3 to, the matrix stays the same.

16. a. $\begin{bmatrix} 250 \\ 185 \end{bmatrix}$

b. $\begin{bmatrix} 0.05 & 0.18 & 0.45 & 0.25 & 0.07 \end{bmatrix}$

c. The number of expected grades (A–E) for students studying Mathematics and Physics.

d. $\begin{bmatrix} 13 & 45 & 113 & 63 & 18 \\ 9 & 33 & 83 & 46 & 13 \end{bmatrix}$

e. 45 students studying Maths are expected to be awarded a B grade.

17. $n = 1,\ m = 3,\ s = 4,\ q = 2$

18. a. Matrix G is of order 3×2 and matrix H is of order 2×1; therefore, GH is of order 3×1. The student's matrix has an order of 3×2.

b. $\begin{bmatrix} 125 \\ 134 \\ 167 \end{bmatrix}$

c. Possible answer:
The student multiplied the first column with the first row, and then the second column with the second row.

d. Possible answer:
Step 1: Determine the order of the product matrix.
Step 2: Multiply the elements in the first row by the elements in the first column.

19. a. Possible answer:
Represent the number of vehicles in a row matrix and the cost for each vehicle in a column matrix, then multiply the two matrices together. The product matrix will have an order of 1×1.

b. [154 000] or $154 000

c. Possible answer:
In this multiplication the number of each type of vehicle is multiplied by the price of each type of vehicle, which is incorrect. For example, the ute is valued at $12 500, but in this multiplication the eight utes sold are multiplied by $4000, $12 500 and $8500 respectively.

20. a. $\begin{bmatrix} 12\,280 \\ 26\,210 \\ 29\,850 \end{bmatrix}$;
Friday $12 280, Saturday $26 210, Sunday $29 850

b. $350 \times 35 + 456 \times 25 + 128 \times 20$

c. No, because you cannot multiply the entry price (3×1) by the number of people (3×3).

4.4 Exam questions

Note: Mark allocations are available with the fully worked solutions online.

1. D

2. D

3. $\begin{bmatrix} 8 & -8 \\ 0 & 24 \end{bmatrix}$

4.5 Inverse matrices and problem-solving with matrices

4.5 Exercise

1. Yes, $AB = \begin{bmatrix} 1 & 0 \\ 0 & 1 \end{bmatrix}$.

2. a. -34; this matrix has an inverse

b. 8; this matrix has an inverse

c. 0; this matrix does not have an inverse

d. -35; this matrix has an inverse

3. a. $\begin{bmatrix} 1 & -2 \\ -2 & 5 \end{bmatrix}$ **b.** $\begin{bmatrix} 1 & -2 \\ -1.5 & 3.5 \end{bmatrix}$ **c.** $\begin{bmatrix} 2 & 1 \\ -3 & -2 \end{bmatrix}$

4. $\det(B) = 0$. We cannot divide by zero; therefore, B^{-1} does not exist.

5. a. $\det(A) = 1,\ \det(B) = 0,\ \det(C) = 0,\ \det(D) = -3.$
Therefore, matrices A and D have inverses.

b. $A^{-1} = \begin{bmatrix} 3 & -1 \\ -5 & 2 \end{bmatrix},\ D^{-1} = \begin{bmatrix} \dfrac{5}{3} & \dfrac{1}{3} \\ 1 & 0 \end{bmatrix}$

6. $x = 2,\ y = -3$

7. a. Matrix X has an order of 1×2. $X = \begin{bmatrix} -2 & 3 \end{bmatrix}$

b. $A = \begin{bmatrix} 4 & 2 \\ 3 & 2 \end{bmatrix}$

8. a. Determine the inverse of $\begin{bmatrix} 4 & 1 \\ 3 & 1 \end{bmatrix}$ and multiply by $\begin{bmatrix} 9 \\ 7 \end{bmatrix}$, using $AX = B$ and $X = A^{-1}B$.

b. $\begin{bmatrix} 1 & -1 \\ -3 & 4 \end{bmatrix}$

c. $x = 2,\ y = 1$

9. $17.50

10. $x = 2,\ y = 1$

11. The determinant $= 0$, so no inverse exists. This means that there is no solution to the simultaneous equations.

12. 7 children and 3 adults

13. $6.40

14. a. Each bracelet costs $4.20 and each necklace costs $9.60.

b. $67.80

15. a. Any one of: did not swap the elements on the diagonal; did not multiply the other elements by -1; or did not multiply the matrix by $\dfrac{1}{\det}$.

b. $\dfrac{1}{6} \begin{bmatrix} 3 & 0 \\ -1 & 2 \end{bmatrix}$

c. The respective order of matrices is 2×2 and 1×2. The number of columns in the first matrix does not equal the number of rows in the second matrix.

d. Possible answers:
Step 1: Find the correct inverse.

Step 2: Multiply $\begin{bmatrix} 5 & 9 \end{bmatrix} \dfrac{1}{6} \begin{bmatrix} 3 & 0 \\ -1 & 2 \end{bmatrix}$.

e. $x = 1, y = 3$

16. a. $\begin{bmatrix} 5 & 7 \\ 6 & 6 \end{bmatrix} \begin{bmatrix} s \\ p \end{bmatrix} = \begin{bmatrix} 156.80 \\ 155.40 \end{bmatrix}$

b. $12.25

c. $K = \begin{bmatrix} 30 & 30 \end{bmatrix}$

d. 2×1

e. $239.70

f. $84.30

17. a. $a =$ number of adult tickets sold
$c =$ number of concession tickets sold
$a + c = 400$ and $15a + 9.5c = 5422.5$

b. $\begin{bmatrix} 1 & 1 \\ 15.00 & 9.50 \end{bmatrix} \begin{bmatrix} a \\ c \end{bmatrix} = \begin{bmatrix} 400 \\ 5422.50 \end{bmatrix}$

c. 295

18. a. $\begin{bmatrix} -\dfrac{4}{19} & -\dfrac{1}{19} & \dfrac{13}{19} \\ -\dfrac{7}{19} & \dfrac{3}{19} & -\dfrac{1}{19} \\ \dfrac{8}{19} & \dfrac{2}{19} & -\dfrac{7}{19} \end{bmatrix}$

b. $\begin{bmatrix} 0 & \dfrac{7}{3} & -\dfrac{8}{3} & \dfrac{10}{3} \\ \dfrac{1}{3} & -\dfrac{7}{9} & \dfrac{11}{9} & -\dfrac{13}{9} \\ 0 & -1 & 1 & -1 \\ -\dfrac{1}{3} & \dfrac{22}{9} & -\dfrac{23}{9} & \dfrac{28}{9} \end{bmatrix}$

19. $a = 2, b = -1, c = 3, d = -2$

4.5 Exam questions

Note: Mark allocations are available with the fully worked solutions online.

1. D

2. B

3. E

4.6 Communications and connections

4.6 Exercise

1. a. $\begin{array}{c} & \begin{matrix} A & B & C \end{matrix} \\ \begin{matrix} A \\ B \\ C \end{matrix} & \begin{bmatrix} 0 & 0 & 1 \\ 0 & 0 & 1 \\ 1 & 1 & 0 \end{bmatrix} \end{array}$

b. $\begin{array}{c} & \begin{matrix} D & E & F & G \end{matrix} \\ \begin{matrix} D \\ E \\ F \\ G \end{matrix} & \begin{bmatrix} 0 & 1 & 1 & 0 \\ 1 & 0 & 1 & 0 \\ 1 & 1 & 0 & 1 \\ 0 & 0 & 1 & 0 \end{bmatrix} \end{array}$

c. $\begin{array}{c} & \begin{matrix} H & I & J & K \end{matrix} \\ \begin{matrix} H \\ I \\ J \\ K \end{matrix} & \begin{bmatrix} 0 & 1 & 0 & 1 \\ 1 & 0 & 1 & 0 \\ 0 & 1 & 0 & 1 \\ 1 & 0 & 1 & 0 \end{bmatrix} \end{array}$

d. $\begin{array}{c} & \begin{matrix} L & M & N & O \end{matrix} \\ \begin{matrix} L \\ M \\ N \\ O \end{matrix} & \begin{bmatrix} 0 & 1 & 0 & 0 \\ 1 & 0 & 1 & 1 \\ 0 & 1 & 0 & 0 \\ 0 & 1 & 0 & 0 \end{bmatrix} \end{array}$

2. Answers may vary. Sample responses are available in the worked solutions in the online resources.

3. There are 2 two-step communications between Karina and Winter: $K - G - W$ and $K - N - W$.

4. The number of two-step connections between Amira and Dev can be obtained from the element in row 4, column 4 or row 4, column 1.
There are 2 two-step connections between Amira and Dev.
There are 2 ways Amira can get a message to Dev via one other person.

5. $\begin{array}{c} & \begin{matrix} A & B & C & D & E \end{matrix} \\ \begin{matrix} A \\ B \\ C \\ D \\ E \end{matrix} & \begin{bmatrix} 0 & 1 & 0 & 0 & 1 \\ 1 & 0 & 1 & 1 & 1 \\ 0 & 1 & 0 & 1 & 0 \\ 0 & 1 & 1 & 0 & 0 \\ 1 & 1 & 0 & 0 & 0 \end{bmatrix} \end{array}$

6. a. 3

b. No

c. They did not communicate with themselves.

d.

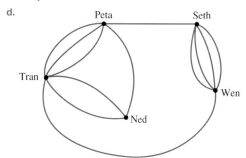

7. a. $\begin{array}{c} & \begin{matrix} W & C & H & K & A \end{matrix} \\ \begin{matrix} W \\ H \\ C \\ K \\ A \end{matrix} & \begin{bmatrix} 0 & 1 & 1 & 1 & 1 \\ 1 & 0 & 0 & 0 & 0 \\ 1 & 0 & 0 & 1 & 1 \\ 1 & 0 & 1 & 0 & 0 \\ 1 & 0 & 1 & 0 & 0 \end{bmatrix} \end{array}$

b. 1

c. Yes. The matrix raised to the power of 3 will provide the number of ways possible.

8. 2

9. a.

b. $\begin{array}{c} & \begin{matrix} B & H & M & N & P \end{matrix} \\ \begin{matrix} B \\ H \\ M \\ N \\ P \end{matrix} & \begin{bmatrix} 3 & 2 & 2 & 2 & 2 \\ 2 & 3 & 2 & 2 & 2 \\ 2 & 2 & 4 & 3 & 1 \\ 2 & 2 & 3 & 4 & 1 \\ 2 & 2 & 1 & 1 & 2 \end{bmatrix} \end{array}$

c. Raise the matrix to the power of 3.

d. Yes. Raise the matrix to the power of 4, as there are five cities in total.

10. a. 3   b. 6

4.6 Exam questions

Note: Mark allocations are available with the fully worked solutions online.

1. A

2. C

3. $\begin{bmatrix} 0 & 1 & 0 & 1 \\ 1 & 0 & 1 & 2 \\ 0 & 1 & 0 & 1 \\ 1 & 2 & 1 & 0 \end{bmatrix}$

4.7 Transition matrices

4.7 Exercise

1.

		From	
		Rain	**No rain**
To	**Rain**	0.3	0.2
	No rain	0.7	0.8

$\begin{bmatrix} 0.3 & 0.2 \\ 0.7 & 0.8 \end{bmatrix}$

2. 40% of trains that start the week in West depot will end up in East depot the next week.
75% of trains that start the week in East depot will end up in West depot the next week.

3.

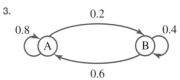

4. 20% of Melburnians move to Canberra each year.
65% of Canberra residents move to Melbourne each year.

5. $\begin{bmatrix} 0.3 & 0.9 \\ 0.7 & 0.1 \end{bmatrix}$

6. After two years, there would be 2775 people in Town J and 4225 people in Town K.

7. $S_5 = \begin{bmatrix} 405 \\ 295 \end{bmatrix}$

8. $S = \begin{bmatrix} 35 \\ 9 \end{bmatrix}$

9. There are 259 birds in Australia and 841 birds in Antarctica at the steady state.

10. a. $T = \begin{bmatrix} 0.7 & 0.35 \\ 0.3 & 0.65 \end{bmatrix}$

b. Sheepies would have 1896 customers after three weeks.

c. At the steady state, 2154 customers shop at Coldes and 1846 customers shop at Sheepies.

4.7 Exam questions

Note: Mark allocations are available with the fully worked solutions online.

1. E

2. 20%

3. $\begin{bmatrix} 300 \\ 780 \\ 300 \\ 620 \end{bmatrix}$

4.8 Review

4.8 Exercise

Multiple choice

1. A   2. C   3. C   4. C
5. A   6. A   7. E   8. C
9. E   10. B

Short answer

11. a. $\begin{bmatrix} 4 \\ 2 \end{bmatrix}$ b. $\begin{bmatrix} -5 \\ -4 \end{bmatrix}$ c. $\begin{bmatrix} 5 \\ 3 \end{bmatrix}$ d. $\begin{bmatrix} 3 & 4.5 \\ 1.5 & 6 \end{bmatrix}$

12. a. 3   b. 1×2

13. a. $\begin{bmatrix} 12.5 \\ 6 \\ 10 \end{bmatrix}$ b. $4000

14. a. i. 1×2 and 2×1; product matrix: 1×1
ii. 1×2 and 2×2; product matrix: 1×2
iii. 2×2 and 2×2; product matrix: 2×2
iv. 3×1 and 1×3; product matrix: 3×3

b. i. $[36]$

ii. $\begin{bmatrix} 32 & 54 \end{bmatrix}$

iii. $\begin{bmatrix} 42 & 101 \\ 55 & 35 \end{bmatrix}$

iv. $\begin{bmatrix} 10 & 6 & 8 \\ 35 & 21 & 28 \\ 40 & 24 & 32 \end{bmatrix}$

15. a. i. 3
ii. -4

b. i. $\begin{bmatrix} 1 & -1 \\ -\dfrac{5}{3} & 2 \end{bmatrix}$

ii. $\begin{bmatrix} -\dfrac{3}{4} & \dfrac{1}{2} \\ -\dfrac{1}{4} & \dfrac{1}{2} \end{bmatrix}$

c. Matrix *M* is not a square matrix; therefore, it does not have an inverse.

16. a. $x = 2$, $y = -1$
b. $x = 7.75$, $y = 5.625$

Extended response

17. a. 8 slices of bread   b. 4 sandwiches
c. $44 \div 5.5 = 8$, $8 \div 4 = 2$ d. 12.8
e. $\begin{bmatrix} 2 & 2 & 1 \end{bmatrix}$

18. a. $\begin{bmatrix} 3890 \\ 2178 \\ 868 \end{bmatrix}$, 3×1 b. $\begin{bmatrix} 62 & 125 & 270 \end{bmatrix}$

c. 1×1 d. 0.95

e. $\begin{bmatrix} 3695.50 \\ 2069.10 \\ 824.60 \end{bmatrix}$

19. a. $\begin{bmatrix} 10 & 9 \\ 8 & 12 \\ 9 & 11 \end{bmatrix}$

  b. $\begin{bmatrix} 45.95 & 25.50 & 18.60 \end{bmatrix}$

  c. i. $\begin{bmatrix} 45.95 & 25.50 & 18.60 \end{bmatrix} \begin{bmatrix} 10 & 9 \\ 8 & 12 \\ 9 & 11 \end{bmatrix}$

  ii. LeisureLand: \$830.90, SportLand: \$924.15

20. a.

		From	
		H	**O**
To	**H**	0.4	0.2
	O	0.6	0.8

  b. $\begin{bmatrix} 0.4 & 0.2 \\ 0.6 & 0.8 \end{bmatrix}$

  c. After three weeks, 1528 people are predicted to be buying the *Herald Moon*.

  d. At the steady state, it is predicted that there will be 1500 people buying the *Herald Moon* and 4500 people buying *The Old Times*.

4.8 Exam questions

Note: Mark allocations are available with the fully worked solutions online.

1. B

2. D

3. C

4. a. 3×2

  b. Add up column 2: $50 + 20 + 40 = 110$

  c. $L = \begin{bmatrix} 50 \\ 20 \\ 40 \end{bmatrix}$

5. E

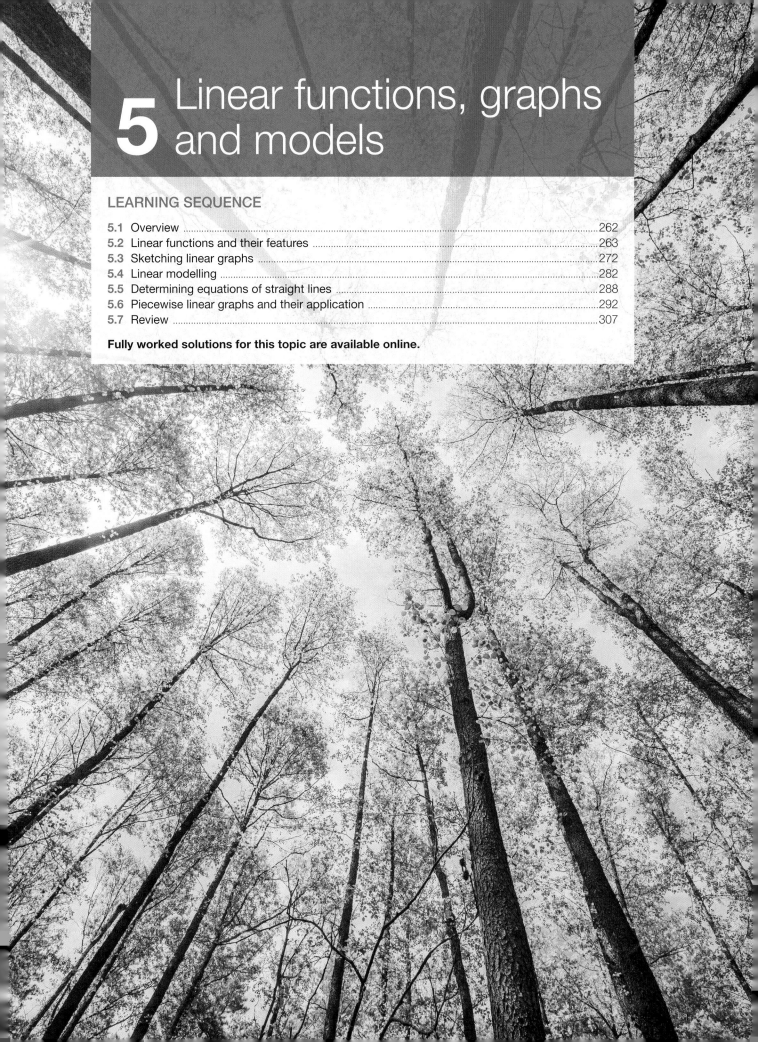

5 Linear functions, graphs and models

Fully worked solutions for this topic are available online.

5.1 Overview

5.1.1 Introduction

Linear equations use one or more variables where one variable is dependent on the other. Almost any situation where there is an unknown quantity can be represented by a linear equation, such as predicting profit or calculating the cost of booking accommodation at a hotel. A useful way to apply linear equations is to make predictions about what will happen in the future. For example, if a linear profit equation is modelled, then this model could be used to predict future profits.

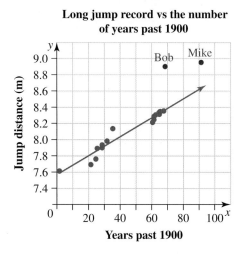

It is always interesting to note that a lot of world records follow a linear trend over time. One event that challenges this is the men's long-jump world record. In 1968, Bob Beamon smashed the record by an amazing 55 cm with a jump of 8.90 m at the Olympics. This jump certainly went against the linear trend. This record stood until 1991, when Mike Powell jumped 8.95 m at the World Championships in Athletics. If we plot the world records from 1900 to 1964 and draw a line of best fit, we can clearly see that Bob's long jump went against the linear trend over the previous 60 years.

KEY CONCEPTS

This topic covers the following key concepts from the VCE Mathematics Study Design:
- the linear function $y = a + bx$, its graph, and interpretation of the parameters, a and b, in terms of initial value and constant rate of change respectively
- graphing linear relations $Ax + By = C$ and equivalent forms
- formulation and analysis of linear models from worded descriptions or relevant data (including simultaneous linear equations in two variables) and their application to solve practical problems including domain of interpretation
- piecewise linear (line segment, step) graphs and their application to modelling practical situations, including tax scales and charges, and payment.

Source: VCE Mathematics Study Design (2023–2027) extracts © VCAA; reproduced by permission.

5.2 Linear functions and their features

5.2.1 Gradient-intercept form

The **gradient-intercept form**, as the name suggests, allows us to easily see the values of the gradient and the y-intercept.

$$y = a + bx$$

y-intercept   gradient

Note: Previously we called the gradient and y-intercept m and c, respectively. In General Mathematics Units 1 & 2 and General Mathematics Units 3 & 4, we will use a and b in the form $y = a + bx$. This notation is used in the field of statistics and is preferred to the form $y = mx + c$ when constructing regression equations.

WORKED EXAMPLE 1 Identifying gradient and intercepts from linear equations

State the gradients and y-intercepts of the following linear equations.

a. $y = 2 + 5x$ **b.** $y = -3 + \dfrac{x}{2}$ **c.** $2x + y = 4$

THINK	WRITE
a. 1. Write the equation. It is in the form $y = a + bx$.	**a.** $y = 2 + 5x$
2. Identify the coefficient of x.	The coefficient of x is 5.
3. Identify the value of a.	The value of a is 2.
4. Write the answer.	The gradient is 5 and the y-intercept is 2.
b. 1. Write the equation. It is in the form $y = a + bx$.	**b.** $y = -3 + \dfrac{x}{2}$
2. Identify the coefficient of x.	x has been multiplied by $\dfrac{1}{2}$, so the coefficient is $\dfrac{1}{2}$.
3. Identify the value of a.	The value of a is -3.
4. Write the answer.	The gradient is $\dfrac{1}{2}$ and the y-intercept is -3.

c.	1. Transpose the equation to be in the form $y = a + bx$.	c.	$2x + y = 4$ $y = 4 - 2x$
	2. Identify the coefficient of x.		The coefficient of x is -2 (the coefficient includes the sign).
	3. Identify the value of a.		The value of a is 4.
	4. Write the answer.		The gradient is -2 and the y-intercept is 4.

5.2.2 The gradient of a linear function

The **gradient** (also known as slope) of a linear function, represented by the letter b, describes how steep the graph is. It also describes how the two variables are related to one another.

The gradient is calculated by dividing the rise (distance up) by the run (distance across).

Suppose a line passes through the points $(1, 4)$ and $(3, 8)$ as shown in the graph.

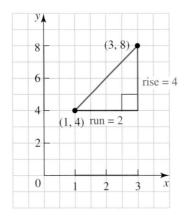

By completing a right-angled triangle, it can be seen that:
- the rise (difference in y-values): $8 - 4 = 4$
- the run (difference in x-values): $3 - 1 = 2$
- to determine the gradient:

$$\text{gradient} = \frac{8 - 4}{3 - 1}$$
$$= \frac{4}{2} = 2$$

Determining the gradient (*b*) of a line passing through two points

To determine the gradient of a straight line that passes through two points. First, identify two clear points (x_1, y_1) and (x_2, y_2) and use one of the following relations:

$$\text{gradient }(b) = \frac{\text{rise}}{\text{run}} = \frac{y_2 - y_1}{x_2 - x_1}$$

Rise: how much the graph has gone up

Run: how much the graph has travelled across

Note: **If the line is sloping upwards (from left to right), the gradient is positive; if it is sloping downwards, the gradient is negative.**

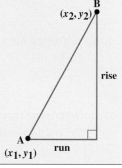

tlvd-3851

WORKED EXAMPLE 2 Determining the gradient of a graph

Determine the values of the gradients of the following graphs.

a.

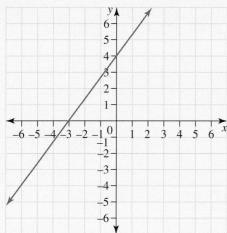

b.

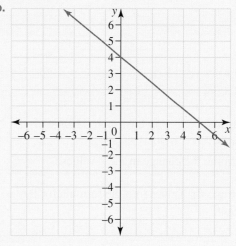

THINK

a. 1. Determine two points on the graph. (Select the *x*- and *y*-intercepts.)

WRITE/DRAW

a.

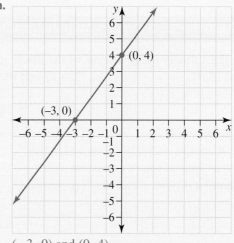

$(-3, 0)$ and $(0, 4)$

2. Determine the rise in the graph (change in y-values).

$$4 - 0 = 4$$

3. Determine the run in the graph (change in x-values).

$$0 - -3 = 3$$

4. Substitute the values into the formula for the gradient.

$$\text{Gradient} = \frac{\text{rise}}{\text{run}}$$
$$= \frac{4}{3}$$

b. 1. Determine two points on the graph. (Select the x- and y-intercepts.)

b.

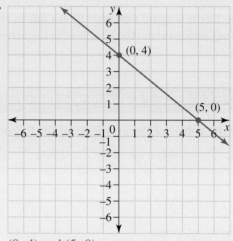

(0, 4) and (5, 0)

2. Determine the rise in the graph (change in y-values).

$$0 - 4 = -4$$

3. Determine the change in the x-values.

$$5 - 0 = 5$$

4. Substitute the values into the formula for the gradient.

$$\text{Gradient} = \frac{\text{rise}}{\text{run}}$$
$$= -\frac{4}{5}$$

WORKED EXAMPLE 3 Determining the gradient between two points

Determine the value of the gradients of the linear graphs that pass through the following points.
a. $(4, 6)$ and $(5, 9)$ **b. $(2, -1)$ and $(0, 5)$**   **c. $(0.5, 1.5)$ and $(-0.2, 1.8)$**

THINK

a. 1. Assign the points.
   Note: It doesn't matter how you assign the points, as long as you're consistent with which point is point 1 and which is point 2.

WRITE

a. Let $(4, 6) = (x_1, y_1)$ and $(5, 9) = (x_2, y_2)$.

2. Write the formula for the gradient and substitute the values.

$$b = \frac{y_2 - y_1}{x_2 - x_1}$$
$$= \frac{9 - 6}{5 - 4}$$
$$= \frac{3}{1}$$

3. Simplify the fraction and write the answer.

The gradient is 3 or $b = 3$.

b. 1. Assign the points.
Note: It doesn't matter how you assign the points, as long as you're consistent with which point is point 1 and which is point 2.

b. Let $(2, -1) = (x_1, y_1)$ and $(0, 5) = (x_2, y_2)$.

2. Write the formula for the gradient and substitute the values.

$$b = \frac{y_2 - y_1}{x_2 - x_1}$$
$$= \frac{5 - -1}{0 - 2}$$
$$= \frac{6}{-2}$$

3. Simplify the fraction and write the answer.

The gradient is -3 or $b = -3$.

c. 1. Assign the points.
Note: It doesn't matter how you assign the points, as long as you're consistent with which point is point 1 and which is point 2.

c. Let $(0.5, 1.5) = (x_1, y_1)$ and $(-0.2, 1.8) = (x_2, y_2)$.

2. Write the formula for the gradient and substitute the values.

$$b = \frac{y_2 - y_1}{x_2 - x_1}$$
$$= \frac{1.8 - 1.5}{-0.2 - 0.5}$$
$$= \frac{0.3}{-0.7}$$

3. Simplify the fraction and write the answer.

The gradient is $-\frac{3}{7}$ or $b = -\frac{3}{7}$.

5.2.3 Identifying features of linear functions

The key features of linear graphs are the gradient and intercepts. The gradient describes the rate of change of a graph and how the two variables are related to one another.

The **x-intercept** tells us what the x-value is when the y-value hits zero. For example, in the case of a car depreciation, the x-intercept tells us when the asset will be worth nothing.

The **y-intercept** tells us the starting point, or initial state of the linear model. For example, in case of a car depreciation, the y-intercept would tell us the starting value of the car.

The intercepts are important because they relate to certain key points of the linear model.

In the graph of $y = 3 + x$, we can see that the x-intercept is at $(-3, 0)$ and the y-intercept is at $(0, 3)$. These points can also be determined algebraically by putting $y = 0$ and $x = 0$ into the equation.

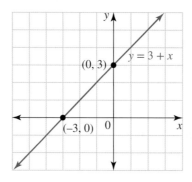

Determining intercepts

To determine the x-intercept, substitute $y = 0$ into your equation and solve for x.

To determine the y-intercept, substitute $x = 0$ into your equation and solve for y.

Standard form

Equations are commonly written in either **standard form** or gradient-intercept form. Writing equations like this makes it easier to identify key features of the linear functions.

The standard form is useful when graphing because it makes it easier to determine the intercepts.

Standard form of a linear equation

$$Ax + By = C$$

where A, B and C are variables and are often integers (whole numbers).

The x-intercept is at $\left(\dfrac{C}{A}, 0\right)$.

The y-intercept is at $\left(0, \dfrac{C}{B}\right)$.

 Resources

 Interactivity Linear graphs (int-6484)

1. **WE1** State the gradients and *y*-intercepts of the following linear equations.

   a. $y = 1 + 2x$

   b. $y = 3 - x$

   c. $y = 4 + \frac{1}{2}x$

   d. $-4x + 4y = 1$

   e. $3x + 2y = 6$

2. Determine the gradients and *y*-intercepts of the following linear equations.

   a. $y = \frac{3x - 1}{5}$

   b. $y = 5(2x - 1)$

   c. $y = \frac{3 - x}{2}$

3. **WE2** Determine the value of the gradient of each of the following graphs.

   a.
   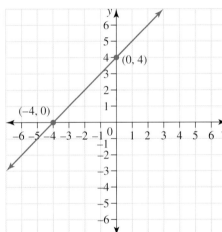

   b.
   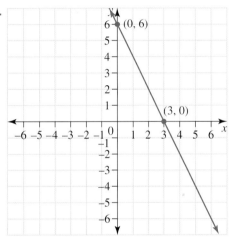

4. Determine the value of the gradient and *y*-intercept of each of the following graphs.

   a.
   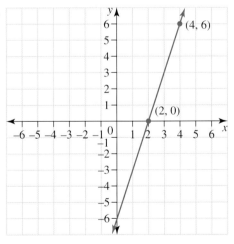

   b.
   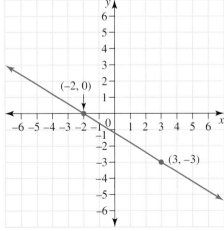

5. **MC** Select the graph with a gradient of $-\dfrac{1}{4}$.

A.
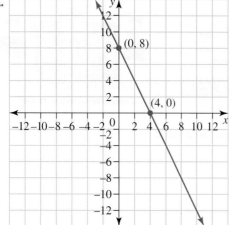

B.
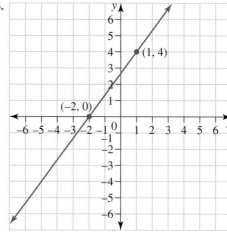

C.
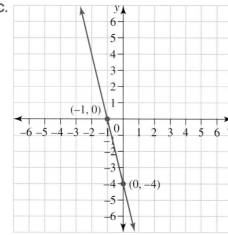

D.
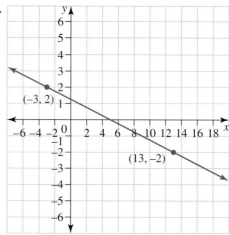

E.
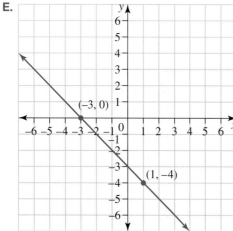

6. **WE3** Determine the values of the gradients of the straight-line graphs that pass through the following points.

   a. (2, 3) and (5, 12)   **b.** (−1, 3) and (2, 7)   **c.** (−0.2, 0.7) and (0.5, 0.9)

7. A line has a gradient of −2 and passes through the points (1, 4) and (a, 8). Calculate the value of a.

8. Write the equations of the linear graphs with the following y-intercepts (a) and gradients (b).

   a. $a = 7$ **b.** $a = -3$ **c.** $a = 2$ **d.** $a = -1$
   $b = 2$ $b = 4$ $b = -1$ $b = 3$

9. Determine the values of the gradients of the straight-line graphs that pass through the following points.

   a. (3, 6) and (2, 9)

   b. (−4, 5) and (1, 8)

   c. (−0.9, 0.5) and (0.2, −0.7)

   d. (1.4, 7.8) and (3.2, 9.5)

   e. $\left(\dfrac{4}{5}, \dfrac{2}{5}\right)$ and $\left(\dfrac{1}{5}, -\dfrac{6}{5}\right)$

   f. $\left(\dfrac{2}{3}, \dfrac{1}{4}\right)$ and $\left(\dfrac{3}{4}, -\dfrac{2}{3}\right)$

10. Write the equations of the linear graphs with the following x- and y-intercepts.

    a. x-intercepts $= 3$
       y-intercepts $= 4$

    b. x-intercepts $= 3$
       y-intercepts $= 9$

    c. x-intercepts $= -2$
       y-intercepts $= 2$

    d. x-intercepts $= -5$
       y-intercepts $= -3$

5.2 Exam questions

Question 1 (1 mark)

MC Which of the following is an accurate sketch of $2x - 3y = 12$?

A.
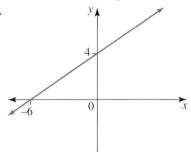

B.
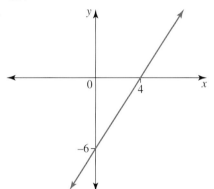

C.
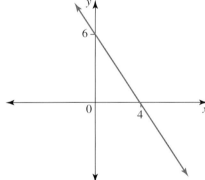

D.
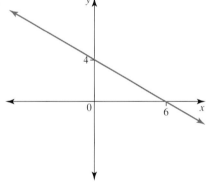

E.
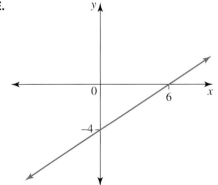

MC The gradient and y-intercept of the graph with equation $3x - 4y + 12 = 0$ are

A. $b = \dfrac{3}{4}$ and $a = 4$ **B.** $b = \dfrac{4}{3}$ and $a = 3$ **C.** $b = \dfrac{4}{3}$ and $a = 4$

D. $b = 3$ and $a = 12$ **E.** $b = \dfrac{3}{4}$ and $a = 3$

Question 3 (1 mark)

MC The gradient of the line connecting the two points $(-6, 2)$ and $(2, 10)$ is

A. 1   **B.** -1 **C.** $\dfrac{4}{3}$ **D.** $-\dfrac{4}{3}$ **E.** 3

More exam questions are available online.

5.3 Sketching linear graphs

LEARNING INTENTION

At the end of this subtopic you should be able to:
- plot a linear graph by using a table of values
- construct a linear graph
- sketch linear graphs using gradient and y-intercepts
- sketch linear graphs using the x- and y-intercepts.

5.3.1 Plotting linear graphs

One way to plot a linear graph is to first draw up a table of values, then plot those coordinates and draw a line through the points.

WORKED EXAMPLE 4 Plotting a linear graph by using a table of values

Plot the graph of $y = 1 + 2x$ by first constructing a table of values with x-values from 0 to 5.

THINK	WRITE
1. To plot, first make a table of values with x-values from 0 to 5.	

x	0	1	2	3	4	5
y						

2. Substitute the x-value into the equation to get the corresponding y-value.

   i. Substitute $x = 0$

$x = 0$
$y = 1 + 2x$
$y = 1 + 2(0)$
$y = 1$

ii. Substitute $x = 1$

$x = 1$
$y = 1 + 2x$
$y = 1 + 2(1)$
$y = 3$

iii. Substitute $x = 2$

$x = 2$
$y = 1 + 2x$
$y = 1 + 2(2)$
$y = 5$

iv. Substitute $x = 3$

$x = 3$
$y = 1 + 3x$
$y = 1 + 2(3)$
$y = 7$

v. Substitute $x = 4$

$x = 4$
$y = 1 + 2x$
$y = 1 + 2(4)$
$y = 9$

vi. Substitute $x = 5$

$x = 5$
$y = 1 + 2x$
$y = 1 + 2(5)$
$y = 11$

3. Fill in the table with the values obtained.

x	0	1	2	3	4	5
y	1	3	5	7	9	11

4. Plot the points on a Cartesian plane, remembering that the x-axis is horizontal and y-axis is vertical.

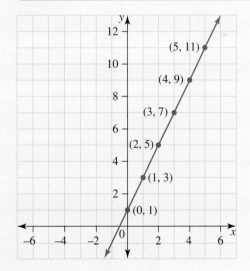

WORKED EXAMPLE 5 Constructing a linear graph

Construct a linear graph that passes through the points $(-1, 2), (0, 4), (1, 6)$ and $(3, 10)$:
a. without technology
b. using CAS.

THINK

a. 1. Using grid paper, rule up the Cartesian plane
   (set of axes) and plot the points.

DRAW/DISPLAY

a.

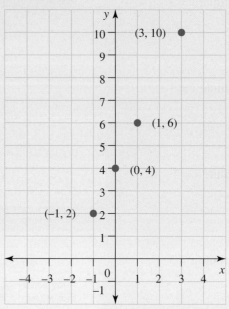

2. Using a ruler, rule a line through the points.

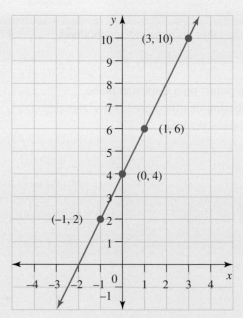

TI	THINK	DISPLAY/WRITE	CASIO	THINK	DISPLAY/WRITE

b. 1. On a Lists & Spreadsheets page, label the first column x and the second column y. Enter the x-coordinates of the given points in the first column, and y-coordinates in the second column.

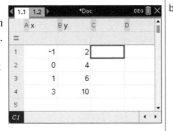

b. 1. On a Statistics screen, label list1 as x and list2 as y. Enter the x-coordinates of the given points in the first column, and y-coordinates in the second column.

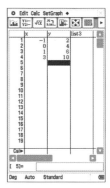

2. On a Data & Statistics page, click the horizontal axis label and select x, then click the vertical axis label and select y.

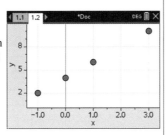

2. Click the G icon and complete the fields as:
Draw: On
Type: LinearR
XList: main\x
YList: main\y
Freq: 1 Then select Set.

3. Press MENU and select:
−4: Analyze
−6: Regression
−2: Show Linear $(a + bx)$

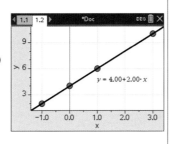

3. Click the y icon.

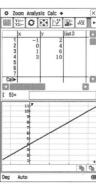

5.3.2 Sketching linear graphs using the gradient and y-intercept

This is useful when you know the starting point of a function and its rate of change. For example, you may only know how much water there is in a tank and the rate at which the water is flowing out.

tlvd-3852

WORKED EXAMPLE 6 Sketching a linear graph using the gradient and y-intercept

Using the gradient and the y-intercept, sketch the graph of each of the following.

a. A linear graph with a gradient of 3 and a y-intercept of 1

b. $y = 4 - 2x$

c. $y = -2 + \dfrac{3}{4}x$

THINK

a. 1. Interpret the gradient.

WRITE/DRAW

a. A gradient of 3 means that for an increase of 1 in the x-value, there is an increase of 3 in the y-value.

2. Write the coordinates of the *y*-intercept.

y-intercept: (0, 1)

3. Determine the *x*- and *y*-values of another point using the gradient.

New *x*-value $= 0 + 1$
$= 1$
New *y*-value $= 1 + 3$
$= 4$
Another point on the graph is (1, 4).

4. Construct a set of axes and plot the two points. Using a ruler, rule a line through the points.

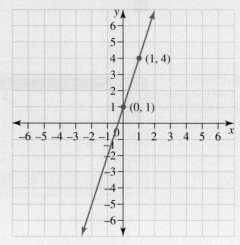

b. 1. Identify the value of the gradient and *y*-intercept.

b. $y = 4 - 2x$ has a gradient of -2 and a *y*-intercept of 4.

2. Interpret the gradient.

Note: Recall that gradient is $\dfrac{\text{rise}}{\text{run}}$.

The gradient of -2 means that for an increase of 1 in the *x*-value, there is a decrease of 2 in the *y*-value.

3. Write the coordinates of the *y*-intercept.

y-intercept: (0, 4)

4. Determine the *x*- and *y*-values of another point using the gradient.

New *x*-value $= 0 + 1$
$= 1$
New *y*-value $= 4 - 2$
$= 2$
Another point on the graph is (1, 2).

5. Construct a set of axes and plot the two points. Using a ruler, rule a line through the points.

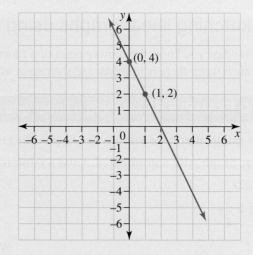

c. 1. Identify the value of the gradient and y-intercept.

c. $y = -2 + \dfrac{3}{4}x$ has a gradient of $\dfrac{3}{4}$ and a y-intercept of -2.

2. Interpret the gradient.

Note: Recall that gradient is $\dfrac{\text{rise}}{\text{run}}$.

A gradient of $\dfrac{3}{4}$ means that for an increase of 4 in the x-value, there is an increase of 3 in the y-value.

3. Write the coordinates of the y-intercept.

y-intercept: $(0, -2)$

4. Determine the x- and y-values of another point using the gradient.

New x-value $= 0 + 4$
$\qquad\qquad\quad = 4$
New y-value $= -2 + 3$
$\qquad\qquad\quad = 1$
Another point on the graph is $(4, 1)$.

5. Construct a set of axes and plot the two points. Using a ruler, rule a line through the points.

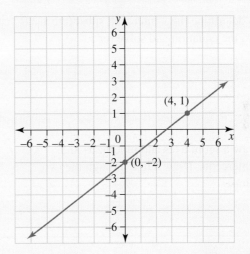

TI	THINK	DISPLAY/WRITE

b. 1. On a Graphs page, complete the entry line for function 1 as: $f1(x) = 4 - 2x$ Then press ENTER.

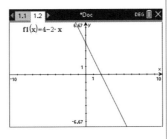

CASIO	THINK	DISPLAY/WRITE

b. 1. On a Graph & Table screen, complete the entry line for equation 1 as: $y1 = 4 - 2x$ Then click the tick box. Click the $ icon to view the graph.

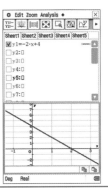

5.3.3 Sketching linear graphs using the x- and y-intercept

When given the x and y-intercepts, plot the coordinates of the intercepts and rule a line through both points.

Determine the values of the x- and y-intercepts for the following linear equations, and sketch their graphs.

a. $3x + 4y = 12$ b. $y = 5x$ c. $3y = 1 + 2x$

THINK | **WRITE/DRAW**

a. 1. To determine the x-intercept, substitute $y = 0$ and solve for x.

a. x-intercept: $y = 0$
$$3x + 4y = 12$$
$$3x + 4 \times 0 = 12$$
$$3x = 12$$
$$\frac{3x}{3} = \frac{12}{3}$$
$$x = 4$$
x-intercept: $(4, 0)$

2. To determine the y-intercept, substitute $x = 0$ into the equation and solve for y.

y-intercept: $x = 0$
$$3x + 4y = 12$$
$$3 \times 0 + 4y = 12$$
$$\frac{4y}{4} = \frac{12}{4}$$
$$y = 3$$
y-intercept: $(0, 3)$

3. Draw a set of axes and plot the x- and y-intercepts. Draw a line through the two points.

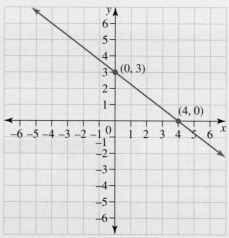

b. 1. To determine the x-intercept, substitute $y = 0$ into the equation and solve for x.

b. x-intercept: $y = 0$
$$y = 5x$$
$$0 = 5x$$
$$x = 0$$
x-intercept: $(0, 0)$

2. To determine the y-intercept, substitute $x = 0$ into the equation and solve for y.

y-intercept: $x = 0$
$$y = 5x$$
$$= 5 \times 0$$
$$= 0$$
y-intercept: $(0, 0)$

3. As the x- and y-intercepts are the same, we need to determine another point on the graph. Substitute $x = 1$ into the equation.

$$y = 5x$$
$$= 5 \times 1$$
$$= 5$$
Another point on the graph is $(1, 5)$.

4. Draw a set of axes. Plot the intercept and the second point. Draw a line through the points.

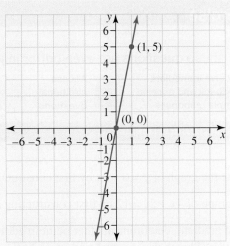

c. 1. To determine the x-intercept, substitute $y = 0$ into the equation.

c. x-intercept: $y = 0$

$$3y = 1 + 2x$$
$$3 \times 0 = 1 + 2x$$
$$0 = 1 + 2x$$

2. Solve the equation for x.

$$0 - 1 = 1 + 2x - 1$$
$$-1 = 2x$$
$$\frac{-1}{2} = \frac{2x}{2}$$
$$x = \frac{-1}{2}$$

x-intercept: $\left(-\dfrac{1}{2}, 0\right)$

3. To determine the y-intercept, substitute $x = 0$ into the equation and solve for y.

y-intercept: $x = 0$

$$3y = 1 + 2 \times 0$$
$$3y = 1$$
$$\frac{3y}{3} = \frac{1}{3}$$
$$y = \frac{1}{3}$$

y-intercept: $\left(0, \dfrac{1}{3}\right)$

4. Draw a set of axes and mark the x- and y-intercepts. Draw a line through the intercepts.

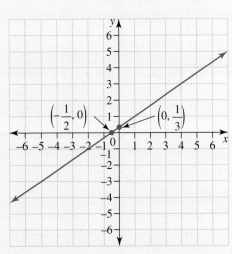

1. **WE4** Plot the graph of $y = 6 - 3x$ by first constructing a table of values, with x-values from 0 to 5.

2. Plot the graph of $4x + 2y = 12$ by first constructing a table of values from $x = 0$ to $x = 4$.

3. Using the gradient, determine another point in addition to the y-intercept that lies on each of the following straight lines. Sketch the graph of each straight line.
   a. Gradient $= 4$; y-intercept $= 3$
   b. Gradient $= -3$; y-intercept $= 1$

4. Using the gradient, determine another point in addition to the y-intercept that lies on each of the following straight lines. Sketch the graph of each straight line.
   a. Gradient $= \dfrac{1}{4}$; y-intercept $= 4$
   b. Gradient $= -\dfrac{2}{5}$; y-intercept $= -2$

5. **WE5** Construct a straight-line graph that passes through the points (2, 5), (4, 9) and (0, 1):
   a. without technology
   b. using CAS.

6. A straight line passes through the following points: (3, 7), (0, a), (2, 5) and (−1, −1). Construct a graph and hence determine the unknown value, a.

7. A straight line passes through the points (2, 5), (0, 9), (−1, 11) and (4, a). Construct a graph of the straight line and hence determine the unknown value, a.

8. A line has a gradient of 5. If it passes through the points (−2, b) and (−1, 7), determine the value of b.

9. **WE6** Using the gradient and the y-intercept, sketch the following linear graphs.
   a. Gradient $= 2$; y-intercept $= 5$
   b. Gradient $= -3$; y-intercept $= 0$
   c. Gradient $= \dfrac{1}{2}$; y-intercept $= 3$

10. **WE7** Determine the values of the x- and y-intercepts for the following linear equations, and sketch their graphs.
    a. $2x + 5y = 20$
    b. $y = 4 + 2x$
    c. $4y = 5 + 3x$

11. Determine the values of the x- and y-intercepts for the following linear equations, and sketch their graphs.
    a. $2x + y = 6$
    b. $y = 9 + 3x$
    c. $2y = 4 + 3x$
    d. $3y - 4 = 5x$

12. The table shows the value of x- and y-intercepts for different linear equations.

Equation	x-intercept	y-intercept
$y = 7 + 2x$	$-\dfrac{7}{2}$	7
$y = 5 + 3x$	$-\dfrac{5}{3}$	5
$y = -1 + 4x$	$\dfrac{1}{4}$	-1
$y = -4 + 2x$	$\dfrac{4}{2} = 2$	-4
$y = 2 + x$	-2	2
$y = 1 + \dfrac{1}{2}x$	$\dfrac{-1}{\frac{1}{2}} = -2$	1
$y = 2 + \dfrac{x}{3}$	$\dfrac{-2}{\frac{1}{3}} = -6$	2

a. Explain how you can determine the x- and y-intercepts for equations of the form shown. State whether this method works for all linear equations.

b. Using your explanation from part **a**, write the x- and y-intercept for the equation $y = a + bx$.

c. A straight line has an x-intercept of $-\dfrac{4}{5}$ and a y-intercept of 4. Write its rule.

5.3 Exam questions

Question 1 (1 mark)

MC The following graph shows a straight line that crosses the x-axis at $(-10, 0)$, passes through the point $(-6, 8)$ and crosses the y-axis at $(0, q)$.

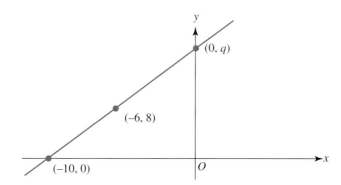

Determine the value of q.
   A. 14   **B.** 16   **C.** 18   **D.** 20   **E.** 22

Question 2 (1 mark)

MC The following graph shows a straight line that passes through the points (6, 6) and (8, 9).

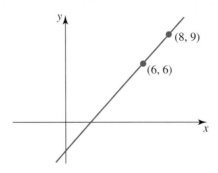

The coordinates of the point where the line crosses the x-axis are

   A. $(-3, 0)$ **B.** $(1, 0)$ **C.** $(1.5, 0)$ **D.** $(2, 0)$ **E.** $(4, 0)$

Question 3 (3 marks)

The equation of a straight line is $y = 4 + \dfrac{1}{2}x$.

   a. State the gradient and the y-intercept.
   b. Determine the x-intercept.
   c. Draw the graph.

More exam questions are available online.

5.4 Linear modelling

> **LEARNING INTENTION**
>
> At the end of this subtopic you should be able to:
> - construct a linear model
> - interpret a linear model
> - identify domains of linear models.

5.4.1 Constructing a linear model

One of the key applications of linear equations is creating a **linear model**. Linear models represent real situations and can be used to make predictions.

When we construct a linear model, we look for two key things:
- the starting state (which is represented by a)
- the rate of change (which is represented by b).

For example, in order to construct a model of the growth of a tree's height over time, it's important to know the starting height and the amount the tree has grown every year.

WORKED EXAMPLE 8 Constructing a linear model

Elle is an occupational therapist who charges an hourly rate of $35 on top of an initial charge of $50. Construct a linear equation to represent Elle's charge, C, for a period of t hours.

THINK	WRITE
1. Identify the constant change and the starting point.	Constant change $= 35$ Starting point $= 50$
2. Construct the equation in terms of C by writing the value of the constant change as the coefficient of the pronumeral (t) that affects the change, and writing the starting point as the y-intercept.	$C = 50 + 35t$

5.4.2 Interpreting a linear model

When we are interpreting a linear model, we look for the same key points as when we are constructing a linear model. We look for the a value, the starting point, and the b value (the rate of change).

tlvd-3853

WORKED EXAMPLE 9 Constructing and evaluating a linear model

A bike tyre with 500 cm³ of air in it is punctured by a nail. After the puncture, the air in the tyre leaks at a rate of 5 cm³/minute.
a. Construct an equation to represent the amount of air, A, in the tyre t minutes after the puncture occurred.
b. State what the value of the gradient in the equation means.
c. Determine the amount of air in the tyre after 12 minutes.
d. By solving your equation from part a, determine how long, in minutes, it will take before the tyre is completely flat (i.e. there is no air left).

THINK	WRITE
a. 1. Identify the constant change and the starting point.	a. Constant change $= -5$ Starting point $= 500$
2. Construct the equation in terms of A by writing the value of the constant change as the coefficient of the pronumeral that affects the change, and writing the starting point as the y-intercept.	$A = 500 - 5t$
b. 1. Identify the value of the gradient in the equation.	b. $A = 500 - 5t$ The value of the gradient is -5.
2. Identify what this value means in terms of the problem.	The value of the gradient represents the rate at which the air is leaking from the tyre. In this case it means that for every minute, the tyre loses 5 cm³ of air.
c. 1. Using the equation found in part a, substitute $t = 12$ and evaluate.	c. $A = 500 - 5t$ $\quad = 500 - 5 \times 12$ $\quad = 440$

2. Write the answer in a sentence.

There is $440\,\text{cm}^3$ of air in the tyre after 12 minutes.

d. 1. When the tyre is completely flat, $A = 0$.

d. $0 = 500 - 5t$

2. Solve the equation for t.

$$0 - 500 = 500 - 5t - 500$$
$$-500 = -5t$$
$$\frac{-500}{-5} = \frac{-5t}{-5}$$
$$100 = t$$

3. Write the answer as a sentence.

The tyre will be flat after 100 minutes.

5.4.3 Domain of a linear model

If we were modelling the height of a child over time, we might come to a growth rate of 10 cm per year. Can you think of how this particular model might be improved?

We need to consider the domain of the linear model — that is, the range of values that the linear model is valid for. If we used the model of growth rate above to make a prediction about how tall a person would be at 30, we would arrive at the prediction that the height of this person will be 340 cm by that time. That's more than half a metre taller than the tallest human who has ever lived, Robert Pershing Wadlow!

He was 2.72 m (8 ft 11.1 in) tall, when he was 22 years old.

From this we can conclude that these equations will only be valid for a certain domain. Anything outside that domain is outside the scope of the linear model. To determine the domain of a linear model, identify all possible x-values.

WORKED EXAMPLE 10 Identifying domains of linear models

Express the following situations as linear models and give their domains.
a. A truck drives across the country for 6 hours at a constant speed of 80 km/h before reaching its destination.
b. The temperature in an ice storage room starts at $-20\,°\text{C}$ and falls at a constant rate of $0.8\,°\text{C}$ per minute for the next 22 minutes.

THINK	WRITE
a. 1. Use pronumerals to represent the information given in the question.	**a.** Let d = the distance travelled by the truck in km. Let t = the time of the journey in hours.
2. Represent the given information as a linear model.	$d = 80t$
3. Determine the domain for which this model is valid.	The model is valid from 0 to 6 hours.

4. Express the domain with the model in algebraic form.	$d = 80t,\ 0 \le t \le 6$	
b. 1. Use pronumerals to represent the information given in the question.	**b.** Let $i =$ the temperature of the ice room. Let $t =$ the time in minutes.	
2. Represent the given information as a linear model.	$i = -20 - 0.8t$	
3. Determine the domain for which this model is valid.	The model is valid from 0 to 22 minutes.	
4. Express the domain with the model in algebraic form.	$i = -20 - 0.8t,\ 0 \le t \le 22$	

5.4 Exercise

Students, these questions are even better in jacPLUS

 Receive immediate feedback and access sample responses

 Access additional questions

 Track your results and progress

Find all this and MORE in jacPLUS ▶

1. **WE8** An electrician charges a call-out fee of $90 plus an hourly rate of $65 per hour. Construct an equation to determine the amount the electrician charges, C, for a period of t hours.

2. An oil tanker is leaking oil at a rate of 250 litres per hour. Initially there was 125 000 litres of oil in the tanker. Construct an equation that represents the amount of oil, A, in litres in the oil tanker t hours after the oil started leaking.

3. A children's swimming pool is being filled with water. The amount of water in the pool at any time can be found using the equation $A = 5 + 20t$, where A is the amount of water in litres and t is the time in minutes.

   a. Explain why this equation can be represented by a straight line.
   b. State the value of the y-intercept and what it represents.
   c. Sketch the graph of $A = 5 + 20t$ on a set of axes.
   d. The pool holds 500 litres. By solving an equation, determine how long it will take to fill the swimming pool. Write your answer correct to the nearest minute.

4. **WE9** A yoga ball is being pumped full of air at a rate of $40\,\text{cm}^3/\text{second}$. Initially there is $100\,\text{cm}^3$ of air in the ball.

   a. Construct an equation that represents the amount of air, A, in the ball after t seconds.
   b. Interpret what the value of the y-intercept in the equation means.
   c. Determine how much air, in cm^3, is in the ball after 2 minutes.
   d. When fully inflated, the ball holds $100\,000\,\text{cm}^3$ of air. Determine how long, in minutes, it takes to fully inflate the ball. Write your answer to the nearest minute.

5. A long-distance runner can run at a rate of 12 km/h. The distance, d, in km they travel from the starting point of a race can be represented by the equation $d = -0.5\,at$.

   a. Write the value of a.
   b. Write the y-intercept. In the context of this problem, explain what this value means.
   c. Determine how far the runner is from the starting point after 30 minutes.
   d. The finish point of this race is 21 km from the starting point. Determine how long, in hours and minutes, it takes the runner to run the 21 km. Give your answer correct to the nearest minute.

6. A large fish tank is being filled with water. After 1 minute the height of the water is 2 cm and after 4 minutes the height of the water is 6 cm. The height of the water in cm, h, after t minutes can be modelled by a linear equation.

   a. Determine the gradient of the graph of this equation.
   b. In the context of this problem, state what the gradient represents.
   c. Using the gradient found in part a, determine the value of the y-intercept. Write your answer correct to 2 decimal places.
   d. State if the fish tank was empty of water before being filled. Justify your answer using calculations.

7. A family deposit $40 in their bank account each week. At the start of the year they have $120 in the account. The amount in dollars, A, that they have in their account after t weeks can be found using the equation $A = a + bt$.

   a. State the values of a and b.
   b. In the context of this problem, state what the y-intercept represents.
   c. Determine the number of weeks it will take the family to save $3000.

8. **WE10** Express the following situations as linear models and give the domains of the models.

   a. Julie works at a department store and is paid $19.20 per hour. She has to work for a minimum of 10 hours per week, but due to her study commitments she can work for no more than 20 hours per week.
   b. The results in a driving test are marked out of 100, with 4 marks taken off for every error made on the course. The lowest possible result is 40 marks.

9. Petrol is being pumped into an empty tank at a rate of 15 litres per minute.

   a. Construct an equation to represent the amount of petrol in litres, P, in the tank after t minutes.
   b. State what the value of the gradient in the equation represents.
   c. If the tank holds 75 litres of petrol, determine the time taken, in minutes, to fill the tank.
   d. The tank had 15 litres of petrol in it before being filled. Write another equation to represent the amount of petrol, P, in the tank after t minutes.
   e. State the domain of the equation formulated in part d.

10. Gert rides to and from work on his bike. The distance and time taken for him to ride home can be modelled using the equation $d = 37 - 22t$, where d is the distance from home in km and t is the time in hours.

    a. Determine the distance, in km, between Gert's work and home.
    b. Explain why the gradient of the line in the graph of the equation is negative.
    c. Solve the equation to determine the time, in hours and minutes, Gert takes to ride home. Write your answer correct to the nearest minute.
    d. State the domain of the equation.
    e. Sketch the graph of the equation.

11. A real estate agent receives a commission of 1.5% on house sales, plus a payment of $800 each month. Their monthly wage can be modelled by the equation $W = a + bx$, where W represents their total monthly wage and x represents the value of their house sales in dollars.
    a. State the values of a and b.
    b. State if there is an upper limit to the domain of the model. Explain your answer.

c. In March the total value of their house sales was $452 000. Determine their monthly wage for March.

d. In September they earned $10 582.10. Determine the value of the house sales they made in September.

12. Monique is setting up a new business selling T-shirts through an online auction site. Her supplier in China agrees to a deal whereby they will supply each T-shirt for $3.50 providing she buys a minimum of 100 T-shirts. The deal is valid for up to 1000 T-shirts.

a. Set up a linear model (including the domain) to represent this situation.

b. Explain what the domain represents in this model.

c. State why there is an upper limit to the domain.

13. A basic mobile phone plan designed for school students charges a flat fee of $15 plus 13 cents per minute of a call. Text messaging is free.

a. Construct an equation that determines the cost, in dollars, for any time spent on the phone, in minutes.

b. In the context of this problem, state what the gradient and y-intercept of the graph of the equation represent.

c. Using CAS, complete the following table to determine the cost at any time, in minutes.

Time (min)	Cost ($)	Time (min)	Cost ($)
5		35	
10		40	
15		45	
20		50	
25		55	
30		60	

5.4 Exam questions

Question 1 (1 mark)

Source: VCE 2017, Further Mathematics Exam 1, Section A, Core, Q12; © VCAA

MC Steven is a wedding photographer.

He charges his clients a fixed fee of $500, plus $250 per hour of photography.

The equation that represents the total amount, $C, Steven charges, for t hours of photography is

A. $C = 250t$ B. $C = 500t$ C. $C = 750t$

D. $C = 500 + 250t$ E. $C = 250 + 500t$

Question 2 (1 mark)

MC The height of a boy, h cm, is related to his age, a years, according to the equation $h = 70.2 + 6.5a$. How old was the boy when he was 161.2 cm tall?

A. 9.8   B. 13   C. 14   D. -14 E. 15.2

▶ **Question 3 (1 mark)**

MC According to my smart phone, a hire car was 15 km away when I called. Its driver undertook that she could average 50 km/h when she drove to pick me up.

What equation can be used to model the distance the car is away from me, d km ($0 \leq d \leq 15$), after time t hours?
- **A.** $d = 50t$
- **B.** $d + 15 = 50t$
- **C.** $d + 50t + 15 = 0$
- **D.** $15 - d = 50t$
- **E.** $d = 50t - 15$

More exam questions are available online.

5.5 Determining equations of straight lines

> **LEARNING INTENTION**
>
> By the end of this subtopic you should be able to:
> - determine the equation of a straight line when the gradient and y-intercept are given
> - determine the equation of a straight line when the gradient and one point are given
> - determine the equation of a straight line when the gradient and two points are given.

5.5.1 Determining the equation of straight lines when given the gradient and y-intercept

When we are given the gradient and y-intercept of a straight line, we can enter these values into the equation $y = a + bx$ to determine the equation of the straight line. Remember that b is equal to the value of the gradient and a is equal to the value of the y-intercept.

For example, if we are given a gradient of 3 and a y-intercept of 6, then the equation of the straight line would be $y = 6 + 3x$.

Sometimes a graph is given and we need to be able to work backwards to determine the equation, and thus work out the initial state and the rate of change of the linear model.

The first strategy we will look at is how we determine the equation of a line when given the gradient of the linear model (slope) and the y-intercept (initial value).

> **WORKED EXAMPLE 11 Determining the equation of a straight line when the gradient and y-intercept are given**
>
> Determine the equation of a straight line with a gradient of 2 and y-intercept of 1.
>
> | THINK | WRITE |
> |---|---|
> | 1. Write the gradient–intercept form of a straight line. | $y = a + bx$ |
> | 2. Substitute the value of the gradient into the equation (in place of b). | $b = 2$
 $y = a + 2x$ |
> | 3. Substitute the value of the y-intercept into the equation (in place of a) and it will give you the answer. | $a = 1$
 $y = 1 + 2x$ |

5.5.2 Determining the equation of straight lines when a gradient and one point are given

When we are given the gradient and one point of a straight line, we need to establish the value of the y-intercept to determine the equation of the straight line. This can be done by substituting the coordinates of the given point into the equation $y = a + bx$ and then solving for a. Remember that b is equal to the value of the gradient, so this can also be substituted into the equation.

> **WORKED EXAMPLE 12 Determining the equation of a straight line when a gradient and one point are given**
>
> Determine the equation of a straight line with a gradient of 2 and passing through point (3, 7).
>
> | THINK | WRITE |
> |---|---|
> | 1. Write the gradient–intercept form of a straight line. | $y = a + bx$ |
> | 2. Substitute the value of the gradient into the equation (in place of b). | Gradient $= b = 2$
 $y = a + 2x$ |
> | 3. Substitute the values of the given point into the equation and solve for a. | (3, 7)
 $7 = a + 2(3)$
 $7 = a + 6$
 $a = 1$ |
> | 4. Substitute the value of a back into the equation and write the answer. | The equation of the straight line is $y = 1 + 2x$. |

5.5.3 Determining the equation of straight lines when two points are given

When we are given two points of a straight line, we can determine the value of the gradient of a straight line between these points as discussed in Section 5.2.2 (by using $b = \dfrac{y_2 - y_1}{x_2 - x_1}$). Once the gradient has been calculated, we can determine the y-intercept by substituting one of the points into the equation $y = a + bx$ and then solving for a.

> **WORKED EXAMPLE 13 Determining the equation of a straight line when given two points**
>
> Determine the equation of the straight line that passes through the points (1, 6) and (3, 0).
>
> | THINK | WRITE |
> |---|---|
> | 1. Write the formula to find the gradient given two points. | $b = \dfrac{y_2 - y_1}{x_2 - x_1}$ |
> | 2. Let one of the given points be (x_1, y_1) and let the other point be (x_2, y_2). | Let $(1, 6) = (x_1, y_1)$ and $(3, 0) = (x_2, y_2)$. |
> | 3. Substitute the values into the equation to determine the value of b. | $b = \dfrac{0 - 6}{3 - 1}$
 $= \dfrac{-6}{2}$
 $= -3$ |

▶

4. Substitute the value of b into the equation $y = a + bx$.
   $$y = a + bx$$
   $$y = a - 3x$$

5. Substitute the values of one of the points into the equation and solve for a.
   Note: The point $(1, 6)$ could also be used.
   $$(3, 0)$$
   $$0 = a - 3(3)$$
   $$0 = a - 9$$
   $$a = 9$$

6. Substitute the value of a back into the equation and write the answer.
   The equation of the straight line is $y = 9 - 3x$.

Note: The equation of a line can also be determined by using CAS and following the instructions in Worked example 5 and inputting the coordinates of the two points.

These three strategies for determining the equation of a straight line can help us to create a linear model when we are given different information. It can also help us to work backwards to develop a linear model based on things we observe and measure.

5.5 Exercise

Students, these questions are even better in jacPLUS

 Receive immediate feedback and access sample responses

 Access additional questions

 Track your results and progress

Find all this and MORE in jacPLUS

1. **WE11** Determine the equation of each straight line given the gradient and y-intercept.
   a. Gradient $= 2$; y-intercept $= -3$
   b. Gradient $= 1$; y-intercept $= 4$
   c. Gradient $= -3$; y-intercept $= 1$
   d. Gradient $= -2$; y-intercept $= -1$

2. **WE12** Determine the equation of each straight line given the gradient and the coordinates of a point.
   a. Gradient $= 1$; point $= (3, 7)$
   b. Gradient $= 3$; point $= (3, 3)$
   c. Gradient $= -2$; point $= (-4, 10)$
   d. Gradient $= \dfrac{1}{2}$; point $= (8, 0)$

3. **WE13** For each of the following, determine the equation of the line that passes through the given points.
   a. $(-3, 4)$ and $(1, 6)$
   b. $(-3, 7)$ and $(0, 7)$
   c. $(2, 8)$ and $(5, 14)$
   d. $(1, -6)$ and $(-3, 10)$

4. **MC** Select which of the following equations represents the line that passes through the points $(3, 8)$ and $(12, 35)$.
   A. $y = 1 + 3x$
   B. $y = 1 - 3x$
   C. $y = -1 + 3x$
   D. $y = 1 + \dfrac{1}{3}x$
   E. $y = -1 + \dfrac{1}{3}x$

5. Steve is looking at data comparing the size of different music venues across the country and the average ticket price at these venues. He calculates the cost to be $y = 15 + 0.04x$, where y is the average ticket price in dollars and x is the capacity of the venue.
   a. Explain what the value of the gradient (b) represents in Steve's equation.
   b. Explain what the value of the y-intercept (a) represents in Steve's equation.
   c. Explain whether the y-intercept is a realistic value for this data.

6. A linear model has been created for the distance a person has walked over a period of time. The person was at a distance of 5 m at 3 minutes and 9 m at 5 minutes. Determine the equation that represents this situation.

7. Abdul invested $50 000 into a term deposit. The investment makes $500 per year. Create a linear model to represent how much the investment is worth, V, as time, t (years), passes.

8. Tommy hires a cinema and invites friends to watch a movie he made. The theatre has a package where it costs $110 to hire the theatre and a $15 ticket is charged per person.

   a. Using C for the cost of hiring the venue and n for the number of friends, write an equation to represent this scenario.
   b. Use the equation to calculate how much it would cost in total for Tommy to invite 12 friends.

9. Yamato, a tree enthusiast, wanted to create a linear model to predict the height of a tree over time. Yamato measured the tree after one year and found that it was 1.5 m tall. He measured the tree two years later and found that it was 4 m tall.

   a. Create a linear model to represent the tree's growth, using H for the tree's height (m) and t for the number of years.
   b. Use the equation to determine how tall the tree was when it was first put into the ground.
   c. Use the equation to predict how tall the tree would be at the start of the fifth year.

10. Snoozy caught a cab and fell asleep while the cab was travelling to their destination. When they woke up, Snoozy noticed the meter read $200. The cab charged a flat fee of $9.50, and 50 cents per km travelled.

   a. Write an equation for the cost of the cab.
   b. Solve to determine how far the cab had travelled before Snoozy woke up.

5.5 Exam questions

Question 1 (1 mark)
Source: VCE 2018, Further Mathematics Exam 1, Module 4, Q1; © VCAA

MC The graph below shows a line intersecting the x-axis at $(4, 0)$ and the y-axis at $(0, 2)$.

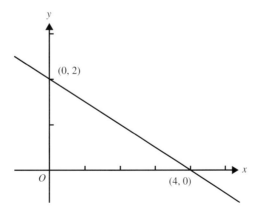

The gradient of this line is

   A. -4 **B.** -2 **C.** $-\dfrac{1}{2}$ **D.** $\dfrac{1}{2}$ **E.** 4

Question 2 (1 mark)

Source: VCE 2018, NH Further Mathematics Exam 1, Module 4, Q1; © VCAA

MC A straight line passes through the point $(-2, 0)$ and $(0, 2)$, as shown in the diagram below.

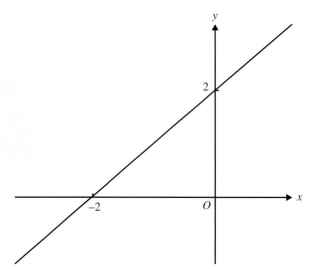

The equation of this straight line is

    A. $y = 2x$ **B.** $y = x + 2$ **C.** $y = 2x + 2$ **D.** $y = x - 2$ **E.** $y = 2x - 2$

Question 3 (1 mark)

MC The amount needed to pay a taxi fare, $\$C$, to travel a distance of x kilometres is given by the rule:

$$C = a + bx$$

To travel a distance of 10 kilometres, the taxi fare is $40.

To travel a distance of 15 kilometres, the taxi fare is $55.

The cost per kilometre, b, is

    A. 2   **B.** 4   **C.** 3   **D.** 2.5   **E.** 3.5

More exam questions are available online.

5.6 Piecewise linear graphs and their application

LEARNING INTENTION

At the end of this subtopic you should be able to:
- evaluate and sketch piecewise linear graphs.

5.6.1 Piecewise linear graphs

A piecewise graph is a graph that is made up of two or more linear equations. Often these are used to represent a situation that involves a transition or a change in a rate. The piecewise graph will be continuous, and the equations of the lines will intersect at each transition point.

Step graphs are formed by two or more linear graphs that have zero gradients. Step graphs have breaks, as shown in the diagram. The end points of each line depend on whether the point is included in the interval.

For example, the interval $-1 < x \le 5$ will have an open end point at $x = -1$, because x does not equal -1 in this case. The same interval will have a closed end point at $x = 5$, because x is less than or equal to 5.

A closed end point means that the x-value is also 'equal to' the value. An open end point means that the x-value is not equal to the value; that is, it is less than or greater than only.

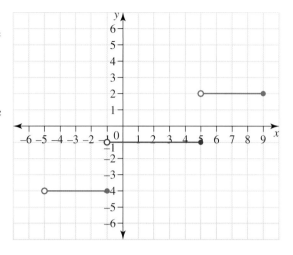

WORKED EXAMPLE 14 Constructing a step graph from the equations

Construct a step graph from the following equations, making sure to take note of the relevant end points.

$$y = 1,\ -3 < x \le 2$$
$$y = 4,\ 2 < x \le 4$$
$$y = 6,\ 4 < x \le 6$$

THINK

1. Construct a set of axes and draw each line within the stated x-intervals.

WRITE/DRAW

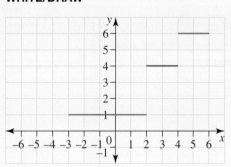

2. Draw in the end points.

For the line $y = 1$:
$-3 < x \le 2$
$x > -3$ is an open circle.
$x \le 2$ is a closed circle.

For the line $y = 4$:
$2 < x \le 4$
$x > 2$ is an open circle.
$x \le 4$ is a closed circle.

For the line $y = 6$:
$4 < x \le 6$
$x > 4$ is an open circle.
$x \le 6$ is a closed circle.

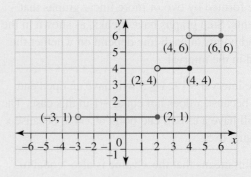

tlvd-3854

WORKED EXAMPLE 15 Evaluating and sketching piecewise linear graph

A piecewise linear graph is constructed from the following linear graphs.

$$y = 1 + 2x, \; x \le a$$
$$y = -1 + 4x, \; x > a$$

a. **By solving the equations simultaneously, determine the point of intersection and hence state the value of a.**

b. **Sketch the piecewise linear graph.**

THINK	WRITE/DRAW
a. 1. Determine the intersection point of the two graphs by solving the equations simultaneously.	a. $y = 1 + 2x$ $y = -1 + 4x$ Solve by substitution: $\quad\quad 1 + 2x = -1 + 4x$ $1 + 2x - 2x = -1 + 4x - 2x$ $\quad\quad\quad\quad 1 = -1 + 2x$ $\quad\quad 1 + 1 = -1 + 2x + 1$ $\quad\quad\quad\quad 2 = 2x$ $\quad\quad\quad\quad x = 1$ Substitute $x = 1$ to determine y: $y = 1 + 2(1)$ $\quad = 3$ The point of intersection is $(1, 3)$. $x = 1$ and $y = 3$
2. The x-value of the point of intersection determines the x-intervals for where the linear graphs meet.	$x = 1$, therefore $a = 1$.

b. 1. Using CAS, sketch the two graphs without taking into account the intervals.

b.

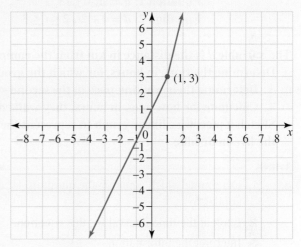

2. Identify which graph exists within the stated x-intervals to sketch the piecewise linear graph.

$y = 1 + 2x$ exists for $x \le 1$.
$y = -1 + 4x$ exists for $x > 1$.
Remove the sections of each graph that do not exist for these values of x.

TI	THINK	DISPLAY/WRITE	CASIO	THINK	DISPLAY/WRITE

a 1. On a Calculator page, press MENU then select:
3: Algebra
1: Solve
Complete the entry line as:solve $(y = 1 + 2x$ and $y = -1 + 4x), \{x, y\}$ then press ENTER.

a 1. On the Main screen, use the template, which is found in Math1 and complete the entry lines as:
$$\begin{cases} y = 1 + 2x \\ y = -1 + 4x \end{cases} | x, y$$
Then press EXE.

2. The answer is shown on the screen.

$x = 1$, hence $a = 1$.

2. The answer is shown on the screen.

$x = 1$, hence $a = 1$.

b 1. On a Graphs page, complete the entry line for function 1 as:
$$f1(x) = \begin{cases} 1 + 2x \,|\, x \leq 1 \\ -1 + 4x \,|\, x > 1 \end{cases}$$
then press ENTER.
Note: The piecewise template can be found by pressing 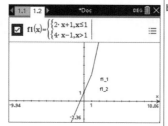 and then select one of them.

b 1. On a Graph & Table screen, complete the entry line for equation 1 as:
$$y1 = \begin{cases} 1 + 2x \,|\, x \leq 1 \\ -1 + 4x \,|\, x > 1 \end{cases}$$
Click the tick box and then click the $ icon.
Note: The piecewise template can be found in the Math3 tab in the Keyboard menu.

WORKED EXAMPLE 16 Evaluating and sketching a piecewise graph

The following two equations represent the distance travelled by a group of students over 5 hours. Equation 1 represents the first section of the hike, when the students are walking at a pace of 4 km/h. Equation 2 represents the second section of the hike, when the students change their walking pace.

$$\text{Equation 1: } d = 4t, \ 0 \leq t \leq 2$$
$$\text{Equation 2: } d = 4 + 2t, \ 2 \leq t \leq 5$$

The variable d is the distance in km from the campsite, and t is the time in hours.
a. Determine the time, in hours, for which the group travelled in the first section of the hike.
b. i. Determine their walking pace in the second section of their hike.
 ii. Calculate for how long, in hours, they walked at this pace.
c. Sketch a piecewise linear graph to represent the distance travelled by the group of students over the five-hour hike.

THINK	WRITE/DRAW
a. 1. Determine which equation the question applies to.	**a.** This question applies to equation 1.
2. Look at the time interval for this equation.	$0 \leq t \leq 2$
3. Interpret the information.	The group travelled for 2 hours.
b. i. 1. Determine which equation the question applies to.	**b. i.** This question applies to equation 2.
2. Interpret the equation. The walking pace is found by the coefficient of t, as this represents the gradient.	$d = 4 + 2t, \ 2 \leq t \leq 5$ The coefficient of t is 2.
3. Write the answer as a sentence.	The walking pace is 2 km/h.
ii. 1. Look at the time interval shown.	**ii.** $2 \leq t \leq 5$
2. Interpret the information and answer the question.	They walked at this pace for 3 hours.
c. 1. Calculate the distance travelled before the change of pace.	**c.** Change after $t = 2$ hours: $d = 4t$ $d = 4 \times 2$ $d = 8$ km

2. Using a calculator, spreadsheet or otherwise, sketch the graph $d = 4t$ between $t = 0$ and $t = 2$.

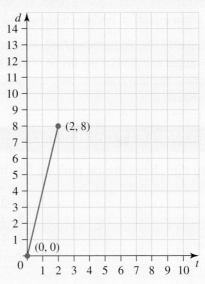

3. Solve the simultaneous equations to determine the point of intersection.

$$4t = 4 + 2t$$
$$4t - 2t = 4 + 2t - 2t$$
$$2t = 4$$
Substitute $t = 2$ into $d = 4t$:
$$d = 4 \times 2$$
$$= 8$$

4. Using CAS, sketch the graph of $d = 4 + 2t$ between $t = 2$ and $t = 5$.

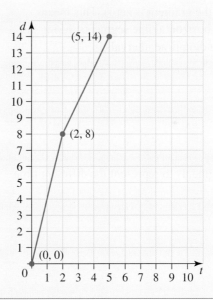

5.6.2 Modelling with piecewise linear and step graphs

Consider the real-life situation of a leaking water tank. For the first 3 hours it leaks at a constant rate of 12 litres per minute; after 3 hours the rate of leakage slows down (decreases) to 9 litres per minute. The water leaks at a constant rate in both situations and can therefore be represented as a linear graph. However, after 3 hours the slope of the line changes because the rate at which the water is leaking changes.

Because there will be multiple different linear equations representing different parts of the situation, we need to be mindful to use the appropriate equation for what we want to find.

A key thing to remember when solving problems with piecewise graphs is to calculate based on the correct corresponding linear equation, because there will be a different equation for each section of the graph.

In Worked example 16, the equation is $d = 4t$ between the start and 2 hours, and $d = 4 + 2t$ between 2 hours and 5 hours. If we wanted to calculate the distance travelled after 4 hours, we would use the second equation, not the first. This is because the first equation only applies up to the 2-hour mark and the second equation applies to all times from 2 to 5 hours.

WORKED EXAMPLE 17 Constructing a step graph to represent the information

The following sign shows the car parking fees in a shopping carpark.

$$0-2\,\text{hours}\ \$1.00$$
$$2-4\,\text{hours}\ \$2.50$$
$$4-6\,\text{hours}\ \$5.00$$
$$6+\,\text{hours}\ \$6.00$$

Construct a step graph to represent this information.

THINK	WRITE/DRAW
1. Draw up a set of axes, labelling the axes in terms of the context of the problem, that is, the time and cost. There is no change in cost during the time intervals, so there is no rate (i.e. the gradient is zero). This means we draw horizontal line segments during the corresponding time intervals.	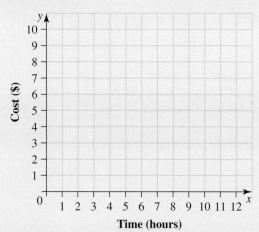
2. Draw segments to represent the different time intervals.	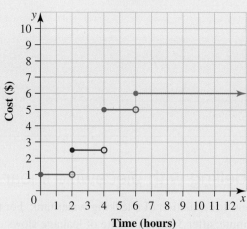

Applying piecewise linear graphs to tax brackets

One application of piecewise linear graphs is in the construction of tax brackets. Depending on how much a person earns, they will be required to pay differing rates of tax, calculated from a piecewise function.

The following table illustrates Australian income tax brackets in 2021–2022:

Taxable income	Tax on this income
0 – $18 200	Nil
$18 201 – $45 000	19 cents for each $1 over $18 200
$45 001 – $120 000	$5 092 plus 32.5 cents for each $1 over $45 000
$120 001 – $180 000	$29 467 plus 37 cents for each $1 over $120 000
$180 001 and over	$51 667 plus 45 cents for each $1 over $180 000

5.6 Exercise

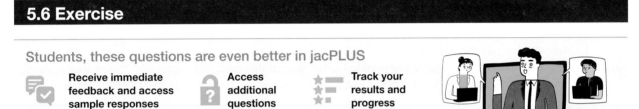

Students, these questions are even better in jacPLUS

Receive immediate feedback and access sample responses

Access additional questions

Track your results and progress

Find all this and MORE in jacPLUS

1. **WE14** Construct a step graph from the following equations, making sure to take note of the relevant end points.

$$y = 3,\ 1 < x \le 4$$
$$y = 1.5,\ 4 < x \le 6$$
$$y = -2,\ 6 < x \le 8$$

2. A step graph is shown below. Write the equations that make up the graph.

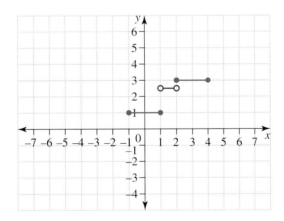

3. **WE15** A piecewise linear graph is constructed from the following linear graphs.

$$y = -3x - 3,\ x \le a$$
$$y = x + 1,\ x \ge a$$

a. By solving the equations simultaneously, find the point of intersection and hence state the value of a.
b. Sketch the piecewise linear graph.

4. Consider the following linear graphs that make up a piecewise linear graph.

$$y = 2x - 3, \ x \le a$$
$$y = 3x - 4, \ a \le x \le b$$
$$y = 5x - 12, \ x \ge b$$

a. Using CAS, sketch the three linear graphs.
b. Determine the two points of intersection.
c. Using the points of intersection, find the values of a and b.
d. Sketch the piecewise linear graph.

5. **WE16** The following two equations represent water being added to a water tank over 15 hours, where w is the water in litres and t is the time in hours.

$$\text{Equation 1: } w = 25t, \ 0 \le t \le 5$$
$$\text{Equation 2: } w = -25 + 30t, \ 5 \le t \le 15$$

a. Determine how many litres of water are in the tank after 5 hours.
b. i. State at what rate the water is being added to the tank after 5 hours.
   ii. Determine for how long the water is added to the tank at this rate.
c. Sketch a piecewise graph to represent the water in the tank at any time, t, over the 15-hour period.

6. The following table shows the costs to hire a plumber.

Time (minutes)	Cost ($)
0–15	45
15–30	60
30–45	80
45–60	110

a. Represent this information on a step graph.
b. Anton hired the plumber for a job that took 23 minutes. Determine how much Anton can expect to be charged for this job.

7. Airline passengers are charged an excess for any luggage that weighs 20 kg or over. The following graph shows these charges for luggage weighing over 20 kg.

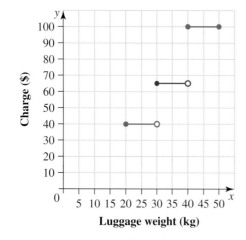

a. Determine how much excess a passenger will be charged for luggage that weighs 31 kg.
b. Nerada checks in her luggage and is charged $40. Determine the maximum excess luggage she could have without having to pay any more.
c. Hilda and Hanz have two pieces of luggage between them. One piece weighs 32 kg and the other piece weighs 25 kg. Explain how they could minimise their excess luggage charges.

8. A car hire company charges a flat rate of $50 plus 75 cents per kilometre up to and including 150 kilometres. An equation to represent this cost, C, in dollars is given as $C = 50 + ak$, $0 \leq k \leq b$, where k is the distance travelled in kilometres.

a. Write the values of a and b.
b. Using CAS, sketch this equation on a set of axes, using appropriate values.

The cost charged for distances over 150 kilometres is given by the equation $C = 87.50 + 0.5k$.

c. Determine the charge in cents per kilometre for distances over 150 kilometres.
d. By solving the two equations simultaneously, determine the point of intersection and hence show that the graph will be continuous.
e. Sketch the equation $C = 87.50 + 0.5k$ for $150 \leq k \leq 300$ on the same set of axes as part b.

9. For the following piecewise graph, identify the equations used.

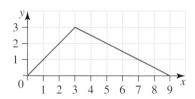

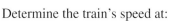_____, $0 \leq x < 3$
_____, $3 \leq x < 9$

10. An N700 series bullet train was running express from Kyoto to Osaka. Its speed is given by the following equations, where v is velocity (km/h) and t is time in minutes.

$$v = 90t, \ 0 \leq t < 3$$
$$v = 270, \ 3 \leq t < 11$$
$$v = 1260 - 90t, \ 11 \leq t < 15$$

Determine the train's speed at:
a. $t = 1$ b. $t = 4$ c. $t = 8$ d. $t = 12$

11. The speed of a model race car along a racetrack is given by the following graph. Determine the speed at $t=1$, $t=2$, $t=5$ and $t=7$.

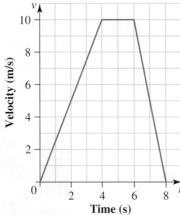

12. Giorno monitored the growth of a tomato plant over the course of a year. The growth over the first year followed two equations, but Giorno forgot what order they were in and where they intersected.

$$H = 5 + 3t$$
$$H = 4t$$

He remembers that the plant grew fastest in the first 6 months. Determine when the growth of the plant started slowing down. (*Hint:* When did the gradient in the piecewise function change?)

13. The temperature of a wood-fired oven, T °C, steadily increases until it reaches 200 °C. Initially the oven has a temperature of 18 °C and it reaches the temperature of 200 °C in 10 minutes.

a. Construct an equation to be used to determine the temperature of the oven during the first 10 minutes. Include the time interval, t, in your answer.

Once the oven has heated up for 10 minutes, a loaf of bread is placed in the oven to cook for 20 minutes. An equation that represents the temperature of the oven during the cooking of the bread is $T = 200$, $a \le t \le b$.

b. i. Write the values of a and b.
   ii. In the context of this problem, state what a and b represent.

After the 20 minutes of cooking, the oven's temperature is lowered. The temperature decreases steadily, and after 30 minutes reaches 60 °C. An equation that determines the temperature of the oven during the last 30 minutes is $T = mt + 340$, $d \le t \le e$.

c. Determine the values of m, d and e.

d. State what m represents in this equation.

e. Using your values from the previous parts, sketch the graph that shows the changing temperature of the wood-fired oven during the 60-minute interval.

14. The amount of money in a savings account over 12 months is shown in the following piecewise graph, where A is the amount of money in dollars and t is the time in months.

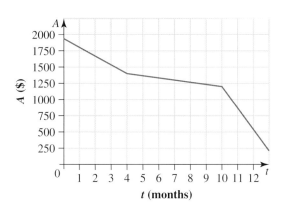

One of the linear graphs that make up the piecewise linear graph is $A = 2000 - 150t$, $0 \leq t \leq a$.

a. Determine the value of a.

b. The equation that intersects with $A = 2000 - 150t$ is given by $A = b - 50t$. If the two equations intersect at the point (4, 1400), show that $b = 1600$.

c. The third equation is given by the rule $A = 4100 - 300t$. By solving a pair of simultaneous equations, determine the time interval for this equation.

d. Using an appropriate equation, determine the amount of money in the account at the end of the 12 months.

15. The following linear equations represent the distance sailed by a yacht from the yacht club during a race, where d is the distance in kilometres from the yacht club and t is the time in hours from the start of the race.

Equation 1: $d = 20t$, $0 \leq t \leq 0.75$

Equation 2: $d = 3.75 + 15t$, $0.75 \leq t \leq 1.25$

Equation 3: $d = 37.5 - 12t$, $1.25 \leq t \leq b$

a. Using CAS, determine the points of intersection.

b. In the context of this problem, explain why equation 3 has a negative gradient.

c. Calculate how far the yacht is from the starting point before it turns and heads back to the yacht club.

d. Determine the duration, to the nearest minute, of the yacht's sailing time for this race. Hence, determine the value of b.

Write your answer correct to 2 decimal places.

16. A small inflatable swimming pool that holds 1500 litres of water is being filled using a hose. The amount of water, A, in litres in the pool after t minutes is shown in the following graph.

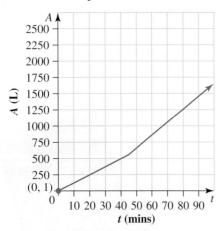

a. Estimate the amount of water, in litres, in the pool after 45 minutes.

b. Determine the amount of water being added to the pool each minute during the first 45 minutes.

After 45 minutes the children become impatient and turn the hose up. The equation $A = -359 + 20t$ determines the amount of water, A, in the pool t minutes after 45 minutes.

c. Using this equation, determine the time taken, in minutes, to fill the pool. Give your answer to the nearest whole minute.

17. Consider the following tax bracket equations.

$$T = 0 \quad \{0 \leq x \leq 18\,200\}$$
$$T = 0.19\,(x - 18\,200) \quad \{18\,200 < x \leq 45\,000\}$$
$$T = 5092 + 0.325\,(x - 45\,000) \quad \{45\,000 < x \leq 120\,000\}$$

where $x =$ income, and $T =$ income tax to be paid.

Determine the income tax to be paid for a person who has earned:

a. $16\,000

b. $40\,000

c. $82\,000

d. $65\,000

18. **WE17** The costs to hire a paddle boat are listed in the table.

Time (minutes)	Hire cost ($)
0–20	15
20–30	20
30–40	25

Construct a step graph to represent the cost of hiring a paddle boat for up to 40 minutes.

19. The postage costs to send parcels from the Northern Territory to Sydney are shown in the following table:

Weight of parcel (kg)	Cost ($)
0–0.5	6.60
0.5–1	16.15
1–2	21.35
2–3	26.55
3–4	31.75
4–5	36.95

a. Represent this information in a step graph.
b. Pammie has two parcels to post to Sydney from the Northern Territory. One parcel weighs 450 g and the other weighs 525 g. Is it cheaper to send the parcels individually or together? Justify your answer using calculations.

5.6 Exam questions

Question 1 (1 mark)

MC The graph below shows the cost, C, of printing n wedding invitations.

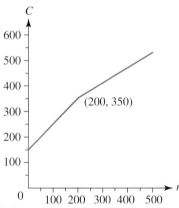

A function that can be used to model this is

$$C = \begin{cases} n + 150 & 0 \leq n < 200 \\ 0.6n + p & 200 \leq n \leq 500 \end{cases}$$

The value of p is
  A. 30
  B. 150
  C. 230
  D. 380
  E. 470

Question 2 (1 mark)

MC The graph for the relationship described below is

$$y = \begin{cases} 2x & \text{if } 0 \le x \le 2 \\ x+2 & \text{if } 2 \le x \le 8 \\ 10 & \text{if } 8 \le x \le 10 \end{cases}$$

A.
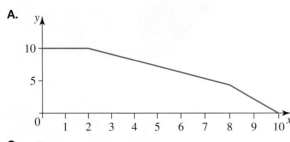

B.
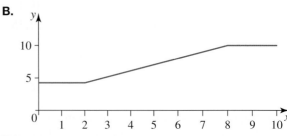

C.
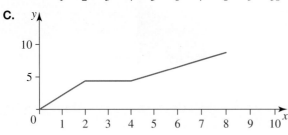

D.
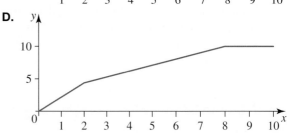

E.
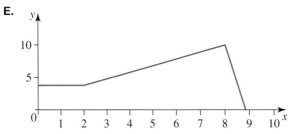

Question 3 (1 mark)

MC The relationship for this graph is

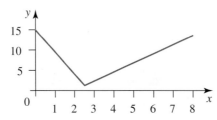

A. $y = \begin{cases} 15 - 5x & \text{if } 0 \le x \le 2.5 \\ -2.5 + 3x & \text{if } 2.5 \le x \le 8 \end{cases}$

B. $y = \begin{cases} 15 - 5x & \text{if } 0 \le x \le 2.5 \\ -2.5 + 2x & \text{if } 2.5 \le x \le 8 \end{cases}$

C. $y = \begin{cases} 15 - 5x & \text{if } 0 \le x \le 3 \\ -2.5 + 2x & \text{if } 3 \le x \le 8 \end{cases}$

D. $y = \begin{cases} 15 - 5x & \text{if } 0 \le x \le 2 \\ -2.5 + 2x & \text{if } 2 \le x \le 8 \end{cases}$

E. $y = \begin{cases} 2x & \text{if } 0 \le x \le 2 \\ x+2 & \text{if } 2 \le x \le 8 \\ 10 & \text{if } 8 \le x \le 10 \end{cases}$

More exam questions are available online.

5.7 Review

5.7.1 Summary

doc-37615

5.7 Exercise

Multiple choice

1. **MC** The gradient of the line passing through the points (4, 6) and (−2, −6) is:
   A. −2
   B. −0.5
   C. 0
   D. 0.5
   E. 2

2. **MC** The *x*- and *y*-intercepts of the linear graph with equation $3x − y = 6$ are:
   A. (2, 6)
   B. (0, 2) and (−6, 0)
   C. (0, 2) and (6, 0)
   D. (2, 0) and (0, −6)
   E. (2, 0) and (0, 6)

3. **MC** The gradient of the graph shown in the following diagram is:

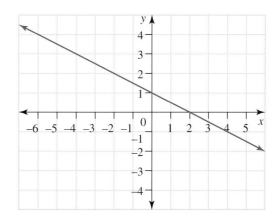

   A. −2
   B. −1
   C. $−\dfrac{1}{2}$
   D. 1
   E. $\dfrac{1}{2}$

4. **MC** Select the sketch of the graph with equation $y = 2 + 4x$.

A.
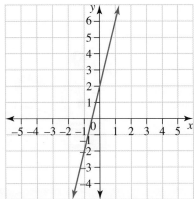

B.
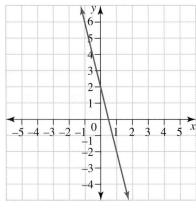

C.
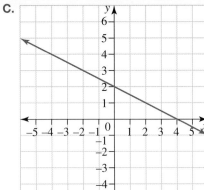

D.
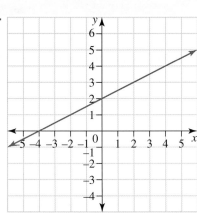

E.
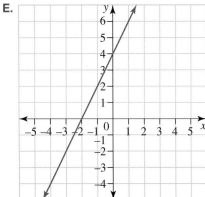

5. **MC** This step graph shows the parking fees for a multilevel carpark in a major city.

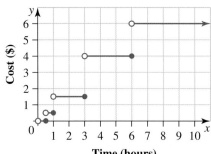

Roberta parked in the carpark and was charged $4.00. If she arrived in the carpark at 9.30 am, the time at which she most likely drove out was:

A. 10.30 am   B. 11.00 pm   C. 12.30 pm   D. 1.30 pm   E. 3.00 pm

6. **MC** This piecewise graph shows the cost of renting a ski equipment package over a period of days.

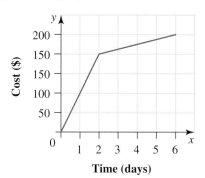

Archibald decides to get the ski equipment package for 3 days. Select which of the following is closest to the amount he would pay for equipment hire.

A. 145   B. 150   C. 163   D. 187   E. 125

7. **MC** A piecewise linear graph is constructed from the following linear equations:
$$y = -5 + 3x, \ x \leq a$$
$$y = -4 + 4x, \ a \leq x \geq b$$
$$y = -5 + 6x, \ x \geq b$$

The values of a and b are respectively:

A. -1 and $\dfrac{1}{2}$.
      B. $\dfrac{1}{2}$ and -1.
      C. $-\dfrac{1}{2}$ and 1.

D. 0 and 1.
      E. -4 and -5.

The following information relates to questions 8 and 9.

An inflated party balloon has a small hole and is slowly deflating. The initial volume of the balloon is $1000 \, \text{cm}^3$ and the balloon loses $5 \, \text{cm}^3$ of air every minute.

8. **MC** If V represents the volume of the balloon in cm^3 and t represents the time in minutes, the equation that represents the volume of the balloon after t minutes is:

A. $V = \dfrac{1000}{5t}$
      B. $V = \dfrac{1000 - 5t}{60}$
      C. $V = 1000 + 5t$

D. $V = 1000 - 5t$
      E. $V = 200 - t$

9. **MC** The time needed for the balloon to lose $650 \, \text{cm}^3$ of air is:

A. 70 minutes.   B. 130 minutes.   C. 270 minutes.   D. 330 minutes.   E. 930 minutes.

10. **MC** The equation of a straight line passing through the points $(-2, 3)$ and $(5, 1)$ is:

A. $y = 2\dfrac{3}{7} - \dfrac{2}{7}x$
      B. $y = -\dfrac{2}{7} + 2\dfrac{3}{7}x$
      C. $y = 2\dfrac{3}{7} + \dfrac{2}{7}x$

D. $y = -2\dfrac{3}{7} + \dfrac{2}{7}x$
      E. $y = 2\dfrac{3}{7} + \dfrac{2}{7}x$

Short answer

11. Sketch the following graphs by finding the x- and y-intercepts.

a. $2x + y = 5$ b. $y - 4x = 8$ c. $4(x + 3y) = 16$ d. $3x + 4y - 10 = 0$

12. Determine the gradients of the lines passing through the following pairs of points.

a. $(3, -2)$ and $(0, 4)$
      b. $(5, 11)$ and $(-2, 18)$

c. $(0.3, 4.1)$ and $(1.2, 5.3)$
      d. $\left(\dfrac{2}{5}, \dfrac{1}{4} \right)$ and $\left(-\dfrac{1}{4}, \dfrac{3}{5} \right)$

13. A line has a gradient of $-\dfrac{3}{4}$. If the line passes through the points $(-a,\ 3)$ and $(-2,\ 6)$, determine the value of a.

14. Complete the following table.

	Equation	Gradient	y-intercept	x-intercept
a	$y = -3 + 5x$	5		
b	$y = 1 + 3x$			
c	$6x - 3y = 9$			
d	$2y + 4x = 8$			
e			5	−5
f		2		2

15. The following two linear equations make a piecewise linear graph.

$$y = 1 - 2x,\ x \le a$$
$$y = 2 - 3x,\ x \ge a$$

a. Solve the equations simultaneously, and hence determine the value of a.
b. Sketch the piecewise linear graph.

Extended response

16. The recommended maximum heart rate during exercise is given by the equation

$$H = 0.85(220 - A)$$

where H is the person's heart rate in beats per minute and A is their age in years.

a. Explain why the maximum heart rate is given by a linear equation.
b. Determine the recommended maximum heart rate for a 25-year-old person. Write your answer correct to the nearest whole number.
c. Determine the gradient and y-intercept of the linear equation.
d. Using your answers from part **c**, sketch the graph that shows the recommended maximum heart rate for persons aged 20 to 70 years.
e. Charlie is working out at the recommended maximum heart rate. His measured heart rate is 162 beats per minute. By solving a linear equation, determine Charlie's age.
f. In the context of this problem, explain why determining the x-intercept would be meaningless.

17. Jerri and Samantha have both entered a 10-km fun run for charity. The distance travelled by Jerri can be modelled by the linear equation

$$d = -0.1 + 6t$$

where d is the distance in km from the starting point and t is time in hours.

a. Determine the time taken for Jerri to run the 10 kilometres. Give your answer correct to the nearest minute.
b. In the context of this problem, explain the meaning of the d-intercept (y-intercept).

The distance Samantha is from the starting point at any time, t hours, can be modelled by the piecewise linear graph

$$d = 4t, \ 0 \le t \le \frac{1}{2}$$

$$d = -2 + 8t, \ \frac{1}{2} \le t \le b$$

c. Calculate how far, in kilometres, Samantha travelled in the first 30 minutes.
d. Determine the speed at which Samantha was travelling in the first 30 minutes.
e. Explain how Samantha's run changed after 30 minutes.
f. i. Determine the value of b.
   ii. Hence, show that Samantha crossed the finishing line ahead of Jerri by 11 minutes.
g. By solving a pair of simultaneous equations, determine:

   i. the time at which Samantha passed Jerri on the run
   ii. the distance from the starting point at which Samantha passed Jerri.

h. Construct two graphs on the same set of axes to represent the distances travelled by Jerri and Samantha in the 10-km race.

18. Trudy is unaware that there is a small hole in the petrol tank of her car. Petrol is leaking out of the tank at a constant rate of 5 mL/min. Trudy has parked her car in a long-term carpark at the airport and gone on a holiday. Initially there are 45 litres of petrol in the tank.

a. In terms of linear graphs, state what the leaking rate and the initial amount of petrol determine.
b. Determine how many litres of petrol leak out of the tank each hour. Explain what assumption is being made about the rate of petrol leaking each hour.
c. Determine how many litres of petrol are lost after four hours.
d. Determine after how many hours there will be 39.75 litres of petrol in the tank.
e. An equation is used to represent the amount of petrol left in the tank, l, after t hours.

   i. Explain why the amount of petrol in the tank would be best modelled by a linear equation.
   ii. Explain why the linear equation will have a negative gradient.
   iii. Write an equation to determine the amount of petrol left in the tank, l, after t hours.

f. Use CAS to aid in sketching the graph of the equation found in part e iii. Clearly label the x- and y-intercepts.
g. Determine how many hours it will take for the petrol tank to become empty.

5.7 Exam questions

▶ **Question 1 (1 mark)**

MC A carpet steam-cleaning company charges $25 to clean one room and $75 to clean six rooms. Which of the given equations is a possible linear model for this steam-cleaning charge (C) if r = number of rooms?

A. $C = 75 + 25r$ **B.** $C = 25 + 75r$ **C.** $C = 15 + 10r$
D. $C = 10 + r$ **E.** $C = 10 + 15r$

Question 2 (1 mark)

MC The gradient of the line shown is

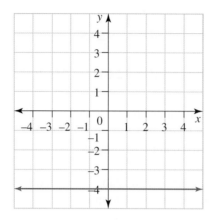

A. undefined.
B. 1
C. −1
D. −4
E. 0

Question 3 (1 mark)

MC A local taxi company uses the following graph to calculate the cost of a taxi trip. Determine the linear model which represents this scenario. Using this linear model, it can be determined that the cost of a 25-km trip will be

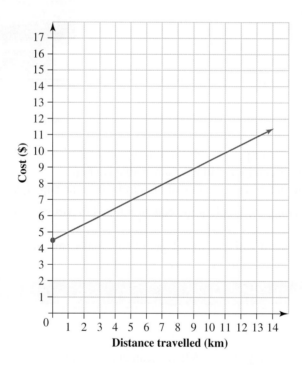

A. $17.00
B. $4.50
C. $25.00
D. $12.50
E. $112.50

Question 4 (5 marks)

In her new job, Josie is paid $20 per hour up to 35 hours in a week and then $30 per hour for any extra hours she works in that week.
  a. How much will Josie be paid if she works 35 hours a week? **(1 mark)**
  b. How much will Josie be paid if she works 36 hours a week? **(1 mark)**
  c. Write an equation relating Josie's income in a week, I, and the number of hours she works, n. **(1 mark)**
  d. Draw a graph of I against n. **(1 mark)**
  e. How many hours did Josie work if she got paid $1000? **(1 mark)**

Question 5 (4 marks)

When his light globes stopped working, Alex changed to halogen globes. His records showed that it cost him 0.7 cents per hour for the 800 hours his original light globes had lasted. Then, the halogen globes cost him 0.5 cents per hour and they lasted 2400 hours.

The relationship showing the total amount he paid per globe, P cents, after h hours from the beginning, is:

$$P = \begin{cases} 0.7h & \text{if } 0 \leq h \leq 800 \\ 160 + 0.5h & \text{if } 800 \leq h \leq 3200 \end{cases}$$

  a. Plot the graph of P against h. **(1 mark)**
  b. How much did running the original globes cost altogether? **(1 mark)**
  c. How much did running both globes cost altogether? **(1 mark)**
  d. After how many hours had he spent $10 on running the globes? **(1 mark)**

More exam questions are available online.

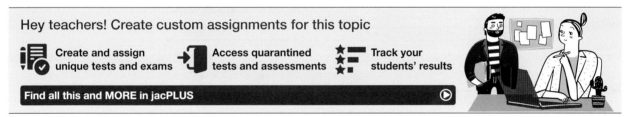

Hey teachers! Create custom assignments for this topic

Create and assign unique tests and exams

Access quarantined tests and assessments

Track your students' results

Find all this and MORE in jacPLUS

Answers

Topic 5 Linear functions, graphs and models

5.2 Linear functions and their features

5.2 Exercise

1. a. Gradient $= 2$; y-intercept $= 1$
   b. Gradient $= -1$; y-intercept $= 3$
   c. Gradient $= \dfrac{1}{2}$; y-intercept $= 4$
   d. Gradient $= 1$; y-intercept $= \dfrac{1}{4}$
   e. Gradient $= -\dfrac{3}{2}$; y-intercept $= 3$

2. a. Gradient $= \dfrac{3}{5}$; y-intercept $= -\dfrac{1}{5}$
   b. Gradient $= 10$; y-intercept $= -5$
   c. Gradient $= -\dfrac{1}{2}$; y-intercept $= \dfrac{3}{2}$

3. a. 1   b. -2

4. a. Gradient $= 3$; y-intercept $= -6$
   b. Gradient $= -\dfrac{3}{5}$; y-intercept $= -\dfrac{6}{5}$

5. D

6. a. 3   b. $\dfrac{4}{3}$ c. $\dfrac{2}{7}$

7. -1

8. a. $y = 7 + 2x$ b. $y = -3 + 4x$ c. $y = 2 - x$
   d. $y = -1 + 3x$

9. a. -3
   b. $\dfrac{3}{5}$
   c. $-\dfrac{12}{11}$
   d. $\dfrac{17}{18}$
   e. $\dfrac{8}{3}$
   f. -11

10. a. $y = 4 - \dfrac{4}{3}x$ b. $y = 9 - 3x$
    c. $y = 2 + x$ d. $y = -3 - \dfrac{3}{5}x$

5.2 Exam questions

Note: Mark allocations are available with the fully worked solutions online.

1. E
2. A
3. E

5.3 Sketching linear graphs

5.3 Exercise

1.

x	0	1	2	3	4	5
y	6	3	0	-3	-6	-9

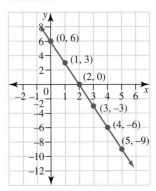

2.

x	0	1	2	3	4
y	6	4	2	0	-2

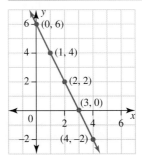

3. a. $(1, 7)$

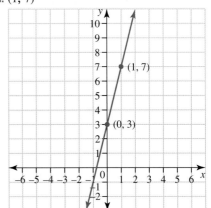

b. $(1, -2)$

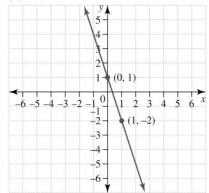

4. a. $(4, 5)$

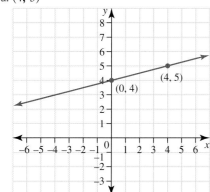

b. $(5, -4)$

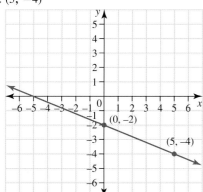

5. a, b.

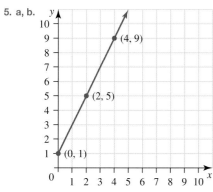

6. $a = 1$

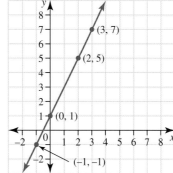

7. $a = 1$

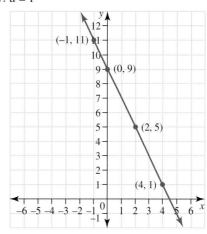

8. $b = 2$

9. a. $(1, 7)$

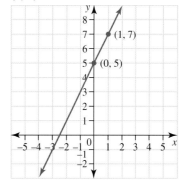

b. $(1, -3)$

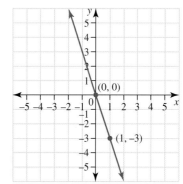

c. $\left(1, 3\dfrac{1}{2}\right)$

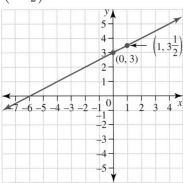

11. a. (3, 0) and (0, 6)

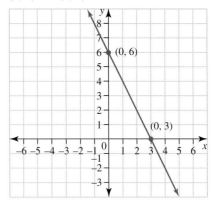

10. a. (10, 0) and (0, 4)

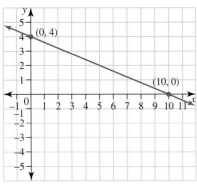

b. (−3, 0) and (0, 9)

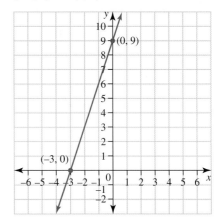

b. (−2, 0) and (0, 4)

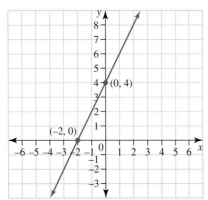

c. $\left(-\dfrac{4}{3}, 0\right)$ and (0, 2)

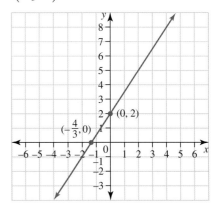

c. $\left(-\dfrac{5}{3}, 0\right)$ and $\left(0, \dfrac{5}{4}\right)$

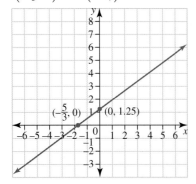

d. $\left(-\dfrac{4}{5}, 0\right)$ and $\left(0, \dfrac{4}{3}\right)$

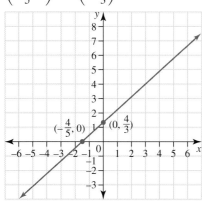

12. a. The y-intercept is the number separate from x (the constant), and the x-intercept is equal to $\dfrac{-y\text{-intercept}}{\text{gradient}}$. This method only works when the equation is in the form $y = a + bx$.

b. y-intercept $= a$, x-intercept $= \dfrac{-a}{b}$

c. $y = 4 + 5x$

5.3 Exam questions

Note: Mark allocations are available with the fully worked solutions online.

1. D

2. D

3. i. gradient $= \dfrac{1}{2}$ or 0.5

   y-intercept $= 4$

   ii. x-intercept $= -8$

   iii.

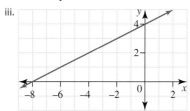

5.4 Linear modelling

5.4 Exercise

1. $C = 90 + 65t$

2. $A = 125\,000 + -250t$

3. a. Both variables in the equation have a power of 1.

   b. y-intercept $= 5$. This represents the amount of water initially in the pool.

c.

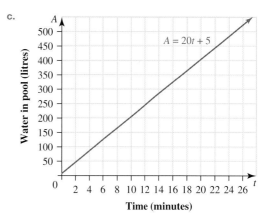

d. 25 minutes

4. a. $A = 100 + 40t$

   b. How much air was initially in the ball

   c. $4900\,\text{cm}^3$

   d. 41 minutes 38 seconds

5. a. 12

   b. y-intercept $= -0.5$. This means that Kirsten starts 0.5 km before the starting point of the race.

   c. 5.5 km

   d. 1 hour, 48 minutes

6. a. $\dfrac{4}{3}$

   b. The increase in the height of the water each minute

   c. 0.67 or $\dfrac{2}{3}$

   d. No, the y-intercept calculated in part **c** is not 0, so there was water in the tank to start with.

7. a. $a = 40$, $b = 120$

   b. The amount of money in the family's account at the start of the year

   c. 72 weeks

8. a. $P = 19.2t$, $10 \le t \le 20$

   b. $R = 100 - 4e$, $0 \le e \le 15$

9. a. $P = 15t$

   b. The additional amount of petrol in the tank each minute

   c. 5 minutes

   d. $P = 15 + 15t$

   e. $0 \le t \le 5$

10. a. 37 km

   b. The distance to Gert's home is reducing as time passes.

   c. 1 hour, 41 minutes

   d. $0 \le t \le 1.682$

e.

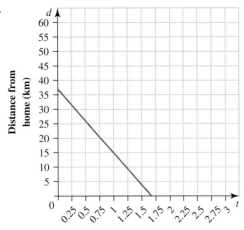

Time (hours)

11. a. $a = 800,\ b = 0.015$

b. No, there is no limit to how much the real estate agent can earn in a month.

c. $7580

d. $652 140

12. a. $C = 3.5t,\ 100 \le t \le 1000$

b. The domain represents the number of T-shirts Monique can buy.

c. There is an upper limit as the deal is valid only up to 1000 T-shirts.

13. a. $C = 15 + 0.13t$

b. The gradient represents the call cost per minute and the y-intercept represents the flat fee.

c.

Time	Cost ($)
5	15.65
10	16.30
15	16.95
20	17.60
25	18.25
30	18.90
35	19.55
40	20.20
45	20.85
50	21.50
55	22.15
60	22.80

5.4 Exam questions

Note: Mark allocations are available with the fully worked solutions online.

1. D

2. C

3. D

5.5 Linear equations and predictions

5.5 Exercise

1. a. $y = -3 + 2x$ **b.** $y = 4 + x$
    c. $y = 1 - 3x$ **d.** $y = -1 - 2x$

2. a. $y = 4 + 4x$ **b.** $y = \dfrac{1}{3} + 3x$
    c. $y = 20 - 2x$ **d.** $y = -4 + \dfrac{1}{2}x$

3. a. $y = \dfrac{11}{2} + \dfrac{1}{2}x$ **b.** $y = 7$
    c. $y = 4 + 2x$ **d.** $y = 18 - 4x$

4. C

5. a. The increase in price of 4 cents for every additional person the venue holds

b. The price of a ticket if a venue has no capacity

c. No, as the smallest venues would still have some capacity

6. $y = 0.5 + 0.5x$

7. $V = 50\,000 + 500t$

8. a. $C = 110 + 15n$

b. It would cost $290 for Tommy to invite 12 friends.

9. a. $H = 0.25 + 1.25t$

b. The tree was 25 cm tall when it was first planted.

c. The tree is predicted to be 6.5 m tall at the start of the fifth year.

10. a. $C = 9.5 + 0.5x$

b. The cab had travelled 381 km before Snoozy woke up.

5.5 Exam questions

Note: Mark allocations are available with the fully worked solutions online.

1. C

2. B

3. C

5.6 Piecewise linear graphs and their application

5.6 Exercise

1.

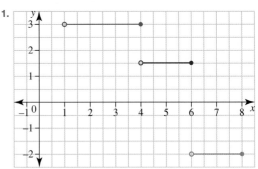

2. $y = 1,\ 1 \le x \le 1;\ y = 2.5,\ 1 < x < 2;\ y = 3,\ 2 \le x \le 4$

3. a. Point of intersection $= (-1, 0)$, $a = -1$

b.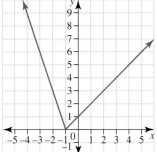

4. a.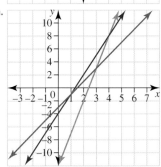

b. $(1, -1)$ and $(4, 8)$

c. $a = 1$ and $b = 4$

d.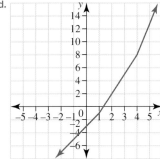

5. a. 125 L

b. i. 30 L/h   **ii.** 10 h

c.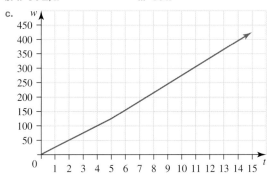

6. a.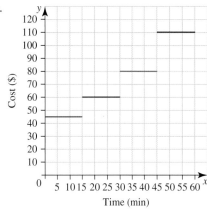

b. $60

7. a. $65

b. 10 kg

c. Place 2 to 3 kg from the 32-kg bag into the 25-kg bag and pay $80 rather than $105.

8. a. $a = 0.75$, $b = 150$

b.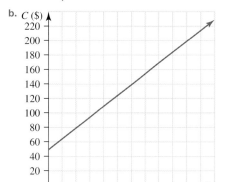

c. 50 cents/km

d. $k = 150$, $c = 162.50$. This means that the point of intersection $(150, 162.5)$ is the point where the charges change. At this point both equations will have the same value, so the graph will be continuous.

e.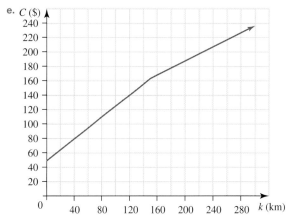

9. $y = x, 0 \leq x < 3$

$y = 4.5 + 0.5x, 3 \leq x < 9$

10. a. 90 km/h   **b.** 270 km/h

    c. 270 km/h   **d.** 180 km/h

11. At $t = 1$, $v = 2.5$ m/s
At $t = 2$, $v = 5$ m/s
At $t = 5$, $v = 10$ m/s
At $t = 7$, $v = 5$ m/s

12. The growth of the plant will slow down at the 5-month mark.

13. a. $T = 18 + 18.2t$, $0 \leq t \leq 10$

    b. i. $a = 10$, $b = 30$

      ii. a is the time the oven first reaches 200 °C and b is the time at which the bread stops being cooked.

    c. $m = \dfrac{-14}{3}$, $d = 30$, $e = 60$

    d. The change in temperature for each minute in the oven

    e.

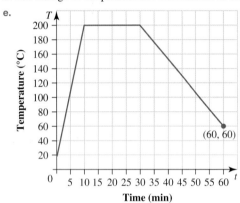

14. a. $a = 4$

    b. $b - 50(4) = 1400$
            $b = 1600$

    c. $10 \leq t \leq 12$

    d. \$500

15. a. $(0.75, 15)$ and $(1.25, 22.5)$

    b. The yacht is returning to the yacht club during this time period.

    c. 22.5 km

    d. 3 hours, 8 minutes; $b = 3.13$

16. a. 540 L   **b.** 12 L/min   **c.** 93 min

17. a. \$0   **b.** \$4142
    c. \$17 117   **d.** \$11 592

18.

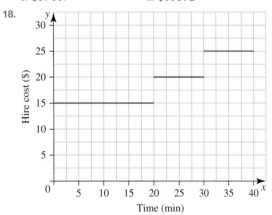

19. a.

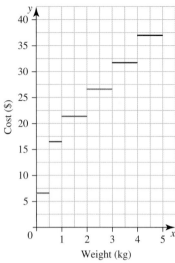

    b. It is cheaper to post them together (\$16.15 together versus \$22.75 individually).

5.6 Exam questions

Note: Mark allocations are available with the fully worked solutions online.

1. C

2. D

3. C

5.7 Review

5.7 Exercise

Multiple choice

  1. E

  2. D

  3. C

  4. A

  5. D

  6. C

  7. A

  8. D

  9. B

10. A

Short answer

11. a. x-intercept: $(2.5, 0)$
       y-intercept: $(0, 5)$

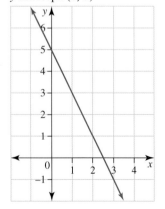

b. x-intercept: $(-2, 0)$

  y-intercept: $(0, 8)$

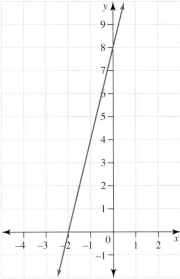

c. x-intercept: $(4, 0)$

  y-intercept: $\left(0, \dfrac{4}{3}\right)$

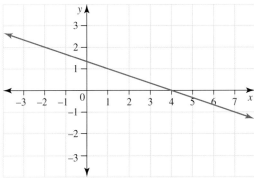

d. x-intercept: $\left(\dfrac{10}{3}, 0\right)$

  y-intercept: $(0, 2.5)$

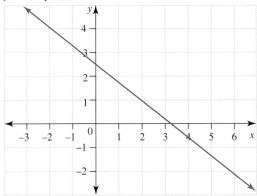

12. a. -2 **b.** -1

  c. $\dfrac{4}{3}$ **d.** $-\dfrac{7}{13}$

13. -2

14.

	Equation	Gradient	y-intercept	x-intercept
a	$y = -3 + 5x$	5	-3	0.6
b	$y = 1 + 3x$	3	1	$-\dfrac{1}{3}$
c	$6x - 3y = 9$	2	-3	1.5
d	$2y + 4x = 8$	-2	4	2
e	$y = 5 + x$	1	5	-5
f	$y = -4 + 2x$	2	-4	2

15. a. 1

b.

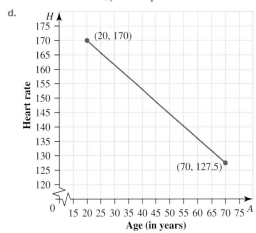

Extended response

16. a. The power of both variables in the equation (H and A) is 1.

  b. 166

  c. Gradient $= -0.85$; y-intercept $= 187$

  d.

  e. 29

  f. At the x-intercept, heart rate $= 0$; therefore, the person would no longer be alive.

17. a. 1 hour, 41 minutes

  b. Jerri started 0.1 km (100 metres) behind the starting line.

  c. 2 km

  d. The gradient of the equation equals the speed; therefore, Samantha was travelling at 4 km/h.

  e. After 30 minutes, Samantha increased her speed from 4 km/h to 8 km/h.

  f. i. 1.5

   ii. Samantha took 1 hour, 30 minutes to run 10 km; Jerri took 1 hour, 41 minutes.
   Difference: $41 - 30$ minutes $= 11$ minutes

  g. i. 0.95 hours (57 minutes)

   ii. 5.6 km

  h.

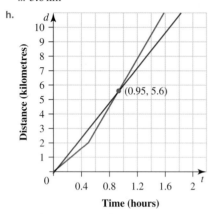

18. a. Leaking rate $=$ gradient; initial petrol $= y$-intercept

  b. 0.3 L/h. It is assumed that the petrol is leaking at a constant rate.

  c. 1.2 L

  d. 17.5 hours

  e. i. Petrol is leaking at a constant rate (the gradient).

   ii. Petrol is leaking from the tank; therefore, the amount of petrol is decreasing.

   iii. $l = 45 - 0.3t$

  f. See the graph at the foot of the page.*

  g. 150 hours

5.7 Exam questions

Note: Mark allocations are available with the fully worked solutions online.

1. C

2. E

3. A

4. i. $700

   ii. $730

   iii. $I = 20n, 0 \leq x < 35$
   $I = -350 + 30n, n > 35$

   iv.

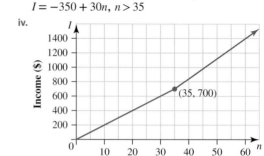

   v. She worked 45 hours.

5. i.

Price (cents) vs Time (hours) graph with points (800, 560) and (3200, 1760).

   ii. $5.60

   iii. $17.60

   iv. 1680 hours

---

*18. f.

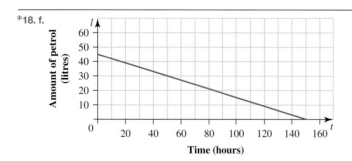

6 Sequences and first-order linear recurrence relations

Fully worked solutions for this topic are available online.

6.1 Overview

6.1.1 Introduction

Humans have been studying and observing patterns since the very origins of mathematical thought.

The study of sequences and series has to do with our observations of patterns in the world. By identifying patterns, we can use them to make predictions that can help aid our decision-making.

Some examples of where sequences can be used are:

- calculating the amount of money in a savings account over time
- predicting the size of a population that is growing exponentially, for example, bacteria in a petri dish, viral spread in a community, population growth in a country
- risk analysis in large financial institutions.

In mathematics, sequences are always ordered, and the links between different terms of sequences can be identified and expressed using mathematical equations. In this chapter, we will specifically be looking at arithmetic and geometric sequences and their applications in modelling growth and decay. This includes constant population growth, flat rate depreciation and simple interest.

KEY CONCEPTS

This topic covers the following key concepts from the VCE Mathematics Study Design:

- the concept of an arithmetic or geometric sequence as a function with the set of non-negative integers as its domain
- tabular and graphical display of sequences, investigation of their behaviour (increasing, decreasing, constant, oscillating, limiting values)
- use of a first-order linear recurrence relation of the form $u_0 = a$, $u_{n+1} = u_n + d$ where a and d are constants, to generate the values of an arithmetic sequence
- use of a first-order linear recurrence relation of the form $u_0 = a$, $u_{n+1} = u_n + d$ where a and d are constants, to model and analyse practical situations involving discrete linear growth or decay such as a simple interest loan or investment, the depreciating value of an asset using the unit cost or flat-rate method
- use of a first-order linear recurrence relation of the form $u_0 = a$, $u_{n+1} = Ru_n$ where a and R are constants, to generate the values of a geometric sequence
- use of a first-order linear recurrence relation of the form $u_0 = a$, $u_{n+1} = Ru_n$ where a and R are constants, to model growth and decay and analyse practical situations involving geometric sequences such as the reducing height of a bouncing ball, reducing balance depreciation, compound interest loans or investments
- generation of the explicit rule, u_n, of an arithmetic or geometric sequence, its use and evaluation, including various practical and financial contexts.

Source: VCE Mathematics Study Design (2023–2027) extracts © VCAA; reproduced by permission.

6.2 Arithmetic sequences

LEARNING INTENTION

At the end of this subtopic you should be able to:
- determine terms in a sequence
- determine the values of the first term and common difference for the arithmetic sequences
- generate values of a sequence
- graphically display an arithmetic sequence
- plot arithmetic sequences to determine terms of the sequence.

6.2.1 Introduction to sequences and terminology

A **sequence** is string of numbers that follows a particular order. In a number sequence, we call each number a **term** of the sequence.

Sequences

In general, mathematical sequences can be displayed as:

First term	Second term	Third term	and so on
$u_0 = a$	u_1	u_2	...

The first term of a mathematical sequence is referred to as u_0 or a.

Arithmetic sequences are generated through repeated addition, and geometric sequences are generated by repeated multiplication.

WORKED EXAMPLE 1 Determining terms in a sequence

Determine the terms in the following sequence.

$$2, 4, 6, 8, 10$$

a. u_1 b. u_3 c. u_5

THINK

a. u_1 means the second term, which is 4.

b. u_3 means the fourth term, which is 8.

c. u_5 means the sixth term. We need to identify the pattern to determine the sixth term.
We apply the pattern to determine the sixth term.

WRITE

a. $u_1 = 4$

b. $u_3 = 8$

c. For each term, the value goes up by 2.
$$u_5 = u_4 + 2$$
$$u_5 = 10 + 2$$
$$u_5 = 12$$

6.2.2 Identifying arithmetic sequences

An **arithmetic sequence** is an example of a sequence generated by first-order linear recurrence relations.

The difference between two consecutive terms in an arithmetic sequence is known as the **common difference**. If the difference is positive, the sequence is increasing. If the common difference is negative, the sequence is decreasing.

Arithmetic sequences

In an arithmetic sequence, the first term is referred to as a and the common difference is referred to as d.

Consider the sequence $3, 7, 11, 15, 19$. There is a constant increase of 4 between each of the consecutive terms. This increase is known as the common difference.

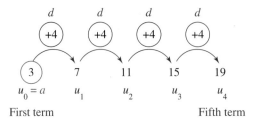

For the sequence above, the first term a is 3 and the common difference d is 4.

To determine the common difference in an arithmetic sequence, we calculate the difference between any term and its previous term.

Calculating common difference in an arithmetic sequence

To calculate the common difference d in an arithmetic sequence:

$$d = \text{any term} - \text{previous term}$$

For example, $d = u_4 - u_3$

$$d = u_7 - u_6$$

WORKED EXAMPLE 2 Determining whether a sequence is arithmetic

Determine which of the following sequences are arithmetic sequences, and for those sequences that are arithmetic, state the values of a and d.

a. $2, 5, 8, 11, 14, \ldots$

b. $4, -1, -6, -11, -16, \ldots$

c. $3, 5, 9, 17, 33, \ldots$

THINK	WRITE
a. 1. Calculate the difference between consecutive terms of the sequence.	**a.** $u_1 - u_0 = 5 - 2$ $\qquad = 3$ $u_2 - u_1 = 8 - 5$ $\qquad = 3$ $u_3 - u_2 = 11 - 8$ $\qquad = 3$ $u_4 - u_3 = 14 - 11$ $\qquad = 3$
2. If the differences between consecutive terms are constant, then the sequence is arithmetic. The first term of the sequence is a and the common difference is d.	The common differences are constant, so the sequence is arithmetic. $a = 2$ and $d = 3$
b. 1. Calculate the difference between consecutive terms of the sequence.	**b.** $u_1 - u_0 = -1 - 4$ $\qquad = -5$ $u_2 - u_1 = -6 - -1$ $\qquad = -6 + 1$ $\qquad = -5$ $u_3 - u_2 = -11 - -6$ $\qquad = -11 + 6$ $\qquad = -5$ $u_4 - u_3 = -16 - -11$ $\qquad = -16 + 11$ $\qquad = -5$
2. If the differences between consecutive terms are constant, then the sequence is arithmetic. The first term of the sequence is a and the common difference is d.	The common differences are constant, so the sequence is arithmetic. $a = 4$ and $d = -5$
c. 1. Calculate the difference between consecutive terms of the sequence.	**c.** $u_1 - u_0 = 5 - 3$ $\qquad = 2$ $u_2 - u_1 = 9 - 5$ $\qquad = 4$ $u_3 - u_2 = 17 - 9$ $\qquad = 8$ $u_4 - u_3 = 33 - 17$ $\qquad = 16$
2. If the differences between consecutive terms are constant, then the sequence is arithmetic.	The common differences are not constant, so the sequence is not arithmetic.

6.2.3 Generating arithmetic sequences using repeated addition

As we've seen above, arithmetic sequences are defined by the pattern of repeatedly adding a number (common difference) to the previous term.

WORKED EXAMPLE 3 Generating terms for an arithmetic sequence

Determine the next four terms in the following arithmetic sequence.

$$5, 8, 11, 14, \ldots$$

THINK	WRITE
1. First, calculate the common difference of the arithmetic sequence.	$u_1 - u_0 = 8 - 5$ $\qquad = 3$ $d = 3$
2. Add the common difference to terms to find the next term.	$u_4 = 14 + 3$ $u_4 = 17$ $u_5 = 17 + 3$ $u_5 = 20$ $u_6 = 20 + 3$ $u_6 = 23$ $u_7 = 23 + 3$ $u_7 = 26$
3. Write the next four terms in the arithmetic sequence.	The next four terms are $17, 20, 23$ and 26.

TI	THINK	DISPLAY/WRITE	CASIO	THINK	DISPLAY/WRITE
1. In a Calculator page, enter the value of the first term, 5, and press ENTER.		1. In the MAIN application, enter the value of the first term, 5, and press EXE.	
2. The common difference for the arithmetic sequence is 3, so type in '+3' and then press ENTER. The next term in the sequence will appear.		2. The common difference for the arithmetic sequence is 3, so type in '+3' and then press EXE. The next term in the sequence will appear.	

3. Continue to press ENTER to generate terms.

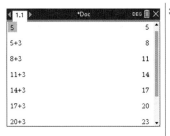

3. Continue to press EXE to generate terms.

4. The next four terms of the sequence can be read from the CAS screen.

The next four terms are 17, 20, 23 and 26.

4. The next four terms of the sequence can be read from the CAS screen.

The next four terms are 17, 20, 23 and 26.

6.2.4 Graphs of arithmetic sequences

Tables of values

When we draw a graph of a mathematical sequence, it helps to first construct a table of values for the sequence. The top row of the table displays the term number of the sequence where the nth term has a term number of $n - 1$ (eg. the 3rd term has a term number of 2), and the bottom row of the table displays the term value.

Term number	0	1	2	...	$n - 1$
Term value					

The data from the table of values can then be used to identify the points to plot in the graph of the sequence.

Drawing graphs of sequences

When we draw a graph of a numerical sequence, the term number is the explanatory variable, so it appears on the x-axis of the graph. The term value is the response value, so it appears on the y-axis of the graph.

We can put the terms of a sequence into a table and graph it to make observations about the sequence.

For the sequence 4, 6, 8, 10, we can draw up a table as follows:

Term number	0	1	2	3
Term value	4	6	8	10

We can draw a graph from this table as shown.

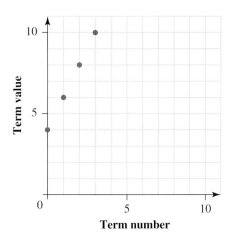

From the graph we could suggest that this sequence is **increasing** at a constant rate.

Because there is a common difference between the terms of an arithmetic sequence, the relationship between the terms is a linear relationship. This means that when we graph the terms of an arithmetic sequence, we can join the points to form a straight line.

When we draw a graph of an arithmetic sequence, we can extend the straight line to determine values of terms in the sequence that haven't yet been determined.

WORKED EXAMPLE 4 Plotting an arithmetic sequence to determine a term of the sequence

An arithmetic sequence is given by the equation $u_n = 7 + 2n$, where $n = 0, 1, 2, 3...$.
a. Create a table of values showing the term number and term value for the first 5 terms of the sequence.
b. Plot the graph of the sequence.
c. Use your graph of the sequence to determine the 12th term ($n = 11$) of the sequence.

THINK

a. 1. Set up a table with the term number in the top row and the term value in the bottom row.

2. Substitute the first 5 values of n into the equation to determine the missing values.

3. Complete the table with the calculated values.

WRITE

a.

Term number	0	1	2	3	4
Term value					

$u_0 = 7 + 2(0)$, when $n = 0$
$= 7 + 2 \times 0$
$= 7 + 0$
$= 7$

$u_1 = 7 + 2(1)$, when $n = 1$
$= 7 + 2 \times 1$
$= 7 + 2$
$= 9$

$u_2 = 7 + 2(2)$, when $n = 2$
$= 7 + 2 \times 2$
$= 7 + 4$
$= 11$

$u_3 = 7 + 2(3)$, when $n = 3$
$= 7 + 2 \times 3$
$= 7 + 6$
$= 13$

$u_4 = 7 + 2(4)$, when $n = 4$
$= 7 + 2 \times 4$
$= 7 + 8$
$= 15$

Term number	0	1	2	3	4
Term value	7	9	11	13	15

b. 1. Use the table of values to identify the points to be plotted.

2. Plot the points on the graph.

b. The points to be plotted are (0, 7), (1, 9), (2, 11), (3, 13) and (4, 15).

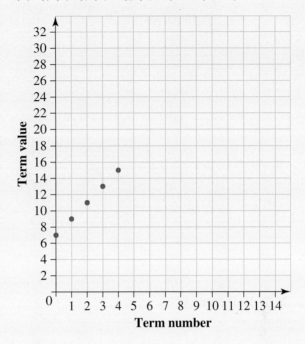

c. 1. Join the points with a straight line and extend the line to cover future values of the sequence.

c.

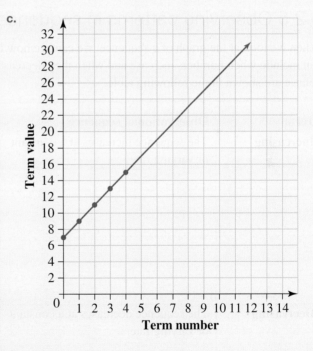

2. Read the required value from the graph (when $n = 11$).

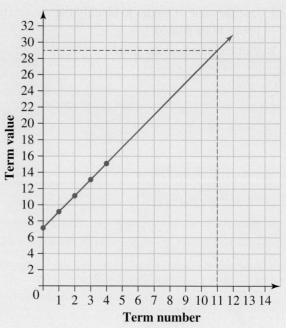

3. Write the answer.

The 12th term ($n = 11$) of the sequence is 29.

6.2.5 Observing patterns in sequences

When we look at the graph of a sequence, we can see how the numbers tend to change in a particular way. This can include when numbers increase and when they decrease. Some key terms to describe the way sequences behave are shown in the following table.

Term	Observation of sequence	Examples	
Increasing	The sequence increases at a constant or varying rate.	This sequence is increasing in a linear pattern (straight line).	This graph is increasing in a non-linear pattern.
Decreasing	The sequence decreases at a constant or varying rate.	This graph is decreasing in a linear pattern (straight line).	This graph is decreasing in a non-linear pattern.

Constant	The sequence values stay the same.	u_n 4 ● ● ● ● 0 n This graph has a constant value of 4.
Oscillating	The sequence increases and decreases within a range of numbers.	u_n 3 ● ● 1 ● ● ● 0 n This graph is oscillating between 1 and 3.

Limiting value	The sequence approaches a value as the number of terms reaches infinity. This is also referred to as convergence.	u_n
		●
		  ●
		   ●
		    ● ● ● ●
		0 n
		This graph has a limiting value. It approaches zero as the number of terms reaches infinity.

WORKED EXAMPLE 5 Identifying features of sequences

Graph and describe the pattern observed in the following sequence.

1, 3, 5, 7, 9 …

THINK

a. 1. First, lay out the sequence in a table with term numbers, n, in one row and term values, u_n, in the other row.

2. Sketch the graph of the sequence, with term numbers, n, on the x-axis and term values, u_n, on the y-axis.

WRITE

a.

Term number	0	1	2	3	4
Term value	1	3	5	7	9

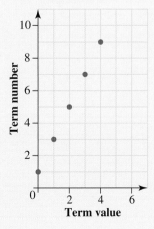

3 Write your conclusion regarding the given graph.	The graph of the sequence is increasing in a straight line.

TI | THINK	DISPLAY/WRITE	CASIO | THINK	DISPLAY/WRITE
1. In a Lists & Spreadsheet page, label the first column term numbers (*n*) and the second column term values (*u*). Enter the values 0 to 4 in the first column.	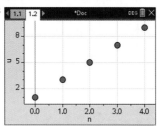	1. In a Statistics screen, label the first column term numbers (*n*) and the second column term values (*u*). Enter the values 0 to 4 in the first column and enter the corresponding term values in the next column.	
2. Open a Data & Statistics page. Click on the horizontal axis label and change it to *n*. Click on the vertical axis label and change it to *u*.		2. Click on the StatGraphs button. Set the following: Type: Scatter XList: main/*n* YList: main/*u* Press "Set" Press the Graph button to see the graph of the sequence.	
3. Make conclusion from the graph.	The graph of the sequence is increasing in a straight line.	3. Make conclusion from the graph.	The graph of the sequence is increasing in a straight line.

  Resources

> **Interactivities** Terms of an arithmetic sequence (int-6261)
>
> Arithmetic sequences (int-6258)

6.2 Exercise

Students, these questions are even better in jacPLUS

 Receive immediate feedback and access sample responses

 Access additional questions

 Track your results and progress

Find all this and MORE in jacPLUS

1. **WE1** Determine the terms in the following sequence.

$$6, \ 8, \ 10, \ 12, \ 14, \ 16$$

   a. u_1　　　　　b. u_3　　　　　c. u_5

2. Determine the terms in the following sequence.

$$2, \ 6, \ 13, \ 17, \ 21, \ 13, \ 6, \ 30$$

   a. u_0　　　b. u_2　　　c. u_5　　　d. u_6

3. Determine u_1, u_2 and u_4 for the following sequences.
   a. $6, 12, 6, 12, 6$　　　　　　b. $3, 6, 9, 12, 15$　　　　　　c. $40, 35, 30, 25, 20$
   d. $1.5, 2.0, 2.5, 3.0, 3.5$　　　e. $64, 32, 16, 8, 4$　　　　　f. $2, 2, 4, 6, 10$

4. **MC** Select the option that gives u_0, u_2 and u_3 for this sequence.

$$5, \ 7, \ 9, \ 11, \ 13$$

   A. $5, 7, 9$　　　B. $5, 9, 11$　　　C. $5, 9, 13$　　　D. $7, 11, 13$　　　E. $7, 9, 11$

5. **WE2** Determine which of the following sequences are arithmetic sequences, and for those sequences that are arithmetic, state the values of a and d.
   a. $23, 68, 113, 158, 203, \ldots$　　b. $3, 8, 23, 68, 203, \ldots$　　c. $\dfrac{1}{2}, \dfrac{3}{4}, 1, \dfrac{5}{4}, \dfrac{3}{2}, \dfrac{7}{4}, \ldots$

6. Determine the missing values in the following arithmetic sequences.
   a. $13, -12, -37, f, -87, \ldots$　　b. $2.5, j, 8.9, 12.1, k, \ldots$　　c. $p, q, r, \dfrac{9}{2}, \dfrac{25}{4}, \ldots$

7. **MC** Select the option that is **not** an arithmetic sequence.
   A. $4, 6, 8, 10$　　B. $1, 3, 5, 7$　　C. $4, 7, 10, 13$　　D. $3, 6, 9, 12$　　E. $2, 4, 8, 16$

8. Determine the values of a and d in the following arithmetic sequences
   a. $2, 4, 6, 8$　　　　　　　　　b. $3, 7, 11, 15$
   c. $23, 20, 17, 14$　　　　　　　d. $12, 11, 10, 9$

9. **WE3** Determine the next four terms in the following arithmetic sequence.

$$4, 10, 16, 22, \ldots$$

10. Determine the first five terms of a sequence with a first term of 12 and a common difference of 5.

11. **MC** Select the sequence that has the features $a = 3$ and $d = 5$.

   A. $3, 6, 9, 12, ...$
   B. $3, 5, 7, 9, ...$
   C. $5, 10, 15, 20, ...$
   D. $3, 8, 13, 18, ...$
   E. $5, 8, 11, 14, ...$

12. Write the first five terms of an arithmetic sequence with a common difference of 9.

13. An arithmetic sequence was constructed where n number of squares were made with matchsticks. One square required 4 matchsticks, two squares required 7 match sticks, and three squares required 10 match sticks. Determine the values of the first term, a, and the common difference, d, for this sequence.

14. For a sequence with $u_2 = 10$ and a common difference of 3, determine the first term.

15. **WE4** An arithmetic sequence is given by the equation $u_n = 5 + 10n$.

   a. Create a table of values showing the term number and term value for the first 5 terms of the sequence.
   b. Plot the graph of the sequence.
   c. Use your graph of the sequence to determine the 9th term of the sequence.

16. An arithmetic sequence is defined by the equation $u_n = 6.4 + 1.6n$.

   a. Create a table of values showing the term number and term value for the first 5 terms of the sequence.
   b. Plot the graph of the sequence.
   c. Use your graph of the sequence to determine the 13th term of the sequence.

17. Namjoon is making a pattern with various sticks.

Design 1	Design 2	Design 3	Design 4

   Create a table and then draw a graph for this sequence and use it to predict how many sticks will be used in the 10th design.

18. Consider the following sequence.

$$1, 4, 7, 10, 13$$

   a. Create a table for the sequence.
   b. Plot a graph for the sequence.
   c. Use the graph to predict the value of the 8th term.
   d. Discuss the limitations of using the graphical method to predict values in a sequence.

19. **WE 5** Graph and describe the pattern observed in the following sequences

   a. 1, 3, 5, 7, 9
   b. 7, 6, 5, 4
   c. 5, 10, 15, 20, 25
   d. 1, 2, 4, 9

20. For the following sequence graphs, describe the shape of the sequence.

a.

b.

c.

d.

21. **MC** Xiaoming sketched a graph for a sequence but didn't know how to describe it. Select from the following how he should describe the graph.

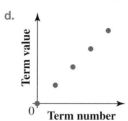

A. The values are constant
B. The values are decreasing
C. The values are oscillating
D. The values are increasing
E. There is a limiting value

22. For the following sequence, sketch a graph and comment on the sequence.

$$3, 6, 12, 24, 48$$

6.2 Exam questions

Question 1 (1 mark)

Determine the value of u_1 in the following sequence.

$$1, -\frac{1}{4}, \frac{1}{16}, \frac{1}{64}, \frac{1}{256}$$

Question 2 (1 mark)

MC For the sequence $-44, -40, -36, \ldots$, the seventh term is
A. -24 **B.** -20 **C.** -18 **D.** -16 **E.** -12

Question 3 (1 mark)

MC Select the three numbers that complete the pattern of the sequence with terms $u_1 = 10$, $u_2 = 6$ and $u_3 = 2$.
A. -2 **B.** $0, -4, -8$ **C.** $1, 0, -1$ **D.** $4, 5, 6$ **E.** $-2, -6, -10$

More exam questions are available online.

6.3 Arithmetic sequence applications

LEARNING INTENTION

At the end of this subtopic you should be able to:
- determine the equation for an arithmetic sequence
- determine the nth term in an arithmetic sequence.

6.3.1 Generating equations for arithmetic sequences

To determine any term of an arithmetic sequence, we need to set up an equation to represent the sequence.

The equation for an arithmetic sequence

An arithmetic sequence can be expressed by the equation

$$u_0 = a, \ u_n = a + nd$$

where u_n is the $(n + 1)$ term, n represents the term number, a is the first term and d is the common difference.

Think of a sequence that starts at 2 and has a common difference of 4. How would we calculate each term individually?

$$u_0 = 2$$
$$u_1 = 2 + 4 = 6$$
$$u_2 = 2 + 2 \times 4 = 10$$
$$u_3 = 2 + 3 \times 4 = 14$$
$$u_4 = 2 + 4 \times 4 = 18$$

This breakdown shows us how we can derive the formula for arithmetic sequences by observing the pattern that emerges as we calculate each term in an arithmetic sequence. Notice that the number that the common difference is multiplied by is always n.

Using this equation, as long as we have the values for a and d, we can construct the equation for any arithmetic sequence.

WORKED EXAMPLE 6 Determining equations that represent arithmetic sequences

Determine the equations that represent the following arithmetic sequences.
a. $3, 6, 9, 12, 15, \ldots$
b. $40, 33, 26, 19, 12, \ldots$

THINK	WRITE
a. 1. Determine the values of a and d.	a. $a = 3$
	$d = u_1 - u_0$
	$= 6 - 3$
	$= 3$

2. Substitute the values for a and d into the formula for an arithmetic sequence.		$\begin{aligned} u_n &= a + nd \\ &= 3 + n \times 3 \\ &= 3 + 3n \end{aligned}$
b. 1. Determine the values of a and d.		**b.** $a = 40$ $\begin{aligned} d &= u_1 - u_0 \\ &= 33 - 40 \\ &= -7 \end{aligned}$
2. Substitute the values for a and d into the formula for an arithmetic sequence.		$\begin{aligned} u_n &= a + nd \\ &= 40 + n \times -7 \\ &= 40 - 7n \end{aligned}$

6.3.2 Determining the nth term in an arithmetic sequence

After an equation has been set up to represent an arithmetic sequence, we can use the equation to determine the value of any term in the sequence. Simply substitute the value of n into the equation to determine the value of that term.

In addition, we can also determine the values of a, d and n for any arithmetic sequence by transposing the equation.

> **Determining values in an arithmetic sequence**
>
> $$a = u_n - nd$$
>
> $$d = \frac{u_n - a}{n}$$
>
> $$n = \frac{u_n - a}{d}$$

tlvd-4466

WORKED EXAMPLE 7 Determining other values of an arithmetic sequence

a. Determine the 15th term of the sequence $2, 8, 14, 20, 26, \ldots$
b. Determine the first term of the arithmetic sequence in which $u_{21} = 1008$ and $d = -8$.
c. Determine the common difference of the arithmetic sequence that has a first term of 12 and an 11th term of 102.
d. An arithmetic sequence has a first term of 40 and a common difference of 12. Determine which term number has a value of 196.

THINK	WRITE
a. 1. As it has a common difference, this is an arithmetic sequence. State the known values.	**a.** $a = 2$, $d = 6$, $n = 14$
2. Substitute the known values into the equation for an arithmetic sequence and solve.	$\begin{aligned} u_n &= a + nd \\ u_{14} &= 2 + 14 \times 6 \\ &= 2 + 84 \\ &= 86 \end{aligned}$
3. Write the answer.	The 15th term of the sequence is 86.
b. 1. State the known values of the arithmetic sequence.	**b.** $d = -8$, $n = 21$, $u_{21} = 1008$

2. Substitute the known values into the equation to determine the first term and solve.

$$a = u_n - nd$$
$$= 1008 - (21) \times (-8)$$
$$= 1008 - (-168)$$
$$= 1008 + 168$$
$$= 1176$$

3. Write the answer.

The first term of the sequence is 1176.

c. 1. State the known values of the arithmetic sequence.

c. $a = 12, \quad n = 10, \quad u_{10} = 102$

2. Substitute the known values into the equation to determine the common difference and solve.

$$d = \frac{u_n - a}{n}$$
$$= \frac{102 - 12}{10}$$
$$= \frac{90}{10}$$
$$= 9$$

3. Write the answer.

The common difference is 9.

d. 1. State the known values of the arithmetic sequence.

d. $a = 40, \quad d = 12, \quad u_n = 196$

2. Substitute the known values into the equation to determine the term number and solve.

$$n = \frac{u_n - a}{d}$$
$$= \frac{196 - 40}{12}$$
$$= 13$$

3. Write the answer.

The 14th term in the sequence has a value of 196.

6.3.3 Applications of arithmetic sequences

When using sequences for applications outside of finding terms, we consider u_n a little differently. For example, in a financial scenario, u_0 is the initial amount of money, d is the amount of money being gained or lost in each time period, and n is the number of time periods that have passed.

WORKED EXAMPLE 8 Application of an arithmetic sequence

Brian deposited $20 per week into his bank account. Determine how much money would be in the account at the end of the fifth week if the account started with $15.

THINK	WRITE
1. First, identify the values of u_0 and d. Brian starts with $15 and the bank account increases by $20 each week	$u_0 = 15$ $d = 20$ $u_n = 15 + 20n$
2. Calculate u_5 to calculate the amount of money in Brian's bank account at the end of the fifth week.	$u_5 = 15 + 20 \times 5$ $u_5 = 115$
3. Write the amount of money in Brian's bank account after five weeks.	At the end of the fifth week, Brian's bank account would have $115 in it.

6.3 Exercise

Students, these questions are even better in jacPLUS

Receive immediate feedback and access sample responses

Access additional questions

Track your results and progress

Find all this and MORE in jacPLUS

1. **WE6** Determine the equations that represent the following arithmetic sequences.
   a. $-1, 3, 7, 11, 15, \ldots$
   b. $1.5, -2, -5.5, -8, -11.5$
   c. $\dfrac{7}{2}, \dfrac{11}{2}, \dfrac{15}{2}, \dfrac{19}{2}, \dfrac{23}{2}, \ldots$

2. Determine the first five terms of the following arithmetic sequences.
   a. $u_n = 5 + 3n$
   b. $u_n = -1 - 7n$
   c. $u_n = \dfrac{1}{3} + \dfrac{2}{3}n$

3. **WE7**
   a. Determine the 20th term of the sequence $85, 72, 59, 46, 33, \ldots$
   b. Determine the first value of the arithmetic sequence in which $u_{70} = 500$ and $d = -43$.

4. **WE7**
   a. Determine the common difference of the arithmetic sequence that has a first term of -32 and an 8th term of 304.
   b. An arithmetic sequence has a first term of 5 and a common difference of 40. Determine which term number has a value of 85.
   c. An arithmetic sequence has a first term of 40 and a common difference of 12. Determine which term number has a value of 196.

5. a. Determine the 15th term of the arithmetic sequence $6, 13, 20, 27, 34, \ldots$
   b. Determine the 20th term of the arithmetic sequence $9, 23, 37, 51, 65, \ldots$
   c. Determine the 30th term of the arithmetic sequence $56, 48, 40, 32, 24, \ldots$
   d. Determine the 55th term of the arithmetic sequence $\dfrac{72}{5}, \dfrac{551}{40}, \dfrac{263}{20}, \dfrac{501}{40}, \dfrac{119}{10}, \ldots$

6. a. Determine the first value of the arithmetic sequence that has a common difference of 6 and a 31st term of 904.
   b. Determine the first value of the arithmetic sequence that has a common difference of $\dfrac{2}{5}$ and a 40th term of -37.2.
   c. Determine the common difference of an arithmetic sequence that has a first value of 564 and a 51st term of 54.
   d. Determine the common difference of an arithmetic sequence that has a first value of -87 and a 61st term of 43.

7. a. An arithmetic sequence has a first value of 120 and a common difference of 16. State the term with a value of 712.
   b. An arithmetic sequence has a first value of 320 and a common difference of 4. State the term with a value of 1160.

8. The graph shows some points of an arithmetic sequence.

   a. Determine the common difference between consecutive terms.
   b. Determine the value of the first term of the sequence.
   c. Determine the value of the 12th term of the sequence.

9. **WE8** Jolyne deposited $40 per week into her bank account. Determine how much money would be in her bank account in the seventh week if the account started with $55 in it.

10. A population of elephants is being tracked in Central Africa. There are 3 calves born every year. Determine how many elephants will be in the herd at the end of the 6th year if there were 7 elephants in the first year.

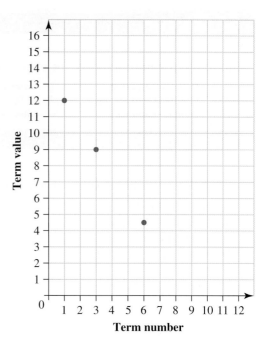

6.3 Exam questions

Question 1 (1 mark)

MC The cost of a television depends on the number that are made. If 50 are made, the cost per television is $250. If 51 are made, the cost is $1.50 less. If 52 are made then the cost is $3 less. Determine the cost if 100 are made.

   A. $175
   B. $173.50
   C. $176.50
   D. $178
   E. $179.50

Question 2 (1 mark)

MC For the information shown in the table, the function has the rule

Term number	0	1	2	3	4
Term value	−5	−3	−1	1	3

   A. $u_n = u(n-1) + 2$
   B. $u_n = 2n - 6$
   C. $u_n = n - 2$
   D. $u_n = -5 + 2n$
   E. $u_n - u_n - 1 = 2$

Question 3 (1 mark)

MC The seating in an arena is arranged in 60 rows. There are 20 seats in the front row, 24 in the second row, 28 in the third row and so on. The number of seats in the back row is

   A. 260   **B.** 220   **C.** 256   **D.** 20   **E.** 32

More exam questions are available online.

6.4 Generate and analyse an arithmetic sequence using a recurrence relation

LEARNING INTENTION

At the end of this subtopic you should be able to:
- use recurrence relations to generate arithmetic sequences
- analyse arithmetic sequences using recurrence relations.

6.4.1 Using a recurrence relation to generate arithmetic sequences

A **first-order linear recurrence relation** is a relation whereby the terms of the sequence depend only on the previous term of the sequence, which means that we need only an initial value to be able to generate all remaining terms of the sequence.

Generating number sequences by using recurrence relations

In an arithmetic sequence recurrence relation, each term is formed by adding or subtracting a common difference, d, to the previous term. We use $u_{n+1} = u_n + d$. The initial value of the sequence is represented as u_0 or a and the next value in the sequence is represented by u_{n+1}.

If the initial value in a recurrence relation changes, then the whole sequence changes. If we are not given an initial value, we cannot determine any terms in the sequence.

If we know the values of a and d in an arithmetic sequence, we can set up a recurrence relation to generate the sequence.

Generating arithmetic sequences using recurrence relations

A recurrence relation representing an arithmetic sequence will be of the form $u_0 = a$, $u_{n+1} = u_n + d$.

WORKED EXAMPLE 9 Generating a recurrence relation to represent an arithmetic sequence

Set up a recurrence relation to represent the arithmetic sequence $-9, -5, -1, 3, 7, \ldots$

THINK	WRITE
1. Determine the common difference by subtracting the first term from the second term.	$d = -5 - -9$ $= -5 + 9$ $= 4$
2. u_0 represents the first term of the sequence.	$u_0 = -9$
3. Set up the recurrence relation with the given information.	$u_0 = -9, \; u_{n+1} = u_n + 4$

6.4.2 Application of arithmetic sequences using recurrence relations

A recurrence relation can be used to generate arithmetic sequences. The worked example below shows the application of arithmetic sequences using recurrence relations.

tlvd-4467

WORKED EXAMPLE 10 Application of arithmetic sequences using recurrence relations

Magnus is stacking coins. In his first stack he stacks 3 coins, and every subsequent stack has 4 more coins than the previous stack.
a. Generate a recurrence relation for this sequence.
b. Generate the equation for this sequence.
c. Calculate the number of coins he would have in his 9th stack.

THINK	WRITE
a. 1. Determine the values of u_0 and d.	**a.** $u_0 = 3$ $d = 4$
2. Use the values to write the recurrence relation.	$u_0 = 3, u_{n+1} = u_n + 4$
b. Substitute the values of u_0 and d to write the equations for this sequence.	**b.** $u_n = a + nd$ $u_n = 3 + 4n$
c. 1. Substitute $n = 8$ to calculate u_8.	**c.** $u_8 = 3 + 4 \times 8$ $u_8 = 3 + 4 \times 8$ $\quad = 35$
2. Write the answer in a sentence.	Magnus would have 35 coins in his 9th stack.

6.4 Exercise

Students, these questions are even better in jacPLUS

 Receive immediate feedback and access sample responses

 Access additional questions

 Track your results and progress

Find all this and MORE in jacPLUS

1. **WE9** Set up a recurrence relation to represent the arithmetic sequence 2, −3, −8, −13, −18.

2. An arithmetic sequence is represented by the recurrence relation $u_0 = -2.2$, $u_{n+1} = u_n + 3.5$. Determine the first 5 terms of the sequence.

3. The 3rd and 4th terms of an arithmetic sequence are -7 and -11.5. Set up a recurrence relation to define the sequence.

4. **WE10** Aryan was stacking chairs. The number of chairs in each stack followed the arithmetic sequence $u_0 = 1$, $u_1 = 4$ and $u_2 = 7$.

a. Generate a recurrence relation for this sequence.
b. Generate the equation for this sequence.
c. Calculate the number of chairs he would have in his 15th stack.

5. An ice shelf is shrinking at a rate of 1200 km^2 per year. When measurements of the ice shelf began, the area of the shelf was $37\,000 \text{ km}^2$.

a. Create a recurrence relation to express the area of the ice shelf after n years.
b. Use your relation to determine the area of the ice shelf after each of the first 6 years.
c. Plot a graph showing the area of the shrinking ice shelf over time.

6.4 Exam questions

Question 1 (1 mark)

MC For the sequence $13, 9, 5, 1, -3, \ldots$, the recurrence relation is
A. $u_{n+1} = u_n - 4$, $u_0 = -3$
B. $u_{n+1} = u_n - 3$, $u_0 = 13$
C. $u_{n+1} = u_n + 4$, $u_0 = 13$
D. $u_{n+1} = u_n + 3$, $u_0 = 13$
E. $u_{n+1} = u_n - 4$, $u_0 = 13$

Question 2 (2 marks)

From the information in the table

Term number	0	2	4	6	8
Term value	3	11	19	27	35

a. write down the recurrence relation   **(1 mark)**
b. write down the rule for the function relating term value and term number.   **(1 mark)**

A ramp is supported by equally spaced vertical struts, as shown. The struts are, in sequence, 30 cm, 39 cm, 48 cm, ...

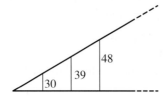

    a. Write down the recurrence relation for the sequence of strut lengths. **(1 mark)**
    b. Write down the rule for the length of the *n*th strut. **(1 mark)**
    c. Calculate the height of the ramp if it finishes at the eighth strut. **(1 mark)**

More exam questions are available online.

6.5 Geometric sequences

> **LEARNING INTENTION**
>
> At the end of this subtopic you should be able to:
> * identify whether sequences are geometric sequences
> * determine equations for geometric sequences
> * determine the values a and R for a geometric sequence
> * graphically display a geometric sequence.

6.5.1 Identifying geometric sequences

A **geometric sequence** is a pattern of numbers whose consecutive terms increase or decrease in the same ratio.

First consider the sequence 1, 3, 9, 27, 81, ... This is a geometric sequence, as each term is obtained by multiplying the preceding term by 3.

Now consider the sequence 1, 3, 6, 10, 15, ... This is not a geometric sequence, as the consecutive terms are not increasing in the same ratio.

Common ratios

The ratio between two consecutive terms in a geometric sequence is known as the **common ratio**.

> **Geometric sequences**
>
> In a geometric sequence, the first term is referred to as $u_0 = a$ and the common ratio is referred to as R.

WORKED EXAMPLE 11 Identifying whether sequences are geometric sequences

Determine which of the following sequences are geometric sequences, and for those sequences that are geometric, state the values of *a* and *R*.

a. $\mathbf{20, 40, 80, 160, 320, \ldots}$

b. $\mathbf{8, 4, 2, 1, \frac{1}{2}, \ldots}$

c. $\mathbf{3, -9, 27, -81, \ldots}$

d. $\mathbf{2, 4, 6, 8, 10, \ldots}$

THINK	WRITE
a. 1. Calculate the ratio $\frac{u_{n+1}}{u_n}$ between all consecutive terms in the sequence.	a. $\frac{u_1}{u_0} = \frac{40}{20} = 2$ $\frac{u_2}{u_1} = \frac{80}{40} = 2$ $\frac{u_3}{u_2} = \frac{160}{80} = 2$ $\frac{u_4}{u_3} = \frac{320}{160} = 2$
2. If the ratios between consecutive terms are constant, then the sequence is geometric. The first term of the sequence is *a* and the common difference is *R*.	The ratios between consecutive terms are all 2, so this is a geometric sequence. $a = 20,\ R = 2$
b. 1. Calculate the ratio $\frac{u_{n+1}}{u_n}$ between all consecutive terms in the sequence.	b. $\frac{u_1}{u_0} = \frac{4}{8} = \frac{1}{2}$ $\frac{u_2}{u_1} = \frac{2}{4} = \frac{1}{2}$ $\frac{u_3}{u_2} = \frac{1}{2}$ $\frac{u_4}{u_3} = \frac{\left(\frac{1}{2}\right)}{1} = \frac{1}{2}$
2. If the ratios between consecutive terms are constant, then the sequence is geometric. The first term of the sequence is *a* and the common difference is *R*.	The ratios between consecutive terms are all $\frac{1}{2}$ so this is a geometric sequence. $a = 8,\ R = \frac{1}{2}$

▶

c. 1. Calculate the ratio $\dfrac{u_{n+1}}{u_n}$ between all consecutive terms in the sequence.

c. $\dfrac{u_1}{u_0} = \dfrac{-9}{3}$

$= -3$

$\dfrac{u_2}{u_1} = \dfrac{27}{-9}$

$= -3$

$\dfrac{u_3}{u_2} = \dfrac{-81}{27}$

$= -3$

2. If the ratios between consecutive terms are constant, then the sequence is geometric. The first term of the sequence is a and the common difference is R.

The ratios between consecutive terms are all -3, so this is a geometric sequence.

$a = 3, R = -3$

d. 1. Calculate the ratio $\dfrac{u_{n+1}}{u_n}$ between all consecutive terms in the sequence.

d. $\dfrac{u_1}{u_0} = \dfrac{4}{2}$

$= 2$

$\dfrac{u_2}{u_1} = \dfrac{6}{4}$

$= \dfrac{3}{2}$

$\dfrac{u_3}{u_2} = \dfrac{8}{6}$

$= \dfrac{4}{3}$

$\dfrac{u_4}{u_3} = \dfrac{10}{8}$

$= \dfrac{5}{4}$

2. If the ratios between consecutive terms are constant, then the sequence is geometric.

All of the ratios between consecutive terms are different, so this is not a geometric sequence.

6.5.2 Generating geometric sequences using repeated multiplication

As we have seen above, geometric sequences are defined by the pattern of repeatedly multiplying the previous term by a number (common ratio).

WORKED EXAMPLE 12 Generating terms for a geometric sequence

Determine the next four terms in the following geometric sequence.

$$2, 6, 18, 54, \ldots$$

THINK

1. First, calculate the common ratio of the geometric sequence.

WRITE

$R = \dfrac{u_1}{u_0} = \dfrac{6}{2}$

$R = 3$

2. Multiply the terms with the common ratio to determine the next term.

$$u_4 = 54 \times 3$$
$$u_4 = 162$$
$$u_5 = 162 \times 3$$
$$u_5 = 486$$
$$u_6 = 486 \times 3$$
$$u_6 = 1458$$
$$u_7 = 1458 \times 3$$
$$u_7 = 4374$$

3. State the next four terms in the geometric sequence.

The next four terms are 162, 486, 1458 and 4374.

TI	THINK	DISPLAY/WRITE	CASIO	THINK	DISPLAY/WRITE
1. In a Calculator page, enter the value of the first term, 2, and press ENTER.		**1.** In the MAIN application, enter the value of the first term, 2, and press EXE.	
2. The common ratio for the arithmetic sequence is 3, so type in '×3' and then press ENTER. The next term in the sequence will appear.		**2.** The common ratio for the geometric sequence is 3, so type in '×3' and then press EXE. The next term in the sequence will appear.	
3. Continue to press ENTER to generate terms.	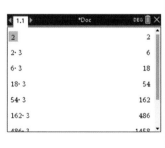	**3.** Continue to press EXE to generate terms.	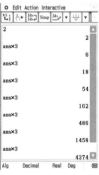
4. The next four terms of the sequence can be read from the CAS screen.	The next four terms are 162, 486, 1458 and 4374.	**4.** The next four terms of the sequence can be read from the CAS screen.	The next four terms are 162, 486, 1458 and 4374.

6.5.3 Graphs of geometric sequences

The shape of the graph of a geometric sequence depends on the value of R.

$R > 1$

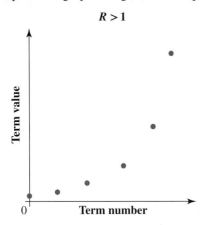

When $R > 1$, the values of the terms increase or decrease at an exponential rate.

$0 < R < 1$

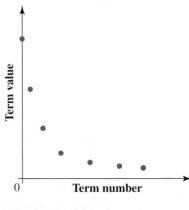

When $0 < R < 1$, the values of the terms converge towards 0.

$-1 < R < 0$

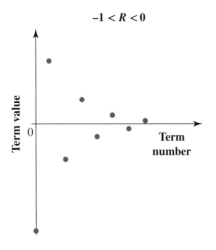

When $-1 < R < 0$, the values of the terms oscillate on either side of 0 but converge towards 0.

$R < -1$

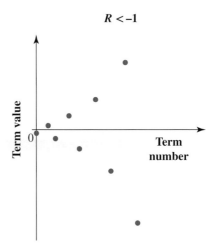

When $R < -1$, the values of the terms oscillate on either side of 0 and move away from the starting value at an exponential rate.

WORKED EXAMPLE 13 Plotting the graph of the geometric sequence by drawing up a table

A geometric sequence is defined by the equation $u_n = 5 \times 2^n$.

a. Set up a table of values showing the term number and term value for the first 5 terms of the sequence.

b. Plot the graph of the sequence.

THINK

a. 1. Set up a table with the term numbers in the top row and the term values in the bottom row.

WRITE

a.

Term number	0	1	2	3	4
Term value					

2. Substitute the first 5 values of the term number into the equation to determine the missing values.

$$u_0 = 5 \times 2^0$$
$$= 5 \times 1$$
$$= 5$$
$$u_1 = 5 \times 2^1$$
$$= 5 \times 2$$
$$= 10$$
$$u_2 = 5 \times 2^2$$
$$= 5 \times 4$$
$$= 20$$
$$u_3 = 5 \times 2^3$$
$$= 5 \times 8$$
$$= 40$$
$$u_4 = 5 \times 2^4$$
$$= 5 \times 16$$
$$= 80$$

3. Complete the table with the calculated values.

Term number	0	1	2	3	4
Term value	5	10	20	40	80

b. 1. Use the table of values to identify the points to be plotted.

b. The points to be plotted are $(0, 5)$, $(1, 10)$, $(2, 20)$, $(3, 40)$ and $(4, 80)$.

2. Plot the points on the graph.

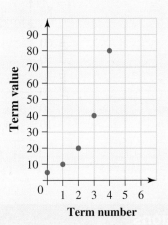

6.5 Exercise

1. WE11 Determine which of the following sequences are geometric sequences, and for those sequences that are geometric, state the values of a and R.

a. 3, 6, 12, 24, 48, ...

b. $\dfrac{1}{2}, \dfrac{5}{4}, \dfrac{25}{8}, \dfrac{125}{16}, ...$

c. 9, 6, 3, 0, −3, ...

d. $\dfrac{1}{2}, \dfrac{1}{5}, \dfrac{2}{25}, \dfrac{4}{125}, ...$

2. Determine the missing values in the following geometric sequences.

   a. $1, 6, c, 216, 1296$ b. $3, g, h, -24, 48$ c. $p, q, s, 300, 1500$

3. State which of the following are geometric sequences. Where applicable, state the first term, u_0, and common ratio, R.

   a. $3, 15, 75, 375, 1875, \ldots$ b. $7, 13, 25, 49, 97, \ldots$ c. $-8, 24, -72, 216, -648, \ldots$

   d. $128, 32, 8, 2, \dfrac{1}{2}, \ldots$ e. $2, 6, 12, 20, 30, \ldots$ f. $3, 3\sqrt{3}, 9, 9\sqrt{3}, 27, \ldots$

4. **WE12** Determine the next four terms in the following geometric sequence.

$$2, 4, 8, 16 \ldots$$

5. **WE13** A geometric sequence is defined by the equation $u_n = 64 \times \left(\dfrac{1}{2}\right)^n$.

   a. Set up a table of values showing the term number and term value for the first 5 terms of the sequence.
   b. Plot the graph of the sequence.

6. A geometric sequence is defined by the equation $u_n = 1.5 \times 3^n$.

   a. Set up a table of values showing the term number and term value for the first 5 terms of the sequence.
   b. Plot the graph of the sequence.

7. Determine the values of the 2nd and 3rd terms of the geometric sequence shown in the following graph.

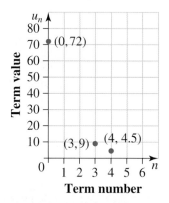

6.5 Exam questions

Question 1 (1 mark)

MC Select the geometric sequence from the following.
   A. $3, 6, 9, 12, \ldots$ **B.** $2, 5, 8, \ldots$ **C.** $8, 11, 14, 17, \ldots$
   D. $3, 9, 27, \ldots$ **E.** $3, 45, 36, 42, \ldots$

Question 2 (1 mark)

MC The rule for the sequence $5, 15, 45, \ldots$ is
   A. $u_n = 3 \times 5^{n-1}$ **B.** $u_n = 5 \times 3^n$ **C.** $u_n = 3 \times 5^n$
   D. $u_n = 5 \times 3^{n-1}$ **E.** $u_n = 5 + 3^{n-1}$

Question 3 (1 mark)

MC The next number in the geometric sequence $4, 6, 9, \ldots$ would be
   A. 13   **B.** 13.5   **C.** 14   **D.** 14.5   **E.** 12.5

More exam questions are available online.

6.6 Geometric sequence applications

LEARNING INTENTION

At the end of this subtopic you should be able to:
- determine the values of a or R in a geometric sequence
- calculate the percentage change
- apply geometric sequences to solve problems.

6.6.1 Determining other values of a geometric sequence

After an equation has been set up to represent a geometric sequence, we can use this equation to determine any term in the sequence. Simply substitute the value of n into the equation to determine the value of that term.

Determining values in a geometric sequence

We can obtain the values a and R for a geometric sequence by transposing the equation.

$$u_0 = a, \quad a = \frac{u_n}{R^n}$$

$$R = \left(\frac{u_n}{a} \right)^{\frac{1}{n}}$$

Note: **The value of n can be calculated using the 'solve' function on a CAS.**

---

tlvd-4468

WORKED EXAMPLE 14 Determining other values of a geometric sequence

a. **Determine the 20th term of the geometric sequence with $a = 5$ and $R = 2$.**
b. **A geometric sequence has a first term of 3 and a 20th term of 1 572 864. Determine the common ratio between consecutive terms of the sequence.**
c. **Determine the first term of a geometric series with a common ratio of 2.5 and a 5th term of 117.1875.**

THINK	WRITE
a. 1. Identify the known values in the question.	a. $a = 5$ $R = 2$ $n = 19$
2. Substitute these values into the geometric sequence formula and solve to determine the missing value.	$u_n = aR^n$ $u_{19} = 5 \times 2^{19}$ $\quad = 2\,621\,440$
3. Write the answer.	The 20th term of the sequence is $2\,621\,440$.
b. 1. Identify the known values in the question.	b. $u_{19} = 1\,572\,864$ $a = 3$ $n = 19$

2. Substitute these values into the formula to calculate the common ratio and solve to determine the missing value.

$$R = \left(\frac{u_n}{a}\right)^{\frac{1}{n}}$$

$$= \left(\frac{1\,572\,864}{3}\right)^{\frac{1}{19}}$$

$$= 524\,288^{\frac{1}{19}}$$

$$= 2$$

3. Write the answer.

The common ratio between consecutive terms of the sequence is 2.

c. 1. Identify the known values in the question.

c. $u_4 = 117.1875$
$R = 2.5$
$n = 4$

2. Substitute these values into the formula to calculate the first term and solve to determine the missing value.

$$a = \frac{u_n}{R^n}$$

$$= \frac{117.1875}{2.5^4}$$

$$= \frac{117.1875}{39.0625}$$

$$= 3$$

3. Write the answer.

The first term of the sequence is 3.

6.6.2 Percentage change

One major use of geometric sequences is calculating sequential percentage changes. We can use the formulas below to work out the common ratios that correspond to different percentage changes.

> **Formula for calculating the common ratio for a *P%* increase or decrease**
>
> The common ratio for a percentage *increase* when each new term is created is given by:
>
> $$R = 1 + \frac{P}{100}$$
>
> The common ratio for a percentage *decrease* when each new term is created is given by:
>
> $$R = 1 - \frac{P}{100}$$

WORKED EXAMPLE 15 Calculating the common ratio from percentage change

Determine the first four terms in each of these geometric sequences for the given percentage changes. Give answers correct to 2 decimal places where appropriate.
a. Starts at 400 and increases by 2%
b. Starts at 100 and decreases by 5%

THINK	WRITE
a. 1. Identify the known values.	**a.** $a = 400, P = 2$
2. Substitute the values of a and P in the formula $R = 1 + \dfrac{P}{100}$.	$R = 1 + \dfrac{2}{100}$ $R = 1.02$ or 102%
3. Start at 400 and multiply by 0.98 to generate the next term.	$400 \xrightarrow{\times 1.02} 408 \xrightarrow{\times 1.02} 416.16 \xrightarrow{\times 1.02} 424.48$
4. Write the answer correct to 2 decimal places.	The first four terms are $400, 392, 384.16$ and 376.48.
b. 1. Identify the known values.	**b.** $a = 100, P = 5$
2. Substitute the values of a and P in the formula $R = 1 - \dfrac{P}{100}$.	$R = 1 - \dfrac{5}{100}$ $R = 0.95$ or 95%
3. Start at 100 and multiply by 0.95 to generate the next term.	$100 \xrightarrow{\times 0.95} 95 \xrightarrow{\times 0.95} 90.25 \xrightarrow{\times 0.95} 85.74$
4. Write the answer correct to 2 decimal places.	The first four terms are $100, 95, 90.25$ and 85.74.

6.6.3 Applications of a geometric sequence

The key aspects to identify in a question that relates to geometric sequences are the initial state, a, and the common ratio, R (usually from a percentage). From there you will be able to derive a general formula for the geometric sequence. Remember to read questions carefully and focus on figuring out what the question is asking of you.

WORKED EXAMPLE 16 Applying a geometric sequence to solve a worded problem

Melbourne's population density is 453 people per square kilometre, which makes it the most populated state capital city in Australia. This is due to many people coming to Melbourne for jobs, education and entertainment from interstate and overseas. In 2011 the population was 3.85 million. The average growth rate for Melbourne was 3.60% over the last 10 years since 2011 according to the Australian Bureau of Statistics.

a. Identify the first term, a, and the common ratio, R, for the geometric sequence for the population growth each year.
b. Determine a formula for the population growth at the start of the nth year. Give the answer correct to 2 decimal places.
c. Calculate the population in 10 years' time. Give the answer correct to 2 decimal places.

THINK	WRITE
a. 1. Identify the known values.	**a.** $a = 3.85$ million, $P = 3.60$
2. Substitute the values of a and P in the formula $R = 1 + \dfrac{P}{100}$.	$R = 1 + \dfrac{3.60}{100}$ $R = 1.036$ or 103.6%

b. 1. Substitute the values of a and R in the formula $u_n = aR^n$.

b. $u_n = aR^n$
$u_n = 3\,850\,000 \times 1.036^n$

2. Write the answer correct to 2 decimal places.

The formula for the population growth at the start of the nth year is $u_n = 3\,850\,000 \times 1.04^n$.

c. 1. Substitute $n = 9$ into the formula for u_n.

c. $u_n = 3\,850\,000 \times 1.036^9$
$= 5\,292\,958.98$

2. Write the answer correct to 2 decimal places.

In 10 years' time the population of Melbourne will be 5.29 million.

6.6 Exercise

Students, these questions are even better in jacPLUS

 Receive immediate feedback and access sample responses

 Access additional questions

 Track your results and progress

Find all this and MORE in jacPLUS

1. **WE14**

   a. Determine the 15th term of the geometric sequence with $a = 4$ and $R = 3$.

   b. A geometric sequence has a first term of 2 and a 12th term of 97 656 250. Determine the common ratio between consecutive terms of the sequence.

   c. Determine the first term of a geometric series with a common ratio of $-\dfrac{1}{2}$ and a 6th term of 13.125.

2. **a.** Determine the 11th term of the geometric sequence with a first value of 1.2 and a common ratio of 4.

   b. A geometric sequence has a first term of -1.5 and a 10th term of 768. Determine the common ratio between consecutive terms of the sequence.

   c. Determine the first term of a geometric series with a common ratio of 0.4 and a 6th term of 6.5536.

3. **a.** Determine the first four terms of the geometric sequence where the 6th term is 243 and the 8th term is 2187.

   b. Determine the first four terms of the geometric sequence where the 3rd term is 331 and the 5th term is 8275.

4. A geometric sequence has a 1st term (u_0) of 200 and a 6th term (u_5) of 2.048. Identify the values of the 2nd, 3rd, 4th and 5th terms.

5. **WE15** Determine the first four terms in each of these geometric sequences for the given percentage changes. Give answers correct to 2 decimal places where appropriate.

   a. Starts at 200 and increases by 4%

   b. Starts at 600 and decreases by 3%

6. Sanka purchased his first home for $670 000 and expects it to appreciate (increase) in value by 8.0% each year. Calculate the value of his house after 10 years.

7. **WE16** The median house price in regional Victoria was $400 600 in 2010. Since then, house prices have increased due to people wanting to move from the city to regional Victoria. Median house prices in regional Victoria have grown by 15% each year.

a. Identify the first term, a, and the common ratio, R, for the geometric sequence for the median house prices each year.

b. Determine the formula for the house price at the start of the nth year.

c. Calculate the median house price at the start of the 15th year.

8. Nuclear medicine procedures help detect and treat diseases by using a small amount of radioactive material called a radiopharmaceutical. Iodine-131 is used to diagnose and treat thyroid cancer and has a half-life (the time taken for 50% of the iodine-131 to decay) of 8 days. A patient is given a single dose of 5 mg of iodine-131 in their system.

a. Identify the first term, a, and the common ratio, R, for the geometric sequence for the radioactive decay.

b. Calculate the amount of iodine-131 remaining after 24 days.

6.6 Exam questions

Question 1 (1 mark)

MC For the sequence $-4, 12, -36, \ldots$ the seventh term is

A. 2916   **B.** -2916 **C.** 972   **D.** -972 **E.** -324

Question 2 (1 mark)

MC Henry's Maths mark in April is 50%. He promises his mother that he will work so that his mark rises by 5% every month. If he is able to keep his promise, he can expect to a have a mark of more than 70% in maths in

A. August   **B.** September   **C.** October
D. November   **E.** December

Question 3 (10 marks)

When Jenny was born, her parents decided to invest $1000 into a term deposit account that she could access once she turned 18. The account received 7% interest per year.

a. State whether this information, when expressed as a sequence, would be represented by an arithmetic sequence or a geometric sequence. Explain your answer. **(1 mark)**

b. Write the rule for this sequence in terms of the number of years of Jenny's life. **(2 marks)**

c. Determine how much money will be in the account after 1 year. **(2 marks)**

d. Determine how much money Jenny will receive from her term deposit when she turns 18. **(2 marks)**

e. Determine how much Jenny would have received if her parents had invested $1500 instead. **(3 marks)**

More exam questions are available online.

6.7 Generate and analyse a geometric sequence using a recurrence relation

LEARNING INTENTION

At the end of this subtopic you should be able to:
- generate a recurrence relation for a geometric sequence
- calculate percentage change with geometric sequences
- model and analyse practical situations.

6.7.1 Using a recurrence relation to generate geometric sequences

If we know the values of a and R in a geometric sequence, we can set up a recurrence relation to generate the sequence.

Recurrence relations

A recurrence relation representing a geometric sequence will be of the form $u_0 = a,\ u_{n+1} = Ru_n$.

WORKED EXAMPLE 17 Setting up a recurrence relation to represent a sequence

Set up a recurrence relation to represent the sequence 10, 5, 2.5, 1.25, 0.625, ...

THINK	WRITE
1. Determine the common ratio by dividing the second term by the first term.	$R = \frac{u_1}{u_0} = \frac{5}{10} = \frac{1}{2}$
2. u_0 represents the first term of the sequence.	$u_0 = 10$
3. Set up the recurrence relation with the given information.	$u_0 = 10,\ u_{n+1} = \frac{1}{2}u_n$

6.7.2 Application of geometric sequences using recurrence relations

A recurrence relation can be used to model a situation. The worked example below shows how a recurrence relation could be used to model the bounce height of a tennis ball.

WORKED EXAMPLE 18 Using a recurrence relation to model a situation

According to the International Federation of Tennis, a tennis ball must meet certain bounce regulations. The test involves dropping a ball vertically from a height of 254 cm and then measuring the rebound height. To meet the regulations, the ball must rebound 135 to 147 cm high, just over half the original distance.

Janine decided to test the ball bounce theory. She dropped a ball from a height of 200 cm. She found that it bounced back up to 108 cm, with the second rebound reaching 58.32 cm and the third rebound reaching 31.49 cm.

a. Set up a recurrence relation to model the bounce height of the ball.
b. Use your relation from part a to estimate the height of the 4th and 5th rebounds, giving your answers correct to 2 decimal places.
c. Sketch the graph of the number of bounces against the height of each bounce.

THINK	WRITE
a. 1. List the known information.	a. $u_0 = 200$, 1st bounce: $u_1 = 108$ cm 2nd bounce: $u_2 = 58.32$ cm 3rd bounce: $u_3 = 31.49$ cm
2. Check if there is a common ratio between consecutive terms. If so, this situation can be modelled using a geometric sequence.	$\dfrac{u_1}{u_0} = \dfrac{108}{200}$ $\quad = 0.54$ $\dfrac{u_2}{u_1} = \dfrac{58.32}{108}$ $\quad = 0.541...$ $\quad \approx 0.54$ There is a common ratio between consecutive terms of 0.54.
3. Set up the equation to represent the geometric sequence.	$a = 200$ $R = 0.54$ $u_0 = a,\ u_{n+1} = R u_n$ $u_0 = 108,\ u_{n+1} = 0.54 u_n$
b. 1. Use the formula from part a to calculate the height of the 4th rebound ($n = 3$).	b. $u_3 = 31.49$ $u_{n+1} = 0.54 u_n$ $u_4 = 0.54 u_3$ $\quad = 0.54 \times 31.49$ $\quad = 17.0046$ $\quad = 17.00$ (correct to 2 decimal places)
2. Use the formula from part a to calculate the height of the 5th rebound ($n = 4$).	$u_{n+1} = 0.54 u_n$ $u_5 = 0.54 u_4$ $\quad = 0.54 \times 17.00$ $\quad = 9.18$

▶

3. Write the answer.

The estimated height of the 4th rebound is 17.00 cm, and the estimated height of the 5th rebound is 9.18 cm.

c. 1. Draw up a table showing the bounce number against the rebound height.

c.

Rebound number	0	1	2	3	4	5
Rebound height (cm)	200	108	58.32	31.49	17.00	9.18

2. Identify the points to be plotted on the graph.

The points to be plotted are (0, 200), (1, 108), (2, 58.32), (3, 31.49), (4, 17.01) and (5, 9.18).

3. Plot the points on the graph.

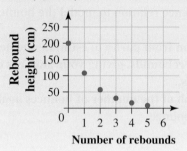

6.7 Exercise

Students, these questions are even better in jacPLUS

 Receive immediate feedback and access sample responses

 Access additional questions

 Track your results and progress

Find all this and MORE in jacPLUS

1. **WE17** Set up a recurrence relation to represent the geometric sequence 2.5, −7.5, 22.5, −67.5, 202.5, ...

2. A geometric sequence is represented by the recurrence relation $u_0 = -4$, $u_{n+1} = -3.5u_n$. Determine the first five terms of the sequence.

3. The 4th (u_3) and 5th (u_4) terms of a geometric sequence are −4 and −1. Set up a recurrence relation to define the sequence.

4. **WE18** Eric decided to test the rebound height of a tennis ball. He dropped a ball from a height of 300 cm and found that it bounced back up to 165 cm, with the second rebound reaching 90.75 cm, and the third rebound reaching 49.91 cm.

   a. Set up a recurrence relation to model the bounce height of the ball.

   b. Use your relation from part **a** to estimate the height of the 4th and 5th rebounds, giving your answers correct to 2 decimal places.

   c. Sketch the graph of the number of bounces against the height of the bounce.

5. Rosanna decided to test the ball rebound height of a basketball. She dropped the basketball from a height of 500 cm and noted that each successive rebound was two-fifths of the previous height.
   a. Set up a recurrence relation to model the bounce height of the ball.
   b. Use your relation from part a to estimate the heights of the first 5 rebounds, correct to 2 decimal places.
   c. Sketch the graph of the first 5 bounces against the rebound height.

6. A bouncing ball rebounds to 70% of its previous height.
   a. Determine the height from which the ball would have to be dropped for the 10th bounce to reach 50 cm in height. Give your answer correct to 1 decimal place.
   b. Define a recurrence relation to determine the height of the ball after n bounces.

6.7 Exam questions

Question 1 (1 mark)

MC The recurrence relation to represent the geometric sequence 8, 16, 32 ... is

A. $u_{n+1} = u_n + 2$ B. $u_{n+1} = u_n - 2$ C. $u_{n+1} = \frac{1}{2} u_n$

D. $u_{n+1} = 2u_{n+1}$ E. $u_{n+1} = 2u_n$

Question 2 (1 mark)

MC A colony of bacteria grows by each organism splitting into two every 5 minutes. If there are 100 bacteria in a dish initially, the number in the dish after half an hour is

A. 800 B. 1600 C. 3200
D. 10 E. 16

Question 3 (1 mark)

MC The 5000 litres of water in a tank is evaporating, being reduced to 4500 litres and 4050 litres in successive months. Determine approximately how much water will be in the tank after 6 months.

A. 2657 litres B. 2657 millilitres C. 2391 litres
D. 2952 litres E. 3281 litres

More exam questions are available online.

6.8 Modelling growth and decay using recurrence relations

LEARNING INTENTION

At the end of this subtopic you should be able to:
- use arithmetic sequences to calculate the simple interest and depreciation cost
- use geometric sequences to calculate the compound interest or reduced balance depreciation.

6.8.1 Linear growth — simple interest

If we have a practical situation involving linear growth or decay in discrete steps, we can generate terms in a sequence using a recurrence relation. A rule can be set up to generate any term in the sequence without having to generate all of the other terms first, as discussed in section 6.4.1.

Rule for the nth term in a sequence involving linear growth

If u_0 is the initial (starting) value, then for linear growth, the value of the nth term in the sequence generated by the recurrence relation is:

$$u_n = u_0 + nd$$

WORKED EXAMPLE 19 Using a first-order linear recurrence relation to calculate simple interest

Boris invests \$6000 in a term deposit with 4% p.a. simple interest.
a. Calculate the amount of interest that is added to the investment each year.
b. Set up a recurrence rule that represents Boris's situation.
c. Use the recurrence rule to calculate the value of Boris's investment after 8 years.

THINK	WRITE
a. 1. The principal, u_0, is 6000.	a. $u_0 = 6000$
2. Determine the interest by calculating 4% of \$6000.	4% of \$6000 $= 0.04 \times 6000$ The interest added to the investment each year is \$240.
b. Take the information from part **a** to write the recurrence rule for linear growth for this investment in the form $u_n = u_0 + nd$.	b. $u_0 = 6000,\ u_n = 6000 + 240n$
c. Substitute $n = 8$ into the equation to calculate the value of this investment after 8 years.	c. $u_8 = 6000 + 240 \times 8$ $u_8 = 7920$ After 8 years, Boris's investment would have a value of \$7920.

6.8.2 Depreciating assets

Many items, such as automobiles or electronic equipment, decrease in value over time as a result of wear and tear. At tax time individuals and companies use depreciation of their assets to offset expenses and to reduce the amount of tax they have to pay.

Unit cost depreciation

Unit cost depreciation is a way of depreciating an asset according to its use. For example, you can depreciate the value of a car based on how many kilometres it has driven. The unit cost is the amount of depreciation per unit of use, which would be 1 kilometre of use in the example of the car.

Future value and write-off value

When depreciating the values of assets, companies will often need to know the **future value** of an item. This is the value of that item at that specific time.

The **write-off value** or scrap value of an asset is the point at which the asset is effectively worthless (i.e. has a value of $0) due to depreciation.

Rule for the nth term in a sequence involving linear decay

If u_0 is the initial (starting) value, then for linear decay, the value of the nth term in the sequence generated by the recurrence relation is:

$$u_n = u_0 - nd$$

---

tlvd-4470

WORKED EXAMPLE 20 Using a first-order linear recurrence relation to calculate depreciation of an asset

Loni purchases a new car for \$25 000 and decides to depreciate it at a rate of \$0.20 per km.
a. Set up a recurrence rule to determine the value of the car after n km of use.
b. Use the recurrence rule to determine the future value of the car after it has 7500 km on its clock.

THINK	WRITE
a. 1. Identify and state the values of u_0 and d.	a. $u_0 = 25\,000$ $d = -0.2$
2. Substitute these values to write a recurrence rule for linear decay in the form $u_n = u_0 - nd$.	$u_0 = 25\,000$ $u_n = 25\,000 - 0.2n$
b. 1. Substitute $n = 7500$ into the recurrence rule determined in part a.	b. $u_{7500} = 25\,000 - 0.2 \times (7500)$ $u_{7500} = 23\,500$
2. Write the answer.	After 7500 km the car will be worth \$23 500.

6.8.3 Geometric growth — compound interest

As covered in Topic 3, compound interest is calculated on the sum of an investment at the start of each compounding period. The amount of interest accrued varies throughout the life of the investment and can be modelled by a geometric sequence using a recurrence relation. A rule can be set up to generate any term in the sequence without having to generate all of the other terms (as discussed in section 6.7.1).

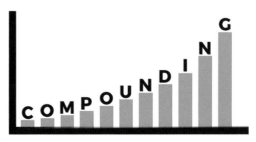

Rule for the *n*th term in a sequence involving geometric growth

If u_0 is the initial (starting) value, then for geometric growth, the value of the *n*th term in the sequence generated by the recurrence relation is:

$$u_n = R^n u_0$$

For geometric growth, $R > 1$ and for geometric decay, $0 < R < 1$.

tlvd-4471

WORKED EXAMPLE 21 Using a recurrence rule to calculate compound interest

Alexis puts $2000 into an investment account that earns compound interest at a rate of 0.5% per month.

a. Set up a recurrence rule that represents Alexis's situation as a geometric sequence, where u_n is the amount in Alexis's account after *n* months. (*Hint:* the common ratio in the geometric sequence equation is equal to $1 + \dfrac{P}{100}$.)

b. Use the recurrence rule to determine the amount in Alexis's account at the end of each of the first 6 months.

c. Calculate the amount in Alexis's account at the end of 15 months.

THINK	WRITE
a. 1. Determine the common ratio by substituting the values in the equation $R = 1 + \dfrac{P}{100}$	a. Common ratio $(R) = 1 + \dfrac{P}{100}$ $R = 1 + \dfrac{0.5}{100}$ $= 1.005$
2. Determine the value of R and the amounts in the account after each of the first two months.	$u_0 = 2000$, $u_{n+1} = 1.005 \times u_n$ $u_1 = 2000 \times 1.005 = 2010$ $u_2 = 2000 \times 1.005^2 = 2020.05$
3. Calculate the common ratio between consecutive terms.	$R = \dfrac{u_1}{u_0}$ $= \dfrac{2010}{2000}$ $= 1.005$
4. Substitute these values into the recurrence rule.	$a = 2000$, $R = 1.005$ $u_n = 2000 \times 1.005^n$
b. 1. Use the recurrence rule from part **a** to determine the values of u_3, u_4 and u_5. Round all values correct to 2 decimal places.	b. $u_3 = 2000 \times 1.005^n$ $= 2000 \times 1.005^3$ $= 2030.150...$ ≈ 2030.15 $u_4 = 2000 \times 1.005^n$ $= 2000 \times 1.005^4$ $= 2040.301...$ ≈ 2040.30

$$u_5 = 2000 \times 1.005^n$$
$$= 2000 \times 1.005^5$$
$$= 2050.502\ldots$$
$$\approx 2050.50$$

2. Write the answer.

The amounts in Alexis's account at the end of each of the first 6 months are $2010, $2020.05, $2030.15, $2040.30, and $2050.50.

c. 1. Use the recurrence rule from part **a** to determine the values of u_{15}, rounding your answer correct to 2 decimal places.

c. $u_{15} = 2000 \times 1.005^{15}$
$$= 2000 \times 1.005^{15}$$
$$= 2155.365\ldots$$
$$\approx 2155.37$$

2. Write the answer.

After 15 months Alexis has $2155.37 in her account.

6.8.4 Geometric decay — reducing balance depreciation

Another method of depreciation is **reducing balance depreciation**. When an item is depreciated by this method, rather than the value of the item depreciating by a fixed amount each year, it depreciates by a percentage of the previous future value of the item.

Due to the nature of reducing balance depreciation, we can represent the sequence of the future values of an item that is being depreciated by a recurrence rule.

WORKED EXAMPLE 22 Using a recurrence rule to calculate depreciation

A hot water system purchased for $1250 is depreciated by the reducing balance method at a rate of 8% p.a.
a. Set up a recurrence rule to determine the value of the hot water system after n years of use.
b. use the recurrence rule to determine the future value of the hot water system after 6 years of use (correct to the nearest cent).

THINK

a. 1. Calculate the common ratio by identifying the value of the item in any given year as a percentage of the value in the previous year. Convert the percentage to a ratio by dividing by 100.

WRITE

a. $100\% - 8\% = 92\%$

Each year the value of the item is 92% of the previous value.

$$92\% = \frac{92}{100}$$
$$= 0.92$$
$$R = 0.92$$

2. Calculate the value of the hot water system after 1 year of use.	$u_0 = 1250$ $u_1 = 1250 \times 0.92$ $= 1150$	
3. Substitute the values of a and R into the recurrence rule.	$u_n = 1250 \times 0.92^n$	
b. 1. Substitute $n = 6$ into the equation determined in part **a**. Give your answer correct to 2 decimal places.	**b.** $u_n = 1250 \times 0.92^n$ $u_6 = 1250 \times 0.92^6$ $= 757.943...$ ≈ 757.94	
2. Write the answer.	After 6 years the book value of the hot water system is $757.94.	

6.8 Exercise

Students, these questions are even better in jacPLUS

 Receive immediate feedback and access sample responses

 Access additional questions

 Track your results and progress

Find all this and MORE in jacPLUS ▶

1. **WE19** Sean puts $10 000 into a term deposit with 3% p.a. simple interest.

   a. Set up a recurrence relation that represents the value of Sean's investment.
   b. Set up an equation that represents Sean's situation as an arithmetic sequence.
   c. Use your equation from part **b** to calculate the value of Sean's investment after 5 years.

2. Oceania got a loan of $6000 that increased at a rate of 8% p.a. simple interest. Determine how much she would have to repay if she waited 3 years before paying off her loan.

3. Grigor puts $1500 into an investment account that earns simple interest at a rate of 4.8% per year.

   a. Set up an equation that represents Grigor's situation as an arithmetic sequence, where u_n is the amount in Grigor's account after n months.
   b. Use your equation from part **a** to determine the amount in Grigor's account after each of the first 6 months.
   c. Calculate the amount in Grigor's account at the end of 18 months.

4. Justine sets up an equation to model the amount of her money in a simple interest investment account after n months. Her equation is $u_n = 8000 + 50n$, where u_n is the amount in Justine's account after n months.

   a. Determine how much Justine invested in the account.
   b. Calculate the annual interest rate of the investment.

5. **WE20** Phillipe purchases a new car for $24 000 and decides to depreciate it at a rate of $0.25 per km.

   a. Set up an equation to determine the value of the car after n km of use.
   b. Use your equation from part **a** to determine the future value of the car after it has travelled 12 000 km.

6. Dougie is in charge of the equipment for his office. He decides to depreciate the value of a photocopier at the rate of x cents for every n copies made. Dougie's equation for the value of the photocopier after n copies is $u_n = 5400 - 0.001n$.

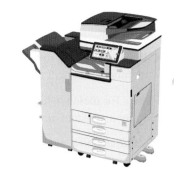

   a. Determine how much the photocopier cost.
   b. Calculate the rate of depreciation per copy made.

7. An employee starts a new job with a $60 000 salary in the first year and the promise of a pay rise of $2500 a year.

   a. Calculate how much her salary will be in her 6th year.
   b. Calculate how long it will take for her salary to reach $85 000.

8. Nadia wants to invest her money and decides to place $90 000 into a credit union account earning simple interest at a rate of 6% per year.

   a. Calculate how much interest Nadia will receive after 1 year.
   b. Calculate the total amount Nadia has in the credit union after n years.
   c. Determine how long Nadia should keep her money invested for if she wants a total of $154 800 returned.

9. Tom bought a car for $23 000, knowing it would depreciate in value by $210 per month.

   a. Calculate the value of the car after 18 months.
   b. Calculate how much the value of the car depreciates by in 3 years.
   c. Determine how many months it will take for the car to be valued at $6200.

10. A confectionary manufacturer introduces a new sweet and produces 50 000 packets of the sweets in the first week. The stores sell them quickly, and in the following week there is demand for 30% more. In each subsequent week the increase in production is 30% of the original production.

   a. Calculate how many packs are manufactured in the 20th week.
   b. Determine in which week the confectionary manufacturer will produce 5 540 000 packs.

11. A canning machine was purchased for a total of $250 000 and is expected to produce 500 000 000 cans before it is written off.

   a. Calculate how much the canning machine depreciates by with each can made.
   b. If the canning machine were to make 40 200 000 cans each year, determine when the machine will theoretically be written off.
   c. Determine when the machine will have a book value of $89 200.

12. The local rugby club wants to increase its membership. In the first year they had 5000 members, and so far they have managed to increase their membership by 1200 members per year.

   a. If the increase in membership continues at the current rate, calculate how many members they will have in 15 years' time.

   Tickets for membership in the first year were $200, and each year the price has risen by a constant amount, with memberships in the 6th year costing $320.

   b. Calculate how much the tickets would cost in 15 years' time.
   c. Determine the total membership income in both the first and 15th years.

13. **WE21** Hussein puts $2500 into an investment that earns compound interest at a rate of 0.3% per month.

   a. Set up an equation that represents Hussein's situation as a geometric sequence, where u_n is the amount in Hussein's account after n months.
   b. Use your equation from part a to determine the amount in Hussein's account at the end of each of the first 6 months.
   c. Calculate the amount in Hussein's account at the end of 15 months.

14. Tim sets up an equation to model the amount of his money in a compound interest investment account after n months. His equation is $u_n = 4500 \times 1.0035^n$, where u_n is the amount in his account after n months.

   a. Determine how much Tim initially invested in the account.
   b. Calculate the annual interest rate of the investment.

15. Jonas starts a new job with a salary of $55\,000 per year and the promise of a 3% pay rise for each subsequent year in the job.

   a. Write an equation to determine Jonas's salary in his nth year in the job.
   b. Determine how much Jonas will earn in his 5th year in the job.

16. **WE22** A refrigerator purchased for $1470 is depreciated by the reducing balance method at a rate of 7% p.a.

   a. Set up an equation to determine the value of the refrigerator after n years of use.
   b. Use your equation from part a to determine the future value of the refrigerator after 8 years of use, correct to the nearest cent.

17. Ivy buys a new oven and decides to depreciate the value of the oven by the reducing balance method. Ivy's equation for the value of the oven after n years is $u_n = 1800 \times 0.925^n$.

   a. Determine how much the oven cost.
   b. Calculate the annual rate of depreciation for the oven.

18. Julio's parents invest $5000 into a college fund on his 5th birthday. The fund pays a compound interest rate of 5.5% p.a. Determine how much the fund will be worth when Julio turns 18.

Question 1 (3 marks)

Jacqui's mother invests $500 for her daughter in a bank account that pays 4% per annum compound interest, adjusted monthly. Jacqui wishes to use the money to buy furniture for $520.

She draws up a table to help her work out when she will have enough.

n	0	1	2	3	4	5
A	$500.00	$501.67	$503.34	$505.02	$506.70	$508.39

    a. Calculate how much will she have after 6 months. **(1 mark)**
    b. Determine how many months it will take for her to have enough to buy the furniture. **(1 mark)**
    c. Determine whether it would be quicker if she invested with simple interest of 0.25% per month.
        Calculate how long the simple interest investment would take to reach her goal. **(1 mark)**

Question 2 (3 marks)

A ball dropped from the top of a 64-metre-high building rebounds to a height of 48 metres on the first bounce, 36 metres on the second bounce and so on. The height of each bounce is $\frac{3}{4}$ of the height of the previous bounce.

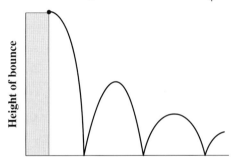

    a. Write the recurrence relation for the sequence of bounces. **(1 mark)**
    b. Write down the rule for the height of the nth bounce. **(1 mark)**
    c. Determine the height of the fifth bounce. **(1 mark)**

Question 3 (3 marks)

A swimming pool pump is purchased for $18 200 with reducing-balance depreciation at 20% per annum.
    a. Calculate the future value of the pump after 3 years. **(1 mark)**
    b. Calculate the accumulated depreciation over the period. **(1 mark)**
    c. Determine after how many full years the book value of the pump will be less than $2500. **(1 mark)**

More exam questions are available online.

6.9 Review

6.9 Exercise

Multiple choice

1. **MC** Select the geometric sequence from the following.
   - **A.** 12, 16, 20, 24, 28
   - **B.** 12, 4.8, 1.92, 0.77, 0.31
   - **C.** 12, 15, 21, 30, 42
   - **D.** 12, 6.2, 3.3, 1.88, 0.42
   - **E.** 12, 14.8, 17.6, 20.4, 23.2

2. **MC** For the sequence $-16, -11.2, -6.4, -1.6, 3.2$, the correct values for a and d are:
   - **A.** $a = -16, d = -11.2$
   - **B.** $a = 4.8, d = 3.2$
   - **C.** $a = -16, d = 4.8$
   - **D.** $a = -19.2, d = 3.2$
   - **E.** $a = 16, d = -4.8$

3. **MC** If $u_0 = 65$, the third term in the recurrence relation $u_{n+1} = 2u_n - 10$ is:
   - **A.** 120
   - **B.** 230
   - **C.** 119
   - **D.** 250
   - **E.** 35

4. **MC** The missing value in the arithmetic sequence of 65, x, 58, 54.5, 51 is:
   - **A.** -3.5
   - **B.** 68.5
   - **C.** 60
   - **D.** 61.5
   - **E.** 62

5. **MC** The common ratio for the sequence 4.8, 14.4, 43.2, 129.6, 388.8 is:
   - **A.** 3
   - **B.** 4.8
   - **C.** 9.6
   - **D.** 81
   - **E.** 4

6. **MC** If $u_3 = 18$, then according to the recurrence relation $u_{n+1} = 0.6u_n + 4.5$, u_4 will equal:
   - **A.** 30
   - **B.** 22.5
   - **C.** 17.4
   - **D.** 10.8
   - **E.** 15.3

7. **MC** The first two numbers of a geometric sequence are 7 and 21. The fourth term is:
   - **A.** 35
   - **B.** 189
   - **C.** 63
   - **D.** 39
   - **E.** 1029

8. **MC** Select the correct recurrence relation for the sequence 45, 11.25, 2.81, 0.70, 0.18.
   - **A.** $u_{n+1} = u_n + 33.75, u_0 = 45$
   - **B.** $u_{n+1} = u_n + 33.75, u_0 = 0.18$
   - **C.** $u_{n+1} = 0.25u_n, u_0 = 33.75$
   - **D.** $u_{n+1} = 0.25u_n, u_0 = 45$
   - **E.** $u_{n+1} = 0.5u_n, u_0 = 45$

9. **MC** Select the graph that represents an arithmetic sequence.

   A.
   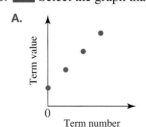

   B.
   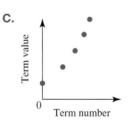

   C.

   D.
   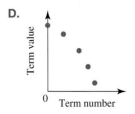

   E.
   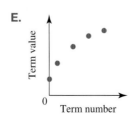

10. **MC** Select the equation that represents a geometric sequence.

    A. $u_n = a + nd$ **B.** $u_{n+1} = Ru_n + d, \; u_0 = a$ **C.** $u_{n+1} = u_n + d, \; u_0 = a$

    D. $u_{n+1} = u_n, \; u_1 = a$ **E.** $u_n = aR$

Short answer

11. Determine the first five terms of each of the following:

    a. Arithmetic sequence: $u_n = 2 + 5n$

    b. Geometric sequence: $u_n = 17 \times 2.2^n$

    c. Recurrence relation: $u_{n+1} = u_n - 6, \; u_0 = 15$

12. Construct a recurrence rule to represent each of the following sequences.

    a. Arithmetic sequence: $3, \; 0.5, \; -2, \; -4.5, \; -7, \; -9.5$

    b. Geometric sequence: $-2, \; -8, \; -32, \; -128, \; -512$

    c. Recurrence relation: $22, \; 143, \; 929.5, \; 6041.75, \; 39\,271.375$

13. Determine the 8th term of the following sequences, correct to 2 decimal places where appropriate.

    a. Arithmetic sequence: $14, \; 18.75, \; 23.5, \; 28.25$

    b. Geometric sequence: $\dfrac{11}{25}, \; \dfrac{33}{50}, \; \dfrac{99}{100}, \; \dfrac{149}{100}, \; \dfrac{223}{100}$

    c. Recurrence relation: $45, \; 28, \; 11, \; -6, \; -23$

14. The graph shows some points of an arithmetic sequence.

    a. Determine the common difference between consecutive terms.

    b. Determine the value of the first term of the sequence.

    c. Determine the equation for this sequence.

    d. Calculate the value of the 9th term.

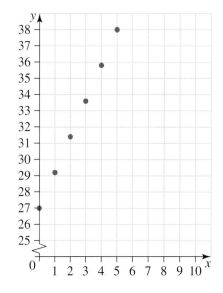

15. A geometric sequence is defined by the equation $u_n = 48 \times \left(\dfrac{1}{2}\right)^n$.

    a. Draw a table of values showing the first 5 terms.

    b. Plot the graph of the sequence.

16. Graph the first 5 terms of an arithmetic sequence that starts with the numbers -4 and 2.

Extended response

17. Chris is saving for his first car. He put $900 into a simple interest savings account that earns 8.2% per year.

    a. Set up an equation that represents Chris's situation as an arithmetic sequence, where u_n is the amount in the account after n months.

    b. Use your equation from part **a** to determine the amount in Chris's account after each of the first 5 months.

    c. Calculate the amount in the savings account at the end of 20 months.

    d. Determine how many months at this interest rate it will take Chris to save $1200.

18. Barry dropped a ball from a second story window, 6 m from the ground. The ball rebounded to two-thirds of the original height.

    a. Set up a recurrence relation to model the bounce of the height of the ball.

    b. Use your relation from part **a** to estimate the height of the 5th and 6th rebounds, giving your answer correct to 2 decimal places.

    c. Sketch the graph of the number of bounces against the height of the bounce.

19. Kane's salary in his first year at his job was $68\,000$. Each year his salary increases by 2.5%.

    a. Write an equation to reflect Kane's salary in his nth year in the job.

    b. Calculate how much he will earn in his 4th year in the job.

Kane decided to put some of the money from his pay rise into a compound interest account. He established the equation $u_n = 1551 \times 1.034^n$ to model the amount of money in the account after n years.

   c. Determine how much Kane originally put in the account and at what interest rate.

   d. Using the equation, determine how much the account will have increased by after 5 years.

20. As part of an experiment, bacteria are grown in a laboratory. On day 7 of the experiment the bacteria count is 10 935. The number of bacteria has tripled each day.

   a. Calculate the amount of bacteria expected on the 9th day.

   b. Determine the original number of bacteria.

21. Sebastien just bought a new gaming computer for $3000. He does some calculations and concludes that the computer depreciates at a flat rate of $250 per year.

   a. Write an equation that represents the depreciation of Sebastien's computer.

   b. Determine how much the computer depreciates by in 5 years.

   c. Determine how long it will take for the computer to reach its write-off value ($0).

22. The number of migrants to Australia is increasing at a rate of 13 000 people per year. Current figures indicate that there are currently 170 000 migrants each year.

   a. Write a recurrence relation to model the growth of migration numbers.

   b. Determine the number of migrants predicted to arrive in Australia in 6 years.

6.9 Exam questions

Question 1 (1 mark)

MC Select the sequence that has no recognisable pattern.
   A. $2, 4, 6, 8, \ldots$ **B.** $2, 8, 14, 20, \ldots$ **C.** $1, 34, 35, 67, \ldots$ **D.** $32, 16, 8, 4, \ldots$ **E.** $89, 82, 75, 68, \ldots$

Question 2 (1 mark)

MC The recursive equation for the sequence $39, 30, 21, 12, \ldots$ is
   A. $u_{n+1} = u_n - 9$ **B.** $u_n = u_{n+1} - 9$ **C.** $u_n = 9u_{n+1}$ **D.** $u_n = u_{n+1} + 9$ **E.** $u_{n+1} = 9$

Question 3 (2 marks)

The number of apples on a tree is seen to increase by 11 each year. If in the first year after the tree was planted it grew 4 apples, determine how many apples it grew in the 4th year.

Question 4 (1 mark)

MC The rule for the sequence $67, 61, 55, 49, \ldots$ is
   A. $u_n = 67 + 6(n-1)$ **B.** $u_n = 67 - 6(n+1)$ **C.** $u_n = 67 - 6n$
   D. $n = 67 - 6(n+1)$ **E.** $u_n = 67 + 6n$

Question 5 (1 mark)

MC The reducing balance depreciation on sporting equipment is 25%. After 3 years, the book value of sporting equipment originally costing $1850 will be
   A. $462.50   **B.** $1387.50   **C.** $780.47   **D.** $1040.63   **E.** $616.67

More exam questions are available online.

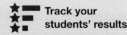

Answers

Topic 6 Sequences and first-order linear recurrence relations

6.2 Arithmetic sequences

6.2 Exercise

1. a. The 2nd term, u_1, is 8.
   b. The 4th term, u_3, is 12.
   c. The 6th term, u_5, is 16.
2. a. The 1st term, u_0, is 2.
   b. The 3rd term, u_2, is 13.
   c. The 6th term, u_5, is 13.
   d. The 7th term, u_6, is 6.
3. a. The 2nd term, u_1, is 12.
      The 3rd term, u_2, is 6.
      The 5th term, u_4, is 6.
   b. The 2nd term, u_1, is 6.
      The 3rd term, u_2, is 9.
      The 5th term, u_4, is 15.
   c. The 2nd term, u_1, is 35.
      The 3rd term, u_2, is 30.
      The 5th term, u_4, is 20.
   d. The 2nd term, u_1, is 2.0.
      The 3rd term, u_2, is 2.5.
      The 5th term, u_4, is 3.5.
   e. The 2nd term, u_1, is 32.
      The 3rd term, u_2, is 16.
      The 5th term, u_4, is 4.
   f. The 2nd term, u_1, is 2.
      The 3rd term, u_2, is 4.
      The 5th term, u_4, is 10.
4. B
5. a. Arithmetic; $a = 23$, $d = 45$
   b. Not arithmetic
   c. Arithmetic; $a = \dfrac{1}{2}$, $d = \dfrac{1}{4}$
6. a. $f = -62$
   b. $j = 5.7$, $k = 15.3$
   c. $p = -\dfrac{3}{4}$, $q = 1$, $r = \dfrac{11}{4}$
7. E
8. a. $a = 2$, $d = 2$ b. $a = 3$, $d = 4$
   c. $a = 23$, $d = -3$ d. $a = 12$, $d = -1$
9. The next four terms are $28, 34, 40$ and 46.
10. The first five terms are $12, 17, 22, 27$ and 32.
11. D
12. The difference between each consecutive term must be identical.
    Please see the worked solution for the sample response.
13. $a = 4$, $d = 3$
14. The first term is 4.

15. a.

Term number	0	1	2	3	4
Term value	15	15	25	35	45

b.

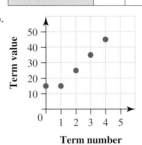

c. 85

16. a.

Term number	0	1	2	3	4
Term value	6.4	8.2	9.6	11.2	12.8

b.

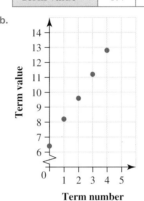

c. 25.6

17.

Design number	1	2	3	4	5
Number of sticks	4	7	10	13	16

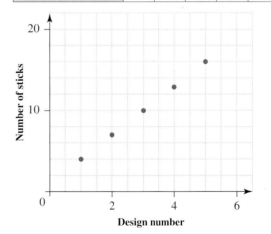

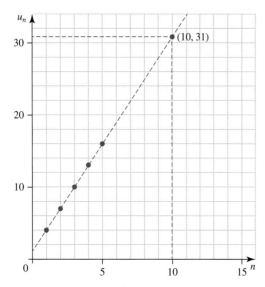

Namjoon will use 31 sticks to create the 10th design.

18. a.

n	1	2	3	4	5
u_n	1	4	7	10	13

b.

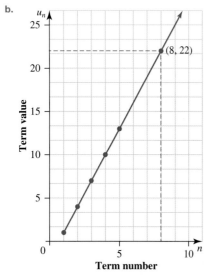

c. Interpreting from the graph, the value of the 8th term is 22.

d. Sample responses are given below.
- It can be tedious and time consuming.
- Could lack accuracy if drawn incorrectly.
- Difficult to determine the values of very high term numbers.

19. a.

Term number	0	1	2	3	4
Term value	1	3	5	7	9

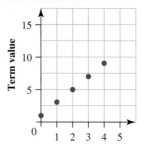

The graph is increasing in a straight line.

b.

Term number	0	1	2	3
Term value	7	6	5	4

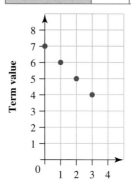

The graph is decreasing in a linear pattern.

c.

Term number	0	1	2	3	4
Term value	5	10	15	20	25

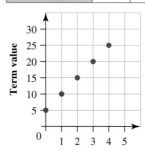

The graph is increasing in a linear pattern.

d.

Term number	0	1	2	3
Term value	1	2	4	9

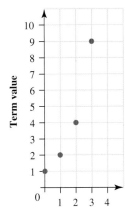

Term number

The graph is increasing in a non-linear pattern.

20. a. The sequence is increasing in a non-linear pattern.
    b. The sequence is decreasing in a linear pattern.
    c. The sequence has a constant value of 2.
    d. The sequence is increasing in a linear pattern.

21. B

22.

Term number	0	1	2	3	4
Term value	3	6	12	24	48

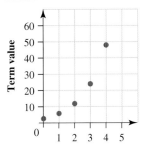

Term number

The graph is increasing in a non-linear pattern.

6.2 Exam questions

Note: Mark allocations are available with the fully worked solutions online.

1. $u_1 = -\dfrac{1}{4}$

2. B

3. E

6.3 Arithmetic sequence applications

6.3 Exercise

1. a. $u_n = -1 + 4n$
   b. $u_n = 1.5 - 3.5n$
   c. $u_n = \dfrac{7}{2} + 2n$

2. a. $5, 8, 11, 14, 17$
   b. $-1, -8, -15, -22, -29$
   c. $\dfrac{1}{3}, 1, \dfrac{5}{3}, \dfrac{7}{3}, 3$

3. a. -162 b. 3467

4. a. 48 b. The 2nd term   c. The 13th term

5. a. 104 b. 275 c. -176 d. $-\dfrac{387}{20}$

6. a. 724 b. -52.8 c. -10.2 d. $\dfrac{13}{6}$

7. a. The 37th term b. The 210th term

8. a. -1.5 b. 13.5 c. -3

9. $335

10. 25 elephants

6.3 Exam questions

Note: Mark allocations are available with the fully worked solutions online.

1. A

2. D

3. C

6.4 Generate and analyse an arithmetic sequence using a recurrence relation

6.4 Exercise

1. $u_{n+1} = u_n - 5, \; u_0 = 2$

2. $-2.2, \; 1.3, \; 4.8, \; 8.3, \; 11.8$

3. $u_{n+1} = u_n - 4.5, \; u_0 = 2$

4. a. $u_0 = 1, \; u_{n+1} = u_n + 3$ b. $u_n = 1 + 3n$
   c. 43 chairs

5. a. $u_0 = 37\,000 \; u_{n+1} = u_n - 1200$
   b. The area of the ice shelf after each of the first 6 years is $35\,800 \text{ km}^2$, $34\,600 \text{ km}^2$, $33\,400 \text{ km}^2$, $32\,200 \text{ km}^2$ and $31\,000 \text{ km}^2$.
   c. See the table at the bottom of the page*
      The points to be plotted are $(1, 35\,800)$, $(2, 34\,600)$, $(3, 33\,400)$, $(4, 32\,200)$, $(5, 31\,000)$ and $(6, 29\,800)$.

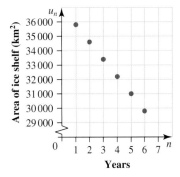

Years

---

*5. c.

Year	1	2	3	4	5	6
Area (km²)	35 800	34 600	33 400	32 200	31 000	29 800

6.4 Exam questions

Note: Mark allocations are available with the fully worked solutions online.

1. E
2. a. $u_0 = 3$, $u_{n+1} = u_n + 4$
   b. $u_n = 4n + 3$
3. a. $u_0 = 30$, $u_{n+1} = u_n + 9$
   b. $u_n = 30 + 9n$
   c. 93 cm

6.5 Geometric sequences

6.5 Exercise

1. a. Geometric; $a = 3$, $R = 2$
   b. Geometric; $a = \frac{1}{2}$, $R = 2\frac{1}{2}$
   c. Not geometric
   d. Geometric; $a = \frac{1}{2}$, $R = \frac{2}{5}$
2. a. $c = 36$
   b. $g = -6$, $h = 12$
   c. $p = 2.4$, $q = 12$, $s = 60$
3. a. Geometric; first term $= 3$, common ratio $= 5$
   b. Not geometric
   c. Geometric; first term $= -8$, common ratio $= -3$
   d. Geometric; first term $= 128$, common ratio $= \frac{1}{4}$
   e. Not geometric
   f. Geometric; first term $= 3$, common ratio $= \sqrt{3}$
4. The next four terms are $32, 64, 128$ and 256.
5. a.

Term number	0	1	2	3	4
Term value	64	32	16	18	14

   b.

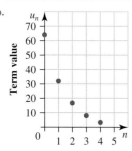

6. a.

Term number	0	1	2	3	4
Term value	1.5	4.5	13.5	40.5	121.5

   b.

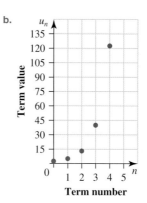

7. 2nd term $= 36$, 3rd term $= 18$

6.5 Exam questions

Note: Mark allocations are available with the fully worked solutions online.

1. D
2. B
3. B

6.6 Geometric sequence applications

6.6 Exercise

1. a. 19 131 876   b. 5   c. -420
2. a. 1 258 291.2   b. -2 c. 640
3. a. $1, 3, 9, 27$ b. $13.24, 66.2, 331, 1655$
4. 2nd term $= 80$, 3rd term $= 32$, 4th term $= 12.8$, 5th term $= 5.12$
5. a. The first four terms are $200, 208, 216.32$ and 224.97.
   b. The first four terms are $600, 583, 564.54$ and 547.60.
6. $1 339 333 or $1.34 million
7. a. $a = 400 600$
      $R = 1.15\%$
   b. $u_n = 400 600 \times 1.15^n$
   c. 3.26 million
8. a. $a = 5.0$ and $R = 0.5$
   b. 0.525 mg

6.6 Exam questions

Note: Mark allocations are available with the fully worked solutions online.

1. B
2. D
3. a. This information would be expressed as a geometric sequence, because a percentage increase in the total amount will be represented by multiplication, not addition.
   b. $u_{n+1} = 1.07 \times u_n$
      where $u_0 = 1000$
   c. $1070
   d. $3379.93
   e. $5069.90

6.7 Generate and analyse a geometric sequence using a recurrence relation

6.7 Exercise

1. $u_{n+1} = -3u_n$, $u_0 = 2.5$
2. -4, 14, -49, 171.5, -600.25
3. $u_{n+1} = \frac{1}{4}u_n$, $u_0 = -256$
4. a. $u_{n+1} = 0.55u_n$, $u_0 = 165$
   b. 4th rebound: 27.45 cm; 5th rebound: 15.10 cm
   c.

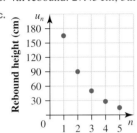

5. a. $u_{n+1} = \frac{2}{5}u_n$, $u_0 = 200$
   b. 1st rebound: 200 cm; 2nd rebound: 80 cm; 3rd rebound: 32 cm; 4th rebound: 12.8 cm; 5th rebound: 5.12 cm
   c.

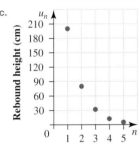

6. a. 17.7 metres
   b. $u_{n+1} = 0.7u_n$, $u_0 = 12.39$

6.7 Exam questions

Note: Mark allocations are available with the fully worked solutions online.
1. E
2. C
3. A

6.8 Modelling growth and decay using recurrence relations

6.8 Exercise

1. a. $u_0 = 10\,000$, $u_{n+1} = u_n + 300$
   b. $u_n = 10\,000 + 300n$
   c. $11\,500
2. $7440
3. a. $u_n = 1500 + 6n$
   b. $1506, $1512, $1518, $1524, $1530 and $1536
   c. $1608
4. a. $8000
   b. 7.5%

5. a. $u_n = 24\,000 - 0.25n$
   b. $21\,000
6. a. $5400   b. 0.1 cents per copy
7. a. $72\,500   b. 10 years
8. a. $5400
   b. $u_n = 90\,000 + 5400n$
   c. 12
9. a. $19\,220   b. $7560   c. 80 months
10. a. 335\,000   b. The 366th week
11. a. 0.05 cents
    b. In the 12th year
    c. After 7 years
12. a. 21\,800
    b. $536
    c. Year 1: 1\,000\,000, Year 15: 11\,684\,800
13. a. $u_n = 2507.5 \times 1.003^n$
    b. $2507.50, $2515.02, $2522.57, $2530.14, $2537.73, $2545.34
    c. $2614.89
14. a. $4500   b. 4.2%
15. a. $u_n = 55\,000 \times 1.03^n$ b. $61\,902.98
16. a. $u_n = 1367.1 \times 0.93^n$ b. $822.59
17. a. $1800   b. 7.5%
18. $10\,028.87

6.8 Exam questions

Note: Mark allocations are available with the fully worked solutions online.
1. a. $510.08
   b. 12 months
   c. No; it would take 16 months.
2. a. $u_0 = 64$, $u_{n+1} = \frac{3}{4}u_n$
   b. $u_n = 64\left(\frac{3}{4}\right)^n$
   c. 20.25 m
3. a. $9318.40   b. $8881.60   c. 9 years

6.9 Review

6.9 Exercise

Multiple choice

1. B
2. C
3. B
4. D
5. A
6. E
7. B
8. D
9. A
10. E

Short answer

11. a. 2, 7, 12, 17, 22

b. 17, 37.4, 82.28, 181.02, 398.24

c. 15, 9, 3, −3, −9

12. a. $u_n = 3 - 2.5n$

b. $u_n = -2 \times 4^n$

c. $u_{n+1} = 6.5\,u_n,\ u_0 = 22$

13. a. 47.25   **b.** 7.52   **c.** −74

14. a. 2.2   **b.** 2.7

c. $u_n = 27 + 2.2n$ **d.** 44.6

15. a.

Term number	0	1	2	3	4
Term value	48	24	12	6	3

b.

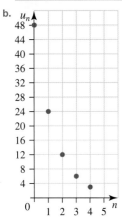

16. $u_0 = -4$

$u_1 = 2$

$d = 2 - -4$

$ = 6$

$u_2 = 2 + 6$

$ = 8$

$u_3 = 8 + 6$

$ = 14$

$u_4 = 14 + 6$

$ = 20$

Term number	0	1	2	3	4
Term value	−4	2	8	14	20

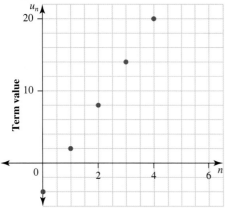

Extended response

17. a. $u_n = 900 + 6.15n$

b. $906.15, $912.30, $918.45, $924.60, $930.75

c. $1023

d. 49 months

18. a. $u_{n+1} = \dfrac{2}{3}u_n,\ u_0 = 400$

b. 79.01 cm, 52.67 cm

c.

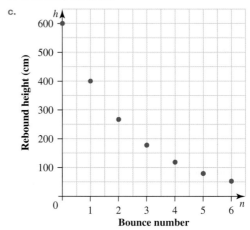

19. a. $u_n = 68\,000 \times 1.025^n$

b. $73\,228.56

c. $1500, 3.4%

d. $272.94

20. a. 98\,415   **b.** 15

21. a. $u_0 = 3000,\ d = -250$

$u_n = u_0 + nd$

$u_n = 3000 - 250n$

b. The computer will be worth $1750 in 5 years. It will have depreciated by $1250.

c. 12 years

22. a. $u_0 = 170\,000,\ u_{n+1} = u_n + 13\,000$

b. 235\,000 people

6.9 Exam questions

Note: Mark allocations are available with the fully worked solutions online.

1. C

2. A

3. $d = 11$

$u_n = 4 + 11n$

$u_3 = 37$

4. C

5. C

7 Financial mathematics extension

Fully worked solutions for this topic are available online.

7.1 Overview

7.1.1 Introduction

When we invest money with a financial institution, the institution uses our money to lend to others, so they pay us interest. On the other hand, when we borrow money from an institution, we are using their money, so they charge us interest. The rates of interest charged, or given, vary and can change at any time.

In this topic, we mainly consider reducing balance loans. These loans are given for specific purposes, such as to buy a home or a car. The financial institution loans the money and the borrower agrees to repay the money, plus interest, over a given period of time. In reducing balance loans, interest is charged, usually every month, on the outstanding balance, and the borrower makes repayments on a regular basis. To reduce the balance, these repayments need to be more than the interest charged for the same period of time. The outstanding balance will continually decrease with each repayment until the loan is paid out. Consequently, the interest charged each month will also decrease throughout the period of the loan.

The theory behind reducing balance loans can be applied to other situations, such as scholarships and superannuation payments. These payments are known as annuities, when the amount of the investment decreases with each payment. Superannuation is probably the most well-known annuity.

KEY CONCEPTS

This topic will introduce concepts from Units 3 and 4 to help you understand various key knowledge points for these units from the Study Design.

- use of a first-order linear recurrence relation to model and analyse (numerically and graphically) the amortisation of a reducing balance loan, including the use of a recurrence relation to determine the value of the loan or investment after *n* payments for an initial sequence from first principles
- use of a table to investigate and analyse the amortisation of a reducing balance loan on a step-by-step basis, the payment made, the amount of interest paid, the reduction in the principal and the balance of the loan
- use of technology with financial modelling functionality to solve problems involving reducing balance loans, such as repaying a personal loan or a mortgage, including the impact of a change in interest rate on repayment amount, time to repay the loan, total interest paid and the total cost of the loan
- use of a first-order linear recurrence relation to model and analyse (numerically and graphically) the amortisation of an annuity, including the use of a recurrence relation to determine the value of the annuity after *n* payments for an initial sequence from first principles
- use of a table to investigate and analyse the amortisation of an annuity on a step-by-step basis, the payment made, the interest earned, the reduction in the principal and the balance of the annuity
- use of technology to solve problems involving annuities including determining the amount to be invested in an annuity to provide a regular income paid, for example, monthly, quarterly
- simple perpetuity as a special case of an annuity that lasts indefinitely.

Source: VCE Mathematics Study Design (2023–2027) extracts © VCAA; reproduced by permission.

7.2 Reducing balance loans modelled using recurrence relations

LEARNING INTENTION

At the end of this subtopic you should be able to:
- create a table to investigate and analyse the amortisation of a reducing balance loan or an annuity, and interpret amortisation tables.

7.2.1 Modelling reducing balance loans with a recurrence relation

When we borrow money from a financial institution, such as a bank, we are using the institution's money, so it charges us interest on the outstanding balance. We need to repay both the amount borrowed (the principal) and the interest charged. We do this by making regular repayments over the period of the loan.

In reducing balance loans:
- interest is charged by the institution per compounding period, normally monthly
- repayments are made by the borrower on a regular basis
- repayments need to be more than the interest for the same period to reduce the amount still owing
- interest is charged on the outstanding balance before the repayment is made
- a **recurrence relation** can be used to model the loan.

Formula for reducing balance loans using the recurrence relation

The general form of the recurrence relation is:

where

$$V_{n+1} = RV_n - d, \quad V_0 = a$$

V_0 = the initial amount borrowed (the principal)

V_n = the balance of the loan after n payments

$R = 1 + \dfrac{r}{100}$, where r is the interest rate per compounding period

d = the payment made per compounding period.

Rounding money

In 'Recursion and financial modelling', all answers should be rounded to the nearest cent (2 decimal places) unless otherwise instructed.

tlvd-4466

WORKED EXAMPLE 1 Modelling reducing balance loans using a recurrence relation

Heidi takes out a loan for $3500 to pay for her business course. The interest rate charged by the financial institution is 9.6% per annum, compounded monthly. She has agreed to make monthly payments of $325.
a. **Write a recurrence relation to describe this situation.**
b. **Calculate the outstanding balance on the loan after Heidi has made:**
   i. **3 payments**
   ii. **6 payments.**
c. **Use your calculator to determine the number of monthly repayments needed to pay off the loan.**

THINK		WRITE	
a.	**1.** State the values of V_0 and d.	**a.**	$V_0 = 3500 \quad d = 325$
	2. Calculate the value of r and R.		$r = \dfrac{9.6}{12}$
			$= 0.8$
			$R = 1 + \dfrac{0.8}{100}$
			$= 1.008$
	3. Write the general form of the recurrence relation and substitute.		$V_{n+1} = RV_n - d, \ V_0 = a$
			$V_{n+1} = 1.008V_n - 325, \ V_0 = 3500$
b. i.	**1.** After 1 payment: V_1	**b. i.**	$V_1 = 1.008 \times 3500 - 325$
			$= 3203$
	2. After 2 payments: V_2		$V_2 = 1.008 \times 3203 - 325$
			$= 2903.62$
	3. After 3 payments: V_3		$V_3 = 1.008 \times 2903.62 - 325$
			$= 2601.85$
	4. Write the answer.		The outstanding balance after 3 payments is \$2601.85.

THINK		WRITE	
b. ii.	*Using a calculator*	**b. ii.**	

3500	3500
$3500 \times 1.008 - 325$	3203
$3203 \times 1.008 - 325$	2903.62
$2901.62 \times 1.008 - 325$	2601.85
$2601.85 \times 1.008 - 325$	2297.67
$2297.67 \times 1.008 - 325$	1991.05
$1991.05 \times 1.008 - 325$	1681.98

THINK	WRITE
b. ii. **1.** On a calculator page, type 3500. **2.** Press ENTER/EXE. **3.** Type $\times 1.008 - 325$. **4.** Press ENTER/EXE 6 times (for 6 payments).	
5. Write the answer.	The outstanding balance after 6 payments is \$1681.98.

THINK		WRITE	
c.	*Using a calculator*	**c.**	

3500	3500
$3500 \times 1.008 - 325$	3203
$3203 \times 1.008 - 325$	2901.62
$2901.62 \times 1.008 - 325$	2601.85
$2601.85 \times 1.008 - 325$	2297.67
$2297.67 \times 1.008 - 325$	1991.05
$1991.05 \times 1.008 - 325$	1681.98
$1681.98 \times 1.008 - 325$	1370.43
$1370.43 \times 1.008 - 325$	1056.40
$1056.40 \times 1.008 - 325$	739.85
$739.85 \times 1.008 - 325$	420.77
$420.77 \times 1.008 - 325$	99.13
$99.13 \times 1.008 - 325$	-225.07

c.
1. On a calculator page, type 3500.
2. Press ENTER/EXE.
3. Type $\times 1.008 - 325$.
4. Press ENTER/EXE until the answer is less than zero (the loan has been repaid and there is no outstanding balance).
5. Count the number of times you pressed ENTER/EXE (or the number of repayments).

| 6. | Write the answer. | After 12 months, the account balance is -225.07, so it would take 12 months to pay off the loan.
 Note: The last payment would only be $99.93 This is calculated by $(325 - 225.07)$ or 99.13×1.008. |

WORKED EXAMPLE 2 Determining the number of monthly repayments needed to pay off a loan

Henry purchased a car for $10 000. He had saved $7600 and needed to borrow the remainder. His financial institution would give him a loan at 8.4% p.a., compounded monthly. Henry has agreed to repay the loan with monthly instalments of $410.

a. Calculate the amount Henry borrowed.
b. Write a recurrence relation to describe this situation.
c. Use your calculator to determine the number of monthly repayments needed to pay off the loan.

THINK	WRITE
a. 1. Subtract savings from the purchase price.	a. $10\,000 - \$7600 = \$2400$
2. Write the answer.	Henry borrowed $2400.
b. 1. State the values of V_0 and d.	b. $V_0 = 2400 \quad d = 410$
2. Calculate the value of r and R.	$r = \dfrac{8.4}{12}$ $= 0.7$ $R = 1 + \dfrac{0.7}{100}$ $= 1.007$
3. Write the general form of the recurrence relation and substitute.	$V_{n+1} = RV_n - d, V_0 = a$ $V_{n+1} = 1.007V_n - 410, V_0 = 2400$

c. *Using a calculator*
   1. On a calculator page, type 2400.
   2. Press ENTER/EXE.
   3. Type $\times 1.007 - 410$.
   4. Press ENTER/EXE until the answer is less than zero (the loan has been repaid and there is no outstanding balance).
   5. Count the number of times you pressed ENTER/EXE (or the number of repayments).

c.

2400	2400
$2400 \times 1.007 - 410$	2006.80
$2006.80 \times 1.007 - 410$	1610.85
$1610.85 \times 1.007 - 410$	1212.12
$1212.12 \times 1.007 - 410$	810.61
$810.61 \times 1.007 - 410$	406.28
$406.28 \times 1.007 - 410$	-0.87

6. Write the answer.			The loan will be repaid in 6 months. *Note:* The final payment will be adjusted for the overpayment of 87 cents. The above calculations have been written correct to 2 decimal places, but kept accurate on the calculator.	

7.2.2 Amortisation of a reducing balance loan using a table

- The process of paying off a loan by regular payments over a period of time is known as amortisation.
- The amortisation of a loan can be tracked on a step-by-step basis by following the payments made, the interest added and the reduction in the principal.
- Remember, the repayments need to cover both the interest and part of the principal, so that the principal is reduced in each period.
- A recurrence relation gives the last column of the amortisation table without showing any of the intermediate steps.

In an amortisation table for reducing balance loans:
- payments remain fixed
- interest is charged on the previous period's balance (normally months)
- principal reduction = payment − interest for the month
- balance of the loan = previous month's balance − principal reduction.

Consider an amortisation table for Worked example 2.
- Payments: $410 per month
- Interest: 8.4% p.a. $= 0.7\%$ per month
- Principal: $2400

Payment number, n	Payment ($)	Interest ($)	Principal reduction ($)	Balance of loan ($)
0				2400
1	410	$0.007 \times 2400 = 16.80$	$410 - 16.80 = 393.20$	$2400 - 393.20 = 2006.80$
2	410	$0.007 \times 2006.80 = 14.05$	$410 - 14.05 = 395.95$	$2006.80 - 395.95 = 1610.85$
3	410	$0.007 \times 1610.85 = 11.28$	$410 - 11.28 = 398.72$	$1610.85 - 398.72 = 1212.13$
4	410	$0.007 \times 1212.13 = 8.48$	$410 - 8.48 = 401.52$	$1212.13 - 401.52 = 810.62$
5	410	$0.007 \times 810.62 = 5.67$	$410 - 5.67 = 404.33$	$810.62 - 404.33 = 406.29$
6	410	$0.007 \times 406.29 = 2.84$	$410 - 2.84 = 407.12$	$406.29 - 407.12 = -0.87$

Note: Calculations may vary by a few cents with rounding to 2 decimal places.

tlvd-4678

WORKED EXAMPLE 3 Analysing the amortisation of a reducing balance loan

Drew and Elise are renovating their kitchen. To complete the renovation, they borrow $4800 from a financial institution at 10.4% p.a., compounded quarterly, with quarterly payments of $1280. The incomplete amortisation table for this loan is below.

a. **Calculate the interest rate per payment period.**
b. **Calculate the interest charged for the first quarter.**
c. **Determine the balance of the loan after two payments have been made.**

d. Calculate the principal reduction that occurs after the third payment.
e. Determine the amount of the final payment to ensure a balance of zero at the end of the loan period.
f. Calculate the total interest paid on this loan.

Payment number, n	Payment ($)	Interest ($)	Principal reduction ($)	Balance of loan ($)
0				4800.00
1	1280.00		1155.20	3644.80
2	1280.00	94.76	1185.24	
3	1280.00	63.95		1243.51
4	1280.00	32.33	1247.67	-4.16

THINK

a. 1. State the annual interest rate.

2. Calculate the quarterly interest rate.

3. Write the answer.

b. 1. Calculate 2.6% of $4800.

2. Write the answer.

c. 1. Calculate the balance after two payments.

2. Write the answer.

d. 1. Calculate principal reduction after the third payment.

2. Write the answer.

e. 1. Last payment is an overpayment of $4.16.

2. Write the answer.

f. 1. Total interest paid is the sum of the interest column.

2. Write the answer.

WRITE

a. 10.4% per year

$= \dfrac{10.4}{4}$% per quarter

$= 2.6$% per quarter

The interest rate is 2.6% per quarter.

b. $2.6\% \times 4800$

$= \dfrac{2.6}{100} \times 4800$

$= 124.80$

Interest charged in the first quarter is $124.80.

c. Previous balance – principal reduction

$= 3644.80 - 1185.24$

$= 2459.56$

Balance owing after 2 payments is $2459.56.

d. Payment – interest charged

$= 1280 - 63.95$

$= 1216.05$

The principal is reduced by $1216.05 after the third payment.

e. Adjusted last payment

$= 1280 - 4.16$

$= 1275.84$

The last payment should be $1275.84.

f. $124.80 + 94.76 + 63.95 + 32.33 = 315.84$

Total interest paid is $315.84.

1. **WE1** Stephen takes out a loan of $4200. The interest rate charged by the financial institution is 10.5% per annum, compounded monthly. He has agreed to make monthly payments of $380.

   a. Write a recurrence relation to describe this situation.

   b. Calculate the outstanding balance on the loan after Stephen has made:

      i. 2 payments   ii. 5 payments.

   c. Use your calculator to determine the number of monthly repayments Stephen needs to make to pay off the loan.

2. Sallyanne took out a personal loan of $2600 for a new TV, with the loan having an interest rate of 15% p.a., compounded monthly. She has agreed to make regular monthly payments of $345.

   a. Write a recurrence relation to describe this situation.

   b. Calculate the outstanding balance on the loan after Sallyanne has made:

      i. 3 payments   ii. 7 payments.

   c. Use your calculator to determine the number of monthly repayments Sallyanne needs to make to completely repay the loan.

3. **WE2** Hazel purchased her first car for $12 500. She had saved $7200 and borrowed the remainder. Her financial institution gave her a loan at 12.6% p.a., compounded monthly. Hazel has agreed to repay the loan with monthly instalments of $575.

   a. Calculate the amount Hazel borrowed.

   b. Write a recurrence relation to describe this situation.

   c. Use your calculator to determine the number of monthly repayments needed to pay off the loan.

4. Leo purchased his first car for $7500. His grandparents gave him $5000 for his birthday and he borrowed the remainder from a bank at an interest rate of 8.4% p.a., compounded monthly. He agreed to make monthly repayments of $300.

   a. Calculate the amount Leo borrowed.

   b. Write a recurrence relation to describe this situation.

   c. Use your calculator to determine the number of monthly repayments needed to pay off the loan.

5. **WE3** Paul and Penny are moving into their first home. To pay for some furniture, they borrowed $4500 from a financial institution at 11.8% p.a., compounded quarterly. They agree to make quarterly payments of $1200.

The incomplete amortisation table for this loan is below.

a. Calculate the interest rate per payment period.
b. Calculate the interest charged for the first quarter.
c. Determine the balance of the loan after two payments have been made.
d. Calculate the principal reduction after the third payment.
e. Determine the amount of the final payment to ensure a balance of zero at the end of the loan period.
f. Calculate the total interest paid on this loan.

Payment number, n	Payment ($)	Interest ($)	Principal reduction ($)	Balance of loan ($)
0				4500.00
1	1200.00		1067.25	3432.75
2	1200.00	101.27	1098.73	
3	1200.00	68.85		1202.87
4	1200.00	35.48	1164.52	38.35

6. **MC** Ben took out a loan for $20 000 for a small start-up business at 16.5% p.a. The contract required that he repay the loan over 5 years with monthly instalments of $421.02. After 8 months Ben still owes:

A. $12 344.05
B. $14 761.22
C. $9653.13
D. $19 001.67
E. $18 774.05

7. Michael wants to purchase a new bike. He has taken out a $7600 loan at 9.6% p.a., compounded quarterly, with quarterly payments of $1375.

The incomplete amortisation table for this loan is below.

a. Calculate the interest rate per payment period.
b. Calculate the interest charged for the first quarter.
c. Determine the balance of the loan after two payments have been made.
d. Calculate the principal reduction after the third payment is made.
e. Complete the last two rows of the table.
f. Michael's final payment needs to be adjusted to ensure a zero balance on the loan. Calculate Michael's final payment.

Payment number, n	Payment ($)	Interest ($)	Principal reduction ($)	Balance of loan ($)
0				7600.00
1	1375.00		1192.60	6407.40
2	1375.00	153.78	1121.22	
3	1375.00	124.47		3935.65
4	1375.00	94.46	1280.54	2655.11
5	1375.00			
6	1375.00			

8. Adam has taken out a loan of $25 000 for a small business venture. The interest rate is 15% per annum, compounded quarterly, with equal quarterly payments of $3250.

a. Calculate the interest rate per quarter.
b. Write a recurrence relation to describe this situation.
c. Calculate the outstanding balance on the loan after Adam has made 3 payments.
d. Calculate the outstanding balance after 18 months.

9. You are Adam's accountant (from question 8). Adam has asked you to prepare a table to show him how his quarterly payments are reducing the balance of his loan. Complete the amortisation table below to illustrate this for the first year of the loan.

Payment number, n	Payment ($)	Interest ($)	Principal reduction ($)	Balance of loan ($)
0				25 000.00
1	3250.00			
2				
3				
4				

10. Jamie took out a personal loan of $3750 to replace her washing machine and dryer. The terms of the loan were 10 monthly payments of $390 with an interest rate of 9% p.a., compounded monthly.

a. Calculate the interest rate per month.
b. Write a recurrence relation to describe this situation.
c. Calculate the outstanding balance on the loan after Jamie has made 2 payments.
d. Determine Jamie's final payment to ensure a zero balance on the loan.

7.2 Exam questions

Question 1 (1 mark)
Source: VCE 2019, Further Mathematics Exam 1, Section A, Q20; © VCAA.

MC Consider the following amortisation table for a reducing balance loan.

Payment number	Payment	Interest	Principal reduction	Balance
0	0.00	0.00	0.00	300 000.00
1	1050.00	900.00	150.00	299 850.00
2	1050.00	899.55	150.45	299 699.55
3	1050.00	899.10	150.90	299 548.65

The annual interest rate for this loan is 3.6%.

Interest is calculated immediately before each payment.

For this loan, the repayments are made
A. weekly.   **B.** fortnightly.   **C.** monthly.   **D.** quarterly.   **E.** yearly.

Question 2 (1 mark)
Source: VCE 2017, Further Mathematics Exam 1, Section A, Core, Q17; © VCAA.

MC The value of a reducing balance loan, in dollars, after n months, V_n, can be modelled by the recurrence relation shown below.

$$V_0 = 26\,000, \qquad V_{n+1} = 1.003\ V_n - 400$$

Determine the value of this loan after five months.
A. $24 380.31   **B.** $24 706.19   **C.** $25 031.10   **D.** $25 355.03   **E.** $25 678.00

▶ **Question 3 (1 mark)**
Source: VCE 2016, Further Mathematics Exam 1, Section A, Q22; © VCAA.

MC The first three lines of an amortisation table for a reducing balance home loan are shown below.

The interest rate for this home loan is 4.8% per annum, compounding monthly.

The loan is to be repaid with monthly payments of $1500.

Payment number, n	Payment ($)	Interest ($)	Principal reduction ($)	Balance of loan ($)
0	0	0	0	250 000.00
1	1500	1000.00	500.00	249 500.00
2	1500			

The amount of payment number 2 that goes towards reducing the principal of the loan is
   A. $486   **B.** $502   **C.** $504   **D.** $996   **E.** $998

More exam questions are available online.

7.3 Solving reducing balance loan problems with technology

LEARNING INTENTION

At the end of this subtopic you should be able to:
- use technology (Finance Solver) to solve problems involving reducing balance loans
- calculate the number of repayments, principal and interest needed to pay off the loan.

7.3.1 Calculating the future value

In the previous subtopic, recurrence relations were used to solve reducing balance loan problems. However, it is very time consuming, particularly if the loan is to be repaid over a long period of time.

For example, if a home loan is to be repaid with monthly payments over a 30-year period, this would mean 360 iterations of the recurrence relation. Instead, the Finance Solver application on your CAS calculator can be used to solve reducing balance loan problems.

Finance Solver has a particular set of fields, similar to those in the following table.

N	The total number of repayment periods
I%	The interest rate per annum
PV	The present value of the loan — this is positive as the money is coming to you.
PMT (Pmt)	The payment per period — this is negative if you are repaying a loan.
FV	The future value of the loan — positive, negative or zero
P/Y (PpY)	The number of payments per year
C/Y (CpY)	The number of compounding periods per year (nearly always the same as P/Y)

Remember:

- **When entering values into Finance Solver, pay particular attention to the sign of the value.**
- **Cash flowing towards you is considered to be positive.**
  - **If you take out a loan, the money is coming to you, so PV is positive.**
- **Cash flowing away from you is considered to be negative.**
  - **If you make a payment, the money is leaving you, so PMT is negative.**
- **If the FV is:**
  - **negative, then there is still money to be paid on the loan**
  - **zero, then the loan has been completely paid out**
  - **positive, then the borrower has overpaid the loan.**
- **Fill in all of the fields except the unknown field. Place your cursor in that field and press ENTER/EXE.**
- **To move between fields, use the TAB key.**
- **Total interest paid equals the total of the payments minus the principal repaid.**

The balance of a loan can be found at any point in time by completing all of the fields except FV, the future value of the loan. Remember the future value of the loan will be negative if money is still owed (i.e. the loan has not been repaid in full) and positive if the loan has been overpaid.

WORKED EXAMPLE 4 Using Finance Solver to determine the future value of the loan

Noah has taken out a \$45 000 loan to purchase equipment for his business. The loan is to be repaid over 10 years with interest charged at 7.5% p.a., compounded monthly. Noah has agreed to make monthly payments of \$535.
Calculate the amount still owing after 6 years. Give your answer to the nearest cent.

THINK

1. Using Finance Solver on your CAS calculator, fill in the appropriate fields.
   N = 72 payments (6 × 12 months)
   I% = 7.5 (annual interest rate)
   PV = 45 000 (amount borrowed)
   PMT = −535 (monthly repayment)
   FV = ??? unknown
   P/Y = 12 (payments per year — monthly)
   C/Y = 12 (compounding periods per year)

WRITE

N	72
I%	7.5
PV	45 000
PMT	−535
FV	
P/Y	12
C/Y	12

2. Press TAB to place the cursor in the FV field, then press ENTER.

   FV = −22 015.63

3. Write the answer.

   The amount of the loan still owing after 6 years is \$22 015.63.

7.3.2 Calculating the payments

The payments required to repay a loan can be found at any point in time by completing all of the fields except PMT. When a loan is repaid, the future value will be zero. Remember, money is leaving, so payments will be negative.

tlvd-4679

WORKED EXAMPLE 5 Calculating the number of repayments to repay a loan

Olivia is considering taking out a 30-year loan of $350 000 to purchase a unit. The bank will lend her the money at 4.5% p.a., with interest compounded monthly.

a. Calculate the amount Olivia will need to pay each month if she is to fully repay the loan in the 30 years.
b. Determine the total amount Olivia would pay over the 30 years.
c. Determine the interest Olivia will pay during this time.

THINK

a. 1. Using Finance Solver on your CAS calculator, fill in the appropriate fields.
N = 360 payments (30 × 12 months)
I% = 4.5 (annual interest rate)
PV = 350 000 (amount borrowed)
PMT = ??? unknown
FV = 0 (loan is repaid in full)
P/Y = 12 (payments per year — monthly)
C/Y = 12 (compounding periods per year)

2. Press TAB to place the cursor in the PMT field, then press ENTER.

3. Write the answer.

b. 1. Total = payments × number of payments

2. Write the answer.

c. 1. Total interest = total payments – principal

2. Write the answer.

WRITE

a.

N	360
I%	4.5
PV	350 000
PMT	
FV	0
P/Y	12
C/Y	12

PMT = −1773.398

The amount of each monthly payment would be $1773.40.

b. $1773.40 × 360 = $638 424

Olivia will pay $638 424 over the 30 years of the loan.

c. $638 424 − $350 000 = $288 424

Olivia will pay $288 424 in interest.

7.3.3 Calculating the present value

The present value of a loan is the amount of money that can be borrowed under certain conditions. This is found by completing all of the fields except PV.

Remember, the present value, which is money coming into the account, will be positive, but payments, which are money leaving the account, will be negative. The future value will be zero, since the loan will be fully repaid over the given period of time.

WORKED EXAMPLE 6 Calculating the present value of a loan

William knows he is earning enough to repay a loan with payments of $1000 per month. The current interest rate for a car loan is 8.2% p.a., compounded monthly. He wishes to repay the loan fully in 5 years.

Calculate the amount William can borrow for a new car.

Give your answer to the nearest dollar.

THINK		WRITE	
1. Using Finance Solver on your CAS calculator, fill in the appropriate fields. N = 60 payments (5 × 12 months) I% = 8.2 (annual interest rate) PV = ??? **unknown** PMT = −1000 (monthly repayment) FV = 0 (loan is repaid in full) P/Y = 12 (payments per year) C/Y = 12 (compounding periods per year)		N	60
		I%	8.2
		PV	
		PMT	−1000
		FV	0
		P/Y	12
		C/Y	12
2. Press TAB to place the cursor in the PV field, then press ENTER.		PV = 49 086.388 45	
3. Write the answer.		The amount William could borrow is $49 086 (to the nearest dollar).	

7.3.4 Calculating time

The time period of a loan depends on the value of N. This is the number of payment periods. The periods may be months, quarters or years. To convert the number of payment periods to years, divide by 12 if the periods are in months, and divide by 4 if the periods are in quarters.

Remember the present value will be positive as money is coming into the account and the payments will be negative as money is leaving the account.

WORKED EXAMPLE 7 Calculating the time period of a loan

A loan of $24 000 is being repaid by quarterly instalments of $2050. The interest rate charged is 9.4% per annum.
a. Determine how long it would take to repay the loan in full.
b. Calculate the last payment needed to ensure a zero balance at the end of this time. Give your answer to the nearest cent.

THINK		WRITE	
a. 1. Using Finance Solver on your CAS calculator, fill in the appropriate fields. N = ??? **unknown** I% = 9.4 (annual interest rate) PV = 24 000 (amount borrowed) PMT = −2050 (monthly repayment) FV = 0 (loan is repaid in full) P/Y = 4 (payments per year — quarters) C/Y = 4 (compounding periods per year)	a.	N	
		I%	9.4
		PV	24 000
		PMT	−2050
		FV	0
		P/Y	4
		C/Y	4
2. Press TAB to place the cursor in the N field, then press ENTER.		N = 13.851 82	
3. Write the answer.		It would take 14 quarters (3.5 years) to repay the loan.	

b. 1. Using Finance Solver on your CAS calculator, fill in the appropriate fields.

N = 14 (number of payments)
I% = 9.4 (annual interest rate)
PV = 24 000 (amount borrowed)
PMT = −2050 (monthly repayment)
FV = ??? unknown
P/Y = 4 (payments per year — quarters)
C/Y = 4 (compounding periods per year)

b.

N	14
I%	9.4
PV	24 000
PMT	−2050
FV	
P/Y	4
C/Y	4

2. Press TAB to place the cursor in the FV field, then press ENTER.

FV = 300.769 80

This is an overpayment of $300.77, since the value is positive.

3. Subtract overpayment from PMT to adjust the last payment.

Adjusted last payment
= $2050 − 300.77
= $1749.23

4. Write the answer.

The final payment is $1749.23.

7.3.5 Calculating the interest rate

Interest rates are set by individual financial institutions and may vary according to the type of loan. The interest rate is found by completing all of the fields except I. The value of I is the interest rate as a percent per annum, or per year.

WORKED EXAMPLE 8 Calculating the interest rate needed to pay off a loan

Charlie and Chloe approached their banks to take out a personal loan of $15 000, to be repaid over a 4-year period. The conditions from Charlie's bank were payments of $370 per month, with interest charged monthly. Chloe's bank asked for payments of $1090 per quarter, with interest adjusted quarterly.
Determine who has the lower interest rate. State the difference.

THINK

Charlie:

1. Using Finance Solver on your CAS calculator, fill in the appropriate fields.

N = 48 (payments: 4 × 12 months)
I% = ??? unknown
PV = 15 000 (amount borrowed)
PMT = −370 (monthly repayment)
FV = 0 (loan is repaid in full)
P/Y = 12 (payments per year — months)
C/Y = 12 (compounding periods per year)

2. Press TAB to place the cursor in the I% field, then press ENTER.

WRITE

N	48
I%	
PV	15 000
PMT	−370
FV	0
P/Y	12
C/Y	12

I = 8.538 886 65

Chloe:

3. Using Finance Solver on your CAS calculator, fill in the appropriate fields.
   N = 16 (payments: 4 × 4 quarters)
   I% = ??? unknown
   PV = 15 000 (amount borrowed)
   PMT = −1090 (quarterly repayment)
   FV = 0 (loan is repaid in full)
   P/Y = 4 (payments per year — quarters)
   C/Y = 4 (compounding periods per year)

N	16
I%	
PV	15 000
PMT	−1090
FV	0
P/Y	4
C/Y	4

4. Press TAB to place the cursor in the I% field, then press ENTER.

   I = 7.323 216

5. State the interest rates, correct to 2 decimal places.

   Charlie: 8.54% p.a.
   Chloe: 7.32% p.a.

6. Write the answer.

   Chloe has the lower interest rate by 1.22%.

7.3 Exercise

Students, these questions are even better in jacPLUS

 Receive immediate feedback and access sample responses

 Access additional questions

 Track your results and progress

Find all this and MORE in jacPLUS

1. **WE4** Zoe has taken out a $28 500 loan to purchase equipment for her business. The loan is to be repaid over 6 years with interest charged at 8.25% p.a., compounded monthly. Zoe has agreed to make monthly payments of $503.
   Calculate the amount still owing after 4 years. Give your answer to the nearest cent.

2. Jack wants to borrow $3750 to purchase a new sound system. He has arranged for a loan at 12.45% p.a., with interest adjusted monthly, to be repaid with monthly payments of $126 over a three-year period.
   Calculate the amount outstanding after 18 months. Give your answer to the nearest cent.

3. **WE5** Charlotte is considering taking out a 30-year loan of $475 000 to purchase a unit. The bank will lend her the money at 4.28% p.a., with interest compounded monthly.

   a. Calculate the amount Charlotte would need to pay each month if she is to repay the loan in 30 years.
   b. Determine the amount Charlotte would pay in total over the 30 years.
   c. Calculate the interest Charlotte will pay during this time.

4. Thomas borrowed $75 800 to purchase a work vehicle. He agreed to repay the loan with monthly instalments for four years, with interest being charged at 9.8% p.a., compounding monthly.

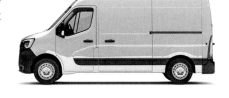

   a. Calculate Thomas's monthly payments. Give your answer to the nearest cent.
   b. Determine the total amount Thomas would pay over the 4 years.
   c. Calculate the interest Thomas would pay during this time.

5. **WE6** Henry knows he is earning enough to repay a loan with payments of $4580 per month. Knowing interest rates may change, he allows for an interest rate of 6% p.a., compounded monthly. He wishes to repay the loan fully in 25 years.
   Determine the amount Henry can borrow. Give your answer to the nearest dollar.

6. Calculate the amount that can be borrowed if it is to be fully repaid in 18 months, with payments of $350 per month and an interest rate of 5.2% per annum, compounded monthly. Give your answer to the nearest hundred dollars.

7. **WE7** A loan of $32 500 is being repaid with quarterly instalments of $1910. The interest rate charged is 6.35% p.a., compounding quarterly.

   a. Determine how long it will take to repay the loan in full.
   b. Calculate the last payment that is required to ensure a zero balance at the end of this time. Give your answer to the nearest cent.

8. Joshua agrees to repay a loan of $17 500 with quarterly instalments of $1535.75. The interest rate, compounding quarterly, is 7.75% per annum.

   a. Determine how long it would take Joshua to repay the loan in full.
   b. Determine what his last payment would need to be to ensure a zero balance at the end of this time. Give your answer to the nearest cent.

9. **WE8** Liam and Amelia approached their banks for a personal loan of $12 500 to be repaid over a 3-year period. The conditions from Liam's bank were payments of $388.55 per month, with interest charged monthly. Amelia's bank asked for payments of $1177.50 per quarter, with interest adjusted quarterly. Determine who has the lower annual interest rate. State the difference.

10. You are comparing two deals for a personal loan of $34 000 for a new car. The reducing balance loan from A-bank requires monthly payments of $850 with the loan repaid in 4 years, with interest adjusted monthly. B-bank is advertising quarterly payments of $2551 for the same length of time, with interest adjusted quarterly.
    Determine which bank has the lower annual interest rate. State the difference.

11. Grace wanted to borrow $315 000 for a home loan and was offered a reducing balance loan over a period of 25 years at 4.9% p.a., with interest adjusted monthly.

    a. Calculate her monthly repayments. Give your answer to the nearest cent.
    b. Calculate the total Grace would repay for this loan.
    c. Calculate the total interest paid on the loan.

12. James borrowed $35 650 to purchase some new equipment for his business. He agreed to repay the loan over 8 years, with quarterly payments and interest charged quarterly at 9.8% p.a.

    a. Calculate the amount James needs to pay quarterly to fully repay the loan in 8 years. Give your answer to the nearest cent.
    b. After 5 years, James decides to pay out the loan. Calculate the amount James needs to pay. Give your answer to the nearest cent.

13. Charlie and Matilda took out a reducing balance loan of $475 000 to purchase their first home. The loan is to be repaid with monthly instalments over a period of 30 years. The interest rate, compounded monthly, is 4.25% p.a.

    a. Calculate their monthly instalments, correct to the nearest cent.
    b. Calculate the amount still owing after 15 years.
    c. They now wish to repay the loan in full in the next 5 years. Determine their new monthly instalments, correct to the nearest cent.

14. Aria has taken out a reducing balance loan of $38 500 to pay for a new kitchen in her unit. The loan is to be repaid over a period of 5 years with an interest rate, compounded per period of time, of 6.2% p.a.

    a. Calculate the payments, to the nearest cent, to fully repay the loan if the payments were made:

       i. quarterly
       ii. monthly
       iii. fortnightly (assume 26 fortnights per year).

    b. Determine which is the better deal.

15. Stuart and Kate have taken out a home loan of $395 000 to purchase their first home. They agreed to repay the loan in full in 25 years, with monthly payments. The interest rate initially was 3.95% p.a., compounded monthly.

    a. Calculate their monthly payments. Give your answer to the nearest cent.
    b. After 10 years, the interest rate was increased to 5.2% p.a. Determine their new monthly payments if they still want to repay the loan in full in 25 years.

7.3 Exam questions

Question 1 (1 mark)
Source: VCE 2018, Further Mathematics Exam 1, Section A, Q22; © VCAA.

MC Adam has a home loan with a present value of $175 260.56.

The interest rate for Adam's loan is 3.72% per annum, compounding monthly.

His monthly repayment is $3200.

The loan is to be fully repaid after five years.

Adam knows that the loan cannot be exactly repaid with 60 repayments of $3200.

To solve this problem, Adam will make 59 repayments of $3200. He will then adjust the value of the final repayment so that the loan is fully repaid with the 60th repayment.

The value of the 60th repayment will be closest to
    A. $368.12   **B.** $2831.88   **C.** $3200.56   **D.** $3557.09   **E.** $3568.12

Question 2 (2 marks)
Source: VCE 2016, Further Mathematics Exam 2, Section A, Q7a; © VCAA.

Ken has borrowed $70 000 to buy a new caravan.

He will be charged interest at the rate of 6.9% per annum, compounding monthly.

For the first year (12 months), Ken will make monthly repayments of $800.

    a. Find the amount that Ken will owe on his loan after he has made 12 repayments. **(1 mark)**
    b. What is the total interest that Ken will have paid after 12 repayments? **(1 mark)**

▶ **Questions 3 (2 marks)**

Source: VCE 2014, Further Mathematics Exam 1, Section B, Module 4, Q5; © VCAA.

MC A bank approves a $90 000 loan for a customer.

The loan is to be repaid fully over 20 years in equal monthly payments.

Interest is charged at a rate of 6.95% per annum on the reducing monthly balance.

To the nearest dollar, the monthly payment will be
    A. $478   **B.** $692   **C.** $695   **D.** $1409   **E.** $1579

More exam questions are available online.

7.4 Annuities

LEARNING INTENTION

At the end of this subtopic you should be able to:
- model annuities using a recurrence relation
- solve annuity problems using Finance Solver
- calculate the principal (V_0) and interest rate (r).

7.4.1 Modelling an annuity using a recurrence relation

An annuity is an investment account that at some future date will provide a regular stream of income for a set period of time. An annuity is similar to a reducing balance loan, but with the payment coming to you from an investment account you already have in place. Annuities are often scholarships, bursaries or retirement funds.

A recurrence relation can be used to model an annuity. Since it uses the same concepts as reducing balance loan problems, the recurrence relation is written in the same form.

Recurrence relation for an annuity

A recurrence relation that can be used to model the value of an annuity after n payments is:

$$V_{n+1} = R V_n - d, \quad V_0 = a$$

where

V_0 = the initial amount invested (the principal)

V_n = the balance of the investment after n payments

$R = 1 + \dfrac{r}{100}$ (r is the interest rate per compounding period)

d = the payment received per compounding period.

---

tlvd-4680

WORKED EXAMPLE 9 Modelling an annuity using a recurrence relation

To study overseas for 3 months, Mia was awarded a scholarship of \$10 250. The scholarship was invested in an annuity that pays 3.96% p.a., compounded monthly. Mia is paid \$3440 per month from the annuity.
a. **Write a recurrence relation to describe this situation.**
b. **Calculate how much money is left in the annuity at the end of 2 months.**
c. **Determine the adjustment made to Mia's last payment to ensure zero balance in the investment account.**

THINK	WRITE
a. 1. State the values of V_0 and d.	a. $V_0 = 10\,250$, $d = 3440$
2. Calculate the value of r and R.	$r = \dfrac{3.96}{12}$ $= 0.33$ $R = 1 + \dfrac{0.33}{100}$ $= 1.0033$
3. Write the general form of the recurrence relation and substitute.	$V_{n+1} = RV_n - d,\ V_0 = a$ $V_{n+1} = 1.0033V_n - 3440,\ V_0 = 10\,250$
b. 1. After 1 payment: V_1	b. $V_1 = 1.0033 \times 10\,250 - 3440$ $= 6843.83$
2. After 2 payments: V_2	$V_2 = 1.0033 \times 6843.83 - 3440$ $= 3426.41$
3. Write the answer.	The balance of the annuity investment account after 2 payments is \$3426.41.
c. 1. After 3 payments: V_3	c. $V_3 = 1.0033 \times 3426.41 - 3440$ $= -2.283\,23$
2. Interpret the result to give a zero balance.	Overdrawn by \$2.28
3. Write the answer.	Last payment will be \$3437.72. (\$3440 − \$2.28).

WORKED EXAMPLE 10 Calculating the balance of an annuity

Peter was granted a sporting scholarship for a 6-month season overseas. The scholarship of \$18 500 was invested in an annuity that paid 4.2% p.a. compounded monthly. Peter received a monthly payment of \$3120 from the annuity.
a. **Write a recurrence relation to describe this situation.**
b. **Use a calculator to determine the balance of the annuity account after:**
   i. **3 months**   ii. **6 months.**
c. **Determine the adjustment made to Peter's last payment to ensure a zero balance in the investment account.**

THINK	WRITE
a. 1. State the values of V_0 and d.	**a.** $V_0 = 18\,500$, $\qquad d = 3120$
2. Calculate the value of r and R.	$r = \dfrac{4.2}{12}$ $= 0.35$ $R = 1 + \dfrac{0.35}{100}$ $= 1.0035$
3. Write the general form of the recurrence relation and substitute.	$V_{n+1} = RV_n - d, \quad V_0 = a$ $V_{n+1} = 1.0035V_n - 3120, V_0 = 18\,500$

b. *Using a calculator*

1. On a calculator page, type 18 500.

2. Press ENTER/EXE.

3. Type × 1.0035 − 3120.

4. Press ENTER/EXE 6 times (for 6 payments).

b.

18 500	18 500
$18\,500 \times 1.0035 - 3120$	15 444.75
$15\,444.75 \times 1.0035 - 3120$	12 378.81
$12\,378.81 \times 1.0035 - 3120$	9302.13
$9302.13 \times 1.0035 - 3120$	6214.69
$6214.69 \times 1.0035 - 3120$	3116.44
$3116.44 \times 1.0035 - 3120$	7.35

5. Write the answers:
  i. after 3 months
  ii. after 6 months.

The balance of the annuity after 3 months is $9302.13.
The balance of the annuity after 6 months is $7.35.

c. 1. Interpret the balance after 6 months.

c. Balance = $7.35 (remaining in account)

2. Adjust the last payment by adding balance.

Payment = $3120 + $7.35
  = $3127.35

3. Write the answer.

Peter's last payment will be $3127.35.

7.4.2 Annuity problems and Finance Solver

The Finance Solver application on your CAS calculator can be used to solve annuity problems. The fields are the same as those used in reducing balance loans but are interpreted differently.

N	The total number of repayment periods
I%	The interest rate per annum
PV	The present value of the investment — this is negative as money went into the account.
PMT (Pmt)	The payment per period — this is positive as money is coming to you.
FV	The future value of the investment — positive, negative or zero
P/Y (PpY)	The number of payments per year
C/Y (CpY)	The number of compounding periods per year (nearly always the same as P/Y)

Remember:

- When entering values into Finance Solver, pay particular attention to the sign of the value.
- Cash flowing towards you is considered to be positive.
  - In an annuity, the money comes to you as a payment, so PMT is positive.
- Cash flowing away from you is considered to be negative.
  - In an annuity, the money is given to a financial institution, so PV is negative.
- If the FV is:
  - negative, the annuity has been overdrawn
  - zero, the annuity has been completely paid out
  - positive, money is still in the annuity account.
- Fill in all of the fields except the unknown field, place your cursor in that field and press ENTER/EXE.
- To move between fields, use the TAB key.

tlvd-4681

WORKED EXAMPLE 11 Solving annuity problems using Finance Solver

Leo has $450 000 in his superannuation fund. He is about to retire and invests his funds in an annuity paying 5.6% p.a. compounded monthly.
a. Determine how long Leo's funds will last if the monthly payment is $4800. Give your answer in years and months, to the nearest month.
b. If Leo wishes the annuity to last 20 years, calculate the amount he could receive each month. Give your answer to the nearest cent.

THINK

a. 1. Using Finance Solver on your CAS calculator, fill in the appropriate fields.
   N = ??? unknown
   I% = 5.6 (annual interest rate)
   PV = −450 000 (amount invested in the annuity)
   PMT = 4800 (monthly payment)
   FV = 0 (annuity balance of zero)
   P/Y = 12 (payments per year — months)
   C/Y = 12 (compounding periods per year)

   2. Press TAB to place the cursor in the N field, then press ENTER.

   3. Write the answer.

b. 1. Using Finance Solver on your CAS calculator, fill in the appropriate fields.
   N = 240 (20 years × 12 for months)
   I% = 5.6 (annual interest rate)
   PV = −450 000 (amount invested in the annuity)
   PMT= ??? unknown
   FV = 0 (annuity balance of zero)
   P/Y = 12 (payments per year — months)
   C/Y = 12 (compounding periods per year)

WRITE

a.

N	
I%	5.6
PV	−450 000
PMT	4800
FV	0
P/Y	12
C/Y	12

N = 123.579

The annuity would last for 124 months, or 10 years and 4 months (to the nearest month).

b.

N	240
I%	5.6
PV	−450 000
PMT	
FV	0
P/Y	12
C/Y	12

2. Press TAB to place the cursor in the PMT field, then press ENTER.

PMT = 3120.9638

3. Write the answer.

For the annuity to last 20 years, the monthly payment would be $3120.96 (to the nearest cent).

7.4 Exercise

Students, these questions are even better in jacPLUS

 Receive immediate feedback and access sample responses

 Access additional questions

Track your results and progress

Find all this and MORE in jacPLUS

1. **WE9** To study overseas for 4 months, Thomas was awarded a scholarship of $8500. The scholarship was invested in an annuity that paid 4.8% p.a., compounded monthly. Thomas was paid $2145 per month from the annuity.

   a. Write a recurrence relation to describe this situation.
   b. Calculate how much money is left in the annuity at the end of 3 months.
   c. Determine the adjustment made to Thomas's last payment to ensure zero balance in the annuity investment account.

2. Lisa is travelling overseas for 3 months after winning a bursary of $6500. She invested the bursary in an annuity that pays 3.24% p.a., compounded monthly. Lisa will be paid $2180 per month from the annuity.

   a. Write a recurrence relation to describe this situation.
   b. Calculate how much money is left in the annuity at the end of 2 months.
   c. Determine the adjustment made to Lisa's last payment to ensure zero balance in the investment account.
   d. Complete the amortisation table below for this annuity.

Payment number, n	Payment ($)	Interest ($)	Annuity reduction ($)	Balance of annuity ($)
0				6500.00
1	2180.00			
2				
3				

3. **WE10** Terry saved for an 8-month overseas trip. To pay for this, Terry invested his savings of $12 500 in an annuity that pays 3.6% p.a. compounded monthly, from which he will receive a monthly payment of $1580.

   a. Write a recurrence relation to describe this situation.
   b. Use a calculator to determine the balance of the annuity account after:

      i. 4 months
      ii. 8 months.

   c. Determine the adjustment made to Terry's last payment to ensure a zero balance in the investment account.

4. Melanie was granted a scholarship to study abroad for 9 months. The scholarship is for $15 250 and has been invested in an annuity that pays 4.5% p.a. compounded monthly. Melanie receives a monthly payment of $1725 from the annuity.

a. Write a recurrence relation to describe this situation.
b. Use a calculator to find the balance of the annuity account after 6 months.
c. Determine the adjustment required to Melanie's last payment to ensure a zero balance in the investment account.

5. Anita has been granted a scholarship of $16 000 to assist with the expenses of attending a sporting institute for one year. The money has been invested in an annuity that pays 6.4% p.a., compounded quarterly. The scholarship pays Anita $4150 per quarter from the annuity.

a. Write a recurrence relation to describe this situation.
b. Use your calculator to determine how much money is left in the annuity at the end of the year.
c. Determine the adjustment made to Anita's last payment to ensure a zero balance in the annuity.
d. Complete the following amortisation table for this annuity.

Payment number, n	Payment ($)	Interest ($)	Annuity reduction ($)	Balance of annuity ($)
0				16 000.00
1	4150.00			
2				
3				
4				

6. **WE11** Sam has $725 000 in his superannuation fund. He is about to retire and invests his funds in an annuity paying 5.25% p.a. compounded monthly.

a. Determine how long Sam's funds will last if the monthly payment is $5500. Give your answer in years and months, to the nearest month.
b. If Sam wishes the annuity to last for 25 years, determine how much he should receive each month. Give your answer to the nearest cent.

7. Georgia has $850 000 in her superannuation fund. She is about to retire and invests her funds in an annuity paying 5.4% p.a. compounded monthly.

a. Georgia wishes the annuity to last for 30 years. Determine how much she should receive each month. Give your answer to the nearest cent.
b. Determine how long Georgia's funds will last if her monthly payment is $5500. Give your answer in years and months, to the nearest month.

8. Patricia has invested her inheritance of $236 000 in an annuity paying 7.3% p.a., compounded quarterly. Determine Patricia's quarterly payments if she wishes the annuity to last for 15 years.

9. Stephen is investigating superannuation funds. He hopes to have enough in a fund to give him a monthly payment of $6200 for 25 years. Currently, his superannuation fund is paying an annuity at the rate of 6.2% p.a., compounded monthly.
Determine how much Stephen would need to have invested in his superannuation fund to achieve his goals. Give your answer to the nearest dollar.

10. A superannuation fund is advertising an annuity paying 7.32% p.a., compounded quarterly. Marion wishes to have a quarterly payment of $13 200 for 20 years.
Determine how much Marion would need to have invested in this annuity to achieve her retirement goals. Give your answer to the nearest dollar.

11. James has invested his superannuation of $560 000 in an annuity paying 8.45% p.a., compounded monthly. He wishes to receive a monthly payment of $5000.
   a. Determine how long this annuity would last. Give your answer to the nearest month.
   b. Determine the balance of the annuity after 5 years. Give your answer to the nearest dollar.
   c. After 5 years, James wishes to increase his monthly payments by $1500 to cover some unexpected expenses. Determine how many more years the annuity will now last. Give your answer to the nearest month.

12. Samantha has received a scholarship of $120 000 to study for another two years. The scholarship has been invested in an annuity paying 5.6% p.a. for the two years, with Samantha receiving regular payments.

   a. Calculate her payments, correct to the nearest cent, if she receives:
      i. quarterly payments, with interest compounded quarterly
      ii. monthly payments, with interest compounded monthly
      iii. weekly payments, with interest compounded weekly (assume 52 weeks in a year).
   b. Determine the payment period that would give Samantha the best value from her scholarship.

13. Billy has just retired and invested his superannuation of $468 000 into an annuity paying 6.48% p.a., compounded monthly.
   a. Determine how much Billy should receive each month if he wishes the annuity to last 20 years. Give your answer to the nearest cent.
   b. After 6 years, Billy received an inheritance of $75 000 and decided to add this to his annuity. Determine how much Billy would now receive each month if he wishes the annuity to last the same period of time.

14. Matilda has invested her life savings of $256 000 in an annuity paying 6.73% p.a., compounded monthly. For the first 5 years, Matilda received a monthly payment of $4000. Due to the economic climate, the annuity's interest rate was then changed to 5.95% p.a., compounded monthly.
If Matilda continued to receive the same monthly payment, determine how long, in total, this annuity would last. Give your answer to the nearest month.

15. Paul and Mary have contributed to their superannuation fund all of their working lives. They have both now retired and their superannuation fund has a balance of $1 645 250. They have invested these funds into an annuity paying 7.24% p.a., compounded monthly. For the first 10 years, they wish to receive a monthly payment of $10 500. After 10 years, the interest rate is increased by 1.25% p.a.
If they had planned for this annuity to last 25 years in total, determine the new monthly payment, correct to the nearest cent.

Question 1 (1 mark)

Source: VCE 2020, Further Mathematics Exam 1, Section A, Q25; © VCAA.

MC The graph below represents the value of an annuity investment, A_n, in dollars, after n time periods.

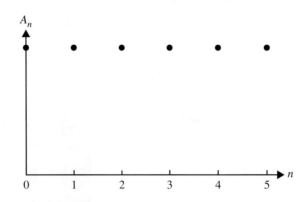

A recurrence relation that could match this graphical representation is

A. $A_0 = 200\,000,\ A_{n+1} = 1.015A_n - 2500$
B. $A_0 = 200\,000,\ A_{n+1} = 1.025A_n - 5000$
C. $A_0 = 200\,000,\ A_{n+1} = 1.03A_n - 5500$
D. $A_0 = 200\,000,\ A_{n+1} = 1.04A_n - 6000$
E. $A_0 = 200\,000,\ A_{n+1} = 1.05A_n - 8000$

Question 2 (1 mark)

Source: VCE 2016, Further Mathematics Exam 1, Section A, Q18; © VCAA.

MC The value of an annuity, V_n, after n monthly payments of $555 have been made can be determined using the recurrence relation

$$V_0 = 100\,000, \quad V_n + 1 = 1.0025V_n - 555$$

The value of the annuity after five payments have been made is closest to

A. $97\,225
B. $98\,158
C. $98\,467
D. $98\,775
E. $110\,224

Question 3 (1 mark)

Source: VCE 2016, Further Mathematics Exam 1, Section A, Q24; © VCAA.

MC Mai invests in an annuity that earns interest at the rate of 5.2% per annum compounding monthly.

Monthly payments are received from the annuity.

The balance of the annuity will be $130\,784.93 after five years.

The balance of the annuity will be $66\,992.27 after 10 years.

The monthly payment that Mai receives from the annuity is closest to

A. $1270
B. $1400
C. $1500
D. $2480
E. $3460

More exam questions are available online.

7.5 Perpetuities

7.5.1 Calculating a perpetuity

A perpetuity is a permanently invested sum of money that provides regular payments that continue forever.
- The funds will last indefinitely, provided the amount paid out is no more than the interest earned on the funds.
- If using the perpetuity formula, the period of the regular payment must be the same as the compounding period of the given interest rate.
- The balance invested does not change, so it will remain the same indefinitely.
- The type of investment used to earn the interest is often called a bond.
- Many scholarships or grants offered by universities or colleges are provided by funds invested as perpetuities.
- Perpetuities are often said to have long-term interest rates.

Perpetuity formula

The perpetuity formula is

$$d = \frac{V_0 r}{100}$$

where

V_0 = the principal, the amount invested in the perpetuity ($)

r = the interest rate earned per period (%)

d = the amount of the regular payment received per period ($).

The perpetuity formula can be transposed to make V_0 or r the subject:

$$V_0 = \frac{100d}{r} \text{ and } r = \frac{100d}{V_0}$$

WORKED EXAMPLE 12 Calculating the annual amount from a perpetuity

A charity has invested \$150 000 to set up a perpetuity to give grants to a local community centre. The funds have been invested in a bond that offers a long-term interest rate of 4.5% p.a. Calculate the amount of the annual grant to the local community centre if the interest is calculated yearly.

THINK	WRITE
1. Write the perpetuity formula.	$d = \dfrac{V_0 r}{100}$
2. State the values of V_0 and r.	$V_0 = 150\,000$ and $r = 4.5$

▶

3. Substitute and calculate the value of d.

$$d = \frac{150\,000 \times 4.5}{100}$$
$$= 6750$$

4. Write the answer.

The annual grant to the community centre will be $6750.

---

tlvd-4682

WORKED EXAMPLE 13 Calculating the amount invested in a perpetuity

A family trust wishes to provide a scholarship of $20 000 per year, in perpetuity, for a talented music student to attend a college. The trust has secured a bond at an interest rate of 8% p.a.

Calculate the amount that needs to be invested if the payments and the interest were paid:

a. **annually**

b. **quarterly.**

THINK	WRITE
a. 1. Write the perpetuity formula to calculate V_0.	**a.** $V_0 = \dfrac{100d}{r}$
2. State the values of d and r.	$d = 20\,000, \quad r = 8$
3. Substitute and calculate the value of V_0.	$V_0 = \dfrac{100 \times 20\,000}{8}$ $= 250\,000$
4. Write the answer.	The amount invested in the perpetuity would be $250 000.
b. 1. Write the perpetuity formula to calculate V_0.	**b.** $V_0 = \dfrac{100\,d}{r}$
2. State the values of d and r. *Remember they need to have the same time period for the formula to be used.*	$d = 5000, \quad r = \dfrac{8}{4} = 2$
3. Substitute and calculate the value of V_0.	$V_0 = \dfrac{100 \times 5000}{2}$ $= 250\,000$
4. Write the answer.	The amount invested in the perpetuity would be $250 000.

7.5.2 Perpetuity problems and Finance Solver

The Finance Solver application on your CAS calculator can be used to solve perpetuity problems.

The fields are the same as those used in annuities. However, the principal that is invested, PV, needs to be known for Finance Solver to be used.

N	1 — it can take any value as the balance never changes.
I%	The interest rate per annum
PV	The present value of the investment — this is negative as money went into the account.
PMT (Pmt)	The payment per period — this is positive as money is coming to you.
FV	The future value of the investment — positive and equal to PV
P/Y (PpY)	The number of payments per year
C/Y (CpY)	The number of compounding periods per year

Remember:

- **When entering values into Finance Solver, pay particular attention to the sign of the value.**
- **The number of payments, N, can take any value since the balance never changes.**
  - **For simplicity, let N = 1.**
- **Cash flowing towards you is considered to be positive.**
  - **In a perpetuity, the money is coming to you as a payment, so PMT is positive.**
- **Cash flowing away from you is considered to be negative.**
  - **In a perpetuity, the money goes into an investment account, so PV is negative.**
- **The amount invested never changes, so FV equals PV but with a positive sign.**
- **In perpetuities, the number of payments per year, P/Y, and the compounding periods per year, C/Y, are not necessarily the same.**
- **Fill in all of the fields except the unknown field, place your cursor in that field and press ENTER/EXE.**
- **To move between fields, use the TAB key.**

WORKED EXAMPLE 14 Calculating the annual amount from a perpetuity using Finance Solver

A charity has invested \$175 000 to set up a perpetuity to give grants to a local sporting team. The funds have been invested in a bond that offers a long-term interest rate of 5.25% p.a. Calculate the amount of the annual grant if the interest is calculated:

a. **yearly**
b. **monthly**
c. **daily (assuming 365 days in a year).**

THINK

a. Interest compounded yearly

   1. Using Finance Solver on your CAS calculator, fill in the appropriate fields.

      N = 1
      I% = 5.25 (annual interest rate)
      PV = −175 000 (amount invested in perpetuity)
      PMT = ??? unknown
      FV = 175 000 (balance always the same)
      P/Y = 1 (payments per year)
      C/Y = 1 (compounding periods)

WRITE

a.

N	1
I%	5.25
PV	−175 000
PMT	
FV	175 000
P/Y	1
C/Y	1

2. Press TAB to place the cursor in the PMT field, then press ENTER.

PMT = 9187.50

3. Write the answer.

With interest compounded yearly, the annual payment would be $9187.50 in perpetuity.

b. Interest compounded monthly

1. Using Finance Solver on your CAS calculator, fill in the appropriate fields.
N = 1
I% = 5.25 (annual interest rate)
PV = −175 000 (amount invested in perpetuity)
PMT = ??? unknown
FV = 175 000 (balance always the same)
P/Y = 1 (payments per year)
C/Y = 12 (compounding periods — months)

b.

N	1
I%	5.25
PV	−175 000
PMT	
FV	175 000
P/Y	1
C/Y	12

2. Press TAB to place the cursor in the PMT field, then press ENTER.

PMT = 9411.8302

3. Write the answer.

With interest compounded monthly, the annual payment would be $9411.83 in perpetuity.

c. Interest compounded daily

1. Using Finance Solver on your CAS calculator, fill in the appropriate fields.
N = 1
I% = 5.25 (annual interest rate)
PV = −175 000 (amount invested in perpetuity)
PMT = ??? unknown
FV = 175 000 (balance always the same)
P/Y = 1 (payments per year)
C/Y = 365 (compounding periods — days)

c.

N	1
I%	5.25
PV	−175 000
PMT	
FV	175 000
P/Y	1
C/Y	365

2. Press TAB to place the cursor in the PMT field, then press ENTER.

PMT = 9432.252 07

3. Write the answer.

With interest compounded daily, the annual payment would be $9432.25 in perpetuity.

WORKED EXAMPLE 15 Calculating the interest rate in a perpetuity

The owner of a company wishes to set up an annual grant of $25 000 for a research centre. He has $450 000 to invest in a perpetuity bond. Determine the interest rate that he would need to find, with interest calculated daily, for this grant to be possible.
Give your answer correct to 1 decimal place.

THINK	WRITE

Interest compounded daily

1. Using Finance Solver on your CAS calculator, fill in the appropriate fields.
   N = 1
   I% = ??? unknown
   PV = −450 000 (amount invested in perpetuity)
   PMT = 25 000 (annual payment)
   FV = 450 000 (balance always the same)
   P/Y = 1 (payments per year)
   C/Y = 365 (compounding periods — days)

N	1
I%	
PV	−450 000
PMT	25 000
FV	450 000
P/Y	1
C/Y	365

2. Press TAB to place the cursor in the I% field, then press ENTER.

   I = 5.407 122

3. Write the answer.

   The interest rate would need to be 5.4% p.a. in perpetuity.

7.5 Exercise

1. **WE12** A charity has invested $75 000 to set up a perpetuity to give grants to a local language centre. The funds have been invested in a bond that offers a long-term interest rate of 6.25% p.a.
   Calculate the amount of the annual grant to the local language centre if the interest is calculated yearly.

2. A football club has invested $200 000 to set up a perpetuity to give grants to a local sporting club. The funds have been invested in a bond that offers a long-term interest rate of 5.82% p.a.

   a. Calculate the amount of the annual grant to the local sporting club if the interest is calculated yearly.

   b. Determine how much the sporting club would receive in total over 10 years.

3. **WE13** A family trust wishes to provide a scholarship of $15 000 per year for a talented art student to attend a college. The trust has secured a bond with a long-term interest rate of 5% p.a. Determine how much needs to be invested if the payments and the interest were paid:

   a. annually   b. quarterly.

4. A benefactor wishes to provide a scholarship for an outstanding Year 11 student to attend a particular college. He wishes to provide $23 400 annually in four equal payments. The benefactor has secured a bond at an interest rate of 6% p.a., with interest compounded quarterly.

   a. Determine the amount of each payment.

   b. Calculate the principal that needs to be invested for the benefactor to meet his commitments.

5. **WE14** A university has invested $1 500 000 to set up a perpetuity to give grants to students in need. The funds have been invested in a bond that offers a long-term interest rate of 4.95% p.a. Calculate the amount of the annual grants available to students if the interest is calculated:

   a. yearly

   b. monthly

   c. daily (assuming 365 days in a year).

6. Mr and Mrs Smith have set up an investment of $350 000 in a perpetuity to assist with their grandchildren's education. The funds have been invested in a bond that offers a long-term interest rate of 7.85% p.a.

   Calculate the amount, to the nearest dollar, that can be provided each year for their grandchildren if the interest is calculated:

   a. quarterly

   b. monthly

   c. daily (assuming 365 days in a year).

7. **WE15** The Penny Company wishes to set up a biannual grant of $12 500 for research. The company has $625 000 to invest in a perpetuity bond. Determine the interest rate they would need to find, with interest calculated daily, for this grant to be possible. Give your answer correct to 2 decimal places.

8. A benefactor has $85 000 to set up a perpetuity as an annual grant to help refugees. The money has been invested in bonds that return 5.8% p.a. compounded monthly.

   a. Calculate the amount of the annual grant. Give your answer correct to the nearest cent.

   b. Determine the interest rate that the benefactor would need to find, with interest calculated daily, for the perpetuity to provide an annual grant of $6750. Give your answer correct to 1 decimal place.

9. A scholar wishes to set up a perpetuity as a scholarship fund for his old school, to be paid annually. He intends to invest $185 000 in bonds that return 6.75% p.a.

   a. Calculate the amount of the scholarship fund, to the nearest dollar, if the interest is compounded monthly.

   b. Calculate the interest rate that would be needed, with interest calculated daily, if he wished to increase the scholarship by $1000 the next year. Give your answer correct to 2 decimal places.

10. The Farthing Company set up a perpetuity for research and development. On average, they anticipate a return of 9.6% p.a., compounded quarterly, from their investment.

   a. Determine how much the Farthing Company needs to invest in the perpetuity bond if they wish to receive $48 000 annually, paid in equal quarterly payments.

   b. The Farthing Company has found a different perpetuity bond paying 9.6% p.a., but with interest compounded daily. Calculate the new quarterly payments.

11. Gemma wishes to give an annual grant of $4600 to a deserving student each year to assist with the cost of their education. To achieve this, she has set up a fund in perpetuity, paying 6.25% p.a., compounded annually.

   a. Determine how much Gemma invested in the fund.

   b. Another fund is offering 6.22% p.a., compounded daily. State whether Gemma would be able to increase her annual grant if she changed to this fund.

12. A rugby club has been given half a million dollars to invest in perpetuity to encourage participation at the local level. The club has found a bond that offers a long-term interest rate of 5.8% p.a., compounded daily.

    a. Determine how much the rugby club can give annually to the local clubs. Give your answer to the nearest dollar.
    b. If the rugby club continues to support the local clubs in this way for the next 15 years, determine how much they would have contributed to the local clubs.

13. A club is considering investing $950 000 in a perpetuity to give grants to needy athletes. They have found two options, both paying 6.2% p.a. Plan A will give quarterly payments with interest calculated monthly. Plan B gives monthly payments with interest calculated daily.
    Over a 2-year period, explain which plan gives the club more funds to give as grants, and by how much.

14. An organisation wishes to support a charity with monthly donations of $2750. They were able to secure a bond offering a long-term interest rate of 6.6% p.a., compounded monthly.

    a. Calculate the amount the organisation needs to invest to support the charity in this way.
    b. Determine by how much the organisation could increase its monthly donations if the bond, for the same amount, offered an interest rate of 6.6% p.a., compounded daily.

15. Penelope, the accountant of a large institution, is in charge of various trusts that have been set up in perpetuity for specific projects. The details of three perpetuities are:

    Trust A: $800 000 providing $12 500 biannually, interest compounded monthly
    Trust B: $800 000 providing $2100 monthly, interest compounded monthly
    Trust C: $800 000 providing $980 per fortnight, interest compounded daily

    Explain which trust is paying the better annual interest rate.

7.5 Exam questions

Question 1 (3 marks)
Source: VCE 2021, Further Mathematics Exam 2, Section A Core, Q6; © VCAA.

Sienna invests $420 000 in a perpetuity from which she will receive a regular monthly payment of $1890.

The perpetuity earns interest at the rate of 5.4% per annum.

a. Determine the total amount, in dollars, that Sienna will receive after one year of monthly payments. **(1 mark)**

b. Write down the value of the perpetuity after Sienna has received one year of monthly payments. **(1 mark)**

c. Let S_n be the value of Sienna's perpetuity after n months.

Complete the recurrence relation, in terms of S_0, S_{n+1} and S_n, that would model the value of this perpetuity over time. Write your answers in the boxes provided. **(1 mark)**

$$S_0 = \boxed{}, \quad S_{n+1} = \boxed{} \times S_n - 1890$$

Source: VCE 2017, Further Mathematics Exam 2, Section A, Core, Q7; © VCAA.

Alex sold his mechanics' business for $360 000 and invested this amount in a perpetuity.

The perpetuity earns interest at the rate of 5.2% per annum.

Interest is calculated and paid monthly.

    a. Determine the monthly payment Alex will receive from this investment. **(1 mark)**

    b. Later, Alex converts the perpetuity to an annuity investment. This annuity investment earns interest at the rate of 3.8% per annum, compounding monthly. For the first four years Alex makes a further payment each month of $500 to his investment. This monthly payment is made immediately after the interest is added.

      After four years of these regular monthly payments, Alex increases the monthly payment. This new monthly payment gives Alex a balance of $500 000 in his annuity after a further two years. Determine the value of Alex's new monthly payment. Round your answer to the nearest cent. **(2 marks)**

▶ **Question 3 (2 marks)**

Source: VCE 2015, Further Mathematics Exam 2, Module 4, Q3; © VCAA.

Jane and Michael decide to set up an annual music scholarship.

To fund the scholarship, they invest in a perpetuity that pays interest at a rate of 3.68% per annum.

The interest from this perpetuity is used to provide an annual $460 scholarship.

    a. Determine the minimum amount they must invest in the perpetuity to fund the scholarship. **(1 mark)**

    b. For how many years will they be able to provide the scholarship? **(1 mark)**

More exam questions are available online.

7.6 Review

7.6.1 Summary

doc-37617

7.6 Exercise

Multiple choice

1. **MC** A loan of $14 000 is taken out over 4 years at 9.75% p.a. (debited fortnightly) on the outstanding balance. The fortnightly repayment needed to repay the loan in full, to the nearest dollar, is:
   A. $135   **B.** $145   **C.** $163   **D.** $170   **E.** $319

2. **MC** Rachel repaid a reducing balance loan of $22 000 in 5 years with quarterly repayments and interest charged quarterly at 8.2% p.a. on the outstanding balance. The total amount of interest she paid at the end of the first quarter is closest to:
   A. $1804   **B.** $451   **C.** $150   **D.** $9000   **E.** $10 000

3. **MC** The number of monthly repayments required to repay a $41 000 reducing balance loan in full, if the repayments are $588.39 and interest is debited monthly at 10.5% p.a., will be closest to:
   A. 500   **B.** 600   **C.** 90   **D.** 100   **E.** 110

4. **MC** A reducing balance loan of $56 000 is repaid by quarterly instalments of $1332.24 over 15 years at an interest rate of 5% p.a. (adjusted quarterly). If, instead, repayments of $1500 per quarter were made throughout the loan (other variables remaining unchanged), the term of the loan would be:
   A. 15 years.   **B.** 14 years.   **C.** $12\frac{3}{4}$ years.   **D.** $12\frac{1}{2}$ years.   **E.** 12 years.

Questions 5–7 refer to the following information.

A reducing balance loan of $24 000 attracting interest at 6.5% p.a. can be repaid over 5 years by either quarterly repayments or fortnightly repayments.

5. **MC** The fortnightly repayment value is between:
   A. $120 and $140.   **B.** $140 and $160.   **C.** $160 and $180.
   D. $180 and $200.   **E.** $200 and $220.

6. **MC** The quarterly repayment value is between:
   A. $1200 and $1220.   **B.** $1250 and $1300.   **C.** $1350 and $1360.
   D. $1400 and $1420.   **E.** $1460 and $1470.

7. **MC** If, instead, a rival institution offered a rate of 5.5% p.a., the quarterly repayment value that would enable the loan to be repaid in full in the same time would be between:
   A. $1350 and $1375.   **B.** $1375 and $1400.   **C.** $1400 and $1425.
   D. $1425 and $1450.   **E.** $1450 and $1500.

8. **MC** Elena has $2000 to invest in her new superannuation fund, which is paying interest of 6.0% p.a. compounded monthly with $360 monthly contributions from her employer. Determine which of the following recurrence relations could be used to model this situation.

A. $V_{n+1} = 1.5V_n + 360, V_0 = 2000$

B. $V_{n+1} = 1.005V_n - 360, V_0 = 2000$

C. $V_{n+1} = 1.5V_n - 360, V_0 = 2000$

D. $V_{n+1} = 1.005V_n + 360, V_0 = 2000$

E. $V_{n+1} = 1.05V_n + 360, V_0 = 2000$

9. **MC** Kerry Green borrows $305 000 to invest in an apartment. She wishes to use the bank's money to purchase the property. If the terms of an interest-only loan are 6.47% p.a. compounded monthly, the monthly repayment is closest to:

A. $1644.45   B. $19 733.50   C. $30 611.62   D. $1644.46   E. none of the above.

10. **MC** Claire is aged 48 and is planning to retire at 65. Her annual salary is $60 000 and her employer superannuation contributions are 10% of her gross monthly income. The superannuation fund has been returning an interest rate of 9.6% p.a., compounded monthly. Claire's current balance is $92 200, which she wants to grow to $800 000. The extra amount that Claire will have to contribute each month to ensure this final balance is achieved is closest to:

A. $0   B. $1990   C. $650   D. $150   E. $240

Short answer

11. Claire has borrowed $16 000 to purchase a new car. The terms of the loan were monthly payments of $340, and an interest rate of 9.6% p.a., charged monthly.

a. Write a recurrence relation to describe this situation.

b. Calculate the outstanding balance on the loan after Claire has made three payments.

12. Tom borrowed $6500 to purchase more equipment for his online business. The financial institution charges interest, compounded monthly, at a rate of 12.24% p.a. Tom agrees to repay the loan with monthly payments of $750.

a. Write a recurrence relation to show the outstanding balance on the loan at any time.

b. Calculate the outstanding balance after Tom has made four payments.

c. Use your calculator to determine the number of monthly payments needed to reduce the loan to under $100.

d. Calculate the last payment needed to ensure a zero balance.

13. a. Complete the amortisation table below for a personal loan of $5000 from a financial institution charging 9.2% p.a., compounded quarterly, with quarterly payments of $1325.

Payment number, n	Payment ($)	Interest ($)	Principal reduction ($)	Balance of loan ($)
0				5000.00
1	1325.00			
2	1325.00			
3	1325.00			
4	1325.00			

b. Calculate the last payment to ensure a zero balance on the loan.

14. Zoe needs to purchase equipment to expand her business. She has taken out a loan of $76 000 from her bank, whose current interest rate is 7.35% p.a., compounded monthly.

a. Calculate Zoe's monthly payments to the nearest cent if she wishes to repay the loan in 5 years.

b. Calculate the interest Zoe paid during the 5 years.

15. Penelope earned a grant, worth $85 000, to continue her research for a year. She has invested the grant in an annuity that pays 10.08% p.a., compounded monthly. She will receive monthly payments of $7475.

    a. Write a recurrence relation to illustrate the outstanding monthly balance of Penelope's annuity.
    b. Calculate the balance of the annuity after 3 months. Give your answer correct to the nearest dollar.

16. A benefactor has set up a scholarship fund for her old college. The money, $550 000, has been placed in a long-term bond. Calculate the annual payment she can give the college if the bond paid 3.52% p.a., compounded annually.

17. A company wishes to establish a fund to give a monthly grant of $8250 for a research and development project. They found a fund that gives a long-term return of 10% p.a. Determine how much the company would need to invest in this fund for it to continue indefinitely.

18. A local bank is advertising home loans for 25 years at 3.95% p.a., compounded monthly. Jack and Jill are looking for their first home together and wish to take out a home loan of $480 000.
    Calculate Jack and Jill's monthly payment from this bank. Give your answer correct to the nearest cent.

19. Anita knows she can afford to repay $3500 per month on a home loan. If the expected interest rate is 4.2% p.a., compounded monthly, and she would like to repay the loan fully in 20 years, determine how much Anita can borrow. Give your answer to the nearest dollar.

20. Paul took out a reducing balance loan of $125 000 for an extension on his home. The loan is to be repaid with monthly instalments over a period of 15 years. The interest rate, compounded monthly, is 4.65% p.a.

    a. Calculate Paul's monthly instalment, correct to the nearest cent.
    b. Determine how much is owing after 10 years.
    c. If Paul now wishes to repay the outstanding balance of the loan in the next 2 years, determine, to the nearest cent, his new monthly instalments.

21. A sporting club has invested $250 000 in a perpetuity fund to give grants to promising athletes. The funds have been invested in a bond offering a long-term interest rate of 9.25% p.a.
    Calculate the amount, to the nearest dollar, the club can give as an annual grant if the interest is compounded:

    a. monthly
    b. daily.

Extended response

22. Mabel has $657 500 in her superannuation fund. She will convert these funds to an annuity when she retires. The annuity from her fund pays 5.8% p.a., compounded monthly.

    a. Determine how long Mabel's funds will last if she receives a monthly payment of $5250.
       Give your answer in years and months, to the nearest month.
    b. Determine the amount of her monthly payments if Mabel wishes the annuity to last 20 years.
       Give your answer to the nearest dollar.

23. You approach your bank about taking out a personal loan of $23 400, repaid over 4 years, to purchase a car.

   a. Your bank asks for fortnightly payments of $275. Calculate the bank's interest rate per annum, correct to 3 decimal places.
   b. Another bank offered the same rate of interest, but charged monthly. Calculate their monthly repayments.
   c. Considering the total is repaid over the 4 years, determine which one would be the better loan.

24. You are comparing the amount payable from various superannuation funds. You have $750 000 to invest. Plan A returns 7.8% p.a. compounded quarterly and Plan B returns 7.8% p.a. compounded daily. You would like both of these to still have a balance of $750 000 at any time.
   Explain which plan gives the better monthly payment and by how much.

25. Jacinta has $525 000 to invest.

   a. Calculate her monthly payment, to the nearest cent:

      i. if she invests in an annuity paying 5.32% p.a., compounded monthly, lasting 25 years
      ii. if she invests in a perpetuity paying 5.32% p.a., compounded daily.

   b. Determine the balance of the two different investments after 15 years.

26. Sally has a reducing balance home loan of $540 000. Interest is charged at 3.85% p.a. monthly for 30 years.

   a. Calculate Sally's monthly payment to the nearest cent.
   b. After 5 years, the interest rate was increased by 0.5% p.a. Sally still wishes to pay off the loan within 30 years. Calculate her new monthly instalment, correct to the nearest cent.
   c. After another 10 years, Sally receives an inheritance of $150 000 that she wants to use to reduce the balance of her loan. She also wants to repay the loan in 20 years instead of the original 30 years. Calculate her new loan repayments, correct to the nearest dollar.

7.6 Exam questions

Question 1 (1 mark)
Source: VCE 2021, Further Mathematics Exam 1, Section A Core, Q18; © VCAA.

MC Deepa invests $500 000 in an annuity that provides an annual payment of $44 970.55

Interest is calculated annually.

The first five lines of the amortisation table are shown below.

Payment number	Payment ($)	Interest ($)	Principal reduction ($)	Balance ($)
0	0.00	0.00	0.00	500 000.00
1	44 970.55	20 000.00	24 970.55	475 029.45
2	44 970.55	19 001.18	25 969.37	449 060.08
3	44 970.55	17 962.40		422 051.93
4	44 970.55	16 882.08	28 088.47	393 963.46

The principal reduction associated with payment number 3 is

A. $17 962.40   **B.** $25 969.37   **C.** $27 008.15   **D.** $28 088.47   **E.** $44 970.55

Question 2 (1 mark)

Source: VCE 2021, Further Mathematics Exam 1, Section A Core, Q19; © VCAA.

MC Deepa invests $500 000 in an annuity that provides an annual payment of $44 970.55

Interest is calculated annually.

The first five lines of the amortisation table are shown below.

Payment number	Payment ($)	Interest ($)	Principal reduction ($)	Balance ($)
0	0.00	0.00	0.00	500 000.00
1	44 970.55	20 000.00	24 970.55	475 029.45
2	44 970.55	19 001.18	25 969.37	449 060.08
3	44 970.55	17 962.40		422 051.93
4	44 970.55	16 882.08	28 088.47	393 963.46

The number of years, in total, for which Deepa will receive the regular payment of $44 970.55 is closest to

A. 12   **B.** 15   **C.** 16   **D.** 18   **E.** 20

Question 3 (3 marks)

Source: VCE 2020, Further Mathematics Exam 2, Section A, Q8; © VCAA.

Samuel has a reducing balance loan.

The first five lines of the amortisation table for Samuel's loan are shown below.

Payment number	Payment ($)	Interest ($)	Principal reduction ($)	Balance ($)
0	0.00	0.00	0.00	320 000.00
1	1600.00	960.00	640.00	319 360.00
2	1600.00	958.08	641.92	318 718.08
3	1600.00	956.15		318 074.23
4	1600.00			

Interest is calculated monthly and Samuel makes monthly payments of $1600.

Interest is charged on this loan at the rate of 3.6% per annum.

    a. Using the values in the amortisation table

       i. calculate the principal reduction associated with payment number 3 **(1 mark)**

      ii. calculate the balance of the loan after payment number 4 is made.

        Round your answer to the nearest cent. **(1 mark)**

b. Let S_n be the balance of Samuel's loan after n months.

Write down a recurrence relation, in terms of S_0, S_{n+1} and S_n, that could be used to model the month-to-month balance of the loan. **(1 mark)**

Question 4 (1 mark)

Source: VCE 2018, Further Mathematics Exam 1, Section A, Q23; © VCAA.

MC Five lines of an amortisation table for a reducing balance loan with monthly repayments are shown below.

Repayment number	Repayment	Interest	Principal reduction	Balance of loan
25	$2200.00	$972.24	$1227.76	$230 256.78
26	$2200.00	$967.08	$1232.92	$229 023.86
27	$2200.00	$961.90	$1238.10	$227 785.76
28	$2200.00	$1002.26	$1197.74	$226 588.02
29	$2200.00	$996.99	$1203.01	$225 385.01

The interest rate for this loan changed immediately before repayment number 28.

This change in interest rate is best described as
  A. an increase of 0.24% per annum.
  B. a decrease of 0.024% per annum.
  C. an increase of 0.024% per annum.
  D. a decrease of 0.0024% per annum.
  E. an increase of 0.00024% per annum.

Question 5 (1 mark)

Source: VCE 2018, Further Mathematics Exam 1, Section A, Core, Q19; © VCAA.

MC Daniel borrows $5000, which he intends to repay fully in a lump sum after one year.

The annual interest rate and compounding period for five different compound interest loans are given below.
  • Loan I: 12.6% per annum, compounding weekly
  • Loan II: 12.8% per annum, compounding weekly
  • Loan III: 12.9% per annum, compounding weekly
  • Loan IV: 12.7% per annum, compounding quarterly
  • Loan V: 13.2% per annum, compounding quarterly

When fully repaid, the loan that will cost Daniel the least amount of money is

  A. Loan I.   **B.** Loan II.   **C.** Loan III.   **D.** Loan IV.   **E.** Loan V.

More exam questions are available online.

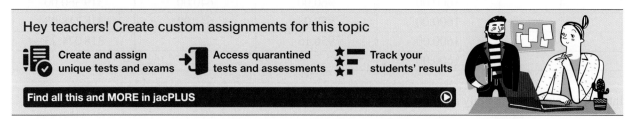

Hey teachers! Create custom assignments for this topic

Create and assign unique tests and exams   Access quarantined tests and assessments   Track your students' results

Find all this and MORE in jacPLUS

Answers

Topic 7 Financial mathematics extension

7.2 Reducing balance loans modelled using recurrence relations

7.2 Exercise

1. a. $V_{n+1} = 1.008\,75V_n - 380, \quad V_0 = 4200$
   b. i. $3510.50   ii. $2453.45

   c. 12 payments
   (*Note:* The last payment would be reduced by $123.12 to give a zero balance.)

2. a. $V_{n+1} = 1.0125V_n - 345, \quad V_0 = 2600$
   b. i. $1650.73   ii. $328.74

   c. 8 payments
   (*Note:* The last payment would be reduced by $12.15 to give a zero balance.)

3. a. $5300
   b. $V_{n+1} = 1.0105V_n - 575, \quad V_0 = 5300$
   c. 10 payments
   (*Note:* The last payment would be reduced by $145.89 to give a zero balance.)

4. a. $2500
   b. $V_{n+1} = 1.007V_n - 300, \quad V_0 = 2500$
   c. 9 payments
   (*Note:* The last payment would be reduced by $114.87 to give a zero balance.)

5. a. 2.95% per quarter   b. $132.75
   c. $2334.02   d. $1131.15
   e. $1238.35   f. $338.35

6. E

7. a. 2.4% per quarter

   b. $182.40

   c. $5186.18

   d. $1250.53

   e.
5	1375.00	63.72	1311.28	1343.83
6	1375.00	32.25	1342.75	1.08

   f. $1376.08

8. a. 3.75% per quarter
   b. $V_{n+1} = 1.0375V_n - 3250, \quad V_0 = 25\,000$
   c. $17\,799.10
   d. $9757.32

9. See the table at the foot of the page.*

10. a. 0.75% per month
    b. $V_{n+1} = 1.0075V_n - 390, \quad V_0 = 3750$
    c. $3023.54
    d. $396.64

7.2 Exam questions

Note: Mark allocations are available with the fully worked solutions online.

1. B   2. A   3. B

7.3 Solving reducing balance loan problems with technology

7.3 Exercise

1. $11\,108.05

2. $2036.11

3. a. $2345.06   b. $844\,221.60   c. $369\,221.60

4. a. $1915.21   b. $91\,930.08   c. $16\,130.08

5. $710\,847

6. $6000

7. a. The loan would be repaid in 20 quarterly payments, a total of 5 years.
   b. $1894.95

8. a. The loan would be repaid in 13 quarterly payments, a total of 3 years and 3 months.
   b. $1535.39

9. Liam has the lower interest rate of 7.45%, lower by 0.3%.

10. B-bank has the lower interest rate of 8.94%, lower by 0.3%.

11. a. $1823.15   b. $546\,945   c. $231\,945

12. a. $1620.19 per quarter   b. $16\,669.86

13. a. $2336.71   b. $310\,619.44   c. $5755.64

14. a. i. $2253.53   ii. $747.90   iii. $344.78
    b. The better deal is paying fortnightly (by a minimum of $52.60).

15. a. $2074.07   b. $2254.31

7.3 Exam questions

Note: Mark allocations are available with the fully worked solutions online.

1. E

2. a. $65\,076.22   b. $4676.22

3. C

*9.

Payment number, n	Payment ($)	Interest ($)	Principal reduction ($)	Balance of loan ($)
0				25 000.00
1	3250	$0.0375 \times 25\,000 = 937.50$	$3250 - 937.50 = 2312.5$	$25\,000 - 2312.5 = 22\,687.50$
2	3250	$0.0375 \times 22\,687.5 = 850.78$	$3250 - 850.78 = 2399.22$	$22\,687.5 - 2399.22 = 20\,288.28$
3	3250	$0.0375 \times 20\,288.28 = 760.81$	$3250 - 760.81 = 2489.19$	$20\,288.28 - 2489.19 = 17\,799.09$
4	3250	$0.0375 \times 17\,700.09 = 667.47$	$3250 - 667.47 = 2582.53$	$17\,799.09 - 2582.53 = 15\,216.56$

7.4 Annuities

7.4 Exercise

1. a. $V_{n+1} = 1.004V_n - 2145, \quad V_0 = 8500$
   b. $2141.63
   c. The annuity still has a balance of $5.20, so the last payment has to be increased by $5.20 to $2150.20.

2. a. $V_{n+1} = 1.0027V_n - 2180, \quad V_0 = 6500$
   b. $2169.26
   c. The annuity has been overdrawn by $4.88, so the last payment has to be decreased by $4.88 to $2175.12.
   d. See the table at the foot of the page.*

3. a. $V_{n+1} = 1.003V_n - 1580, \quad V_0 = 12500$
   b. i. $6302.18   ii. $29.65
   c. The annuity still has a balance of $29.65, so Terry's last payment will have to be increased by $29.65 to $1609.65.

4. a. $V_{n+1} = 1.00375V_n - 1725, \quad V_0 = 15250$
   b. $5148.84
   c. The annuity still has a balance of $12.55, so Melanie's last payment will have to be increased by $12.55 to $1737.55.

5. a. $V_{n+1} = 1.016V_n - 4150, \quad V_0 = 16000$
   b. $46.17
   c. The annuity still has a balance of $46.17, so the last payment will have to be increased by $46.17 to $4196.17.
   d. See the table at the foot of the page.*

6. a. 16 years and 5 months   b. $4344.55

7. a. $4773.01   b. 22 years and 1 month

8. $6504.65 per quarter

9. $944 284

10. $552 243

11. a. 18 years and 6 months
    b. $481 442
    c. 8 years and 9 months

12. a. i. $15 960.32
        ii. $5296.87
        iii. $1220.29
    b. Quarterly

13. a. $3483.77   b. $4164.04

14. 6 years and 7 months

15. $15 201.59

7.4 Exam questions

Note: Mark allocations are available with the fully worked solutions online.

1. B
2. C
3. C

7.5 Perpetuities

7.5 Exercise

1. $4687.50
2. a. $11 640   b. $116 400
3. a. $300 000   b. $300 000
4. a. $5850   b. $390 000
5. a. $74 250   b. $75 957.93   c. $76 113.10
6. a. $28 294   b. $28 485   c. $28 579
7. 3.96% p.a.
8. a. $5063.19   b. 7.6% p.a.
9. a. $12 881   b. 7.24% p.a.
10. a. $500 000   b. $12 143.54
11. a. $73 600
    b. $4722.88, an increase of $122.88 from the original fund
12. a. $29 855   b. $447 825
13. Plan A would give $314.80 more for grants.

---

*2. d.

Payment number, n	Payment ($)	Interest ($)	Annuity reduction ($)	Balance of annuity ($)
0				6500.00
1	2180.00	$0.0027 \times 6500 = 17.55$	$2180 - 17.55 = 2162.45$	$6500 - 216.45 = 4337.55$
2	2180.00	$0.0027 \times 4337.55 = 11.71$	$2180 - 11.71 = 2168.29$	$4337.55 - 2168.29 = 2169.26$
3	2180.00	$0.0027 \times 2169.26 = 5.86$	$2180 - 5.86 = 2174.14$	$2169.26 - 2174.14 = -4.88$

*5. d.

Payment. number, n	Payment ($)	Interest ($)	Annuity reduction ($)	Balance of annuity ($)
0				16 000.00
1	4150	$0.016 \times 16\,000 = 256$	$4150 - 256 = 3894$	$16\,000 - 3894 = 12\,106$
2	4150	$0.016 \times 12\,106 = 193.70$	$4150 - 193.70 = 3956.30$	$12\,106 - 3956.3 = 8149.70$
3	4150	$0.016 \times 8149.7 = 130.39$	$4150 - 130.39 = 4019.61$	$8149.70 - 4019.61 = 4130.09$
4	4150	$0.016 \times 4130.09 = 66.08$	$4150 - 66.08 = 4083.92$	$4130.09 - 4083.02 = 46.17$

14. a. $500\,000

b. $2757.33, an increase of $7.33 per month.

15. Trust C

7.5 Exam questions

Note: Mark allocations are available with the fully worked solutions online.

1. a. $22\,680

b. $420\,000

c. $S_n = 1.0045 \times S_n - 1890$

2. a. $1560 b. $805.65

3. a. $12\,500 b. 27 years

7.6 Review

7.6 Exercise

Multiple choice

1. C

2. B

3. E

4. C

5. E

6. D

7. B

8. D

9. D

10. D

Short answer

11. a. $V_{n+1} = 1.008V_n - 340, \quad V_0 = 16\,000$

b. $15\,358.90

12. a. $V_{n+1} = 1.0102V_n - 750, \quad V_0 = 6500$

b. $3723.07

c. Nine

d. $839.58

13. a. See the table at the foot of the page.*

b. $1315.45

14. a. $1517.47 b. $15\,048.20

15. a. $V_{n+1} = 1.0084V_n - 7475, \quad V_0 = 85\,000$

b. $64\,546

16. $19\,360

17. $990\,000

18. $2520.38

19. $567\,656

20. a. $965.85 b. $51\,618.96 c. $2256.51

21. a. $24\,131 b. $24\,225

Extended response

22. a. 16 years and 1 month

b. $4635

23. a. 10.307% p.a.

b. $596.94

c. The fortnightly loan would be the better choice as you would pay $54.12 less over the 4 years.

24. Plan B is the better plan by an extra $46.70 per month.

25. a. i. $3167.77 ii. $2332.50

b. i. $294\,302.21 ii. $525\,000

26. a. $294\,302.21 b. $2666.83 c. $3755

7.6 Exam questions

Note: Mark allocations are available with the fully worked solutions online.

1. C

2. B

3. a. i. $643.85 ii. $317\,428.45

b. $S_0 = 320000, S_{n+1} = 1.003 \times S_n - 1600$

4. A

5. D

*13. a.

Payment number, n	Payment ($)	Interest ($)	Annuity reduction ($)	Balance of annuity ($)
0				5000.00
1	1325.00	$0.023 \times 5000 = 115$	$1325 - 115 = 1210$	$5000 - 1210 = 3790$
2	1325.00	$0.023 \times 3790 = 87.17$	$1325 - 87.17 = 1237.83$	$3790 - 1237.83 = 2552.17$
3	1325.00	$0.023 \times 2552.17 = 58.70$	$1325 - 58.70 = 1266.30$	$2552.17 - 1266.30 = 1285.87$
4	1325.00	$0.023 \times 1285.87 = 29.58$	$1325 - 29.58 = 1295.42$	$1285.87 - 1295.42 = -9.55$

8 Investigating relationships between two numerical variables

LEARNING SEQUENCE

Fully worked solutions for this topic are available online.

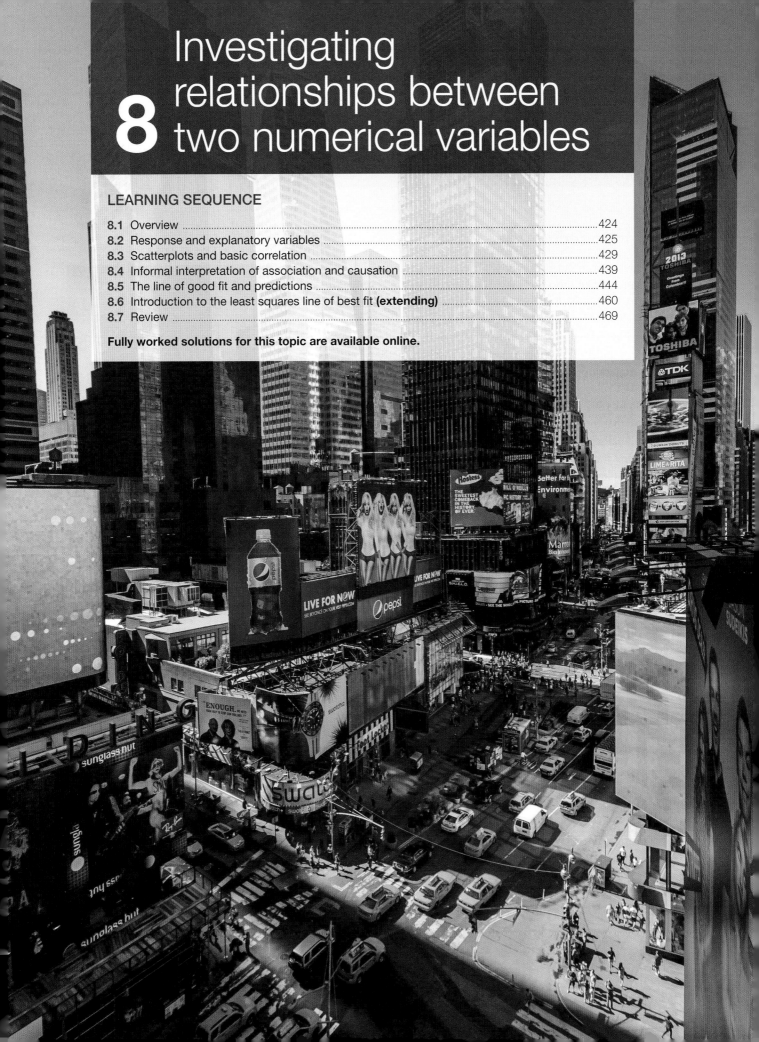

8.1 Overview

8.1.1 Introduction

Karl Pearson, born in London in 1857, was an influential mathematician and biostatistician. He was educated privately at University College School in London before moving to King's College, Cambridge, in 1876 to study mathematics. Pearson then travelled to Germany to study physics at the University of Heidelberg under GH Quincke and Kuno Fischer. In 1884 Pearson was appointed professor of applied mathematics and mechanics at University College, London. His work there sparked his interest in statistics. In 1892 he published *The Grammar of Science*, in which he argued that the scientific method is essentially descriptive rather than explanatory. He then made similar arguments about statistics, highlighting the importance of quantification for biology, medicine and social sciences.

Through his mathematical work and institution building, Pearson played a leading role in the creation of modern statistics. The basis for his statistical mathematics came from the long tradition of work on the method of least square approximation to estimate quantities from repeated astronomical and geodetic measurements using probability theory. As a statistician, Pearson emphasised measuring correlations and fitting curves to data (Pearson's correlation coefficient), and later went on to develop the chi-squared distribution (test of goodness of fit of observed data).

A 23-year-old Albert Einstein started the Olympia Academy study group in 1902 with his friends Maurice Solovine and Conrad Habicht. The first reading suggestion made was Pearson's *The Grammar of Science*. This book covered a number of themes that were later to become part of Einstein's and other scientists' theories.

KEY CONCEPTS

This topic covers the following key concepts from the VCE Mathematics Study Design:
- response and explanatory variables
- scatterplots and their use in identifying and qualitatively describing the association between two numerical variables in terms of direction, form and strength
- informal interpretation of association and causation
- use of a line of good fit by eye to make predictions, including the issues of interpolation and extrapolation
- interpretation of a line of good fit, its intercept and slope in the context of the data.

Source: VCE Mathematics Study Design (2023–2027) extracts ©VCAA; reproduced by permission.

8.2 Response and explanatory variables

8.2.1 Identifying response and explanatory variables

To investigate the relationship between two numeric variables, it is necessary to construct a scatterplot. However, before we can construct a scatterplot, the **response variable (RV)** and **explanatory variable (EV)** must be identified.

An EV is the variable that is altered or observed to change, which can affect or explain changes in the RV. The EV is always plotted along the horizontal (x) axis of a scatterplot.

The RV is what we expect to change in response to variations in the EV. It is expected that you would only see a change in the RV after a change has occurred in the EV. The RV is always plotted along the vertical (y) axis of a scatterplot.

Consider the variables *training time,* which represents the amount of time spent in training in hours per week, and *finish time,* which represents the amount of time taken to complete a marathon. A logical conclusion to make is that as the training time increases, an athlete would likely become faster and therefore their finish time would decrease.

Thus, it can be seen that training time, when altered, will likely result in a change in the finish time. This makes training time the explanatory variable (EV), and since finish time responds to changes in training time, it will be the response variable (RV). When plotting this data in a scatterplot, training time will be placed on the horizontal axis and finish time on the vertical axis.

In situations where it does not seem clear which is the EV and the RV, just remember that when we conduct research, we seek to predict the RV using values of the EV. Thus, through careful reading of how the research question is phrased, we can determine which variable is which.

Response and explanatory variables

- **The explanatory variable (EV) can explain changes in the response variable. It is placed on the horizontal axis of a scatterplot.**
- **The response variable (RV) changes in response to the explanatory variable. The RV is predicted using values of the EV. It is placed on the vertical axis of a scatterplot.**

tlvd-4814

WORKED EXAMPLE 1 Identifying explanatory and response variables

Determine the explanatory and response variable in each case.

a. When researching housing, a reasearcher compares the sale price (*price*) and distance from the city centre (*distance*).

b. Data is collected comparing the amount of caffeine in the bloodstream (*dosage*) and the reaction time of an athlete (*time*).

c. A scientist wants to create a model to predict the height of an adult male athlete (*height*) based on their arm span (*span*).

THINK

a. Houses that are further out from the city centre tend to be cheaper than those in inner city suburbs. This would mean that price responds to changes in the distance from the city.

b. Caffeine is known to improve athletes' performance in a number of areas, including reaction time. Having more caffeine in the system is likely to reduce reaction time. This means reaction time responds to changes in the dosage of caffeine.

c. It may not be clear whether height or arm span depends on the other, but since we are predicting the height it will be the response variable.

WRITE

a. The explanatory variable (EV) is distance.
The response variable (RV) is price.

b. The explanatory variable (EV) is dosage.
The response variable (RV) is time.

c. The explanatory variable (EV) is span.
The response variable (RV) is height.

8.2 Exercise

Students, these questions are even better in jacPLUS

  Receive immediate feedback and access sample responses

 Access additional questions

 Track your results and progress

Find all this and MORE in jacPLUS

1. **WE1** Identify the explanatory variable (EV) and response variable (RV) in each of the following scenarios. The two variables are shown in brackets for each question.

   a. The results of a student's exam (*percentage*) and the number of hours spent studying for the exam (*time*)

   b. A scatterplot of the data of a person's age (*age*) and their annual income (*income*)

2. Identify the explanatory variable (EV) and response variable (RV) in each of the following scenarios. The two variables are shown in brackets for each question.

   a. An investigation is being conducted on the life expectancy (*age*) and the median income (*income*) of workers in a country.

   b. A model is being created to explore the time spent exercising each week (*hours*) and the resting heart rate (*BPM*) of athletes in various sports.

3. Identify the explanatory variable (EV) and response variable (RV) in each of the following scenarios. The two variables are shown in brackets for each question.

   a. A diabetes researcher wants to investigate the relationship between the weight of a person in kilograms (*weight*) and their health by measuring the circumference of their waist in cm (*waist*).

   b. It is important for fishermen to not catch fish that are too small. Thus, data is recorded to predict the weight of a fish in kg (*weight*) from its length (*length*) so fishermen can quickly determine whether a fish is too small and should be thrown back.

4. A researcher is conducting a study on sleep and is interested in the relationship between the amount of screen time each day (screen time; hours per day) and the amount of sleep the person has each night (sleep; hours per night).

   Decide whether screen time is the response or explanatory variable. Justify your answer.

5. An office manager is trying to predict the amount of coffee to order for all the staff in the building. They know that during heavy periods, where large amounts of after-hours work are required, much more coffee is consumed. The manager plans to track the office workers' sleeping patterns (sleep; hours per night) in order to predict the amount of coffee required each day (coffee; cups per day).

   Decide whether coffee (cups per day) is the response or explanatory variable. Justify your answer.

Question 1 (1 mark)

Source: VCE 2019, Further Mathematics, Exam 2, Q4; © VCAA.

The relative humidity (%) at 9 am and 3 pm on 14 days in November is shown in the table below.

Relative humidity (%)	
9 am	**3 pm**
100	87
99	75
95	67
63	57
81	57
94	74
96	71
81	62
73	53
53	54
57	36
77	39
51	30
41	32

Note: Data: Australian Government, Bureau of Meteorology, www.bom.gov.au

A least squares line is to be fitted to the data with the aim of predicting the relative humidity at 3 pm (*humidity 3 pm*) from the relative humidity at 9 am (*humidity 9 am*). Name the explanatory variable.

Question 2 (1 mark)

Source: VCE 2018, Further Mathematics, Exam 2, Q2; © VCAA.

The congestion level in a city can be recorded as the percentage increase in travel time due to traffic congestion in peak periods (compared to non-peak periods). This is called the percentage congestion level.

The percentage congestion levels for the morning and evening peak periods for 19 large cities are plotted on the scatterplot below. Name the response variable.

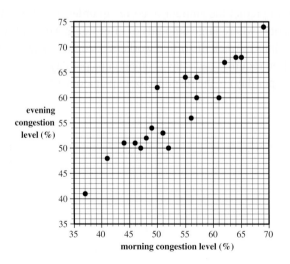

Source: VCE 2014, Further Mathematics, Exam 2, Q2; © VCAA.

The scatterplot below shows the *population* and *area* (in square kilometres) of a sample of inner suburbs of a large city.

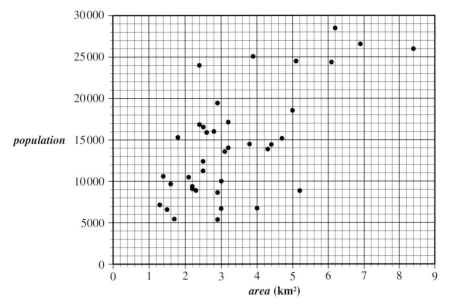

State the response variable.

More exam questions are available online.

8.3 Scatterplots and basic correlation

LEARNING INTENTION

At the end of this subtopic you should be able to:
• represent data in scatterplots
• use a scatterplot to describe an observed association between two numerical variables in terms of strength, direction and form
• determine the strength of association using Pearson's product-moment correlation coefficient.

8.3.1 Scatterplots and correlations

A common way to interpret bivariate data is through the use of a **scatterplot**. Scatterplots provide a visual display of data and can be used to draw **correlations** and **causations** between two variables.

On a scatterplot, each point represents a single combination of the explanatory and response variables for a single person, place or case. Values for the explanatory variable are placed along the horizontal axis, and values for the response variable on the vertical axis. A coordinate on the scatterplot will have the value of the EV listed first and the value of the RV second.

Consider the explanatory variable training time (hours) and response variable finish time (hours), which represents the amount of time taken to complete a marathon. The coordinate $(21, 3)$ represents the result of a single person who trains 21 hours a week and finishes a marathon in 3 hours.

Scatterplots

- **The explanatory variable (EV) is placed on the horizontal axis of a scatterplot.**
- **The response variable (RV) is placed on the vertical axis of a scatterplot.**
- **For each coordinate on the scatterplot (x, y), x represents a value of the EV and y represents a value of the RV.**

Correlation

When interpreting a scatterplot, the correlation provides an insight into the relationship between the two variables. The correlation is a measure of the strength of the linear relationship between the two variables. There are three classifications for the correlation of data:

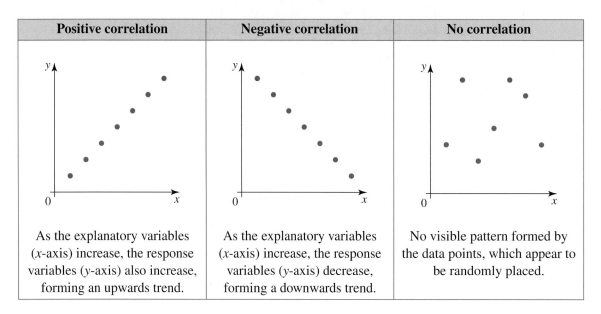

Positive correlation	Negative correlation	No correlation
As the explanatory variables (x-axis) increase, the response variables (y-axis) also increase, forming an upwards trend.	As the explanatory variables (x-axis) increase, the response variables (y-axis) decrease, forming a downwards trend.	No visible pattern formed by the data points, which appear to be randomly placed.

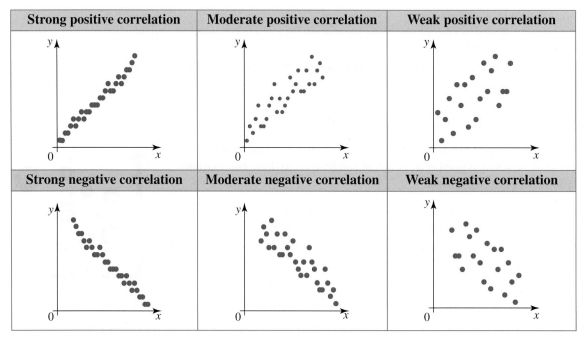

A local cafe recorded the number of ice-creams sold per day as well as the daily maximum temperature for 12 days.

Temp (°C)	36	32	28	26	30	24	19	25	33	35	37	34
No. of ice-creams sold	162	136	122	118	134	121	65	124	140	154	156	148

a. Identify the response and explanatory variables.
b. Represent the data in a scatterplot.
c. Discuss the strength of the relationship between the variables.

THINK

a. Consider which variable does not rely on the other. This will be the explanatory variable.

b. Select a reasonable scale for each variable that covers the full range of the data set. Plot the given points, remembering that the explanatory variable should be represented on the x-axis and the response variable should be represented on the y-axis.

WRITE/DRAW

a. Explanatory variable = temperature
   Response variable = number of ice-creams sold

b.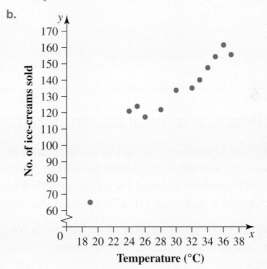

c. Look at the pattern of the data points. Do they form a linear pattern? Are they progressing in a similar direction, either positive or negative? How strong is the correlation between the variables?

c. There is a linear relationship between the two variables. As the temperature increases so does the number of ice-creams sold. The correlation between the variables is strong. Therefore, this graph could be described as having a strong positive correlation.

Note: The lowest point on the scatterplot could be considered a potential outlier.

TI	THINK	DISPLAY/WRITE	CASIO	THINK	DISPLAY/WRITE

b. 1. On a Lists & Spreadsheet page, label the first column as *temp* and the second column as *icecreams*. Enter the temperature values in the first column and the corresponding ice-cream sales values in the second column.

b. 1. On a Statistics screen, relabel list 1 as *temp* and list 2 as *icecream*. Enter the temperature values in the first column and the corresponding ice-cream sales values in the second column.

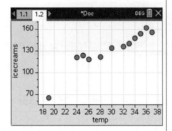

2. On a Data & Statistics page, click on the horizontal axis label and select *temp*. Click on the vertical axis label and select *icecreams*.

2. Click the G icon, and complete the fields as:
Draw: On Type: Scatter X
List: main\temp
Y List: main\icecream
Press Set.

3. Click the y icon.

8.3.2 Pearson's product-moment correlation coefficient

The strength of a linear association can be observed from a scatterplot of the data, but it is not always easy to determine the strength by eye alone. In order to determine exactly how strong this association is, we can use **Pearson's product-moment correlation coefficient**, r, which measures the strength of a linear trend and associates it with a numerical value between -1 and $+1$.

A value of either -1 or $+1$ indicates a perfect linear correlation, while a result closer to zero indicates no correlation between the variables. The following scale is a guide when using r to describe the strength of a linear relationship.

Type of association	r-value
Perfect positive	$r = 1$
Strong positive	$0.75 \le r < 1$
Moderate positive	$0.50 \le r < 0.75$
Weak positive	$0.25 \le r < 0.5$
No association	$-0.25 < r < 0.25$
Weak negative	$-0.50 < r \le -0.25$
Moderate negative	$-0.75 < r \le -0.50$
Strong negative	$-1 < r \le -0.75$
Perfect negative	$r = -1$

In general, it is rare to use the full name for r. Most commonly it is referred to just as the correlation coefficient.

Pearson's product-moment correlation coefficient formula

$$r = \frac{1}{n-1} \sum_{i=1}^{n} \left(\frac{x_i - \bar{x}}{s_x} \right) \left(\frac{y_i - \bar{y}}{s_y} \right)$$

where

n is the number of pieces of data in the data set

x_i is an x-value (explanatory variable)

y_i is a y-value (response variable)

s_x is the standard deviation of the x-values

s_y is the standard deviation of the y-values

$\bar{x}$ is the mean of the x-values

$\bar{y}$ is the mean of the y-values.

As this formula can be difficult to work with, once the raw data has been gathered we will use CAS to generate Pearson's product-moment correlation coefficient value.

We go into more depth on how to calculate the value of r in subtopic 8.4.

8.3.3 Describing the association displayed in a scatterplot

To describe the association shown in a scatterplot, we must comment on the following three patterns that appear in all scatterplots:
- Direction (and outliers)
- Strength
- Form

If there is no association seen in a scatterplot, then it is not necessary to comment on any of the above.

Direction

The association seen in a scatterplot can have either positive direction or negative direction. An outlier is a point that is located in an isolated space well away from the rest of the data.

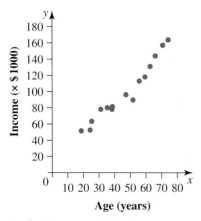

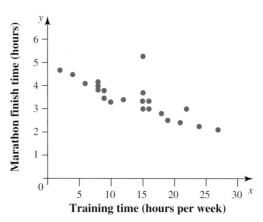

A positive association between age and income can be seen. Specifically, this means that as age increases, so does income. There are no outliers shown in the scatterplot.

A negative association between training time and finish time can be seen. Specifically, this means that as training time increases, finish time decreases. This scatterplot also has an outlier that occurs at $(15, 5.3)$.

Strength

The strength of a relationship will be either strong, moderate, or weak. To determine the strength of the association seen in a scatterplot, we use the value of r, the correlation coefficient, as shown in the table in 8.3.2.

Form

The form of the association seen in a scatterplot will be either linear or non-linear. A linear association can be seen if the points follow along a path that traces out a straight line. The form will be non-linear if the path of the points curves.

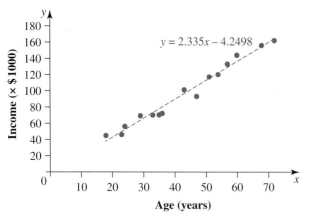

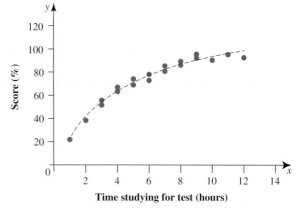

A linear association between age and income can be seen from the straight line that fits through the data.

When comparing time studying against score, studying past a certain point does not yield any significant benefit. That is because test scores can only go up to 100%. Thus, the data levels off and appears to form a curved path. The form of this association is non-linear.

Describing the association of a scatterplot

- **The direction will be: positive or negative.**
- **The strength will be: strong, moderate or weak. This depends on the value of r.**
- **The form will be: linear (straight line) or non-linear (curved line).**
- **There may be an outlier if there is a data point that does not appear to fit the association in the scatterplot.**

WORKED EXAMPLE 3 Describing the association in a scatterplot

Use the scatterplot to describe the association between actual temperature and apparent temperature in terms of strength, direction, and form.

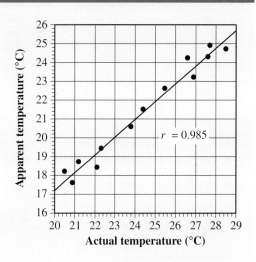

THINK

Given that $r = 0.985$ is positive and above 0.75, we can see that the association is positive and strong.

The data follows along a straight line, telling us the form is linear.

WRITE

The association is strong, positive and linear.

8.3 Exercise

1. **WE1** A survey was conducted to record how long it takes to eat a pizza and the time of day during which it is eaten. Identify the explanatory and response variables.

2. A study recorded the amount of data needed on a phone plan and the time spent using phone apps. Identify the explanatory and response variables.

3. For each of the following scenarios, identify the explanatory and response variables.

   a. The age of people (in years) and the number of star jumps they can complete in one minute
   b. The cost of purchasing various quantities of chocolate
   c. The number of songs stored on a media player and the memory capacity used
   d. The growth rate of bacteria in a laboratory and the quantity of food supplied

4. **WE2** The scatterplot shown has been established.

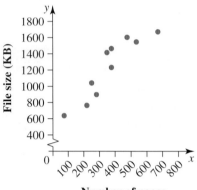

   a. State the response variable.
   b. Describe the relationship between these variables.

5. With reference to the data in the following table:

Time (minutes)	2	4	6	8	10	12	14	16	18	20
Weight that can be held (kg)	55	52	46	33	28	25	19	20	17	12

   a. identify the response and explanatory variables
   b. represent the data in a scatterplot
   c. identify the type of correlation, if any, that is evident from the scatterplot of these two variables.

6. The weights and heights of a random sample of people were collected, with the following table displaying the collected data.

Height (cm)	140	145	150	155	160	165	170	175	180	185
Weight (kg)	58	62	66	70	75	77	78	80	88	90

   a. Identify the explanatory and response variables.
   b. Using a reasonable scale, plot the data.

7. **WE3** Describe any association shown in each scatterplot in terms of strength, direction and form.

a.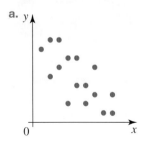

b.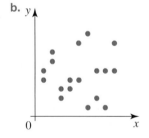

c.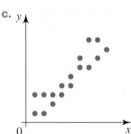

8. Suggest a combination of explanatory variable and response variable that may produce each of the following correlation trends:

   a. negative correlation
   b. no correlation.

9. Describe the association seen in the scatterplot below in terms of strength, direction and form. Identify if there are any outliers.

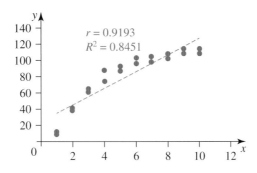

10. Discuss the association seen in the scatterplot below in terms of strength, direction and form. Identify if there are any outliers.

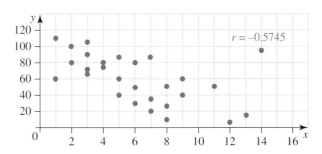

8.3 Exam questions

Question 1 (1 mark)

Source: VCE 2016, Further Mathematics Exam 1, Section A, Q12; © VCAA.

MC There is a strong positive association between a country's Human Development Index and its carbon dioxide emissions.

From this information, it can be concluded that
  A. increasing a country's carbon dioxide emissions will increase its Human Development Index.
  B. decreasing a country's carbon dioxide emissions will increase its Human Development Index.
  C. this association must be a chance occurrence and can be safely ignored.
  D. countries that have higher Human Development Indices tend to have higher levels of carbon dioxide emissions.
  E. countries that have higher Human Development Indices tend to have lower levels of carbon dioxide emissions.

▶ **Question 2 (1 mark)**

Source: VCE 2015, Further Mathematics Exam 2, Q4; © VCAA.

The table below shows male life expectancy (*male*) and female life expectancy (*female*) for a number of countries in 2013. A scatterplot has been constructed from this data.

Life expectancy (in years) in 2013	
Male	*Female*
80	85
60	62
73	80
70	71
70	78
78	83
77	80
65	69
74	77
70	78

The correlation coefficient, *r*, is equal to 0.9496.

Use the scatterplot to describe the association between *male* life expectancy and *female* life expectancy in terms of strength, direction and form.

▶ **Question 3 (1 mark)**

MC Which of the following scatterplots shows weak negative correlation?

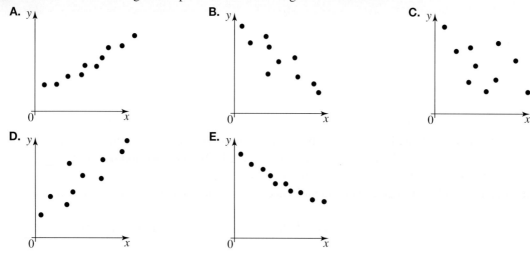

More exam questions are available online.

8.4 Informal interpretation of association and causation

8.4.1 Causation and coefficient of determination

The measure of how much the change in one variable is caused by the other variable is referred to as causation. It is important to note that a strong linear relationship between two variables does not necessarily mean that a change in one variable will cause a change in the other. Other factors often need to be considered. When there is a clear explanatory and response variable, the **coefficient of determination**, r^2, can be calculated to explore the impact a change in one variable may have on the other.

For example, if a data set generated an r-value of 0.9, indicating a very strong linear relationship, the r^2-value would be 0.81. This indicates that 81% of the variation in the *response variable* can be explained by the variation in the *explanatory variable*, and 19% can be explained by other factors.

Interpreting the coefficient of determination

- **The coefficient of determination is found by evaluating r^2.**
- **The coefficient of determination tells us the percentage of variation in the response variable that is explained by variation in the explanatory variable.**

It is also important to understand that while a strong association tells us that a high percentage of the change in the RV is explained by the change in the EV, this does not mean there is causation between the two variables. High correlation does not mean one specific variable definitely causes the response in the other variable, as there may be other factors involved.

An example of this is common responses where two variables appear to be correlated, such as the number of TVs per home and life expectancy in a country. These are both positively correlated to income as the population of a wealthy nation can afford more TVs and better health care.

Confounding is similar but involves interaction between a number of different variables, for example heart disease is affected by diet, exercise, family history and income. All these variables affect the likelihood of heart disease, and it is difficult to examine any one variable individually. Finally, it is always possible to have a high level of correlation purely through coincidence, meaning that while we see a strong association, there is no causation between the two variables at all.

WORKED EXAMPLE 4 Calculating and interpreting the coefficient of determination

Use the data from Worked example 2 to answer the following questions.
a. Use CAS to calculate Pearson's product-moment correlation coefficient correct to 4 decimal places.
b. Determine the coefficient of determination for this situation correct to 4 decimal places. Write your interpretation that could be drawn from this information.

THINK	WRITE
a. Enter the data values into a CAS to generate the r-value.	a. $r = 0.9355$

▶

| TI | THINK | DISPLAY/WRITE | CASIO | THINK | DISPLAY/WRITE |

a. 1. On a Lists & Spreadsheet page, label the first column as *temp* and the second column as *icecreams*. Enter the temperature values in the first column and the corresponding ice-cream sales values in the second column.

a. 1. On a Statistics screen, relabel list 1 as *temp* and list 2 as *icecream*. Enter the temperature values in the first column and the corresponding ice-cream sales values in the second column.

2. On a Calculator page, press MENU then select:
6: Statistics
1: Stat Calculations
3: Linear Regression $(a + bx)$
Complete the fields as:
X List: temp Y List: icecreams, then select OK.

2. Select:
- Calc
- Regression
- Linear Reg
Complete the fields as:
X List: main\temp
Y List: main\icecream, then select OK.

3. Pearson's product-moment correlation coefficient can be read from the screen.

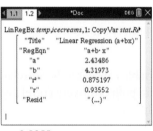

$r = 0.9355$

3. Pearson's product-moment correlation coefficient can be read from the screen.

$r = 0.9355$

b. 1. The coefficient of determination can be read from the screen.

$r^2 = 0.8752$, meaning that 87.52% of the variation in ice-cream sales can be explained by the variation in temperature.

b. 1. The coefficient of determination can be read from the screen.

$r^2 = 0.8752$, meaning that 87.52% of the variation in ice-cream sales can be explained by the variation in temperature.

Assumptions made when calculating the correlation coefficient

When we calculate the correlation coefficient, either using CAS or the formula, we are assuming that:
- the data is numeric
- the data is linear (straight)
- there are no outliers.

If these assumptions are not correct, such as the data being non-linear (curved) instead of straight, then the values of r and r^2 will be inaccurate and will not properly represent the data. This means that any conclusions made about strength or variation will be invalid. This will require other techniques to be applied to the data to analyse the association of the scatterplot (these techniques are not covered in this chapter).

8.4 Exercise

1. **WE4** The following table outlines the cost of an annual magazine subscription along with the number of magazine issues per year.

No. of issues per year	7	9	10	6	8	4	4	5	11	9	10	5	11	3	7	12	7	6	12
Subscription cost ($)	34	40	52	38	50	25	28	40	55	55	45	28	65	24	38	55	50	33	59

   a. Use CAS to determine Pearson's product-moment correlation coefficient for this data correct to 4 decimal places. Describe what this tells you about the strength of the linear relationship between the variables.
   b. Calculate the coefficient of determination correct to 4 decimal places. State what causations could be drawn from this information. Determine what other factors might contribute to this result.

2. After assessing a series of bivariate data, a coefficient of determination value of 0.52 was calculated.

   a. State what this value tells you about the strength of relationship between the two variables.
   b. Referring to the variables as x and y, state the causation that could be suggested.
   c. Explain why we can't use the coefficient of determination to draw exact conclusions.

3. Use your understanding of Pearson's product-moment correlation coefficient to explain what the following results indicate.

   a. $r = 0.68$ b. $r = -0.97$ c. $r = -0.1$ d. $r = 0.30$

4. Calculate the value of the coefficient of determination in the following scenarios and interpret the meaning behind the result. Give your answers correct to 4 decimal places.

   a. A survey found that the correlation between a child's diet and their health is $r = 0.8923$.
   b. The correlation between global warming and the amount of water in the ocean was found to be $r = 0.9997$.

5. Consider the data table below.

Height (cm)	140	145	150	155	160	165	170	175	180	185
Weight (kg)	58	62	66	70	75	77	78	80	88	90

   Calculate Pearson's product-moment correlation coefficient and the coefficient of determination. Give your answers correct to 3 decimal places.

6. The coefficient of determination for a data set was found to be 0.5781. Calculate the percentage of variation that can be explained by other factors.

7. A series of data looked at the amount of time rugby teams spent warming up before a match and the number of wins. The correlation coefficient is 0.86. Draw your conclusions from this.

8. A survey asked random people for their house number and the combined age of the household members. The following data was collected.

House no.	Total age of household
14	157
65	23
73	77
58	165
130	135
95	110
54	94
122	25
36	68

House no.	Total age of household
101	53
57	64
34	120
120	180
159	32
148	48
22	84
9	69

a. Using the house number as the explanatory variable, plot this data.
b. Comment on the resulting scatterplot.
c. Determine Pearson's product-moment correlation coefficient and the coefficient of determination. Give your answers correct to 4 decimal places.
d. Draw your conclusions from these values.
e. Determine the percentage of variation that could be contributed by other factors.

9. A class of Year 11 students were asked to record the amount of time in hours that they spent on a History assignment and the mark out of 100 that they received for the assignment.

Time spent (hours)	Mark (%)
2	72
0.5	52
1.5	76
2.5	82
0.25	36
2	73
2.5	84
2.5	80
2	74
0.5	48

Time spent (hours)	Mark (%)
0.75	58
1.5	69
1	62
2	78
3	90
3.5	94
1	70
3	92
2.5	88
3	97

a. Identify the explanatory and response variables.
b. Draw a scatterplot to represent this data.
c. Comment on the direction and correlation of the data points.
d. Explain why the data is not perfectly linear.
e. Using the data table, calculate Pearson's product-moment correlation coefficient and the coefficient of determination. Give your answers correct to 3 decimal places.
f. State what these values suggest about the relationship between a student's assignment mark and the time spent on it.

10. Use CAS to design a data set that meets the following criteria:
   • Contains 10–15 data points
   • Produces a negative trend
   • Has a coefficient of determination between 0.25 and 0.75

▶ **Question 1 (1 mark)**

Source: VCE 2018, Further Mathematics, Exam 1, Question 9; © VCAA.

MC The scatterplot below displays the *resting pulse rate*, in beats per minute, and the *time spent exercising*, in hours per week, of 16 students. A least squares line has been fitted to the data.

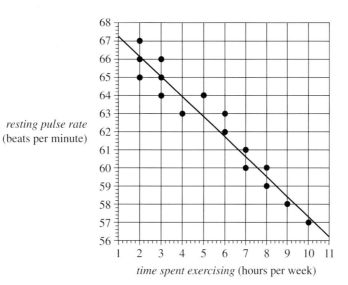

time spent exercising (hours per week)

The coefficient of determination is 0.8339.

The correlation coefficient *r* is closest to

    A. −0.913

    B. −0.834

    C. −0.695

    D. 0.834

    E. 0.913

▶ **Question 2 (1 mark)**

Source: VCE 2017, Further Mathematics Exam 2, Section A, Core, Q3; © VCAA.

The *number of male moths* caught in a trap set in a forest and the *egg density* (eggs per square metre) in the forest are shown in the table below.

Number of male moths	35	37	45	49	65	74	77	86	95
Egg density (eggs per square metre)	471	635	664	997	1350	1100	2010	1640	1350

The correlation coefficient is $r = 0.862$.

Determine the percentage of the variation in *egg density* in the forest explained by the variation in the *number of male moths* caught in the trap.

Round your answer to 1 decimal place.

▶ **Question 3 (1 mark)**

Source: VCE 2017, Further Mathematics Exam 1, Section A, Core, Q12; © VCAA

MC Data collected over a period of 10 years indicated a strong, positive association between the number of stray cats and the number of stray dogs reported each year ($r = 0.87$) in a large, regional city.

A positive association was also found between the population of the city and both the number of stray cats ($r = 0.61$) and the number of stray dogs ($r = 0.72$).

During the time that the data was collected, the population of the city grew from 34 564 to 51 055.

From this information, we can conclude that:
- **A.** if cat owners paid more attention to keeping dogs off their property, the number of stray cats reported would decrease.
- **B.** the association between the number of stray cats and stray dogs reported cannot be causal because only a correlation of +1 or –1 shows causal relationships.
- **C.** there is no logical explanation for the association between the number of stray cats and stray dogs reported in the city, so it must be a chance occurrence.
- **D.** because larger populations tend to have both a larger number of stray cats and stray dogs, the association between the number of stray cats and the number of stray dogs can be explained by a common response to a third variable, which is the increasing population size of the city.
- **E.** more stray cats were reported because people are no longer as careful about keeping their cats properly contained on their property as they were in the past.

More exam questions are available online.

8.5 The line of good fit and predictions

LEARNING INTENTION

At the end of this subtopic you should be able to:
- apply the equation of a line of good fit by eye to the data
- use the model to make predictions, including interpolation and extrapolation
- interpret the regression line equation.

8.5.1 Lines of good fit by eye

In order to model the association between two numeric variables shown in a scatterplot, it is helpful to add a line that describes the behaviour of the points. This is called a **regression line** or regression equation, and can be used to make predictions. There are many different methods to determine a regression equation and we explore a few of them in this topic.

Sometimes the data for a practical problem may not be in the form of a perfect linear relationship, but the data can still be modelled by an approximate linear relationship.

When we are given a scatterplot representing data that appears to be approximately represented by a linear relationship, we can draw a **line of good (best) fit** by eye so that approximately half of the data points are on either side of the line of good fit.

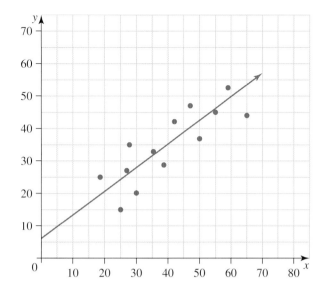

After drawing a line of good fit, the equation of the line can be determined by picking two points on the line and determining the equation, as demonstrated in Worked example 5.

Creating a line of good fit when given only two points

In some instances, we may be given only two points of data in a data set. For example, we may know how far someone had travelled after 3 and 5 hours of their journey without being given other details. In these instances we can make a line of good fit using these two values to estimate other possible values that might fit into the data set.

Although this method can be useful, it is much less reliable than drawing a line of good fit by eye, as we do not know how typical these two points are of the data set. Also, when we draw a line of good fit through two points of data that are close together in value, we are much more likely to have an inaccurate line for the rest of the data set.

tlvd-4815

WORKED EXAMPLE 5 Determining the equation of the line of good fit using two points

The following table and scatterplot represent the relationship between the test scores in Mathematics and Physics for ten Year 11 students. A line of good fit by eye has been drawn on the scatterplot.

| Test score in Mathematics | 65 | 43 | 72 | 77 | 50 | 37 | 68 | 89 | 61 | 48 |
| Test score in Physics | 58 | 46 | 78 | 83 | 35 | 51 | 61 | 80 | 55 | 62 |

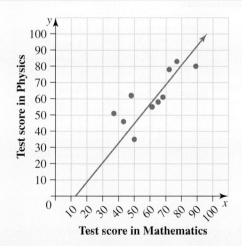

Choose two appropriate points that lie on the line of good fit and determine the equation for the line.

THINK	**WRITE**
1. Look at the scatterplot and pick two points that lie on the line of good fit.	Two points that lie on the line of good fit are $(40, 33)$ and $(80, 81)$.
2. Calculate the value of the gradient between the two points.	Let $(40, 33) = (x_1, y_1)$. Let $(80, 81) = (x_2, y_2)$. $m = \dfrac{y_2 - y_1}{x_2 - x_1}$ $\quad= \dfrac{81 - 33}{80 - 40}$ $\quad= 1.2$
3. Substitute the value of b into the equation $y = a + bx$.	$y = a + 1.2x$
4. Substitute the values of one of the points into the equation and solve for a.	$(40, 33)$ $\quad 33 = a + 1.2 \times 40$ $\quad 33 = a + 48$ $33 - 48 = a$ $\quad\quad a = -15$
5. Substitute the value of a back into the equation and write the answer.	The line of good fit for the data is $y = -15 + 1.2x$.

TI | THINK	**DISPLAY/WRITE**	**CASIO | THINK**	**DISPLAY/WRITE**
1. Select two points that lie on the line of good fit.	Two points are $(40, 33)$ and $(80, 81)$.	1. Select two points that lie on the line of good fit.	Two points are $(40, 33)$ and $(80, 81)$.
2. On a Lists & Spreadsheets page, label the first column x and the second column y. Enter the x-coordinates of the chosen points in the first column, and y-coordinates in the second column.		2. On a Statistics screen, label list1 as x and list2 as y. Enter the x-coordinates of the chosen points in the first column, and y-coordinates in the second column.	
3. Press MENU then select: 6: Statistics 1: Stat Calculations 3: Linear Regression $(a + bx)$… Complete the fields as: X List: x Y List: y then select OK.		3. Select: - Calc - Regression - Linear Reg Complete the fields as: XList: main\x YList: main\y then select OK.	

4.	Interpret the output shown on the screen.	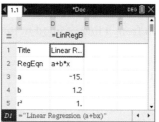 The equation is given in the form $y = a + bx$, where $a = -15$ and $b = 1.2$.	4.	Interpret the output shown on the screen.	 The equation is given in the form $y = a + bx$, where $b = 1.2$ and $a = -15$.
5.	Write the answer.	The line of good fit is $y = -15 + 1.2x$	5.	Write the answer.	The line of good fit is $y = -15 + 1.2x$

Note: If you use CAS, there are shortcuts you can take to find the equation of a straight line given two points. Refer to the CAS instructions available in eBookPLUS.

8.5.2 Making predictions: Interpolation and extrapolation

Interpolation

When we use **interpolation**, we are making a prediction from a line of good fit that appears within the parameters of the original data set.

If we plot our line of good fit on the scatterplot of the given data, then interpolation will occur between the first and last points of the scatterplot.

The following diagram shows the interval for interpolation when using the line of good fit to make a prediction of the test scores in Physics (RV) using a value of the test scores in Mathematics (EV).

For the situation where a value of the RV is used to predict the EV, the region for interpolation and extrapolation would look like the following diagram:

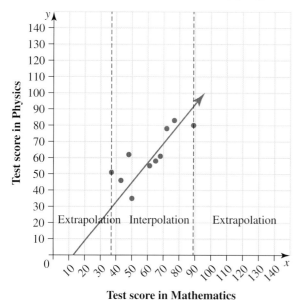

Test score in Mathematics

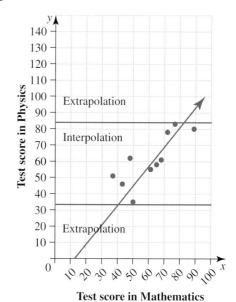

Test score in Mathematics

Extrapolation

When we use **extrapolation**, we are making a prediction from a line of good fit that appears outside the parameters of the original data set.

If we plot our line of good fit on the scatterplot of the given data, then extrapolation will occur before the first point or after the last point of the scatterplot.

tlvd-4816

WORKED EXAMPLE 6 Making predictions of interpolation or extrapolation

Flowers with a diameter of 5–17 cm were measured and the number of petals on each flower was documented. A regression equation of $N = 0.41 + 1.88d$, where N is the number of petals and d is the diameter of the flower (in cm), was established.

a. Identify the explanatory variable.
b. Determine the number of petals that would be expected on a flower with a diameter of 15 cm. Round to the nearest whole number.
c. State whether the value found in part b is an example of interpolated or extrapolated data.
d. A flower with 35 petals is found. Use the equation to predict the diameter of the flower correct to 1 decimal place.
e. State whether part d is an example of interpolated or extrapolated data.

THINK	WRITE
a. Consider the format of the equation. The variable on the right-hand side will be the explanatory variable.	a. Explanatory variable = flower diameter
b. 1. Using the equation, substitute 15 in place of d.	b. $N = 0.41 + 1.88d$ $= 0.41 + 1.88 \times 15$ $= 28.61$
2. Round to the nearest whole value.	29 petals
c. Consider the data range given in the opening statement.	c. 15 cm is inside the data range, so this is interpolation, not extrapolation.
d. 1. Using the equation, substitute 35 in place of N.	d. $35 = 0.41 + 1.88d$
2. Transpose the equation to solve for d.	$d = \dfrac{35 - 0.41}{1.88}$ $= 18.40$
3. Round to 1 decimal place.	$= 18.4$ (correct to 1 decimal place)
e. Consider the data range given in the opening statement.	e. 18.4 cm is outside the data range, so this is an example of extrapolated data.

8.5.3 Reliability of predictions

The more pieces of data there are in a set, the better the line of good fit you will be able to draw. More data points allow more reliable predictions. However, other factors should also be considered. Interpolation closer to the centre of a data set will be more reliable than interpolation closer to the edge. A strong correlation between the points of data will give a more reliable line of good fit. This is shown when all of the points appear close to the line of good fit. The more points that appear far away from the line of good fit, the less reliable other predictions will be.

By comparison, extrapolation cannot be considered reliable as we are assuming that the association shown in the scatterplot continues beyond the data. In the line of good fit shown in 8.5.2, it appears that if the test score in Mathematics falls below 10%, then the test score in Physics would be below 0%, which is not possible. For most sets of data, there will be a point where predictions made through extrapolation end up being nonsensical. Thus, extrapolation is considered to be an unreliable prediction in most cases.

WORKED EXAMPLE 7 Making predictions using the line of good fit

The following data represent the air temperature (°C) and depth of snow (cm) at a popular ski resort.

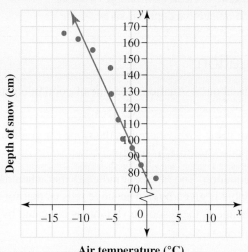

Air temperature (°C)

Air temperature (°C)	−4.5	−2.3	−8.9	−11.0	−13.3	−6.2	−0.4	1.5	−3.7	−5.4
Depth of snow (cm)	111.3	95.8	155.6	162.3	166.0	144.7	84.0	77.2	100.5	129.3

The line of good fit for this data set has been calculated as $y = 84 - 7.2x$.
a. Use the line of good fit to estimate the depth of snow if the air temperature is −6.5 °C.
b. Use the line of good fit to estimate the depth of snow if the air temperature is 25.2 °C.
c. Comment on the reliability of your estimations in parts a and b.

THINK	WRITE
a. 1. Enter the value of x into the equation for the line of good fit.	a $x = -6.5$ $y = 84 - 7.2x$ $= 84 - 7.2 \times -6.5$
2. Evaluate the value of y.	$= 130.8$
3. Write the answer.	The depth of snow if the air temperature is −6.5 °C will be approximately 130.8 cm.
b. 1. Enter the value of x into the equation for the line of good fit.	b. $x = 25.2$ $y = 84 - 7.2x$ $= 84 - 7.2 \times 25.2$
2. Evaluate the value of y.	$= -97.4$ (to 1 decimal point).
3. Write the answer.	The depth of snow if the air temperature is 25.2 °C will be approximately −97.4 cm.

c. Relate the answers back to the original data to check their reliability.

c. The estimate in part **a** was made using interpolation, with the point being comfortably located within the parameters of the original data. The estimate appears to be consistent with the given data and as such is reliable. The estimate in part **b** was made using extrapolation, with the point being located well outside the parameters of the original data. This estimate is clearly unreliable, as there cannot be a negative depth of snow.

8.5.4 Interpreting the regression line equation

Often data is collected to make informed decisions or predictions about a situation. The line of good fit equation from a scatterplot can be used for this purpose.

Remember that the equation for the line of good fit is in the form $y = a + bx$, where b is the slope (or gradient), a is the y-intercept, and x and y refer to the explanatory and response variables respectively.

Two important pieces of information can be attained from this equation. The intercept indicates that when the explanatory variable is equal to 0, the value of the response variable is given by a. The slope tells us that for each increment of 1 unit of change in the explanatory variable, the change in the response variable is indicated by the value of the slope, b.

Interpreting the intercept and slope

- **Interpretation of the intercept (a): On average, when the value of the *explanatory variable* is 0, it is predicted that the value of the *response variable* will be a.**
- **Interpretation of the slope (b): On average, for each *1 unit* increase in the *explanatory variable*, the *response variable* will *increase/decrease by b units*.**

Note: **A negative slope indicates that the response variable will decrease with an increase in the explanatory variable.**

---

tlvd-4817

WORKED EXAMPLE 8 Interpreting the line of good fit

The line of good fit is $y = 62 - 8x$.
a. Identify the y-intercept.
b. For each unit of change in the explanatory variable, determine how much the response variable changes.
c. Explain what your answer to part b tells you about the direction of the line.

THINK	WRITE
a. Consider the equation in the form $y = a + bx$. Identify the value that represents a.	**a.** y-intercept $= 62$
b. The change in the response variable due to the explanatory variable is reflected in the slope. Identify the b value in the equation.	**b.** $b = -8$
c. A positive b value indicates a positive trending line, while a negative b value indicates a negative trending line.	**c.** As the b value is negative, the trend of the line is negative.

1. Steve is looking at data comparing the size of different music venues across the country and the average ticket price at these venues. After plotting the data in a scatterplot, he calculates a line of good fit as $y = 15 + 0.04x$, where y is the average ticket price in dollars and x is the capacity of the venue.

   a. State what the value of the gradient (b) represents in Steve's equation.
   b. State what the value of the y-intercept represents in Steve's equation.
   c. State whether the y-intercept is a realistic value for this data.

2. **WE5** A sports scientist is looking at data comparing the heights of athletes and their performance in the high jump. The following table and scatterplot represent the data they have collected. A line of good fit by eye has been drawn on the scatterplot.

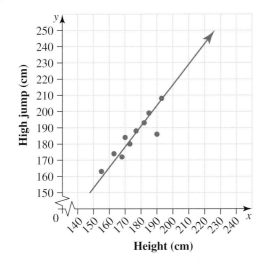

Height (cm)

Height (cm)	168	173	155	182	170	193	177	185	163	190
High jump (cm)	172	180	163	193	184	208	188	199	174	186

   Choose two appropriate points that lie on the line of good fit and determine the equation for the line.

3. Nidya is analysing the data from question **2**, but a clerical error means that she only has access to two points of data: $(170, 184)$ and $(177, 188)$.

   a. Determine Nidya's equation for the line of good fit, rounding all decimal numbers to 2 places.
   b. Add Nidya's line of good fit to the scatterplot of the data.
   c. Comment on the similarities and differences between the two lines of good fit.

4. The following table and scatterplot shows the age and height of a field of sunflowers planted at different times throughout summer.

Age of sunflower (days)	63	71	15	33	80	22	55	47	26	39
Height of sunflower (cm)	237	253	41	101	264	65	218	182	82	140

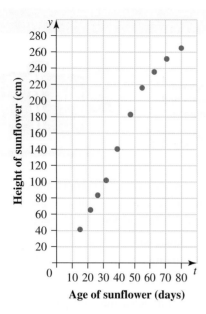

a. Xavier draws a line of good fit by eye that goes through the points $(10, 16)$ and $(70, 280)$. Draw the line of good fit on the scatterplot and comment on his choice of line.

b. Calculate the equation of the line of good fit using the two points that Xavier selected.

c. Patricia draws a line of good fit by eye that goes through the points $(10, 18)$ and $(70, 258)$. Draw the line of good fit on the scatterplot and comment on her choice of line.

d. Calculate the equation of the line of good fit using the two points that Patricia selected.

e. Explain why the value of the y-intercept is not 0 in either equation.

5. Olivia is analysing historical figures for the prices of silver and gold. The price of silver (per ounce) at any given time (x) is compared with the price of gold (per gram) at that time (y). She asks her assistant to note down the points she gives him and to create a line of good fit from the data.

On reviewing her assistant's notes, she has trouble reading the handwriting. The only complete pieces of information she can make out are one of the points of data $(16, 41.5)$ and the gradient of the line of good fit (2.5).

a. Use the gradient and data point to determine the equation of the line of good fit.

b. Use your answer from part a to answer the following questions.

   i. Calculate the price of a gram of gold if the price of silver is $25 per ounce.
   ii. Calculate the price of an ounce of silver if the price of gold is $65 per gram.
   iii. Calculate the price of a gram of gold if the price of silver is $11 per ounce.
   iv. Calculate the price of an ounce of silver if the price of gold is $28 per gram.

6. A student is calculating the equation of a straight line passing through the points $(-2, 5)$ and $(3, 1)$. Their working is shown below.

$$\frac{y_2 - y_1}{x_2 - x_1} = \frac{1 - 5}{-2 - 3} \qquad\qquad y = \frac{4}{5}x + c$$

$$= \frac{-4}{-5} \qquad\qquad\qquad 1 = \frac{4}{5} \times 3 + c$$

$$= \frac{4}{5} \qquad\qquad\qquad c = 1 - \frac{12}{5}$$

$$\qquad\qquad\qquad\qquad\qquad = \frac{-7}{5}$$

$$\qquad\qquad\qquad\qquad y = \frac{4}{5}x - \frac{7}{5}$$

a. Identify the error in the student's working.

b. Calculate the correct equation of the straight line passing through these two points.

7. A government department is analysing the population density and crime rate of different suburbs to see if there is a connection. The following table and scatterplot display the data that has been collected so far.

Population density (persons per km²)	3525	2767	4931	3910	1572	2330	2894	4146	1968	5337
Crime rate (per 1000 people)	185	144	279	227	65	112	150	273	87	335

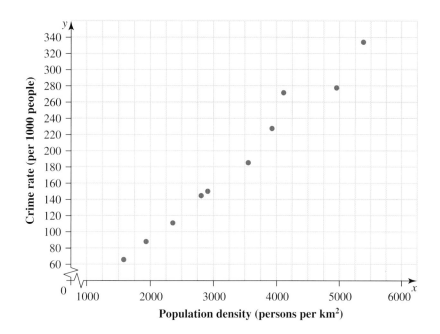

a. Draw a line of good fit on the scatterplot of the data.
b. Choose two points from the line of good fit and determine the equation of the line.
c. State what the value of the x-intercept means in terms of this problem.
d. Explain whether the x-intercept value is realistic. Explain your answer.

8. A marine biologist is studying the lives of sea turtles. They collect data comparing the number of sea turtle egg nests and the number of survivors from those nests.
The following table and scatterplot display the data they have collected.

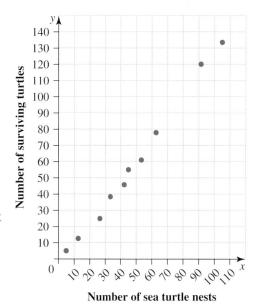

Number of sea turtle nests	45	62	12	91	27	5	53	33	41	105
Number of surviving turtles	55	78	13	120	25	5	61	39	46	133

a. The marine biologist draws a line of good fit for the data that goes through the points $(0, -5)$ and $(100, 127)$. Determine the equation for the line of good fit.
b. State what the gradient of the line represents in terms of the problem.
c. State what the y-intercept represents in terms of the problem. State whether this value is realistic.

d. Use the equation to answer the following questions.

    i. Estimate how many turtles you would expect to survive from 135 nests.

    ii. Estimate how many nests would need to be used to have 12 surviving turtles.

e. Comment on the reliability of your answers to part **d**.

9. A straight line passes through the points $(-2, 2)$ and $(-2, 6)$.

a. State the gradient of the line.

b. Determine the equation of this line.

10. **WE7** The owner of an ice-cream parlour has collected data relating the outside temperature to ice-cream sales.

Outside temperature (°C)	23.4	27.5	26.0	31.1	33.8	22.0	19.7	24.6	25.5	29.3
Ice-cream sales	135	170	165	212	204	124	86	144	151	188

A line of good fit for this data has been calculated as $y = -77 + 9x$.

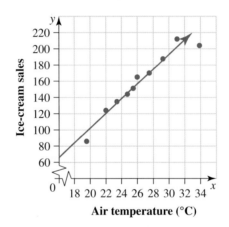

Air temperature (°C)

a. Use the line of good fit to estimate ice-cream sales if the air temperature is 27.9 °C.

b. Use the line of good fit to estimate ice-cream sales if the air temperature is 15.2 °C.

c. Comment on the reliability of your answers to parts **a** and **b**.

11. Georgio is comparing the cost and distance of various long-distance flights, and after drawing a scatterplot, he creates an equation for a line of good fit to represent his data. Georgio's line of good fit is $y = 55 + 0.08x$, where y is the cost of the flight and x is the distance of the flight in kilometres.

a. Estimate the cost of a flight between Melbourne and Sydney (713 km) using Georgio's equation.

b. Estimate the cost of a flight between Melbourne and Broome (3121 km) using Georgio's equation.

c. All of Georgio's data came from flights of distances between 400 km and 2000 km. Comment on the suitability of using Georgio's equation for shorter and longer flights than those he analysed.
Suggest other factors that might affect the cost of these flights.

12. Mariana is a scientist and is collecting data measuring lung capacity (in L) and time taken to swim 25 metres (in seconds). Unfortunately a spillage in her lab causes all of her data to be erased apart from the records of a person with a lung capacity of 3.5 L completing the 25 metres in 55.8 seconds and a person with a lung capacity of 4.8 L completing the 25 metres in 33.3 seconds.

   a. Use the remaining data to construct an equation for the line of good fit relating lung capacity (x) to the time taken to swim 25 metres (y).
   Give any numerical values correct to 2 decimal places.
   b. State what the value of the gradient represents in the equation.
   c. Use the equation to estimate the time it would take people with the following lung capacities to swim 25 metres.

      i. 3.2 litres
      ii. 4.4 litres
      iii. 5.3 litres

   d. Comment on the reliability of creating the equation from Mariana's two remaining data points.

13. An AFL coach is analysing data comparing the kicking efficiency (x) with the handball efficiency (y) of different AFL players. The data is shown in the following table.

Kicking efficiency (%)	75.3	65.6	83.1	73.9	79.0	84.7	64.4	72.4	68.7	80.2
Handball efficiency (%)	84.6	79.8	88.5	85.2	87.1	86.7	78.0	81.3	82.4	90.3

   a. A line of good fit for the data goes through the points $(66, 79)$ and $(84, 90)$. Determine the equation for the line of good fit for this data.
   Give any numerical values correct to 2 decimal places.
   b. Use the equation from part a and the figures for kicking efficiency to create a table for the predicted handball efficiency of the same group of players.

Kicking efficiency (%)	75.3	65.6	83.1	73.9	79.0	84.7	64.4	72.4	68.7	80.2
Predicted handball efficiency (%)										

   c. Comment on the differences between the predicted handball efficiency and the actual handball efficiency.

14. Chenille is comparing the price of new television sets versus their aggregated review scores (out of 100). The following table and scatterplot display the data she has collected.

RRP of television set ($)	799	1150	2399	480	640	999	1450	1710	3500	1075
Aggregated review score (%)	75	86	93	77	69	81	88	90	96	84

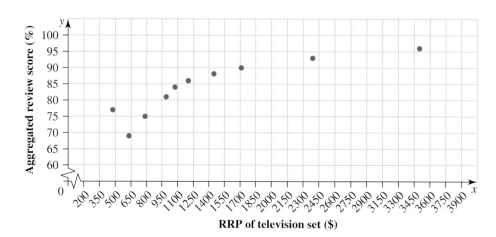

a. Using CAS, a spreadsheet or otherwise, calculate the equations of the straight lines that pass through each of the following pairs of points.

  i. $(400, 80)$ and $(3400, 97)$
  ii. $(300, 78)$ and $(3200, 95)$
  iii. $(400, 75)$ and $(2400, 95)$
  iv. $(430, 67)$ and $(1850, 95)$

b. Draw these lines on the scatterplot of the data.
c. Explain which line you think is the most appropriate line of good fit for the data.

15. Karyn is investigating whether the salary of the leading actors/actresses in movies has any impact on the box office receipts of the movie.
The following table and scatterplot display the data that Karyn has collected.

Actor/actress salary ($m)	0.2	3.5	8.0	2.5	10.0	6.5	1.6	14.0	4.7	7.5
Box office receipts ($m)	135	72	259	36	383	154	25	330	98	232

a. Explain why a line of good fit for the data would never go through the point of data $(0.2, 135)$.
b. Draw a line of good fit on the scatterplot.
c. Use two points from your line of good fit to determine the equation for this line.
d. Explain what the value of the gradient means in the context of this problem.
e. Calculate the expected box office receipts for films where the leading actor/actress is paid the following amounts.

  i. $1.2 million
  ii. $11 million
  iii. $50 000
  iv. $20 million

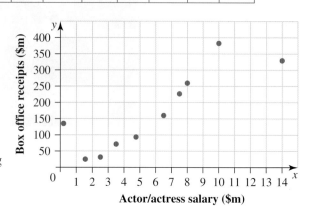

f. Comment on the reliability of your answers to part **e**.

16. Giant sequoias are the world's largest trees, growing up to 100 or more metres in height and 10 or more metres in diameter. Throughout their lifetime they continue to grow in size, with the largest of them among the fastest growing organisms that we know of.

Sheila is examining the estimated age and diameter of giant sequoias. The following table and scatterplot show the data she has collected.

Estimated age (years)	450	1120	330	1750	200	1400	630	800	980	2050
Diameter (cm)	345	485	305	560	240	525	390	430	465	590

a. The line of good fit shown on the scatterplot passes through the points $(0, 267)$ and $(2000, 627)$. Determine the equation for the line of good fit.

b. Using a spreadsheet, CAS or otherwise, calculate the average (mean) age of the trees in Sheila's data set.

c. Using a spreadsheet, CAS or otherwise, calculate the average (mean) diameter of the trees in Sheila's data set.

d. Subtract the y-intercept from your equation in part a from the average age calculated in part b, and divide this total by the average diameter calculated in part c.

e. State how the answer in part d compares to the gradient of the equation calculated in part a.

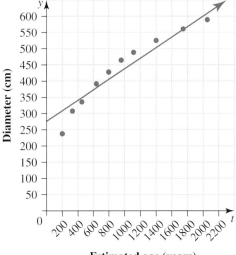

17. **WE8** The least squares regression equation for a line is $y = -1.837 + 1.701x$.

a. Identify the y-intercept.

b. For each unit of change in the explanatory variable, determine how much the response variable changes.

c. State what your answer to part b tells you about the direction of the line.

18. The least squares regression equation for a line is $y = 105.90 - 1.476x$.

a. Identify the y-intercept.

b. For each unit of change in the explanatory variable, determine how much the response variable changes.

c. State what your answer to part b tells you about the direction of the line.

19. Answer the following questions for the equation $y = 60 - 5x$.

a. Identify the y-intercept.

b. For each unit of change in the explanatory variable, determine how much the response variable changes.

c. State whether the trend of the data is positive or negative.

d. Calculate the value of y when $x = 40$.

20. **WE6** A brand of medication for babies bases the dosage on the age (in months) of the child. The regression equation for this situation is $M = 0.157 + 0.312A$, where M is the amount of medication in mL and A is the age in months.

a. Identify the explanatory variable.

b. Calculate the amount of medication required for a child aged 6 months.

c. Determine the age of a child who requires 2.5 mL of the medication. Give your answer correct to 1 decimal place.

21. A survey of the nightly room rate for Sydney hotels and their proximity to the Sydney Harbour Bridge produced the regression equation $C = 281.92 - 50.471d$, where C is the cost of a room per night in dollars and d is the distance to the bridge in kilometres.

   a. Identify the response variable.

   b. Based on this equation, calculate the cost of a hotel room 2.5 km from the bridge. Give your answer correct to the nearest cent.

   c. Determine the distance of a hotel room from the bridge if the cost of the room was $115. Give your answer correct to 2 decimal places.

22. Answer the following questions for the equation $I = 0.43 + 1.1s$, where I is the number of insects caught and s is the area of a spider's web in cm^2.

   a. Identify the response variable.

   b. For each unit of change in the explanatory variable, determine how much the response variable changes.

   c. State whether the trend of the data is positive or negative.

   d. Determine how many insects are likely to be caught if the area of the spider's web is 60 cm^2. Give your answer correct to the nearest whole number.

8.5 Exam questions

Question 1 (1 mark)

Source: VCE 2021, Further Mathematics Exam 1, Section A Core, Q11; © VCAA.

MC The table below shows the *weight*, in kilograms, and the *height*, in centimetres, of 10 adults.

Weight (kg)	Height (cm)
59	173
67	180
69	184
84	195
64	173
74	180
76	192
56	169
58	164
66	180

A least squares line is fitted to the data.

The least squares line enables an adult's *weight* to be predicted from their *height*.

The number of times that the predicted value of an adult's *weight* is greater than the actual value of their *weight* is

   A. 3

   B. 4

   C. 5

   D. 6

   E. 7

Question 2 (3 marks)

Source: VCE 2020, Further Mathematics Exam 2, Section A, Q5; © VCAA.

The scatterplot below shows *body density*, in kilograms per litre, plotted against *waist measurement*, in centimetres, for 250 men.

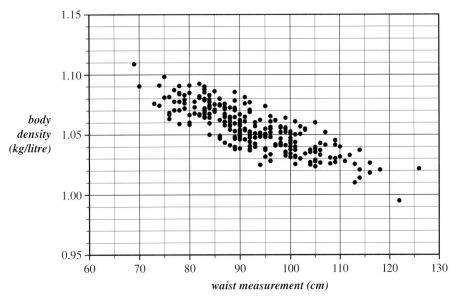

Data: RW Johnson, 'Fitting percentage of body fat to simple body measurements', *Journal of statistics education*, 4:1, 1996, https://doi.org/10.1080/10691898.1996.11910505

When a least squares line is fitted to the scatterplot, the equation of this line is:

$$body\ density = 1.195 - 0.001512 \times waist\ measurement$$

a. Use the equation of this least squares line to predict the body density of a man whose waist measurement is 65 cm. **(1 mark)**
Round your answer to 2 decimal places.

b. When using the equation of this least squares line to make the prediction in **part a.**, state whether you are extrapolating or interpolating. **(1 mark)**

c. Interpret the slope of this least squares line in terms of a man's *body density* and *waist measurement*. **(1 mark)**

Question 3 (3 marks)

Source: VCE 2019, Further Mathematics, Exam 2, Section A, Q5; © VCAA.

The scatterplot below shows the atmospheric pressure, in hectopascals (hPa), at 3 pm (*pressure 3 pm)* plotted against the atmospheric pressure, in hectopascals, at 9 am (*pressure 9 am*) for 23 days in November 2017 at a particular weather station.

A least squares line has been fitted to the scatterplot as shown.

The equation of this line is

$$pressure\ 3\ pm = 111.4 + 0.8894 \times pressure\ 9\ am$$

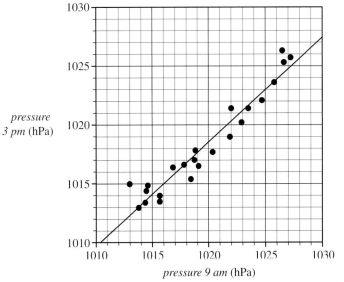

Data: Australian Government, Bureau of Meteorology,

a. Interpret the slope of this least squares line in terms of the atmospheric pressure at this weather station at 9 am and at 3 pm. **(1 mark)**

b. Use the equation of the least squares line to predict the atmospheric pressure at 3 pm when the atmospheric pressure at 9 am is 1025 hPa. **(1 mark)**

Round your answer to the nearest whole number.

c. Is the prediction made in **part b.** an example of extrapolation or interpolation? **(1 mark)**

More exam questions are available online.

8.6 Introduction to the least squares line of best fit (extending)

LEARNING INTENTION

At the end of this subtopic you should be able to:
- determine the equation of the least squares line.

8.6.1 Least squares regression

Sometimes, in order to make a quick prediction, a line of good fit can be drawn by eye; however, in most other situations it is necessary to be more accurate. When there are no outliers in a scatterplot, we can generate an equation using the **least squares regression line**.

This line minimises the vertical distances between the data points and the line of good fit. It is called the least squares regression line because if we took the squares of these vertical distances, this line would represent the smallest possible sum of all of these squares.

We have already used this method to determine the equation of the line between two points in Worked example 5. The only difference is that rather than using two points, we enter in all data points provided in the data.

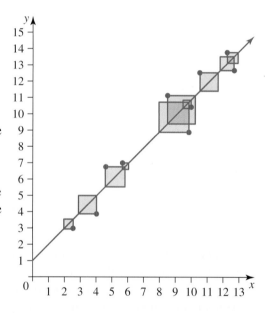

Least squares regression line

The equation for the least squares regression line takes the form $y = a + bx$

where

y is the response variable

x is the explanatory variable

b is the gradient or slope of the line

a is the y-intercept.

The following data give the height (in cm) and weight (in kg) of 12 students in a class.

Height (cm)	151	172	183	166	156	167	159	188	149	161	153	165
Weight (kg)	48	78	82	63	54	68	61	85	46	62	58	76

a. Calculate the least squares regression equation. Give all values to 3 significant figures.
b. Use the answer to a to predict the weight of a student who is 190 cm tall.
c. Use the answer to a to predict the height of a student who weighs 60 kg.
d. State whether the answers to b and c are reliable.

THINK

a. In this case we will assume that height is the EV and weight is the RV.
Enter the data into CAS and generate the equation of the least squares line. Write the equation using the variables *height* and *weight*.

b. Using the equation, substitute 190 for the height.

c. Using the equation, substitute 60 for the weight.

d. State which prediction was made using values within the data.
A height of 190 cm is above any value in the table, so it was extrapolation.
A weight of 60 kg falls between the highest and lowest weights in the table.

WRITE

a. The slope is 0.997.
The intercept is −98.6.
The least squares equation is:
$weight = -98.6 + 0.997 \times height$

b. $weight = -98.6 + 0.997 \times 190$
$= 90.83 \, kg$

c. $60 = -98.6 + 0.997 \times height$
$height = 159 \, cm$

d. The prediction in **b** was extrapolation and thus not reliable.
The prediction in **c** was interpolation and thus reliable.

TI | THINK   **DISPLAY/WRITE**

a. 1. On a Lists & Spreadsheet page, label the first column as *height* and the second column as *weight*. Enter the height values in the first column and the corresponding weight values in the second column.

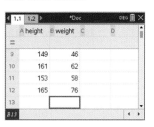

2. On a Calculator page, press MENU, then select:
6: Statistics
1: Stat Calculations
4: Linear Regression (a+bx)
Complete the fields as:
X List: height
Y List: weight, then select OK.

CASIO | THINK   **DISPLAY/WRITE**

a. 1. On a Statistics screen, relabel list1 as *height* and list2 as *weight*. Enter the height values in the first column and the corresponding weight values in the second column.

2. Select:
• Calc
• Regression
• Linear Reg
Complete the fields as:
XList: main\height
YList: main\weight
then select OK.

3. The value of a and b can be read from the screen, allowing for the equation of the least squares regression line to be formed.

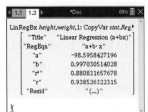

$a = -98.6$ and $b = 0.997$
$weight = -98.6 + 0.997h$

3. The value of a and b can be read from the screen, allowing for the equation of the least squares regression line to be formed.

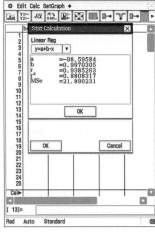

$a = -98.6$ and $b = 0.997$
$weight = -98.6 + 0.997h$

8.6 Exercise

1. **WE9** A researcher investigating the proposition that tall mothers have tall sons measures the heights of 12 mothers and their adult sons. The results are shown below.

Height of mother (cm)	Height of son (cm)
185	188
155	157
171	172
169	173
170	174
175	180

Height of mother (cm)	Height of son (cm)
158	159
156	150
168	172
169	175
179	180
173	190

a. State which variable is the response variable.
b. Draw a scatterplot and a line of good fit.
c. Determine the equation of the line of good fit, expressing the equation in terms of height of mother (M) and height of son (S). Give values correct to 4 significant figures.

2. Data on the daily sales of gumboots and the maximum daily temperature were collected.

Temp (°C)	Daily sales (no. of pairs)
17	2
16	3
12	8
10	16
14	7
17	3
18	2
22	1

Temp (°C)	Daily sales (no. of pairs)
23	1
19	2
17	3
15	3
12	12
15	9
20	1

a. Draw a scatterplot of this data.
b. Determine the equation of the line of good fit, expressed in terms of temperature (T) and daily sales (D). Give values correct to 4 significant figures.
c. Determine Pearson's product-moment correlation coefficient and the coefficient of determination. Give your answers correct to 4 significant figures.
d. Interpret these values in the context of the data.

3. a. Use the data given to draw a scatterplot and a line of good fit by eye.

x	1	2	3	4	5	6	7	8	9	10
y	35.3	35.9	35.7	36.2	37.3	38.6	38.4	39.1	40.0	41.1

b. Determine the equation of the line of good fit and use it to predict the value of y when $x = 15$. Give your answers correct to 4 significant figures.

4. Use the data given below to complete the following questions.

x	1	2	3	4	5	6	7	8	9	10
y	4	1	2	3	5	5	3	6	8	7

a. Draw a scatterplot and a line of good fit by eye.
b. Determine the equation of the line of good fit. Give values correct to 2 significant figures.
c. Predict the value of y when $x = 20$.
d. Predict the value of x when $y = 9$. Give your answers correct to 2 decimal places.

5. While camping a mathematician estimated that:

$$\text{number of mosquitos around fire} = 10.2 + 0.5 \times \text{temperature of the fire (°C)}$$

a. Determine the number of mosquitoes that would be expected if the temperature of the fire was 240 °C. Give your answer correct to the nearest whole number.
b. Determine the temperature of the fire if there were only 12 mosquitoes in the area.
c. Identify some factors that could affect the reliability of this equation.

6. The following scatterplot has the line of good fit equation of $c = 13.33 + 2.097m$, where c is the number of new customers each hour and m is the number of market stalls.

   a. Using the line of good fit, interpolate the data to find the number of new customers expected if there are 30 market stalls.

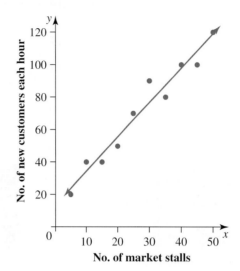

No. of market stalls

   b. Use the formula to extrapolate the number of market stalls required in order to expect 150 new customers.
   c. Explain why part **a** is an example of interpolation, while part **b** demonstrates extrapolation.

7. Use the data from the table to complete the following questions.

x	10	11	12	13	14	15	16	17	18	19
y	22	18	20	15	17	11	11	7	9	8

   a. Draw a scatterplot and a line of good fit by eye.
   b. Determine the equation of the line of good fit. Give values correct to 4 significant figures.
   c. Extrapolate the data to predict the value of y when $x = 23$.
   d. Determine the assumptions made when extrapolating data.

8. Data on people's average monthly income and the amount of money they spend at restaurants was collected.

Average monthly income ($\times \$1000$)	Money spent at restaurants per month (\$)	Average monthly income ($\times \$1000$)	Money spent at restaurants per month (\$)
2.8	150	4.1	600
2.5	130	3.5	360
3.0	220	2.9	175
3.1	245	3.6	350
2.2	100	2.7	185
4.0	400	4.2	620
3.7	380	3.6	395
3.8	200		

a. Draw a scatterplot of this data on your calculator.
b. Determine the equation of the line of good fit in terms of average monthly income in thousands of dollars (I) and money spent at restaurants in dollars (R). Give values correct to 4 significant figures.
c. Extrapolate the data to predict how much a person who earns \$5000 a month might spend at restaurants each month.
d. Explain why part c is an example of extrapolation.
e. A person spent \$265 eating out last month. Estimate their monthly income, giving your answer to the nearest \$10. State whether this an example of interpolation or extrapolation.

9. Data on students' marks in Geography and Music were collected.

Geography	Music
65	91
80	57
72	77
61	89
99	51
54	76
39	62
66	87

Geography	Music
78	88
89	64
84	90
73	45
68	60
57	79
60	69

a. State whether there is an obvious explanatory variable in this situation.
b. Draw a scatterplot of this data on your calculator, using the marks in Geography as the explanatory variable.
c. Determine the equation of the line of good fit. Give values correct to 4 significant figures.
d. Based on your equation, if a student received a mark of 85 for Geography, predict the mark (to the nearest whole number) they would receive for Music.
e. State how confident you feel about making predictions for this data. Explain your response.
f. Calculate Pearson's product-moment correlation coefficient for this data. Explain how you can use this value to evaluate the reliability of your data.

10. For three months, an athlete has been wearing an exercise-tracking wristband that records the distance they walk and the number of calories they burn. The graph shows their weekly totals. The regression-line equation for this data is $y = 14\,301 + 115.02x$.

a. Identify the response variable in this situation.
b. Rewrite the equation in terms of the explanatory and response variables.
c. Using the equation for the regression line, determine the number of calories burned if the athlete walked 50 km in a week. State whether this is an example of interpolation or extrapolation. Explain your response.
d. Due to an injury, in one week the athlete only walked 10 km. Use the data to determine the number of calories this distance would burn. State whether this is an example of interpolation or extrapolation. Explain your response.

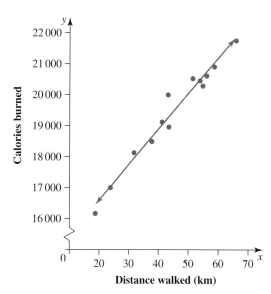

e. Pearson's product-moment correlation coefficient for this data is 0.9678. Explain how you can use this value to evaluate the reliability of the data.

f. List at least two other factors that could influence this data set.

8.6 Exam questions

Question 1 (1 mark)

Source: VCE 2019, Further Mathematics Exam 1, Section A, Q13; © VCAA.

MC The time, in minutes, that Liv ran each day was recorded for nine days.

These times are shown in the table below.

Day number	1	2	3	4	5	6	7	8	9
Time (minutes)	22	40	28	51	19	60	33	37	46

The time series plot below was generated from this data.

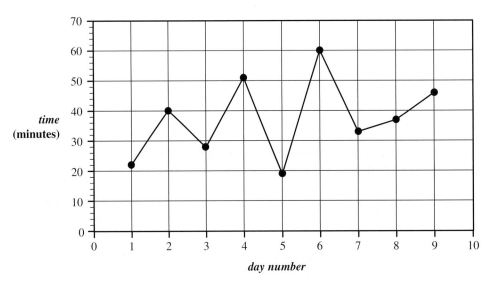

A least squares line is to be fitted to the time series plot shown.

The equation of this least squares line, with day number as the explanatory variable, is closest to

A. *day number* $= 23.8 + 2.29 \times time$

B. *day number* $= 28.5 + 1.77 \times t$

C. *time* $= 23.8 + 1.77 \times day\ number$

D. *time* $= 23.8 + 2.29 \times day\ number$

E. *time* $= 28.5 + 1.77 \times day\ number$

Question 2 (1 mark)

Source: VCE 2016, Further Mathematics Exam 1, Section A, Q9; © VCAA.

MC The scatterplot below shows life expectancy in years (*life expectancy*) plotted against the Human Development Index (*HDI*) for a large number of countries in 2011.

A least squares line has been fitted to the data and the resulting residual plot is shown.

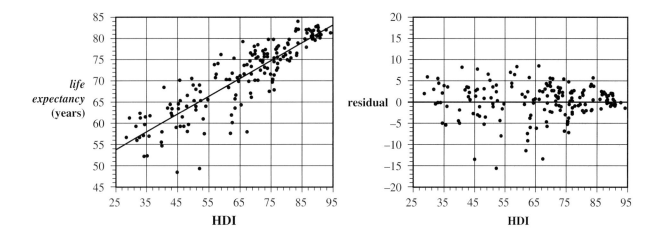

The equation of this least squares line is

$$life\ expectancy = 43.0 + 0.422\ HDI$$

The coefficient of determination is $r^2 = 0.875$

Given the information above, determine which one of the following statements is **not** true.
- **A.** The value of the correlation coefficient is close to 0.94
- **B.** 12.5% of the variation in life expectancy is not explained by the variation in the Human Development Index.
- **C.** On average, life expectancy increases by 43.0 years for each 10-point increase in the Human Development Index.
- **D.** Ignoring any outliers, the association between life expectancy and the Human Development Index can be described as strong, positive and linear.
- **E.** Using the least squares line to predict the life expectancy in a country with a Human Development Index of 75 is an example of interpolation.

Question 3 (3 marks)

Source: VCE 2016, Further Mathematics Exam 2, Core, Q3; © VCAA.

The data in the table below shows a sample of actual temperatures and apparent temperatures recorded at a weather station. A scatterplot of the data is also shown.

The data will be used to investigate the association between the variables *apparent temperature* and *actual temperature*.

Apparent temperature (°C)	Actual temperature (°C)
24.7	28.5
24.3	27.6
24.9	27.7
23.2	26.9
24.2	26.6
22.6	25.5
21.5	24.4
20.6	23.8
19.4	22.3
18.4	22.1
17.6	20.9
18.7	21.2
18.2	20.5

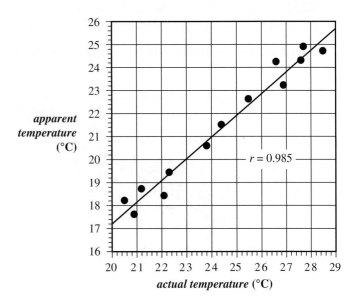

a. i. Determine the equation of the least squares line that can be used to predict the *apparent temperature* from the *actual temperature*.

Round your answers to 2 significant figures. **(1 mark)**

apparent temperature = ☐ + ☐ × actual temperature

ii. Interpret the intercept of the least squares line in terms of the variables *apparent temperature* and *actual temperature*. **(1 mark)**

b. The coefficient of determination for the association between the variables *apparent temperature* and *actual temperature* is 0.97

Interpret the coefficient of determination in terms of these variables. **(1 mark)**

More exam questions are available online.

8.7 Review

8.7.1 Summary

doc-37618

Hey students! Now that it's time to revise this topic, go online to:

Access the topic summary	**Review your** results	**Watch teacher-led** videos	**Practise exam** questions

Find all this and MORE in jacPLUS

8.7 Exercise

Multiple choice

1. a. **MC** The regression-line equation for the following graph is closest to:

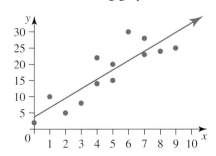

A. $y = 3.8 + 2.9x$ **B.** $y = -3.8 - 2.9x$ **C.** $y = -3.8 + 2.9x$

D. $y = 3.8 - 2.9x$ **E.** $y = 2.9 + 3.8x$

b. **MC** The type of correlation shown in the graph in part **a**, would best be described as:

  A. weak, positive correlation.

  B. moderate, positive correlation.

  C. strong, positive correlation.

  D. no correlation.

  E. moderate, negative correlation.

2. **MC** Select the type of correlation indicated by an r-value of 0.64.

  A. Strong, positive correlation

  B. Strong, negative correlation

  C. Moderate, positive correlation

  D. Moderate, negative correlation

  E. Weak, positive correlation

3. **MC** A gardener tracks a correlation coefficient of 0.79 between the growth rate of his trees and the amount of fertiliser used. Select what the gardener can conclude from this result.

  A. An increase in tree growth increases the use of fertiliser.

  B. An increase in the use of fertiliser increases the health of the trees.

  C. The growth rate of the trees is influenced by the amount of fertiliser used.

  D. The growth rate of the trees influences the quality of the fertiliser used.

  E. There is no correlation between the growth rate of the trees and the amount of fertiliser used.

4. **MC** Select the scatterplot that best demonstrates a line of good fit.

A.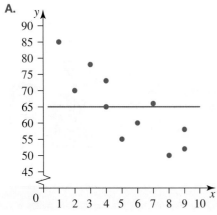

B.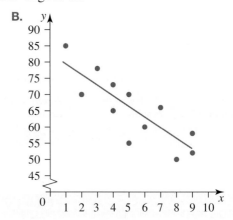

C.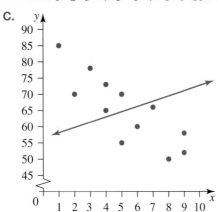

D.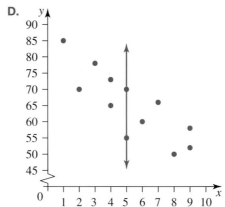

E.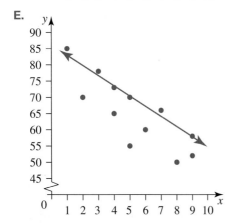

5. **MC** When $y = 0.54 + 15.87x$ and $x = 2.5$, the value of y is:

A. 18.91   B. 40.215   C. 39.135   D. 6.888   E. 34.019

6. **MC** For the sample data set shown, select which of the following is an example of interpolating data.

x	1	5	15	25
y	10	16	18	22

A. Calculating the value of x when $y = -7$
B. Calculating the value of y when $x = 17$
C. Calculating the value of x when $y = 27$
D. Calculating the value of y when $x = 37$
E. Calculating the value of x when $y = 5$

7. **MC** The graph for the line of best fit equation $y = 85 - 4x$ is most likely to be:

A.

B.

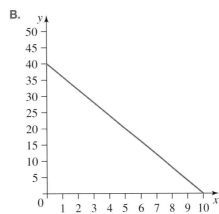

C.

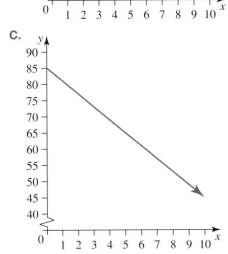

D.

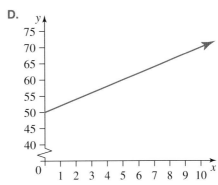

E.

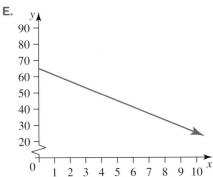

8. **MC** A series of data points recorded a coefficient of determination value of 0.82. Calculate Pearson's product-moment coefficient.

A. 82%   B. 0.18   C. 0.67   D. 0.91   E. 18%

9. **MC** For the data set from question **6**, the line of best fit equation is:

A. $y = 10 + x$
B. $y = 0.435 + 11.456x$
C. $y = 0.876 + 0.936x$
D. $y = 11.456 + 0.435x$
E. $y = 0.936 + 0.876x$

10. **i.** For each of the following graphs, describe the strength of correlation between the explanatory and response variables.

a.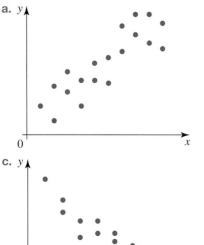

b.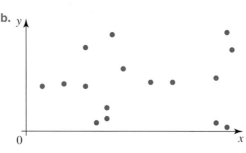

c.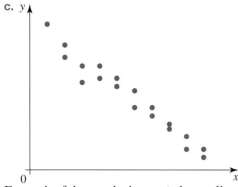

ii. For each of the graphs in part **i**, draw a line of good fit where possible.

11. Identify the explanatory and response variable for each of the following scenarios.

   a. In a junior Science class, students plot the time taken to boil various quantities of water.
   b. Extra buses are ordered to transport a number of students to the school athletics carnival.

12. Use the following data to complete this question.

x	10	9	8	7	6	5	4	3	2	1
y	6	10	4	11	13	18	15	19	21	26

   a. Plot the data on a scatterplot.
   b. Comment on the direction and strength of the data.
   c. Determine Pearson's product-moment correlation coefficient and the coefficient of determination for the data.
   d. Use you answer from part **c** to further discuss the relationship between the variables.

13. Pearson's product-moment correlation coefficient for a scatterplot was found to be -0.7564.

   a. Determine the value of the coefficient of determination.
   b. State what these values indicate to you about the strength of the relationship between the two variables.

14. During an interview investigating the link between the sales of healthy snack foods (functional foods) and the increasing consumer demand for these products, an advertising expert made the following comment:

   'There is a correlation but it's not causation … our increasing need for healthy food and our laziness has resulted in mass innovation of functional foods.'

   Explain why they might have stated that there is no causative link between the sales of healthy foods and laziness.

Extended response

15. Data on 15 people's shoe size and the length of their hair was collected.

Shoe size	Length of hair (cm)
6	9
8	14
7	12
8	1
9	7
6	8
7	5
12	22
8	15
9	8
10	18
12	4
7	5
9	9
11	3

a. Draw a scatterplot of this data (length of hair is the response variable).
b. Determine the equation of the line of good fit.
c. Determine Pearson's product-moment correlation coefficient and the coefficient of determination for the data.
d. State your conclusions from this data.

16. An independent agency test-drove a random sample of current model vehicles and measured their fuel tank capacity against the average fuel consumption. Along with the following scatterplot, a regression equation of $y = 0.6968 + 0.1119x$ was established.

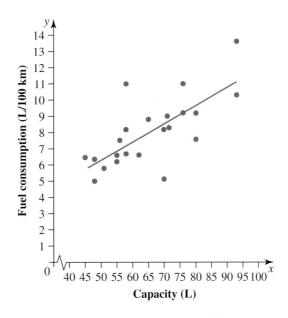

a. Identify the response variable in this situation.
b. Rewrite the equation in terms of the explanatory and response variables.

c. It is often said that smaller vehicles are more economical. Determine, correct to 2 decimal places, the fuel consumption of a vehicle that had a 40-litre fuel tank.

d. State whether your answer to part c is an example of interpolation or extrapolation. Explain your response.

e. Calculate, correct to the nearest whole number, the tank size of a vehicle that had a fuel consumption rate of 10.2 L per 100 km.

f. Pearson's product-moment correlation coefficient for this data is 0.516. Suggest how you can use this value to evaluate the reliability of the data.

g. List at least two other factors that could influence the data.

17. The weight of top brand running shoes was tracked against the recommended retail price, and the results were recorded in the following scatterplot.

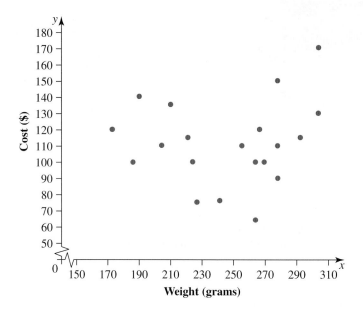

a. Identify the explanatory variable for this situation.

b. Describe the relationship between these two variables.

c. The coefficient of determination for this data is $r^2 = 0.018\,72$. State your conclusions from this result.

d. Identify two external factors that could explain the distribution of the data points.

18. The Bureau of Meteorology records data such as maximum temperatures and solar exposure on a daily and monthly basis. The following data table, for the Botanical Gardens in Melbourne, shows the monthly average amount of solar energy that falls on a horizontal surface and the monthly average maximum temperature.

(*Note:* The data values have been rounded to the nearest whole number.)

Month	Jan	Feb	Mar	Apr	May	Jun	Jul	Aug	Sep	Oct	Nov	Dec
Average solar exposure (MJ)	25	21	17	11	8	6	7	10	13	18	21	24
Average max daily temp. (°C)	43	41	34	33	24	19	24	24	28	32	25	40

a. Identify the explanatory and response variables for this situation.

b. Using CAS, plot the data on a scatterplot.

c. Describe the trend of the data.

d. Calculate Pearson's product-moment coefficient and coefficient of determination for this data. State what these values tell you about the reliability of the data.

e. Plot the regression line for this data and write the equation in terms of the variables.

f. Using your equation, calculate the amount of solar exposure for a monthly maximum temperature of 37 °C.

g. Extrapolate the data to find the average maximum temperature expected for a month that recorded an average solar exposure of 3 MJ.

h. Explain why part g is an example of extrapolation.

8.7 Exam questions

Question 1 (1 mark)

Source: VCE 2021, Further Mathematics Exam 1, Section A Core, Q10; © VCAA.

MC Oscar walked for nine consecutive *days*. The *time*, in minutes, that Oscar spent walking on each day is shown in the table below.

Day	1	2	3	4	5	6	7	8	9
Time	46	40	45	34	36	38	39	40	33

A least squares line is fitted to the data.

The equation of this line predicts that on day 10 the time Oscar spends walking will be the same as the time he spent walking on

A. day 3
B. day 4
C. day 6
D. day 8
E. day 9

Question 2 (3 marks)

Source: VCE 2020, Further Mathematics Exam 2, Q4; © VCAA.

The *age*, in years, *body density*, in kilograms per litre, and *weight*, in kilograms, of a sample of 12 men aged 23 to 25 years are shown in the table below.

Age (years)	Body density (kg/litre)	Weight (kg)
23	1.07	70.1
23	1.07	90.4
23	1.08	73.2
23	1.08	85.0
24	1.03	84.3
24	1.05	95.6
24	1.07	71.7
24	1.06	95.0
25	1.07	80.2
25	1.09	87.4
25	1.02	94.9
25	1.09	65.3

a. What percentage of the variation in *body density* can be explained by the variation in *weight*? Round your answer to the nearest percentage. **(1 mark)**

b. A least squares line is to be fitted to the data with the aim of predicting body *density* from *weight*.

   i. Name the explanatory variable for this least squares line. **(1 mark)**

   ii. Determine the slope of this least squares line. Round your answer to three significant figures. **(1 mark)**

Question 3 (1 mark)

Source: VCE 2018, Further Mathematics Exam 1, Q8; © VCAA.

MC The scatterplot below displays the *resting pulse rate*, in beats per minute, and the *time spent exercising*, in hours per week, of 16 students. A least squares line has been fitted to the data.

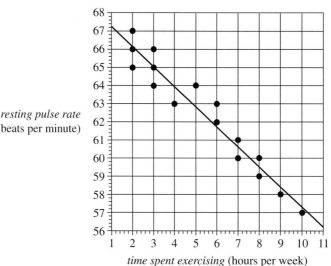

The equation of this least squares line is closest to

  A. *resting pulse rate* $= 67.2 - 0.91 \times$ *time spent exercising*

  B. *resting pulse rate* $= 67.2 - 1.10 \times$ *time spent exercising*

  C. *resting pulse rate* $= 68.3 - 0.91 \times$ *time spent exercising*

  D. *resting pulse rate* $= 68.3 - 1.10 \times$ *time spent exercising*

  E. *resting pulse rate* $= 67.32 - 1.10 \times$ *time spent exercising*

Question 4 (1 mark)

Source: VCE 2018, Further Mathematics Exam 1, Q10; © VCAA.

MC In a study of the association between a person's *height*, in centimetres, and *body surface area*, in square metres, the following least squares line was obtained.

$$body\ surface\ area = -1.1 + 0.019 \times height$$

Which one of the following is a conclusion that can be made from this least squares line?

  A. An increase of $1\ m^2$ in *body surface area* is associated with an increase of 0.019 cm in *height*.

  B. An increase of 1 cm in *height* is associated with an increase of $0.019\ m^2$ in *body surface area*.

  C. The correlation coefficient is 0.019

  D. A person's *body surface area*, in square metres, can be determined by adding 1.1 cm to their *height*.

  E. A person's *height*, in centimetres, can be determined by subtracting 1.1 from their *body surface area*, in square metres.

Source: VCE 2018, Further Mathematics Exam 1, Q14; © VCAA.

MC A least squares line is fitted to a set of bivariate data.

Another least squares line is fitted with response and explanatory variables reversed.

Which one of the following statistics will **not** change in value?

    A. the residual values

    B. the predicted values

    C. the correlation coefficient r

    D. the slope of the least squares line

    E. the intercept of the least squares line

More exam questions are available online.

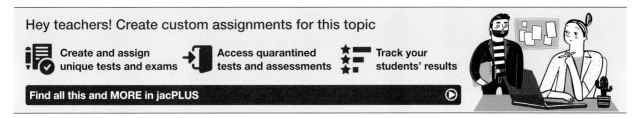

Answers

Topic 8 Investigating relationships between two numerical variables

8.2 Response and explanatory variables

8.2 Exercise

1. a. EV: Time
   RV: Percentage
   b. EV: Age
   RV: Income
2. a. EV: Income
   RV: Age
   b. EV: Hours
   RV: BPM
3. a. EV: Weight
   RV: Waist
   b. EV: Length
   RV: Weight
4. EV
5. RV

8.2 Exam questions

Note: Mark allocations are available with the fully worked solutions online.
1. Humidity at 9 am
2. Evening congestion level (%)
3. Population

8.3 Scatterplots and basic correlation

8.3 Exercise

1. Explanatory variable = time of day
   Response variable = time taken to eat a pizza
2. Explanatory variable = time spent using phone apps
   Response variable = amount of data required
3. a. Explanatory variable = age
   Response variable = number of star jumps
   b. Explanatory variable = quantity of chocolate
   Response variable = cost
   c. Explanatory variable = number of songs
   Response variable = memoryused
   d. Explanatory variable = food supplied
   Response variable = growth rate
4. a. File size
   b. Strong positive correlation
5. a. Explanatory variable = time
   Response variable = weight that can be held(kg)
   b.

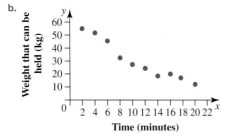

c. Strong negative correlation
6. a. Explanatory variable = height
   Response variable = weight
   b.

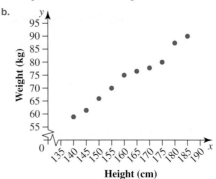

7. a. Weak, negative, linear association
   b. No association
   c. Moderate, positive, linear association
8. Various possible answers. For example:
   a. Loss of money over time
   b. Temperature and number of shoes owned
9. Strong, positive, non-linear
10. Moderate, negative and linear with an outlier

8.3 Exam questions

Note: Mark allocations are available with the fully worked solutions online.
1. D
2. Strong, positive and linear
3. C

8.4 Informal interpretation of association and causation

8.4 Exercise

1. a. $r = 0.8947$, which indicates a strong, positive, linear association.
   b. $r^2 = 0.8005$, which suggests that 80% of the variation in subscription costs can be explained by the variation in the number of issues per year.
   Other factors might include the number of coloured pages, weight of postage, amount of advertising in each issue.
2. a. It indicates a positive linear association between the x-and y-variables, when $r = 0.7211$, and a negative linear association when $r = -0.7211$.
   b. It could be suggested that 52% of the variation in y can be explained by the variation in x.
   c. The coefficient of determination provides information about the strength of the data rather than the causation.
3. a. Moderate, positive, linear association
   b. Strong, negative, linear association
   c. No linear association
   d. Weak, positive, linear association

4. a. $r^2 = 0.7962$. 79.62% of the variation in a child's health can be explained by the variation in the child's diet.

 b. $r^2 = 0.9994$. 99.94% of the variation in the amount of water in the ocean can be explained by the variation in global warming.

5. $r = 0.989$

 $r^2 = 0.978$

6. 42.19%

7. A strong positive relationship between the amount of time spent warming up and the number of matches won

8. a.

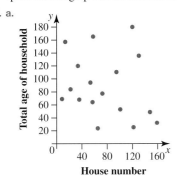

House number

 b. The data points appear random, indicating no correlation.

 c. $r = -0.2135$

 $r^2 = 0.0456$

 d. There is no relationship between the house number and the age of the household.

 e. 95.44%

9. a. Explanatory variable = time spent
 Response variable = mark

 b.

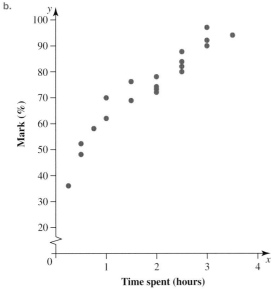
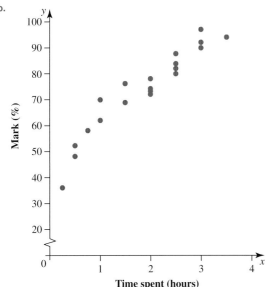

Time spent (hours)

 c. Strong, positive, linear correlation

 d. Each person's understanding of the topic is different and their study habits are unique. Therefore 1 hour spent on the assignment does not guarantee a consistent result. Individual factors will also influence the assignment mark.

 e. $r = 0.952$

 $r^2 = 0.906$

 f. There is a strong relationship between the time spent on an assignment and the resulting grade. As the time spent increased, so did the mark.

10. Various answers are possible. For example data set, see the table on the bottom of the page.*

8.4 Exam questions

Note: Mark allocations are available with the fully worked solutions online.

1. A

2. 74.3%

3. D

8.5 The line of good fit and predictions

8.5 Exercise

1. a. The increase in price for every additional person the venue holds

 b. The price of a ticket if a venue has no capacity

 c. No, as the smallest venues would still have some capacity.

2. $y = -33.75 + 1.25x$

3. a. $y = 86.86 + 0.57x$

 b.

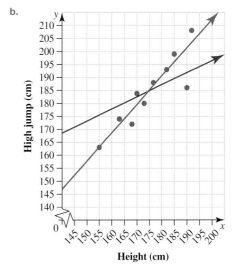

Height (cm)

*10.

x	1	2	2	3	4	5	5	6	7	7	8	9	10	12	14
y	30	25	18	11	18	17	10	11	16	8	6	4	12	10	7

c. Nidya's line of good fit is not a good representation of the data. In this instance having only two points of data to create the line of good fit was not sufficient.

4. a.

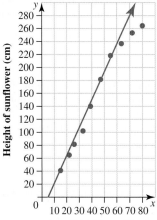

Age of sunflower (days)

Xavier's line is closer to the values above the line than those below it, and there are more values below the line than above it, so this is not a great line of good fit.

b. $y = -28 + 4.4x$

c.

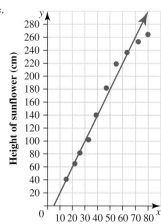

Age of sunflower (days)

Patricia's line is more appropriate as the data points lie on either side of the line and the total distance of the points from the line appears to be minimal.

d. $y = -22 + 4x$

e. The line of good fit does not approximate the height for values that appear outside the parameters of the data set, and the y-intercept lies well outside these parameters.

5. a. $y = 1.5 + 2.5x$
   b. i. $64.00 ii. $25.40
   iii. $29.00 iv. $10.60

6. a. The student did not assign the x- and y-values for each point before calculating the gradient, and they mixed up the values.
   b. $y = -\dfrac{4}{5}x + \dfrac{17}{5}$

7. a. Lines of good fit will vary but should split the data points on either side of the line and minimise the total distance from the points to the line.
   b. Answers will vary.
   c. The amount of crime in a suburb with 0 people
   d. No; if there are 0 people in a suburb there should be no crime.

8. a. $y = 1.32x - 5$
   b. For each additional nest the number of surviving turtles increases by 1.32.
   c. The y-intercept represents the number of surviving turtles from 0 nests. This value is not realistic as you cannot have a negative amount of turtles.
   d. i. 173   ii. 13
   e. The answer to d i was made using extrapolation, so it is not as reliable as the answer to part d ii, which was made using interpolation. However, due to the nature of the data in question, we would expect this relationship to continue and for both answers to be quite reliable.

9. a. The gradient is undefined.
   b. $x = -2$

10. a. 174 ice-creams
    b. 60 ice-creams
    c. The estimate in part a is reliable as it was made using interpolation. It is located within the parameters of the original data set, and it appears consistent with the given data.
    The estimate in part b is unreliable as it was made using extrapolation and is located well outside the parameters of the original data set.

11. a. $112
    b. $305
    c. All estimates outside the parameters of Georgio's original data set (400 km to 2000 km) will be unreliable, with estimates further away from the data set being more unreliable than those closer to the data set.
    Other factors that might affect the cost of flights include air taxes, fluctuating exchange rates and the choice of airlines for various flight paths.

12. a. $y = 116.37 - 17.31x$
    b. For each increase of 1 L of lung capacity, the swimmer will take less time to swim 25 metres.
    c. i. 61.0 seconds   ii. 40.2 seconds
       iii. 24.6 seconds
    d. As Mariana has only two data points and we have no idea of how typical these are of the data set, the equation for the line of good fit and the estimates established from it are all very unreliable.

13. a. $y = 38.67 + 0.61x$
    b. See the table on the bottom of the page.*

*13. b.

Kicking efficiency (%)	75.3	65.6	83.1	73.9	79.0	84.7	64.4	72.4	68.7	80.2
Predicted handball efficiency (%)	84.6	78.7	89.4	83.7	86.9	90.3	78.0	82.8	80.6	87.6

c. The predicted and actual handball efficiencies are very similar in values. A couple of the results are identical, and only a couple of the results are significantly different.

14. a. i. $y = 77.7333 + 0.0057x$
    ii. $y = 76.242 + 0.0059x$
    iii. $y = 71 + 0.01x$
    iv. $y = 58.5211 + 0.0197x$

b.
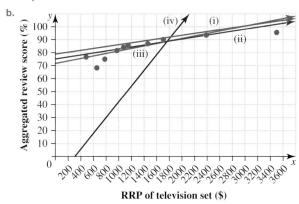

RRP of television set ($)

c. Line iii is the most appropriate line of good fit for this data

15. a. This point of data is clearly an outlier in terms of the data set.
    b. Lines of good fit will vary but should split the data points on either side of the line and minimise the total distance from the points to the line.
    c. Answers will vary.
    d. The increase in box office taking per $1 m increase in the leading actor/actress salary
    e. Answers will vary.
    f. The answers to parts iii and iv are considerably less reliable than the answers to parts i and ii, as they are created using extrapolation instead of interpolation.

16. a. $y = 0.18x + 267$
    b. 971 years
    c. 433.5 cm in diameter
    d. 0.446
    e. The answers are very similar.

17. a. -1.837
    b. 1.701
    c. Positive trend

18. a. 105.9
    b. -1.476
    c. Negative trend

19. a. 60
    b. -5
    c. Negative
    d. -140

20. a. Age in months
    b. 2.029 mL
    c. 7.5 months old

21. a. Cost per night
    b. $155.74
    c. 3.31 km

22. a. Number of insects caught
    b. 1.1
    c. Positive
    d. 66

8.5 Exam questions

Note: Mark allocations are available with the fully worked solutions online.

1. D

2. a. 1.10
   b. Extrapolating
   c. On average, body density decreases by 0.001 512 kg/L for each 1 cm increase in waist measurement.

3. a. On average, for each 1 hPa increase in pressure at 9 am, the pressure at 3 pm increases by 0.8894 hPa.
   b. Pressure at 3 pm $= 111.4 + 08894 \times 1025$
      $= 1023.035$
      $= 1023$ hpa
   c. Interpolation
      The value 1025 hPa is within the data range for pressure at 9 am.

8.6 Introduction to the least squares line of best fit (extending)

8.6 Exercise

1. a. Response variable $=$ height of son

b.
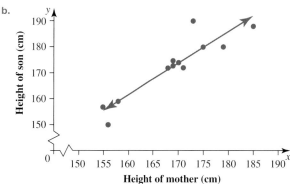

Height of mother (cm)

c. $S = -33.49 + 1.219M$

2. a.
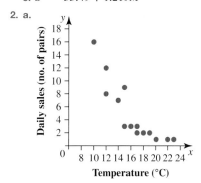

Temperature (°C)

b. $D = 22.50 - 1.071T$

c. $r = -0.8621$

$r^2 = 0.7432$

d. There is a strong negative relationship between the number of gumboots sold and the temperature. The data indicates that 74% of the sales can be explained by the temperature; therefore, 26% of sales are due to other factors.

3. a.
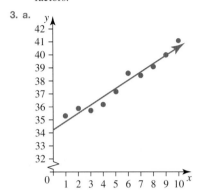

b. $y = 34.23 + 0.6412x$. When $x = 15$, $y = 43.85$.

4. a.
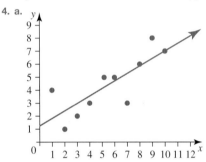

b. $y = 1.2 + 0.58x$

c. 12.8

d. 13.45

5. a. 130

b. $3.6\,°C$

c. The location of the fire, air temperature, proximity to water, etc.

6. a. 76

b. 65

c. Part **a** looks at data within the original data set range, while part **b** predicts data outside of the original data set range of 0–125 new customers each hour.

7. a.
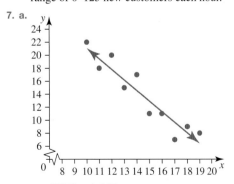

b. $y = 37.70 - 1.648x$

c. -0.204

d. It is assumed the data will continue to behave in the same manner as the data originally supplied.

8. a.
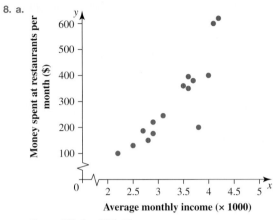

b. $R = -459.8 + 229.5I$

c. $687.70

d. Part **c** asks you to predict data outside of the original data set range.

e. $3160, interpolation

9. a. There is no obvious explanatory variable.

b.
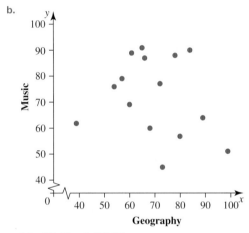

c. $M = 87.63 - 0.2195G$

d. 69

e. Not very confident. The graph does not indicate a strong correlation between the two variables.

f. $r = -0.2172$. This indicates very weak correlation between the data, which supports the view that conclusions cannot be drawn from this data.

10. a. Calories burned

b. Calories burned $= 14\,301 + 115.02 \times$ distance walked

c. $20\,052$. Interpolation, as this data is inside the original range.

d. $15\,451.2$. Extrapolation, as the explanatory variable provided is outside the original data range.

e. An r value of 0.9678 indicates a very strong positive linear relationship, showing that the relationship between the two variables is very strong and can be used to draw conclusions.

f. Examples: speed of walking, difficulty of walking surface, foods eaten.

8.6 Exam questions

Note: Mark allocations are available with the fully worked solutions online.

1. E

2. C

3. a. i. Apparent temperature $= -1.7 + 0.94 \times$ actual temperature

   ii. On average, when the actual temperature is 0 °C, the apparent temperature is −1.7 °C.

   b. 97% of the variation in apparent temperature can be explained by the variation in actual temperature.

8.7 Review

8.7 Exercise

Multiple choice

1. a. A   b. B

2. C

3. C

4. B

5. B

6. B

7. C

8. D

9. D

Short answer

10. i. a. Moderate positive correlation

    b. No correlation

    c. Strong negative correlation

    ii. a.

    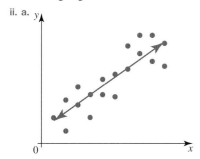

    b. No line of good fit possible

    c.

    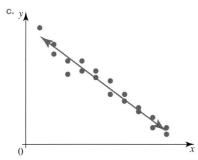

11. a. Explanatory variable = quantity of water
    Response variable = time

    b. Explanatory variable = number of students
    Response variable = number of buses required

12. a.

    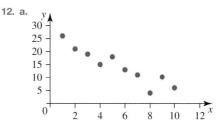

    b. Strong negative correlation

    c. $r = -0.9329$; $r^2 = 0.8703$

    d. Pearson's product-moment correlation coefficient and the coefficient of determination confirm a strong relationship between the two variables.

13. a. 0.5721

    b. There is a strong negative relationship between the variables. The coefficient of determination suggests that 57% of the variation in the y-variable is due to changes in the x-variable, and 43% is due to other factors.

14. Although there appears to be a link between the laziness of people and the increase in sales of healthy foods, there are also many other possible factors besides laziness. Based on this observation alone, the cause of an increase in sales of healthy foods cannot be concluded to be due to laziness.

Extended response

15. a.

    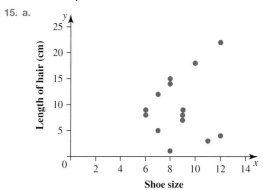

    b. $y = 4.0386 + 0.6157x$

    c. $r = 0.2055$; $r^2 = 0.04221$

    d. There is no association between the two variables.

16. a. Average fuel consumption

    b. Average fuel consumption $= 0.6968 + 0.1119 \times$ fuel tank capacity

    c. 5.17

    d. Extrapolation, as the x-value is outside of the original data range

    e. 85 L

    f. This value indicates a moderate relationship between the variables. Therefore, the data can be used, but other factors should also be considered.

    g. Various answers are possible; for example, the manner in which a person drives the vehicle, weather conditions and road conditions.

17. a. Weight (grams)

    b. No correlation

c. This supports the view that there is no correlation between the variables. Based on this value, no conclusions can be made from the data.

d. Various answers are possible; for example, popularity of the shoe or desired profits.

18. a. Explanatory variable = average solar exposure
Response variable = maximum daily temperature

b.

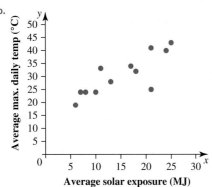

c. Strong positive correlation

d. $r = 0.8242$; $r^2 = 0.6793$
These values indicate a strong relationship between the two variables. The coefficient of determination suggests that nearly 70% of the maximum daily temperature can be explained by the amount of solar exposure.

e.

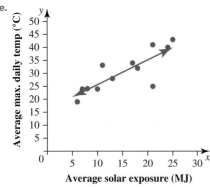

Maximum daily temperature $= 16.232 + 0.9515$
$\times$ average solar exposure

f. 22 MJ

g. 19 °C

h. An x-value of 3 MJ is outside the original data set.

8.7 Exam questions

Note: Mark allocations are available with the fully worked solutions online.

1. B

2. a. 29%

b. i. Body density is being predicted from weight; therefore, weight is the explanatory variable.

ii. Slope $= -0.001\,12$ correct to 3 significant figures.

3. D

4. B

5. C

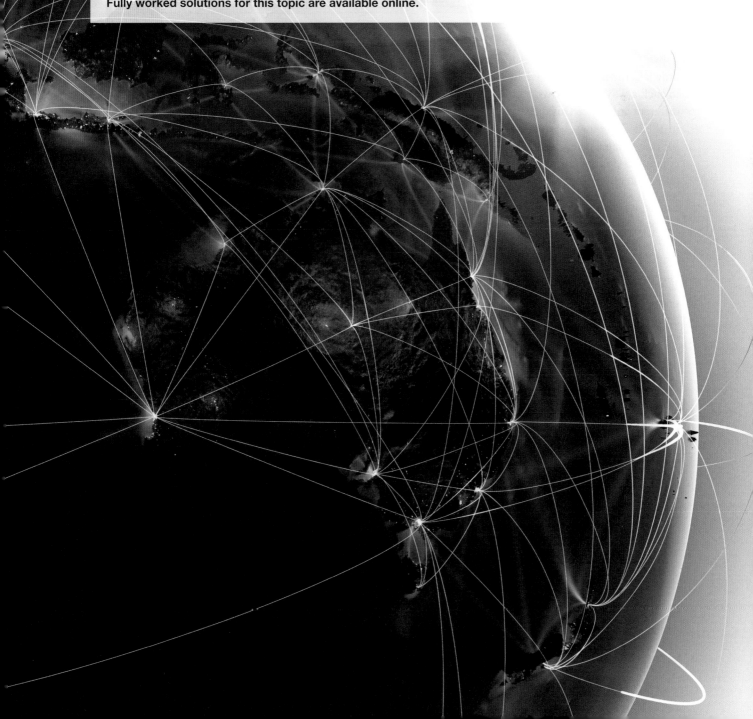

9 Graphs and networks

LEARNING SEQUENCE

Fully worked solutions for this topic are available online.

9.1 Overview

9.1.1 Introduction

Just like matrices, networks are used to show how things are connected. The idea of networks and graph theory is usually credited to Leonhard Euler's 1736 work, *Seven Bridges of Königsberg*. This work carried on with the analysis initiated by Leibniz.

One of the most famous problems in graph theory is the 'four colour problem', which poses the question: 'Is it true that any map drawn in the plane may have its regions coloured with four colours, in such a way that any two regions having a common border have different colours?' This question was first posed by Francis Guthrie in 1852. There have been many failed attempts to prove this. The 'four colour problem' remained unsolved for more than a century, until in 1969 Heinrich Heesch published a method for solving the problem using a computer. Computers allowed networks to be used to solve problems that previously took too long due to the multitude of combinations.

Procedures called algorithms are applied to networks to calculate maximum and minimum values. This study of constructed networks belongs to a field of mathematics called operational research. This developed rapidly during and after World War II, when mathematicians, industrial technicians and members of the armed services worked together to improve military operations.

In more recent times, graph theory and networks have been used to deliver mail, land people on the moon, organise train timetables and improve the flow of traffic. Graph theory and networks have also been applied to a wide range of disciplines from social networks where they are used to examine the structure of relationships and social identities, to biological networks, which analyse molecular networks.

KEY CONCEPTS

This topic covers the following key concepts from the VCE Mathematics Study Design:
- the introduction to the notations, conventions and representations of types and properties of graphs, including edge, loop, vertex, the degree of a vertex, isomorphic and connected graphs and the adjacency matrix
- description of graphs in terms of faces (regions), vertices and edges, and the application of Euler's formula for planar graphs
- connected graphs: walks, trails, paths, cycles and circuits with practical applications
- weighted graphs and networks, and an introduction to the shortest path problem (solution by inspection only) and its practical application
- trees and minimum spanning trees, greedy algorithms, and their use to solve practical problems.

Source: VCE Mathematics Study Design (2023–2027) extracts © VCAA; reproduced by permission.

9.2 Definitions and terms

LEARNING INTENTION

At the end of this subtopic you should be able to:
- define *edge*, *loop*, *vertex* and *the degree of a vertex*
- identify properties and types of graphs, including isomorphic and connected graphs
- identify characteristics of and construct adjacency matrices.

9.2.1 Graphs

The mathematician Leonhard Euler (1707–83) is usually credited with being the founder of graph theory. He famously used it to solve a problem known as the 'Seven Bridges of Königsberg'.

For a long time it had been pondered whether it was possible to travel around the European city of Königsberg (now called Kaliningrad) in such a way that the seven bridges would only have to be crossed once each.

In the branch of mathematics known as graph theory, diagrams involving points and lines are used as a planning and analysis tool for systems and connections. Applications of graph theory include business efficiency, transportation systems, design projects, building and construction, food chains and communications networks.

Leonhard Euler

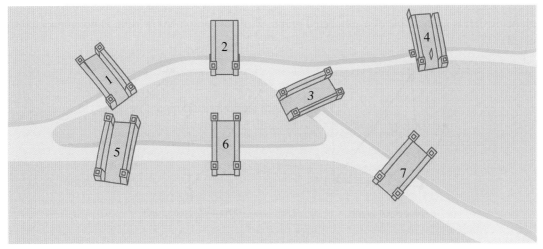

Bridges of Königsberg

9.2.2 Types and properties of graphs

A **graph** is a series of points and lines that can be used to represent the connections that exist in various settings.

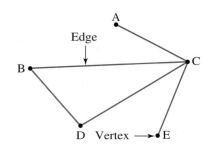

> In a graph, the lines are called edges and the points are called vertices (or nodes), with each edge joining a pair of vertices.

Although edges are often drawn as straight lines, they don't have to be.

When vertices are joined by an edge, they are known as 'adjacent' vertices. Note that the edges of a graph can intersect without there being a vertex. The graph above has five edges and five vertices.

Simple graphs

> A simple graph is one in which pairs of vertices are connected by at most one edge.

The graphs below are examples of simple graphs.

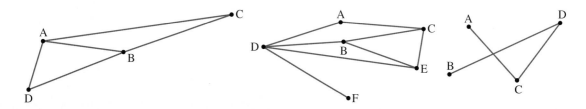

Connected graphs

> If it is possible to reach every vertex of a graph by moving along the edges, it is called a connected graph; otherwise, it is a disconnected graph.

The graph below left is an example of a connected graph, while the graph below right is not connected (it is disconnected).

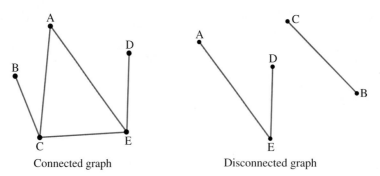

Connected graph   Disconnected graph

A bridge

> A bridge is an edge that disconnects the graph when removed.

> *Note:* A graph can have more than one bridge.

In the following graph, the edge AB is a bridge, because if it is removed, the graph would become disconnected. Can you identify another edge that would be classified as a bridge?

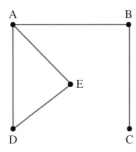

Consider the road map shown.

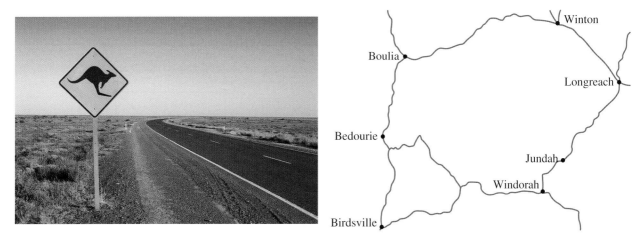

This map can be represented by the following graph.

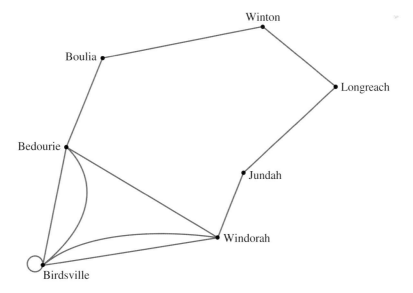

As there is more than one route connecting Birdsville to Windorah and Birdsville to Bedourie, each route is represented by an edge in the graph. In this case, we say there are multiple edges. Also, as it is possible to travel along a road from Birdsville that returns without passing through another town, this is represented by an edge. When this happens, the edge is called a loop.

A loop

A loop is an edge that connects a vertex to itself.

Note: A loop is only counted as one edge.

Multiple edge

A multiple edge occurs when two vertices are connected by more than one edge.

Since graphs can be used to model practical real-life situations, it becomes clear that two types of graphs are needed. Sometimes it may only be possible to move along an edge in one direction, say a one-way street, or it could be possible to move along an edge in both directions.

So far, the graphs you have seen are all examples of **undirected graphs** where we assume you can move along an edge in both directions. Graphs that only allow movement in a specified direction are called **directed graphs** or **digraphs**, and these edges will contain an arrow indicating the direction of travel allowed.

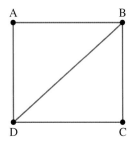

An undirected graph
(travel along an edge in any direction)

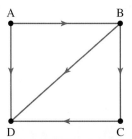

A directed graph
(arrows show the direction you must travel)

tlvd-4818

WORKED EXAMPLE 1 Drawing a graph to represent the possible ways of travelling

The diagram represents a system of paths and gates in a large park. Draw a graph to represent the possible ways of travelling to each gate in the park.

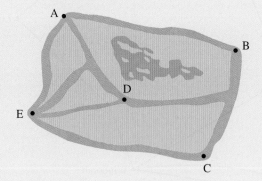

THINK	WRITE/DRAW
1. Identify, draw and label all possible vertices.	Represent each gate as a vertex.

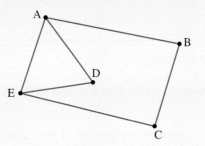

2. Draw edges to represent all the direct connections between the identified vertices.	Direct pathways exist for A–B, A–D, A–E, B–C, C–E and D–E.

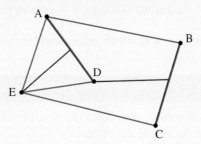

3. Identify all the other unique ways of connecting vertices.	Other unique pathways exist for A–E, D–E, B–D and C–D.

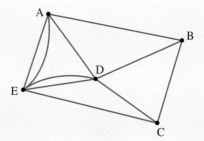

4. Draw the final graph.

9.2.3 The degree of a vertex

When analysing the situation represented by a graph, it can be useful to consider the number of edges that are directly connected to a particular vertex. This is referred to as the **degree** of the vertex and is given the notation $\deg(V)$, where V represents the vertex.

The degree of a vertex

The degree of a vertex in an undirected graph is the number of edges directly connected to that vertex, *except* that a loop at a vertex contributes twice to the degree of that vertex.

In the graph, $\deg(A) = 2$, $\deg(B) = 2$, $\deg(C) = 5$, $\deg(D) = 2$, $\deg(E) = 5$ and $\deg(F) = 2$.

Notice how the loop at vertex E counts twice to the degree of the vertex. The degree of a vertex may be odd or even. The above graph has four vertices of even degree and two vertices of odd degree.

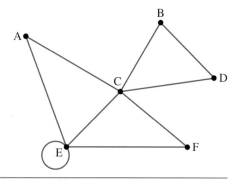

The handshaking theorem

The handshaking theorem states that the sum of the degrees of the vertices of a graph is twice the number of edges.

$$\text{sum of the degrees of the vertices} = 2 \times \text{number of edges}$$

This can also be expressed as:

$$\text{number of edges} = \frac{1}{2} \text{ sum of the degrees of the vertices}$$

Note that this applies even if multiple edges and loops are present.

WORKED EXAMPLE 2 Demonstrating the handshaking theorem

For the graph in the diagram, show that the number of edges is equal to half the sum of the degree of the vertices.

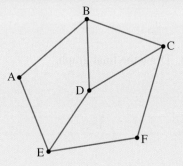

THINK	WRITE
1. Identify the degree of each vertex.	$\deg(A) = 2, \deg(B) = 3, \deg(C) = 3,$ $\deg(D) = 3, \deg(E) = 3$ and $\deg(F) = 2$
2. Calculate the sum of the degrees for the graph.	The sum of the degrees for the graph $= 2 + 3 + 3 + 3 + 3 + 2$ $= 16$
3. Count the number of edges for the graph.	The graph has the following edges: A–B, A–E, B–C, B–D, C–D, C–F, D–E, E–F. The graph has 8 edges.
4. Write the answer as a sentence.	The total number of edges in the graph is therefore half the sum of the degrees.

9.2.4 Isomorphic graphs

Consider the following graphs.

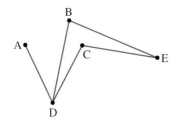 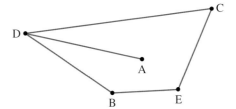

For the two graphs, the connections for each vertex can be summarised as shown in the table.

Although the graphs don't look exactly the same, they could represent exactly the same information. Such graphs are known as **isomorphic graphs**.

Vertex	Connections		
A	D		
B	D	E	
C	D	E	
D	A	B	C
E	B	C	

Isomorphic graphs

Isomorphic graphs have the same number of vertices and edges, with corresponding vertices having identical degrees and connections.

WORKED EXAMPLE 3 Identifying isomorphic graphs

Confirm whether the following two graphs are isomorphic.

Graph 1

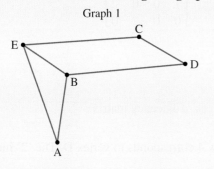

Graph 2

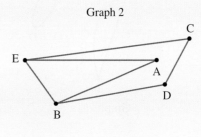

THINK	WRITE
1. Identify the degree of the vertices for each graph.	

Graph	A	B	C	D	E
Graph 1	2	3	2	2	3
Graph 2	2	3	2	2	3

2. Identify the number of edges for each graph.

Graph	Edges
Graph 1	6
Graph 2	6

3. Identify the vertex connections for each graph.

Vertex	Connections		
A	B	E	
B	A	D	E
C	D	E	
D	B	C	
E	A	B	C

4. Comment on the two graphs.

The two graphs are isomorphic as they have the same number of vertices and edges, with corresponding vertices having identical degrees and connections.

9.2.5 Adjacency matrices

Matrices are often used when working with graphs. A matrix that represents the number of edges that connect the vertices of a graph is known as an adjacency matrix.

Adjacency matrices

Each column and row of an adjacency matrix corresponds to a vertex of the graph, and the numbers indicate how many edges connect them.

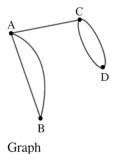

$$\begin{bmatrix} 0 & 2 & 1 & 0 \\ 2 & 0 & 0 & 0 \\ 1 & 0 & 0 & 2 \\ 0 & 0 & 2 & 0 \end{bmatrix}$$

Graph   Adjacency matrix

In the adjacency matrix, column 3 corresponds to vertex C and row 4 corresponds to vertex D. The '2' indicates the number of edges joining these two vertices.

$$\begin{array}{c} \\ A \\ B \\ C \\ D \end{array} \begin{array}{cccc} A & B & \textcircled{C} & D \\ \begin{bmatrix} 0 & 2 & 1 & 2 \\ 2 & 0 & 0 & 0 \\ 1 & 0 & 0 & 2 \\ 0 & 0 & 2 & 0 \end{bmatrix} \end{array}$$

9.2.6 Characteristics of adjacency matrices

Adjacency matrices are square matrices with n rows and columns, where n is equal to the number of vertices in the graph.

$$\text{Column:} \quad 1 \quad 2 \ \dots \ n-1 \quad n \qquad \text{Row}$$

$$\begin{bmatrix} 0 & 2 & \dots & 1 & 0 \\ 2 & 0 & \dots & 0 & 0 \\ \vdots & \vdots & \dots & \vdots & \vdots \\ 1 & 0 & \dots & 0 & 2 \\ 0 & 0 & \dots & 2 & 0 \end{bmatrix} \quad \begin{matrix} 1 \\ 2 \\ \vdots \\ n-1 \\ n \end{matrix}$$

Adjacency matrices are symmetrical around the leading diagonal.

$$\begin{bmatrix} 0 & 2 & 1 & 0 \\ 2 & 0 & 0 & 0 \\ 1 & 0 & 0 & 2 \\ 0 & 0 & 2 & 0 \end{bmatrix}$$

Any non-zero value in the leading diagonal will indicate the existence of a loop.

$$\begin{bmatrix} 0 & 2 & 1 & 2 \\ 2 & \boxed{1} & 0 & 0 \\ 1 & 0 & 0 & 2 \\ 2 & 0 & 2 & 0 \end{bmatrix}$$

The '1' indicates that one loop exists at vertex B:

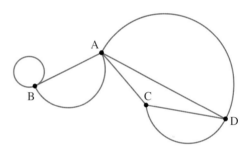

A row consisting of all zeros indicates an isolated vertex (a vertex that is not connected to any other vertex).

$$\begin{bmatrix} 0 & 0 & 0 & 1 \\ \boxed{0 \quad 0 \quad 0 \quad 0} \\ 0 & 0 & 0 & 1 \\ 1 & 0 & 1 & 0 \end{bmatrix}$$

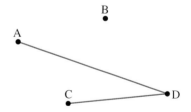

WORKED EXAMPLE 4 Constructing an adjacency matrix from a graph

Construct the adjacency matrix for the given graph.

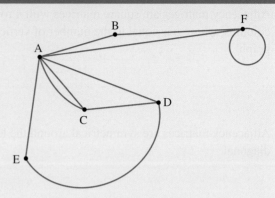

THINK

WRITE

1. Draw a table with rows and columns for each vertex of the graph.

	A	B	C	D	E	F
A						
B						
C						
D						
E						
F						

2. Count the number of edges that connect vertex A to the other vertices and record these values in the corresponding space for the first row of the table.

	A	B	C	D	E	F
A	0	1	2	1	1	1

3. Repeat step 2 for all the other vertices.

	A	B	C	D	E	F
A	0	1	2	1	1	1
B	1	0	0	0	0	1
C	2	0	0	1	0	0
D	1	0	1	0	1	0
E	1	0	0	1	0	0
F	1	1	0	0	0	1

4. Display the numbers as a matrix.

$$\begin{bmatrix} 0 & 1 & 2 & 1 & 1 & 1 \\ 1 & 0 & 0 & 0 & 0 & 1 \\ 2 & 0 & 0 & 1 & 0 & 0 \\ 1 & 0 & 1 & 0 & 1 & 0 \\ 1 & 0 & 0 & 1 & 0 & 0 \\ 1 & 1 & 0 & 0 & 0 & 1 \end{bmatrix}$$

 Resources

 Interactivity The adjacency matrix (int-6466)

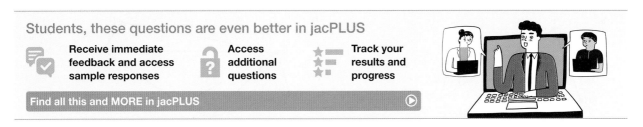

1. **WE1** The diagram shows the plan of a floor of a house. Draw a graph to represent the possible ways of travelling between each room on the floor.

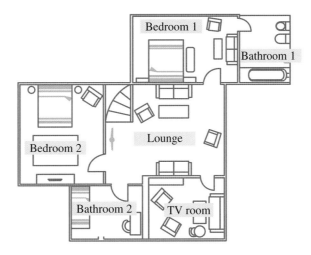

2. Draw a graph to represent the following tourist map.

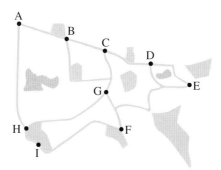

3. **WE2** For each of the following graphs, verify that the number of edges is equal to half the sum of the degree of the vertices.

a.

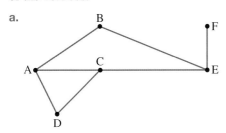

b.

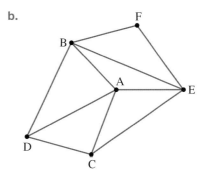

4. For each of the following graphs, verify that the number of edges is equal to half the sum of the degree of the vertices.

a.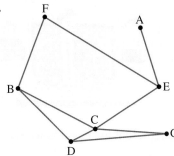

b.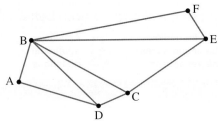

5. Identify the degree of each vertex in the following graphs.

a.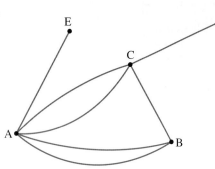

b.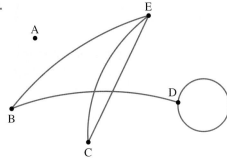

c.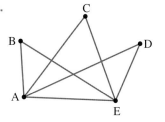

d.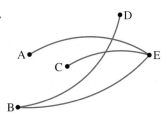

6. **WE3** Confirm whether the following pairs of graphs are isomorphic.

a.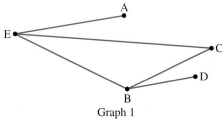

Graph 1   Graph 2

b.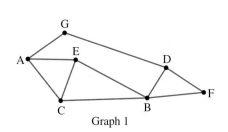

Graph 1

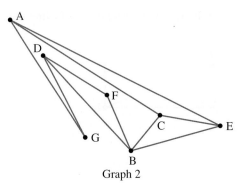

Graph 2

c.

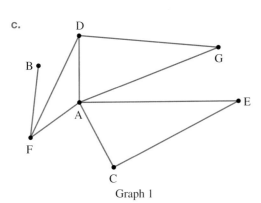

Graph 1

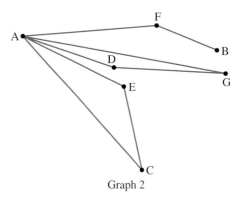

Graph 2

d.

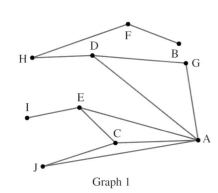

Graph 1

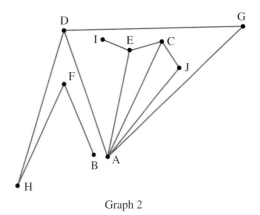

Graph 2

7. Explain why the following pairs of graphs are not isomorphic.

a.

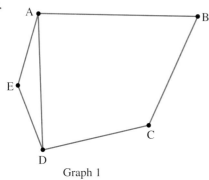

Graph 1

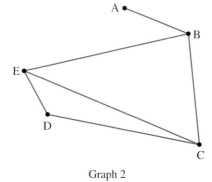

Graph 2

b.

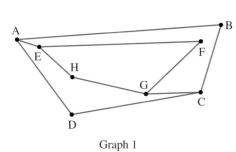

Graph 1

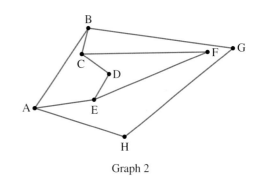

Graph 2

8. Identify pairs of isomorphic graphs from the following.

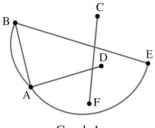

Graph 1

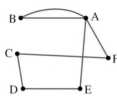

Graph 2

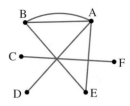

Graph 3

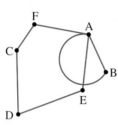

Graph 4

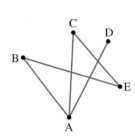

Graph 5

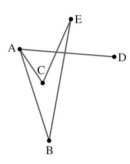

Graph 6

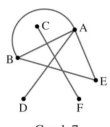

Graph 7

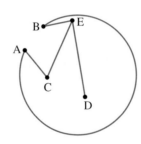

Graph 8

9. **WE4** Construct adjacency matrices for the following graphs.

a.

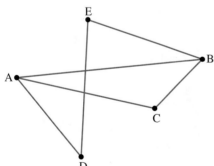

b.

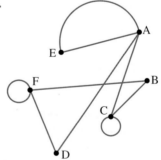

c.

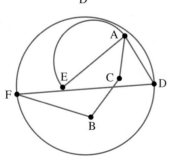

d.

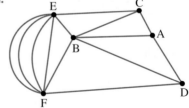

10. Draw graphs to represent the following adjacency matrices.

a. $\begin{bmatrix} 0 & 1 & 0 & 1 \\ 1 & 0 & 0 & 1 \\ 0 & 0 & 0 & 1 \\ 1 & 1 & 1 & 0 \end{bmatrix}$

b. $\begin{bmatrix} 1 & 0 & 2 & 0 \\ 0 & 0 & 0 & 1 \\ 2 & 0 & 0 & 2 \\ 0 & 1 & 2 & 0 \end{bmatrix}$

c. $\begin{bmatrix} 0 & 1 & 2 & 0 & 0 \\ 1 & 1 & 0 & 0 & 1 \\ 2 & 0 & 0 & 1 & 0 \\ 0 & 0 & 1 & 1 & 1 \\ 0 & 1 & 0 & 1 & 1 \end{bmatrix}$

d. $\begin{bmatrix} 2 & 0 & 1 & 1 & 0 \\ 0 & 0 & 0 & 0 & 0 \\ 1 & 0 & 1 & 0 & 2 \\ 1 & 0 & 0 & 0 & 1 \\ 0 & 0 & 2 & 1 & 0 \end{bmatrix}$

11. Construct the adjacency matrices for each of the graphs shown in question **15**.

12. Complete the following adjacency matrices.

a. $\begin{bmatrix} 0 & 0 & \\ 0 & 2 & 2 \\ 1 & & 0 \end{bmatrix}$

b. $\begin{bmatrix} 2 & 1 & & 0 \\ & 0 & & \\ 0 & 1 & 0 & 1 \\ & 2 & & 0 \end{bmatrix}$

c. $\begin{bmatrix} 0 & & 1 & & 0 \\ 0 & 0 & & & 0 \\ & 0 & 0 & 0 & 2 \\ 1 & 0 & 0 & 0 & 1 \\ 0 & & & 1 & 0 \end{bmatrix}$

d. $\begin{bmatrix} 0 & 0 & 0 & 1 & 0 \\ 0 & 0 & 0 & 1 & 0 \\ & 0 & 0 & 0 & 1 \\ & & & 0 & 0 \\ 0 & & & 0 & 1 \end{bmatrix}$

13. Draw a graph of:
    a. a simple, connected graph with 6 vertices and 7 edges
    b. a simple, connected graph with 7 vertices and 7 edges, where one vertex has degree 3 and five vertices have degree 2
    c. a simple, connected graph with 9 vertices and 8 edges, where one vertex has degree 8.

14. By indicating the passages with edges and the intersections and passage endings with vertices, draw a graph to represent the maze shown.

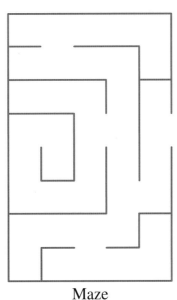

Maze

15. Complete the following table for the graphs shown.

	Simple	Connected
Graph 1	Yes	Yes
Graph 2		
Graph 3		
Graph 4		
Graph 5		

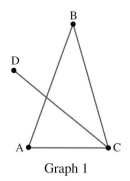

Graph 1

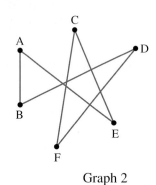

Graph 2

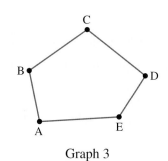

Graph 3

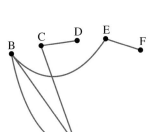

Graph 4

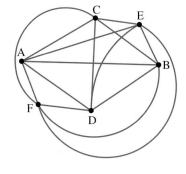

Graph 5

16. A round robin tournament occurs when each team plays all other teams once only.
At the beginning of the school year, five schools play a round robin competition in table tennis.

a. Draw a graph to represent the games played.
b. State what the total number of edges in the graph indicates.

17. The diagram shows the map of some of the main suburbs of Beijing.

a. Draw a graph to represent the shared boundaries between the suburbs.
b. State which suburb has the highest degree.
c. State the type of graph.

18. Jetways Airlines operates flights in South-East Asia.

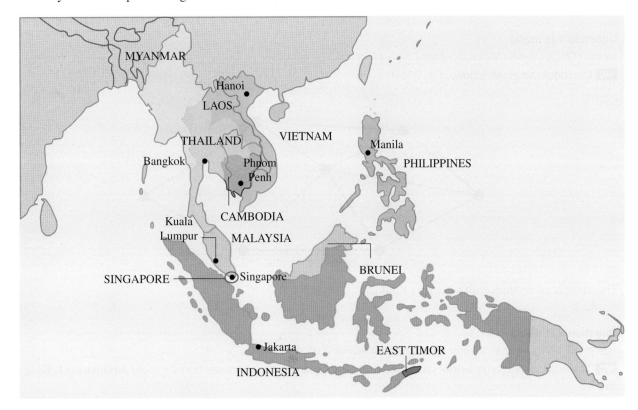

The table indicates the number of direct flights per day between key cities.

From: To:	Bangkok	Manila	Singapore	Kuala Lumpur	Jakarta	Hanoi	Phnom Penh
Bangkok	0	2	5	3	1	1	1
Manila	2	0	4	1	1	0	0
Singapore	5	4	0	3	4	2	3
Kuala Lumpur	3	1	3	0	0	3	3
Jakarta	1	1	4	0	0	0	0
Hanoi	1	0	2	3	0	0	0
Phnom Penh	1	0	3	3	0	0	0

a. Draw a graph to represent the number of direct flights.
b. State whether this graph would be considered directed or undirected. Explain why.
c. State the number of ways you can travel from:

  i. Phnom Penh to Manila
  ii. Hanoi to Bangkok.

▶ **Question 1 (1 mark)**

Source: VCE 2021, Further Mathematics Exam 1, Section B, Module 2, Q1; © VCAA.

MC Consider the graph below.

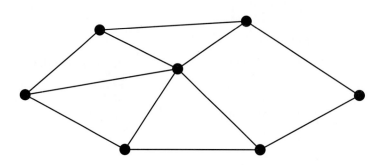

The number of vertices with a degree of 3 is

A. 1   **B.** 2   **C.** 3   **D.** 4   **E.** 5

▶ **Question 2 (1 mark)**

Source: VCE 2020, Further Mathematics Exam 1, Section B, Module 2, Q8; © VCAA.

MC The adjacency matrix below shows the number of pathway connections between four landmarks: J, K, L and M.

$$\begin{array}{c@{}c} & \begin{array}{cccc} J & K & L & M \end{array} \\ \begin{array}{c} J \\ K \\ L \\ M \end{array} & \left[\begin{array}{cccc} 1 & 3 & 0 & 2 \\ 3 & 0 & 1 & 2 \\ 0 & 1 & 0 & 2 \\ 2 & 2 & 2 & 0 \end{array}\right] \end{array}$$

A network of pathways that could be represented by the adjacency matrix is

A.

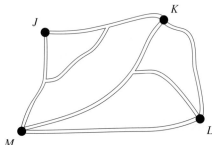

B.

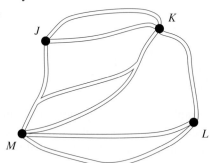

C.

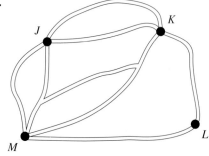

D.

E.

▶ **Question 3 (1 mark)**

Source: VCE 2017, Further Mathematics Exam 1, Section B, Module 2, Q1; © VCAA.

MC Which one of the following graphs contains a loop?

A.

B.

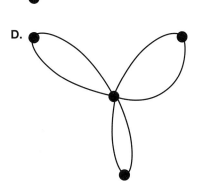

C.

D.

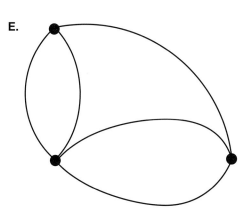

E.

More exam questions are available online.

9.3 Planar graphs

As indicated in Subtopic 9.2, graphs can be drawn with intersecting edges. However, in many applications intersections may be undesirable. Consider a graph of an underground railway network. In this case, intersecting edges would indicate the need for one rail line to be in a much deeper tunnel, which could add significantly to construction costs.

In some cases it is possible to redraw graphs so that they have no intersecting edges. When a graph can be redrawn in this way, it is known as a **planar graph**.

For example, in the graph shown below, it is possible to redraw one of the intersecting edges so that it still represents the same information.

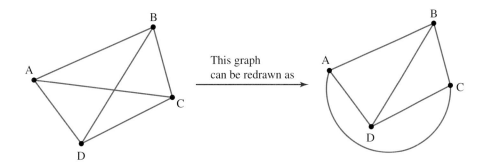

This graph can be redrawn as

WORKED EXAMPLE 5 Redrawing a graph to make it planar

Redraw the graph so that it has no intersecting edges.

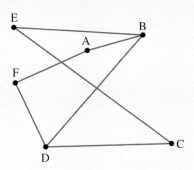

THINK

1. List all connections in the original graph.

2. Draw all vertices and any section(s) of the graph that have no intersecting edges.

3. Draw any further edges that don't create intersections. Start with edges that have the fewest intersections in the original drawing.

WRITE/DRAW

Connections: AB; AF; BD; BE; CD; CE; DF

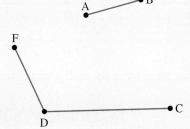

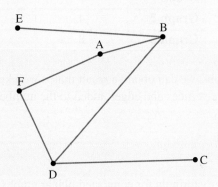

4. Identify any edges yet to be drawn and redraw so that they do not intersect with the other edges.

Connections: ~~AB~~; ~~AF~~; ~~BD~~; ~~BE~~; ~~CD~~; CE; ~~DF~~

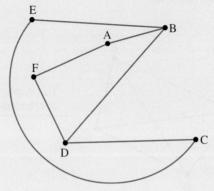

Note: Although a graph may not appear to be planar, by modifying the placement of some edges, the graph may become planar.

9.3.1 Euler's formula

In all planar graphs, the edges and vertices create distinct areas referred to as **faces** (regions). The planar graph shown has five faces *including the area around the outside*.

Consider the following group of planar graphs.

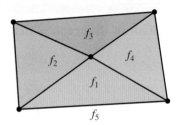

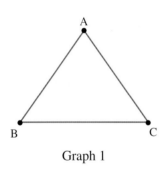

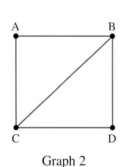

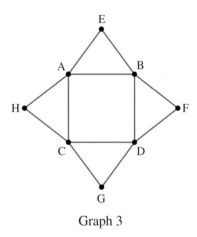

| | Graph 1 | Graph 2 | Graph 3 |

The number of vertices, edges and faces for each graph is summarised in the following table.

	Vertices	Edges	Faces
Graph 1	3	3	2
Graph 2	4	5	3
Graph 3	8	12	6

For each of these graphs, we can obtain a result that is well known for any connected planar graph: the difference between the vertices and edges added to the number of faces will always equal 2.

Graph 1: $3 - 3 + 2 = 2$

Graph 2: $4 - 5 + 3 = 2$

Graph 3: $8 - 12 + 6 = 2$

This is known as Euler's formula for connected planar graphs.

Euler's formula

For any connected planar graph:

$$v - e + f = 2$$

where v is the number of vertices, e is the number of edges and f is the number of faces (regions).

WORKED EXAMPLE 6 Applying Euler's formula

Determine the number of faces in a connected planar graph of 7 vertices and 10 edges.

THINK	WRITE
1. Substitute the given values into Euler's formula.	$v - e + f = 2$ $7 - 10 + f = 2$
2. Solve the equation for the unknown value.	$7 - 10 + f = 2$ $\qquad f = 2 - 7 + 10$ $\qquad f = 5$
3. Write the answer as a sentence.	There will be 5 faces in a connected planar graph with 7 vertices and 10 edges.

 Resources

Interactivities Planar graphs (int-6467)
Euler's formula (int-6468)

9.3 Exercise

1. **WE5** Redraw the following graphs so that they have no intersecting edges.

a.

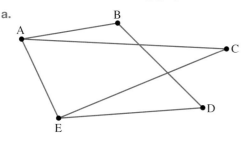

b.

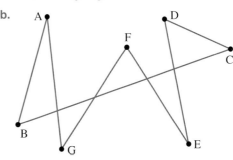

2. Select which of the following are planar graphs.

a. **A.**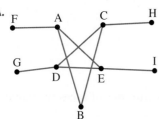

B.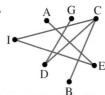

C.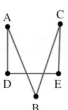

D.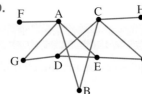

b. **A.**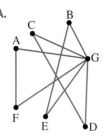

B.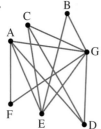

C.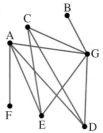

D.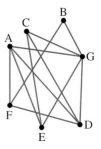

3. Redraw the following graphs to show that they are planar.

a.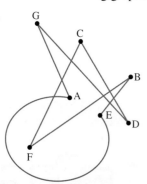

b.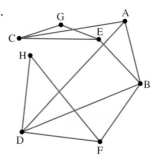

4. Identify which of the following graphs are *not* planar.

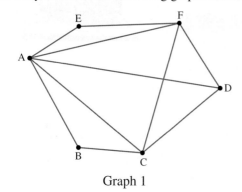

Graph 1

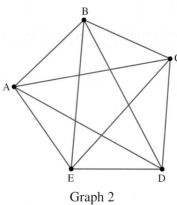

Graph 2

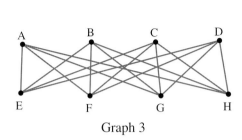

Graph 3

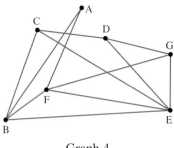

Graph 4

5. **WE6** State the number of faces for a connected planar graph of:

   a. 8 vertices and 10 edges
   b. 11 vertices and 14 edges.

6. **a.** For a connected planar graph of 5 vertices and 3 faces, state the number of edges.
   b. For a connected planar graph of 8 edges and 5 faces, state the number of vertices.

7. For each of the following planar graphs, identify the number of faces.

   a.

   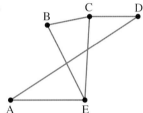

   b.

   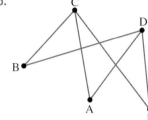

   c.

   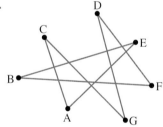

   d.

   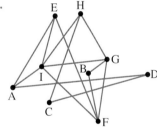

8. Construct a connected planar graph with:

   a. 6 vertices and 5 faces
   b. 11 edges and 9 faces.

9. Use the following adjacency matrices to draw graphs that have no intersecting edges.

   a.
   $$\begin{bmatrix} 0 & 1 & 1 & 1 & 0 \\ 1 & 0 & 1 & 1 & 0 \\ 1 & 1 & 0 & 0 & 1 \\ 1 & 1 & 0 & 0 & 1 \\ 0 & 0 & 1 & 1 & 0 \end{bmatrix}$$

   b.
   $$\begin{bmatrix} 0 & 0 & 1 & 1 & 0 \\ 0 & 0 & 0 & 1 & 1 \\ 1 & 0 & 0 & 0 & 0 \\ 1 & 1 & 0 & 0 & 1 \\ 0 & 1 & 0 & 1 & 0 \end{bmatrix}$$

10. For the graphs you drew in question 9:

    i. identify the number of enclosed faces
    ii. identify the maximum number of additional edges that can be added to maintain a simple planar graph.

11. a. Use the planar graphs shown to complete the given table.

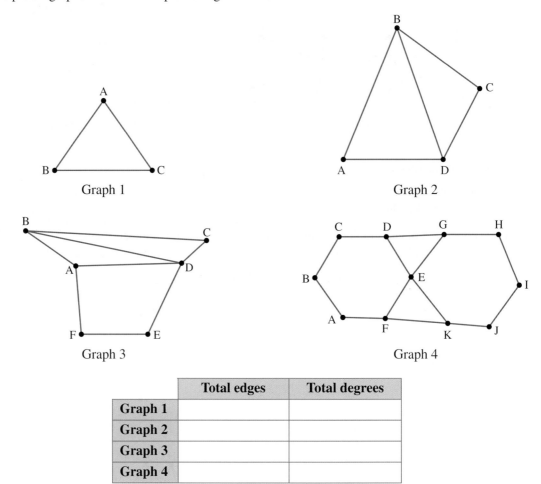

	Total edges	Total degrees
Graph 1		
Graph 2		
Graph 3		
Graph 4		

b. State the pattern that is evident from the table.

12. a. Use the planar graphs shown to complete the given table.

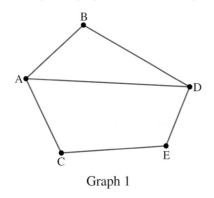

Graph 1

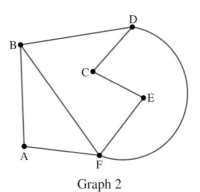

Graph 2

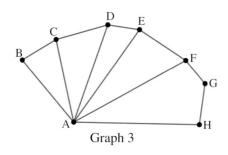

Graph 3

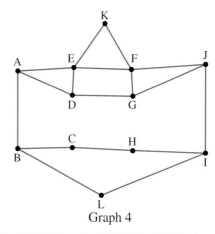

Graph 4

	Total vertices of even degree	Total vertices of odd degree
Graph 1		
Graph 2		
Graph 3		
Graph 4		

b. Comment on any pattern evident from the table.

13. Represent the following 3-dimensional shapes as planar graphs.

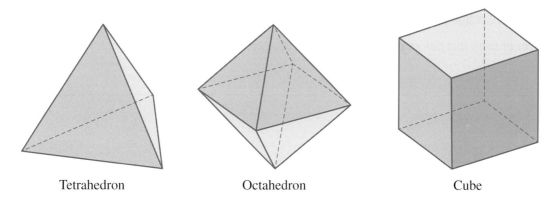

Tetrahedron   Octahedron   Cube

14. A section of an electric circuit board is shown in the diagram.

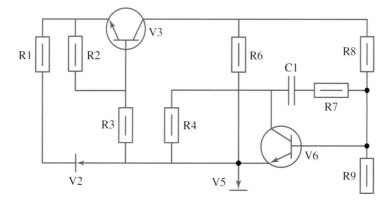

a. Draw a graph to represent the circuit board, using vertices to represent the labelled parts of the diagram.

b. State whether it is possible to represent the circuit board as a planar graph.

15. The table displays the most common methods of communication for a group of people.

	Email	WhatsApp	SMS
Adam	Ethan, Liam	Ethan, Liam	Ethan
Michelle		Sophie, Emma, Ethan	Sophie, Emma
Liam	Adam		
Sophie		Michelle, Chloe	Michelle, Chloe
Emma	Chloe	Chloe, Ethan, Michelle	Chloe, Ethan
Ethan		Emma, Adam, Michelle	Emma
Chloe	Emma, Sophie	Emma, Sophie	Emma, Sophie

a. Display the information for the entire table in a graph.
b. State who would be the best person to introduce Chloe and Michelle.
c. Display the WhatsApp information in a separate graph.
d. If Liam and Sophie began communicating through WhatsApp, determine how many faces the graph from part c would have.

9.3 Exam questions

▶ **Question 1 (1 mark)**

Source: VCE 2020, Further Mathematics Exam 1, Section B, Module 2, Q1; © VCAA.

MC A connected planar graph has seven vertices and nine edges.

The number of faces that this graph will have is

A. 1
B. 2
C. 3
D. 4
E. 5

▶ **Question 2 (1 mark)**

Source: VCE 2016, Further Mathematics Exam 1, Section B, Module 2, Q5; © VCAA.

MC Consider the planar graph below.

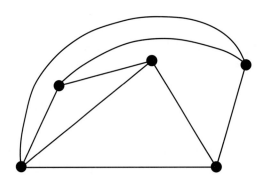

Which one of the following graphs can be redrawn as the planar graph above?

A.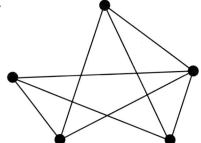

B.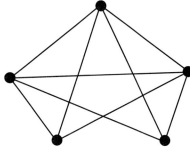

C.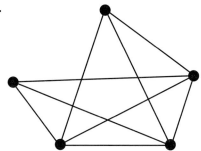

D.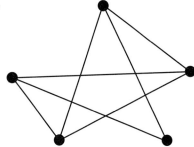

E.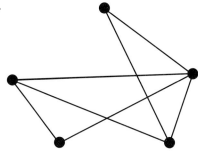

Question 3 (1 mark)

Source: VCE 2013, Further Mathematics, Exam 1, Section B, Module 5, Q7; © VCAA.

MC A connected graph consists of five vertices and four edges.

Consider the following five statements.
- The graph is planar.
- The graph has more than one face.
- All vertices are of even degree.
- The sum of the degrees of the vertices is eight.
- The graph cannot have a loop.

How many of these statements are always true for such a graph?
- **A.** 1
- **B.** 2
- **C.** 3
- **D.** 4
- **E.** 5

More exam questions are available online.

9.4 Connected graphs

9.4.1 Traversing connected graphs

Many applications of graphs involve an analysis of movement around a network. These include fields such as transport, communications and utilities, to name a few.

Movement through a simple connected graph is described in terms of starting and finishing at specified vertices by travelling along the edges. This is usually done by listing the labels of the vertices visited in the correct order.

In more complex graphs, edges may also have to be indicated, as there may be more than one connection between vertices.

The definitions of the main terms used when describing movement across a network are as follows.

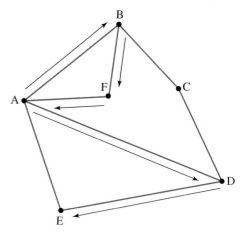

Route: ABFADE

Moving around a graph

Walk: Any route taken through a network, including routes that repeat edges and vertices

Trail: A walk with no repeated edges

Path: A walk with no repeated vertices and no repeated edges

Cycle: A path beginning and ending at the same vertex

Circuit: A trail beginning and ending at the same vertex

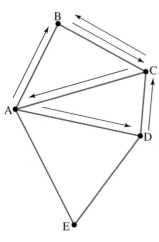

Walk: ABCADCB

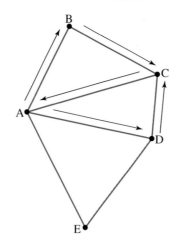

Trail: ABCADC

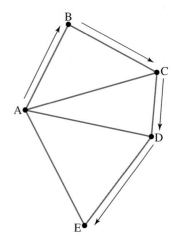

Path: ABCDE

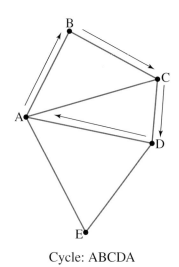

Cycle: ABCDA

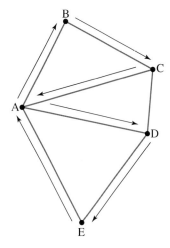

Circuit: ABCADEA

tlvd-4820

WORKED EXAMPLE 7 Identifying different routes in a graph

In the following network, identify two different routes:
one cycle and one circuit.

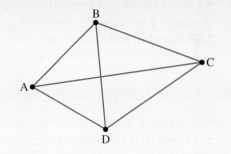

THINK

1. For a cycle, identify a route that doesn't repeat a vertex and doesn't repeat an edge, apart from the start/finish.

2. For a circuit, identify a route that doesn't repeat an edge and ends at the starting vertex.

WRITE

Cycle: ABDCA

Circuit: ADBCA

9.4.2 Eulerian trails and circuits

In some practical situations, it is most efficient if a route travels along each edge only once. Examples include parcel deliveries and council garbage collections. If it is possible to travel a network using each edge only once, the route is known as an **Euler trail** or **Euler circuit**.

Euler trails and circuits

An Euler trail is a trail in which every edge is used once.

An Euler circuit is an Euler trail that starts and ends at the same vertex.

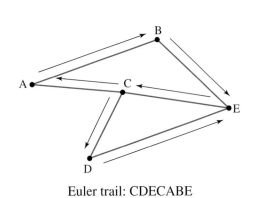

Euler trail: CDECABE

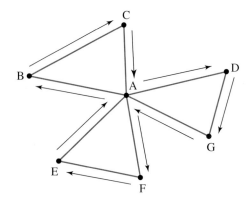

Euler circuit: ABCADGAFEA

Note that in the examples shown, the vertices for the Euler circuit are of even degree, and there are 2 vertices of odd degree for the Euler trail.

Existence conditions for Euler trails and circuits

- **If all of the vertices of a connected graph have an even degree, then an Euler circuit exists. An Euler circuit can begin at any vertex.**
- **If exactly 2 vertices of a connected graph have an odd degree, then an Euler trail exists. An Euler trail will start at one of the odd degree vertices and finish at the other.**

9.4.3 Hamiltonian paths and cycles

In other situations, it may be more practical if all vertices can be reached without using all of the edges of the graph. For example, if you wanted to visit a selection of the capital cities of Europe, you wouldn't need to use all the available flight routes.

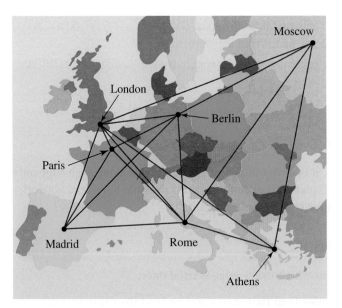

Hamiltonian paths and cycles

A Hamiltonian path is a path that passes through every vertex exactly once.

A Hamiltonian cycle is a Hamiltonian path that starts and finishes at the same vertex.

Hamiltonian paths and Hamiltonian cycles reach all vertices of a network once without necessarily using all of the available edges.

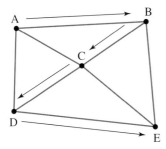

Hamiltonian path: ABCDE

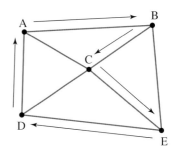

Hamiltonian cycle: ABCEDA

WORKED EXAMPLE 8 Describing an Eulerian trail and Hamiltonian path

Identify an Euler trail and a Hamiltonian path in the following graph.

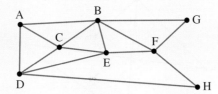

THINK

1. For an Euler trail to exist, there must be exactly 2 vertices with an odd degree.

2. Identify a route that uses each edge once. Begin the route at one of the odd-degree vertices and finish at the other.

3. Identify a route that reaches each vertex once.

4. Write the answer.

WRITE/DRAW

$\deg(A) = 3, \deg(B) = 5, \deg(C) = 4, \deg(D) = 4, \deg(E) = 4,$
$\deg(F) = 4, \deg(G) = 2, \deg(H) = 2$
As there are only two odd-degree vertices, an Euler trail must exist.

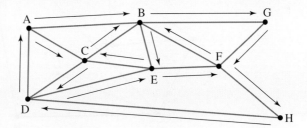

Euler trail: ABGFHDEFBECDACB

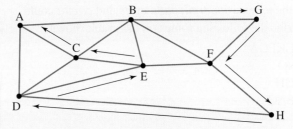

Hamiltonian path: BGFHDECA

Euler trail: ABGFHDEFBECDACB (other answers exist).
Hamiltonian path: BGFHDECA (other answers exist).

9.4 Exercise

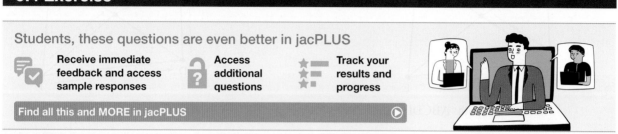
1. **WE7** In the following network, identify two different routes: one cycle and one circuit.

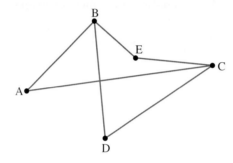

2. In the following network, identify three different routes: one path, one cycle and one circuit.

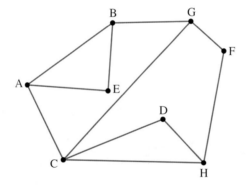

3. State which of the terms *walk*, *trail*, *path*, *cycle* and *circuit* could be used to describe the following routes on the graph shown.

    a. AGHIONMLKFGA
    b. IHGFKLMNO
    c. HIJEDCBAGH
    d. FGHIJEDCBAG

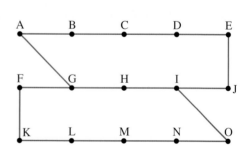

4. Use the following graph to identify the indicated routes.

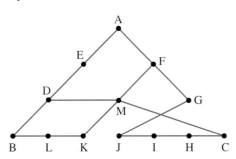

a. A path commencing at M, including at least 10 vertices and finishing at D
b. A trail from A to C that includes exactly 7 edges
c. A cycle commencing at M that includes 10 edges
d. A circuit commencing at F that includes 7 vertices

5. **WE8** Identify an Euler trail and a Hamiltonian path in each of the following graphs.

a.
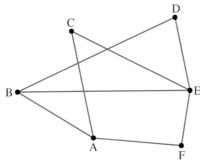

b.
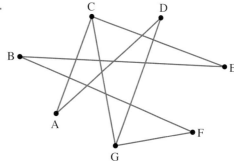

6. Identify an Euler circuit and a Hamiltonian cycle in each of the following graphs, if they exist.

a.
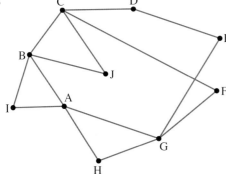

b.
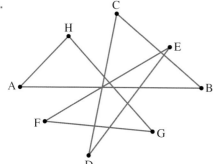

7. Consider the following graphs.

i.
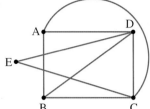

ii.
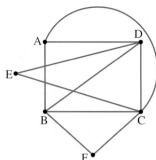

iii.
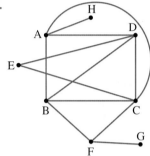

iv.
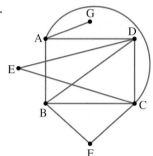

a. Identify which of the graphs have an Euler trail.
b. Identify the Euler trails found.
c. Identify which of the graphs have a Hamiltonian cycle.
d. Identify the Hamiltonian cycles found.
e. Construct adjacency matrices for each of the graphs.
f. Explain how these might assist with making decisions about the existence of Euler trails and circuits, and Hamiltonian paths and cycles.

8. In the following graph, if an Euler trail commences at vertex A, identify the vertices at which it could finish.

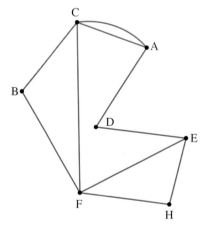

9. In the following graph, identify at which vertices a Hamiltonian path could finish if it commences by travelling from:

   a. B to E

   b. E to A.

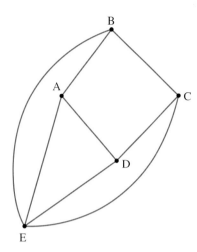

10. In the following graph, other than from G to F, identify the 2 vertices between which you must add an edge in order to create a Hamiltonian path that commences from:

    a. vertex G

    b. vertex F.

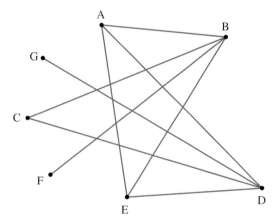

11. On the map shown, a school bus route is indicated in yellow. The bus route starts and ends at the school indicated.

    a. Draw a graph to represent the bus route.
    b. Students can catch the bus at stops that are located at the intersections of the roads marked in yellow. Determine whether it is possible for the bus to collect students by driving down each section of the route only once. Explain your answer.
    c. If roadworks prevent the bus from travelling along the sections indicated by the Xs, determine whether it would be possible for the bus to still collect students on the remainder of the route by travelling each section only once.
    Explain your answer.

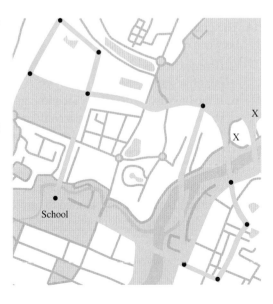

12. The map of an orienteering course is shown. Participants must travel to each of the nine checkpoints along any of the marked paths.

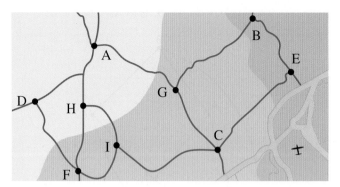

   a. Draw a graph to represent the possible ways of travelling to each checkpoint.
   b. State the degree of checkpoint H.
   c. If participants must start and finish at A, and visit every other checkpoint only once, identify two possible routes they could take.
   d. i. If participants can decide to start and finish at any checkpoint, and the paths connecting D and F, H and I, and A and G are no longer accessible, it is possible to travel the course by moving along each remaining path only once. Explain why.
      ii. Identify the two possible starting points.

13. a. Use the following graph to complete the table to identify all of the Hamiltonian cycles commencing at vertex A.

	Hamiltonian cycle
1.	ABCDA
2.	
3.	
4.	
5.	
6.	

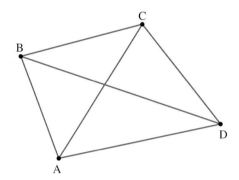

   b. State whether any other Hamiltonian cycles are possible.

14. The graph shown outlines the possible ways a tourist bus can travel between eight locations.

   a. If vertex A represents the second location visited, list the possible starting points.
   b. If the bus also visited each location only once, state which of the starting points listed in part **a** could not be correct.
   c. If the bus also needed to finish at vertex D, list the possible paths that could be taken.

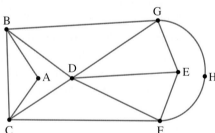

   d. If instead the bus company decided to operate a route that travelled to each connection only once, state the possible starting and finishing points.
   e. If instead the company wanted to travel to each connection only once and finish at the starting point, state which edge of the graph would need to be removed.

9.4 Exam questions

▶ **Question 1 (1 mark)**

Source: VCE 2021, Further Mathematics Exam 1, Section B, Module 2, Q5; © VCAA.

MC Consider the following five statements about the graph shown:
- The graph is planar.
- The graph contains a cycle.
- The graph contains a bridge.
- The graph contains an Eulerian trail.
- The graph contains a Hamiltonian path.

How many of these statements are true?

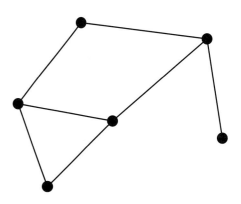

A. 1   **B.** 2   **C.** 3   **D.** 4   **E.** 5

▶ **Question 2 (1 mark)**

Source: VCE 2020, Further Mathematics Exam 1, Section B, Module 2, Q2; © VCAA.

MC Consider the graph below.

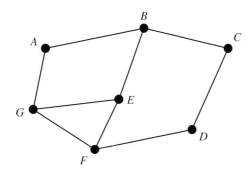

Which one of the following is not a Hamiltonian cycle for this graph?

A. *ABCDFEGA*   **B.** *BAGEFDCB*   **C.** *CDFEGABC*

D. *DCBAGFED*   **E.** *EGABCDFE*

▶ **Question 3 (1 mark)**

Source: VCE 2019, Further Mathematics Exam 1, Section B, Module 2, Q2; © VCAA.

MC Consider the graph below.

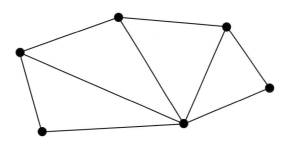

The minimum number of extra edges that are required so that an Eulerian circuit is possible in this graph is

A. 0   **B.** 1   **C.** 2   **D.** 3   **E.** 4

More exam questions are available online.

9.5 Weighted graphs and networks, and trees

LEARNING INTENTION

At the end of this subtopic you should be able to:
- identify the properties of weighted graphs and networks
- solve the shortest path problem
- identify the properties of trees and minimum spanning trees
- apply greedy algorithms to identify the minimum spanning tree.

9.5.1 Weighted graphs

In many applications using graphs, it is useful to attach a value to the edges. These values could represent the length of the edge in terms of time or distance, or the costs involved with moving along that section of the path. Such graphs are known as **weighted graphs**.

Weighted graphs can be particularly useful as analysis tools. For example, they can help determine how to travel through a network in the shortest possible time.

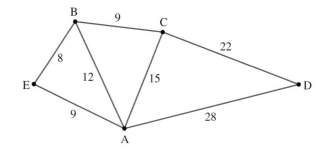

tlvd-4821

WORKED EXAMPLE 9 Identifying the shortest path between two vertices

The graph represents the distances in kilometres between eight locations.
Identify the shortest distance to travel from A to D that goes to all vertices.

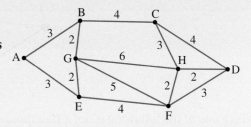

THINK

1. Identify the Hamiltonian paths that connect the two vertices.

2. Calculate the total distances for each path to determine the shortest distance.

3. Write the answer as a sentence.

WRITE

Possible paths:
- **a.** ABGEFHCD
- **b.** ABCHGEFD
- **c.** AEGBCHFD
- **d.** AEFGBCHD
- **e.** AEFHGBCD

- **a.** $3+2+2+4+2+3+4 = 20$
- **b.** $3+4+3+6+2+4+3 = 25$
- **c.** $3+2+2+4+3+2+3 = 19$
- **d.** $3+4+5+2+4+3+2 = 23$
- **e.** $3+4+2+6+2+4+4 = 25$

The shortest distance from A to D that travels to all vertices is 19 km.

9.5.2 Trees

A **tree** is a simple connected graph with no circuits. As such, any pair of vertices in a tree is connected by a unique path, and the number of edges is always 1 less than the number of vertices.

A tree with *n* vertices has *n* − 1 edges.

Spanning trees are sub-graphs (graphs that are formed from part of a larger graph) that include all of the vertices of the original graph. In practical settings, they can be very useful in analysing network connections.

For example, a *minimum* spanning tree for a weighted graph can identify the lowest-cost connections. Spanning trees can be obtained by systematically removing any edges that form a circuit, one at a time.

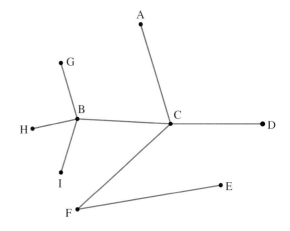

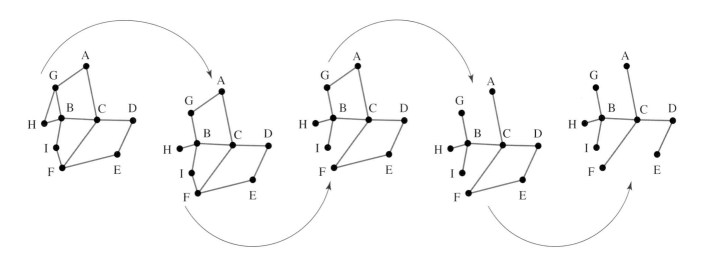

There may be more than one spanning tree in a weighted graph.

9.5.3 Greedy algorithms and minimum spanning trees

An **algorithm** is a finite set of steps or instructions for performing a computation or for solving a problem. Many algorithms are designed to find a solution to a given problem that either minimises or maximises the value of some parameter. A **greedy algorithm** is an algorithm that seems to make the best choice at each step. Although a greedy algorithm will find a solution to a problem, it will not necessarily find the optimal solution every time. We call the algorithm *greedy* whether or not it finds an optimal solution.

Consider the following directed graph, where the aim is to determine the largest product of numbers.

A greedy algorithm will try and make what appears to be the 'best' choice at *each step*. That is:

At step 1, there is only one number to select, so the number 4 will be selected.

At step 2, the algorithm could choose 5 or 10. A greedy algorithm will select 10, as it is the larger of the two numbers.

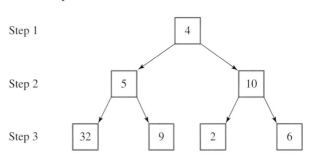

As the number 10 was chosen in step 2, the possible choices at step 3 are 2 or 6. A greedy algorithm will select 6, as it is the larger of the two numbers.

The following diagrams give a visual representation of how a greedy algorithm selected its numbers, compared to the optimal solution. The greedy algorithm found a solution with a product of 240 ($4 \times 10 \times 6$), which, upon inspection, is not the largest product. The largest product of numbers (and the optimal solution) is 640 ($4 \times 5 \times 32$).

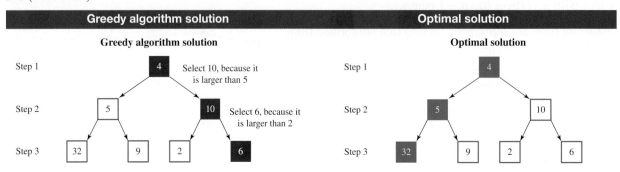

A **minimum spanning** tree is a spanning tree with the lowest total weighting. Prim's algorithm and **Kruskal's algorithm** are examples of greedy algorithms that find a minimum spanning tree. When applied correctly, these greedy algorithms will *always* find the optimal solution. The proof that Prim's algorithm and Kruskal's algorithm produce a minimum spanning tree for every connected weighted graph is beyond the scope of this course.

9.5.4 Prim's algorithm

Prim's algorithm is a set of logical steps that can be used to identify the minimum spanning tree for a weighted connected graph.

> **Prim's algorithm for determining a minimum spanning tree**
>
> **Step 1: Select the edge that has the lowest weight. If two or more edges are the lowest, choose any of these.**
>
> **Step 2: Looking at the two vertices included so far, select the smallest edge leading from either vertex. If two or more edges are the smallest, choose any of these.**
>
> **Step 3: Look at all vertices included so far and select the smallest edge leading from any included vertex. If two or more edges are the smallest, choose any of these.**
>
> **Step 4: Continue step 3 until all vertices are connected or until there are $n - 1$ edges in the spanning tree. Remember to *not* select an edge that would create a circuit.**

tlvd-4822

> **WORKED EXAMPLE 10 Using Prim's algorithm to determine a minimum spanning tree**
>
> a. **Use Prim's algorithm to identify the minimum spanning tree of the graph shown.**
> b. **Determine the length of the minimum spanning tree.**

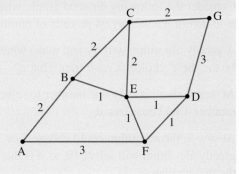

THINK

a. 1. Draw the vertices of the graph and include the edge with the lowest weight.

Note: Since BE, EF, DE and DF all have a weight of 1, choose any of these edges.

2. Look at the two included vertices B and E. Select the smallest edge leading from B or E.

Note: You can select edge ED or EF as they both have a weight of 1.

3. Look at the vertices included so far (B, E and D). Select the smallest edge leading from any of these vertices.

Note: You can select EF or DF as they both have a weight of 1.

4. Look at the vertices included so far (B, E, D and F). Select the smallest edge leading from any of these vertices.

Note: You *cannot* select edge DF as it will create a circuit. You must now select an edge with a weight of 2 (the next lowest). Choose either edge AB, BC or CE.

5. Repeat this process until all vertices are connected. When selecting edges, do not create any circuits.

b. 1. The length of the minimum spanning tree can be determined by adding up the weight of the edges that create the minimum spanning tree.

DRAW

a.

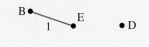

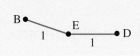

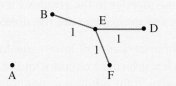

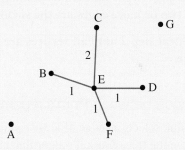

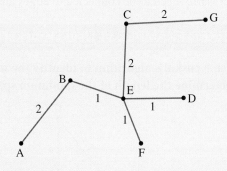

b. Length $= 1 + 1 + 1 + 2 + 2 + 2 = 9$

Note: In Worked example 10, it is possible to draw more than one minimum spanning tree. An example of a different minimum spanning tree is:

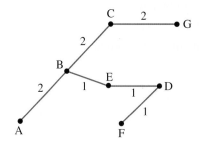

The length of this minimum spanning tree is still 9. Determine how many different minimum spanning trees are possible.

9.5.5 Kruskal's algorithm

Kruskal's algorithm is a set of logical steps that can be used to identify the minimum spanning tree for a connected weighted graph.

Kruskal's algorithm for finding a minimum spanning tree

Step 1: Choose an edge in the graph with the lowest weight and add it to the minimum spanning tree. (If two or more edges are the smallest, choose any of these.)

Step 2: Find the next edge of lowest weight.
Check if it forms a circuit with the spanning tree formed so far. If a circuit is not formed, add this edge to the minimum spanning tree; otherwise, disregard it.
(If two or more edges are the smallest, choose any of these.)

Step 3: Repeat step 2 until all vertices are connected or until there are ($n - 1$) edges in the spanning tree.

The difference between Prim's algorithm and Kruskal's algorithm

Prim's algorithm selects an edge of lowest weight that must be *connected* to the edges already included in the minimum spanning tree. Kruskal's algorithm builds up the minimum spanning tree piece by piece. The edges do not necessarily need to be connected to edges already included in the minimum spanning tree.

WORKED EXAMPLE 11 Using Kruskal's algorithm to determine a minimum spanning tree

a. **Use Kruskal's algorithm to identify the minimum spanning tree of the graph shown.**
b. **Determine the length of the minimum spanning tree.**

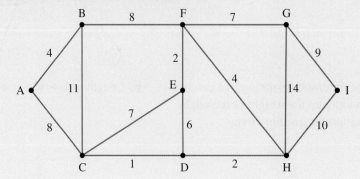

THINK

DRAW

a. 1. Draw the vertices of the graph. Identify the edge with the lowest weight and include this edge in the minimum spanning tree. Edge CD has the lowest weight, 1.

a.

2. The next edge of lowest weight has a weight of 2 (edges DH and EF).
Select edge EF and check if adding this edge to the minimum spanning tree creates a circuit. Adding this edge would not create a circuit, so include it in the minimum spanning tree.
Note: Edge DH could have been selected instead.

3. The next edge of lowest weight has a weight of 2 (edge DH).
Check if adding this edge to the minimum spanning tree creates a circuit.
Adding this edge would not create a circuit, so include it in the minimum spanning tree.

4. The next edge of lowest weight has a weight of 4 (edges AB and FH).
Select edge AB and check if adding this edge to the minimum spanning tree creates a circuit. Adding this edge would not create a circuit, so include it in the minimum spanning tree.

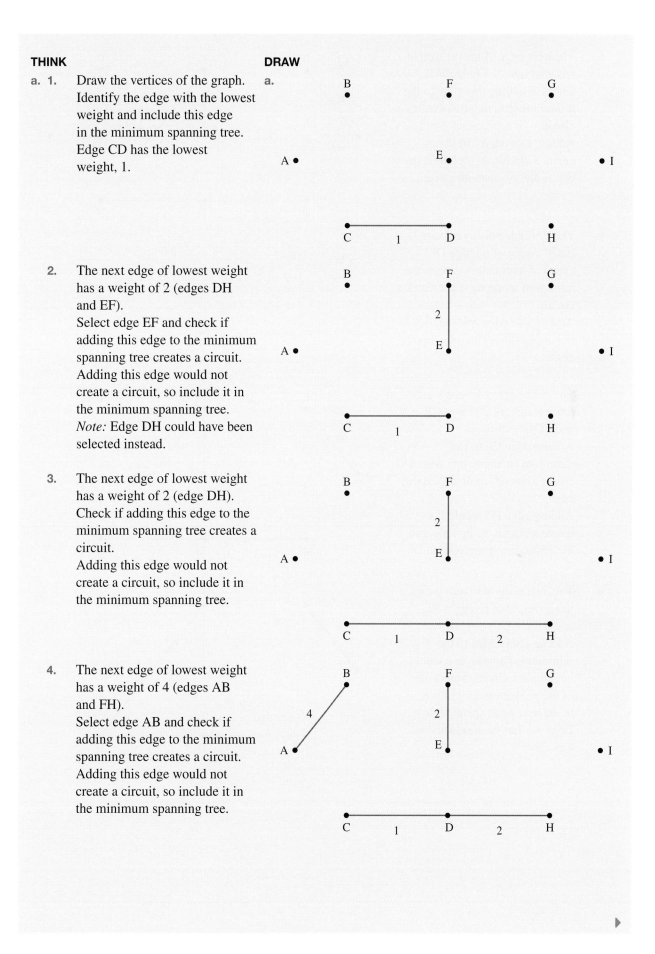

5. The next edge of lowest weight
   has a weight of 4 (edge FH).
   Check if adding this edge to the
   minimum spanning tree creates a
   circuit.
   Adding this edge would not
   create a circuit, so include it in
   the minimum spanning tree.

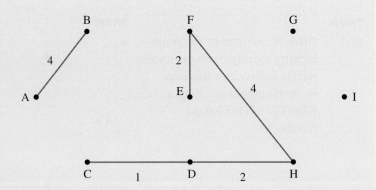

6. The next edge of lowest weight
   has a weight of 6 (edge DE).
   Check if adding this edge to the
   minimum spanning tree creates a
   circuit.
   Adding this edge will create a
   circuit, so do *not* include this
   edge.

7. The next edge of lowest weight
   has a weight of 7 (edges CE
   and FG).
   Adding edge CE to the
   minimum spanning tree would
   create a circuit, so disregard this
   edge.
   Adding edge FG would not
   create a circuit, so include it in
   the minimum spanning tree.

8. The next edge of lowest weight
   has a weight of 8 (edges AC
   and BF).
   Adding either edge to the
   minimum spanning tree would
   not create a circuit. Select *one*
   of these edges and include it
   in the minimum spanning tree.
   Edge AC has been selected in
   this case.

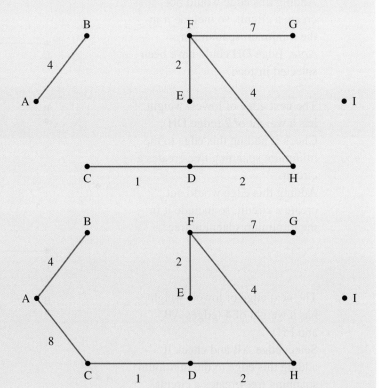

9. The next edge of lowest weight has a weight of 9 (edge GI). Check if adding this edge to the minimum spanning tree creates a circuit.

Adding this edge would not create a circuit, so include it in the minimum spanning tree.

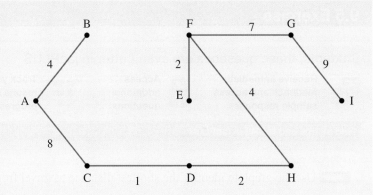

10. At this stage all vertices have been reached, so we have the completed minimum spanning tree.

Also, since the original graph has 9 vertices, the minimum spanning tree should contain 8 edges.

The minimum spanning tree is:

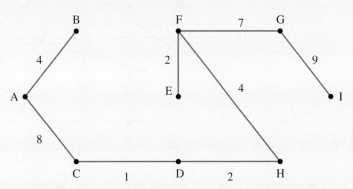

Note: Other minimum spanning trees are possible for this problem.

b. The length of the minimum spanning tree can be determined by adding up the weights of the edges that create the minimum spanning tree.

b. Length $= 1 + 2 + 2 + 4 + 4 + 7 + 8 + 9$
$= 37$

 Resources

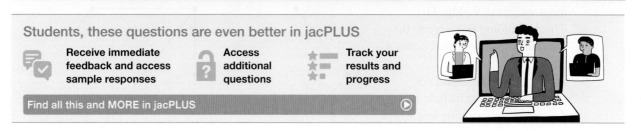
1. **WE9** Use the graph to identify the shortest distance to travel from A to D that goes to all vertices.

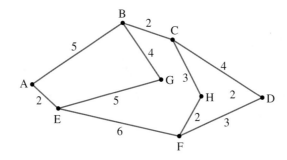

2. Use the graph to identify the shortest distance to travel from A to I that goes to all vertices.

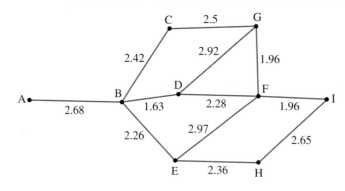

3. Draw three spanning trees for each of the following graphs.

a.

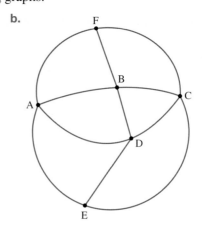

b.

4. A truck starts from the main distribution point at vertex A and makes deliveries at each of the other vertices before returning to A.
   Determine the shortest route the truck can take.

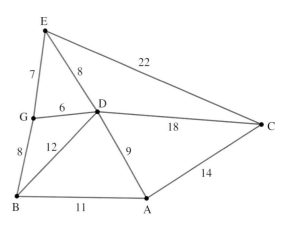

5. For the following trees:
   i. add the minimum number of edges to create an Euler trail
   ii. identify the Euler trail created.

   a.

   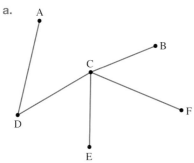

   b.

   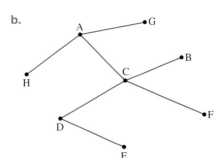

   c.

   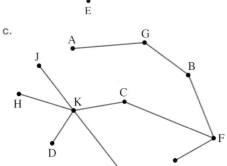

   d.

   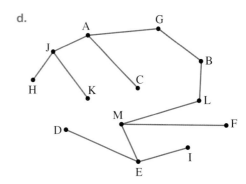

6. **WE10** Consider the graph shown.
   a. Use Prim's algorithm to identify the minimum spanning tree.
   b. State the length of the minimum spanning tree.

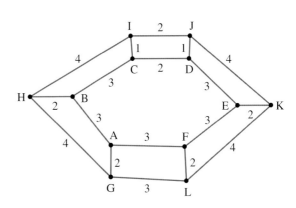

7. Use Prim's algorithm to identify the minimum spanning tree of the graph shown.

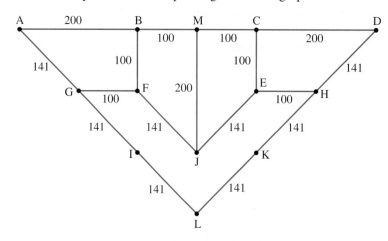

8. Identify the minimum spanning tree for each of the following graphs.

a.

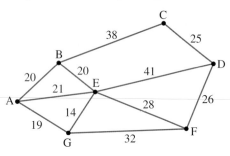

b.

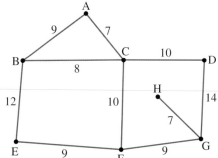

9. Draw diagrams to show the steps you would follow when using Prim's algorithm to identify the minimum spanning tree for the following graph.

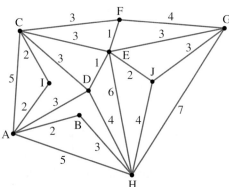

10. **WE11** For each of the following graphs:
   i. use Kruskal's algorithm to identify the minimum spanning tree
   ii. state the length of the minimum spanning tree.

a.

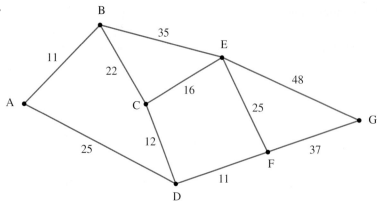

b.

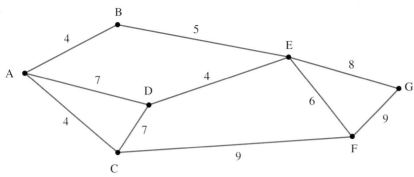

c.

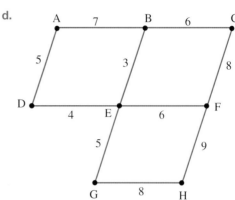

d.

e.

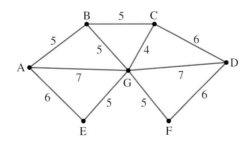

11. Consider the graph shown.

a. If an edge with the highest weighting is removed, identify the shortest Hamiltonian path.
b. If the edge with the lowest weighting is removed, identify the shortest Hamiltonian path.

12. Consider the graph shown.
   a. Identify the longest and shortest Hamiltonian paths.
   b. State the minimum spanning tree for this graph.

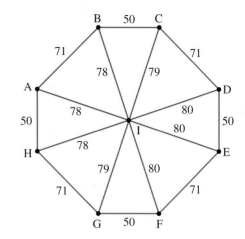

13. The weighted graph represents the costs incurred by a salesperson when moving between the locations of various businesses.
   a. Identify the cheapest way of travelling from A to G.
   b. Identify the cheapest way of travelling from B to G.
   c. If the salesperson starts and finishes at E, state the cheapest way to travel to all vertices.

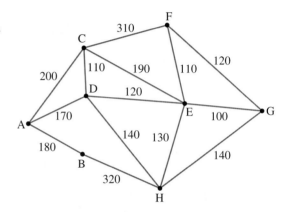

14. The organisers of the Tour de Vic bicycle race are using the following map to plan the event.

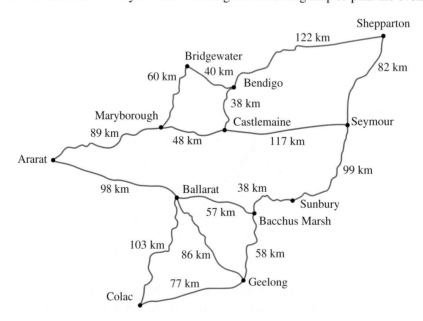

   a. Draw a weighted graph to represent the map.
   b. If they wish to start and finish in Geelong, determine the shortest route that can be taken that includes a total of nine other locations exactly once, two of which must be Ballarat and Bendigo.
   c. Draw the minimum spanning tree for the graph.
   d. If the organisers decide to use the minimum spanning tree as the course, determine the shortest possible distance if each location had to be reached at least once.

15. A mining company operates in several locations in Western Australia and the Northern Territory, as shown on the map.

Flights operate between selected locations, and the flight distances (in km) are shown in the following table.

	A	B	C	D	E	F	G	H	I
A		1090		960			2600		2200
B	1090		360	375	435				
C		360							
D	960	375							
E		435							
F							1590	1400	
G	2600					1590		730	
H						1400	730		220
I	2200							220	

a. Show this information as a weighted graph.

b. State whether a Hamiltonian path exists. Explain your answer.

c. Identify the shortest distance possible for travelling to all sites the minimum number of times if you start and finish at:
   i. location A
   ii. location G.

d. Draw the minimum spanning tree for the graph.

Question 1 (1 mark)

Source: VCE 2020, Further Mathematics Exam 1, Section B, Module 2, Q3; © VCAA.

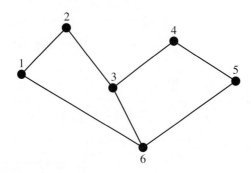

MC Which one of the following is not a spanning tree for the network above?

A.

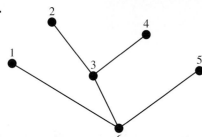

B.

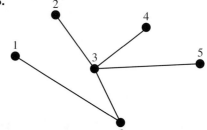

C.

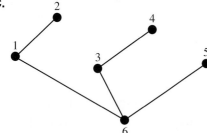

D.

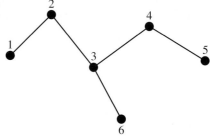

E.

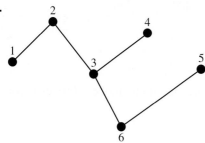

Question 2 (1 mark)

Source: VCE 2020, Further Mathematics Exam 1, Section B, Module 2, Q5; © VCAA.

MC The network below shows the distances, in metres, between camp sites at a camping ground that has electricity.

The vertices *A* to *I* represent the camp sites.

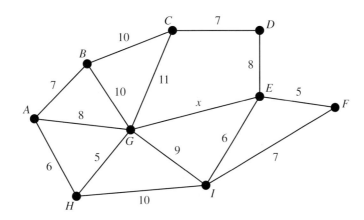

The minimum length of cable required to connect all the camp sites is 53 m.

The value of *x*, in metres, is at least

   A. 5   **B.** 6   **C.** 8   **D.** 9   **E.** 11

Question 3 (1 mark)

Source: VCE 2019, Further Mathematics Exam 1, Section B, Module 2, Q5; © VCAA.

MC The following diagram shows the distances, in metres, along a series of cables connecting a main server to seven points, *A* to *G*, in a computer network.

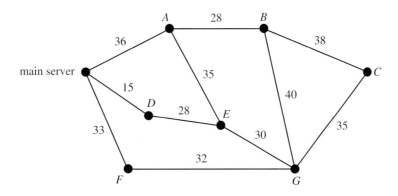

The minimum length of cable, in metres, required to ensure that each of the seven points is connected to the main server directly or via another point is

   A. 175   **B.** 203   **C.** 208   **D.** 221   **E.** 236

More exam questions are available online.

9.6 Exercise

Multiple choice

1. **MC** The minimum number of edges in a connected graph with eight vertices is:

   A. 5   **B.** 6   **C.** 7   **D.** 8   **E.** 9

2. **MC** Select the graph that does not have an Euler trail.

A.

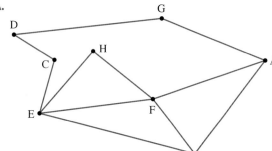

B.

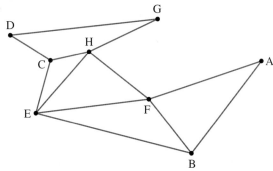

C.

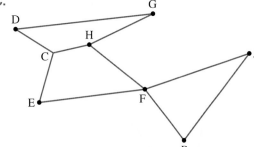

D.

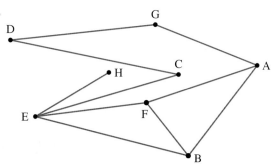

E.

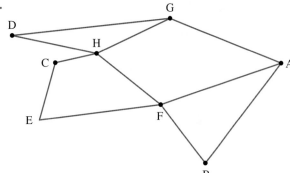

3. **MC** A connected graph with 9 vertices has 10 faces. The number of edges in the graph is:

   A. 15   **B.** 16   **C.** 17   **D.** 18   **E.** 19

4. **MC** Select the graph that is a spanning tree for the following graph.

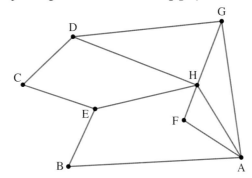

A.

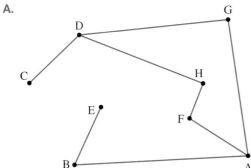

B.

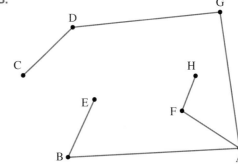

C.

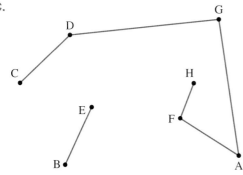

D.

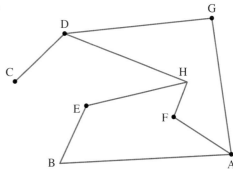

E.

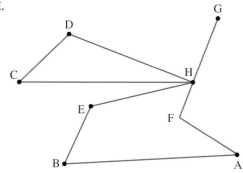

5. **MC** Select the length of the minimum spanning tree of the following graph.

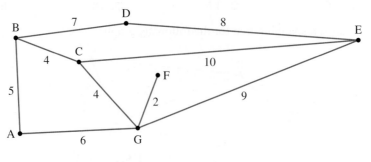

A. 33   **B.** 26   **C.** 34   **D.** 31   **E.** 30

6. **MC** An Euler circuit can be created in the following graph by adding an edge between the vertices:

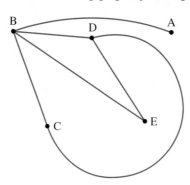

A. A and D.   **B.** A and B.   **C.** A and C.   **D.** B and C.   **E.** A and E.

7. **MC** The adjacency matrix that represents the following graph is:

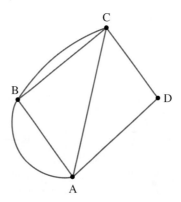

A. $\begin{bmatrix} 0 & 2 & 2 & 2 \\ 2 & 0 & 2 & 0 \\ 2 & 2 & 0 & 1 \\ 2 & 0 & 1 & 0 \end{bmatrix}$ **B.** $\begin{bmatrix} 0 & 1 & 1 & 0 \\ 1 & 0 & 1 & 0 \\ 1 & 1 & 0 & 1 \\ 1 & 0 & 1 & 0 \end{bmatrix}$ **C.** $\begin{bmatrix} 0 & 2 & 1 & 1 \\ 2 & 1 & 2 & 0 \\ 1 & 2 & 1 & 1 \\ 1 & 0 & 1 & 1 \end{bmatrix}$

D. $\begin{bmatrix} 1 & 2 & 1 & 1 \\ 2 & 0 & 2 & 0 \\ 1 & 2 & 1 & 1 \\ 1 & 0 & 1 & 0 \end{bmatrix}$ **E.** $\begin{bmatrix} 0 & 2 & 1 & 1 \\ 2 & 0 & 2 & 0 \\ 1 & 2 & 0 & 1 \\ 1 & 0 & 1 & 0 \end{bmatrix}$

8. **MC** The number of faces in the planar graph shown is:

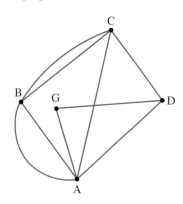

A. 6   B. 7   C. 8   D. 9   E. 10

9. **MC** A Hamiltonian cycle for the graph shown is:

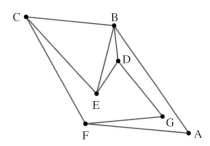

A. ABCEDGFA   B. ABDGFCEA   C. ABDGFCEDEBCFA
D. ABDGFCECFA   E. ABDGFA

10. **MC** In the graph shown, select the number of edges that are bridges.

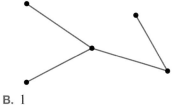

A. 0   B. 1   C. 2
D. 3   E. 4

Short answer

11. a. Identify whether the following graphs are planar or not planar.

i.

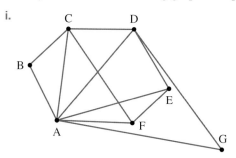

ii.

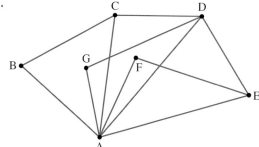

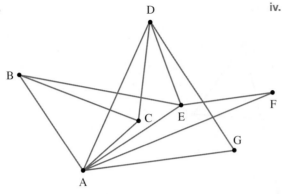

iii.

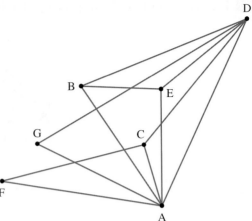

iv.

b. Redraw the graphs that are planar without any intersecting edges.

12. Complete the following adjacency matrices.

a. $\begin{bmatrix} 1 & 1 & 0 & 1 \\ & 0 & & 0 \\ & 3 & 1 & \\ & & 1 & 0 \end{bmatrix}$

b. $\begin{bmatrix} 0 & 1 & 2 & 1 & & \\ & 0 & & 0 & 1 \\ & 2 & 0 & 2 & \\ & & 2 & 2 & \\ 1 & & 3 & & 0 \end{bmatrix}$

c. $\begin{bmatrix} 0 & & 1 & 3 & 1 & \\ 2 & 0 & & & 1 & \\ & 3 & 0 & 2 & & \\ & 1 & & 2 & 2 & 1 \\ & & 3 & & 3 & 1 \\ 2 & 0 & 1 & & & 0 \end{bmatrix}$

d. $\begin{bmatrix} 0 & & & & 0 & & \\ 2 & 0 & & 1 & & & \\ 1 & 2 & 0 & 1 & 1 & 0 & \\ 3 & & & 0 & & & 1 \\ & 2 & & 0 & 0 & & 3 \\ 1 & 1 & & 2 & 0 & 0 & 2 \\ 0 & 0 & 1 & & & & 0 \end{bmatrix}$

13. Identify which of the following graphs are:

   i. simple   **ii.** planar.

a.

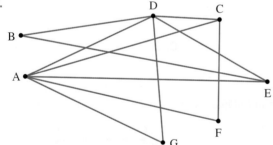

b.

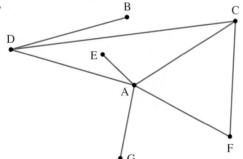

c.

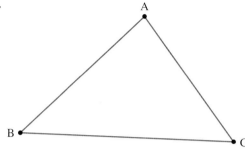

d.

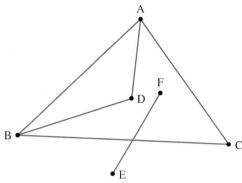

e.

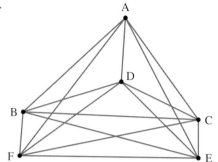

f.

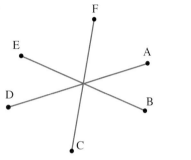

g.

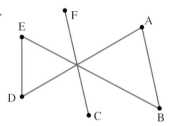

h.

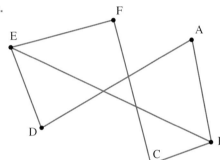

14. Identify which of the following graphs are isomorphic.

a.

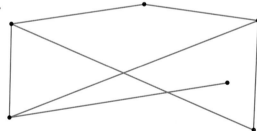

b.

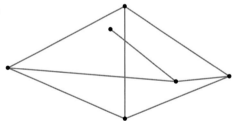

c.

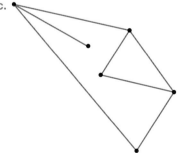

d.

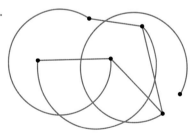

15. For each of the following graphs:

   i. add the minimum number of edges required to create an Euler trail
   ii. state the Euler trail created.

   a.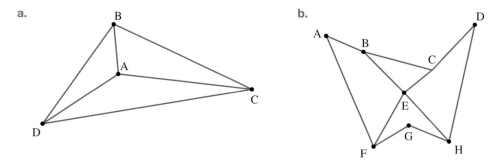

   b.

16. a. Determine the shortest distance from start to finish in the following graph.

   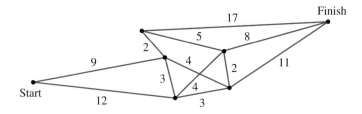

   b. Calculate the total length of the shortest Hamiltonian path from start to finish.
   c. Draw the minimum spanning tree for this graph.

Extended response

17. The flying distances between the capital cities of Australian mainland states and territories are listed in the following table.

	Adelaide	Brisbane	Canberra	Darwin	Melbourne	Perth	Sydney
Adelaide		2055	1198	3051	732	2716	1415
Brisbane	2055		1246	3429	1671	4289	982
Canberra	1198	1246		4003	658	3741	309
Darwin	3051	3429	4003		3789	4049	4301
Melbourne	732	1671	658	3789		3456	873
Perth	2716	4289	3741	4049	3456		3972
Sydney	1415	982	309	4301	873	3972	

   a. Draw a weighted graph to show this information.
   b. If technical problems are preventing direct flights from Melbourne to Darwin and from Melbourne to Adelaide, calculate the shortest way of flying from Melbourne to Darwin.
   c. If no direct flights are available from Brisbane to Perth or from Brisbane to Adelaide, calculate the shortest way of getting from Brisbane to Perth.
   d. Draw the minimum spanning tree for the graph and state its total distance.

18. The diagram shows the streets in a suburb of a city with a section of underground tunnels shown in dark blue. Weightings indicate distances in metres. The tunnels are used for utilities such as electricity, gas, water and drainage.

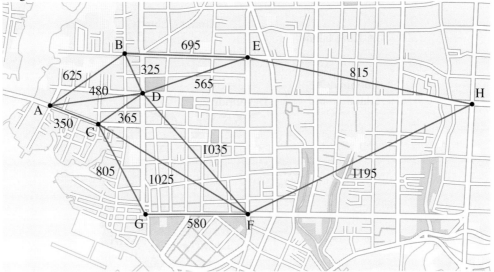

a. i. If the gas company wishes to run a pipeline that minimises its total length but reaches each vertex, calculate the total length required.

   ii. Draw a graph to show the gas lines.

b. If drainage pipes need to run from H to A, calculate the shortest path they can follow. Determine how long this path will be in total.

c. A single line of cable for a computerised monitoring system needs to be placed so that it starts at D and reaches every vertex once. Calculate the minimum length possible and determine the path it must follow.

d. If a power line has to run from D so that it reaches every vertex at least once and finishes back at the start, determine what path it must take to be a minimum.

19. A brochure for a national park includes a map showing the walking trails and available camping sites.

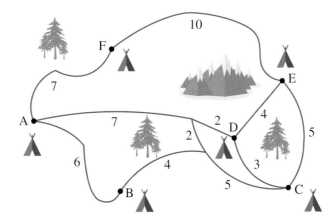

a. Draw a weighted graph to represent all the possible ways of travelling to the camp sites.

b. Draw the adjacency matrix for the graph.

c. Determine whether it is possible to walk a route that travels along each edge exactly once. Explain your answer, and indicate the path if it is possible.

d. If the main entrance to the park is situated at A, calculate the shortest way to travel to each campsite and return to A.

20. A cruise ship takes passengers around Tasmania between the seven locations marked on the map.

The sailing distances between locations are indicated in the table.

	Hobart	Bruny I.	Maria I.	Flinders I.	Devonport	Robbins I.	King I.
Hobart	–	65 km	145 km	595 km	625 km	–	–
Bruny I.	65 km	–	130 km	–	–	715 km	–
Maria I.	145 km	130 km	–	450 km	–	–	–
Flinders I.	595 km	–	450 km	–	330 km	405 km	465 km
Devonport	625 km	–	–	330 km	–	265 km	395 km
Robbins I.	–	715 km	–	405 km	265 km	–	120 km
King I.	–	–	–	465 km	395 km	120 km	–

a. Draw a weighted graph to represent all possible ways of travelling to the locations.
b. Calculate the shortest route from Hobart to Robbins Island.
c. Calculate the shortest way of travelling from Hobart to visit each location only once.
d. Calculate the shortest way of sailing from King Island, visiting each location once and returning to King Island.

9.6 Exam questions

▶ **Question 1 (2 marks)**
Source: VCE 2021, Further Mathematics Exam 2, Section B, Module 2, Q2; © VCAA.

George lives in Town *G* and Maggie lives in Town *M*.

The diagram below shows the network of main roads between Town *G* and Town *M*.

The vertices *G, H, I, J, K, L, M, N* and *O* represent towns.

The edges represent the main roads. The numbers on the edges indicate the distances, in kilometres, between adjacent towns.

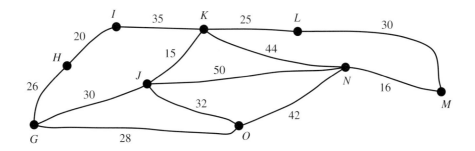

a. What is the shortest distance, in kilometres, between Town G and Town M? **(1 mark)**

b. George plans to travel to Maggie's house. He will pass through all the towns shown above. George plans to take the shortest route possible. Which town will George pass through twice? **(1 mark)**

Question 2 (4 marks)

▶ **Question 2 (4 marks)**

Source: VCE 2020, Further Mathematics Exam 2, Section B, Module 2, Q3; © VCAA.

A local fitness park has 10 exercise stations: M to V.

The edges on the graph below represent the tracks between the exercise stations.

The number on each edge represents the length, in kilometres, of each track.

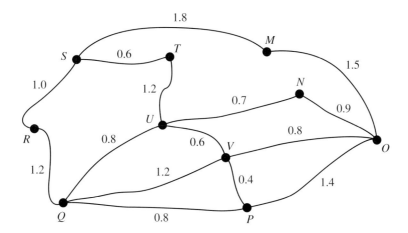

The Sunny Coast cricket coach designs three different training programs, **all starting at exercise station S**.

Training program number	Training details
1	The team must run to exercise station O.
2	The team must run along all tracks just once.
3	The team must visit each exercise station and return to exercise station S.

a. What is the shortest distance, in kilometres, covered in training program 1? **(1 mark)**

b. i. What mathematical term is used to describe training program 2? **(1 mark)**

  ii. At which exercise station would training program 2 finish? **(1 mark)**

c. To complete training program 3 in the minimum distance, one track will need to be repeated.

Complete the following sentence by filling in the boxes provided.

This track is between exercise station _____ and exercise station _____. **(1 mark)**

▶ **Question 3 (3 marks)**

Source: VCE 2019, Further Mathematics Exam 2, Section B, Module 2, Q1; © VCAA.

Fencedale High School has six buildings. The network below shows these buildings represented by vertices. The edges of the network represent the paths between the buildings.

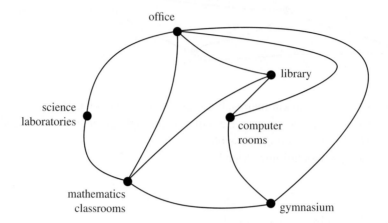

a. Which building in the school can be reached directly from all other buildings? **(1 mark)**

b. A school tour is to start and finish at the office, visiting each building only once.

  i. What is the mathematical term for this route? **(1 mark)**

  ii. Draw in a possible route for this school tour on the diagram below. **(1 mark)**

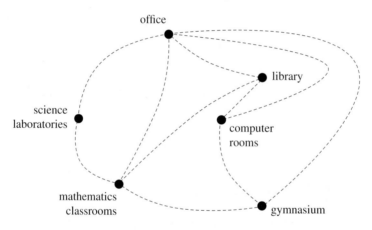

▶ **Question 4 (1 mark)**

Source: VCE 2019, Further Mathematics Exam 1, Section B, Module 2, Q4; © VCAA.

MC Two graphs, labelled Graph 1 and Graph 2, are shown below.

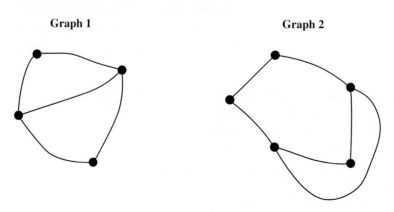

Graph 1   **Graph 2**

Which one of the following statements is **not** true?
  A. Graph 1 and Graph 2 are isomorphic.
  B. Graph 1 has five edges and Graph 2 has six edges.
  C. Both Graph 1 and Graph 2 are connected graphs.
  D. Both Graph 1 and Graph 2 have three faces each.
  E. Neither Graph 1 nor Graph 2 are complete graphs.

Question 5 (1 mark)

Source: VCE 2017, Further Mathematics Exam 1, Section B, Module 2, Q6; © VCAA.

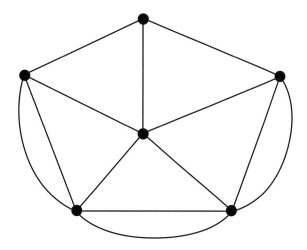

MC An Eulerian trail for the graph above will be possible if only one edge is removed.

In how many different ways could this be done?
  A. 1
  B. 2
  C. 3
  D. 4
  E. 5

More exam questions are available online.

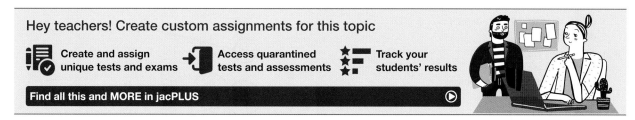

Answers

Topic 9 Graphs and networks

9.2 Definitions and terms

9.2 Exercise

1.

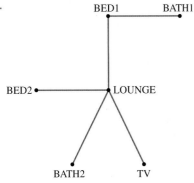

2.

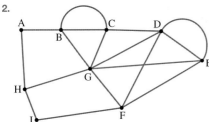

3. a. Edges $= 7$; degree sum $= 14$
   a. Edges $= 10$; degree sum $= 20$

4. a. Edges $= 9$; degree sum $= 18$
   b. Edges $= 9$; degree sum $= 18$

5. a. $\deg(A) = 5$; $\deg(B) = 3$; $\deg(C) = 4$; $\deg(D) = 1$; $\deg(E) = 1$

   b. $\deg(A) = 0$; $\deg(B) = 2$; $\deg(C) = 2$; $\deg(D) = 3$; $\deg(E) = 3$

   c. $\deg(A) = 4$; $\deg(B) = 2$; $\deg(C) = 2$; $\deg(D) = 2$; $\deg(E) = 4$

   d. $\deg(A) = 1$; $\deg(B) = 2$; $\deg(C) = 1$; $\deg(D) = 1$; $\deg(E) = 3$

6. a. The graphs are isomorphic.
   b. The graphs are isomorphic.
   c. The graphs are not isomorphic.
   d. The graphs are isomorphic.

7. a. Different degrees and connections
   b. Different connections

8. The isomorphic pairs are graphs 2 and 4, and graphs 5 and 6.

9. a. $\begin{bmatrix} 0 & 1 & 1 & 1 & 0 \\ 1 & 0 & 1 & 0 & 1 \\ 1 & 1 & 0 & 0 & 0 \\ 1 & 0 & 0 & 0 & 1 \\ 0 & 1 & 0 & 1 & 0 \end{bmatrix}$

   b. $\begin{bmatrix} 0 & 0 & 1 & 1 & 2 & 0 \\ 0 & 0 & 1 & 0 & 0 & 1 \\ 1 & 1 & 1 & 0 & 0 & 0 \\ 1 & 0 & 0 & 0 & 0 & 1 \\ 2 & 0 & 0 & 0 & 0 & 0 \\ 0 & 1 & 0 & 1 & 0 & 1 \end{bmatrix}$

   c. $\begin{bmatrix} 0 & 0 & 1 & 1 & 2 & 0 \\ 0 & 0 & 1 & 0 & 0 & 1 \\ 1 & 1 & 0 & 0 & 0 & 0 \\ 1 & 0 & 0 & 0 & 0 & 3 \\ 2 & 0 & 0 & 0 & 0 & 0 \\ 0 & 1 & 0 & 3 & 0 & 0 \end{bmatrix}$

   d. $\begin{bmatrix} 0 & 1 & 1 & 1 & 0 & 0 \\ 1 & 0 & 1 & 1 & 1 & 1 \\ 1 & 1 & 0 & 0 & 1 & 0 \\ 1 & 1 & 0 & 0 & 0 & 1 \\ 0 & 1 & 1 & 0 & 0 & 4 \\ 0 & 1 & 0 & 1 & 4 & 0 \end{bmatrix}$

10. a.

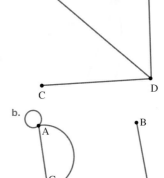

   b.

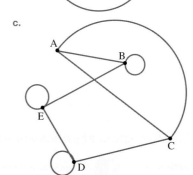

   c.

   d.

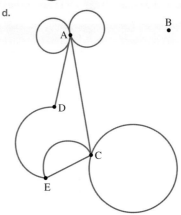

11. Graph 1:

$$\begin{bmatrix} 0 & 1 & 1 & 0 \\ 1 & 0 & 1 & 0 \\ 1 & 1 & 0 & 1 \\ 0 & 0 & 1 & 0 \end{bmatrix}$$

Graph 2:

$$\begin{bmatrix} 0 & 1 & 0 & 0 & 1 & 0 \\ 1 & 0 & 0 & 1 & 0 & 0 \\ 0 & 0 & 0 & 0 & 1 & 1 \\ 0 & 1 & 0 & 0 & 0 & 1 \\ 1 & 0 & 1 & 0 & 0 & 0 \\ 0 & 0 & 1 & 1 & 0 & 0 \end{bmatrix}$$

Graph 3:

$$\begin{bmatrix} 0 & 1 & 0 & 0 & 1 \\ 1 & 0 & 1 & 0 & 0 \\ 0 & 1 & 0 & 1 & 0 \\ 0 & 0 & 1 & 0 & 1 \\ 1 & 0 & 0 & 1 & 0 \end{bmatrix}$$

Graph 4:

$$\begin{bmatrix} 0 & 2 & 1 & 0 & 0 & 0 \\ 2 & 0 & 0 & 0 & 1 & 0 \\ 1 & 0 & 0 & 1 & 0 & 0 \\ 0 & 0 & 1 & 0 & 0 & 0 \\ 0 & 1 & 0 & 0 & 0 & 1 \\ 0 & 0 & 0 & 0 & 1 & 0 \end{bmatrix}$$

Graph 5:

$$\begin{bmatrix} 0 & 1 & 1 & 1 & 1 & 1 \\ 1 & 0 & 1 & 1 & 1 & 1 \\ 1 & 1 & 0 & 1 & 1 & 1 \\ 1 & 1 & 1 & 0 & 1 & 1 \\ 1 & 1 & 1 & 1 & 0 & 1 \\ 1 & 1 & 1 & 1 & 1 & 0 \end{bmatrix}$$

12. a. $\begin{bmatrix} 0 & 0 & 1 \\ 0 & 2 & 2 \\ 1 & 2 & 0 \end{bmatrix}$ **b.** $\begin{bmatrix} 2 & 1 & 0 & 0 \\ 1 & 0 & 1 & 2 \\ 0 & 1 & 0 & 1 \\ 0 & 2 & 1 & 0 \end{bmatrix}$

c. $\begin{bmatrix} 0 & 0 & 1 & 1 & 0 \\ 0 & 0 & 0 & 0 & 0 \\ 1 & 0 & 0 & 0 & 2 \\ 1 & 0 & 0 & 0 & 1 \\ 0 & 0 & 2 & 1 & 0 \end{bmatrix}$ **d.** $\begin{bmatrix} 0 & 0 & 0 & 1 & 0 \\ 0 & 0 & 0 & 1 & 0 \\ 0 & 0 & 0 & 0 & 1 \\ 1 & 1 & 0 & 0 & 0 \\ 0 & 0 & 1 & 0 & 1 \end{bmatrix}$

13. Answers will vary. Possible answers are shown.

a.

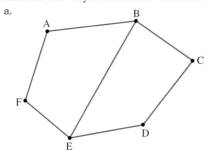

b.

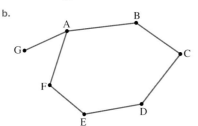

c.

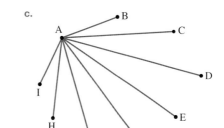

14.

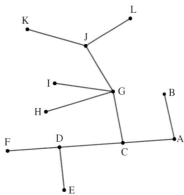

15.

	Simple	Connected
Graph 1	Yes	Yes
Graph 2	Yes	Yes
Graph 3	Yes	Yes
Graph 4	No	Yes
Graph 5	No	Yes

16. a.

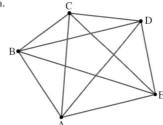

b. Total number of games played

17. a.

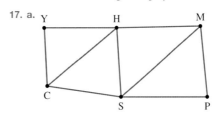

b. Huairou and Shunyi
c. Simple connected graph

18. a.

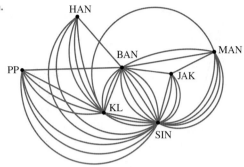

b. Directed, as it would be important to know the direction of the flight

   i. 10   **ii.** 7

9.2 Exam questions

Note: Mark allocations are available with the fully worked solutions online.

1. E

2. E

3. B

9.3 Planar graphs

9.3 Exercise

1. a.

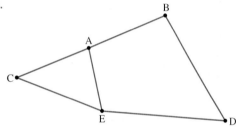

b.

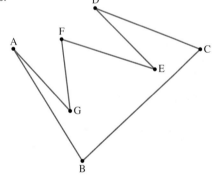

2. a. All of them   **b.** All of them

3. a.

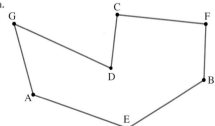

b.

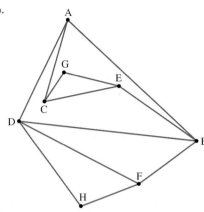

4. Graph 3

5. a. 4

   b. 5

6. a. 6

   b. 5

7. a. 3

   b. 3

   c. 2

   d. 7

8. a.

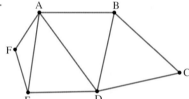

b.

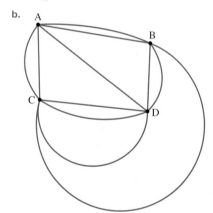

9. a.

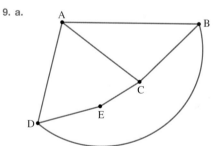

b.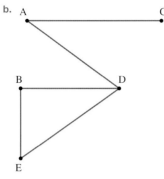

10. a. i. 3   **ii.** 2

   b. i. 1   **ii.** 4

11. a.

	Total edges	**Total degrees**
Graph 1	3	6
Graph 2	5	10
Graph 3	8	16
Graph 4	14	28

   b. Total degrees $= 2 \times$ total edges

12. a.

	Total vertices of even degree	**Total vertices of odd degree**
Graph 1	3	2
Graph 2	4	2
Graph 3	4	4
Graph 4	6	6

   b. No clear pattern is evident.

13.

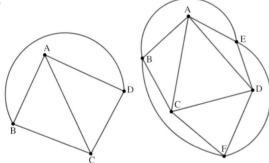

Tetrahedron   Octahedron

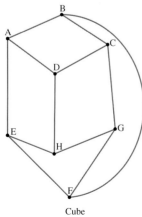

Cube

14. a.

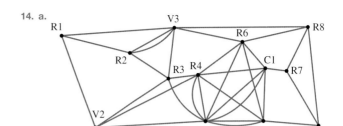

   b. No

15. a.

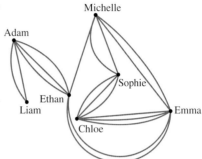

   b. Sophie or Emma

   c.

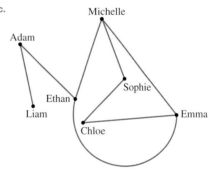

   d. 4

9.3 Exam questions

Note: Mark allocations are available with the fully worked solutions online.

1. D

2. A

3. C

9.4 Connected graphs

9.4 Exercise

1. Cycle: ABECA (others exist)
Circuit: BECDB (others exist)

2. Path: ABGFHDC (others exist)
Cycle: DCGFHD (others exist)
Circuit: AEBGFHDCA (others exist)

3. a. Walk, trail and circuit

   b. Walk, trail and path

   c. Walk, trail, path, cycle and circuit

   d. Walk and trail

4. a. MCHIJGFAED   **b.** AEDBLKMC

   c. MDEAFGJIHCM   **d.** FMCHIJGF

5. a. Euler trail: AFEDBECAB; Hamiltonian path: BDECAF

b. Euler trail: GFBECGDAC;
Hamiltonian path: BECADGF

6. a. Euler circuit: AIBAHGFCJBCDEGA;
Hamiltonian cycle: none exists

b. Euler circuit: ABCDEFGHA (others exist);
Hamiltonian cycle: HABCDEFGH (others exist)

7. a. Graphs i, ii and iv

b. Graph i: ACDABDECB (others exist)
Graph ii: CFBCEDBADCA (others exist)
Graph iv: CFBCEDCADBAH (others exist)

c. Graphs i and ii

d. Graph i: CEDABC
Graph ii: CEDABFC

e. i. $\begin{bmatrix} 0 & 1 & 1 & 1 & 0 \\ 1 & 0 & 1 & 1 & 0 \\ 1 & 1 & 0 & 1 & 1 \\ 1 & 1 & 1 & 0 & 1 \\ 0 & 0 & 1 & 1 & 0 \end{bmatrix}$

ii. $\begin{bmatrix} 0 & 1 & 1 & 1 & 0 & 0 \\ 1 & 0 & 1 & 1 & 0 & 1 \\ 1 & 1 & 0 & 1 & 1 & 1 \\ 1 & 1 & 1 & 0 & 1 & 0 \\ 0 & 0 & 1 & 1 & 0 & 0 \\ 0 & 1 & 1 & 0 & 0 & 0 \end{bmatrix}$

iii. $\begin{bmatrix} 0 & 1 & 1 & 1 & 0 & 0 & 0 & 1 \\ 1 & 0 & 1 & 1 & 0 & 1 & 0 & 0 \\ 1 & 1 & 0 & 1 & 1 & 1 & 0 & 0 \\ 1 & 1 & 1 & 0 & 1 & 0 & 0 & 0 \\ 0 & 0 & 1 & 1 & 0 & 0 & 0 & 0 \\ 0 & 1 & 1 & 0 & 0 & 0 & 1 & 1 \\ 0 & 0 & 0 & 0 & 0 & 1 & 0 & 0 \\ 1 & 0 & 0 & 0 & 0 & 0 & 0 & 0 \end{bmatrix}$

iv. $\begin{bmatrix} 0 & 1 & 1 & 1 & 0 & 0 & 1 \\ 1 & 0 & 1 & 1 & 0 & 1 & 0 \\ 1 & 1 & 0 & 1 & 1 & 1 & 0 \\ 1 & 1 & 1 & 0 & 1 & 0 & 0 \\ 0 & 0 & 1 & 1 & 0 & 0 & 0 \\ 0 & 1 & 1 & 0 & 0 & 0 & 0 \\ 1 & 0 & 0 & 0 & 0 & 0 & 0 \end{bmatrix}$

f. The presence of Euler trails and circuits can be identified using the adjacency matrix to check the degree of the vertices. The presence of Hamiltonian paths and cycles can be identified by using the adjacency matrix to check the connections between vertices.

8. E

9. a. A or C   **b.** B or D

10. a. G to C   **b.** F to E

11. a.

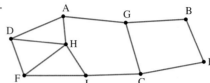

b. Yes, because the degree of each intersection or corner point is an even number.

c. Yes, because the degree of each remaining intersection or corner point is still an even number.

12. a.

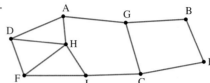

b. 4

c. i. ADHFICEBGA   **ii.** AHDFICEBGA

d. i. Yes, because two of the checkpoints have odd degree.
  ii. H and C

13. a.

	Hamiltonian cycle
1.	ABCDA
2.	ABDCA
3.	ACBDA
4.	ACDBA
5.	ADBCA
6.	ADCBA

b. Yes, commencing on vertices other than A

14. a. B, C, D, F or G   **b.** B or C

c. None possible   **d.** D or E

e. D to E

9.4 Exam questions

Note: Mark allocations are available with the fully worked solutions online.

1. D   **2.** D   **3.** C

9.5 Weighted graphs and networks, and trees

9.5 Exercise

1. 21

2. 20.78

3. a.

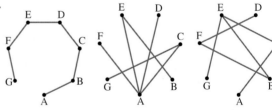

Other possibilities exist.

b.

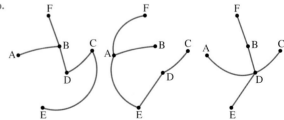

Other possibilities exist.

4. ABGEDCA or ACDEGBA (length 66)

5. a. i.

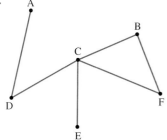

ii. ADCBFCE or ADCFBCE

b. i.

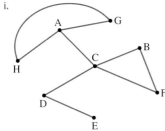

ii. AHGACBFCDE or similar

c. i.

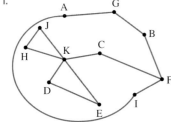

ii. KDEKHJKCFIAGBF or similar

d. i.

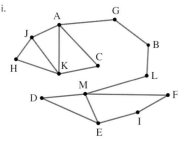

ii. EDMEIFMLBGACKHJKA or similar

6. a.

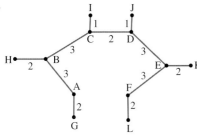

b. 24

7.

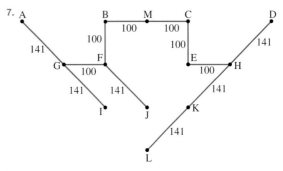

8. a.

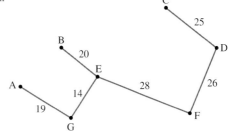

b.

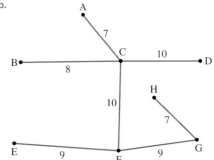

9. Step 1 **Step 2**

Step 3 **Step 4**

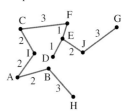

10. a. i.

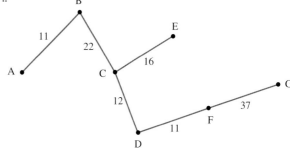

ii. 109

b. i.

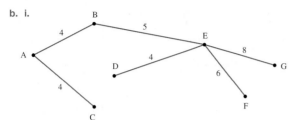

ii. 31

c. i.

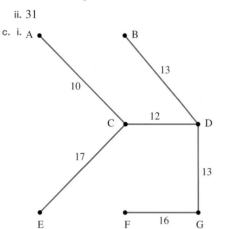

ii. 81

d. i.

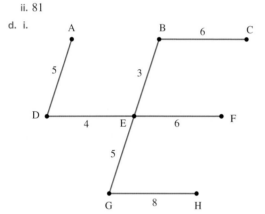

ii. 37

e. i.

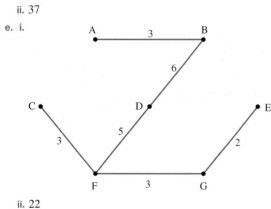

ii. 22

11. **a.** FDCGBAE (other solutions exist)

   b. FDCBAEG (other solutions exist)

12. **a.** Longest: IFEDCBAHG (or similar variation of the same values)

   Shortest: IAHGFEDCB (or similar variation of the same values)

b.

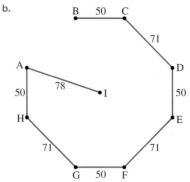

13. **a.** ADEG

   b. BHG

   c. EGFCDABHE

14. **a.**

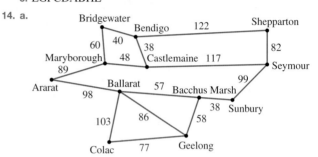

b. 723 km

c.

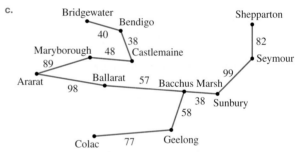

d. 859 km

15. **a.**

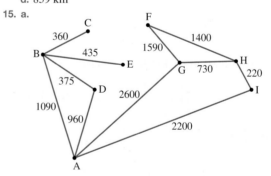

b. No; C and E are both only reachable from B.

c. i. 12 025

   ii. 12 025

d.

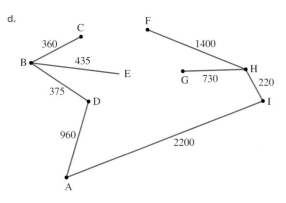

9.5 Exam questions

Note: Mark allocations are available with the fully worked solutions online.

1. B

2. D

3. B

9.6 Review

9.6 Exercise

Multiple choice

1. C   **2.** D   **3.** C   **4.** B

5. E   **6.** A   **7.** E   **8.** A

9. A   **10.** E

Short answer

11. a.   i. Planar   ii. Planar

     iii. Planar   iv. Planar

   b.   i.

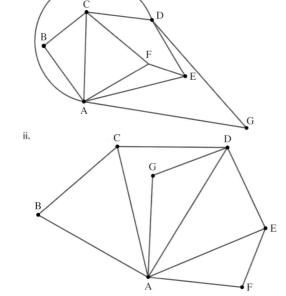

ii.

iii.

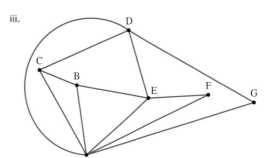

iv.

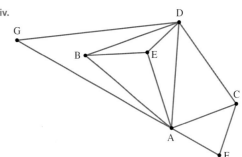

12. a. $\begin{bmatrix} 1 & 1 & 0 & 1 \\ 1 & 0 & 3 & 0 \\ 0 & 3 & 1 & 1 \\ 1 & 0 & 1 & 0 \end{bmatrix}$

b. $\begin{bmatrix} 0 & 1 & 2 & 1 & 1 \\ 1 & 0 & 2 & 0 & 1 \\ 2 & 2 & 0 & 2 & 3 \\ 1 & 0 & 2 & 2 & 2 \\ 1 & 1 & 3 & 2 & 0 \end{bmatrix}$

c. $\begin{bmatrix} 0 & 2 & 1 & 3 & 1 & 2 \\ 2 & 0 & 3 & 1 & 1 & 0 \\ 1 & 3 & 0 & 2 & 3 & 1 \\ 3 & 1 & 2 & 2 & 2 & 1 \\ 1 & 1 & 3 & 2 & 3 & 1 \\ 2 & 0 & 1 & 1 & 1 & 0 \end{bmatrix}$

d. $\begin{bmatrix} 0 & 2 & 1 & 3 & 0 & 1 & 0 \\ 2 & 0 & 2 & 1 & 2 & 1 & 0 \\ 1 & 2 & 0 & 1 & 1 & 0 & 1 \\ 3 & 1 & 1 & 0 & 0 & 2 & 1 \\ 0 & 2 & 1 & 0 & 0 & 0 & 3 \\ 1 & 1 & 0 & 2 & 0 & 0 & 2 \\ 0 & 0 & 1 & 1 & 3 & 2 & 0 \end{bmatrix}$

13. a. Simple, planar   **b.** Simple, planar

     c. Simple, planar   **d.** Simple, planar

     e. Simple   **f.** Simple, planar

     g. Simple, planar   **h.** Simple, planar

14. Graphs **a** and **d** are isomorphic.

15. a. i. 3

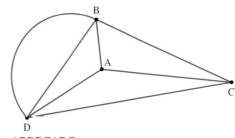

ii. ABDBCADC

b. **i.**

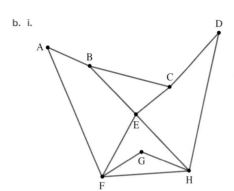

ii. BAFEHGFHDCEBC

16. a. 23
  b. 34

c.

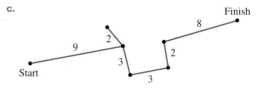

Extended response

17. a. See the figure at the foot of the page.*
  b. Via Canberra (4661 km)
  c. Via Sydney (4954 km)
  d. See the figure at the foot of the page.*

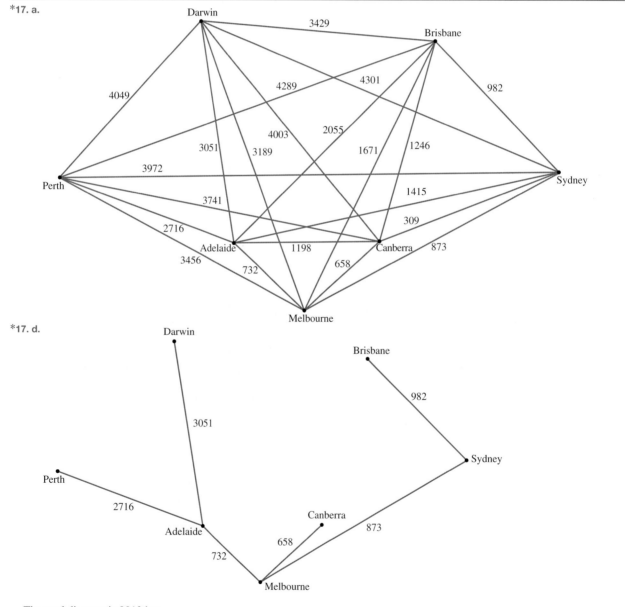

17. a.

17. d.

The total distance is 9012 km.

18. a. i. 3805 m
   ii. See the figure at the foot of the page.*
  b. HEDA, 1860 m
  c. 4905 m, DFGCABEH
  d. DEHFGCABD, 5260 m

19. a.

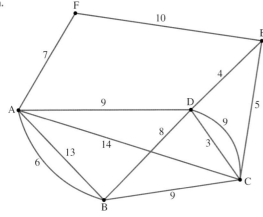

b.
$$\begin{bmatrix} 0 & 2 & 1 & 1 & 0 & 1 \\ 2 & 0 & 1 & 1 & 0 & 0 \\ 1 & 1 & 0 & 2 & 1 & 0 \\ 1 & 1 & 2 & 0 & 1 & 0 \\ 0 & 0 & 1 & 1 & 0 & 1 \\ 1 & 0 & 0 & 0 & 1 & 0 \end{bmatrix}$$

c. No, as there are more than two vertices of odd degree.
d. AFEDCBA(39)

20. a. See the figure at the foot of the page.*
  b. Hobart–Bruny–Robbins (780 km)
  c. Hobart–Bruny–Robbins–King–Devonport–Flinders–Maria (2075 km)
  d. King–Devonport–Flinders–Maria–Hobart–Bruny–Robbins–King

*18. a. ii.

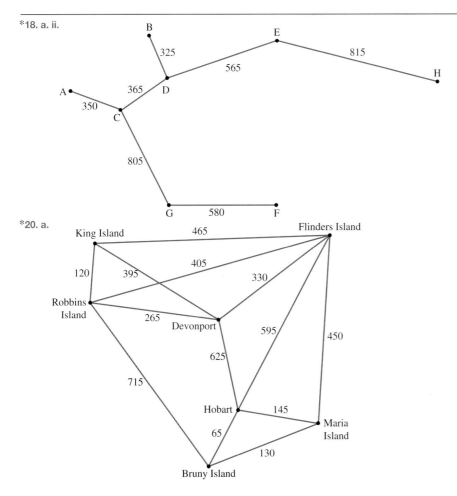

*20. a.

9.6 Exam questions

Note: Mark allocations are available with the fully worked solutions online.

1. a. 86 km   **b.** Town K

2. a. 3.2 km

   b. i. Eulerian trail

     ii. Eulerian trails start and finish at vertices with an odd degree. The training program starts at S, with a degree of 3, and will finish at P, also with a degree of 3.

   c. This track is between exercise station S and exercise station T.

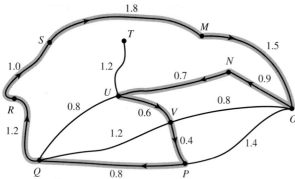

3. a. The office

   b. i. Hamiltonian cycle

     ii.

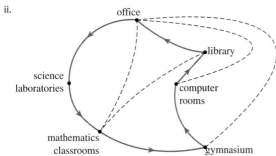

4. A

5. E

10 Variation

Fully worked solutions for this topic are available online.

10.1 Overview

10.1.1 Introduction

Even though a lot of variables in mathematics are linearly related, this is certainly not always the case. Depending on what the scatterplot of the data shows, the data can also be modelled by relationships that are not linear. These can be reciprocal $\left(\dfrac{1}{x}\right)$, square (x^2) or logarithmic relationships $(\log_{10} x)$, to name a few.

The Richter scale, developed in the 1930s, assigns a magnitude number to quantify the size of an earthquake. The Richter scale is a base 10 logarithmic scale. The logarithmic scale allows for the data to cover a huge range of values for different earthquakes. An earthquake that registers 5.0 on the Richter scale has a shaking amplitude 10 times greater than an earthquake that registered 4.0 at the same distance.

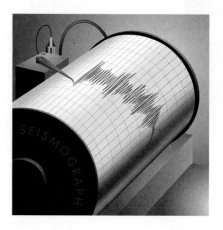

Another well-known use of logarithmic transformation is the pH scale. The pH scale tells us the acidity or basicity of an aqueous solution. Solutions with a pH less than 7 are classified as acidic and solutions with a pH greater than 7 are basic. A solution of pH 2 is 10 times more acidic than a solution of pH 3 and 100 times more acidic than a solution of pH 4. Measurements on the pH scale are very important for industries such as medicine, biology, chemistry, agriculture, food science nutrition, engineering and oceanography.

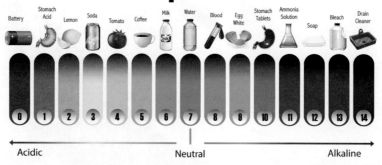

The pH Scale

KEY CONCEPTS

This topic covers the following key concepts from the VCE Mathematics Study Design:
- numerical, graphical, and algebraic approaches to direct and inverse variation
- transformation of data to linearity to establish relationships between variables, for example y and x^2, or y and $\dfrac{1}{x}$ or y and $\log_{10}(x)$
- modelling of given non-linear data using the relationships $y = kx^2 + c$, $y = kx$ where $k > 0$ and $y = k\log_{10}(x) + c$ where $k > 0$.

Source: VCE Mathematics Study Design (2023–2027) extracts © VCAA; reproduced by permission.

10.2 Direct and inverse variation

10.2.1 Direct variation

Frequently in mathematics we deal with investigating how changes that occur in one quantity cause changes in another quantity. Understanding how these quantities vary in relation to each other allows the development of equations that provide mathematical models for determining all possible values in the relationship.

Direct variation involves quantities that are proportional to each other. If two quantities vary directly, then doubling one doubles the other. In direct variation, as one value increases, so does the other; likewise, as one decreases, so does the other. This produces a linear graph that passes through the origin.

Quantities proportional to each other

If two quantities are directly proportional, we say that they 'vary directly' with each other. This can be written as $y \propto x$.

The proportion sign, $\propto$, is equivalent to '$= k$', where k is called the constant of proportionality or the constant of variation. The constant of variation, k, is equal to the ratio of y to x for any data pair.

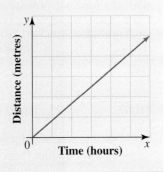

Another way to put this is that k is the rate at which y varies with x, otherwise known as the gradient. This means that $y \propto x$ can be written as $y = kx$.

tlvd-4938

WORKED EXAMPLE 1 Calculating the value of constant of proportionality

The cost of apples purchased at the supermarket varies directly with their mass, as shown in the graph. Calculate the constant of proportionality and use it to write a rule connecting cost, C, and mass, m.

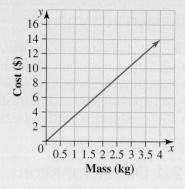

THINK

1. Apply the gradient formula to calculate k.

2. Substitute the value of k into the equation that relates the two variables (that is, in the form $y = kx$).

WRITE

$$k = \frac{\text{rise}}{\text{run}}$$
$$= \frac{14}{4}$$
$$= 3.5$$

Here the variables are cost ($) and mass (kg), so:
$$C = km$$
$$= 3.5m$$

10.2.2 Inverse variation

If two quantities vary inversely, then increasing one variable decreases the other.

Inverse variation produces the graph of a hyperbola.

Statement for inverse variation

The statement 'x varies inversely with y' can be written as $y \propto \dfrac{1}{x}$ or $y = \dfrac{k}{x}$.

k is called the proportionality constant or the constant of variation.

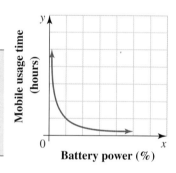

WORKED EXAMPLE 2 Determining the rule for inverse variation

The time taken to complete a task is inversely proportional to the number of workers as shown in the table.

Determine the rule that relates the time to complete the task with the number of workers.

Number of workers (n)	1	2	3	4	6
Hours to complete (T)	12	6	4	3	2

THINK

1. Write the statement for inverse proportion using the variables from the question.

WRITE

$$T \propto \frac{1}{n} \rightarrow T = \frac{k}{n}$$

2. Select a pair of coordinates from the table and substitute them into the equation to solve for k.

$$(n, T) \rightarrow T = \frac{k}{n}$$
$$(1, 12) \rightarrow 12 = \frac{k}{1}$$
$$\therefore k = 12$$

3. Write the equation.

$$T = \frac{12}{n}$$

 Resources

Interactivity Direct, inverse and joint variation (int-6490)

10.2 Exercise

Students, these questions are even better in jacPLUS

 Receive immediate feedback and access sample responses

 Access additional questions

 Track your results and progress

Find all this and MORE in jacPLUS

---

1. **WE1** The cost of a wedding reception varies directly with the number of people attending is shown in the graph.
Calculate the constant of proportionality and use it to write a rule connecting cost, C, and the number of people attending, N.

2. The distance travelled by a vehicle, d (in kilometres) is directly proportional to the time, t (in hours). Use the information in the table to write a rule connecting distance and time.

Time (h)	1	2	3	4
Distance (km)	90	180	270	360

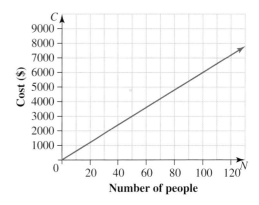

3. Evaluate whether y is directly proportional to x in each of the following tables.

a.
x	0.7	1.2	4	4.1
y	2.8	4.8	12	16.4

b.
x	0.4	1.5	2.2	3.4
y	0.84	3.15	4.62	7.14

4. The amount of interest earned by an investment is proportional to the amount of money invested.
   a. Use the pronumerals I for the amount of interest and A for the amount of money invested, together with the proportionality sign ($\propto$), to write this in mathematical shorthand.
   b. Write your answer to part a using an equals sign and a constant, k.
   c. If an investment of \$30 000 earns \$15 000 interest, find k.
   d. Write an equation connecting I and A.
   e. Apply the equation to find the interest earned by an investment of \$55 000.
   f. Apply the equation to calculate the investment needed to earn \$75 000 in interest.

5. If $K \propto m$ and $K = 27.9$ when $m = 6.2$:
   a. calculate the constant of proportionality
   b. determine the rule connecting K and m
   c. calculate the value of K when $m = 72$
   d. calculate the value of m when $K = 450$.

6. **WE2** The time taken to travel 100 km is inversely proportional to the speed, as shown in the following table. Determine the rule that relates the time taken to travel 100 km (T) with the speed (s).

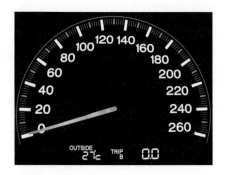

Speed (km/h)	100	50	25	10	5	1
Time (h)	1	2	4	10	20	100

7. A box of chocolates contains 20 pieces that are divided equally among family members.

   a. Copy and complete the table to show the number of sweets that each receives for various numbers of relatives.

Family members	20	10		4	
Number of chocolates	1		4		20

   b. Show this information in a graph.
   c. Write a rule connecting the number of family members (n) with the number of chocolates each receives (C).

8. A company making electronic parts finds that the number sold depends on the price. If the price is higher, fewer parts are sold.
   Market research gives the following expected sales results.

Price ($)	1	5	20	50	100	200	400
Number sold (thousands)	400	80	20	8	4	2	1

   a. Draw a graph of the number sold versus the price.
   b. Use the information to write an equation that connects the price, P, with the number sold, n.
   c. Apply the equation to predict the number sold if the price is $25.
   d. Apply the equation to calculate the price to sell 250 000.

9. The percentage of harmful bacteria, P, found to be present in a sample of cooked food at certain temperatures, t, is shown in the following table.

Food temperature, t (°C)	10	20	40	50	80	100
Harmful bacteria percentage, P	80	40	20	16	10	8

   a. Draw a graph of this information.
   b. Determine an equation connecting the quantities.
   c. Apply the equation to predict the percentage of harmful bacteria present when the food temperature is 25 °C.
   d. Apply the equation to calculate the temperature required to ensure that the food contains no more than 3.2% of harmful bacteria.

Question 1 (1 mark)

MC Identify which of these sets of variables is an example of direct proportion.
- **A.** The number of hours spent studying and the mark received on a test
- **B.** A person's height and weight
- **C.** The number of hours exercising and weight
- **D.** The number of hours worked and the pay received at a constant hourly rate
- **E.** The amount of money in an account and the length of time

Question 2 (1 mark)

MC State which of the following correctly describes the characteristics of inverse proportion.
- **A.** As one variable increases, the other must decrease.
- **B.** • As one variable increases, the other must also increase.
  - The graph relating the two variables must be a hyperbola.
- **C.** As one variable increases, the other must also increase.
- **D.** The variables must not be related.
- **E.** • As one variable increases, the other must decrease.
  - The graph relating the two variables must be a hyperbola.

Question 3 (3 marks)

Explain, using the figure, why y and x are directly proportional.

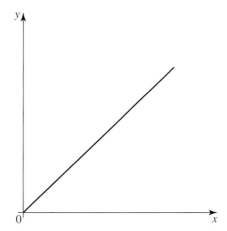

More exam questions are available online.

10.3 Data transformations

10.3.1 Linearising data

We have seen that when quantities have a relationship that is directly proportional, their graphs are straight lines. When we have a linear rule obtained from investigating a direct relationship, it is easy to identify any values that don't match the graph.

If a data set displays non-linear behaviour, we can often apply a mathematical approach to **linearise** it, meaning we transform it into a straight line. This is known as a data transformation, and it can often be achieved through a process of squaring one of the coordinates, taking the reciprocal or applying a logarithmic transformation.

10.3.2 Transforming data with x^2

When the relationship between two variables appears to be quadratic or parabolic in shape, we can often transform the non-linear relation to a linear relation by plotting the y-values against x^2-values instead of x-values.

For example, when we graph the points from the table below, we obtain a typical parabolic shape.

x	0	1	2	3	4
y	1	2	5	10	17

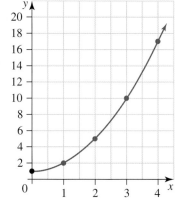

If instead the y-values are plotted against the square of the x-values, the graph becomes linear.

x	0	1	2	3	4
x^2	0	1	4	9	16
y	1	2	5	10	17

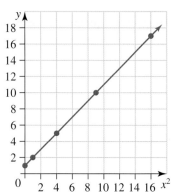

WORKED EXAMPLE 3 Transforming data with x^2

Redraw the graph shown by plotting y-values against x^2-values.

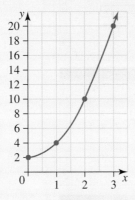

THINK	WRITE
1. Construct a table of values for the points shown in the graph.	

x	0	1	2	3
y	2	4	10	20

2. Add a row for calculating the values of x^2.

x	0	1	2	3
x^2	0	1	4	9
y	2	4	10	20

3. Plot the points using the x^2- and y-values.

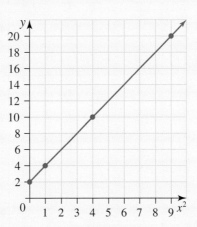

TI	THINK	DISPLAY/WRITE	CASIO	THINK	DISPLAY/WRITE

1. In a Lists & Spreadsheet page, label the first column as x, the second column as xsq (x-square) and the third column as y.
Enter the x-values into the first column and the y-values into the third column.

1. In a Statistics screen, label list1 as x, list2 as xsq (x-square), and list 3 as y. Enter the x-values into the first column and the y-values into the third column.

▶

2. In the function cell below *xsq*, complete the entry line as:
   $= x^2$
   then press ENTER.
   Select the variable reference for *x* when prompted, then select OK.

2. In the function cell below the *xsq* column, complete the entry line as x^2, then press EXE.

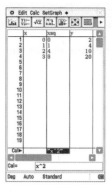

3. In a Data & Statistics page, click on the label of the horizontal axis and select *xsq*. Click on the label of the vertical axis and select *y*. Press MENU, then select:
   4: Analyze
   6: Regression
   1: Show Linear ($mx + b$)
   The graph is displayed on the screen.

3. Select:
   • SetGraph
   • Setting…
   Complete the fields as:
   Draw: On
   Type: Scatter
   XList: main*xsq*
   YList: main*y*
   Freq: 1
   then select Set.

4. Click the *y* icon, then select:
   • Calc
   • Regression
   • Linear Reg
   Complete the fields as:
   XList: main*xsq*
   YList: main*y*
   Select OK, then select OK again.

5. The graph is displayed on the screen.

10.3.3 Transforming data with $\frac{1}{x}$

For data that is not linearised by an x^2 transformation, other adjustments to the scale of the x-axis can be made. Another common transformation to use is $\frac{1}{x}$, called the reciprocal of x. We use a $\frac{1}{x}$ transformation when working with values that start large and get smaller, tapering off as they approach a certain value. An example of this is shown in the following worked example.

WORKED EXAMPLE 4 Transforming data with $\frac{1}{x}$

Redraw the graph shown by plotting y-values against $\frac{1}{x}$-values. Comment on whether the graph is linearised.

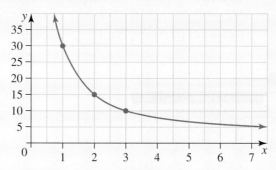

THINK	WRITE
1. Construct a table of values for the points shown in the graph.	

x	1	2	3
y	30	15	10

2. Add a row for calculating the values of $\frac{1}{x}$.

x	1	2	3
$\frac{1}{x}$	1	$\frac{1}{2}$	$\frac{1}{3}$
y	30	15	10

3. Plot the points using the $\frac{1}{x}$- and y-values.

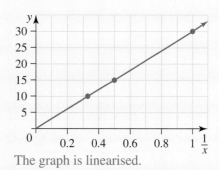

The graph is linearised.

**TI	THINK**	**DISPLAY/WRITE**	**CASIO	THINK**	**DISPLAY/WRITE**

1. In a Lists & Spreadsheet page, label the first column as *x*, the second column as *recipx* (reciprocal *x*) and the third column as *y*.
Enter the *x*-values into the first column and the *y*-values into the third column.

1. In a Statistics screen, label list 1 as *x*, list 2 as *recipx* (reciprocal *x*) and list 3 as *y*.
Enter the *x*-values into the first column and the *y*-values into the third column.

2. In the function cell below *recipx*, complete the entry line as:
$$= \frac{1}{x}$$
then press ENTER.
Select the variable reference for *x* when prompted, then select OK.

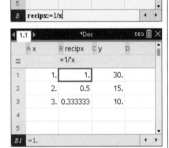

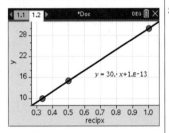

2. In the function cell below the *recipx* column, complete the entry line as:
$$\frac{1}{x}$$
then press EXE.

3. In a Data & Statistics page, click on the label of the horizontal axis and select *recipx*. Click on the label of the vertical axis and select *y*. Press MENU, then select:
4: Analyze
6: Regression
1: Show Linear (*mx* + *b*)
The graph is displayed on the screen.

3. Select:
- SetGraph
- Setting…
Complete the fields as:
Draw: On
Type: Scatter
XList: main\recipx
YList: main\y
Freq: 1
then select Set.

4. Click the *y* icon, then select:
- Calc
- Regression
- Linear Reg
Complete the fields as:
XList: main\recipx
YList: main\y
Select OK, then select OK again.

	5.	The graph is displayed on the screen.	

10.3.4 Transforming data with logarithms

The final common transformation that we will look at in this topic is the logarithmic transformation, using **logarithms** in base 10, $\log_{10}(x)$. This transformation is common when working with data values that get increasingly bigger.

Note: There are logarithms with numerical bases other than 10. Calculators default to using base 10 logarithms, and we use only base 10 logarithms in this topic for simplicity.

WORKED EXAMPLE 5 Transforming data with $\log_{10}(x)$

Redraw the graph shown by plotting y-values against $\log_{10}(x)$ values.

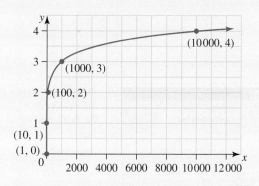

THINK

1. Construct a table of values for the points shown in the graph.

2. Add a row for calculating the values of $\log_{10}(x)$.

WRITE

x	1	10	100	1000	10 000
y	0	1	2	3	4

x	1	10	100	1000	10 000
$\log_{10}(x)$	0	1	2	3	4
y	0	1	2	3	4

3. Plot the points using the $\log_{10}(x)$- and y-values.

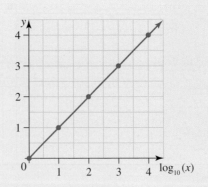

TI | THINK	DISPLAY/WRITE	CASIO | THINK	DISPLAY/WRITE
1. In a Lists & Spreadsheet page, label the first column as x, the second column as $\log x$ (the logarithm of x) and the third column as y. Enter the x-values into the first column and the y-values into the third column.	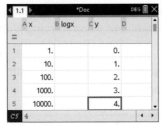	1. In a Statistics screen, label list 1 as x, list 2 as $\log x$ (logarithm of x), and list 3 as y. Enter the x-values into the first column and the y-values into the third column.	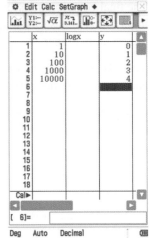
2. In the function cell below $\log x$, complete the entry line as: $= \log(x)$ then press ENTER. Select the variable reference for x when prompted, then select OK.		2. In the function cell below the $\log x$ column, complete the entry line as: $\log(x)$ then press EXE.	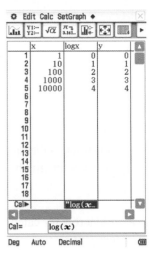

3. In a Data & Statistics page, click on the label of the horizontal axis and select log x. Click on the label of the vertical axis and select y. Press MENU, then select:
4: Analyze
6: Regression
1: Show linear $(mx + b)$
The graph is displayed on the screen

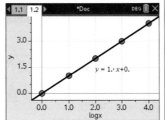

3. Select
   - Set graph
   - Setting
   Complete the fields as:
   Draw: On
   Type: Scatter
   XList: main\log x
   YList: main\y
   Freq: 1
   then select Set.

4. Select:
   - Calc
   - Regression
   - Linear Reg
   Complete the fields as:
   XList: main\logx
   YList: main\y
   Select OK, then select OK again.

5. The graph is displayed on the screen.

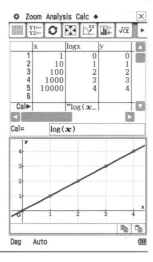

 Resources

Interactivity Linearising data (int-6491)
Transforming to linearity (int-6253)

1. **WE3** Redraw the graphs shown by plotting y-values against x^2-values.

a.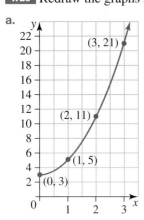

b.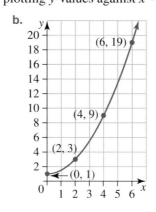

c.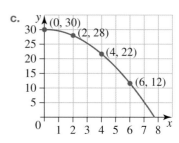

2. i. Redraw each of these graphs by plotting the y-values against x^2-values.
   ii. Identify which graphs are linearised by the x^2 transformation.

a.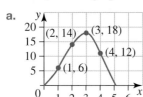

b.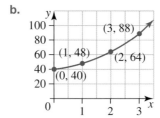

c.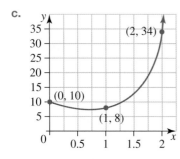

d.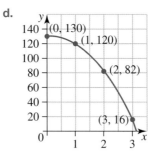

3. For each table of values shown:

   i. sketch a graph of the original values
   ii. construct a table of values to show the original y-values against the x^2-values
   iii. sketch a graph of the transformed values
   iv. write a comment that compares the transformed graph with the original one.

a.

x	0	1	2	3
y	20	18	12	2

b.

x	1	3	5	7	9	11	15
y	4	22	56	106	172	254	466

c.

x	9	10	11	12	13	14	15
y	203	275	315	329	323	303	275

d.

x	0	1	4	9	16	25	36	49	64
y	16	17	18	19	20	21	22	23	24

4. **WE4** Redraw the graphs shown by plotting y-values against $\dfrac{1}{x}$-values. Comment on whether each graph is linearised.

a. 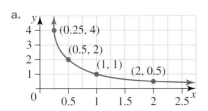 **b.**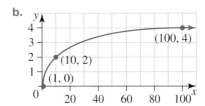

5. **i.** Redraw each of these graphs by plotting the y-values against $\dfrac{1}{x}$-values.

   ii. Identify which of the graphs are linearised by the $\dfrac{1}{x}$ transformation.

a. 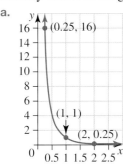 **b.**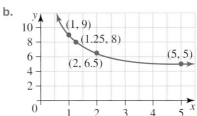

6. For each table of values shown:

   i. sketch a graph of the original values

   ii. construct a table of values to show $\dfrac{1}{x}$-values against the original y-values

   iii. sketch a graph of the transformed values
   iv. write a comment that compares the transformed graph with the original one.

a.

x	2	4	6	8	10
y	40	25	18	17	16

c.

x	0	1	2	3	4	5	6
y	60	349	596	771	844	785	564

b.

x	1	2	5	10	20	25
y	40	27.5	20	17.5	16.25	16

d.

x	0	1	4	6	8	9	10	11
y	72	66	48	36	24	18	12	6

7. **WE5** Redraw the graphs shown by plotting y-values against $\log_{10}(x)$ values.

a.
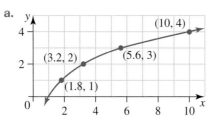

b.
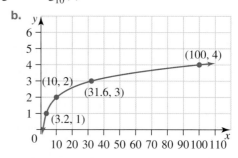

8. Redraw the graphs in Question **4** by plotting the y-values against $\log_{10}(x)$ values. Comment on whether each graph is linearised by the $\log_{10}(x)$ transformation.

9. a. Use the rule $y = \dfrac{1}{x^2}$ to complete the following table of values. For the second row $\left(\dfrac{1}{x}\right)$, give your answers correct to 1 decimal place.

x	0.25	0.5	1	1.25	2	2.5	4	5
$\dfrac{1}{x}$								
y								

b. Draw graphs of the original data and the transformed data.

c. Comment on the effectiveness of this transformation for linearising this data.

10. A science experiment measured the distance an object travels when dropped from a height of 5 m. The results are shown in the table.

Time (s)	0.2	0.4	0.6	0.8	1.0
Distance (cm)	20	78	174	314	490

a. Draw a graph to represent the data.

b. Select an appropriate data transformation to linearise the data and show the resultant table of values and graph.

c. Draw a graph to show the actual height of the object for the time period shown in the table.

d. Explain whether the same data transformation would linearise the data this time.

11. a. Use CAS to complete the details of the table correct to 3 decimal places.

x	10	20	30	40	50	60	70	80
y	4	5.2	5.9	6.4	6.8	7.1	7.4	7.6
$\dfrac{1}{x}$								
$\log(x)$								

b. Use CAS to draw the two data transformations.

c. Explain which transformation gives the better linearisation of the data.

12. Measurements of air pressure are important for making predictions about changes in the weather. Because air pressure also changes with altitude, any predictions made must also take into account the height at which any measurements were taken.

Altitude (m)	Air pressure (kPa)
0	100
300	90
1500	75
2500	65
3000	60
3500	55
4000	50
5000	45
5500	40
6000	35
8000	25
9000	18
12 000	12
15 000	10

The table gives a series of air pressure measurements taken at various altitudes.

a. Sketch a graph to represent the data.
b. Apply an appropriate data transformation to linearise the data and show the resultant table of values and graph.
c. Comment on whether the transformation results in a relationship that puts all of the points in a straight line, and explain what this would mean.

13. Data comparing consumption of electricity with the maximum daily temperature in a country is shown in the table.

Max. temp. (°C)	21.2	22.6	25.7	30.1	35.1	38.7	37.1	32.7	27.9	23.9	21.7	39.6
Consumption (GWh)	3800	3850	3950	4800	5900	6600	6200	5200	4300	3900	3825	7200

a. Explain which variable should be used for the x-axis.
b. Use CAS to:

   i. sketch the graph of the original data.

   ii. redraw the graph using both an x^2 and a $\dfrac{1}{x}$ transformation

   iii. comment on the effect of both transformations.

14. A university student started going to the gym recently. The table shows the amount of weight the student has been able to bench press at the start of every month.

Bench press weight	30	56	71	82	90
Month	1	2	3	4	5

a. Sketch a graph with the amount of weight on the y-axis and the month on the x-axis.
b. Apply an appropriate data transformation to linearise the data and show the resultant table of values and graph.

Question 1 (1 mark)

MC State which of the following scatterplots would benefit most from an $x \rightarrow x^2$ transformation.

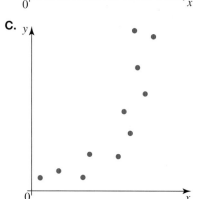

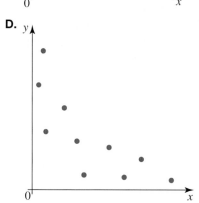

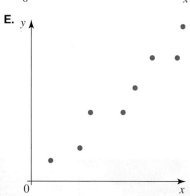

D.

E.

Question 2 (1 mark)

MC The table shows the corresponding values of x and y such that $y = a + \dfrac{b}{x}$.

x	1	2
y	6	4

The respective values of a and b are

    A. 3, 3   **B.** 2, 4   **C.** 2, −4   **D.** 1, 6   **E.** 3, 10

▶ **Question 3 (3 marks)**

The cost of printing advertising flyers is partly constant and varies partly as the area to be printed. If it costs 45 cents to print one flyer in a square pattern of side length 10 cm and 90 cents to print one flyer in a square pattern of side length 21 cm, find

    a. the fixed cost per flyer   **(1 mark)**

    b. the equation relating the cost of printing one flyer, C, to the side length of the square pattern to be printed, l **(1 mark)**

    c. the cost of printing 1000 flyers in a square pattern of side length 14.8 cm.   **(1 mark)**

More exam questions are available online.

10.4 Orders of magnitude

LEARNING INTENTION

At the end of this subtopic you should be able to:
- calculate orders of magnitude difference between two values
- convert between units of measure that range over multiple orders of magnitude
- explain the concept of a logarithmic (base 10) scale.

10.4.1 Using orders of magnitude

Orders of magnitude have close ties to base 10 logarithms, which we use extensively in this topic.

Power of 10	Basic numeral	Order of magnitude
10^{-4}	0.0001	-4
10^{-3}	0.001	-3
10^{-2}	0.01	-2
10^{-1}	0.1	-1
10^{0}	1	0
10^{1}	10	1
10^{2}	100	2
10^{3}	1000	3
10^{4}	10 000	4

As you can see from the table, the exponents of the powers of 10 are equal to the orders of magnitude.

Orders of magnitude

An increase of 1 order of magnitude means an increase in the basic numeral by a multiple of 10. Similarly, a decrease of 1 order of magnitude means a decrease in the basic numeral by a multiple of 10.

We use orders of magnitude to compare different values and to check that estimates we make are reasonable. For example, if we are given the mass of two objects as 1 kg and 10 kg, we can say that the difference in mass between the two objects is 1 order of magnitude, as one of the objects has a mass 10 times greater than the other.

Identify the order of magnitude that expresses the difference in distance between 3.5 km and 350 km.

THINK	WRITE
1. Calculate the difference in size between the two distances by dividing the larger distance by the smaller distance.	$\dfrac{350}{3.5} = 100$
2. Express this number as a power of 10.	$100 = 10^2$
3. The exponent of the power of 10 is equal to the order of magnitude.	The order of magnitude that expresses the difference in distance is 2.

10.4.2 Scientific notation and orders of magnitude

When working with orders of magnitude, it can be helpful to express numbers in scientific notation. If two numbers in scientific notation have the same coefficient, that is, if the numbers in the first part of the scientific notation (between 1 and 10) are the same, then we can easily determine the order of magnitude by calculating the difference in value between the exponents of the powers of 10.

Calculate by how many orders of magnitude the following distances differ.
Distance A: 2.6×10^{-3} km
Distance B: 2.6×10^2 km

THINK	WRITE
1. Check that the coefficients of both numbers are the same in scientific notation.	In scientific notation, both numbers have a coefficient of 2.6.
2. Determine the order of magnitude difference between the numbers by subtracting the exponent of the smaller power of 10 (-3) from the exponent of the larger power of 10 (2).	$2 - -3 = 5$
3. Write the answer.	The distances differ by an order of magnitude of 5.

10.4.3 Logarithmic scales

Earthquakes are measured by seismometers, which record the amplitude of the seismic waves of the earthquake. There are large discrepancies in the size of earthquakes, so rather than using a traditional scale to measure their amplitude, a **logarithmic scale** is used.

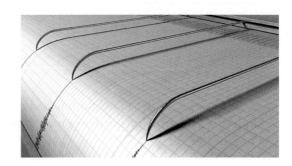

Logarithmic scales

A logarithmic scale represents numbers using a log (base 10) scale. This means that if we express all of the numbers in the form 10^a, the logarithmic scale will represent these numbers as a.

Notations for logarithmic scales

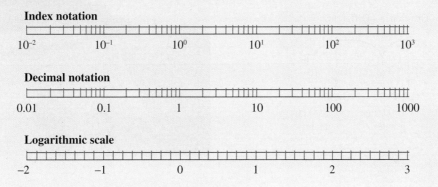

In the diagram above are three possible representations of the same scale. Note how the index notation and decimal notation are related to the logarithmic scale.

This means that for every increase of 1 in the magnitude in the scale, the amplitude or power of the earthquake is increasing by a multiple of 10. This allows us to plot earthquakes of differing sizes on the same scale.

The Richter scale was designed in 1934 by Charles Richter and is the most widely used method for measuring the magnitude of earthquakes.

Consider the following table of historical earthquake data.

World earthquake data			
Year	Location	Magnitude on the Richter scale	Amplitude of earthquake
2021	Australia (Wood's Point, Victoria)	5.9	$10^{5.9}$
2011	Japan (Tohoku)	9.0	10^9
2010	New Zealand (Christchurch)	7.1	$10^{7.1}$
2008	China (Sichuan)	7.9	$10^{7.9}$
2004	Indonesia (Indian Ocean)	9.1	$10^{9.1}$

If we were to plot this data on a linear scale, the amplitude of the largest earthquake would be 1585 times bigger ($10^{9.1}$ compared to $10^{5.9}$) than the size of the smallest earthquake. This would create an almost unreadable graph. By using a logarithmic scale the graph becomes easier for us to read and interpret.

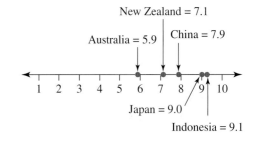

However, this scale doesn't highlight the real difference between the amplitudes of the earthquakes, which only becomes clear when these values are calculated.

The difference in amplitude between an earthquake of magnitude 1 and an earthquake of magnitude 2 on the Richter scale is 1 order of magnitude, or $10^1 = 10$ times.

Let's consider the difference between the 2021 Australian earthquake and the 2010 New Zealand earthquake. According to the Richter scale, the difference in magnitude is 1.2, which means that the real difference in the amplitude of the two earthquakes is $10^{1.2}$. When evaluated, $10^{1.2} = 15.85$, which indicates that the New Zealand earthquake was more than 15 times more powerful than the Australian earthquake.

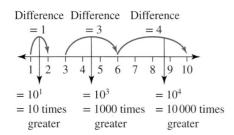

WORKED EXAMPLE 8 Evaluating the order of magnitude difference in amplitude

Using the information from the world earthquake data, compare the real amplitude of the Australian earthquake to that of the Japanese earthquake, which resulted in a tsunami that damaged the Fukushima power plant.

THINK	WRITE
1. Identify the order of magnitude difference between Japan's earthquake and Australia's earthquake.	$9 - 5.9$ $= 3.1$
2. Express the order of magnitude difference in real terms by displaying it as a power of 10.	$10^{3.1}$
3. Evaluate to express the difference in amplitude.	$= 1258.93$
4. Interpret the result and write the answer.	Japan experienced an earthquake that was nearly 1300 times larger than the earthquake in Australia.

As you can see from Worked example 8, very large numbers are involved when dealing with magnitudes of earthquakes. It would be challenging to represent an increase in amplitude of 1300 times while also having a scale accurate enough to accommodate smaller changes. Using the logarithmic scale enables such diversity in numbers to be represented on the same scale.

Logarithmic scales are also used in measuring pH levels. On the pH scale 7 is considered neutral, values from 6 to 0 indicate an increase in acidity levels, and values from 8 to 14 indicate an increase in alkalinity.

1. **WE6** Identify the order of magnitude that expresses the difference between 0.3 metres and 3000 metres.

2. The Big Lobster in South Australia is a 4000 kg sculpture of a lobster. If a normal lobster has a mass of 4 kg, calculate by what order of magnitude the mass of the Big Lobster is greater than the mass of a normal lobster.

3. **WE7** The weight of a brown bear in the wild is 3.2×10^2 kg, and the weight of a cuddly teddy bear is 3.2×10^{-1} kg. Calculate by how many orders of magnitudes the weights of the bears differ.

4. A cheetah covers 100 m in 7.2 seconds. It takes a snail a time 3 orders of magnitude greater than the cheetah to cover this distance. Express the time it takes the snail to cover 100 m in seconds.

5. **MC** A virus has an approximate mass of 1×10^{-20} kg. Select the object that has a mass 10 000 000 000 times smaller than that of the virus.

   A. A hydrogen atom (mass $= 1 \times 10^{-27}$ kg)
   B. An electron (mass $= 1 \times 10^{-30}$ kg)
   C. A bacterium (mass $= 1 \times 10^{-15}$ kg)
   D. An ant (mass $= 1 \times 10^{-6}$ kg)
   E. A grain of fine sand (mass $= 1 \times 10^{-9}$ kg)

6. **WE8** Many of the earthquakes experienced in Australia are between 3 and 5 in magnitude on the Richter scale. Compare the experience of a magnitude 3 earthquake to that of a magnitude 5 earthquake, expressing the difference in amplitude as a basic numeral.

7. A soft drink has a pH of 5, whereas lemon juice has a pH of 2.

   a. Identify the order of magnitude difference between the acidity of the soft drink and the lemon juice.
   b. Determine how many times more acidic the juice is than the soft drink.

8. **MC** Water is considered neutral and has a pH of 7. Select the liquid that is either 10 000 times more acidic or alkaline than water.

   A. Soapy water, pH 12
   B. Detergent, pH 10
   C. Orange juice, pH 3
   D. Vinegar, pH 2
   E. Battery acid, pH 1

9. The largest ever recorded earthquake was in Chile in 1960 and had a magnitude of 9.5 on the Richter scale. In 2011, an earthquake of magnitude 9.0 occurred in Japan.

  a. Determine the difference in magnitude between the two earthquakes.
  b. Reflect this difference in real terms by calculating the difference in amplitude between the two earthquakes, giving your answer as a basic numeral correct to 2 decimal places.

10. Diluted sulfuric acid has a pH of 1, making it extremely acidic. If its acidity was reduced by 1000 times, determine what pH would it register.

11. A tennis player has a tennis ball with a mass of 5×10^1 grams. At practice, the player's coach sets a series of exercises using a 5×10^3 gram medicine ball.

  a. Determine the order of magnitude difference between the mass of the two balls.
  b. Determine the mass of each ball written as basic numerals.

12. The size of a hydrogen atom is 1×10^{-10} m and the size of a nucleus is 1×10^{-15} m.

  a. Calculate by how many orders of magnitude the atom and nucleus differ in size.
  b. Determine how many times larger the atom is than the nucleus.

13. The volume of a 5000 mL container was reduced by 1 order of magnitude.

  a. Express this statement in scientific notation.
  b. Calculate the new volume of the container.

14. The distance between Zoe's house and Gwendolyn's house is 25 km, which is 2 orders of magnitude greater than the distance from Zoe's house to the school.

  a. Express the distance from Zoe's house to the school in scientific notation.
  b. Determine the distance from Zoe's house to the school.

10.4 Exam questions

Question 1 (1 mark)

MC The order of magnitude of the number of seconds in a day is

A. 2　　　　B. 5　　　　C. 24　　　　D. 60　　　　E. 86 400

Question 2 (1 mark)

MC An earthquake registers 3.0 on the Richter scale. A second earthquake registers 6.0. Select how many times bigger the amplitude of the second earthquake is than the amplitude of the first earthquake.

A. 1000　　　　B. 100　　　　C. 10　　　　D. 6　　　　E. 2

Question 3 (1 mark)

MC An order of magnitude estimate of the number of cells in the human body (100 000 000 000 000) is

A. 100　　　　B. 15　　　　C. 14　　　　D. 13　　　　E. 10

More exam questions are available online.

10.5 Non-linear data modelling

10.5.1 Modelling non-linear data

Non-linear data relationships can be represented with straight line graphs, and variation statements can be used to establish the rules that connect them.

The **constant of variation**, k, will be the gradient of the straight line.

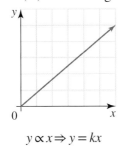

$$y \propto x \Rightarrow y = kx$$
Straight line through origin

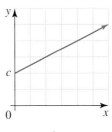

$$y = kx + c$$
Straight line with y-intercept at c

WORKED EXAMPLE 9 Determining the rule for an x^2 relationship

Determine the rule for the transformed data shown in the table.

Original data				
x	0	1	2	3
y	2	4	10	20

Transformed data				
x	0	1	2	3
x^2	0	1	4	9
y	2	4	10	20

THINK	WRITE
1. Identify the variation statement from the labels of the axes of the straight line graph.	The straight line graph indicates that y varies directly with x^2: $y \propto x^2$, which indicates that the rule will be of the form $y = kx^2$.
2. Determine k by calculating the gradient of the straight line.	$k = \dfrac{y_2 - y_1}{x_2 - x_1}$ Using $(9,\ 20)$ and $(0,\ 2)$: $k = \dfrac{20 - 2}{9 - 0}$ $\quad = \dfrac{18}{9}$ $\quad = 2$
3. Identify the value of the y-intercept, c.	The straight line graph intercepts the y-axis at $y = 2$.
4. Substitute the values of k and c in $y = kx^2 + c$ to determine the rule for the transformed data.	$y = kx^2 + c$ $y = 2x^2 + 2$

10.5.2 Modelling with $y = \dfrac{k}{x} + c$

We know that when x and y vary inversely, the rules $y \propto \dfrac{1}{x}$ or $y = \dfrac{k}{x}$ apply, and the graph will be hyperbolic in shape.

As the constant of variation, k, increases, the graph stretches out away from the horizontal axis and parallel to the vertical axis.

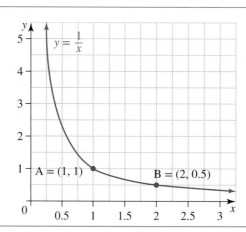

$k = 1 \Rightarrow y = \dfrac{1}{x}$

| $k = 2 \Rightarrow y = \dfrac{2}{x}$ | 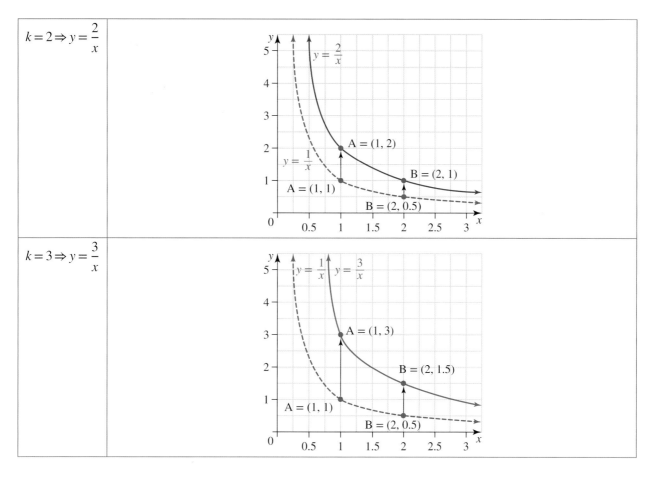 |

For inverse relationships of the type $y = \dfrac{k}{x} + c$, as the value of c increases, every point on the graph translates (moves) parallel to the horizontal axis by a distance of c.

| $c = 0 \Rightarrow y = \dfrac{1}{x}$ | 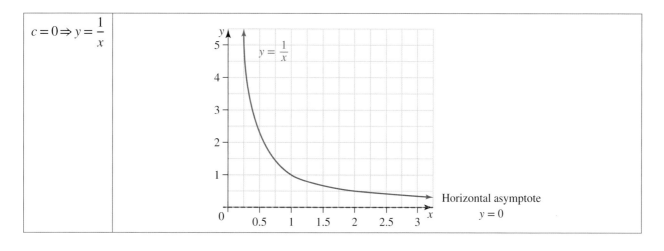 |

$c = 1 \Rightarrow y = \dfrac{1}{x} + 1$	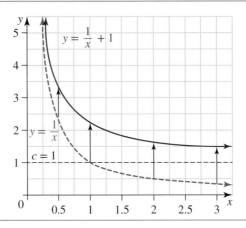
$c = 2 \Rightarrow y = \dfrac{1}{x} + 2$	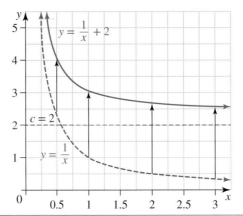

In addition, no y-coordinate on the graph will ever actually equal the value of c, because $\dfrac{1}{x}$ will never equal zero. The line $y = c$ is known as the **horizontal asymptote**.

tlvd-4940

WORKED EXAMPLE 10 Constructing the graph of an inverse relationship

Construct the graph of $y = \dfrac{2}{x} + 3$, indicating the position of the horizontal asymptote and the coordinates for when $x = 1$ and $x = 2$.

THINK	WRITE
1. Identify the value of the horizontal asymptote from the general form of the equation, $y = \dfrac{k}{x} + c$.	$y = \dfrac{2}{x} + 3$, so the horizontal asymptote is $c = 3$.
2. Substitute the values into the rule for the coordinates required.	When $x = 1$, $y = \dfrac{2}{1} + 3 = 5$, and when $x = 2$, $y = \dfrac{2}{2} + 3 = 4$. The required coordinates are $(1, \ 5)$ and $(2, \ 4)$.

3. Mark the asymptote and the required coordinates on the graph.

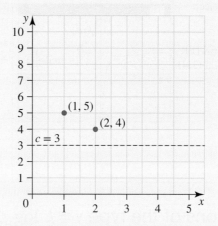

4. Complete the graph by drawing the hyperbola.

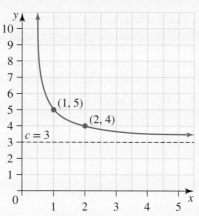

TI	THINK	DISPLAY/WRITE	CASIO	THINK	DISPLAY/WRITE
1. On a Graphs page, complete the entry line for function 1 as: $f1(x) = \dfrac{2}{x} + 3$ then press ENTER.	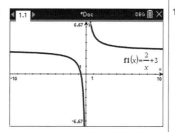	1. In a Graph & Table screen, complete the entry line for $y1$ as: $y1 = \dfrac{2}{x} + 3$ Select the tick box, then press the $ icon.	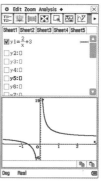
2. Press MENU, then select: 5: Trace 1: Graph Trace Type '1', then press ENTER twice to mark the point $(1, 5)$ on the graph.	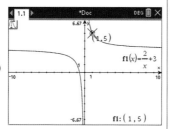	2. Select: • Analysis • Trace Type '1', then select OK. Press EXE to mark the point $(1, 5)$ on the graph.	

3. Press MENU, then select:
   5: Trace
   1: Graph Trace
   Type '2', then press ENTER
   twice to mark the point $(2, 4)$
   on the graph.

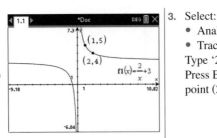

3. Select:
   - Analysis
   - Trace
   Type '2', then select OK.
   Press EXE to mark the
   point $(2, 4)$ on the graph.

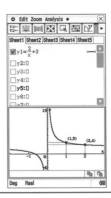

10.5.3 Functions of the type $y = k \log_{10}(x) + c$

Logarithms can be very useful when dealing with various calculations in mathematics, as they are the inverse of exponentials. Log transformations can also be very useful for linearising some data types.

Base 10 logarithms, written '$\log_{10}$', are usually an inbuilt function in calculators and are obtained using the 'log' button. We can work between exponential and logarithmic functions by using the relationship $y = A^x \Leftrightarrow \log_A(y) = x$. The following tables show some values for the functions $y = 10^x$ and $y = \log_{10}(x)$.

x	0	1	2	3
$y = 10^x$	$10^0 = 1$	$10^1 = 10$	$10^2 = 100$	$10^3 = 1000$

x	1	10	100	1000
$y = \log_{10}(x)$	$\log_{10}(1) = 0$	$\log_{10}(10) = 1$	$\log_{10}(100) = 2$	$\log_{10}(1000) = 3$

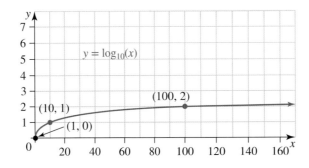

For functions of the type $y = k \log_{10}(x) + c$, as the value of k increases, the graph stretches out, moving parallel to the vertical axis (and therefore moving further away from the horizontal axis).

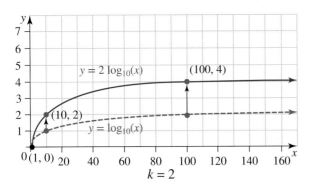

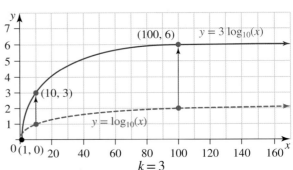

As the value of c increases, every point on the graph translates (moves) parallel to the horizontal axis by a distance of c. This will also change the x-intercept.

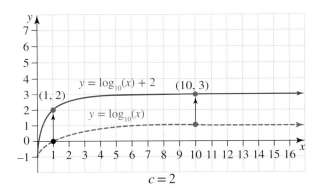

$$c = 2$$

WORKED EXAMPLE 11 Constructing the graph of a logarithmic relationship

Construct the graph of $y = 5\log_{10}(x) + 2$, indicating the coordinates for when $x = 1$ and $x = 10$. Use CAS to determine the x-intercept correct to 1 decimal place.

THINK	WRITE
1. Substitute the values into the rule for the coordinates required.	When $x = 1$, $y = 5\log_{10}(1) + 2$ $= 2$ and when $x = 10$, $y = 5\log_{10}(10) + 2$ $= 7$ The required coordinates are $(1, 2)$ and $(10, 7)$.
2. Use CAS to determine the x-axis intercept by solving the equation equal to zero.	Solve: $5\log_{10}(x) + 2 = 0$ $\Rightarrow x = 10^{\frac{-2}{5}} \approx 0.4$
3. Construct the graph with the indicated coordinates.	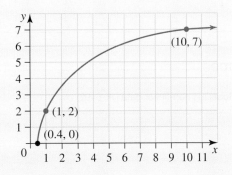

TI | THINK	DISPLAY/WRITE	CASIO | THINK	DISPLAY/WRITE
1. On a Graphs page, complete the entry line for function 1 as: $f1(x) = 5\ \log_{10}(x) + 2$ then press ENTER.	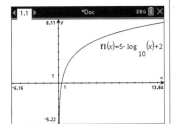	1. In a Graph & Table screen, complete the entry line for $y1$ as: $y1 = 5\ \log_{10}(x) + 2$ Select the tick box, then press the \$ icon.	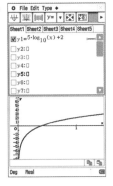

2. Press MENU, then select:
5: Trace
1: Graph Trace
Type '1', then press ENTER twice to mark the point (1, 2) on the graph.

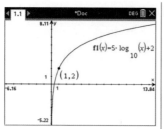

2. Select:
• Analysis
• Trace
Type '1', then select OK. Press EXE to mark the point (1, 2) on the graph.

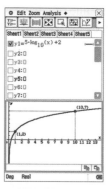

3. Press MENU, then select:
5: Trace
1: Graph Trace
Type '10', then press ENTER twice to mark the point (10, 7) on the graph.

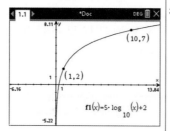

3. Select:
• Analysis
• Trace
Type '10', then select OK. Press EXE to mark the point (10, 7) on the graph.

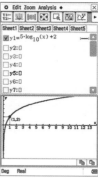

4. Press MENU, then select:
6: Analyze Graph
1: Zero
When prompted for the lower bound, move the cursor to the left of the x-intercept and click. When prompted for the upper bound, move the cursor to the right of the x-intercept and click.

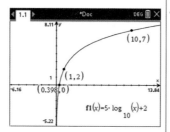

4. Select:
• Analysis
• G-Solve
• Root
then press EXE to mark the x-intercept on the graph.

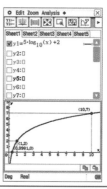

 Resources

Interactivity Modelling non-linear data (int-6492)

1. **WE9** Determine the rule for the transformed data shown in the table.

Original data				
x	0	1	2	3
y	3	6	15	30

Transformed data				
x	0	1	2	3
x^2	0	1	4	9
y	3	6	15	30

2. Determine the rule in the form $y = kx^2 + c$ that relates the variables in the following tables.

a.

x	0	1	2	3
y	25	23	17	7

b.

x	0	1	2	3	4	5	6	7
y	10	10.4	11.6	13.6	16.4	20	24.4	29.6

3. For each of the following applications:

   i. write a formula to represent the relationship
   ii. calculate the required values.

   a. According to Hooke's law, the distance, d, that a spring is stretched by a hanging object varies directly with the object's mass, m. If a 6-kg mass stretches a spring by 40 cm, determine the distance a spring is stretched when the mass hanging from it is 15 kg.

   b. The length of a radio wave, L, is inversely proportional to its frequency, f. If a radio wave has a length of 15 m when the frequency is 600 kHz, calculate the length of a radio wave that has a frequency of 1600 kHz.

   c. The stopping distance of a car, d, after the brakes have been applied varies directly as the square of the speed, s. If a car travelling 100 km/h can stop in 40 m, determine how fast can a car go and still stop in 25 m.

4. a. Apply the rule $y = 2x^2 + 5$ to complete the table of values shown.

x	0	1	2	3	4	5
y						

b. Construct a graph of the table.

c. Apply an x^2 transformation and redraw the graph using the transformed data.

d. Determine the gradient of the transformed data.

5. **WE10** Construct the graph of $y = \dfrac{4}{x} + 5$, indicating the position of the horizontal asymptote and the coordinates for when $x = 0.5$ and $x = 8$.

6. Determine the rule in the form $y = \dfrac{k}{x} + c$ that relates the variables in the following tables.

a.

x	1	2	3	4	6	12
y	18	12	10	9	8	7

b.

x	1	2	4	5
y	16	6	1	0

7. Consider the inverse relation $y = \dfrac{k}{x} + c$ with the graph shown.

a. Use the coordinates $(1, 14)$ and $(3, 6)$ to calculate the values of k and c.

b. Complete the table shown.

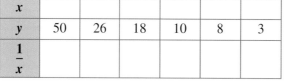

x						
y	50	26	18	10	8	3
$\dfrac{1}{x}$						

c. Sketch a graph of y against $\dfrac{1}{x}$ for the points in the table.

d. Determine the gradient of the transformed data.

8. Consider the table shown.

x	0	1	2	3	4	5
y	6	6.75	9	12.75	18	24.75

Given that y varies directly with x^2:

a. sketch a straight-line representation of the relationship

b. determine the rule for the straight line graph.

9. **WE11** Construct the graph of $y = 4 \log_{10}(x) + 4$, indicating the coordinates for when $x = 1$ and $x = 10$. Use CAS to determine the x-intercept correct to 2 decimal places.

10. Determine the rule in the form $y = k \log_{10}(x) + c$ that relates the variables in the following tables.

a.

x	1	10	100
y	1	3	5

b.

x	10	100	1000
y	1	7	13

11. A study of a marsupial mouse population on an isolated island finds that it changes according to the rule $P = 10 \log_{10}(t) + 600$, for $t \geq 1$, where P is the total population and t is the time in days for the study.

   a. Complete the table.

t	1	10	100
P			

   b. Sketch a graph of the changing population over this time period.
   c. Determine the population on the 1000th day of the study.

12. The relationship between the speed of a car, s km/h, and its exhaust emissions, E g/1000 km, is shown in the following table.

s	50	55	60	65	70	75	80	85
E	1260	1522.5	1810	2122.5	2460	2822.5	3210	3622.5

   a. Sketch a straight-line representation of the relationship.
   b. Determine the rule for the straight-line graph.

13. The maximum acceleration possible for various objects of differing mass is shown in the following table.

Mass (kg)	0.5	3	10	15	20	30	75
Acceleration (m/s^2)	30	5	1.5	1	0.75	0.5	0.2

   a. Describe the type of variation that is present for this data.
   b. Sketch a graph of the relationship.
   c. Determine the rule for the graph.

14. The frequency, f, of the vibrations of the strings of a musical instrument will vary inversely with their length, l.

   a. A string in Katya's cello is 62 cm long and vibrates 5.25 times per second. Complete the following table, giving your answers correct to 1 decimal place.

String length (cm)	70	65	60	55	50	45	40
Vibrations per second							

   b. Sketch a graph of the table of values.
   c. Use CAS to draw a graph of a $\frac{1}{x}$ transformation.
   d. Comment on the effect of the transformation.

15. The graph shown is of the form $y = k \log_{10}(x) + c$.

    a. Use CAS to determine the values of k and c.
    b. Redraw the graph with k and c increased by adding 2 to their values. Indicate the coordinates on the graph that correspond to the x-values of 1, 10 and 100.

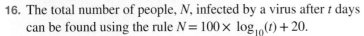

16. The total number of people, N, infected by a virus after t days can be found using the rule $N = 100 \times \log_{10}(t) + 20$.

    a. Giving your answers to the nearest whole number, calculate the total number of people infected after:

    i. 1 day   ii. 2 days   iii. 10 days

    b. Sketch a graph of $N = 100 \times \log_{10}(t) + 20$.
    c. Use CAS to calculate how many days it takes to reach double the number of infected people compared to when $t = 10$.

17. Use CAS to investigate graphs of the form $y = \log_{10}(x - b)$.

    a. Sketch graphs for:
    i. $y = \log_{10}(x - 1)$ ii. $y = \log_{10}(x - 2)$ iii. $y = \log_{10}(x - 3)$.

    b. Explain what happens to the graphs when the value of b is changed.

18. Use CAS to investigate relationships of the form $y = 10 \log_{10}(x + 1) + 2$.

    a. Complete the table for the given values.

x	9	99	999
y			

    b. Sketch a graph for the values in the table.
    c. Complete the table of values for a $k \log_{10}(x)$ transformation. Give answers correct to 4 decimal places.

x	9	99	999
$\log_{10}(x)$			
y			

    d. Sketch a graph of the transformed data.   e. Comment on the transformed graph.

10.5 Exam questions

Question 1 (1 mark)

MC It is known that y varies directly with the square of x. If $y = 300$ when $x = 25$, state which of the following is a correct equation connecting the variables.

A. $y = 300 x^2$ **B.** $y = x^2$ **C.** $y = 0.48 x^2$ **D.** $y = 25 x^2$ **E.** $y = 12 x^2$

Question 2 (1 mark)

MC It is known that y varies inversely with x. If $y = 50$ when $x = 2$, state which of the following is a correct equation connecting the variables.

A. $y = \dfrac{50}{x}$ **B.** $y = \dfrac{2}{x}$ **C.** $y = \dfrac{25}{x}$ **D.** $y = \dfrac{100}{x}$ **E.** $y = \dfrac{1}{x}$

Question 3 (1 mark)

MC It is known that y varies directly with the square of x. If $y = 90$ when $x = 5$, state which of the following is a correct equation connecting the variables.

A. $y = 3.6 x^2$ **B.** $y = 90 x^2$ **C.** $y = 18 x^2$ **D.** $y = 5 x^2$ **E.** $y = x^2$

More exam questions are available online.

10.6 Review

10.6.1 Summary

doc-37620

10.6 Exercise

Multiple choice

1. **MC** The graph shows the connection between the number of revolutions of a bicycle wheel and the distance travelled. State which of the following is true.

   A. $k = 2.2$

   B. $\dfrac{r}{d} = 2.2$ for any point on the graph

   C. The units for k are revolutions per metre.

   D. $r = 2.2d$

   E. $k = 22$

2. **MC** If b is directly proportional to c^2, select the constant of variation if $b = 72$ when $c = 12$.

   A. 0.5   **B.** 2   **C.** 6   **D.** 36   **E.** 144

3. **MC** If $y \propto \dfrac{1}{x}$ and $y = 12$ when $x = 50$, then when $x = 25$:

   A. $y = 6$

   B. $y = 18$

   C. $y = 12.5$

   D. $y = 24$

   E. $y = 11.5$

Graph of distance travelled versus number of revolutions

(y-axis: Distance travelled (m), values 200 to 2200; x-axis: Number of revolutions, values 200 to 1000; axes labelled d and r)

4. **MC** When an amount of gas is enclosed in a container, the pressure, P, is inversely proportional to the volume, V. State which of the following is **not** true.

   A. P varies inversely with V.   **B.** As V increases, P decreases.

   C. If V is halved, P is doubled.   **D.** $P \propto \dfrac{1}{V}$

   E. $\dfrac{P}{V} = k$

5. **MC** If y varies with the square of x and $y = 100$ when $x = 2$, then the constant of variation is:

   A. 400

   B. 200

   C. 100

   D. 50

   E. 25

6. The difference in magnitude between two earthquakes measuring 5.9 and 8.2 on the Richter scale is:

A. 199.5   **B.** 1.1   **C.** 2.3   **D.** 9.2   **E.** 20.0

7. **MC** A liquid with a pH level of 9 is diluted to a pH level of 4. Calculate how many times more acidic the liquid is now.

A. 1 000 000 000   **B.** 1 00 000   **C.** 10 000   **D.** 1000   **E.** 100

8. **MC** The graph shown follows the general rule $y = kx^2 + c$. A possible rule for the graph would be:

A. $y = 4x^2 + 2$

B. $y = \dfrac{3x^2}{4} + 2$

C. $y = 0.2x^2 + 2$

D. $y = \dfrac{x^2}{4} + 2$

E. $y = 0.5x^2 + 2$

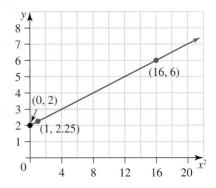

9. **MC** A possible rule for the graph shown would be:

A. $y = \log_{10}(x)$
B. $y = 2\log_{10}(x) + 2$
C. $y = 3\log_{10}(x) + 2$
D. $y = 2\log_{10}(x) + 3$
E. $y = 2\log_{10}(x)$

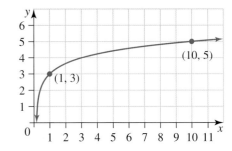

10. **MC** The graph with the rule $y = \dfrac{x^2}{4} + 1$ is:

A.
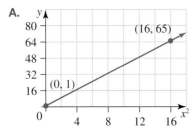

B.
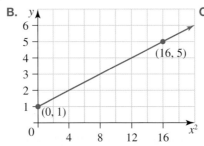

C.
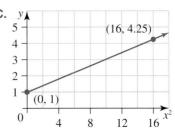

D.
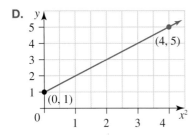

E.
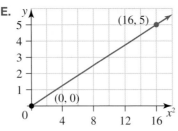

11. The distance, *d* metres, that an object has fallen from rest after a time, *t* seconds, is given in the table.

Time, *t* (seconds)	0	1	2	3	4	5
Distance, *d* (metres)	0	4.9	19.6	44.1	78.4	122.5

    a. State the type of variation in this case.

    b. Determine the rule that relates the distance, *d*, and the time, *t*.

    c. Apply the rule to find the distance fallen in 12 seconds.

    d. Apply the rule to calculate the time taken to fall a distance of 25 metres, correct to 2 decimal places.

12. The time, *t* hours, to complete a 100-kilometre journey varies inversely with the traveller's speed, *v* (m/s).

    a. Write a proportionality statement connecting *t* and *v*.

    b. Write a rule connecting *t* and *v* using a constant of variation, *k*.

    c. Calculate the time to complete the 100 km at 50 km/h.

    d. Calculate the constant of variation.

    e. State the rule connecting *t* and *v*.

13. The intensity, *I* (lux), of the light from a light bulb is inversely proportional to the square of the distance, *d* (m), from the bulb. At a distance of 1.0 m, the intensity is 5.0 lux, as shown in the diagram.

    a. Write a variation statement using a '$\propto$' sign for the relationship between *I* and *d*.

    b. Calculate the constant of variation and hence write a rule for the relationship.

    c. Apply the rule to calculate the intensity of the light at a distance of 3.5 m from the bulb. Give your answer correct to 2 decimal places.

    d. Calculate the distance at which the intensity is 0.05 lux.

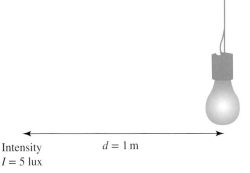

Intensity
I = 5 lux

d = 1 m

14. A call-centre employee is paid by the hour. The table shows the number of hours worked in a 5-day week and the wages earned each day. The wages earned vary directly with the number of hours worked.

Day	Monday	Tuesday	Wednesday	Thursday	Friday
Hours worked (*h*)	7.0	6.5	9.0	7.0	8.5
Wages (*w*)	$119.00	$110.50	$153.00	$119.00	$144.50

    a. Write an equation for the relationship between the variables.

    b. Calculate the constant of variation, *k*.

    c. Calculate the value of $\dfrac{\text{wages}}{\text{hours}}$ for each day.

    d. Evaluate what your answer to part c indicates.

15. Consider the data in the table.

x	3.5	6.1	9.7	11.2
y	7.35	12.81	21.34	23.52

    a. Determine why *y* is not directly proportional to *x*.

    b. One *y*-value can be changed so that *y* is directly proportional to *x*. Determine the required change.

Extended response

16. Johannes Kepler formulated three laws of planetary motion in the early seventeenth century. The second of these stated that for any planetary system, $R^3 \propto T^2$, where R is the radius of orbit and T is the time taken for the orbit (the period).

  a. Copy and complete the following table to test Kepler's Second Law.

Planet	Orbital radius, R (m)	R^3	Orbital period, T (s)	T^2	$\dfrac{R^3}{T^2}$
Mercury	5.79×10^{10}		7.60×10^{6}		
Venus	1.08×10^{11}		1.94×10^{7}		
Earth	1.49×10^{11}		3.16×10^{7}		
Mars	2.28×10^{11}		5.94×10^{7}		
Jupiter	7.78×10^{11}		3.74×10^{8}		
Saturn	1.43×10^{12}		9.30×10^{8}		
Uranus	2.87×10^{12}		2.66×10^{9}		
Neptune	4.50×10^{12}		5.20×10^{9}		

  b. Explain how the data supports Kepler's Second Law.

17. The distance from Earth to the Sun is approximately 1×10^{11} m, whereas the distance to the nearest star is 1×10^{16} m.

  a. Determine by how many orders of magnitude these distances differ.
  b. Express the difference in distance as a basic numeral.
  c. The diameter of the Sun is 1 391 684 km, and the Earth has a diameter of 12 742 km. Express the diameter of the Earth as a percentage of the diameter of the Sun. Give your answer correct to 3 significant figures.

18. Water is being poured into a cone of height 100 cm and radius 20 cm. When the water has reached a height, h, in the vessel, the radius of the surface is r.

  a. Determine the rule that relates the height and the radius of the vessel, and use this rule to calculate the height when the radius is 4 cm.

  Calculate the height, h, when $r = 4$ cm.

  b. Complete the table by calculating the volume of water in the cone at various radii.

r (cm)	4	6	8	10	15	20
Volume, $V = \dfrac{\pi r^2 h}{3}$ (cm³)						

  c. Using CAS, draw a graph of V against r.
  d. Explain whether V varies directly as r.
  e. Complete the table and draw a graph for an r^3 transformation.

r (cm)	4	6	8	10	15	20
r^3 (cm³)						
Volume, $V = \dfrac{\pi r^2 h}{3}$ (cm³)						

  f. Determine the rule that connects V and r^3.

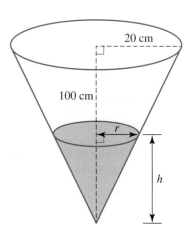

10.6 Exam questions

Question 1 (1 mark)

MC State by how many orders of magnitude 1 gigabyte and 1 megabyte differ.

A. 1 **B.** 2 **C.** 3 **D.** 4 **E.** 5

Question 2 (1 mark)

MC Given the graph of p against q shown, select the true statement from the following.

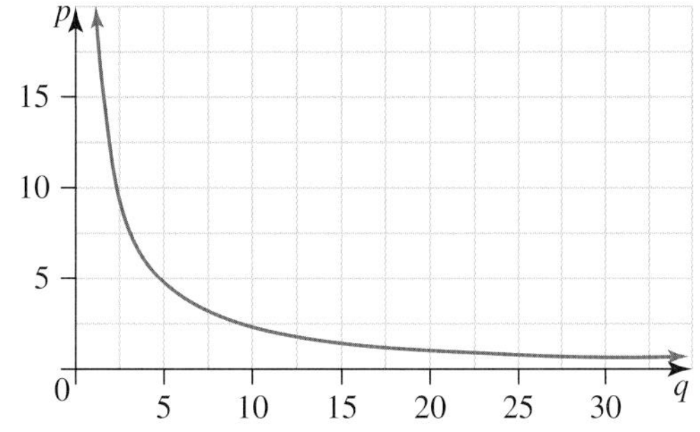

A. p varies inversely as q
B. p varies directly as q
C. p varies as q^2
D. p varies inversely as q^2
E. There is no relationship between q and p.

Question 3 (1 mark)

MC It is known that y varies inversely with x. If $y = 75$ when $x = 2$, state which of the following is a correct equation connecting the variables.

A. $y = \frac{75}{x}$

B. $y = \frac{2}{x}$

C. $y = \frac{37.5}{x}$

D. $y = \frac{150}{x}$

E. $y = \frac{1}{x}$

Question 4 (1 mark)

MC State which of these sets of variables is an example of direct proportion.

A. The temperature and the number of ice creams sold
B. A person's age and height
C. The amount of fuel in a car's tank and the distance that can be travelled at a constant rate
D. The price of a daily train ticket and the length of time travelled
E. The number of hours spent watching TV and the number of hours spent studying

The time it takes to travel from Melbourne to Sydney depends on the average speed of travel. The values of time, t (hours), and average speed, v (kilometres per hour), are shown in the table.

v (km/h)	100	200	300	400	500	600	700	800
t (h)	8.80	4.40	2.93	2.20	1.76	1.47	1.26	1.10

   a. Sketch the graph of time taken against speed. **(1 mark)**

   b. Select the true statements from the following. **(1 mark)**

     i. The graph is a straight line.

     ii. As the speed increases, the time taken decreases.

     iii. As the speed increases, the time taken increases.

     iv. When the speed is doubled, the time taken is halved.

   c. Write the variation relationship between t and v. **(1 mark)**

   d. Determine what graph should be plotted to find the relationship between t and v. **(1 mark)**

   e. Determine the relationship between t and v. **(1 mark)**

More exam questions are available online.

Answers

Topic 10 Variation

10.2 Direct and inverse variation

10.2 Exercise

1. $k = 60$, $C = 60N$

2. $d = 90t$

3. a. No, $y \neq kx$. **b.** Yes, $y = 2.1x$.

4. a. $I \propto A$ **b.** $I = kA$ **c.** $k = \dfrac{1}{2}$

 d. $I = \dfrac{1}{2}A$ **e.** $\$27\,500$ **f.** $\$150\,000$

5. a. $k = 4.5$ **b.** $K = 4.5m$ **c.** 324
 d. 100

6. $T = \dfrac{100}{s}$

7. a.

Family members	20	10	5	4	1
Number of chocolates	1	2	4	5	20

b.

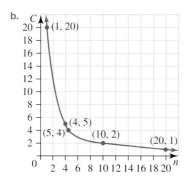

c. $C = \dfrac{20}{n}$

8. a.

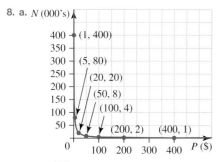

b. $n = \dfrac{400}{P}$

c. 16

d. $\$0.0016$

9. a.

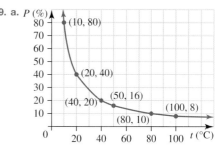

b. $P = \dfrac{800}{t}$

c. 32%

d. 250 °C

10.2 Exam questions

Note: Mark allocations are available with the fully worked solutions online.

1. D

2. E

3. As x increases, y also increases.
The graph relating y and x goes through the origin.
The graph relating y and x is linear.
As these variables satisfy all three conditions, y and x are directly proportional.

10.3 Data transformations

10.3 Exercise

1. a.

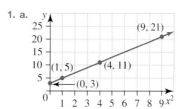

b.

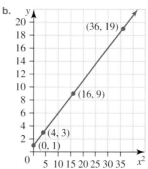

c.

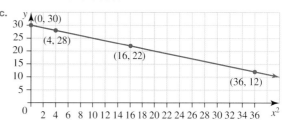

2. a. i.

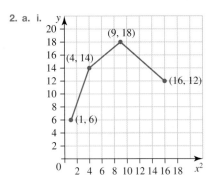

ii. Not linearised

b. i.

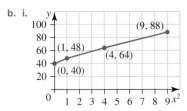

ii. Linearised (not perfectly)

c. i.

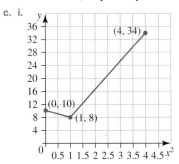

ii. Not linearised

d. i.

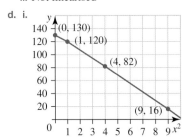

ii. Linearised (not perfectly)

3. a. i.

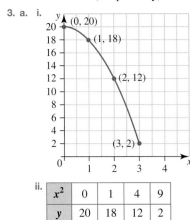

ii.

x^2	0	1	4	9
y	20	18	12	2

iii.

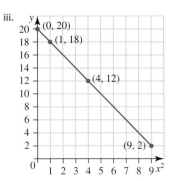

iv. The transformed data has been linearised.

b. i.

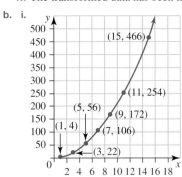

ii.

x^2	1	9	25	49	81	121	225
y	4	22	56	106	172	254	466

iii.

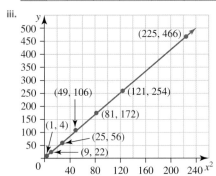

iv. The transformed data has been linearised.

c. i.

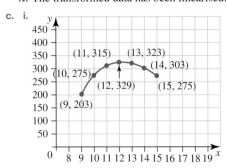

ii.

x^2	81	100	121	144	169	196	225
y	203	275	315	329	323	303	275

iii.

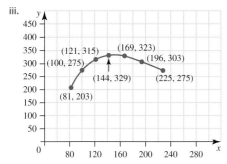

iv. The transformed data has not been linearised.

d. i.

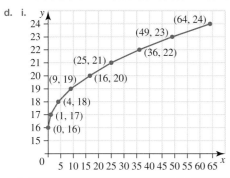

ii.

x^2	0	1	16	81	256	625	1296	2401	4096
y	16	17	18	19	20	21	22	23	24

iii.

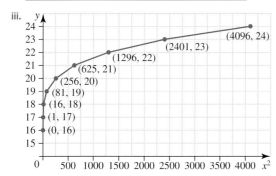

iv. The transformed data has not been linearised.

4. a.

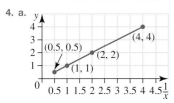

Linearised

b. i.

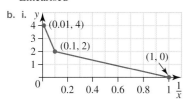

ii. Not linearised

5. a. i.

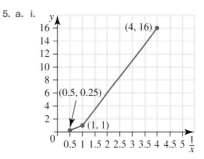

ii. Not linearised (but more linear than the original)

b. i.

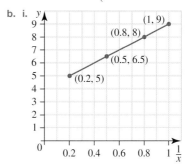

ii. Linearised

6. a. i.

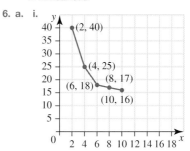

ii.

$\dfrac{1}{x}$	$\dfrac{1}{2}$	$\dfrac{1}{4}$	$\dfrac{1}{6}$	$\dfrac{1}{8}$	$\dfrac{1}{10}$
y	40	25	18	17	16

iii.

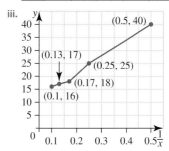

iv. The transformed data has been made more linear.

b. i.

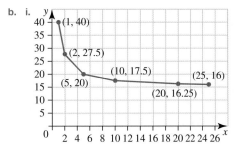

ii.

$\dfrac{1}{x}$	1	$\dfrac{1}{2}$	$\dfrac{1}{5}$	$\dfrac{1}{10}$	$\dfrac{1}{20}$	$\dfrac{1}{25}$
y	40	27.5	20	17.5	16.25	16

iii. See the graph at the bottom of the page.*

iv. The transformed data has been linearised.

c. i.

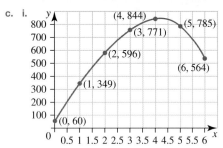

ii.

$\dfrac{1}{x}$	Undefined	1	$\dfrac{1}{2}$	$\dfrac{1}{3}$	$\dfrac{1}{4}$	$\dfrac{1}{5}$	$\dfrac{1}{6}$
y	60	349	596	771	844	785	564

iii.

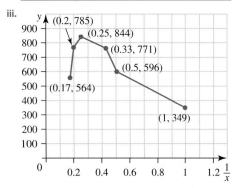

iv. The transformed data has not been linearised.

d. i.

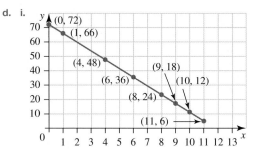

ii.

$\dfrac{1}{x}$	Undefined	1	$\dfrac{1}{4}$	$\dfrac{1}{6}$	$\dfrac{1}{8}$	$\dfrac{1}{9}$	$\dfrac{1}{10}$	$\dfrac{1}{11}$
y	72	66	48	36	24	18	12	6

iii.

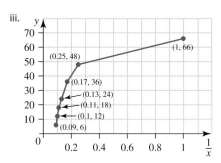

iv. The original data was linear but the transformed data is not.

7. a.

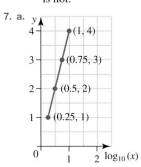

b.

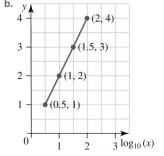

***6. b. iii.**

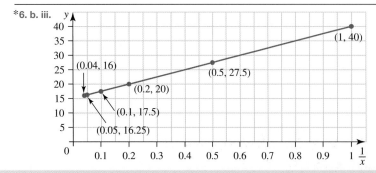

8. a. See the graph at the bottom of the page.*
This graph is not linearised by the $\log_{10}(x)$ transformation.

b.

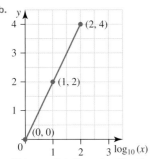

This graph is linearised by the $\log_{10}(x)$ transformation.

9. a.

x	0.25	0.5	1	1.25	2	2.5	4	5
$\dfrac{1}{x}$	4	2	1	0.8	0.5	0.4	0.3	0.2
y	16	4	1	0.64	0.25	0.16	0.0625	0.04

b. Original data:

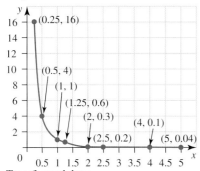

Transformed data:

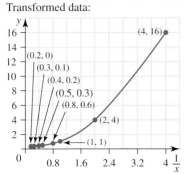

c. This transformation is not very effective at linearising this data.

10. a.

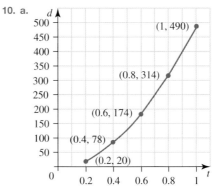

b.

Time^2 (s^2)	0.04	0.16	0.36	0.64	1.0
Distance (cm)	20	78	174	314	490

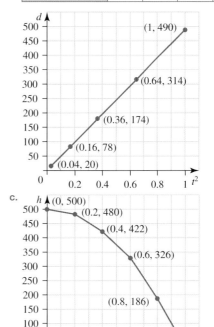

c.

d. The same data transformation will work in this case as the original data is parabolic in its shape, so the transformed data will look like this.

*8. a.

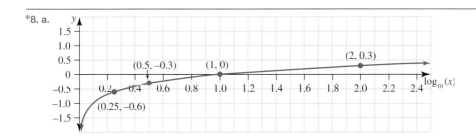

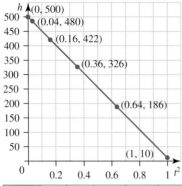

11. a.

x	10	20	30	40	50	60	70	80
y	4	5.2	5.9	6.4	6.8	7.1	7.4	7.6
$\dfrac{1}{x}$	0.10	0.05	0.033	0.025	0.02	0.017	0.014	0.013
$\log(x)$	1.00	1.30	1.48	1.60	1.70	1.78	1.85	1.90

Altitude (m)	Air pressure (kPa)	$\dfrac{1}{\text{Altitude (m)}}$
0	100	Undefined
300	90	0.003 33
1500	75	0.000 66
2500	65	0.000 40
3000	60	0.000 33
3500	55	0.000 29
4000	50	0.000 25
5000	45	0.000 20
5500	40	0.000 18
6000	35	0.000 17
8000	25	0.000 13
9000	18	0.000 11
12 000	12	0.000 08
15 000	10	0.000 06

b.

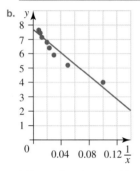 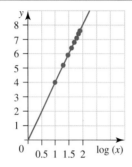

c. When using the calculator, the $\log(x)$ transformation gives the best linearisation of the data.

12. a.

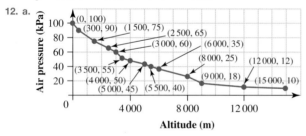

b.

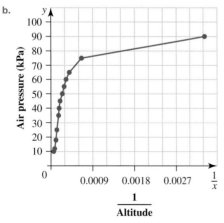

c. The $\dfrac{1}{x}$ transformation doesn't put all the data in a straight line. Most of the values seem to be more linear, but the lower altitude values curve sharply. This would seem to indicate that an alternative transformation (other than $\dfrac{1}{x}$ or x^2) is needed to linearise the data.

13. a. As it is likely that the consumption of electricity would be influenced by the maximum temperature, the temperature should be used for the x-axis.

b. i.

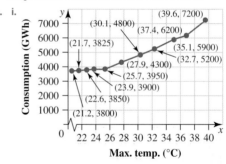

ii. x^2 transformation:

See graph at the bottom of the page*

$\dfrac{1}{x}$ transformation:

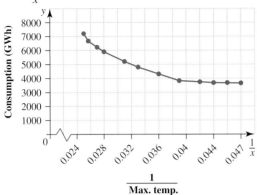

$$\dfrac{1}{\text{Max. temp.}}$$

iii. Both transformations have a linearising effect, but neither appears to be substantially better than the original data.

14. a.

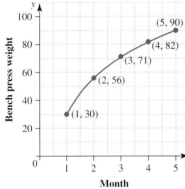

b.

x (months)	1	2	3	4	5
$\log_{10}(x)$	0	0.30	0.48	0.60	0.70
y (weight)	30	56	71	82	90

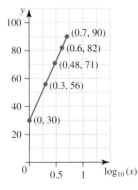

10.3 Exam questions

Note: Mark allocations are available with the fully worked solutions online.

1. C

2. B

3. a. 31.80 cents   b. $31.80 + 0.132l^2$ c. $607.13

10.4 Orders of magnitude

10.4 Exercise

1. 4

2. 3

3. 3

4. 7200 seconds

5. B

6. An earthquake of magnitude 5 is 100 times stronger than an earthquake of magnitude 3.

7. a. 3   b. 1000

8. C

9. a. 0.5   b. 3.16

10. 4

11. a. 2
    b. Tennis ball: 50 g; medicine ball: 5000 g

12. a. 5   b. 100 000

13. a. $5 \times 10^3 \times 10^{-1}$ b. 500 mL

14. a. 2.5×10^{-1} km b. 0.25 km or 250 m

10.4 Exam questions

Note: Mark allocations are available with the fully worked solutions online.

1. B   2. A   3. C

10.5 Non-linear data modelling

10.5 Exercise

1. $y = 3x^2 + 3$

2. a. $y = -2x^2 + 25$ b. $y = 0.4x^2 + 10$

3. a. i. $d = \dfrac{40}{6} m$ ii. 100 cm

   b. i. $L = \dfrac{9000}{f}$ ii. 5.625 m

   c. i. $d = 0.004 s^2$ ii. 79 km/h

*13. b. ii.

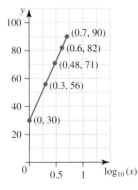

(Max. temp.)²

4. a.

x	0	1	2	3	4	5
y	5	7	13	23	37	55

b.

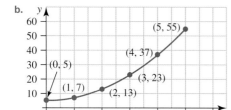

c.

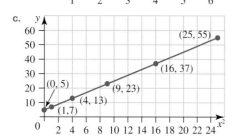

d. 2

5.

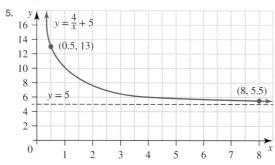

6. a. $y = \dfrac{12}{x} + 6$　　　**b.** $y = \dfrac{20}{x} - 4$

7. a. $k = 12$ and $c = 2$

b.

x	0.25	0.5	0.75	1.5	2	12
y	50	26	18	10	8	3
$\dfrac{1}{x}$	4	2	1.33	0.67	0.5	0.08

c. See the graph at the bottom of the page.*

d. 12

8. a.

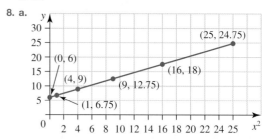

b. $y = 0.75x^2 + 6$

9. See the graph at the bottom of the page*

10. a. $y = 2\log_{10}(x) + 1$　　　**b.** $y = 6\log_{10}(x) - 5$

11. a.

t	1	10	100
P	600	610	620

b. See the graph at the bottom of the page*

c. 630

12. a.

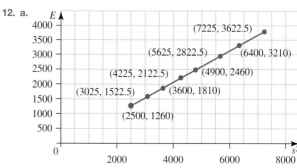

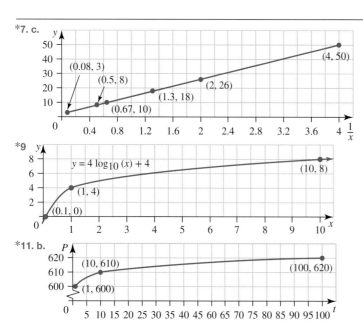

*7. c.

*9

*11. b.

b. $E = 0.5s^2 + 10$

13. a. Inverse variation

b.
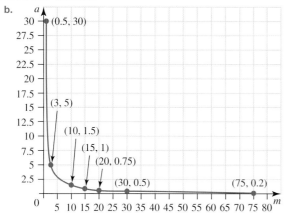

c. $a = \dfrac{15}{m}$

14. a.

String length (cm)	70	65	60	55	50	45	40
Vibrations per second	4.7	5.0	5.4	5.9	6.5	7.2	8.1

b.
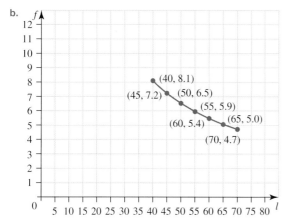

c.
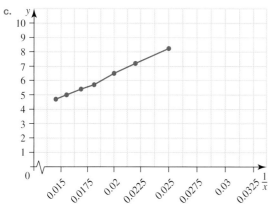

d. The transformed data is linearised.

15. a. $k = 2$ and $c = 3$

b.
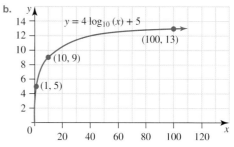

16. a. i. 20   ii. 50   iii. 120

b.
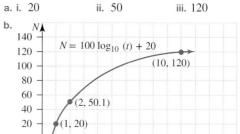

c. 158 days

17. a. i.
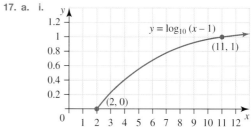

ii.
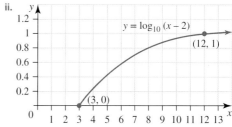

iii.
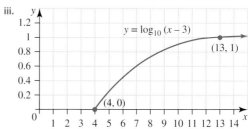

b. As the value of b increases, the graph moves an equivalent distance from the y-axis. The x-axis intercept is one more than the value of b.

18. a.

x	9	99	999
y	12	22	32

b.

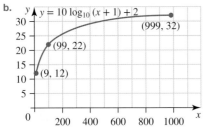

c.

x	9	99	999
$\log_{10}(x)$	0.9542	1.9956	2.9995
y	12	22	32

d.

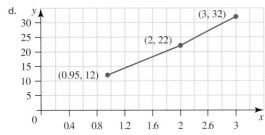

e. The transformed data has been linearised.

10.5 Exam questions

Note: Mark allocations are available with the fully worked solutions online.

1. C
2. D
3. A

10.6 Review

10.6 Exercise

Multiple choice

1. A
2. A
3. D
4. E
5. E
6. C
7. B
8. D
9. D
10. B

Short answer

11. **a.** Direct variation **b.** $d = 4.9t^2$
    c. 705.6 m **d.** 2.26 s

12. **a.** $t \propto \dfrac{1}{v}$ **b.** $t = \dfrac{k}{v}$
    c. 2 hours **d.** $k = 100$
    e. $t = \dfrac{100}{v}$

13. **a.** $I \propto \dfrac{1}{d^2}$ **b.** $k = 5, I = \dfrac{5}{d^2}$
    c. 0.41 lux **d.** 10 m

14. **a.** $w = kh$
    b. $k = 17$
    c. See table at the bottom of the page*
    d. There is direct variation between the two variables: wages and hours worked.

15. **a.** $k \neq \dfrac{y}{x}$ for all values in the table.
    b. Change the y-value from 21.34 to 20.37.

Extended response

16. **a.** See table at the bottom of the page*
    b. $\dfrac{R^3}{T^2}$ is approximately the same value for all the planets in the table.

17. **a.** 5 **b.** 100 000 **c.** 0.916%

---

*14. **c.**

Day	Monday	Tuesday	Wednesday	Thursday	Friday
Wages/hour (w/h)	$17	$17	$17	$17	$17

*16. **a.**

Planet	Orbital radius, R (m)	R^3	Orbital period, T (s)	T^2	$\dfrac{R^3}{T^2}$
Mercury	5.79×10^{10}	1.94×10^{32}	7.60×10^6	5.78×10^{13}	3.36×10^{18}
Venus	1.08×10^{11}	1.26×10^{33}	1.94×10^7	3.76×10^{14}	3.35×10^{18}
Earth	1.49×10^{11}	3.31×10^{33}	3.16×10^7	9.99×10^{14}	3.31×10^{18}
Mars	2.28×10^{11}	1.19×10^{34}	5.94×10^7	3.53×10^{15}	3.37×10^{18}
Jupiter	7.78×10^{11}	4.71×10^{35}	3.74×10^8	1.40×10^{17}	3.36×10^{18}
Saturn	1.43×10^{12}	2.92×10^{36}	9.30×10^8	8.65×10^{17}	3.38×10^{18}
Uranus	2.87×10^{12}	2.36×10^{37}	2.66×10^9	7.08×10^{18}	3.33×10^{18}
Neptune	4.50×10^{12}	9.11×10^{37}	5.20×10^9	2.70×10^{19}	3.37×10^{18}

18. a. $h = 5r$, so
$r = 4 \Rightarrow h = 20$ cm

b. See table at the bottom of the page*

c.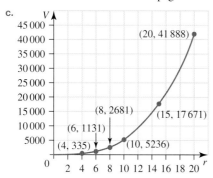

d. No

e. See table at the bottom of the page*

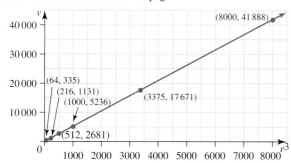

f. $V = 5.24r^3$

4. C

5. a. See figure at the bottom of the page.*

b. ii, iv

c. $t \propto \dfrac{1}{v}$

d. t against $\dfrac{1}{v}$

e. $t = \dfrac{880}{v}$

10.6 Exam questions

Note: Mark allocations are available with the fully worked solutions online.

1. C

2. A

3. D

---

*18. b.

r (cm)	4	6	8	10	15	20
Volume, $V = \dfrac{\pi r^2 h}{3}$ (cm³)	335	1131	2681	5236	17 671	41 888

*18. e.

r (cm)	4	6	8	10	15	20
r^3 (cm³)	64	216	512	1000	3375	8000
Volume, $V = \dfrac{\pi r^2 h}{3}$ (cm³)	335	1131	2681	5236	17 671	41 888

*5. a.

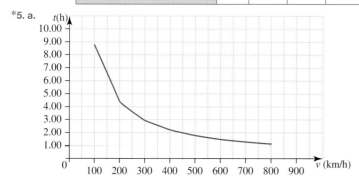

11 Space and measurement

Fully worked solutions for this topic are available online.

11.1 Overview

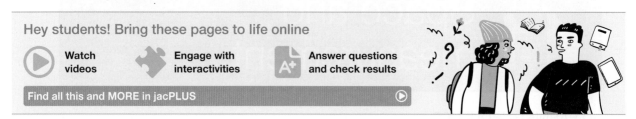

11.1.1 Introduction

When using formulas related to circles, we continually use π. We simply use 3.14 or push the button on our calculator, but what is π really and where did it come from?

π goes back to work done by Archimedes over 2000 years ago when calculators and computers didn't exist. He was able to find π to 99.9% accuracy, without using decimal places. The techniques he used to do this helped build the foundations of calculus.

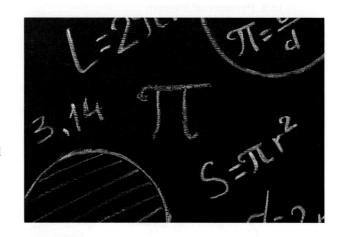

π is the length of the circumference of a circle with a diameter of 1 unit. Archimedes did not know the circumference of a circle, so how did he work out its value? A possible method is to use a square of side length 1 and hence a circle of diameter 1 as shown. Trigonometry shows that the value of π is between the outer square perimeter and the inner square perimeter. Thus, $2.8 < \pi < 4$. If you take the centre of these two values, you calculate π to be equal to 3.4.

Archimedes didn't actually use squares; he started with hexagons (six-sided polygons) to find the range of values that π is between. He then continued his work with 12-, 24-, 48- and 96-sided polygons, and stopped when he found $3\frac{10}{71} < \pi < 3\frac{1}{7}$. Decimals weren't invented until 250 BC, let alone any spreadsheet application that could easily perform these repeat calculations. So Archimedes had to spend a lot of time working through his formulas using fractions. The midpoint between $3\frac{10}{71}$ and $3\frac{1}{7}$ is 3.141 85, which is over 99.9% accurate. This is an amazing achievement considering it was done over 2000 years ago.

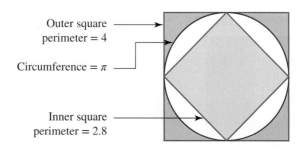

Outer square perimeter = 4

Circumference = π

Inner square perimeter = 2.8

KEY CONCEPTS

This topic covers the following key concepts from the VCE Mathematics Study Design:
- units of measurement of length, angle, area, volume and capacity
- exact and approximate answers, scientific notation, significant figures and rounding
- perimeter and areas of triangles, quadrilaterals, circles including arcs and sectors and composite shapes, and practical applications
- volumes and surface areas of solids (spheres, cylinders, pyramids and prisms and composite objects) and practical applications, including simple applications of Pythagoras' theorem in three dimensions.

Source: VCE Mathematics Study Design (2023–2027) extracts © VCAA; reproduced by permission.

11.2 Scientific notation, significant figures and rounding

LEARNING INTENTION

At the end of this subtopic you should be able to:
- apply scientific notation to simplify very large or small numbers
- use significant figures to simplify a number by rounding
- compare exact and approximate answers.

11.2.1 Scientific notation

Scientific notation is used to simplify very large numbers, such as the mass of the Earth $(5.972 \times 10^{24}\,\text{kg})$, or very small numbers, such as the mass of an electron $(9.109\,382\,91 \times 10^{-31}\,\text{kg})$.

Scientific notation

A number written in scientific notation is in the form $a \times 10^{b}$, where a is a real number between 1 and 10 and b is an integer.

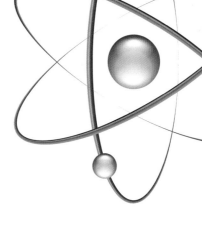

Scientific notation uses multiplications of 10. For example, look at the following pattern:

$$5 \times 10 = 50$$
$$5 \times 10 \times 10 = 500$$
$$5 \times 10 \times 10 \times 10 = 5000$$

Notice how the number of zeros in the answer increases in proportion to the number of times 5 is multiplied by 10. Scientific notation can be used here to simplify the repetitive multiplication.

$$5 \times 10^{1} = 50$$
$$5 \times 10^{2} = 500$$
$$5 \times 10^{3} = 5000$$

There are four key steps to write a basic numeral using scientific notation.

Step	Instructions	Example 1: 7256	Example 2: 0.008 923
1	Identify the first non-zero value of the original number.	7	8
2	Write that digit, followed by a decimal point and all remaining digits.	7.256	8.923
3	Multiply the decimal by 10.	7.256×10	8.923×10
4	Count the number of places the decimal point is moved. The power of the base value 10 will reflect the movement of the decimal point. If the decimal point is moved to the left, the power will be positive. If moved to the right, the power will be negative.	7.256×10^{3}	8.923×10^{-3}

To convert a scientific notation value to basic numerals, use the following steps.

Step 1 Look to see if the power is positive or negative.
For example: 2.007×10^5 (positive power) and 9.71×10^{-4} (negative power)

Step 2a If positive, rewrite the number without the decimal point, adding zeros **behind** the last number to fill in the necessary number of place values. In this case, move the decimal point 5 to the right as the power is +5.
For example: $2.007 \times 10^5 = 200\,700$

Step 2b If negative, rewrite the number without the decimal point, adding zeros **in front of** the last number to fill in the necessary number of place values. In this case, move the decimal point 4 to the left as the power is −4.
For example: $9.71 \times 10^{-4} = 0.000\,971$

Step 3 Double check by counting the number of places the decimal point has moved. This should match the value of the power of 10.

Note: When using CAS or a scientific calculator, you may be presented with an answer such as 3.19E−4. This is an alternative form of scientific notation; 3.19E−4 means 3.19×10^{-4}.

WORKED EXAMPLE 1 Writing numbers using scientific notation

Rewrite these numbers using scientific notation.
a. **640 783**
b. **0.000 005 293**

THINK	WRITE
a. 1. Identify the first digit (6). Rewrite the full number with the decimal place moved directly after this digit.	a. 6.40 783
2. Following the new decimal number, multiply by a base of 10.	$6.40\,783 \times 10$
3. Count the number of places the decimal point has moved, with this becoming the power of the base 10. Here the decimal point has moved 5 places. As the decimal point has moved to the left, the power will be positive.	$6.40\,783 \times 10^5$
b. 1. Identify the first non-zero digit (5). Rewrite the number with the decimal place moved to directly after this digit.	b. 5.293
2. Place the decimal point after the 5 and multiply by a base of 10.	5.293×10
3. Identify the the value of the power of 10 by counting the number of places the decimal point has moved. Here the decimal point has moved 6 places. As the decimal point has moved to the right, the power will be negative.	5.293×10^{-6}

WORKED EXAMPLE 2 Converting scientific notation to a basic numeral

Rewrite these numbers as basic numerals.
a. 2.5×10^{-11} m (the size of a helium atom)

b. 3.844×10^8 m (the distance from Earth to the Moon)

THINK	WRITE
a. 1. First, note that the power is negative, indicating the number will begin with a zero. In front of the 2, record 11 zeros.	a. 0 000 000 000 025
2. Put a decimal point between the first two zeros.	0.000 000 000 025 m
b. 1. The power of this example is positive. Rewrite the number without the decimal point.	b. 3844
2. Add the necessary number of zeros to move the decimal point 8 places to the right, as indicated by the power of 10.	384 400 000 m

11.2.2 Significant figures and rounding

Significant figures are a method of simplifying a number by rounding it to a base 10 value. Questions relating to significant figures will require a number to be written correct to x number of significant figures. To complete this rounding, the relevant significant figure(s) needs to be identified.

Consider the number 123.456 789.

This value has 9 significant figures, as there are 9 numbers that tell us something about the particular place value in which they are located. The most significant of these values is the number 1, as it indicates the overall value of this number is in the hundreds. If we were asked to round this value to 1 significant figure, we would round the number to the nearest hundred, which in this case would be 100. If we were rounding to 2 significant figures, we would round the number to the nearest 10, which is 120.

Rounding this value to 6 significant figures means the first 6 significant figures need to be acknowledged, 123.456. However, as the number following the 6th significant figure is 5 or more, the corresponding value needs to round up, therefore making the final answer 123.457.

Rounding rules

If the number after the required number of significant figures is 5 or more, round up. If this number is 4 or below, leave it as is.

Zeros

Zeros present an interesting challenge when evaluating significant figures and are best explained using examples. 4056 contains 4 significant figures. The zero is considered a significant figure as there are non-zero numbers on either side of it.

4000 contains 1 significant figure. The zeros are ignored as they are place holders and may have been rounded.

4000.0 contains 5 significant figures. In this situation the zeros are considered important due to the zero after the decimal point. A zero after the decimal point indicates the numbers before it are precise.

0.004 contains 1 significant figure. As with 4000, the zeros are place holders.

0.0040 contains 2 significant figures. The zero following the 4 implies the value is accurate to this degree.

Rounding with significant figures

- **When rounding with significant figures, the first non-zero digit is the first significant figure. All remaining digits are also significant unless the number is a whole number ending in zeros.**
- **When the number is a whole number ending in zeros, the final zero or trailing zeros in the whole number are not significant.**

WORKED EXAMPLE 3 Rounding with significant figures

With reference to the following values:
1. **identify the number of significant figures**
2. **round correct to 3 significant figures.**
    a. **19 080**   b. **0.000 076 214**

THINK	WRITE
a. i. This is a whole number ending in a zero. The final zero in this number is not significant. Therefore, the 1, 9, 0 and 8 are the significant figures.	a. 19 080 has 4 significant figures.
ii. Round the number to the third significant figure. It is important to consider the number that follows it. In this case, as the following number is 5 or more, the value of the third significant figure needs to be rounded up by 1.	Rounded to 3 significant figures, 19 080 = 19 100.
b. i. The first significant figure is the 7 since it is the first non-zero digit. All remaining digits are also significant.	b. 0.000 076 214 has 5 significant figures.
ii. The third significant figure is 2; however, it is important to consider the next value as it may require additional rounding. In this case the number following the 3rd significant number is below 5, meaning no additional rounding needs to occur.	Rounded to 3 significant figures, 0.000 076 214 = 0.000 076 2.

11.2.3 Exact and approximate answers

Exact answers are answers that are *not* rounded in any way. They maintain the absolute accuracy of a particular number and capture the entire mathematics of a problem. Exact answers come in many forms such as whole numbers, fractions, decimals and radicals.

Approximate answers also come in many forms and are a result of either making approximations in your working or rounding off. While providing rounded answers is very common in mathematics, it reduces the accuracy of answers.

A comparison of exact and approximate answers

- Solving the equation $3x = 2$ for x gives the *exact* solution $x = \frac{2}{3}$.

We could approximate the value of $\frac{2}{3}$ by rounding it to a particular amount of decimal places to give an *approximate* answer of $x = 0.7$ (correct to 1 decimal place).

- When using Pythagoras' theorem to find the length of a side of a triangle, we need to take the square root of a number. In this situation, we can give an exact answer by leaving the radical symbol in our answer or use a calculator to find an approximate answer.

If $c^2 = 1^2 + 2^2$, the *exact* answer is $c = \sqrt{5}$.

We can use a calculator to give an *approximate* answer, say $c = 2.24$ (correct to 2 decimal places).

Exact and approximate answers

When completing calculations in mathematical problems, pay attention to any rounding instructions. If an *exact* answer is required, it is strongly recommended you do *not* use a calculator to obtain the decimal answer. Instead, use arithmetic to find the solution to a problem or ensure your calculator is working in EXACT arithmetic.

For example, a musician has organised a concert at the local hall. The musician is charging $18.50 a ticket, and on the night 210 people have purchased tickets. To get an idea of the revenue, The musician does a quick estimate of the money made on ticket sales by rounding the values to 1 significant figure.

$$18.50 \times 210 \approx 20 \times 200$$
$$\approx \$4000$$

Note: The use of the approximate equals sign $\approx$ indicates the values used are no longer exact. Therefore, the resulting answer will be an approximation.

When compared to the exact answer, the approximate answer gives a reasonable evaluation of her revenue.

$$18.50 \times 210 = \$3885$$

WORKED EXAMPLE 4 Comparing exact and approximate answers

a. Calculate $42.6 \times 59.7 \times 2.2$, rounding the answer correct to 1 decimal place.
b. Redo the calculation by rounding the original values correct to 1 significant figure.
c. Comparing your two answers, state whether the approximate value would be considered a reasonable result.

THINK	WRITE
a. 1. Complete the calculation. The answer contains 3 decimal places; however, the required answer only needs 1.	a. $42.6 \times 59.7 \times 2.2$ $= 5595.084$
2. Look at the number following the first decimal place (8). As it is 5 or more, additional rounding needs to occur, rounding the 0 up to a 1.	≈ 5595.1
b. 1. Round each value correct to 1 significant figure. Remember to indicate the rounding by using the approximate equals sign ($\approx$).	b. $42.6 \times 59.7 \times 2.2$ $\approx 40 \times 60 \times 2$
2. Complete the calculation.	≈ 4800
c. There is a sizeable difference between the approximate and exact answers, so in this instance the approximate answer would not be considered a reasonable result.	c. No

 Resources

Interactivity Scientific notation (int-6456)

11.2 Exercise

1. **WE1** The Great Barrier Reef stretches 2 600 000 m in length. Rewrite this number using scientific notation.

2. The wavelength of red light is approximately 0.000 000 55 metres. Rewrite this number using scientific notation.

3. Rewrite the following values using scientific notation.
   a. 7319
   b. 0.080 425
   c. 13 000 438
   d. 0.000 260
   e. 92 630 051
   f. 0.000 569 2

4. **WE2** The thickness of a DNA strand is approximately 3×10^{-9} m. Rewrite this value as a basic numeral.

5. The universe is thought to be approximately 1.38×10^{10} years old. Rewrite this value as a basic numeral.

6. Rewrite the following values as basic numerals.
   a. 1.64×10^{-4}
   b. 2.3994×10^{-8}
   c. 1.4003×10^{9}
   d. 8.6×10^{5}

7. **WE3** The distance around the Earth's equator is approximately 40 075 000 metres.
   a. Identify the number of significant figures in this value.
   b. Round this value correct to 4 significant figures.

8. The average width of a human hair is 1.2×10^{-3} cm.
   a. Rewrite this value as a basic numeral.
   b. Identify the number of significant figures in the basic numeral.
   c. Round your answer to part **a** correct to 1 significant figure.

9. For each of the following values:
   i. identify the number of significant figures
   ii. round the value correct to the number of significant figures specified in the brackets.

   a. 1901 (2)
   b. 0.001 47 (2)
   c. 21 400 (1)
   d. 0.094 250 (3)
   e. 1.080 731 (4)
   f. 400.5 (3)

10. **WE4** Answer the following questions.

   a. Calculate $235.47 + 1952.99 - 489.73$, rounding the answer correct to 1 decimal place.
   b. Redo the calculation by rounding the original numbers correct to 1 significant figure.
   c. Comparing your two answers, state whether the approximate answer would be considered a reasonable result.

11. Bruce is planning a flight around Victoria in his light aircraft. On average the plane burns 102 litres of fuel an hour, and Bruce estimates the trip will require 37 hours of flying time.

   a. Manually calculate the amount of fuel required by rounding each value correct to 1 significant figure.
   b. Use a calculator to determine the exact amount of fuel required.

12. The space shuttle *Discovery* completed 39 space missions in its lifetime, travelling a total distance of 238 539 663 km.

   a. State how many significant figures are in the total distance travelled.
   b. Round this value correct to 2 significant figures.
   c. Convert your answer to part b to scientific notation.

13. The diameter of the Darling Harbour Ferris Wheel in Sydney, New South Wales is 110 m.

   a. Using the equation $C = \pi d$, determine the circumference of the wheel correct to 2 decimal places.
   b. Redo the calculation by rounding the diameter and the value of π correct to 2 significant figures.
   Determine how this changes your answer. State whether it would be considered a reasonable approximation.

14. The Robinson family want to lay instant lawn in their backyard. The dimensions of the rectangular backyard are 12.6 m by 7.8 m.

   a. If each square metre of lawn costs $12.70, estimate the cost for the lawn by rounding each number correct to the nearest whole number before completing your calculations.
   b. Compare your answer to part a to the actual cost of the lawn.
   c. State whether you could have used another rounding strategy to improve your estimate.

15. An animal park uses a variety of ventilated boxes to safely transport their animals. The lid of the box used for small animals measures 76 cm long by 20 cm wide. Along the lid of the box are 162 ventilation holes, each of which has a radius of 1.2 cm.

   a. Using the formula $A = \pi r^2$, calculate the area of one ventilation hole, providing the exact answer.
   b. Round your answer to part a correct to 2 decimal places.
   c. Determine the amount of surface area that remains on the lid after the ventilation holes are removed. Round your answer to the nearest whole number.
   d. The company that manufactures the boxes prefers to work in millimetre measurements. Convert your remaining surface area to millimetres squared (*Note*: 1 cm^2 = 100 mm^2.)
   e. Record your answer to part d in scientific notation.

▶ **Question 1 (1 mark)**

MC Select which of these numbers is expressed with 3 significant figures.

A. 250   **B.** 25   **C.** 2.5   **D.** 0.0250   **E.** 0.025

▶ **Question 2 (1 mark)**

MC A film of gold on a computer chip measures 5.14×10^{-3} metres wide, 7.32×10^{-3} metres long and 2.56×10^{-6} metres thick.

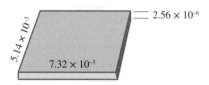

The volume of gold on the chip, in scientific notation, is closest to

A. $4 \times 10^{-11} \text{m}^3$

B. $1.5 \times 10^{-11} \text{m}^3$

C. $1.5 \times 10^{-53} \text{m}^3$

D. $4 \times 10^{-53} \text{m}^3$

E. $9.63 \times 10^{-11} \text{m}^3$

▶ **Question 3 (3 marks)**

Amy is calculating the surface area of Earth in square kilometres using the formula $A = 4\pi r^2$ as an approximation. She knows that Earth is not really spherical and that the radius measurement in kilometres depends on where it is measured. Also, π can be used with different levels of accuracy.

| Value of π | Surface area of Earth | | Average value |
	Radius at equator 6378.1 km	Radius at poles 6356.8 km	
3.141 59	511 201 530.5	507 792 863	509 497 196.8
3.141 592 654	511 201 962.3	507 793 291.9	509 497 627.1
π from computer	511 201 962.3	507 793 291.9	509 497 627.1
3.142 857 143 $\left(\pi \approx \dfrac{22}{7}\right)$	511 407 720.8	507 997 678.4	509 702 699.6

a. State which value of π in the table is likely to be most accurate. **(1 mark)**

b. Determine the most accurate average value of the surface area. **(1 mark)**

c. Write the most accurate average value of the surface area in scientific notation. **(1 mark)**

More exam questions are available online.

11.3 Pythagoras' theorem

LEARNING INTENTION

At the end of this subtopic you should be able to:
- apply Pythagoras' theorem to calculate an unknown side length in a right-angled triangle
- apply Pythagoras' theorem to calculate missing side lengths of three-dimensional objects.

11.3.1 Review of Pythagoras' theorem

Even though the theorem that describes the relationship between the side lengths of right-angled triangles bears the name of the famous Greek mathematician Pythagoras, thought to have lived around 550 BC, evidence exists in some of humanity's earliest relics that it was known and used much earlier than that.

Phythagoras' theorem

The side lengths of any right-angled triangle are related according to the rule $a^2 + b^2 = c^2$, where c represents the hypotenuse (the longest side), and a and b represent the other two side lengths.

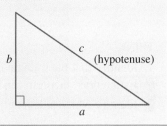

The **hypotenuse** is always the side length that is opposite the right angle. **Pythagoras' theorem** can be used to find an unknown side length of a triangle when the other two side lengths are known.

WORKED EXAMPLE 5 Using Pythagoras' theorem to calculate the hypotenuse

Calculate the unknown side length in the right-angled triangle shown.

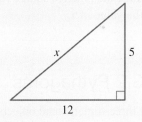

THINK	WRITE
1. Identify that the triangle is right-angled so Pythagoras' theorem can be applied.	$a^2 + b^2 = c^2$
2. Identify which side length is the hypotenuse.	x is opposite the right angle, so it is the hypotenuse; $a = 12$ and $b = 5$.
3. Substitute the known values into the theorem and simplify.	$12^2 + 5^2 = x^2$ $144 + 25 = x^2$ $169 = x^2$
4. Take the positive square root of both sides to obtain the value of x.	$\sqrt{169} = x$ $13 = x$

Note: We always take the positive square root when using Pythagoras' theorem because the length of a side of a triangle must be a positive number.

If the length of the hypotenuse and one other side length are known, the other side length can be found by subtracting the square of the known side length from the square of the hypotenuse.

WORKED EXAMPLE 6 Using Pythagoras' theorem to calculate a shorter side

Calculate the length of the unknown side in the right-angled triangle shown.

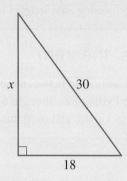

THINK	WRITE
1. Identify that the triangle is right-angled so Pythagoras' theorem can be applied.	$a^2 + b^2 = c^2$
2. Identify which side length is the hypotenuse.	30 is opposite the right angle, so it is the hypotenuse; $a = x$ and $b = 18$.
3. Substitute the known values into the theorem and simplify.	$x^2 + 18^2 = 30^2$ $x^2 = 30^2 - 18^2$ $= 576$
4. Take the positive square root of both sides to obtain the value of x.	$x = \sqrt{576}$ $= 24$

11.3.2 Pythagoras' theorem in three dimensions

Many three-dimensional objects contain right-angled triangles that can be modelled with two-dimensional drawings. Using this method we can calculate missing side lengths of three-dimensional objects.

---

tlvd-5041

WORKED EXAMPLE 7 Applying Pythagoras' theorem in three dimensions

Calculate the maximum length of a metal rod that would fit into a rectangular crate with dimensions $1\,\text{m} \times 1.5\,\text{m} \times 0.5\,\text{m}$.

THINK	WRITE
1. Draw a diagram of a rectangular box with a rod in it, labelling the dimensions.	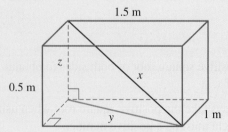

2. Draw in a right-angled triangle that has the metal rod as one of the sides, as shown in pink. The length of y in this right-angled triangle is not known (see step 3).
Draw in another right-angled triangle, as shown in green, to calculate the length of y (see step 4).

3. Draw the right-angled triangle containing the rod and use Pythagoras' theorem to calculate the length of the rod (x).

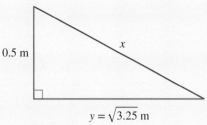

0.5 m

x

$y = \sqrt{3.25}$ m

$c^2 = a^2 + b^2$

$x^2 = (\sqrt{3.25})^2 + 0.5^2$

$\quad = 3.25 + 0.25$

$\quad = 3.5$

$x = \sqrt{3.5}$

4. Calculate the length of y using Pythagoras' theorem. Calculate the exact value of y.

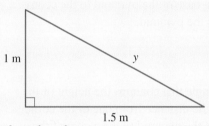

1 m

y

1.5 m

$c^2 = a^2 + b^2$

$y^2 = 1.5^2 + 1^2$

$\quad = 3.25$

$y = \sqrt{3.25}$

5. Write the answer in a sentence.

The maximum length of the metal rod is $\sqrt{3.5}$ m (≈ 1.87 m correct to 2 decimal places).

WORKED EXAMPLE 8 Applying Pythagoras' theorem in three dimensions

A square pyramid has a base length of 30 metres and a slant edge of 65 metres. Determine the height of the pyramid, giving your answer correct to 1 decimal place.

THINK

1. Draw a diagram to represent the situation. Add a point in the centre of the diagram below the apex of the pyramid.

WRITE

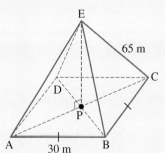

2. Determine the exact diagonal distance across the base of the pyramid by using Pythagoras' theorem.

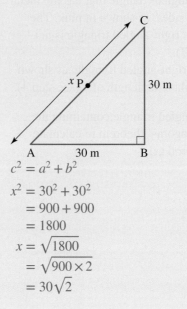

$$c^2 = a^2 + b^2$$
$$x^2 = 30^2 + 30^2$$
$$= 900 + 900$$
$$= 1800$$
$$x = \sqrt{1800}$$
$$= \sqrt{900 \times 2}$$
$$= 30\sqrt{2}$$

3. Calculate the exact distance from one of the corners on the base of the pyramid to the centre of the base of the pyramid.

$$AP = \frac{1}{2} AC$$
$$= \frac{1}{2} \times 30\sqrt{2}$$
$$= 15\sqrt{2}$$

4. Draw the triangle that contains the height of the pyramid and the distance from one of the corners on the base of the pyramid to the centre of the base of the pyramid.

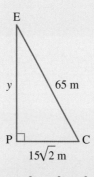

5. Use Pythagoras' theorem to calculate the height of the pyramid, rounding your answer to 1 decimal place.

$$c^2 = a^2 + b^2$$
$$65^2 = y^2 + \left(15\sqrt{2}\right)^2$$
$$4225 = y^2 + 450$$
$$y^2 = 4225 - 450$$
$$= 3775$$
$$y = \sqrt{3775}$$
$$\approx 61.4$$

6. Write the answer in a sentence.

The height of the pyramid is 61.4 metres (correct to 1 decimal place).

TI | THINK	DISPLAY/WRITE	CASIO | THINK	DISPLAY/WRITE
Ensure your Calculation Mode in the Settings is set to EXACT.		Ensure your calculator is set to Standard.	
1. On a Calculator page, press MENU, then select: 3: Algebra 1: Solve Complete the entry line as: solve $(ac^2 = 30^2 + 30^2, ac) \| ac > 0$ then press ENTER. The length of AC can be read from the screen.	 $AC = 30\sqrt{2}$	1. On the Main screen, complete the entry line as: solve $(ac^2 = 30^2 + 30^2, ac) \| ac > 0$ then press EXE. The length of AC can be read from the screen.	 $AC = 30\sqrt{2}$
2. Complete the entry line as: $\dfrac{30\sqrt{2}}{2}$ then press ENTER. The length of AP can be read from the screen.	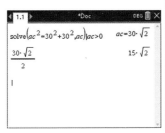 $AP = 15\sqrt{2}$	2. Complete the entry line as: $\dfrac{30\sqrt{2}}{2}$ then press EXE. The length of AP can be read from the screen.	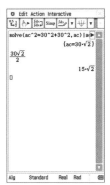 $AP = 15\sqrt{2}$
3. Complete the entry line as: solve $(65^2 = ep^2 + (15\sqrt{2})^2, ep) \| ep > 0$ then press ENTER. The length of EP can be read from the screen.	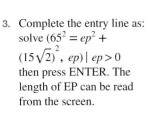 $EP = 5\sqrt{151}$	3. Complete the entry line as: solve $(65^2 = ep^2 + (15\sqrt{2})^2, ep) \| ep > 0$ then press EXE. The length of EP can be read from the screen.	 $EP = 5\sqrt{151}$
4. Use the up arrow to highlight the previous answer and then press ENTER to copy and paste it on a new entry line.	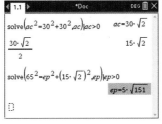	4. Select: • Action • Transformation • approx. then select ans from the Math 1 tab on the Keyboard menu and press EXE.	

5. Press MENU, then select:
   2: Number
   1: Convert to Decimal
   then press ENTER.

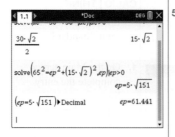

5. State the answer, rounded to 1 decimal place.
   The height of the pyramid is 61.4 metres.

6. State the answer, rounded to 1 decimal place.
   The height of the pyramid is 61.4 metres.

 Resources

 Interactivity Pythagoras' theorem (int-6473)

11.3 Exercise

Students, these questions are even better in jacPLUS

Receive immediate feedback and access sample responses

Access additional questions

Track your results and progress

Find all this and MORE in jacPLUS ▶

1. **WE5** Calculate the unknown side length in each of the right-angled triangles shown, giving your answers correct to 1 decimal place.

a.

x 6
9

b.

x 14
25

2. Show that a right-angled triangle with side lengths of 40 cm and 96 cm will have a hypotenuse of length 104 cm.

3. Evaluate the unknown side lengths in the right-angled triangles shown, giving your answers correct to 2 decimal places.

a.
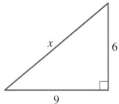
6
3

b.
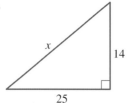
8.5
11

c.
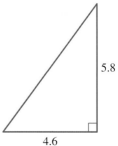
5.8
4.6

d.

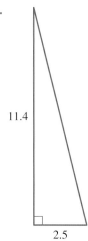

11.4

2.5

e.

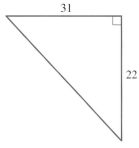

31

22

f.

8

33

4. **WE6** Calculate the length of the unknown side in each of the right-angled triangles shown, giving your answers correct to 1 decimal place.

a.

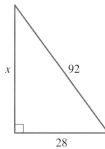

x 92

28

b.

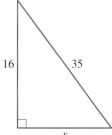

16   35

x

5. Show that if a right-angled triangle has a hypotenuse of 24 cm and a side length of 19.2 cm, the third side length will be 14.4 cm.

6. Evaluate the length of the unknown side in each of the right-angled triangles shown. Give your answers correct to 2 decimal places.

a.

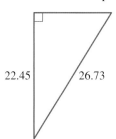

22.45   26.73

b.

18.35   23.39

c.

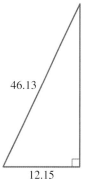

46.13

12.15

d.

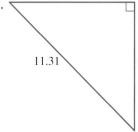

11.31   8

e.

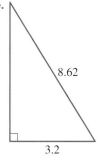

8.62

3.2

f.

3.2   3.62

7. Calculate the length of the unknown side in the following diagram, giving your answer correct to 2 decimal places.

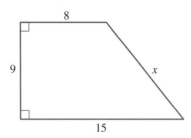

8. **WE7** Calculate the maximum length of a metal rod that would fit into a rectangular crate with dimensions 1.2 m × 83 cm × 55 cm.

9. Determine whether a metal rod of length 2.8 metres would fit into a rectangular crate with dimensions 2.3 m × 1.2 m × 0.8 m.

10. Calculate, correct to 2 decimal places, the lengths of the shorter sides of a right-angled isosceles triangle that has a hypotenuse of length:
    a. 20 cm
    b. 48 cm
    c. 5.5 cm
    d. 166 cm.

11. **WE8** A square pyramid has a base length of 25 metres and a slant edge of 45 metres. Determine the height of the pyramid, giving your answer correct to 1 decimal place.

12. Determine which of the following square pyramids has the greater height.
    Pyramid 1: base length of 18 metres and slant edge of 30 metres
    Pyramid 2: base length of 22 metres and slant edge of 28 metres

13. Calculate the length of the longest metal rod that can fit diagonally into each of the boxes shown.

    a.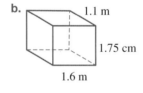
    b.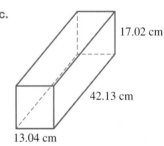
    c.

14. A friend wants to pack an umbrella into their suitcase.
    a. If the suitcase measures 89 cm × 21 cm × 44 cm, determine whether their 1-m umbrella will fit.
    b. Give the length of the longest object that will fit in the suitcase.

15. Determine the value of the pronumeral x, correct to 2 decimal places, in the following diagram.

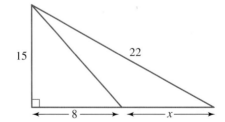

16. A cable joins the top of two vertical towers that are 12 metres high. One of the towers is at the bottom of a hill and the other is at the top. The horizontal distance between the towers is 35 metres and the base of the upper tower is 60 metres above ground level.

Calculate the minimum length of cable required to join the top of the towers. Give your answer correct to 2 decimal places.

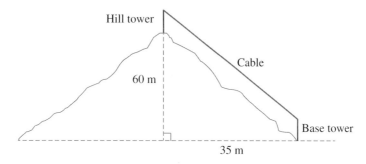

17. A semi-trailer carries a container that has the following internal dimensions: length 14.5 m, width 2.4 m and height 2.9 m. Give your answers to the following questions correct to 2 decimal places.

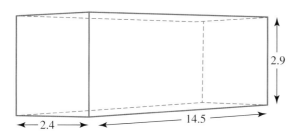 

a. Calculate the length of the longest object that can be placed on the floor of the container.
b. Calculate the length of the longest object that can be placed in the container if only one end is placed on the floor.
c. If a rectangular box with length 2.4 m, width 1.2 m and height 0.8 m is placed on the floor at one end so that it fits across the width of the container, calculate the length of the longest object that can now be placed inside if it touches the floor adjacent to the box.

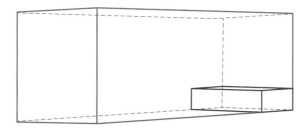

18. An ultralight aircraft is flying at an altitude of 1000 metres and a horizontal distance of 10 kilometres from its landing point.

a. If the aircraft travels in a straight line from its current position to its landing point, calculate how far it travels correct to the nearest metre. (Assume the ground is level.)
b. If the aircraft maintained the same altitude for a further 4 kilometres, calculate the straight-line distance from the new position to the same landing point, correct to the nearest metre.

c. From the original starting point, the pilot mistakenly follows a direct line to a point on the ground that is 2.5 kilometres short of the correct landing point. They realise their mistake when the aircraft is at an altitude of 400 metres and a horizontal distance of 5.5 kilometres from the correct landing point. The pilot then follows a straight-line path to the correct landing point.

Calculate the total distance travelled by the aircraft from its starting point to the correct landing point, correct to the nearest metre.

11.3 Exam questions

▶ **Question 1 (1 mark)**

Source: VCE 2020, Further Mathematics Exam 1, Section B, Module 3, Q3; © VCAA.

MC Two trees stand on horizontal ground.

A 25 m cable connects the two trees at point A and point B, as shown in the diagram below.

Point A is 45 m above the ground and point B is 30 m above the ground.

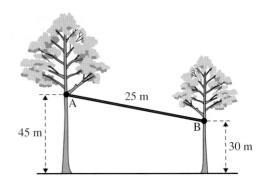

The horizontal distance, in metres, between point A and point B is
- **A.** 10
- **B.** 15
- **C.** 20
- **D.** 30
- **E.** 35

▶ **Question 2 (1 mark)**

Source: VCE 2017, Further Mathematics Exam 1, Section B, Module 3, Q2; © VCAA.

MC A right-angled triangle, *XYZ*, has side lengths $XY = 38.5$ cm and $YZ = 24.0$ cm, as shown in the diagram below.

The length of *XZ*, in centimetres, is closest to
- **A.** 24.8
- **B.** 30.1
- **C.** 38.8
- **D.** 45.4
- **E.** 62.5

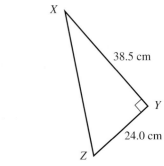

▶ **Question 3 (1 mark)**

Source: VCE 2017, Further Mathematics Exam 1, Section B, Module 3, Q6; © VCAA.

MC A hemispherical bowl of radius 10 cm is shown in the diagram below.

The bowl contains water with a maximum depth of 2 cm.

The radius of the surface of the water, in centimetres, is
- **A.** 2
- **B.** 6
- **C.** 8
- **D.** 9
- **E.** 10

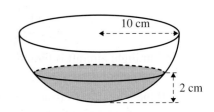

More exam questions are available online.

11.4 Perimeter and area of polygons and triangles

LEARNING INTENTION

At the end of this subtopic you should be able to:
- calculate the perimeter and area of a polygon (including quadrilaterals, circles and arcs)
- calculate the perimeter and area of triangles (including using Heron's formula).

11.4.1 Perimeter and area of standard shapes

Units of length are used to describe the distance between any two points.

The standard unit of length in the metric system is the metre. The most commonly used units of length are the millimetre (mm), centimetre (cm), metre (m) and kilometre (km).

Converting units of length

The following chart is useful when converting from one unit of length to another.

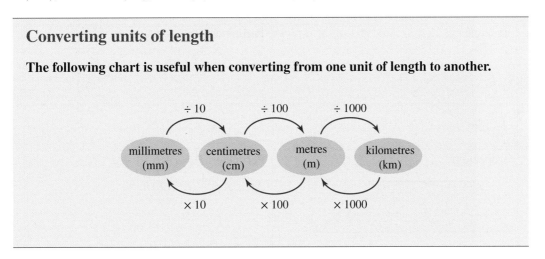

Units of **area** are named by the side length of the square that encloses that amount of space. For example, a square metre is the amount of space enclosed by a square with a side length of 1 metre.

Converting units of area

The following chart is useful when converting between units of area.

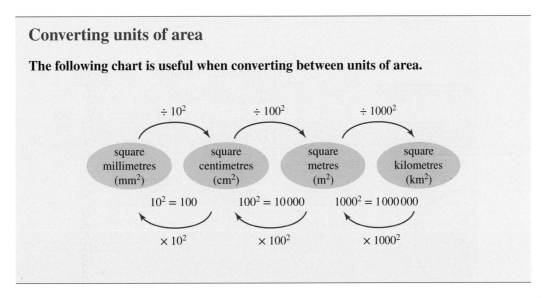

You should be familiar with the methods and units of measurement used for calculating the **perimeter** (distance around an object) and area (two-dimensional space taken up by an object) of standard **polygons** and other shapes. These are summarised in the following table.

Shape	Perimeter and area
Square 	Perimeter: $P = 4l$ Area: $A = l^2$
Rectangle 	Perimeter: $P = 2l + 2w$ Area: $A = lw$
Triangle 	Perimeter: $P = a + b + c$ Area: $A = \dfrac{1}{2}bh$
Trapezium 	Perimeter: $P = a + b + c + d$ Area: $A = \dfrac{1}{2}(a + b)h$
Parallelogram 	Perimeter: $P = 2a + 2b$ Area: $A = bh$
Rhombus 	Perimeter: $P = 4a$ Area: $A = \dfrac{1}{2}xy$
Circle 	Circumference (perimeter): $C = 2\pi r$ $\quad = \pi D$ Area: $A = \pi r^2$

Note: The approximate value of π is 3.14. However, when calculating **circumference** and area, always use the π button on your calculator and make rounding off to the required number of decimal places your final step.

WORKED EXAMPLE 9 Calculating the perimeter and area of a polygon

Calculate the perimeter and area of the shape shown in the diagram.

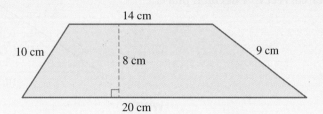

THINK	WRITE
1. Identify the shape.	Trapezium
2. Identify the components for the perimeter formula and evaluate.	$P = 10 + 20 + 14 + 9$ $\quad = 53$
3. Write the perimeter including the units.	$P = 53$ cm
4. Identify the components for the area formula and evaluate.	$A = \dfrac{1}{2}(a+b)h$ $\quad = \dfrac{1}{2}(20+14)8$ $\quad = \dfrac{1}{2} \times 34 \times 8$ $\quad = 136$
5. Write the answer with the correct unit.	$A = 136$ cm^2

11.4.2 Heron's formula

Heron's formula is a way of calculating the area of a triangle if you are given all three side lengths. It is named after Hero of Alexandria, who was a Greek engineer and mathematician.

> **Heron's formula**
>
> **Step 1: Calculate s, the value of half of the perimeter of the triangle:**
>
> $$s = \frac{a+b+c}{2}$$
>
> **Step 2: Use the following formula to calculate the area of the triangle:**
>
> $$A = \sqrt{s(s-a)(s-b)(s-c)}$$

Use Heron's formula to calculate the area of the triangle shown. Give your answer correct to 1 decimal place.

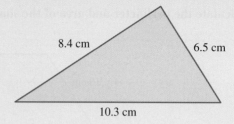

8.4 cm   6.5 cm

10.3 cm

THINK	WRITE

1. Calculate the value of s.

$$s = \frac{a+b+c}{2}$$
$$= \frac{6.5 + 8.4 + 10.3}{2}$$
$$= \frac{25.2}{2}$$
$$= 12.6$$

2. Use Heron's formula to calculate the area of the triangle correct to 1 decimal place.

$$A = \sqrt{s(s-a)(s-b)(s-c)}$$
$$= \sqrt{12.6(12.6-6.5)(12.6-8.4)(12.6-10.3)}$$
$$= \sqrt{12.6 \times 6.1 \times 4.2 \times 2.3}$$
$$= \sqrt{742.4676}$$
$$\approx 27.2$$

3. Write the area with the correct unit.

$$A = 27.2 \text{ cm}^2$$

 Resources

Interactivities Conversion of units of area (int-6269)
Area and perimeter (int-6474)
Using Heron's formula to find the area of a triangle (int-6475)

11.4 Exercise

Students, these questions are even better in jacPLUS

 Receive immediate feedback and access sample responses

 Access additional questions

 Track your results and progress

Find all this and MORE in jacPLUS

In the following questions, assume all measurements are in centimetres unless otherwise indicated.

1. **WE9** Calculate the perimeter and area of the shape shown in the diagram.

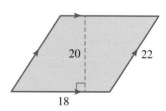

20   22

18

2. Calculate the circumference and area of the shape shown in the diagram, giving your final answers correct to 2 decimal places.

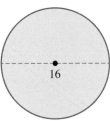

3. Calculate the perimeter and area of each of the following shapes, giving answers correct to 2 decimal places where appropriate.

a.

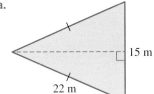

b.

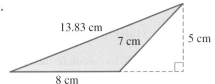

c.

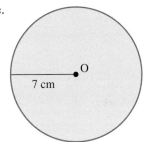

d.

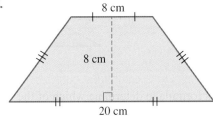

4. Correct to 2 decimal places, calculate the circumference and area of:

a. a circle of radius 5 cm
b. a circle of diameter 18 cm.

5. Calculate the perimeter and area of a parallelogram with side lengths of 12 cm and 22 cm, and a perpendicular distance of 16 cm between the short sides.

6. Calculate the area of a rhombus with diagonals of 11.63 cm and 5.81 cm.

7. Calculate the area of the shaded region shown in the diagram, giving your answer correct to 2 decimal places.

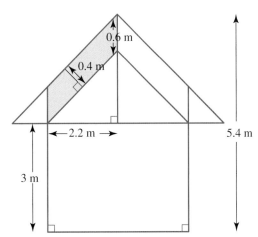

8. Calculate the perimeter of the large triangle shown in the diagram and hence calculate the shaded area.

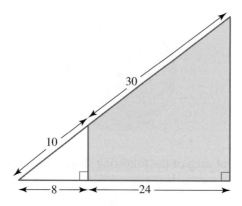

9. A circle has an area of 3140 cm^2. Calculate its radius correct to 2 decimal places.

10. **WE10** Use Heron's formula to calculate the area of the triangle shown.

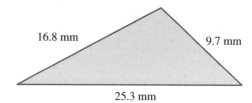

11. Use Heron's formula to determine which of the following triangles has the largest area.
Triangle 1: side lengths of 10.6 cm, 13.5 cm and 16.2 cm
Triangle 2: side lengths of 10.8 cm, 14.2 cm and 24.6 cm
Triangle 3: side lengths of 12.1 cm, 12.6 cm and 12.7 cm

12. The Bayview Council wants to use the triangular park beside the beach to host a special Anzac Day barbecue. However, council rules stipulate that public areas can be used for such purposes only if the area chosen is over 350 m^2 in size. The sides of the triangular park measure 23 metres, 28 metres and 32 metres.
Calculate the area of the park using Heron's formula and determine whether it is of a suitable size to host the barbecue.

13. A rectangle has a side length that is twice as long as its width. If it has an area of 968 cm^2, find the length of its diagonal correct to 2 decimal places.

14. A window consists of a circular metal frame 2 cm wide and two straight pieces of metal that divide the inner region into four equal segments, as shown in the diagram.

a. If the window has an inner radius of 30 cm, calculate, correct to 2 decimal places:

  i. the outer circumference of the window
  ii. the total area of the circular metal frame.

b. If the area of the metal frame is increased by 10% by reducing the size of the inner radius, calculate the circumference of the new inner circle.

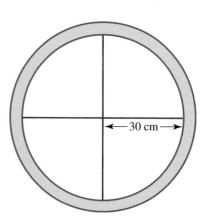

15. A semicircular section of a running track consists of eight lanes that are 1.2 m wide. The innermost line of the first lane has a total length of 100 m.

    a. Calculate how much further someone in lane 8 will run around the curve from the start line to the finish line.

    b. Determine the total area of the curved section of the track.

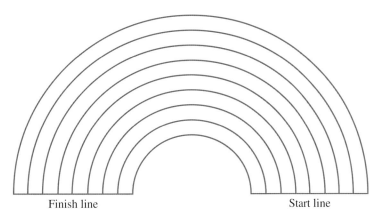

Finish line   Start line

16. A paved area of a garden courtyard forms an equilateral triangle with a side length of 20 m. It is paved using a series of identically sized blue and white triangular pavers as shown in the diagram.

    a. Calculate the total area of the paving correct to 2 decimal places.

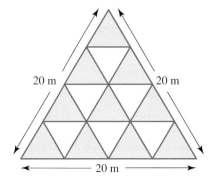

20 m   20 m

20 m

    b. If the pattern is continued by adding two more rows of pavers, calculate the new perimeter and area of the paving correct to 2 decimal places.

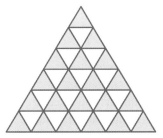

    c. After the additional two rows are added, the architects decide to add two rows of rectangular pavers to each side. Each rectangular paver has a length that is twice the side length of a triangular paver, and a width that is half the side length of a triangular paver.
If this was done on each side of the triangular paved area, calculate the perimeter and area of the paving.

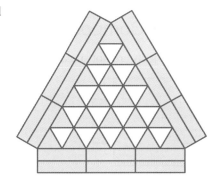

▶ **Question 1 (1 mark)**

MC The perimeter of this figure is
- **A.** 34.44 cm
- **B.** 68.88 cm
- **C.** 12.4 cm
- **D.** 24.8 cm
- **E.** 8.0 cm

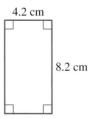

▶ **Question 2 (1 mark)**

MC The area of this figure is

- **A.** 62.5 cm^2
- **B.** 90 cm^2
- **C.** 126 cm^2
- **D.** 35 cm^2
- **E.** 150 cm^2

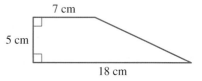

▶ **Question 3 (1 mark)**

MC A right-angled triangle is inside a rectangle, with measurements shown in centimetres.

The shaded area is

- **A.** 240 cm^2
- **B.** 25 cm^2
- **C.** 95 cm
- **D.** 190 cm^2
- **E.** 215 cm^2

More exam questions are available online.

11.5 Perimeter and area of composite shapes and sectors

> **LEARNING INTENTION**
>
> At the end of this subtopic you should be able to:
> - calculate the perimeter and area of composite shapes
> - calculate the perimeter and area of an annulus and a sector
> - solve problems by applying the perimeter and area formulas.

11.5.1 Composite shapes

Many objects are not standard shapes but are combinations of them. For example:

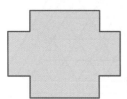

To calculate the areas of composite shapes, split them into standard shapes, calculate the individual areas of the standard shapes and sum the answers together.

To calculate the perimeters of composite shapes, it is often easiest to calculate each individual side length and to then calculate the total, rather than applying any specific formula.

tlvd-5043

WORKED EXAMPLE 11 Calculating the area of a composite shape

Calculate the area of the object shown correct to 2 decimal places.

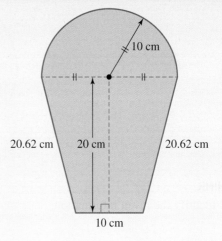

THINK	WRITE
1. Identify the given information.	The shape is a combination of a trapezium and a semicircle.
2. Find the area of each component of the shape.	Area of trapezium: $A = \dfrac{1}{2}(a+b)h$
	$= \dfrac{1}{2}(10+20)20$
	$= 300 \text{ cm}^2$
	Area of semicircle: $A = \dfrac{1}{2}\pi r^2$
	$= \dfrac{1}{2}\pi(10)^2$
	$= 50\pi \text{ cm}^2$
3. Sum the areas of the components.	Total area: $300 + 50\pi \approx 457.08$
4. State the answer.	The area of the shape is 457.08 cm^2.

Some composite shapes do have specific formulas.

11.5.2 Annulus

The area between two circles with the same centre is known as an **annulus**. The area of an annulus is calculated by subtracting the area of the inner circle from the area of the outer circle.

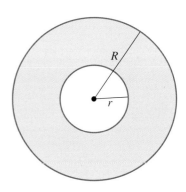

Area of an annulus

Area of annulus = area of outer circle − area of inner circle

$$A = \pi R^2 - \pi r^2$$
$$= \pi(R^2 - r^2)$$

Calculate the area of the annulus shown in the diagram correct to 1 decimal place.

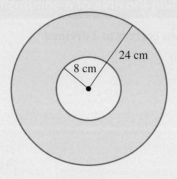

24 cm

8 cm

THINK	WRITE
1. Identify the given information.	The area shown is an annulus. The radius of the outer circle is 24 cm. The radius of the inner circle is 8 cm.
2. Substitute the information into the formula and simplify.	$A = \pi R^2 - \pi r^2$ $\quad = \pi(R^2 - r^2)$ $\quad = \pi(24^2 - 8^2)$ $\quad = 512\pi$ $\quad \approx 1608.5$
3. Write the answer with the correct unit.	The shaded area is 1608.5 cm^2.

11.5.3 Sectors

Sectors are fractions of a circle. Because there are 360 degrees in a whole circle, the area of a sector can be found using $A = \dfrac{\theta}{360} \times \pi r^2$, where θ is the angle between the two radii that form the sector.

The perimeter of a sector is a fraction of the circumference of the related circle plus two radii:

$$P = \left(\dfrac{\theta}{360} \times 2\pi r \right) + 2r$$

$$= 2r \left(\dfrac{\theta}{360}\pi + 1 \right)$$

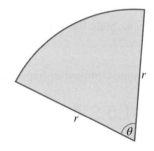

r

r

θ

Perimeter and area of a sector

$$\text{Area of a sector} = \dfrac{\theta}{360} \times \pi r^2$$

$$\text{Perimeter of a sector} = 2r \left(\dfrac{\theta}{360}\pi + 1 \right)$$

11.5.4 Arc length

The length of the circumference between two points A and B on a circle is known as an *arc*. To calculate the arc length of a circle, we must determine what fraction of the total circumference the arc represents.

The angle $\theta°$ shown is a fraction of $360°$.

The length l of arc AB will be the same fraction of the circumference of the circle, $2\pi r$:

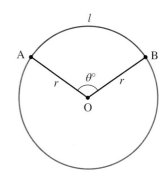

$$l = \frac{\theta}{360} \times 2\pi r$$

$$l = \frac{\pi r \theta}{180}$$

(where θ is measured in degrees).

WORKED EXAMPLE 13 Calculating the arc length, perimeter and area of a sector

Consider the following diagram.

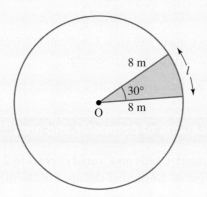

Calculate:
a. the area of the shaded sector
b. the perimeter of the sector
c. the arc length, l.
Give your answers correct to 2 decimal places.

THINK	WRITE
a. 1. Write the known information from the diagram.	**a.** The radius is 8 m. The angle between the two radii that form the arc is $30°$.
2. Substitute the values of r and θ into the area of a sector formula and simplify.	Area of a sector $= \dfrac{\theta}{360} \times \pi r^2$ $= \dfrac{30}{360} \times \pi \times (8)^2$ $= \dfrac{16}{3}\pi$ ≈ 16.75
3. Write the answer with the correct unit.	The area of the sector is 16.75 m^2.

b. 1. Substitute the values of r and θ into the perimeter of a sector formula and simplify.

$$\text{Perimeter of a sector} = 2r\left(\frac{\theta}{360}\pi + 1\right)$$

$$= 2 \times 8 \times \left(\frac{30}{360} \times \pi + 1\right)$$

$$\approx 20.19$$

2. Write your answer with the correct unit.

The perimeter of the sector is 20.19 m.

c. 1. Substitute the values of r and θ into the arc length formula and simplify.

$$\text{Arc length} = \frac{\pi r \theta}{180}$$

$$= \frac{8 \times 30 \times \pi}{180}$$

$$= \frac{4\pi}{3}$$

$$\approx 4.19$$

2. Write your answer with the correct unit.

The arc length is 4.19 m.

11.5.5 Applications

Calculations for perimeter and area have many and varied applications, including in building and construction, painting and decorating, real estate, surveying and engineering.

When dealing with these problems, it is often useful to draw diagrams to represent the given information.

WORKED EXAMPLE 14 Applications of perimeter and area

Calculate the total area of the triangular sails on a yacht correct to 2 decimal places, if the apex of one sail is 4.3 m above its base length of 2.8 m, and the apex of the other sail is 4.6 m above its base of length of 1.4 m.

THINK

1. Draw a diagram of the given information.

WRITE

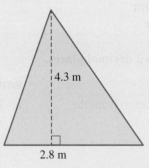
4.3 m
2.8 m

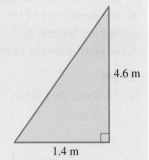
4.6 m
1.4 m

2. Identify the formulas required from the given information.

For each sail, use the formula for area of a triangle:

$$A = \frac{1}{2}bh$$

---

3. Substitute the information into the required formulas for each area and simplify.

Sail 1: $A = \dfrac{1}{2}bh$

$\qquad = \dfrac{1}{2} \times 2.8 \times 4.3$

$\qquad = 6.02$

Sail 2: $A = \dfrac{1}{2}bh$

$\qquad = \dfrac{1}{2} \times 1.4 \times 4.6$

$\qquad = 3.22$

4. Add the areas of each of the required parts.

Area of sail $= 6.02 + 3.22$
$\qquad\qquad\quad = 9.24$

5. Write the answer with the correct unit.

The total area of the sails is 9.24 m^2.

11.5 Exercise

Students, these questions are even better in jacPLUS

Receive immediate feedback and access sample responses

Access additional questions

Track your results and progress

Find all this and MORE in jacPLUS

1. **WE11** Calculate the area of the object shown.

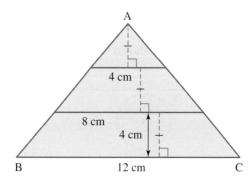

2. Calculate the perimeter and area of the object shown correct to 1 decimal place.

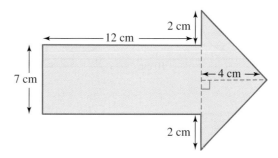

3. A circle of radius 8 cm is cut out from a square of side length 20 cm. Determine how much of the area of the square remains. Give your answer correct to 2 decimal places.

4. a. Calculate the perimeter of the shaded area inside the rectangle shown in the diagram correct to 2 decimal places.

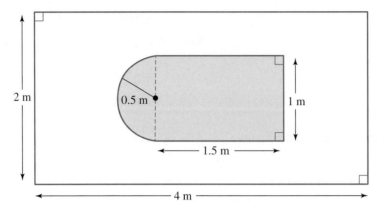

b. If the darker area inside the rectangle is removed, calculate the remaining area.

5. a. Calculate the shaded area in the diagram.

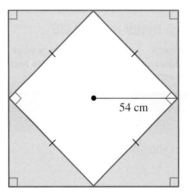

b. Calculate the unshaded area inside the square shown in the diagram, giving your answer correct to 2 decimal places.

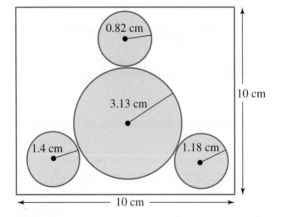

6. **WE12** Calculate the area of the annulus shown in the diagram correct to 2 decimal places.

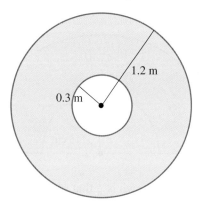

7. **WE13** Calculate the area, perimeter and arc length of the shaded sector shown in the diagram correct to 2 decimal places.

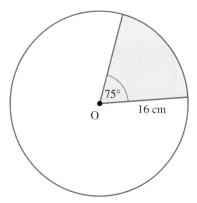

8. Calculate the area and perimeter of the shaded region shown in the diagram correct to 2 decimal places.

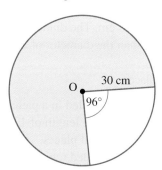

9. Calculate the area and perimeter of the shaded region shown in the diagram correct to 2 decimal places.

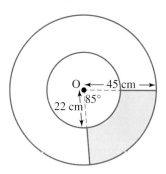

10. Calculate the perimeter and area of the shaded region in the half-annulus formed by 2 semicircles shown in the diagram. Give your answers correct to the nearest whole number.

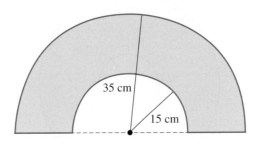

11. The area of the inner circle in the diagram shown is $\frac{1}{9}$ that of the annulus formed by the two outer circles. Calculate the area of the inner circle correct to 2 decimal places given that the units are in centimetres.

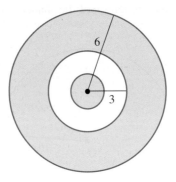

12. In the diagram, the smallest circle has a diameter of 5 cm and the others have diameters that are progressively 2 cm longer than the one immediately before.
    Calculate the area that is shaded green, correct to 2 decimal places.

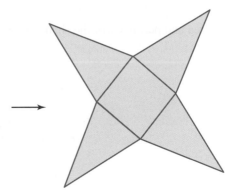

13. **WE14** Calculate the area of glass in a table that consists of three glass circles. The largest circle has a diameter of 68 cm. The diameters of the other two circles are 6 cm and 10 cm less than the diameter of the largest circle.
    Give your answer correct to 2 decimal places.

14. Part of the floor of an ancient Roman building was tiled in a pattern in which four identical triangles form a square with their bases. If the triangles have a base length of 12 cm and a height of 18 cm, calculate the perimeter and area they enclose, correct to 2 decimal places.
    (That is, calculate the perimeter and area of the shaded region shown on the right.)

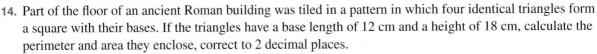

15. The vertices of an equilateral triangle of side length 2 metres touch the edge of a circle of radius 1.16 metres, as shown in the diagram. Calculate the area of the unshaded region correct to 2 decimal places.

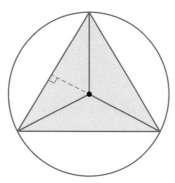

16. A circle of radius 0.58 metres sits inside an equilateral triangle of side length 2 metres so that it touches the edges of the triangle at three points. If the circle represents an area of the triangle to be removed, determine how much area would remain once this was done.

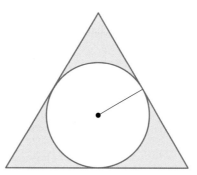

17. An annulus has an inner radius of 20 cm and an outer radius of 35 cm. Two sectors are to be removed. If one sector has an angle at the centre of 38° and the other has an angle of 25°, calculate the remaining area. Give your answer correct to 2 decimal places.

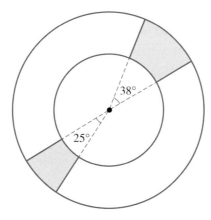

18. A trapezium is divided into five identical triangles of equal size with dimensions as shown in the diagram. Calculate the area and perimeter of the shaded region.

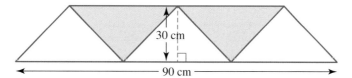

11.5 Exam questions

▶ **Question 1 (1 mark)**

Source: VCE 2021, Further Mathematics Exam 1, Section B, Module 3, Q6; © VCAA.

MC A child's toy has the following design

The area of the shaded region, in square centimetres, is closest to

    A. 13
    B. 27
    C. 31
    D. 45
    E. 113

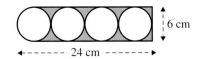

▶ **Question 2 (1 mark)**

Source: VCE 2020, Further Mathematics Exam 1, Section B, Module 3, Q2; © VCAA.

MC A flag consists of three different coloured sections: red, white and blue.

The flag is 3 m long and 2 m wide, as shown in the diagram.

The blue section is an isosceles triangle that extends to half the length of the flag.

The area of the blue section, in square metres, is

    A. 0.75
    B. 1.5
    C. 2
    D. 3
    E. 6

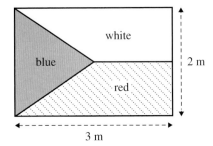

▶ **Question 3 (1 mark)**

Source: VCE 2020, Further Mathematics Exam 1, Section B, Module 3, Q5; © VCAA.

MC The pie chart shown displays the results of a survey.

Eighty per cent of the people surveyed selected 'agree'.

Twenty per cent of the people surveyed selected 'disagree'.

The radius of the pie chart is 16 mm.

The area of the sector representing 'agree', in square millimetres, is closest to

    A. 80
    B. 161
    C. 483
    D. 643
    E. 804

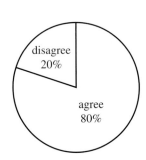

More exam questions are available online.

11.6 Volume

11.6.1 Volume and capacity

The amount of space that is taken up by any solid or three-dimensional object is known as its **volume**. Volume is expressed in cubic units of measurement, such as cubic metres (m^3) or cubic centimetres (cm^3). The relationships between commonly used cubic units are shown below.

Converting units of volume

The following chart is useful when converting between units of volume.

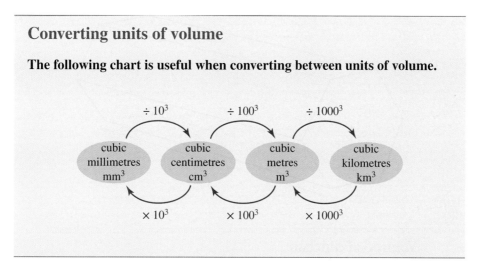

The amount of liquid an object can hold is known as its **capacity**. The relationships between commonly used units of capacity are shown as follows.

Converting units of capacity

The following chart is useful when converting between units of capacity.

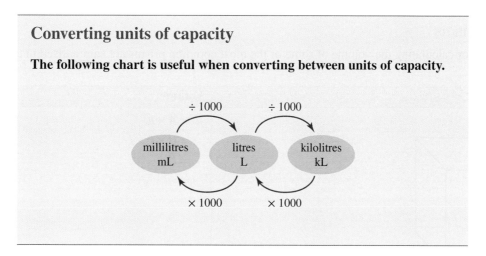

You can convert between volume and capacity by using the following information.

Converting between volume and capacity

- $1\,cm^3 = 1\,mL$
- $1\,L = 1000\,cm^3$

Many standard objects have formulas that can be used to calculate their volume. If the centre point of the top of the solid is directly above the centre point of its base, the object is called a 'right solid'. If the centre point of the top is not directly above the centre point of the base, the object is an 'oblique solid'.

Prisms

If a solid object has identical ends that are joined by flat surfaces, and the object's cross-section is a polygon and is the same along its length, the object is a **prism**. The volume of a prism is calculated by taking the product of the base area and its height (or length). A cylinder is not a prism.

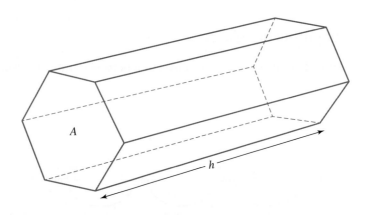

Volume of a prism

$$V = A \times h, (A = \text{base area})$$

Common prisms

The formulas for calculating the volume of some of the most common prisms are summarised in the following table.

Prism	Volume
Cube	$V = A \times h$ $\quad = (l \times l) \times l$ $\quad = l^3$

Prism	Volume

Rectangular
 	
 h w l
	$V = (A \times h)$
	    $= (l \times w) \times h$
	    $= l \times w \times h$

Triangular
 	
 h l b
	$V = A \times h$
	    $= \left(\dfrac{1}{2}bh\right) \times l$
	    $= \dfrac{1}{2}bhl$

Note: These formulas apply to both right prisms and oblique prisms, as long as you remember that the height of an oblique prism is its perpendicular height (the distance between the top and the base).

tlvd-5044

WORKED EXAMPLE 15 Calculating the volume of a triangular prism

Calculate the volume of a triangular prism with length $l = 12$ cm, triangle base length $b = 6$ cm and triangle height $h = 4$ cm.

THINK	WRITE
1. Identify the given information.	Triangular prism, $l = 12$ cm, $b = 6$ cm, $h = 4$ cm
2. Substitute the information into the appropriate formula for the solid object and evaluate.	$V = \dfrac{1}{2}bhl$ $\quad = \dfrac{1}{2} \times 6 \times 4 \times 12$ $\quad = 144$
3. Write the answer.	The volume is 144 cm^3.

11.6.2 Cylinders

A **cylinder** is a solid object with ends that are identical circles and a cross-section that is the same along its length (like a prism). As a result, it has a curved surface along its length.

The volume of a cylinder is calculated by taking the product of the base area and the height.

> **Volume of a cylinder**
>
> $$V = \text{base area} \times \text{height}$$
> $$= \pi r^2 h$$

WORKED EXAMPLE 16 Calculating the volume of a cylinder

Calculate the volume of a cylinder of radius 10 cm and height 15 cm correct to the nearest cubic centimetre.

THINK	WRITE
1. Identify the given information.	Cylinder, $r = 10$ cm, $h = 15$ cm
2. Substitute the information into the appropriate formula for the solid object and evaluate.	$V = \pi r^2 h$ $= \pi \times 10 \times 10 \times 15$ ≈ 4712.39
3. Write the answer with the correct unit.	The volume is 4712 cm^3 correct to the nearest cubic centimetre.

11.6.3 Cones

A **cone** is a solid object that is similar to a cylinder in that it has one end that is circular, but different in that at the other end it has a single vertex.

It can be shown that if you have a cone and a cylinder with identical circular bases and heights, the volume of the cylinder will be three times the volume of the cone. (The proof of this is beyond the scope of this course.)

The volume of a cone can therefore be calculated by using the formula for a comparable cylinder and dividing by three.

> **Volume of a cone**
>
> $$V = \frac{1}{3} \times \text{base area} \times \text{height}$$
> $$= \frac{1}{3} \pi r^2 h$$

Calculate the volume of a cone of radius 20 cm and a height of 36 cm correct to 1 decimal place.

THINK	WRITE
1. Identify the given information.	Given: a cone with $r = 20$ cm and $h = 36$ cm
2. Substitute the information into the appropriate formula for the solid object and evaluate.	$V = \dfrac{1}{3}\pi r^2 h$ $= \dfrac{1}{3} \times \pi \times 20 \times 20 \times 36$ $\approx 15\,079.6$
3. State the answer.	The volume is $15\,079.6 \text{ cm}^3$.

11.6.4 Pyramids

A **pyramid** is a solid object whose base is a polygon and whose sides are triangles that meet at a single point. The most famous examples are the pyramids of Ancient Egypt, which were built as tombs for the pharaohs.

A pyramid is named after the shape of its base. For example, a hexagonal pyramid has a hexagon as its base polygon. The most common pyramids are square pyramids and triangular pyramids. As with cones, the volume of a pyramid can be calculated by using the formula of a comparable prism and dividing by three.

Volume of a pyramid

Square pyramid	Triangular pyramid
$V = \dfrac{1}{3} \times \textbf{base area} \times \textbf{height}$ $= \dfrac{1}{3}l^2 h$	$V = \dfrac{1}{3} \times \textbf{base area} \times \textbf{height}$ $= \dfrac{1}{3}\left(\dfrac{1}{2}ab\right)h$ $= \dfrac{1}{6}abh$

Calculate the volume of a pyramid that is 75 cm tall and has a rectangular base with dimensions 45 cm by 38 cm.

THINK	WRITE
1. Identify the given information.	Given: a pyramid with a rectangular base of 45×38 cm and a height of 75 cm
2. Substitute the information into the appropriate formula for the solid object and evaluate.	$V = \dfrac{1}{3} l \, wh$ $= \dfrac{1}{3} \times 45 \times 38 \times 75$ $= 42\,750$
3. Write the answer with the correct unit.	The volume is $42\,750 \text{ cm}^3$.

11.6.5 Spheres

A **sphere** is a solid object that has a curved surface such that every point on the surface is the same distance (the radius of the sphere) from a central point.

The formula for calculating the volume of a sphere has been attributed to the ancient Greek mathematician Archimedes.

Volume of a sphere

$$V = \frac{4}{3}\pi r^3$$

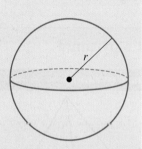

Calculate the volume of a sphere of radius 63 cm correct to 1 decimal place.

THINK	WRITE
1. Identify the given information.	Given: a sphere of $r = 63$ cm
2. Substitute the information into the appropriate formula for the solid object and evaluate.	$V = \dfrac{4}{3}\pi r^3$ $= \dfrac{4}{3} \times \pi \times 63 \times 63 \times 63$ $= 1\,047\,394.4$
3. Write the answer with the correct unit.	The volume is $1\,047\,394.4 \text{ cm}^3$.

11.6.6 Volumes of composite solids

As with calculations for perimeter and area, when a solid object is composed of two or more standard shapes, we need to identify each part and add their volumes to evaluate the overall volume.

WORKED EXAMPLE 20 Calculating the volume of a composite solid

Calculate the volume of an object that is composed of a hemisphere (half a sphere) of radius 15 cm that sits on top of a cylinder of height 45 cm, correct to 1 decimal place.

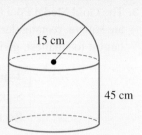

THINK	WRITE
1. Identify the given information.	Given: a hemisphere with $r = 15$ cm and a cylinder of $r = 15$ cm and $h = 45$ cm
2. Substitute the information into the appropriate formula for each component of the solid object and evaluate.	Hemisphere: $V = \dfrac{1}{2}\left(\dfrac{4}{3}\pi r^3\right)$ $= \dfrac{1}{2} \times \left(\dfrac{4}{3} \times \pi \times 15 \times 15 \times 15\right)$ $= 2250\pi$ Cylinder: $V = \pi r^2 h$ $= \pi \times 15 \times 15 \times 45$ $= 10\,125\pi$
3. Add the volume of each component.	Composite object: $V = 2250\pi + 10\,125\pi$ $\approx 38\,877.2$
4. Write the answer with the correct unit.	The volume is $38\,877.2$ cm^3.

  Resources

Interactivity Volume (int-6476)

11.6 Exercise

Students, these questions are even better in jacPLUS

 Receive immediate feedback and access sample responses

 Access additional questions

 Track your results and progress

Find all this and MORE in jacPLUS

---

1. **WE15** Calculate the volume of a triangular prism with length $l = 2.5$ m, triangle base length $b = 0.6$ m and height $h = 0.8$ m.

2. Giving answers correct to the nearest cubic centimetre, calculate the volume of a prism that has:
   a. a base area of 200 cm² and a height of 1.025 m
   b. a rectangular base 25.25 cm by 12.65 cm and a length of 0.42 m
   c. a right-angled triangular base with one side length of 48 cm, a hypotenuse of 73 cm and a length of 96 cm
   d. a height of 1.05 m and a trapezium-shaped base with parallel sides that are 25 cm and 40 cm long and 15 cm apart.

3. The Gold Medal Pool Company sells three types of above-ground swimming pools, with base shapes that are square, rectangular or circular. Use the information in the table to list the volumes of each type from largest to smallest, giving your answers in litres.

Type	Depth	Base dimensions
Square pool	1.2 m	Length: 3 m
Rectangular pool	1.2 m	Length: 4.1 m Width: 2.25 m
Circular pool	1.2 m	Diameter: 3.3 m

4. A builder uses the floor plan of the house they are building to calculate the amount of concrete they need to order for the foundations supporting the brick walls.

   a. The foundation needs to go around the perimeter of the house with a width of 600 mm and a depth of 1050 mm. Calculate how many cubic metres of concrete are required.
   b. The builder also wants to order the concrete required to pour a rectangular slab 3 m by 4 m to a depth of 600 mm. Calculate how many cubic metres of extra concrete they should order.

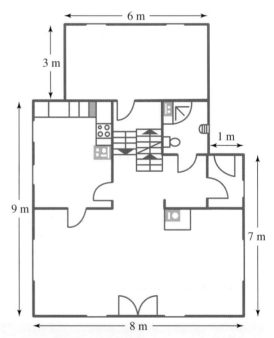

5. **WE16** Giving your answer correct to the nearest cubic centimetre, calculate the volume of a cylinder of radius 22.5 cm and a height of 35.4 cm.

6. Calculate the volume of a cylinder that has:
   a. a base circumference of 314 cm and a height of 0.625 m, giving your answer correct to the nearest cubic centimetre
   b. a height of 425 cm and a radius that is three-quarters of its height, giving your answer correct to the nearest cubic metre.

7. A company manufactures skylights in the shape of a cylinder with a hemispherical lid.
   When they are fitted onto a house, three-quarters of the length of the cylinder is below the roof. If the cylinder is 1.5 metres long and has a radius of 30 cm, calculate the volume of the skylight that is above the roof and the volume that is below it.
   Give your answers correct to the nearest cubic centimetre.

8. The outer shape of a washing machine is a rectangular prism with a height of 850 mm, a width of 595 mm and a depth of 525 mm. Inside the machine, clothes are washed in a cylindrical stainless steel drum that has a diameter of 300 mm and a length of 490 mm.

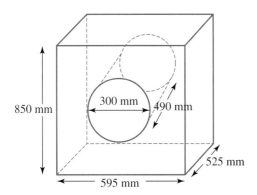

a. Calculate the maximum volume of water, in litres, that the stainless steel drum can hold.
b. Calculate the volume of the washing machine, in cubic metres, after subtracting the volume of the stainless steel drum.

9. **WE17** Calculate the volume of a cone of radius 30 cm and a height of 42 cm correct to 1 decimal place.

10. Calculate the volume of a cone that has:

a. a base circumference of 628 cm and a height of 0.72 m, correct to the nearest whole number
b. a height of 0.36 cm and a radius that is two-thirds of its height, correct to 3 decimal places.

11. Calculate the exact volumes of the solid objects shown in the following diagrams.

a.

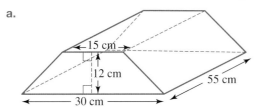

b.

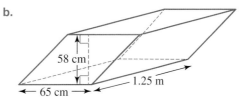

c.

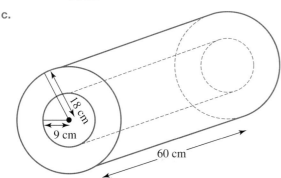

d.

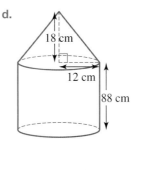

12. **WE18** Calculate the volume of a pyramid that is 2.025 m tall and has a rectangular base with dimensions 1.05 m by 0.0745 m, correct to 4 decimal places.

13. Calculate the volume of a pyramid that has:

a. a base area of 366 cm² and a height of 1.875 m
b. a rectangular base 18.45 cm by 26.55 cm and a length of 0.96 m
c. a height of 3.6 m, a triangular base with one side length of 1.2 m and a perpendicular height of 0.6 m.

14. The diagram shows the dimensions for a proposed house extension. Calculate the volume of insulation required in the roof if it takes up an eighth of the overall roof space.

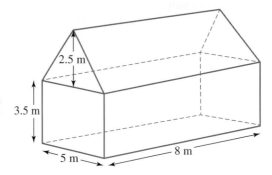

15. **WE19** Calculate the volume of a sphere of radius 0.27 m correct to 4 decimal places.

16. Calculate the radius, correct to the nearest whole number, of a sphere that has:
    a. a volume of 248 398.88 cm³
    b. a volume of 4.187 m³.

17. **WE20** Calculate the volume of an object that is composed of a hemisphere (half a sphere) of radius 1.5 m that sits on top of a cylinder of height 2.1 m. Give your answer correct to 2 decimal places.

18. Calculate the volume, correct to 2 decimal places where appropriate, of an object that is composed of:
    a. a square pyramid of height 48 cm that sits on top of a cube of side length 34 cm
    b. a cone of height 75 cm that sits on top of a 60-cm-tall cylinder of radius 16 cm.

19. A tank holding liquid petroleum gas (LPG) is cylindrical in shape with hemispherical ends. If the tank is 8.7 metres from the top of the hemisphere at one end to the top of the hemisphere at the other, and the cylindrical part of the tank has a diameter of 1.76 metres, calculate the volume of the tank to the nearest litre.

20. The glass pyramid in the courtyard of the Louvre Museum in Paris has a height of 22 m and a square base with side lengths of 35 m.

    a. Calculate the volume of the glass pyramid in cubic metres.
    b. A second glass pyramid at the Louvre Museum is called the Inverted Pyramid as it hangs upside down from the ceiling. If its dimensions are one-third of those of the larger glass pyramid, calculate its volume in cubic centimetres.

21. A wheat farmer needs to purchase a new grain silo and has the choice of two sizes. One is cylindrical with a conical top, and the other is cylindrical with a hemispherical top. Use the dimensions shown in the following diagrams to determine which silo holds the greatest volume of wheat and by how much.

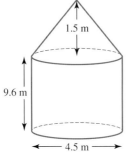

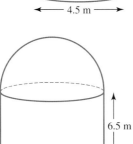

22. Tennis balls are spherical with a diameter of 6.7 cm. They are sold in packs of four in cylindrical canisters whose internal dimensions are 26.95 cm long with a diameter that is 5 mm greater than that of a ball. The canisters are packed vertically in rectangular boxes; each box is 27 cm high and will fit exactly eight canisters along its length and exactly four along its width.

   a. Calculate the volume of free space that is in a canister containing four tennis balls. Give your answer correct to 2 decimal places.
   b. Calculate the volume of free space that is in a rectangular box packed full of canisters. Give your answer correct to 2 decimal places.

23. Using CAS or otherwise, compare the volumes of cylinders that are 50 cm tall but have different radii.

   a. Tabulate the results for cylinders with radii of 10, 20, 40, 80, 160 and 320 cm.
   b. Graph your results.
   c. Use your graph to estimate the volume of a cylinder that is 50 cm tall and has a radius of:

      i. 100 cm   ii. 250 cm.

24. Using CAS or otherwise, compare the volumes of square pyramids that are 20 m tall but have different base lengths.

   a. Tabulate the results for pyramids with base lengths of 5, 10, 15, 20, 25 and 30 m.
   b. Graph your results.
   c. Use your graph to estimate the volume of a square pyramid that is 20 m tall and a base length of:

      i. 9 m   ii. 14 m.

Question 1 (1 mark)

Source: VCE 2019, Further Mathematics Exam 1, Section B, Module 3, Q7; © VCAA.

MC A can of dog food is in the shape of a cylinder. The can has a circumference of 18.85 cm and a volume of 311 cm^3.

The height of the can, in centimetres, is closest to

    A. 2.8   **B.** 3.0   **C.** 6.0   **D.** 11.0   **E.** 16.5

Question 2 (1 mark)

Source: VCE 2016, Further Mathematics Exam 1, Section B, Module 3, Q5, © VCAA.

MC A water tank in the shape of a cylinder with a hemispherical top is shown.

The volume of water that this tank can hold, in cubic metres, is closest to

    A. 80

    B. 88

    C. 96

    D. 105

    E. 121

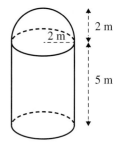

Question 3 (1 mark)

Source: VCE 2015, Further Mathematics Exam 1, Section B, Module 2, Q6, © VCAA.

MC A cylindrical block of wood has a diameter of 12 cm and a height of 8 cm.

A hemisphere is removed from the top of the cylinder, 1 cm from the edge, as shown.

The volume of the block of wood, in cubic centimetres, after the hemisphere has been removed is closest to

    A. 452   **B.** 606   **C.** 643

    D. 1167   **E.** 1357

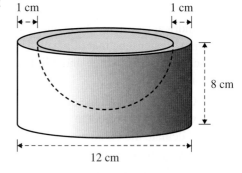

More exam questions are available online.

11.7 Surface area

LEARNING INTENTION

At the end of this subtopic you should be able to:
- calculate the surface areas of solids
- apply the formula for surface area to composite objects.

11.7.1 Using nets to calculate surface area

The **surface area** of a solid object is equal to the combined total of the areas of each individual surface that forms it. Some objects have specific formulas for the calculation of the total surface area, whereas others require the calculation of each individual surface in turn. Surface area is particularly important in design and construction when considering how much material is required to make a solid object.

In manufacturing, it could be important to make an object with the smallest amount of material that is capable of holding a particular volume. Surface area is also important in aerodynamics, as the greater the surface area, the greater the potential air resistance or drag.

Nets

The **net** of a solid object is a pattern or plan for its construction. Each surface of the object is included in its net. Therefore, the net can be used to calculate the total surface area of the object.

For example, the net of a cube will have six squares, whereas the net of a triangular pyramid (or tetrahedron) will have four triangles.

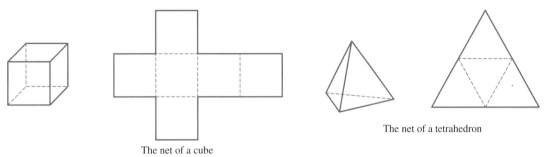

The net of a cube

The net of a tetrahedron

tlvd-5045

WORKED EXAMPLE 21 Using a net to calculate the total surface area

Calculate the surface area of the prism shown by first drawing its net.

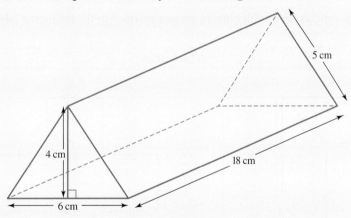

THINK

1. Identify the prism and each surface in it.

2. Redraw the given diagram as a net, making sure to check that each surface is present.

WRITE

The triangular prism consists of two identical triangular ends, two identical rectangular sides and one rectangular base.

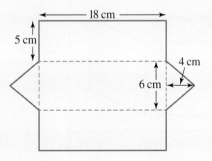

3. Calculate the area of each surface identified in the net.

Triangular ends:
$$A = 2 \times \left(\frac{1}{2}bh\right)$$
$$= 2 \times \left(\frac{1}{2} \times 6 \times 4\right)$$
$$= 24$$
Rectangular sides:
$$A = 2 \times (l\,w)$$
$$= 2 \times (18 \times 5)$$
$$= 180$$
Rectangular base:
$$A = l\,w$$
$$= 18 \times 6$$
$$= 108$$

4. Add the component areas and write the answer with the correct unit.

Total surface area:
$$= 24 + 180 + 108$$
$$= 312 \text{ cm}^2$$

11.7.2 Surface area formulas

The surface area formulas for common solid objects are summarised in the following table.

Object	Surface area
Cube	
	SA $= 6l^2$
Rectangular prism	
	SA $= 2lw + 2lh + 2wh$

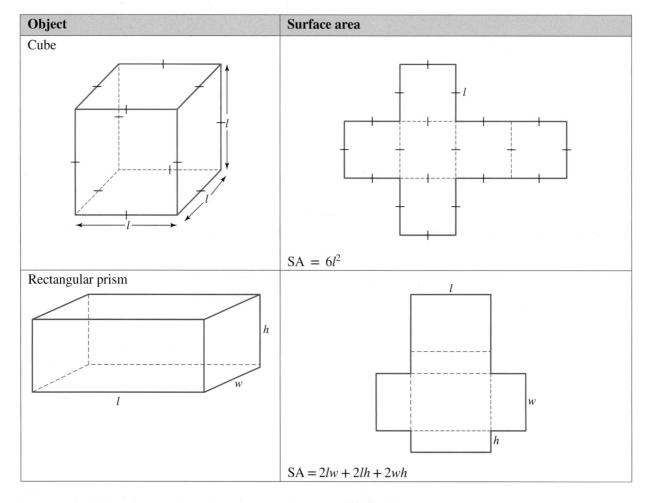

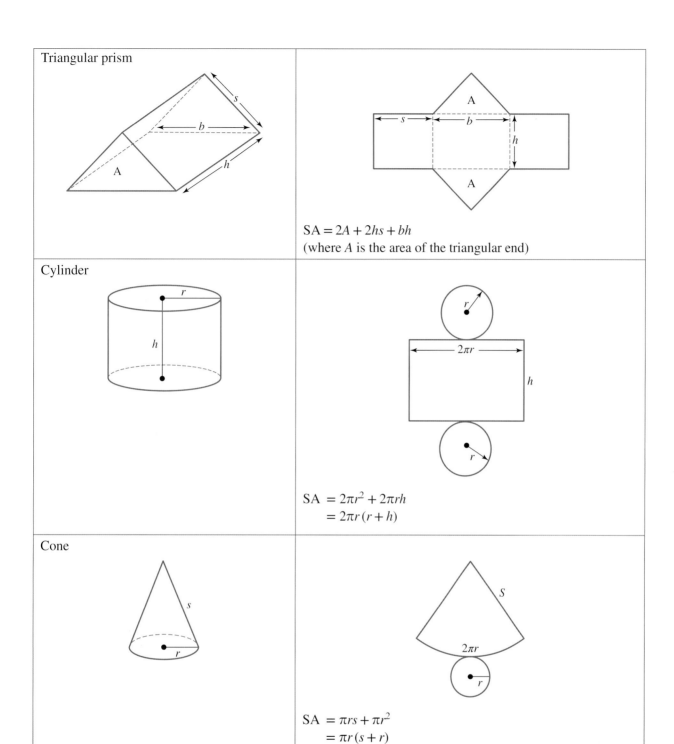

Triangular prism	
	$SA = 2A + 2hs + bh$ (where A is the area of the triangular end)
Cylinder	
	$\begin{aligned} SA &= 2\pi r^2 + 2\pi rh \\ &= 2\pi r\,(r + h) \end{aligned}$
Cone	
	$\begin{aligned} SA &= \pi rs + \pi r^2 \\ &= \pi r\,(s + r) \end{aligned}$ (including the circular base)

(continued)

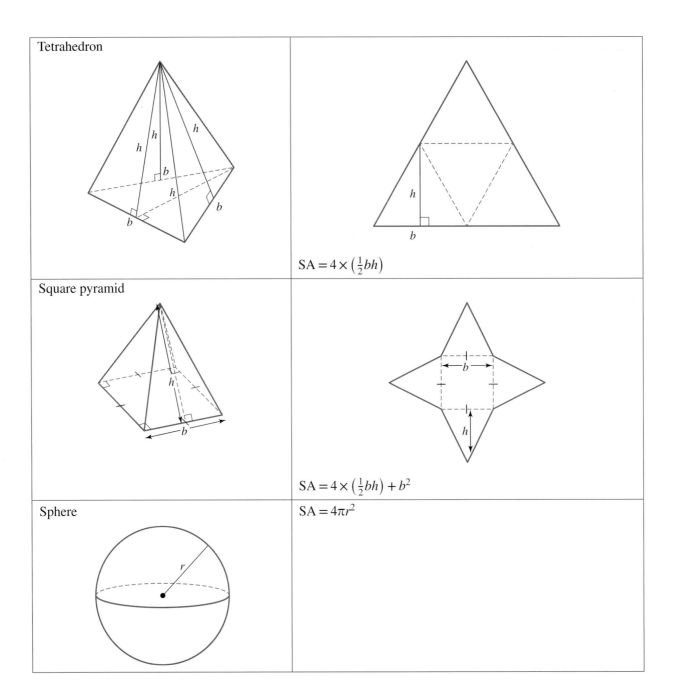

Tetrahedron	$SA = 4 \times \left(\frac{1}{2}bh\right)$
Square pyramid	$SA = 4 \times \left(\frac{1}{2}bh\right) + b^2$
Sphere	$SA = 4\pi r^2$

WORKED EXAMPLE 22 Using formulas to determine the total surface area

Calculate the surface area of the object shown by selecting an appropriate formula. Give your answer correct to 1 decimal place.

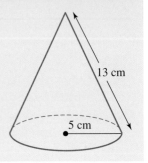

13 cm

5 cm

THINK	WRITE
1. Identify the object and the appropriate formula.	Given the object is a cone, the formula is $SA = \pi rs + \pi r^2$.
2. Substitute the given values into the formula and evaluate.	$$\begin{aligned} SA &= \pi rs + \pi r^2 \\ &= \pi r(s + r) \\ &= \pi \times 5(5 + 13) \\ &= \pi \times 90 \\ &\approx 282.7 \end{aligned}$$
3. Write the answer with the correct unit.	The surface area of the cone is 282.7 cm².

11.7.3 Surface areas of composite solids

For composite solids, be careful to include only those surfaces that form the outer part of the object.

For example, if a solid consisted of a pyramid on top of a cube, the internal surface highlighted in blue would not be included.

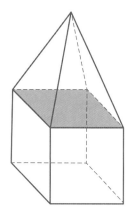

WORKED EXAMPLE 23 Calculating the total surface area of a composite solid

Calculate the surface area of the object shown correct to 2 decimal places.

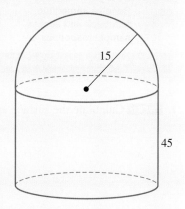

THINK	WRITE
1. Identify the components of the composite solid.	The object consists of a hemisphere that sits on top of a cylinder.

2. Substitute the given values into the formula for each surface of the object and evaluate. Give exact answers, rounding the final answer only.

Hemisphere:
$$SA = \frac{1}{2}(4\pi r^2)$$
$$= \frac{1}{2}(4 \times \pi \times 15^2)$$
$$= 450\pi$$
Cylinder (no top):
$$SA = \pi r^2 + 2\pi rh$$
$$= \pi \times 15^2 + 2 \times \pi \times 15 \times 45$$
$$= 1575\pi$$

3. Add the area of each surface to obtain the total surface area.

Total surface area: $SA = 450\pi + 1575\pi$
$$= 2025\pi$$

4. Use a calculator to determine the value of 2025π correct to 2 decimal places and write the answer with correct units.

The total surface area of the object is 6361.73 cm².

 Resources

 Interactivity Surface area (int-6477)

11.7 Exercise

Students, these questions are even better in jacPLUS

 Receive immediate feedback and access sample responses

 Access additional questions

 Track your results and progress

Find all this and MORE in jacPLUS

1. **WE21** Calculate the surface area of the prism shown by first drawing its net.

24 cm

32 cm   16 cm

2. Calculate the surface area of the tetrahedron shown by first drawing its net.

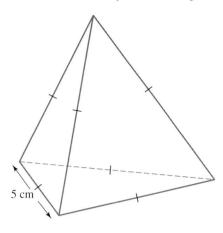

5 cm

3. `WE22` Calculate the surface areas of the objects shown by selecting appropriate formulas.

a.

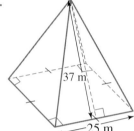

37 m

25 m

b.

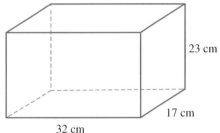

23 cm

17 cm

32 cm

4. Calculate (correct to 2 decimal places where appropriate) the surface area of:

a. a pyramid formed by four equilateral triangles with a side length of 12 cm

b. a sphere with a radius of 98 cm

c. a cylinder with a radius of 15 cm and a height of 22 cm

d. a cone with a radius of 12.5 cm and a slant height of 27.2 cm.

5. Calculate (correct to 2 decimal places where appropriate) the total surface area of:

a. a rectangular prism with dimensions 8 cm by 12 cm by 5 cm

b. a cylinder with a base diameter of 18 cm and a height of 20 cm

c. a square pyramid with a base length of 15 cm and a vertical height of 18 cm

d. a sphere of radius 10 cm.

6. A prism is 25 cm high and has a trapezoidal base whose parallel sides are 8 cm and 12 cm long respectively, and are 10 cm apart.

a. Draw the net of the prism.

b. Calculate the total surface area of the prism.

7. `WE23` Calculate the surface area of the object shown correct to 2 decimal places.

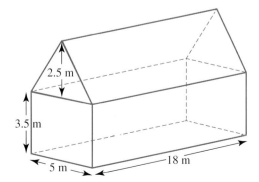

2.5 m

3.5 m

5 m

18 m

8. Calculate the surface area of the object shown correct to 2 decimal places.

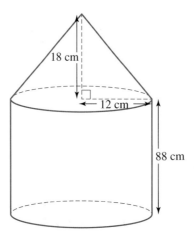

9. A hemispherical glass ornament sits on a circular base that has a radius of 5 cm.
   a. Calculate its total surface area to the nearest square centimetre.
   b. If an artist attaches it to an 8-cm-tall cylindrical stand with the same circumference, calculate the new total surface area of the combined object that is created. Give your answer to the nearest square centimetre.

10. An ice-cream shop sells two types of cones. One is 6.5 cm tall with a radius of 2.2 cm. The other is 7.5 cm tall with a radius of 1.7 cm. By first calculating the slant height of each cone correct to 2 decimal places, determine which cone (not including any ice-cream) has the greater surface area and by how much.

11. The top of a church tower is in the shape of a square pyramid that sits on top of a rectangular prism base that is 1.1 m high. The pyramid is 6 m high with a base length of 4.2 m.
    Calculate the total external surface area of the top of the church tower if the base of the prism forms the ceiling of a balcony. Give your answer correct to 2 decimal places.

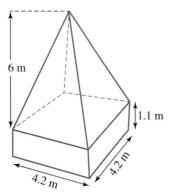

12. A cylindrical plastic vase is 66 cm high and has a radius of 12 cm. The centre has been hollowed out so that there is a cylindrical space with a radius of 9 cm that goes to a depth of 48 cm and ends in a hemisphere, as shown in the diagram. Giving your answers to the nearest square centimetre:
    a. calculate the area of the external surfaces of the vase
    b. calculate the area of the internal surface of the vase.

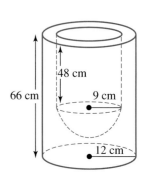

13. A dumbbell consists of a cylindrical tube that is 28 cm long with a diameter of 3 cm, and two pairs of cylindrical discs that are held in place by two locks. The larger discs have a diameter of 12 cm and a width of 2 cm, and the smaller discs are the same thickness with a diameter of 9 cm.

Calculate the total area of the exposed surfaces of the discs when they are held in position as shown in the diagram. Give your answer to the nearest square centimetre.

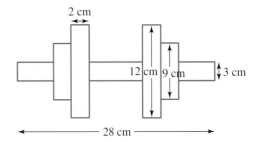

14. A staircase has a section of red carpet down its centre strip. Each of the nine steps is 16 cm high, 25 cm deep and 120 cm wide. The red carpet is 80 cm wide and extends from the back of the uppermost step to a point 65 cm beyond the base of the lower step.

a. Calculate the area of the red carpet.
b. If all areas of the front and top of the stairs that are not covered by the carpet are to be painted white, calculate the area to be painted.

15. A rectangular swimming pool is 12.5 m long, 4.3 m wide and 1.5 m deep. If all internal surfaces are to be tiled, calculate the total area of tiles required.

16. A quarter-pipe skateboard ramp has a curved surface that is one-quarter of a cylinder with a radius of 1.5 m. If the surface of the ramp is 2.4 m wide, calculate the total surface area of the front, back and sides.

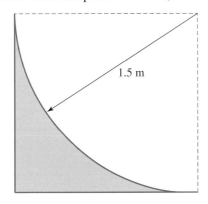

17. Using CAS or otherwise, compare the surface areas of cones that have a slant height of 120 cm but different radii.

   a. Tabulate the results for cones with radii of 15, 30, 60, 120 and 240 cm.
   b. Graph your results.
   c. Use your graph to estimate the surface area of a cone that has a slant height of 120 cm and a radius of:

      i. 100 cm   ii. 200 cm.

18. Using CAS or otherwise, compare the surface areas of cylinders that have a height that is twice the length of their radius.

   a. Tabulate the results for cylinders with radii of 5, 10, 15, 20, 25 and 30 cm.
   b. Graph your results.
   c. Use your graph to estimate the surface area of one of these cylinders that has a height twice the length of its radius and a radius of:

      i. 27 cm   ii. 13 cm.

11.7 Exam questions

Question 1 (1 mark)

Source: VCE 2020, Further Mathematics Exam 1, Section B, Module 3, Q8; © VCAA.

`MC` A cake is in the shape of a rectangular prism, as shown in the diagram.

The cake is cut in half to create two equal portions.

The cut is made along the diagonal, as represented by the dotted line.

The total surface area, in square centimetres, of one portion of the cake is

   A. 132   **B.** 180   **C.** 192   **D.** 212   **E.** 264

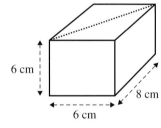

Question 2 (1 mark)

Source: VCE 2019, Further Mathematics Exam 1, Section B, Module 3, Q3; © VCAA.

`MC` An ice cream dessert is in the shape of a hemisphere. The dessert has a radius of 5 cm.

The top and the base of the dessert are covered in chocolate.

The total surface area, in square centimetres, that is covered in chocolate is closest to

   A. 52   **B.** 157   **C.** 236   **D.** 314   **E.** 942

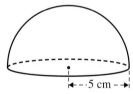

Question 3 (1 mark)

`MC` The surface area of the triangular prism, including the base, is

   A. 1260 m^2
   B. 1710 m^2
   C. 1770 m^2
   D. 2040 m^2
   E. 3600 m^2

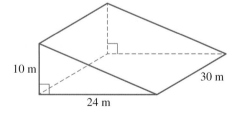

More exam questions are available online.

11.8 Review

11.8.1 Summary

doc-37621

Hey students! Now that it's time to revise this topic, go online to:

Access the topic summary

Review your results

Watch teacher-led videos

A⁺ **Practise exam** questions

Find all this and MORE in jacPLUS

11.8 Exercise

Multiple choice

1. **MC** Select which group of three numbers in the form (a, b, c) would be the side lengths of a right-angled triangle.

   A. 6, 24, 25
   B. 13, 14, 15
   C. 7, 24, 25
   D. 9, 12, 16
   E. 6, 9, 12

2. **MC** If a right-angled isosceles triangle has a hypotenuse of length 32 units, the other sides will be closest to:

   A. 21.54   B. 21.55   C. 22.62   D. 16   E. 22.63

3. **MC** An equilateral triangle with a side length of 4 units will have a height closest to:

   A. 3.46   B. 4.47   C. 4   D. 3.47   E. 4.48

4. **MC** A trapezium has a height of 8 cm and an area of 148 cm². Its parallel sides could be:

   A. 11 cm and 27 cm   B. 12 cm and 24 cm   C. 12 cm and 26 cm
   D. 12 cm and 25 cm   E. 11 cm and 25 cm

5. **MC** A circle has a circumference of 75.4 cm. Its area is closest to:

   A. 440 cm²   B. 475 cm²   C. 461 cm²   D. 448 cm²   E. 452 cm²

6. **MC** A cylinder with a volume of 1570 cm³ and a height of 20 cm will have a diameter that is closest to:

   A. 5 cm   B. 12 cm   C. 15 cm   D. 10 cm   E. 18 cm

7. **MC** A cone with a surface area of 2713 cm² and a diameter of 24 cm will have a slant-height that is closest to:

   A. 58 cm   B. 48 cm   C. 60 cm   D. 46 cm   E. 64 cm

8. **MC** A hemisphere with a radius of 22.5 cm will have a volume and total surface area respectively that are closest to:

   A. 47 689 cm³ and 4764 cm²
   B. 47 689 cm³ and 6358 cm²
   C. 23 845 cm³ and 6359 cm²
   D. 23 856 cm³ and 4771 cm²
   E. 47 688 cm³ and 6359 cm²

9. **MC** A square pyramid with a volume of 500 cm³ and a vertical height of 15 cm will have a surface area that is closest to:

   A. 416 cm²   B. 492 cm²   C. 359 cm²   D. 316 cm²   E. 635 cm²

10. **MC** An open rubbish skip is in the shape of a trapezoidal prism with the dimensions indicated in the diagram.
    The external surface area (m^2) and volume (m^3) respectively are:
    A. 8.1 and 19.8
    B. 19.8 and 8.1
    C. 17.75 and 8.1
    D. 8.1 and 26.3
    E. 26.3 and 8.1

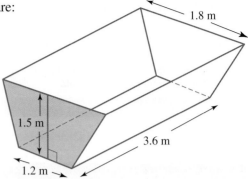

Short answer

11. Calculate the length of wire required to support the mast of a yacht if the mast is 10.8 m long and the support wire is attached to the horizontal deck at a point 3.6 m from the base of the mast.

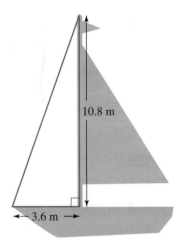

12. The supporting strut of a streetlight must be attached so that its ends are an equal distance from the top of the pole. If the strut is 1.2 m long, determine how far the ends are from the top of the pole.

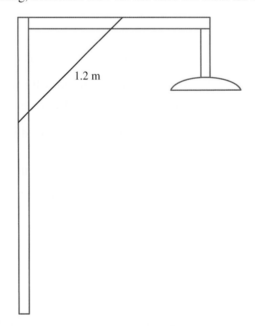

13. A surveyor is measuring a building site and wants to check that the guidelines for the foundations are square (i.e. at right angles). The surveyor places a marker 3600 mm from a corner along one line, and another marker 4800 mm from the corner along the other line. Calculate how far apart the markers must be for the lines to be square.

14. The side pieces of train carriages are made from rectangular sheets of pressed metal of length 25 metres and height 2.5 metres. Rectangular sections for the doors and windows are cut out. The dimensions of the spaces for the doors are 2 metres high by 2.5 metres wide.
The window spaces are 3025 millimetres wide by 900 millimetres high. Each sheet must have spaces cut for two doors and three windows.

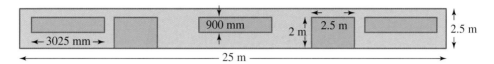

a. Calculate the total area of pressed metal that remains once the sections for the doors and windows have been removed.
b. A thin edging strip is placed around each window and around the top and sides of the door opening. Calculate the total length of edging required.

15. A semicircular arch sits on two columns as shown in the diagram. The outer edges of the columns are 7.6 metres apart and the inner edges are 5.8 metres apart. The width of each column is three-quarters the width of the arch, and the arch overhangs the columns by one-eighth of its width on each edge. The face of the arch (the shaded area) is to be tiled.
Calculate the area the tiles will cover, correct to 2 decimal places.

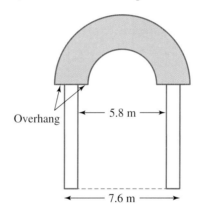

16. A circular pond is placed in the middle of a rectangular garden that is 15 m by 8 m.

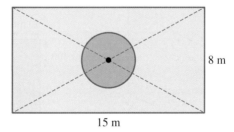

If the radius of the pond is a quarter of the distance from the centre to the corner of the garden, calculate:

a. the circumference of the pond, correct to 3 decimal places
b. the area of the garden, not including the pond, correct to 1 decimal place
c. the volume of water in the pond, correct to the nearest litre, if it is filled to a depth of 850 mm.

17. When viewed from above, a swimming pool can be seen as a rectangle with a semicircle at each end, as shown in the diagram below. The area around the outside of the pool extending 1.5 metres from the edge is to be paved.

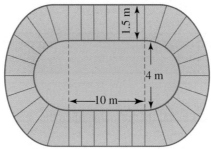

a. Calculate the paved area around the pool, correct to 2 decimal places.

b. If the pool is to be filled to a depth of 900 mm in the semicircular sections and 1500 mm in the rectangular section, calculate the total volume of water in the pool to the nearest litre.

18. A rectangular piece of glass with side lengths 1000 mm and 800 mm has its corners removed for safety, as shown in the diagrams.

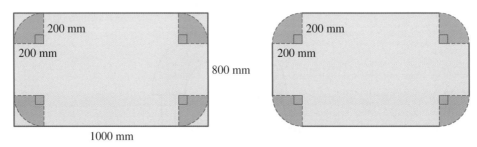

a. Calculate the surface area of the glass after the corners have been removed, correct to 1 decimal place.

b. Calculate the perimeter of the glass after the corners have been removed, correct to 1 decimal place.

19. A piece of timber has the dimensions 400 mm by 400 mm by 1800 mm. The top corners of the piece of timber are removed along its length. The cuts are made at an angle a distance of 90 mm from the corners, so the timber that is removed forms two triangular prisms.

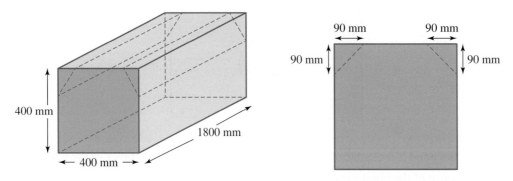

a. Calculate the volume of the piece of timber in cubic metres after the corners are removed. Give your answer correct to 3 decimal places.

b. Calculate the surface area of the piece of timber after the corners are removed.

c. Calculate the total volume and surface area of the two smaller pieces of timber that are cut from the corners, assuming they each remain as one piece.

20. Two tunnels run under a bend in a river. One runs in a straight line from point A to point C, and the other runs in a straight line from point D to point E. Points A and D are 5 km apart, as are points C and E. Point B is 10 km from both D and E, and BD is perpendicular to BE. An access tunnel GF is to be constructed between the midpoints of AC and DE.

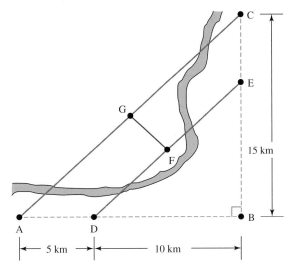

a. Calculate the lengths of AC and DE correct to 2 decimal places.
b. Calculate the length of GF correct to 2 decimal places.
c. The tunnels running from A to C and from D to E are cylindrical in shape with an outer diameter of 40 metres. Calculate the volume of material that was removed to create the two tunnels.
d. The trucks used to remove the excavated material during the construction of the tunnels can carry a maximum of 85 m³. Determine how many truck loads were required to make the two tunnels. Round your answer correct to the nearest whole number.
e. The inner walls of the tunnels are formed of concrete that is 3.5 metres thick. Calculate the total volume of concrete used for the tunnels, correct to 1 decimal place.
f. The inner surface of the concrete in the tunnels is sprayed with a sealant to prevent water seeping through. Calculate the total area that is sprayed with the sealant, correct to 1 decimal place.

11.8 Exam questions

Question 1 (1 mark)

Source: VCE 2021, Further Mathematics Exam 1, Section B, Module 3, Q5; © VCAA.

MC A cone and a cylinder both have a radius of r centimetres.

The height of the cone is 12 cm.

If the cylinder and the cone have the same volume, then the height of the cylinder, in centimetres, is

A. 4
B. 6
C. 8
D. 12
E. 36

Source: VCE 2021, Further Mathematics Exam 2, Section B, Module 3, Q1; © VCAA.

The game of squash is played with a spherical ball that has a radius of 2 cm.

    a. Show that the volume of one squash ball, rounded to 2 decimal places, is 33.51 cm^3.   **(1 mark)**

    Squash balls may be sold in cube-shaped boxes.

    Each box contains one ball and has a side length of 4.1 cm, as shown the diagram below.

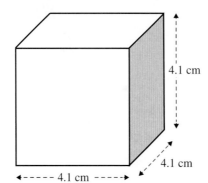

    b. Calculate the empty space, in cubic centimetres, that surrounds the ball in the box.
    Round your answer to 2 decimal places.   **(1 mark)**

    c. Calculate the total surface area, in square centimetres, of one box.   **(1 mark)**

    d. Retail shops store the cube-shaped boxes in a space within a display unit.

    The space has a length of 17.0 cm and a width of 12.5 cm. Due to the presence of a shelf above, there is a maximum height of 8.5 cm available. This is shown in the diagram below.

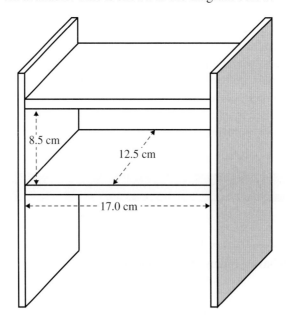

    Calculate the maximum number of cube-shaped boxes that can fit into the space within the display unit.   **(1 mark)**

Question 3 (5 marks)

Source: VCE 2020, Further Mathematics Exam 2, Section B, Module 3, Q1; © VCAA.

Khaleda manufactures a face cream. The cream comes in a cylindrical container.

The area of the circular base is 43 cm². The container has a height of 7 cm, as shown in the diagram below.

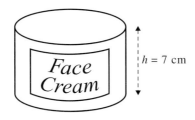

a. Write the volume of the container in cubic centimetres. **(1 mark)**

b. Write a calculation that shows that the radius of the cylindrical container, rounded to 1 decimal place, is 3.7 cm. **(1 mark)**

c. What is the total surface area of the container, in square centimetres, including the base and the lid? Round your answer to the nearest square centimetre. **(1 mark)**

d. The diagram below shows the dimensions of a shelf that will display the containers.

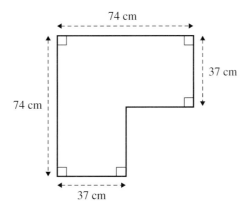

What is the perimeter of the shelf, in centimetres? **(1 mark)**

e. The shelf will display the containers in a single layer. Each container will stand upright on the shelf. What is the maximum number of containers that can fit on the shelf? **(1 mark)**

Question 4 (1 mark)

Source: VCE 2019, Further Mathematics Exam 1, Section B, Module 3, Q8; © VCAA.

MC Four identical circles of radius r are drawn inside a square, as shown in the diagram below.

The region enclosed by the circles has been shaded in the diagram.

The shaded area can be found using

A. $4r^2 - 2\pi$

B. $4r^2 - \pi r^2$

C. $4r - \pi r^2$

D. $2r^2 - \pi r^2$

E. $2r - 2\pi r$

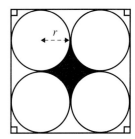

Question 5 (1 mark)

Source: VCE 2018, Further Mathematics Exam 1, Section B, Module 3, Q8; © VCAA.

MC A cone with a radius of 2.5 cm is shown in the diagram below.

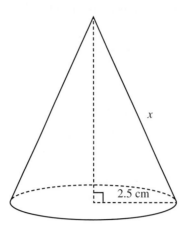

The slant edge, x, of this cone is also shown.

The volume of this cone is 36 cm^3.

The surface area of this cone, including the base, can be found using the rule *surface area* $= \pi r(r + x)$.

The total surface area of this cone, including the base, in square centimetres, is closest to

    A. 20   **B.** 42   **C.** 63   **D.** 67   **E.** 90

More exam questions are available online.

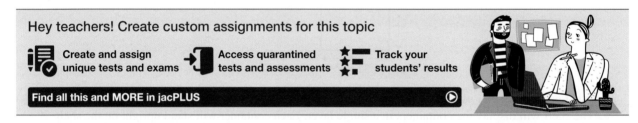

Answers

Topic 11 Space and measurement

11.2 Scientific notation, significant figures and rounding

11.2 Exercise

1. 2.6×10^6
2. 5.5×10^{-7}
3. a. 7.319×10^3 b. 8.0425×10^{-2}
   c. $1.300\,043\,8 \times 10^7$ d. 2.6×10^{-4}
   e. $9.263\,005\,1 \times 10^7$ f. 5.692×10^{-4}
4. $0.000\,000\,003$
5. $13\,800\,000\,000$
6. a. $0.000\,164$ b. $0.000\,000\,023\,994$
   c. $1\,400\,300\,000$ d. $860\,000$
7. a. 5   b. $40\,080\,000$ m
8. a. 0.0012 cm   b. 2   c. 0.001 cm
9. a. i. 4   ii. 1900
   b. i. 3   ii. 0.0015
   c. i. 3   ii. 20 000
   d. i. 5   ii. 0.0943
   e. i. 7   ii. 1.081
   f. i. 4   ii. 401
10. a. 1698.7
    b. 1700
    c. Yes, the answers are very close.
11. a. 4000 litres   b. 3774 litres
12. a. 9
    b. $240\,000\,000$ km
    c. 2.4×10^8 km
13. a. 345.58 m
    b. 341 m; this is a reasonable approximation.
14. a. $1352
    b. Actual cost $= \$1248.16$; there is a significant difference in the cost.
    c. Multiply then round.
15. a. 1.44π cm^2 b. 4.52 cm^2
    c. 788 cm^2 d. 78 800 mm^2
    e. 7.88×10^4 mm^2

11.2 Exam questions

Note: Mark allocations are available with the fully worked solutions online.
1. D
2. E
3. a. π from the computer
   b. $509\,497\,627.1$ km^2
   c. $5.094\,976\,271 \times 10^8$ km^2

11.3 Pythagoras' theorem

11.3 Exercise

1. a. 10.8   b. 28.7
2. $40^2 + 96^2 = 104^2$
3. a. 6.71   b. 13.90   c. 7.40
   d. 11.67   e. 38.01   f. 33.96
4. a. 87.6   b. 31.1
5. $24^2 - 19.2^2 = 14.4^2$
6. a. 14.51   b. 14.50   c. 44.50
   d. 7.99   e. 8.00   f. 1.69
7. 11.40
8. 1.56 m
9. No, the maximum length rod that could fit would be 2.71 m long.
10. a. 14.14 cm   b. 33.94 cm
    c. 3.89 cm   d. 117.38 cm
11. 41.4 m
12. Pyramid 1 has the greatest height.
13. a. 30.48 cm   b. 2.61 cm   c. 47.27 cm
14. a. Yes   b. 1.015 m
15. 8.09
16. 69.46 m
17. a. 14.70 m   b. 14.98 m   c. 13.82 m
18. a. $10\,050$ m   b. 6083 m   c. $10\,054$ m

11.3 Exam questions

Note: Mark allocations are available with the fully worked solutions online.
1. C
2. D
3. B

11.4 Perimeter and area of polygons and triangles

11.4 Exercise

1. Perimeter $= 80$ cm, area $= 360$ cm^2
2. Circumference $= 50.27$ cm, area $= 201.06$ cm^2
3. a. Perimeter $= 59$ m, area $= 155.12$ m^2
   b. Perimeter $= 28.83$ cm, area $= 20$ cm^2
   c. Perimeter $= 43.98$ cm, area $= 153.94$ cm^2
   d. Perimeter $= 48$ cm, area $= 112$ cm^2
4. a. Circumference $= 31.42$ cm, area $= 78.54$ cm^2
   b. Circumference $= 56.55$ cm, area $= 254.47$ cm^2
5. Perimeter $= 68$ cm, area $= 192$ cm^2
6. 33.79 cm^2
7. Area $= 1.14$ m^2
8. Perimeter $= 96$ units, area $= 360$ units^2
9. 31.61 cm
10. 47.9 mm^2
11. Triangle 1 has the largest area.
12. The area of the park is 313.8 m^2, so it is not of a suitable size to host the barbecue.

13. 49.19 cm

14. a. i. 201.06 cm   ii. 389.56 cm^2

   b. 187.19 cm

15. a. 26.39 m   b. 1104.73 m^2

16. a. 173.21 m^2

   b. Perimeter = 90 m, area = 389.71 m^2

   c. Perimeter = 120 m, area = 839.7 m^2

11.4 Exam questions

Note: Mark allocations are available with the fully worked solutions online.

1. D

2. A

3. E

11.5 Perimeter and area of composite shapes and sectors

11.5 Exercise

1. 72 cm^2

2. Perimeter = 48.6 cm, area = 106 cm^2

3. 198.94 cm^2

4. a. 5.57 m   b. 6.11 m^2

5. a. 5831.62 cm^2 b. 56.58 cm^2

6. 4.24 cm^2

7. Perimeter = 52.94 cm, area = 167.55 cm^2, arc length = 20.94 cm

8. Perimeter = 198.23 cm, area = 2073.45 cm^2

9. Perimeter = 145.40 cm, area = 1143.06 cm^2

10. Perimeter = 197 cm, area = 1571 cm^2

11. 9.42 cm^2

12. 25.13 cm^2

13. 9292.83 cm^2

14. Perimeter = 151.76 cm, area = 576 cm^2

15. 2.50 m^2

16. 0.67 m^2

17. 2138.25 cm^2

18. Area = 900 cm^2, perimeter = 194.16 cm

11.5 Exam questions

Note: Mark allocations are available with the fully worked solutions online.

1. B

2. B

3. D

11.6 Volume

11.6 Exercise

1. 0.6 m^3

2. a. 20 500 cm^3 b. 13 415 cm^3

   c. 126 720 cm^3 d. 51 188 cm^3

3. Rectangular pool: 11 070 litres
   Square pool: 10 800 litres
   Circular pool: 10 263.58 litres

4. a. 25.2 m^3 b. 7.2 m^3

5. 56 301 cm^3

6. a. 490 376 cm^3 b. 136 m^3

7. Volume above = 162 578 cm^3,
   Volume below = 318 086 cm^3

8. a. 34.64 litres   b. 0.231 m^3

9. 39 584.1 cm^3

10. a. 753 218 cm^3 b. 0.022 cm^3

11. a. 14 850 cm^3 b. 471 250 cm^3

    c. 14 580π cm^3 d. 13 536π cm^3

12. 0.0528 m^3

13. a. 22 875 cm^3 b. 15 675.12 cm^3 c. 0.432 m^3

14. 6.25 m^3

15. 0.0824 m^3

16. a. 39 cm   b. 1 m

17. 21.91 m^3

18. a. 57 800 cm^3 b. 68 361.05 cm^3

19. 19 739 litres

20. a. 8983.3 m^3 b. 332 716 049.4 cm^3

21. The hemispherical-topped silo holds 37.35 m^3 more.

22. a. 467.27 cm^3 b. 9677.11 cm^3

23. a.

Cylinder radius (cm)	Volume (cm^3)
10	15 708
20	62 832
40	251 327
80	1 005 310
160	4 021 239
320	16 084 954

b.

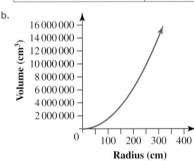

c.

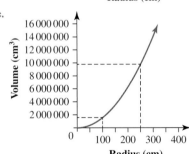

   i. Approximately 1 600 000 cm^3

   ii. Approximately 9 800 000 cm^3

24. a.

Base length (cm)	Volume (cm^3)
5	166.7
10	666.7
15	1500.0
20	2666.7
25	4166.7
30	6000.0

b.

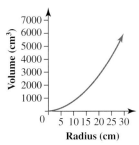

c.

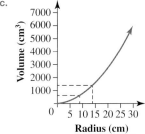

i. Approximately 550 cm^3

ii. Approximately 1300 cm^3

11.6 Exam questions

Note: Mark allocations are available with the fully worked solutions online.

1. D

2. A

3. B

11.7 Surface area

11.7 Exercise

1.

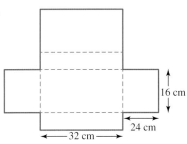

3328 cm^2

2.

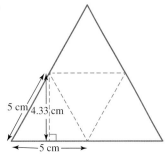

43.3 cm^2

3. a. 2475 m^2 b. 3342 cm^2

4. a. 249.42 cm^2 b. 120 687.42 cm^2
   c. 3487.17 cm^2 d. 1559.02 cm^2

5. a. 392 cm^2 b. 1639.91 cm^2
   c. 810 cm^2 d. 1256.64 cm^2

6. a.

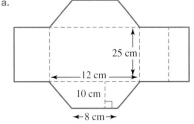

b. 1210 cm^2

7. 390.94 m^2

8. 7902.86 cm^2

9. a. 236 cm^2 b. 487 cm^2

10. The cone with height 6.5 cm and radius 2.2 cm has the greater surface area by 6.34 cm^2.

11. 71.90 m^2

12. a. 5627 cm^2 b. 3223 cm^2

13. 688 cm^2

14. a. 34 720 cm^2 b. 14 760 cm^2

15. 104.15 m^2

16. 10.22 m^2

17. a.

Radius	Surface area (cm^2)
15	6361.7
30	14 137.2
60	33 929.2
120	90 477.9
240	271 433.6

b.

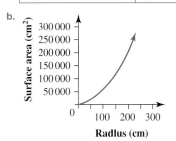

c.

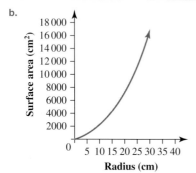

i. Approximately 60 000 cm²

ii. Approximately 210 000 cm²

18. a.

Radius (cm)	Surface area (cm²)
5	471.24
10	1884.96
15	4241.15
20	7539.82
25	11 780.97
30	16 964.60

b.

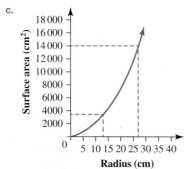

c.

i. Approximately 14 000 cm²

ii. Approximately 3200 cm²

11.7 Exam questions

Note: Mark allocations are available with the fully worked solutions online.

1. C

2. C

3. D

11.8 Review

11.8 Exercise

Multiple choice

1. C	2. E	3. A	4. D	5. E
6. D	7. C	8. D	9. A	10. B

Short answer

11. 11.38 m

12. 0.849 m

13. 6000 mm

14. a. 44.33 m²   b. 36.55 m

15. 12.63 m²

16. a. 13.352 m   b. 105.8 m²   c. 12 058 L

Extended response

17. a. 55.92 m²   b. 71 310 L

18. a. 765 663.7 mm²

   b. 3256.6 mm

19. a. 0.273 m³

   b. 2 994 005 mm²

   c. Volume = 14 580 000 mm³
      Surface area = 1 122 405 mm²

20. a. 21.21 km and 14.14 km   b. 3.54 km

   c. 44 422 120 m³   d. 522 613

   e. 249 cm²   f. 3 664 824.9 m²

11.8 Exam questions

Note: Mark allocations are available with the fully worked solutions online.

1. A

2. a. $V = \dfrac{4}{3} \times \pi \times 2^3 \approx 33.51$ cm³

   b. 35.41 cm³

   c. 100.86 cm²

   d. 24 boxes

3. a. 301 cm³

   b. $r = \sqrt{\dfrac{43}{\pi}} \approx 3.7$

   c. 249 cm²

   d. 296 cm

   e. 75 containers

4. B

5. D

12 Applications of trigonometry

Fully worked solutions for this topic are available online.

12.1 Overview

12.1.1 Introduction

Trigonometry is the branch of mathematics that describes the relationship between angles and side lengths of triangles. Triangles were first studied in the second millennium BCE by Egyptians and Babylonians. The study of trigonometric functions began in Hellenistic (Greek) mathematics. This understanding enabled early explorers to plot the stars and navigate the seas.

Architecture and engineering rely on the formation of triangles for support structures. The construction of the Golden Gate Bridge (shown in the photo) involved hundreds of thousands of trigonometric calculations.

Music theory involves sound waves, which travel in a repeating wave pattern. This repeating pattern can be represented graphically by sine and cosine functions. This graphical representation of music allows computers to create sound, so sound engineers can adjust volume and pitch, and make different sound effects.

The electricity that is sent to our house requires an understanding of trigonometry. Power companies use alternating current (AC) to send electricity over long distances. The alternating current has a sinusoidal (sine wave) behaviour.

Consider the way the video game *Mario* uses trigonometry. When you see Mario smoothly glide over road blocks, he doesn't really jump straight along the *y*-axis; it is a slightly curved or parabolic path. Trigonometry helps Mario to jump over these obstacles, demonstrating one of the many areas where trigonometry is used.

KEY CONCEPTS

This topic covers the following key concepts from the VCE Mathematics Study Design:
- the use of trigonometric ratios and Pythagoras' theorem to solve practical problems involving a right-angled triangle in two dimensions, including the use of angles of elevation and depression
- the use of the sine rule, including the ambiguous case, the cosine rule, as a generalisation of Pythagoras' theorem, and their application to solving practical problems involving non-right-angled triangles, including three-figure (true) bearings in navigation.

Source: VCE Mathematics Study Design (2023–2027) extracts © VCAA; reproduced by permission.

12.2 Trigonometric ratios

12.2.1 Trigonometric ratios

A ratio of the lengths of two sides of a right-angled triangle is called a **trigonometric ratio**. The three most common trigonometric ratios are **sine**, **cosine** and **tangent**. They are abbreviated as sin, cos and tan respectively. Trigonometric ratios are used to calculate the unknown length or acute angle size in right-angled triangles.

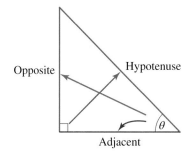

It is important to identify and label the features in a right-angled triangle. The labelling convention of a right-angled triangle is as follows:

The longest side of a right-angled triangle is always called the hypotenuse and is opposite the right angle. The other two sides are named in relation to the reference angle, θ. The opposite side is opposite the reference angle, and the adjacent side is next to the reference angle.

The sine ratio

The sine ratio is used when we want to calculate an unknown value given two out of the three following values: opposite, hypotenuse and reference angle.

The sine ratio of θ is written as $\sin(\theta)$ and is defined as follows.

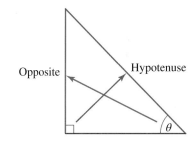

> **Sine ratio**
>
> $$\sin(\theta) = \frac{\text{opposite}}{\text{hypotenuse}} \quad \text{or} \quad \sin(\theta) = \frac{\text{O}}{\text{H}}$$

The inverse sine function is used to determine the value of the unknown reference angle given the lengths of the hypotenuse and opposite side.

> **Sine ratio for calculating an angle**
>
> $$\theta = \sin^{-1}\left(\frac{\text{O}}{\text{H}}\right)$$

Calculate the value of the pronumeral x correct to 2 decimal places.

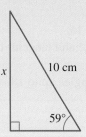

THINK

WRITE

1. Label all the given information on the triangle.

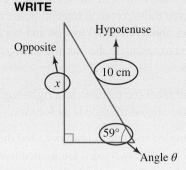

2. Since we have been given the combination of opposite, hypotenuse and the reference angle θ, we need to use the sine ratio. Substitute the given values into the ratio equation.

$$\sin(\theta) = \frac{O}{H}$$
$$\sin(59°) = \frac{x}{10}$$

3. Rearrange the equation to make the unknown the subject and solve. Make sure your calculator is in degree mode.

$$x = 10\sin(59°)$$
$$= 8.57$$
The opposite side length is 8.57 cm.

TI | THINK	DISPLAY/WRITE	CASIO | THINK	DISPLAY/WRITE
1. Ensure your calculator is in DEGREE mode. On the Calculator page, press MENU, then select: 3 : Algebra 1 : Solve Complete the entry line as: solve $\left(\sin(59) = \dfrac{x}{10}, x\right)$ then press ENTER.	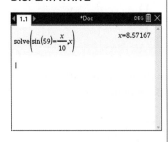	1. Ensure your calculator is in DEGREE mode. On the Main screen, complete the entry line as: solve $\left(\sin(59) = \dfrac{x}{10}, x\right)$ then press EXE.	
2. The answer appears on the screen.	$x = 8.57$ cm (to 2 decimal places)	2. The answer appears on the screen.	$x = 8.57$ cm (to 2 decimal places)

Calculate the value of the unknown angle, θ, correct to 2 decimal places.

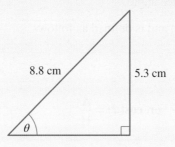

THINK

1. Label all the given information on the triangle.

WRITE/DRAW

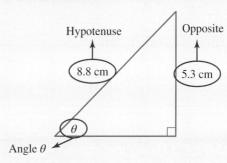

2. Since we have been given the combination of opposite, hypotenuse and the reference angle θ, we need to use the sine ratio. Substitute the given values into the ratio equation.

$$\sin(\theta) = \frac{O}{H}$$
$$= \frac{5.3}{8.8}$$

3. To calculate the angle θ, we need to use the inverse sine function. Make sure your calculator is in degree mode.

$$\theta = \sin^{-1}\left(\frac{5.3}{8.8}\right)$$
$$= 37.03°$$

TI	THINK	DISPLAY/WRITE	CASIO	THINK	DISPLAY/WRITE
1. Ensure the calculator is in DEGREE mode. On the Calculator page, press MENU, then select: 3 : Algebra 1 : Solve Complete the entry line as: $\text{solve}\left(\sin(\theta) = \frac{5.3}{8.8}, \theta\right)$ $\|0 < \theta < 90$ then press ENTER. *Note:* The θ symbol can be found by pressing the $\boxed{\pi\blacktriangleright}$ button. Alternatively, use a letter to represent the angle.		1. Ensure your calculator is in DEGREE mode. On the Main screen, complete the entry line as: $\text{solve}\left(\sin(\theta) = \frac{5.3}{8.8}, \theta\right)$ $\|0 < \theta < 90$ then press EXE. *Note:* The θ symbol can be found in the Math2 tab in the Keyboard menu. Alternatively, use a letter to represent the angle.	
2. The answer appears on the screen.	$\theta = 37.03°$ (to 2 decimal places)	2. The answer appears on the screen.	$\theta = 37.03°$ (to 2 decimal places)

12.2.2 The cosine ratio

The cosine ratio is used when we want to calculate an unknown value given two out of the three following values: adjacent, hypotenuse and reference angle.

The cosine ratio of θ is written as $\cos(\theta)$ and is defined as follows.

Cosine ratio

$$\cos(\theta) = \frac{\text{adjacent}}{\text{hypotenuse}} \quad \text{or} \quad \cos(\theta) = \frac{A}{H}$$

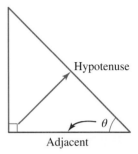

The inverse cosine function is used to calculate the value of the unknown reference angle when given lengths of the hypotenuse and adjacent side.

Cosine ratio for calculating an angle

$$\theta = \cos^{-1}\left(\frac{A}{H}\right)$$

WORKED EXAMPLE 3 Using the cosine ratio to calculate an unknown side length

Calculate the value of the pronumeral y correct to 2 decimal places.

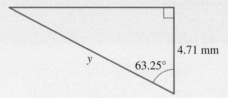

THINK	WRITE
1. Label all the given information on the triangle.	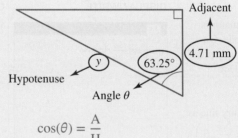
2. Since we have been given the combination of adjacent, hypotenuse and the reference angle θ, we need to use the cosine ratio. Substitute the given values into the ratio equation.	$\cos(\theta) = \dfrac{A}{H}$ $\cos(63.25°) = \dfrac{4.71}{y}$
3. Rearrange the equation to make the unknown the subject and solve. Make sure your calculator is in degree mode.	$y = \dfrac{4.71}{\cos(63.25°)}$ $= 10.46$ The length of the hypotenuse is 10.46 mm.

Calculate the value of the unknown angle, θ, correct to 2 decimal places.

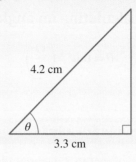

4.2 cm

θ

3.3 cm

THINK	WRITE

1. Label all the given information on the triangle.

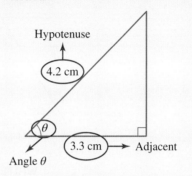

Hypotenuse

4.2 cm

θ

3.3 cm → Adjacent

Angle θ

2. Since we have been given the combination of adjacent, hypotenuse and the reference angle θ, we need to use the cosine ratio. Substitute the given values into the ratio equation.

$$\cos(\theta) = \frac{adjacent}{hypotenuse} = \frac{A}{H}$$

$$\cos(\theta) = \frac{3.3}{4.2}$$

3. To calculate the angle θ, we need to use the inverse cosine function. Make sure your calculator is in degree mode.

$$\theta = \cos^{-1}\left(\frac{3.3}{4.2}\right)$$

$$= 38.21°$$

12.2.3 The tangent ratio

The tangent ratio is used when we want to calculate an unknown value given two out of the three following values: opposite, adjacent and reference angle.

The tangent ratio of θ is written as $\tan(\theta)$ and is defined as follows.

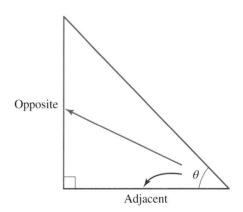

Opposite

Adjacent

θ

Tangent ratio

$$\tan(\theta) = \frac{\textbf{opposite}}{\textbf{adjacent}} \quad \text{or} \quad \tan(\theta) = \frac{\textbf{O}}{\textbf{A}}$$

The inverse tangent function is used to calculate the value of the unknown reference angle given the lengths of the adjacent and opposite sides.

Tangent ratio for calculating an angle

$$\theta = \tan^{-1}\left(\frac{\text{O}}{\text{A}}\right)$$

tlvd-5048

WORKED EXAMPLE 5 Using the tangent ratio to calculate an unknown side length

Calculate the value of the pronumeral x correct to 2 decimal places.

THINK	WRITE
1. Label all the given information on the triangle.	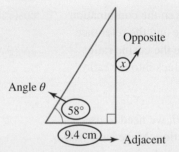
2. Since we have been given the combination of opposite, adjacent and the reference angle θ, we need to use the tangent ratio. Substitute the given values into the ratio equation.	$\tan(\theta) = \dfrac{\text{O}}{\text{A}}$ $\tan(58°) = \dfrac{x}{9.4}$
3. Rearrange the equation to make the unknown the subject and solve. Make sure your calculator is in degree mode.	$x = 9.4\ \tan(58°)$ $x = 15.04$ The opposite side length is 15.04 cm.

Determine the value of the unknown angle, θ, correct to 2 decimal places.

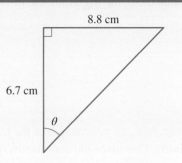

THINK	WRITE
1. Label all the given information on the triangle.	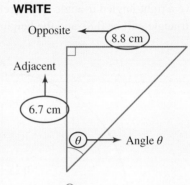
2. Since we have been given the combination of opposite, adjacent and the reference angle θ, we need to use the tangent ratio. Substitute the given values into the ratio equation.	$\tan(\theta) = \dfrac{O}{A}$ $= \dfrac{8.8}{6.7}$
3. To calculate the angle θ, we need to use the inverse tangent function. Make sure your calculator is in degree mode.	$\theta = \tan^{-1}\left(\dfrac{8.8}{6.7}\right)$ $= 52.72°$

12.2.4 The unit circle

If we draw a circle of radius 1 in the Cartesian plane with its centre located at the origin, then we can locate the coordinates of any point on the circumference of the circle using right-angled triangles.

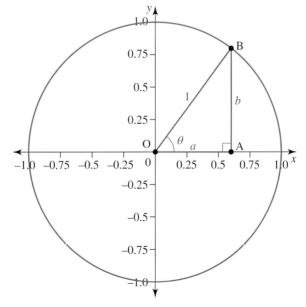

As the radius of the circle in this diagram is 1, the length of the hypotenuse is 1 and the coordinates of B can be found using the trigonometric ratios.

$$\cos(\theta) = \frac{A}{H} \text{ and } \sin(\theta) = \frac{O}{H}$$
$$= \frac{a}{1} \qquad\qquad = \frac{b}{1}$$
$$= a \qquad\qquad\quad = b$$

Therefore, the base length of the triangle, a, is equal to $\cos(\theta)$, and the height of the triangle, b, is equal to $\sin(\theta)$. This gives the coordinates of B as $(\cos(\theta), \sin(\theta))$.

For example, if we have a right-angled triangle with a reference angle of 30° and a hypotenuse of length 1, then the base length of the triangle will be 0.87 and the height of the triangle will be 0.5, as shown in the following triangle.

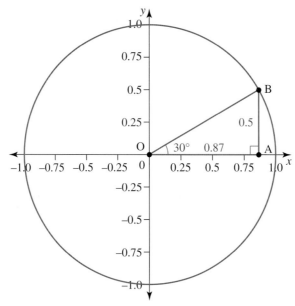

Similarly, if we calculate the value of $\cos(30°)$ and $\sin(30°)$, we get 0.87 and 0.5 respectively.

We can actually extend this definition to any point B on the unit circle as having the coordinates $(\cos(\theta), \sin(\theta))$, where θ is the angle measured in an anticlockwise direction.

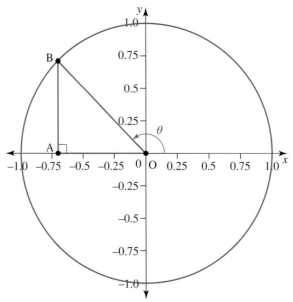

Extending sine and cosine to 180°

We can place any right-angled triangle with a hypotenuse of 1 in the unit circle so that one side of the triangle lies on the positive x-axis. The following diagram shows a triangle with base length 0.64, height 0.77 and reference angle 50°.

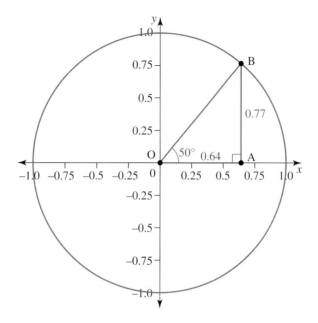

The coordinates of point B in this triangle are (0.64, 0.77) or (cos(50°), sin(50°)). Now reflect the triangle in the y-axis as shown in the following diagram.

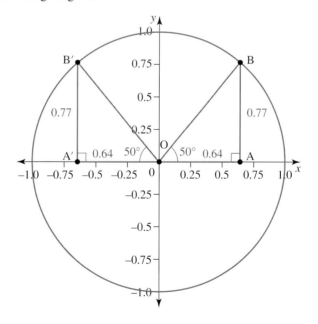

We can see that the coordinates of point B′ are (−0.64, 0.77) or (− cos(50°), sin(50°)).

We have previously determined the coordinates of any point B on the circumference of the unit circle as (cos(θ), sin(θ)), where θ is the angle measured in an anticlockwise direction. In this instance, the value of θ = 180° − 50 = 130°.

Therefore, the coordinates of point B′ are (cos(130°), sin(130°)).

This discovery can be extended when we place any right-angled triangle with a hypotenuse of length 1 inside the unit circle.

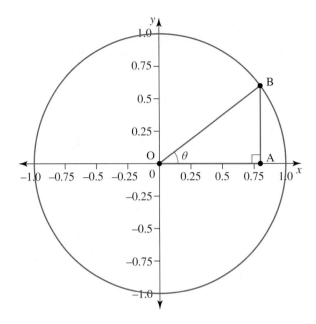

As previously determined, the point B has coordinates $(\cos(\theta),\ \sin(\theta))$.

Reflecting this triangle in the y-axis gives:

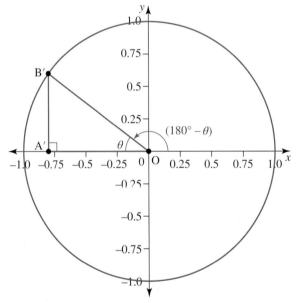

So the coordinates of point B′ are $(-\cos(\theta),\ \sin(\theta))$. We also know that the coordinates of B′ are $(\cos(180 - \theta),\ \sin(180 - \theta))$ from the general rule about the coordinates of any point on the unit circle.

Equating the two coordinates for B′ gives us the following equations:

$$-\cos(\theta) = \cos(180 - \theta)$$
$$\sin(\theta) = \sin(180 - \theta)$$

So, to calculate the values of the sine and cosine ratios for angles up to 180°, we can use the following definitions.

Definition of sine and cosine for angles up to 180°

$$\cos(\theta) = -\cos(180 - \theta)$$
$$\sin(\theta) = \sin(180 - \theta)$$

Remember that if two angles sum to 180°, then they are supplements of each other. So if we are calculating the sine or cosine of an angle between 90° and 180°, then start by determining the supplement of the given angle.

tlvd-5049

WORKED EXAMPLE 7 Calculating sine and cosine for angles up to 180°

Calculate the values of:

a. sin (140°)   **b. cos (160°)**

giving your answers correct to 2 decimal places.

THINK	WRITE
a. 1. Calculate the supplement of the given angle.	a. $180° - 140° = 40°$
2. Calculate the sine of the supplement angle correct to 2 decimal places.	$sin(40°) = 0.642\,787\ldots$ $= 0.64$ (to 2 decimal places)
3. The sine of an obtuse angle is equal to the sine of its supplement.	$sin(140°) = sin(40°)$ $= 0.64$ (to 2 decimal places)
b. 1. Calculate the supplement of the given angle.	b. $180° - 160° = 20°$
2. Calculate the cosine of the supplement angle correct to 2 decimal places.	$cos(20°) = 0.939\,692\ldots$ $= 0.94$ (to 2 decimal places)
3. The cosine of an obtuse angle is equal to the negative cosine of its supplement.	$cos(160°) = -cos(20°)$ $= -0.94$ (to 2 decimal places)

TI	THINK	DISPLAY/WRITE	CASIO	THINK	DISPLAY/WRITE
1. Ensure your calculator is in DEGREE mode. On the Calculator page, complete the entry line as: sin(140) then press ENTER.		1. Ensure your calculator is in DEGREE mode. On the Main screen, complete the entry line as: sin(140) then press EXE.	
2. The answer appears on the screen.	sin(140) = 0.64 (to 2 decimal places)	2. The answer appears on the screen.	sin(140) = 0.64 (to 2 decimal places)

SOH–CAH–TOA

Trigonometric ratios are relationships between the sides and angles of a right-angled triangle.

In solving trigonometric ratio problems for sine, cosine and tangent, we need to:

1. determine which ratio to use
2. write the relevant equation
3. substitute values from given information
4. make sure the calculator is in degree mode
5. solve the equation for the unknown lengths, or use the inverse trigonometric function to determine an unknown angle.

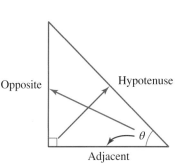

To assist in remembering the trigonometric ratios, the mnemonic **SOH–CAH–TOA** has been developed.

Shortcut to remembering the trigonometric ratios

SOH–CAH–TOA stands for:
- **Sine is Opposite over Hypotenuse**
- **Cosine is Adjacent over Hypotenuse**
- **Tangent is Opposite over Adjacent.**

on Resources

Interactivity Trigonometric ratios (int-2577)

12.2 Exercise

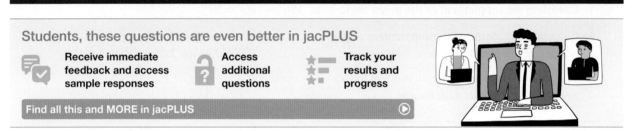

Students, these questions are even better in jacPLUS

Receive immediate feedback and access sample responses

Access additional questions

Track your results and progress

Find all this and MORE in jacPLUS

1. **WE1** Calculate the value of the pronumeral x correct to 2 decimal places.

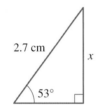

2.7 cm

x

53°

2. Calculate the value of the pronumeral x.

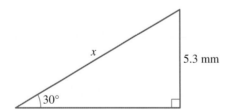

x

5.3 mm

30°

3. **WE2** Calculate the value of the unknown angle, θ, correct to 2 decimal places.

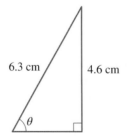

6.3 cm

4.6 cm

θ

4. Calculate the value of the unknown angle, θ, correct to 2 decimal places.

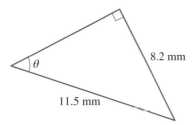

5. **WE3** Calculate the value of the pronumeral y correct to 2 decimal places.

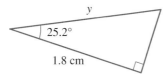

6. Calculate the value of the pronumeral y correct to 2 decimal places.

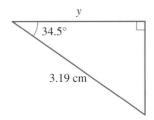

7. **WE4** Calculate the value of the unknown angle, θ, correct to 2 decimal places.

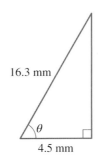

8. Calculate the value of the unknown angle, θ, correct to 2 decimal places.

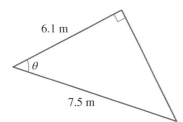

9. **WE5** Calculate the value of the pronumeral x correct to 2 decimal places.

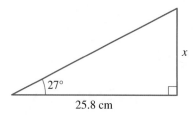

10. Calculate the value of the pronumeral y correct to 2 decimal places.

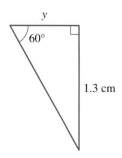

11. **WE6** Calculate the value of the unknown angle, θ, correct to 2 decimal places.

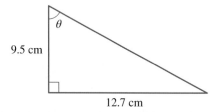

12. Calculate the value of the unknown angle, θ, correct to 2 decimal places.

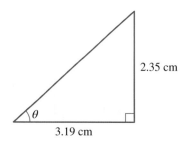

13. **WE7** Calculate the values of:

    a. $\sin(125°)$ b. $\cos(152°)$
    giving your answers correct to 2 decimal places.

14. Calculate the values of:

    a. $\sin(99.2°)$ b. $\cos(146.7°)$
    giving your answers correct to 2 decimal places.

15. Calculate the value of the pronumeral x correct to 2 decimal places.

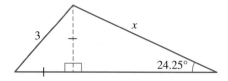

16. Calculate the value of the unknown angle, θ, correct to 2 decimal places.

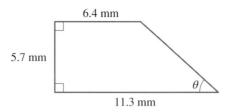

17. A kitesurfer has a kite of length 2.5 m and strings of length 7 m as shown. Calculate the values of the angles θ and α, correct to 2 decimal places.

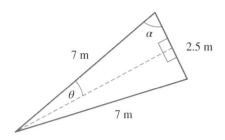

18. A stuntman is to be catapulted into the air in a capsule at an angle of 64.3° to the ground. Assuming the stuntman travels in a straight path, calculate the horizontal distance he will have covered, correct to 2 decimal places, if he reached a height of 20 metres.

19. A yacht race follows a triangular course as shown below. Calculate, correct to 1 decimal place:

a. the distance of the final leg, y
b. the total distance of the course.

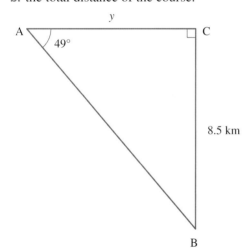

20. A railway line rises for 300 metres at a uniform slope of 6° with the horizontal. Calculate the distance travelled by the train, correct to the nearest metre.

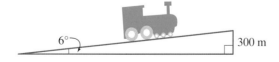

21. A truss is used to build a section of a roof. If the vertical height of the truss is 1.5 metres and the span (horizontal distance between the walls) is 8 metres wide, calculate the pitch of the roof (its angle with the horizontal) correct to 2 decimal places.

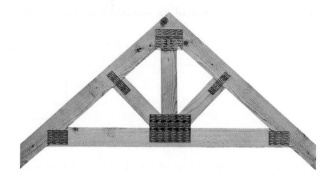

22. If 3.5 metres of Christmas lights are attached directly from the tip of the bottom branch to the top of a Christmas tree at an angle of 35.2° from the ground, calculate the height of the Christmas tree, correct to 2 decimal places.

23. A 2.5 m ladder is placed against a wall. The base of the ladder is 1.7 m from the wall.
    a. Calculate the angle, correct to 2 decimal places, that the ladder makes with the ground.
    b. Calculate how far the ladder reaches up the wall, correct to 2 decimal places.

24. A school is building a wheelchair ramp of length 4.2 m to be inclined at an angle of 10.5°.
    a. Calculate the horizontal length of the ramp, correct to 1 decimal place.
    b. The ramp is too steep at 10.5°. Instead, the vertical height of the ramp needs to be 0.5 m and the ramp inclined at an angle of 5.7°. Calculate the new length of the wheelchair ramp, correct to 2 decimal places.

25. A play gym for monkeys is constructed at a zoo. A rope is tied from a tree branch 1.6 m above the ground to another tree branch 2.5 m above the ground. The monkey swings along the rope, which makes an angle of 11.87° to the horizontal. Determine how far apart the trees are, correct to 2 decimal places.

26. A dog training obstacle course ABCDEA is shown in the diagram. Point B is vertically above point D. Calculate the total length of the obstacle course in metres, giving your answer correct to 2 decimal places.

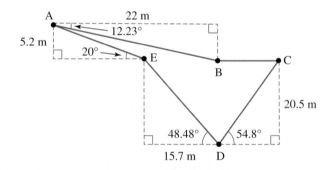

12.2 Exam questions

Question 1 (1 mark)

Source: VCE 2014, Further Mathematics Exam 1, Section B, Module 2, Q1; © VCAA.

MC The top of a ladder that is 4.50 m long rests 3.25 m up a wall, as shown in the diagram.

The angle, θ, that the ladder makes with the wall is closest to

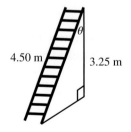

    A. 36°
    B. 44°
    C. 46°
    D. 50°
    E. 54°

Question 2 (1 mark)

MC The value of x in the diagram is

A. 11.4 cm

B. 20 cm

C. 24.4 cm

D. 9.8 cm

E. 8 cm

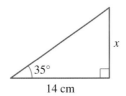

Question 3 (1 mark)

MC A diagonal cut 45 cm long is made in a rectangular piece of cardboard 30 cm wide.

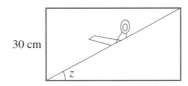

Calculate the angle, z, that the cut makes with the longer side of the cardboard.

A. 33.7°

B. 48.6°

C. 56.3°

D. 48.2°

E. 41.8°

More exam questions are available online.

12.3 Angles of elevation and depression, and bearings

LEARNING INTENTION

At the end of this subtopic you should be able to:
- apply angles of elevation and depression to calculate distance
- determine a bearing from one point to another
- apply trigonometry in bearings problems.

12.3.1 Angles of elevation and depression

An **angle of elevation** is the angle between a horizontal line from the observer to an object that is above the horizontal line.

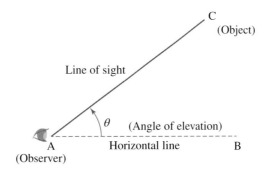

An **angle of depression** is the angle between a horizontal line from the observer to an object that is below the horizontal line.

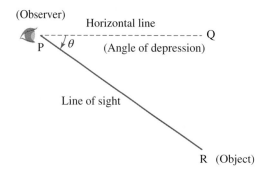

(Observer)
Horizontal line
Q
P θ (Angle of depression)
Line of sight
R (Object)

We use angles of elevation and depression to locate the positions of objects above or below the horizontal (reference) line. Angles of elevation and angles of depression are equal as they are alternate angles.

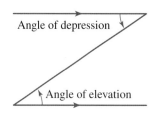

Angle of depression

Angle of elevation

WORKED EXAMPLE 8 Applying an angle of depression

The angle of depression from a scuba diver at the water's surface to a hammerhead shark on the sea floor of the Great Barrier Reef is 40°. The depth of the water is 35 m. Calculate the horizontal distance from the scuba diver to the shark, correct to 2 decimal places.

THINK	WRITE
1. Draw a diagram to represent the information.	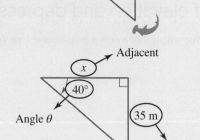 x 40° 35 m
2. Label all the given information on the triangle.	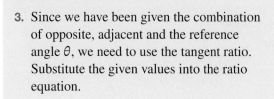 Adjacent x 40° Angle θ 35 m Opposite
3. Since we have been given the combination of opposite, adjacent and the reference angle θ, we need to use the tangent ratio. Substitute the given values into the ratio equation.	$\tan(\theta) = \dfrac{O}{A}$ $\tan(40°) = \dfrac{35}{x}$

4.	Rearrange the equation to make the unknown the subject and solve. Make sure your calculator is in degree mode.	$x = \dfrac{35}{\tan(40°)}$ $= 41.71$ The horizontal distance from the scuba diver to the shark is 41.71 m.

12.3.2 Bearings

Bearings are used to locate the positions of objects or the direction of a journey on a two-dimensional plane.

The four main directions or standard bearings of a directional compass are known as **cardinal points**. They are North (N), South (S), East (E) and West (W).

There are two types of bearings: **conventional (compass) bearings** and **true bearings**.

In this course, however, we will only use true bearings.

True bearings (three-figure bearings)

True bearings are measured in a clockwise direction from north around to the required direction. They are written with all three digits of the angle stated.

If the angle measured is less than 100°, place a zero in front of the angle. For example, if the angle measured is 20° from the north, the true the bearing is 020°.

A three-figure bearing can sometimes be followed by a capital T to indicate that it is a true bearing; however, this is not compulsory. For example, a true bearing of 150° could also be written as 150°T.

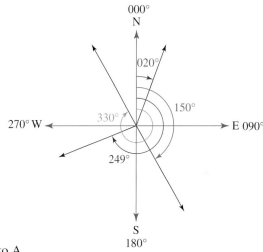

Bearings from A to B

The bearing from A to B is **not** the same as the bearing from B to A.

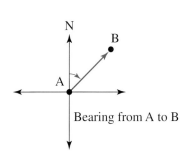

Bearing from A to B

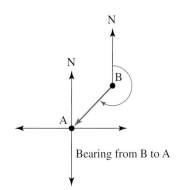

Bearing from B to A

When determining the bearing from a point to another point, it is important to follow the instructions and draw a diagram. Always draw the centre of the compass at the starting point of the direction requested.

When a problem asks to calculate the bearing of B from A, mark in north and join a directional line to B to work out the bearing. Returning to where you came from requires a change in bearing of 180°.

WORKED EXAMPLE 9 Determining a bearing from one point to another

Calculate the true bearing from:
a. **Town A to Town B**
b. **Town B to Town A.**

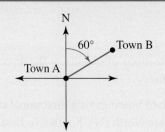

THINK	WRITE
a. 1. To calculate the bearing from Town A to Town B, make sure the centre of the compass is marked at town A. The angle is measured clockwise from north to the bearing line at Town B.	a. 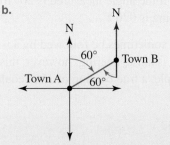
	The angle measured from north is 60°.
2. A true bearing is written with all three digits of the angle.	The true bearing from Town A to Town B is 060°.
b. 1. To calculate the bearing from Town B to Town A, make sure the centre of the compass is marked at Town B. The angle is measured clockwise from north to the bearing line at Town A.	b.
	The angle measured from north is 60° + 180° = 240°.
2. A true bearing is written with all three digits of the angle.	The true bearing from Town B to Town A is 240°.

12.3.3 Using trigonometry in bearings problems

As the four cardinal points (N, E, S, W) are at right angles to each other, we can use trigonometry to solve problems involving bearings.

When solving a bearings problem with trigonometry, always start by drawing a diagram to represent the problem. This will help you to identify what information you already have, and determine which trigonometric ratio to use.

WORKED EXAMPLE 10 Using trigonometry in bearings problems

A boat travels for 25 km on a true bearing of 310°.
a. Calculate how far north the boat travels, correct to 2 decimal places.
b. Calculate how far west the boat travels, correct to 2 decimal places.

THINK	WRITE
a. 1. Draw a diagram of the situation, remembering to label the compass points as well as all of the given information.	a. 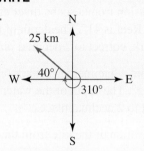
2. Identify the information you have in respect to the reference angle, as well as the information you need.	Reference angle $= 40°$ Hypotenuse $= 25$ Opposite $= ?$
3. Determine which of the trigonometric ratios to use.	$\sin(\theta) = \dfrac{O}{H}$
4. Substitute the given values into the trigonometric ratio and solve for the unknown.	$\sin(40°) = \dfrac{O}{25}$ $25 \ \sin(40°) = O$ $\qquad O = 16.069 \ldots$ $\qquad = 16.07 \text{ (to 2 decimal places)}$
5. Write the answer in a sentence.	The boat travels 16.07 km north.
b. 1. Use your diagram from part **a** and identify the information you have in respect to the reference angle, as well as the information you need.	b. Reference angle $= 40°$ Hypotenuse $= 25$ Adjacent $= ?$
2. Determine which of the trigonometric ratios to use.	$\cos(\theta) = \dfrac{A}{H}$
3. Substitute the given values into the trigonometric ratio and solve for the unknown.	$\cos(40°) = \dfrac{A}{25}$ $25 \ \cos(40°) = A$ $\qquad A = 19.151 \ldots$ $\qquad = 19.15 \text{ (to 2 decimal places)}$
4. Write the answer in a sentence.	The boat travels 19.15 km west.

 Resources

Interactivity Bearings (int-6481)

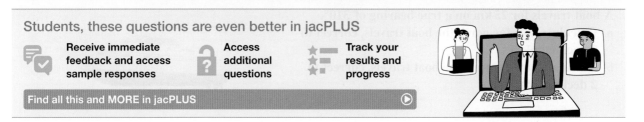
1. **WE8** The angle of depression from a scuba diver at the water's surface to a hammerhead shark on the sea floor of the Great Barrier Reef is 41°. The depth of the water is 32 m. Calculate the horizontal distance from the scuba diver to the shark, correct to 2 decimal places.

2. The angle of elevation from a hammerhead shark on the sea floor of the Great Barrier Reef to a scuba diver at the water's surface is 35°. The depth of the water is 33 m. Calculate the horizontal distance from the shark to the scuba diver, correct to 2 decimal places.

3. Calculate the angle of elevation of the kite from the ground, correct to 2 decimal places.

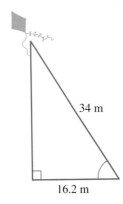

34 m

16.2 m

4. Calculate the angle of depression from the boat to the treasure at the bottom of the sea, correct to 2 decimal places.

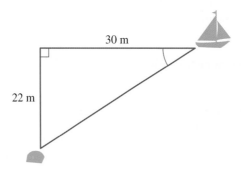

30 m

22 m

5. A crocodile is fed on a 'jumping crocodile tour' on the Adelaide River. The tour guide dangles a piece of meat on a stick at an angle of elevation of 60° from the boat, horizontal to the water. If the stick is 2 m long and held 1 m above the water, calculate the vertical distance the crocodile has to jump out of the water to get the meat, correct to 2 decimal places.

6. A ski chair lift operates from the Mt Buller village and has an angle of elevation of 45° to the top of the Federation ski run. If the vertical height is 707 m, calculate the ski chair lift length, correct to 2 decimal places.

7. State each of the following as a true bearing.

a.
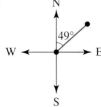

b.
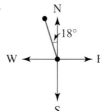

c.
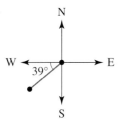

d.
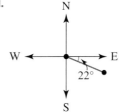

8. State each of the following as a true bearing.

a.
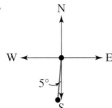

b.
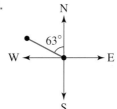

c.
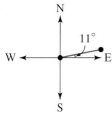

d.
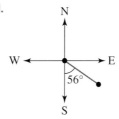

9. **WE9** In the diagram shown, calculate the true bearing from:
   a. Town A to Town B
   b. Town B to Town A.

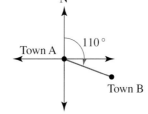

10. In the diagram shown, calculate the true bearing from:
    a. Town A to Town B
    b. Town B to Town A.

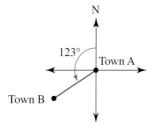

11. For the diagram shown, calculate the true bearing of:

a. B from A
b. C from B
c. A from C.

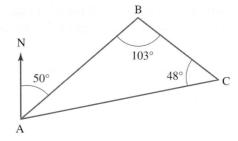

12. **WE10** A boat travels for 36 km on a true bearing of 155°.

a. Calculate how far south the boat travels, correct to 2 decimal places.
b. Calculate how far east the boat travels, correct to 2 decimal places.

13. A boat travels north for 6 km, west for 3 km, then south for 2 km. Calculate the boat's true bearing from its starting point. Give your answer in decimal form to 1 decimal place.

14. A student uses an inclinometer to measure an angle of elevation of 50° from the ground to the top of Uluru. If the student is standing 724 m from the base of Uluru, determine the height of Uluru correct to 2 decimal places.

15. A tourist looks down from the Eureka Tower's Edge on the Skydeck to see people below on the footpath. If the angle of depression is 88° and the people are 11 m from the base of the tower, calculate how high up the tourist is standing in the glass cube, correct to the nearest metre.

16. A student uses an inclinometer to measure the height of his house. The angle of elevation is 54°. He is 1.5 m tall and stands 7 m from the base of the house. Calculate the height of the house correct to 2 decimal places.

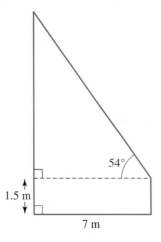

17. A tourist 1.72 m tall is standing 50 m away from the base of the Sydney Opera House. The Opera House is 65 m tall. Calculate the angle of elevation, to the nearest degree, from the tourist to the top of the Opera House.

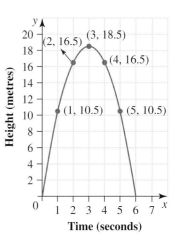

18. A parachutist falls from a vertical distance of 5000 m while travelling over a horizontal distance of 150 m. Calculate the angle of depression of the descent correct to 2 decimal places.

19. Air traffic controllers in two control towers, which are both 87 m high, spot a plane at an altitude of 500 metres. The angle of elevation from tower A to the plane is 5° and from tower B to the plane is 7°. Calculate the distance between the two control towers correct to the nearest metre.

20. An AFL footballer takes a set shot at goal, with the graph showing the path that the ball took as it travelled towards the goal.
If the footballer's eye level is at 1.6 metres, calculate the angle of elevation from his eyesight to the ball, correct to 2 decimal places, after:

a. 1 second
b. 2 seconds
c. 3 seconds
d. 4 seconds
e. 5 seconds.

12.3 Exam questions

Question 1 (1 mark)
Source: VCE 2019, Further Mathematics Exam 1, Section B, Module 3, Q2; © VCAA.

MC Town B is located on a bearing of 60° from Town A.

The diagram that could illustrate this is

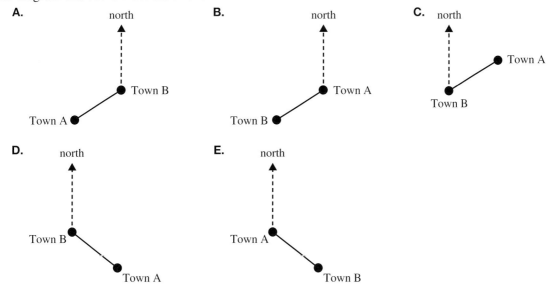

Question 2 (1 mark)

MC If the angle of elevation of the top of a cliff from the deck of a ship is 56°, determine which of the following is true.

    A. The angle of depression of the cliff is 56° from the deck of the ship.
    B. The angle of depression of the cliff is 34° from the deck of the ship.
    C. The angle of elevation of the deck of a ship is 56° from the cliff.
    D. The angle of depression of the deck of a ship is 56° from the cliff.
    E. The angle of depression of the deck of a ship is 34° from the cliff.

Question 3 (3 marks)

A yacht race starts at point S. The yachts sail for 6 km on a true bearing of 030° to a buoy at point B. They then turn to a bearing of 190° and sail to the finish at point F, 10 km away.

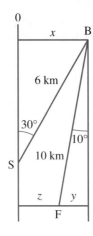

    a. Calculate how far east of S buoy B is.   **(1 mark)**
    b. Calculate how far east of buoy B the finish, F, is.   **(1 mark)**
    c. Calculate how far east of the start, S, the finish, F, is.   **(1 mark)**

More exam questions are available online.

12.4 The sine rule

LEARNING INTENTION

At the end of this subtopic you should be able to:
- apply the sine rule in non-right-angled triangles to determine the unknown angle and side length
- determine when to use the ambiguous case of the sine rule.

12.4.1 The sine rule

The **sine rule** can be used to determine the side length or angle in non-right-angled triangles.

To help us solve non-right-angled triangle problems, the labelling convention of a non-right-angled triangle, ABC, is as follows:

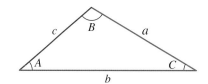

Angle A is opposite side length a.

Angle B is opposite side length b.

Angle C is opposite side length c.

The largest angle will always be opposite the longest side length, and the smallest angle will always be opposite to the smallest side length.

Formulating the sine rule

We can divide an acute non-right-angled triangle into two right-angled triangles as shown in the following diagrams.

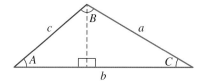 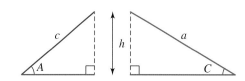

If we apply trigonometric ratios to the two right-angled triangles we get:

$$\frac{h}{c} = \sin(A) \text{ and } \frac{h}{a} = \sin(C)$$
$$h = c\sin(A) \quad h = a\sin(C)$$

Equating the two expressions for h gives:

$$c\sin(A) = a\sin(C)$$

$$\frac{a}{\sin(A)} = \frac{c}{\sin(C)}$$

In a similar way, we can split the triangle into two using side a as the base, giving us:

$$\frac{b}{\sin(B)} = \frac{c}{\sin(C)}$$

This gives us the sine rule.

The sine rule

$$\frac{a}{\sin(A)} = \frac{b}{\sin(B)} = \frac{c}{\sin(C)}$$

We can apply the sine rule to determine all of the angles and side lengths of a triangle if we are given either:
- 2 side lengths and an angle opposite one of the given sides.
- 1 side length and 2 angles.

tlvd-5051

WORKED EXAMPLE 11 Using the sine rule to determine a side length

Calculate the value of the unknown length x correct to 2 decimal places.

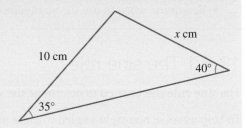

THINK

1. Label the triangle with the given information, using the conventions for labelling.
   Angle A is opposite to side a.
   Angle B is opposite to side b.

2. Substitute the known values into the sine rule.

3. Rearrange the equation to make x the subject and solve.
   Make sure your calculator is in degree mode.

4. Write the answer in a sentence.

WRITE

$$\frac{a}{\sin(A)} = \frac{b}{\sin(B)}$$

$$\frac{x}{\sin(35°)} = \frac{10}{\sin(40°)}$$

$$\frac{x}{\sin(35°)} = \frac{10}{\sin(40°)}$$

$$x = \frac{10 \, \sin(35°)}{\sin(40°)}$$

$$x = 8.92$$

The unknown side length x is 8.92 cm.

WORKED EXAMPLE 12 Using the sine rule to determine an angle

A non-right-angled triangle has values of side $b = 12.5$, angle $A = 25.3°$ and side $a = 7.4$. Calculate the value of angle B correct to 2 decimal places.

THINK

1. Draw a non-right-angled triangle, labelling with the given information.
   Angle A is opposite to side a.
   Angle B is opposite to side b.

WRITE

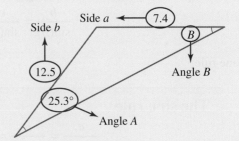

2. Substitute the known values into the sine rule.

$$\frac{a}{\sin(A)} = \frac{b}{\sin(B)}$$

$$\frac{7.4}{\sin(25.3°)} = \frac{12.5}{\sin(B)}$$

3. Rearrange the equation to make $\sin(B)$ the subject and solve.
Make sure your calculator is in degree mode.

$$\frac{7.4}{\sin(25.3°)} = \frac{12.5}{\sin(B)}$$

$$7.4 \sin(B) = 12.5 \sin(25.3°)$$

$$\sin(B) = \frac{12.5 \sin(25.3°)}{7.4}$$

$$B = \sin^{-1}\left(\frac{12.5 \sin(25.3°)}{7.4}\right)$$

$$= 46.21°$$

4. Write the answer.

Angle B is 46.21°.

12.4.2 The ambiguous case of the sine rule

When we are given two sides lengths of a triangle and an acute angle opposite one of these side lengths, there are two different triangles we can draw. So far we have only dealt with triangles that have all acute angles; however, it is also possible to draw triangles with an obtuse angle.

For example, take the triangle ABC, where $a = 12$, $c = 8$ and $C = 35°$.

There are two triangles that satisfy these conditions, as shown in the diagrams below.

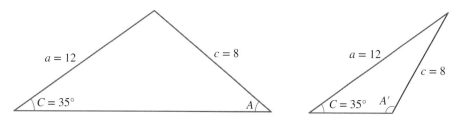

When we solve this for angle A in the first triangle, we get an acute angle as shown:

$$\frac{8}{\sin(35°)} = \frac{12}{\sin(A)}$$

$$8 \sin(A) = 12 \sin(35°)$$

$$\sin(A) = \frac{12 \sin(35°)}{8}$$

$$A = \sin^{-1}\left(\frac{12 \sin(35°)}{8}\right)$$

$$= 59.36°$$

In the second triangle, the size of the obtuse angle is the supplement of the acute angle calculated previously.

$$A' = 180° - A$$

$$= 180° - 59.36°$$

$$= 120.64°$$

Determining when we can use the ambiguous case

The case where two different triangles can represent the same information is known as the *ambiguous case of the sine rule*.

The ambiguous case of the sine rule occurs when the smaller known side is opposite the known angle.

The ambiguous case does *not* apply if the given angle is obtuse or a right angle; then there is only one such triangle.

WORKED EXAMPLE 13 Applying the ambiguous case of the sine rule

Determine the two possible values of angle A for triangle ABC, given $a = 15$, $c = 8$ and $C = 25°$. Give your answers correct to 2 decimal places.

THINK

1. Draw a non-right-angled triangle, labelling with the given information.
   Angle A is opposite to side a.
   Angle C is opposite to side c.
   Note that two triangles can be drawn, with angle A being either acute or obtuse, since the smaller known side is opposite the known angle.

2. Substitute the known values into the sine rule.

3. Rearrange the equation to make $\sin(A)$ the subject and solve.
   Make sure your calculator is in degree mode.
   The calculator will only give the acute angle value.

4. Solve for the obtuse angle A'.

5. Write the answer in a sentence.

WRITE

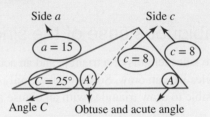

Side a Side c

$a = 15$ $c = 8$ $c = 8$

$C = 25°$ A' A

Angle C Obtuse and acute angle

$$\frac{a}{\sin(A)} = \frac{c}{\sin(C)}$$

$$\frac{15}{\sin(A)} = \frac{8}{\sin(25°)}$$

$$\frac{15}{\sin(A)} = \frac{8}{\sin(25°)}$$

$$15 \sin(25°) = 8 \sin(A)$$

$$\sin(A) = \frac{15 \sin(25°)}{8}$$

$$A = \sin^{-1}\left(\frac{15 \sin(25°)}{8}\right)$$

$$= 52.41°$$

$$A' = 180° - A$$
$$= 180° - 52.41°$$
$$= 127.59°$$

The two possible values for A are $52.41°$ and $127.59°$.

TI| THINK	DISPLAY/WRITE	CASIO| THINK	DISPLAY/WRITE
1. Ensure your calculator is in DEGREE mode. On the Calculator page, press MENU, then select: 3: Algebra 1: Solve Complete the entry line as: $solve\left(\dfrac{15}{\sin(a)} = \dfrac{8}{\sin(25)}, a\right)$ $\|0 < a < 180$ then press ENTER.	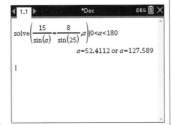	1. Ensure your calculator is in DEGREE mode. On the Main screen, complete the entry line as: $solve\left(\dfrac{15}{\sin(A)} = \dfrac{8}{\sin(A)}, A\right)$ $\|0 < A < 180$ then press EXE.	
2. The answer appears on the screen.	The two possible values for A are 52.41° and 127.59°.	2. The answer appears on the screen.	The two possible values for A are 52.41° and 127.59°.

 Resources

 Interactivities   The sine rule (int-6275)

Solving non-right-angled triangles (int-6482)

12.4 Exercise

Students, these questions are even better in jacPLUS

 Receive immediate feedback and access sample responses

 Access additional questions

 Track your results and progress

Find all this and MORE in jacPLUS

1. **WE11** Calculate the value of the unknown length x correct to 2 decimal places.

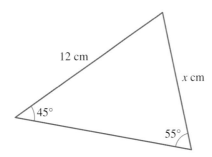

2. Calculate the value of the unknown length x correct to 2 decimal places.

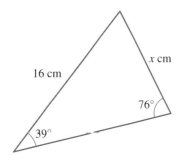

3. Calculate the value of the unknown length x correct to 2 decimal places.

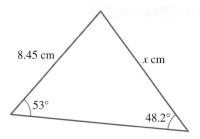

4. Calculate the value of the unknown length x correct to the nearest cm.

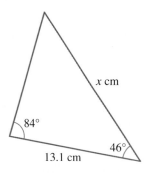

5. **WE12** A non-right-angled triangle has values of side $b = 10.5$, angle $A = 22.3°$ and side $a = 8.4$. Calculate the value of angle B correct to 1 decimal place.

6. A non-right-angled triangle has values of side $b = 7.63$, angle $A = 15.8°$ and side $a = 4.56$. Calculate the value of angle B correct to 1 decimal place.

7. For the triangle ABC shown, calculate the acute value of θ correct to 1 decimal place.

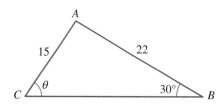

8. **WE13** Calculate the two possible values of angle A for triangle ABC, given $a = 8$, $c = 6$ and $C = 43°$. Give your answers correct to 2 decimal places.

9. Calculate the two possible values of angle A for triangle ABC, given $a = 7.5$, $c = 5$ and $C = 32°$. Give your answers correct to 2 decimal places.

10. If triangle ABC has values $b = 19.5$, $A = 25.3°$ and $a = 11.4$, calculate both possible angle values of B correct to 2 decimal places.

11. Calculate all the side lengths, correct to 2 decimal places, for the triangle ABC, given $a = 10.5$, $B = 60°$ and $C = 72°$.

12. Calculate the acute and obtuse angles that have a sine value of approximately 0.573 58. Give your answer to the nearest degree.

13. Part of a roller-coaster track is in the shape of an isosceles triangle, ABC, as shown in the following triangle. Calculate the track length AB correct to 2 decimal places.

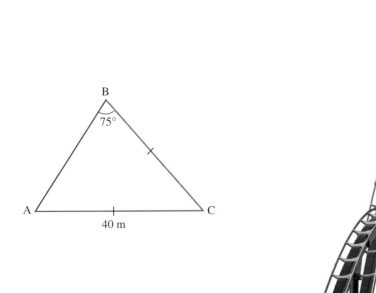

14. The shape of a water slide follows the path of PY and YZ in the diagram shown.

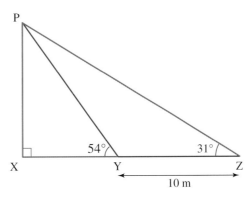

Calculate, correct to 2 decimal places:

a. the total length of the water slide

b. the height of the water slide, PX.

15. Calculate the two unknown angles shown in the diagram below, correct to 1 decimal place.

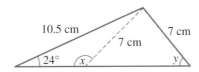

16. In the triangle ABC, $a = 11.5$ m, $c = 6.5$ m and $C = 25°$.

a. Draw the two possible triangles with this information.

b. Calculate the two possible values of angle A and hence the two possible values of angle B. Give all values correct to 2 decimal places.

17. At a theme park, the pirate ship swings back and forth on a pendulum. The centre of the pirate ship is secured by a large metal rod that is 5.6 metres in length. If one of the swings covers an angle of 122°, determine the distance between the point where the rod meets the ship at the two extremes of the swing. Give your answer correct to 2 decimal places.

18. Andariel went for a ride on her dune buggy in the desert. She rode east for 6 km, then turned 125° to the left for the second stage of her ride. After 5 minutes riding in the same direction, she turned to the left again, and from there travelled the 5.5 km straight back to her starting position. Determine how far Andariel travelled in the second section of her ride, correct to 2 decimal places.

12.4 Exam questions

Question 1 (1 mark)

Source: VCE 2017, Further Mathematics Exam 1, Section B, Module 3, Q7; © VCAA.

MC A triangle ABC has:
- one side, $\overline{AB}$, of length 4 cm
- one side, $\overline{BC}$, of length 7 cm
- one angle, $\angle ACB$, of 26°.

Which one of the following angles, correct to the nearest degree, could **not** be another angle in triangle ABC?

A. 24°   **B.** 50°   **C.** 104°   **D.** 130°   **E.** 144°

Question 2 (1 mark)

Source: VCE 2016, Further Mathematics Exam 1, Section B, Module 3, Q6; © VCAA.

MC Marcus is on the opposite side of a large lake from a horse and its stable. The stable is 150 m directly east of the horse. Marcus is on a bearing of 170° from the horse and on a bearing of 205° from the stable.

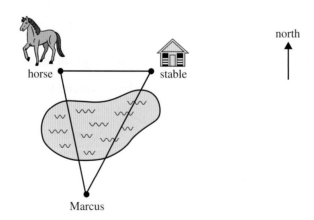

The straight-line distance, in metres, between Marcus and the horse is closest to

A. 45   **B.** 61   **C.** 95   **D.** 192   **E.** 237

Question 3 (1 mark)

MC In the diagram shown, the angle(s) opposite the side with length 5 units is/are

A. $38.68°$
B. $51.32°$
C. $141.32°$
D. $111.32°$ and $68.68°$
E. $38.68°$ and $141.32°$

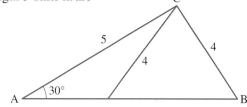

More exam questions are available online.

12.5 The cosine rule

LEARNING INTENTION

At the end of this subtopic you should be able to:
- apply the cosine rule in non-right-angled triangles to determine the unknown angle or side length.

12.5.1 Formulating the cosine rule

The **cosine rule**, like the sine rule, is used to calculate the length or angle in a non-right-angled triangle. We use the same labelling conventions for non-right-angled triangles as when using the sine rule.

As with the sine rule, the cosine rule is derived from a non-right-angled triangle being divided into two right-angled triangles, where the base side lengths are equal to $(b - x)$ and x.

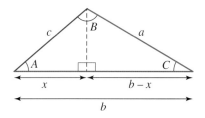   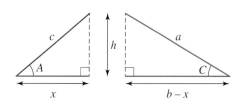

Using Pythagoras' theorem we get:

$$c^2 = x^2 + h^2 \qquad a^2 = (b-x)^2 + h^2$$
$$h^2 = c^2 - x^2 \quad and \quad h^2 = a^2 - (b-x)^2$$

Equating the two expressions for h^2 gives:

$$c^2 - x^2 = a^2 - (b-x)^2$$
$$a^2 = (b-x)^2 + c^2 - x^2$$
$$a^2 = b^2 - 2bx + c^2$$

Substituting the trigonometric ratio $x = c\cos(A)$ from the right-angled triangle into the expression, we get:

$$a^2 = b^2 - 2b\,(c\cos(A)) + c^2$$
$$= b^2 + c^2 - 2bc\cos(A)$$

This is known as the cosine rule, and we can interchange the pronumerals to get the equations below.

> ### The cosine rule to determine a side length
>
> **Use one of these rules to determine a side length.**
>
> $$a^2 = b^2 + c^2 - 2bc\cos(A)$$
> $$b^2 = a^2 + c^2 - 2ac\cos(B)$$
> $$c^2 = a^2 + b^2 - 2ab\cos(C)$$

We can apply the cosine rule to determine all of the angles and side lengths of a triangle if we are given either:
- 3 side lengths or
- 2 side lengths and the included angle.

The cosine rule can also be transposed to give the following equations.

> ### The cosine rule to determine an angle
>
> **Use one of these rules to determine the size of an angle.**
>
> $$\cos(A) = \frac{b^2 + c^2 - a^2}{2bc}$$
> $$\cos(B) = \frac{a^2 + c^2 - b^2}{2ac}$$
> $$\cos(C) = \frac{a^2 + b^2 - c^2}{2ab}$$

WORKED EXAMPLE 14 Using the cosine rule to determine a side length

tlvd-5052

Determine the value of the unknown length x correct to 2 decimal places.

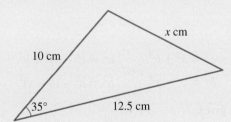

THINK

1. Draw the non-right-angled triangle, labelling it with the given information.
   Angle A is opposite to side a.
   If three side lengths and one angle are given, always label the angle as A and the opposite side as a.

WRITE

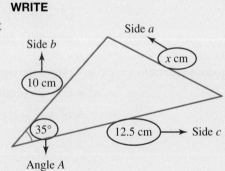

2. Substitute the known values into the cosine rule.

$$a^2 = b^2 + c^2 - 2bc\cos(A)$$

$$x^2 = 10^2 + 12.5^2 - 2 \times 10 \times 12.5\cos(35°)$$

3. Solve for x.
   Make sure your calculator is in degree mode.

$$x^2 = 51.462$$
$$x = \sqrt{51.462}$$
$$\approx 7.17$$

4. Write the answer in a sentence.

The unknown length x is 7.17 cm.

TI	THINK	DISPLAY/WRITE	CASIO	THINK	DISPLAY/WRITE

TI | THINK

1. Ensure your calculator is in DEGREE mode.
   On the Calculator page, press MENU, then select:
   3: Algebra
   1: Solve
   Complete the entry line as:
   solve($x^2 = 10^2 + 12.5^2 - 2 \times 10 \times 12.5\cos(35)$, x)
   then press ENTER.

DISPLAY/WRITE

2. Reject the negative solution as lengths can only be positive.

$x = 7.17$ cm (to 2 decimal places)

CASIO | THINK

1. Ensure your calculator is in DEGREE mode.
   On the Main screen, complete the entry line as:
   solve($x^2 = 10^2 + 12.5^2 - 2 \times 10 \times 12.5\cos(35)$, x)
   then press EXE.

DISPLAY/WRITE

2. Reject the negative solution as lengths can only be positive.

$x = 7.17$ cm (to 2 decimal places)

WORKED EXAMPLE 15 Using the cosine rule to determine an angle

A non-right-angled triangle ABC has values $a = 7$, $b = 12$ and $c = 16$. Determine the magnitude of angle A correct to 2 decimal places.

THINK

1. Draw the non-right-angled triangle, labelling with the given information.

WRITE

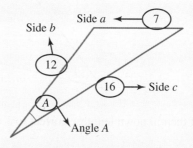

2. Substitute the known values into the cosine rule.

$$a^2 = b^2 + c^2 - 2bc\cos(A)$$

$$7^2 = 12^2 + 16^2 - 2 \times 12 \times 16\cos(A)$$

3. Rearrange the equation to make $\cos(A)$ the subject and solve.
   Make sure your calculator is in degree mode.

$$\cos(A) = \frac{12^2 + 16^2 - 7^2}{2 \times 12 \times 16}$$

$$A = \cos^{-1}\left(\frac{12^2 + 16^2 - 7^2}{2 \times 12 \times 16}\right)$$

$$\approx 23.93°$$

4. Write the answer in a sentence.

The magnitude of angle A is 23.93°.

Note: In the example above, it would have been quicker to substitute the known values directly into the transposed cosine rule for cos(A).

on Resources

Interactivity   The cosine rule (int-6276)

12.5 Exercise

Students, these questions are even better in jacPLUS

Receive immediate feedback and access sample responses

Access additional questions

Track your results and progress

Find all this and MORE in jacPLUS

1. **WE14** Calculate the value of the unknown length x correct to 2 decimal places.

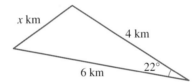

2. Calculate the value of the unknown length x correct to 2 decimal places.

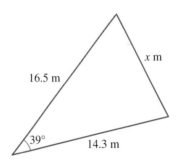

3. Calculate the value of the unknown length x correct to 1 decimal place.

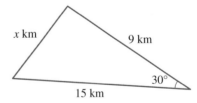

4. Calculate the value of the unknown length x correct to 2 decimal places.

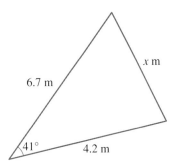

5. **WE15** A non-right-angled triangle ABC has values $a = 8$, $b = 13$ and $c = 17$. Determine the magnitude of angle A correct to 2 decimal places.

6. A non-right-angled triangle ABC has values $a = 11$, $b = 9$ and $c = 5$. Determine the magnitude of angle A correct to 2 decimal places.

7. For triangle ABC, determine the magnitude of angle A correct to 2 decimal places, given $a = 5$, $b = 7$ and $c = 4$.

8. For triangle ABC with $a = 12$, $B = 57°$ and $c = 8$, calculate the side length b correct to 2 decimal places.

9. Determine the largest angle, correct to 2 decimal places, between any two legs of the following sailing course.

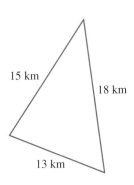

10. A triangular paddock has sides of length 40 m, 50 m and 60 m. Determine the magnitude of the largest angle between the sides, correct to 2 decimal places.

11. A triangle has side lengths of 5 cm, 7 cm and 9 cm. Determine the size of the smallest angle correct to 2 decimal places.

12. ABCD is a parallelogram. Calculate the length of the diagonal AC correct to 2 decimal places.

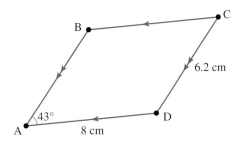

13. An orienteering course is shown in the following diagram. Calculate the total distance of the course correct to 2 decimal places.

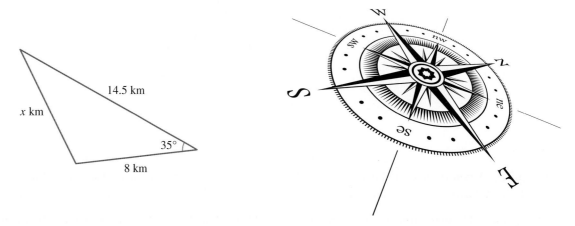

14. Britney is mapping out a new running path around her local park. She is going to run west for 2.1 km before turning 105° to the right and running another 3.3 km. From there, she will run in a straight line back to her starting position.
Determine how far will Britney run in total. Give your answer correct to the nearest metre.

15. A cruise boat is travelling to two destinations. To get to the first destination, it travels for 4.5 hours at a speed of 48 km/h. From there, it takes a 98° turn to the left and travels for 6 hours at a speed of 54 km/h to reach the second destination. The boat then travels directly back to the start of its journey.
Determine how long this leg of the journey will take if the boat is travelling at 50 km/h. Give your answer correct to the
nearest minute.

16. Two air traffic control towers are 180 km apart. At the same time, they both detect a plane, P. The plane is 100 km from Tower A at the bearing shown in the diagram.

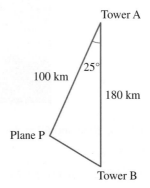

Calculate the distance of the plane from Tower B correct to 2 decimal places.

Question 1 (1 mark)

Source: VCE 2013, Further Mathematics Exam 1, Section B, Module 2, Q2; © VCAA.

MC The distances from a kiosk to points A and B on opposite sides of a pond are found to be 12.6 m and 19.2 m respectively.

The angle between the lines joining these points to the kiosk is $63°$.

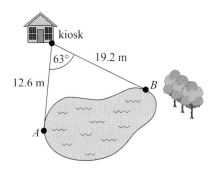

The distance, in m, across the pond between points A and B can be found by evaluating

A. $\dfrac{1}{2} \times 12.6 \times 19.2 \times \sin(63°)$

B. $\dfrac{19.2 \times \sin(63°)}{12.6}$

C. $\sqrt{12.6^2 + 19.2^2}$

D. $\sqrt{12.6^2 + 19.2^2 - 2 \times 12.6 \times 19.2 \times \cos(63°)}$

E. $\sqrt{s(s-12.6)(s-19.2)(s-63)}$, where $s = \dfrac{1}{2}(12.6 + 10.2 + 63)$

Question 2 (1 mark)

MC The towns Alferton (A) and Betaton (B) are 80 km apart. Gamerton (G) is 60 km from Alferton, on a bearing of $050°$. Betaton is due east of Alferton.

The distance between Gamerton and Betaton is

A. 131.73 km **B.** 79.52 km **C.** 61.88 km **D.** 51.44 km **E.** 51.42 km

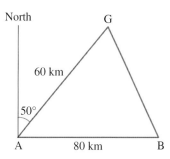

Question 3 (1 mark)

MC The smallest angle of the triangle is θ.

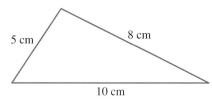

Select the equation that would give the correct value of θ.

A. $\cos(\theta) = \dfrac{5^2 + 10^2 - 8^2}{2 \times 5 \times 10}$

B. $\cos(\theta) = \dfrac{8^2 + 10^2 - 5^2}{2 \times 8 \times 10}$

C. $\cos(\theta) = \dfrac{5^2 + 8^2 - 10^2}{2 \times 5 \times 10}$

D. $\cos(\theta) = \dfrac{8^2 + 10^2 - 5^2}{2 \times 10 \times 5}$

E. $\cos(\theta) = \dfrac{8^2 + 5^2 - 10^2}{2 \times 5 \times 10}$

More exam questions are available online.

12.6 Area of triangles

12.6.1 Area of triangles

You should be familiar with calculating the area of a triangle using the rule: **area** $= \dfrac{1}{2} \boldsymbol{bh}$, where b is the base length and h is the perpendicular height. However, for many triangles we are not given the perpendicular height, so this rule cannot be directly used.

Take the triangle ABC as shown.

If h is the perpendicular height of this triangle, then we can calculate the value of h using the sine ratio:

$$\sin(A) = \frac{h}{c}$$

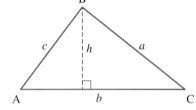

Transposing this equation gives $h = c \sin(A)$, which we can substitute into the rule for the area of the triangle to give the following.

Area of a triangle

$$\textbf{Area} = \frac{1}{2} bc \sin(A)$$

Note: We can label any sides of the triangle a, b and c, and this formula can be used as long as we have the length of two sides of a triangle and know the value of the included angle.

---

tlvd-5053

WORKED EXAMPLE 16 Calculating the area of a triangle using two sides and the included angle

Calculate the areas of the following triangles. Give both answers correct to 2 decimal places.

a.

7 cm

63°

5 cm

b. A triangle with sides of length 8 cm and 7 cm, and an included angle of 55°

THINK	**WRITE**

a. 1. Label the vertices of the triangle.

a.

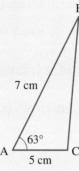

2. Write down the known information.

$b = 5$ cm
$c = 7$ cm
$A = 63°$

3. Substitute the known values into the formula to calculate the area of the triangle.

$$\text{Area} = \frac{1}{2}bc\sin(A)$$
$$= \frac{1}{2} \times 5 \times 7 \times \sin(63°)$$
$$= 15.592\ldots$$
$$= 15.59 \text{ (to 2 decimal places)}$$

4. Write the answer, remembering to include the units.

The area of the triangle is 15.59 cm^2 correct to 2 decimal places.

b. 1. Draw a diagram to represent the triangle.

b.

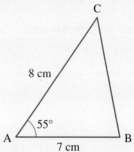

2. Write down the known information.

$b = 8$ cm
$c = 7$ cm
$A = 55°$

3. Substitute the known values into the formula to calculate the area of the triangle.

$$\text{Area} = \frac{1}{2}bc\sin(A)$$
$$= \frac{1}{2} \times 8 \times 7 \times \sin(55°)$$
$$= 22.936\ldots$$
$$= 22.94 \text{ (to 2 decimal places)}$$

4. Write the answer, remembering to include the units.

The area of the triangle is 22.94 cm^2 correct to 2 decimal places.

12.6.2 Heron's formula

As shown in Topic 11, we can also use Heron's formula to calculate the area of a triangle.

To use Heron's formula, we need to know the length of all three sides of the triangle.

Heron's formula for the area of a triangle

$$\text{Area} = \sqrt{s\,(s-a)\,(s-b)\,(s-c)}$$

where $s = \dfrac{a+b+c}{2}$

and a, b, c are the side lengths of the triangle.

WORKED EXAMPLE 17 Calculating the area of a triangle using Heron's formula

Calculate the area of a triangle with sides of 4 cm, 7 cm and 9 cm, giving your answer correct to 2 decimal places.

THINK	WRITE
1. Write down the known information.	$a = 4$ cm $b = 7$ cm $c = 9$ cm
2. Calculate the value of s (the semi-perimeter).	$s = \dfrac{a+b+c}{2}$ $= \dfrac{4+7+9}{2}$ $= \dfrac{20}{2}$ $= 10$
3. Substitute the values into Heron's formula to calculate the area.	$\text{Area} = \sqrt{s(s-a)(s-b)(s-c)}$ $= \sqrt{10(10-4)(10-7)(10-9)}$ $= \sqrt{10 \times 6 \times 3 \times 1}$ $= \sqrt{180}$ $= 13.416\ldots$ ≈ 13.42
4. Write the answer, remembering to include the units.	The area of the triangle is 13.42 cm^2 correct to 2 decimal places.

Determining which formula to use

In some situations you may have to perform some calculations to determine either a side length or angle size before calculating the area. This may involve using the sine or cosine rule. The following table should help if you are unsure what to do.

Given	What to do	Example
Base length and perpendicular height	Use area $= \frac{1}{2}bh$.	8 cm 13 cm
Two side lengths and the included angle	Use area $= \frac{1}{2}bc\sin(A)$.	5 cm 104° 9 cm
Three side lengths	Use Heron's formula: area $= \sqrt{s(s-a)(s-b)(s-c)}$, where $s = \dfrac{a+b+c}{2}$	22 mm 12 mm 19 mm
Two angles and one side length	Use the sine rule to determine a second side length, and then use area $= \frac{1}{2}bc\sin(A)$. *Note:* The third angle may have to be calculated.	75° 29° 9 cm
Two side lengths and an angle opposite one of these lengths	Use the sine rule to calculate the angle opposite the other length, then determine the final angle before using area $= \frac{1}{2}bc\sin(A)$. *Note:* Check if the ambiguous case is applicable.	14 cm 72° 12 cm

on Resources

Interactivity Area of triangles (int-6483)

12.6 Exercise

1. **WE16** Calculate the area of the following triangle correct to 2 decimal places.

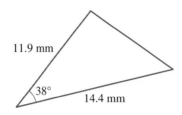

11.9 mm

38°

14.4 mm

2. Calculate the area of a triangle with sides of length 14.3 mm and 6.5 mm, and an inclusive angle of 32°. Give your answer correct to 2 decimal places.

3. A triangle has one side length of 8 cm and an adjacent angle of 45.5°. If the area of the triangle is 18.54 cm², calculate the length of the other side that encloses the 45.5° angle, correct to 2 decimal places.

4. The smallest two sides of a triangle are 10.2 cm and 16.2 cm respectively, and the largest angle of the same triangle is 104.5°. Calculate the area of the triangle correct to 2 decimal places.

5. **WE17** Calculate the area of a triangle with sides of 11 cm, 12 cm and 13 cm, giving your answer correct to 2 decimal places.

6. Calculate the area of a triangle with sides of 22.2 mm, 13.5 mm and 10.1 mm, giving your answer correct to 2 decimal places.

7. Calculate the areas of the following triangles, correct to 2 decimal places where appropriate.

   a.

   11.7 cm

   19.4 cm

   b.

   9.1 mm

   10.7 mm

   12.4 mm

   c.

   31.2 cm

   38°

   22.5 cm

   d.

   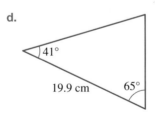

   41°

   19.9 cm

   65°

8. Calculate the areas of the following triangles, correct to 2 decimal places where appropriate.

   a. Triangle ABC, given $a = 12$ cm, $b = 15$ cm, $c = 20$ cm
   b. Triangle ABC, given $a = 10.5$ mm, $b = 11.2$ mm and $C = 40°$
   c. Triangle DEF, given $d = 19.8$ cm, $e = 25.6$ cm and $D = 33°$
   d. Triangle PQR, given $p = 45.9$ cm, $Q = 45.5°$ and $R = 67.2°$

9. A triangular field is defined by three trees, each of which sits in one of the corners of the field, as shown in the following diagram.

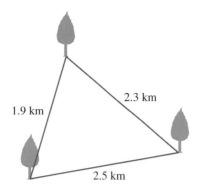

Calculate the area of the field in km², correct to 3 decimal places.

10. A triangle ABC has values $a = 11$ cm, $b = 14$ cm and $A = 31.3°$. Answer the following correct to 2 decimal places.
   a. Calculate the sizes of the other two angles of the triangle.
   b. Calculate the other side length of the triangle.
   c. Calculate the area of the triangle.

11. A triangle has side lengths of $3x$, $4x$ and $5x$. If the area of the triangle is 121.5 cm², use any appropriate method to determine the value of the pronumeral x.

12. A triangular-shaped piece of jewellery has two side lengths of 8 cm and an area of 31.98 cm². Use trial and error to calculate the length of the third side correct to 1 decimal place.

13. A triangle has two sides of length 9.5 cm and 13.5 cm, and one angle of 40.2°. Calculate all three possible areas of the triangle correct to 2 decimal places.

14. A BMX racing track encloses two triangular sections, as shown in the following diagram. Calculate the total area that the race track encloses to the nearest m².

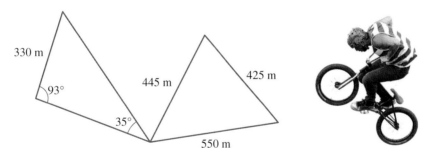

15. Calculate the area of the following shape correct to 2 decimal places.

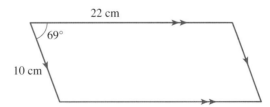

16. A dry field is in the shape of a quadrilateral, as shown in the following diagram.

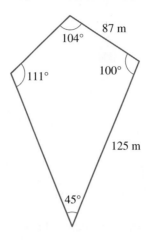

Determine the amount of grass seed needed to cover the field in 1 mm of grass seed. Give your answer correct to 2 decimal places.

12.6 Exam questions

Question 1 (1 mark)

MC A triangle has side lengths of 24 cm, 26 cm and 20 cm. The area of the triangle is closest to
A. 322 cm²
B. 2791 cm²
C. 51 975 cm²
D. 228 cm²
E. 53 cm²

Question 2 (1 mark)

MC The area of the triangle shown is
A. 71.11 cm²
B. 53.13 cm²
C. 42.03 cm²
D. 11.93 cm²
E. 142.22 cm²

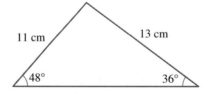

Question 3 (1 mark)

MC The area of the triangular sheet of cardboard shown in the diagram is
A. 20.88 sin(50°)
B. 20.88 sin(30°)
C. 20.88 sin(100°)
D. 10.44 sin(50°)
E. 10.44 sin(100°)

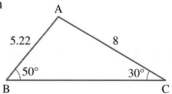

More exam questions are available online.

12.7 Review

📄 12.7.1 Summary

doc-37622

12.7 Exercise

Multiple choice

1. **MC** The length of side x in the triangle shown can be calculated using:

   A. $37\cos(25°)$

   B. $37\sin(25°)$

   C. $\dfrac{37}{\cos(25°)}$

   D. $\dfrac{37}{\sin(25°)}$

   E. $\dfrac{\cos(25°)}{37}$

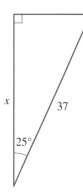

2. **MC** The length of side x in the triangle shown, correct to the nearest metre, is:

   A. 9

   B. 14

   C. 19

   D. 23

   E. 27

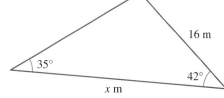

3. **MC** The magnitude of angle A in the triangle shown, correct to the nearest degree, is:

   A. 22°

   B. 30°

   C. 34°

   D. 56°

   E. 57°

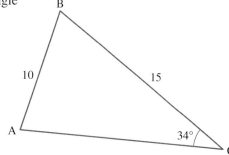

4. **MC** The largest angle in the triangle shown can be calculated using:

A. $\cos^{-1}\left(\dfrac{8^2 + 3^2 - 5^2}{2 \times 3 \times 8}\right)$

B. $\cos^{-1}\left(\dfrac{8^2 + 5^2 - 3^2}{2 \times 5 \times 8}\right)$

C. $\cos^{-1}\left(\dfrac{5^2 + 3^2 - 8^2}{2 \times 3 \times 5}\right)$

D. $\cos^{-1}\left(\dfrac{5^2 + 3^2 - 8^2}{2 \times 3 \times 8}\right)$

E. $\cos^{-1}\left(\dfrac{8^2 - 3^2 - 5^2}{2 \times 3 \times 5}\right)$

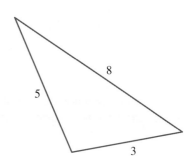

5. **MC** The acute and obtuse angles that have a sine of approximately 0.529 92, correct to the nearest degree, are respectively:

A. 31° and 149°   B. 32° and 58°   C. 31° and 59°   D. 32° and 148°   E. 58° and 148°

6. **MC** Using Heron's formula, the area of a triangle with sides 4.2 cm, 5.1 cm and 9 cm is:

A. 5.2 cm²   B. 9.2 cm²   C. 13.7 cm²   D. 18.3 cm²   E. 28.2 cm²

7. **MC** The locations of Jo's, Nelson's and Sammy's homes are shown in the diagram. Jo's home is due north of Nelson's home.

The true bearings of Jo's and Nelson's homes from Sammy's home are respectively:

A. 030° and 090°
B. 060° and 090°
C. 030° and 180°
D. 060° and 180°
E. 210° and 270°

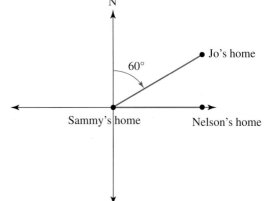

8. **MC** A boy is standing 150 m away from the base of a building. His eye level is 1.65 m above the ground. He observes a hot air balloon hovering above the building at an angle of elevation of 30°.
If the building is 20 m high, the distance the hot air balloon is above the top of the building is closest to:

A. 64 m   B. 65 m   C. 66 m   D. 67 m   E. 68 m

9. **MC** The area of the triangle shown, correct to 2 decimal places, is:

A. 2.20 cm²
B. 2.54 cm²
C. 3.36 cm²
D. 4.41 cm²
E. 5.07 cm²

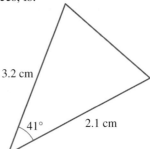

10. **MC** A unit of cadets walked from their camp for 7.5 km on a true bearing of 064°. They then travelled on a truc bearing of 148° until they came to a signpost that indicated they were 14 km in a straight line from their camp.

The true bearing from the signpost back to their camp is closest to:

A. 032°   **B.** 064°   **C.** 096°   **D.** 296°   **E.** 328°

Short answer

11. a. Calculate the value of the unknown length x correct to 2 decimal places.

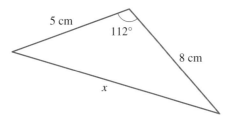

     b. Calculate the value of x correct to the nearest degree.

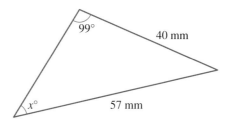

12. In the triangle ABC, $a = 18.5$ m, $c = 12.6$ m and $C = 31.35°$.

     a. Draw the two possible triangles with this information.

     b. Determine, correct to 2 decimal places, the two possible values of A and hence the two possible values of B.

13. a. Calculate the area of the triangle shown correct to 2 decimal places.

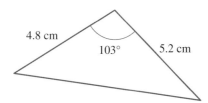

     b. Calculate the area of the triangle shown correct to 2 decimal places.

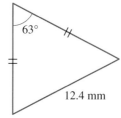

14. ABCDEF is a regular hexagon with the point B being due north of A. Determine the true bearing of the point:

a. C from B
b. D from C
c. F from E
d. E from B.

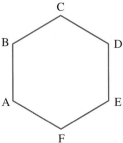

15. Three immunity idols are hidden on an island for a TV reality game. Idol B is 450 m on a bearing of 072° from idol A. Immunity idol C is 885 m on a true bearing of 105° from idol A.

a. Calculate the distance between immunity idols B and C.
b. Calculate the triangular area between immunity idols A, B and C that contestants need to search to find all three idols.

16. a. Calculate the value of the unknown length x correct to 2 decimal places.

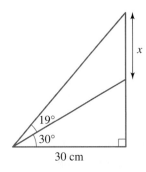

b. Calculate the values of x and y, correct to 2 decimal places where appropriate.

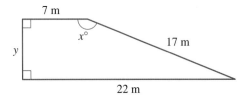

Extended response

17. Three treasure chests are buried on an island. Treasure chest B is 412 m on a true bearing of 073° from treasure chest A. Treasure chest C is 805 m on a true bearing of 108° from treasure chest A.

a. Draw a diagram to represent the information.
b. Show that angle A is 35°.
c. Calculate the distance between treasure chests B and C correct to 1 decimal place.
d. Calculate the triangular area between treasure chests A, B and C correct to 1 decimal place.
e. A treasure hunter misreads the information as 'Treasure chest B is 412 m on a true bearing of 078° from treasure chest A' rather than 'Treasure chest B is 412 m on a true bearing of 073° from treasure chest A.'

   i. Draw a diagram to represent the misread information.
   ii. Determine how far the treasure hunter has travelled from the actual position of treasure chest B to his incorrect location of treasure chest B, correct to 1 decimal place.
   iii. Calculate the true bearing from his incorrect location of treasure chest B to the actual location of treasure chest B.

18. A dog kennel is placed in the corner of a triangular garden at point C. The kennel is positioned 30.5 m at an angle of 32.8° from one corner of the backyard fence (A) and 20.8 m from the other corner of the backyard fence (B).

a. Draw a diagram to represent the information.
b. Calculate, correct to 1 decimal place:

   i. the shortest distance between the kennel and the backyard fence
   ii. the length of the backyard fence between points A and B.

c. Using Heron's formula, calculate the triangular area between the kennel and the two corners of the backyard fence, correct to 1 decimal place.

19. A stained glass window frame consisting of five triangular sections is to be made in the shape of a regular pentagon with a side length of 30 cm.

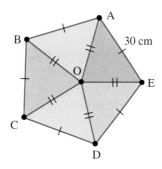

a. Show that $\angle AOB = 72°$.
b. Calculate $\angle ABO$.
c. Use the sine rule to calculate the length of OB, correct to 1 decimal place.
d. Calculate the total length of frame required to construct the window, correct to 1 decimal place.
e. Three of the triangular panels must have coloured glass. Calculate the total area of coloured glass required correct to the nearest cm^2.

20. A triangular flag ABC has a printed design with a circle touching the sides of a flag and a 1-metre vertical line as shown in the diagram.

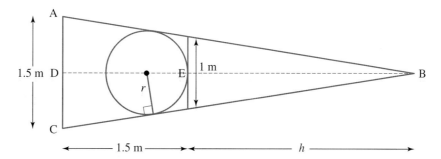

a. Calculate h, the horizontal distance from the vertex point of the flag B to the vertical line at point E, correct to 1 decimal place.
b. Calculate $\angle ACB$ correct to 2 decimal places.
c. Calculate r, the radius of the circle, correct to 2 decimal places.
d. The printed circle design in the flag is to be coloured yellow. Calculate the area of the circle correct to 2 decimal places.
e. Calculate the total area of the flag correct to 2 decimal places.

Question 1 (1 mark)
Source: VCE 2021, Further Mathematics Exam 1, Section B, Module 3, Q8; © VCAA.

MC Rod and Lucia went on a bushwalk.

They walked 1400 m from the car park to reach a lookout that was directly east of the car park.

From the lookout, Rod returned to the car park via a cafe and Lucia returned to the car park via a swimming hole.
- The bearing of the swimming hole from the lookout is 290°.
- The bearing of the cafe from the lookout is 240°.
- The swimming hole is 950 m from the car park.
- The cafe is 700 m from the car park.
- The swimming hole is closer to the lookout than it is to the car park.

In relation to the total distance each of them individually walked from the lookout back to the car park, which one of the following statements is true?
- **A.** Rod and Lucia walked the same distance.
- **B.** Rod walked 467 m further than Lucia, to the nearest metre.
- **C.** Rod walked 717 m further than Lucia, to the nearest metre.
- **D.** Lucia walked 924 m further than Rod, to the nearest metre.
- **E.** Lucia walked 1174 m further than Rod, to the nearest metre.

Question 2 (3 marks)
Source: VCE 2016, Further Mathematics Exam 2, Module 3, Q4; © VCAA.

During a game of golf, Salena hits a ball twice, from P to Q and then from Q to R.

The path of the ball after each hit is shown in the diagram below.

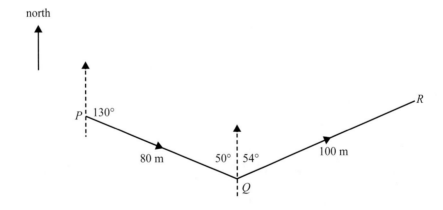

After Salena's first hit, the ball travelled 80 m on a bearing of 130° from point P to point Q.

After Salena's second hit, the ball travelled 100 m on a bearing of 054° from point Q to point R.
- **a.** Another ball is hit and travels directly from P to R.
  Use the cosine rule to find the distance travelled by this ball.   **(2 marks)**
  Round your answer to the nearest metre.
- **b.** What is the bearing of R from P?
  Round your answer to the nearest degree.   **(1 mark)**

Question 3 (2 marks)

Source: VCE 2016, Further Mathematics Exam 2, Module 3, Q2; © VCAA.

Salena practises golf at a driving range by hitting golf balls from point *T*.

The first ball that Salena hits travels directly north, landing at point *A*.

The second ball that Salena hits travels 50 m on a bearing of 030°, landing at point *B*.

The diagram below shows the positions of the two balls after they have landed.

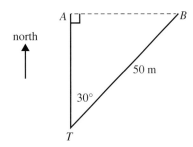

a. How far apart, in metres, are the two golf balls? **(1 mark)**

b. A fence is positioned at the end of the driving range.
The fence is 16.8 m high and is 200 m from the point *T* .

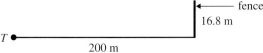

What is the angle of elevation from *T* to the top of the fence?
Round your answer to the nearest degree. **(1 mark)**

Question 4 (1 mark)

Source: VCE 2014, Further Mathematics Exam 1, Section B, Module 2, Q7; © VCAA.

MC A cross-country race is run on a triangular course. The points *A*, *B* and *C* mark the corners of the course, as shown below.

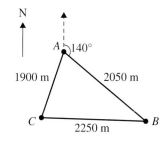

The distance from *A* to *B* is 2050 m.

The distance from *B* to *C* is 2250 m.

The distance from *A* to *C* is 1900 m.

The bearing of *B* from *A* is 140°.

The area within the triangular course *ABC*, in square metres, can be calculated by evaluating

A. $\sqrt{3100 \times 1200 \times 1050 \times 850}$

B. $\sqrt{3100 \times 2250 \times 2050 \times 1900}$

C. $\sqrt{6200 \times 4300 \times 4150 \times 3950}$

D. $\frac{1}{2} \times 2050 \times 2250 \times \sin(140°)$

E. $\frac{1}{2} \times 2050 \times 2250 \times \sin(40°)$

Source: VCE 2014, Further Mathematics Exam 2, Module 2, Q2; © VCAA.

The chicken coop has two spaces, one for nesting and one for eating.

The nesting and eating spaces are separated by a wall along the line *AX*, as shown in the diagrams below.

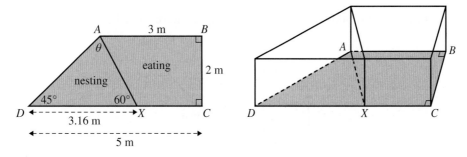

$DX = 3.16\,\text{m}$, $\angle ADX = 45°$ and $\angle AXD = 60°$.

a. Write down a calculation to show that the value of θ is 75°. **(1 mark)**

b. The sine rule can be used to calculate the length of the wall *AX*.
Fill in the missing numbers below. **(1 mark)**

$$\frac{AX}{\sin \boxed{}^{\circ}} = \frac{\boxed{}}{\sin \boxed{}^{\circ}}$$

c. What is the length of *AX*?
Write your answer in metres, correct to two decimal places. **(1 mark)**

d. Calculate the area of the floor of the nesting space, *ADX*.
Write your answer in square metres, correct to one decimal place. **(1 mark)**

e. The height of the chicken coop is 1.8 m.
Wire mesh will cover the roof of the eating space.
The area of the walls along the lines *AB*, *BC* and *CX* will also be covered with wire mesh.
What total area, in square metres, will be covered by wire mesh?
Write your answer, correct to the nearest square metre. **(2 marks)**

More exam questions are available online.

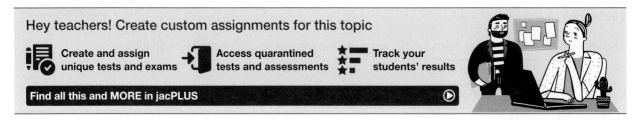

Answers

Topic 12 Applications of trigonometry

12.2 Trigonometric ratios

12.2 Exercise

1. $x = 2.16$ cm
2. $x = 10.6$ mm
3. $\theta = 46.90°$
4. $\theta = 45.48°$
5. $y = 1.99$ cm
6. $y = 2.63$ cm
7. $\theta = 73.97°$
8. $\theta = 35.58°$
9. $x = 13.15$ cm
10. $y = 0.75$ cm
11. $\theta = 53.20°$
12. $\theta = 36.38°$
13. a. 0.82   b. −0.88
14. a. 0.99   b. −0.84
15. $x = 5.16$
16. $\theta = 49.32°$
17. $\theta = 10.29°$, $\alpha = 79.71°$
18. 9.63 m
19. a. 7.4 km   b. 27.2 km
20. 2870 m
21. 20.56°
22. 2.02 m
23. a. 47.16°   b. 1.83 m
24. a. 4.1 m   b. 5.03 m
25. 4.28 m
26. 100.94 m

12.2 Exam questions

Note: Mark allocations are available with the fully worked solutions online.
1. B
2. D
3. E

12.3 Angles of elevation and depression, and bearings

12.3 Exercise

1. 36.81 m
2. 47.13 m
3. 61.54°
4. 36.25°
5. 2.73 m
6. 999.85 m
7. a. 049°   b. 342°   c. 231°   d. 112°

8. a. 185°   b. 297°   c. 079°   d. 124°
9. a. 110°   b. 290°
10. a. 237°   b. 057°
11. a. 050°   b. 127°   c. 259°
12. a. 32.63 km   b. 15.21 km
13. 323.1°
14. 862.83 m
15. 315 m
16. 11.13 m
17. 52°
18. 88.28°
19. 1357 m
20. a. 83.59°   b. 82.35°   c. 79.93°
    d. 74.97°   e. 60.67°

12.3 Exam questions

Note: Mark allocations are available with the fully worked solutions online.
1. A
2. D
3. a. 3 km   b. 1.74 km   c. 1.26 km

12.4 The sine rule

12.4 Exercise

1. $x = 10.36$ cm
2. $x = 10.38$ cm
3. $x = 9.05$ cm
4. $x = 17$ cm
5. $B = 28.3°$
6. $B = 27.1°$
7. $\theta = 47.2°$
8. $A = 65.41°$ or $114.59°$
9. $A = 52.64°$ or $127.36°$
10. $B = 46.97°$ or $133.03°$
11. $b = 12.24$, $c = 13.43$
12. 35° and 145°
13. 20.71 m
14. a. 23.18 m   b. 10.66 m
15. $x = 142.4°$, $y = 37.6°$
16. a.

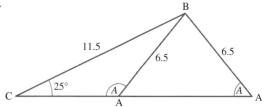

b. $A = 48.39°$ or $131.61°$; $B = 106.61°$ or $23.39°$
17. 9.80 m
18. 5.91 km

12.4 Exam questions

Note: Mark allocations are available with the fully worked solutions online.

1. E
2. E
3. E

12.5 The cosine rule

12.5 Exercise

1. $x = 2.74$ km
2. $x = 10.49$ m
3. $x = 8.5$ km
4. $x = 4.48$ m
5. $A = 26.95°$
6. $A = 99.59°$
7. $A = 44.42°$
8. $b = 10.17$
9. $79.66°$
10. $82.82°$
11. $33.56°$
12. 13.23 cm
13. 31.68 km
14. 8822 m
15. 7 hours, 16 minutes
16. 98.86 km

12.5 Exam questions

Note: Mark allocations are available with the fully worked solutions online.

1. D
2. D
3. B

12.6 Area of triangles

12.6 Exercise

1. 52.75 mm^2
2. 24.63 mm^2
3. 6.50 cm
4. 79.99 cm^2
5. 61.48 cm^2
6. 43.92 mm^2
7. a. 113.49 cm^2 b. 47.45 mm^2
   c. 216.10 cm^2 d. 122.48 cm^2
8. a. 89.67 cm^2 b. 37.80 mm^2
   c. 247.68 cm^2 d. 750.79 cm^2
9. 2.082 km^2
10. a. $B = 41.39°$, $C = 107.31°$
    b. $c = 20.21$ cm
    c. 73.51 cm^2
11. $x = 4.5$ cm
12. 11.1 cm
13. 41.39 cm^2, 61.41 cm^2 and 59.12 cm^2

14. $167\ 330$ m^2
15. 205.39 cm^2
16. 8.14 m^3

12.6 Exam questions

Note: Mark allocations are available with the fully worked solutions online.

1. D
2. A
3. C

12.7 Review

12.7 Exercise

Multiple choice

1. A   2. E   3. E   4. C   5. D
6. A   7. B   8. E   9. A   10. D

Short answer

11. a. 10.91 cm   b. $44°$

12. a.

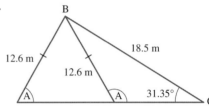

b. $A = 49.81°$, $B = 98.84°$ or $A = 130.19°$, $B = 18.46°$

13. a. 12.16 cm^2 b. 62.73 mm^2
14. a. $060°$ b. $120°$
    c. $240°$ d. $120°$
15. a. 563.67 m   b. $108\ 451$ m^2
16. a. 17.19 cm
    b. $x = 151.93°$, $y = 8$ m

Extended response

17. a.

b. $A = 180 - (73 + 72)$
   $= 35°$
c. 523.8 m
d. $95\ 116.2$ m^2
e. i.

ii. 35.9 m
iii. $345.6°$T

18. a.

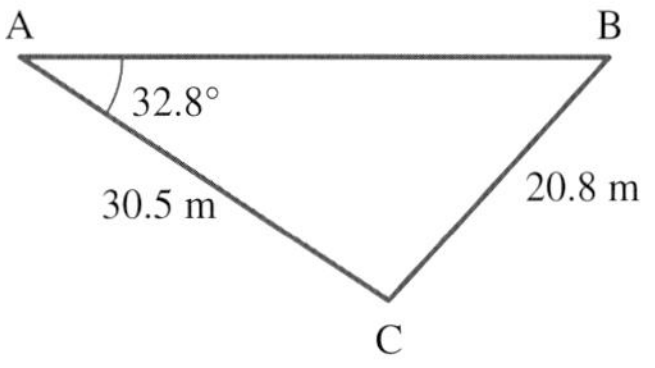

Triangle with vertices A (top left), B (top right), C (bottom). Angle at A is 32.8°. Side from A to C is 30.5 m, side from B to C is 20.8 m.

b. i. 16.5 m **ii.** 38.3 m

c. 316.1 m^2

19. a. $\dfrac{360}{5} = 72°$ **b.** 54° **c.** 25.5 cm

d. 277.5 cm **e.** 928 cm^3

20. a. 3 m **b.** 80.54° **c.** 0.59 m

d. 1.09 m^2 **e.** 3.38 m^2

12.7 Exam questions

Note: Mark allocations are available with the fully worked solutions online.

1. B

2. a. 142 m **b.** 087°

3. a. 25 m **b.** 5°

4. A

5. a. Angles in a triangle add to 180°.

$\theta = 180 - (45 + 60) = 75°$

b. $\dfrac{AX}{\sin\left(45°\right)} = \dfrac{3.16}{\sin\left(75°\right)}$

c. 2.31 m

d. 3.2 m^2

e. 17 m^2

13 Similarity and scale

Fully worked solutions for this topic are available online.

13.1 Overview

13.1.1 Introduction

The construction industry is constantly growing. You only need to look at the cranes on buildings in any major city in the world. Some of these buildings are used as homes; others are hospitals, schools, offices and places of leisure.

To construct buildings and make sure they are functional, designers must apply the principles of mathematics. These principles involve scaled drawings in the form of plans. A plan is a drawing of the object to be built, reduced in size in a way that all the measurements correspond to the actual object.

Architects often use a different set of scales than engineers, surveyors or furniture designers. This depends on the size of what is being designed, as well as the complexity of the design.

Scale is not just used for plans; it can also be used to create a scale model of the design. A scale model is generally a physical representation of an object that maintains accurate relationships between all important aspects of the model. The scale model allows you to see some behaviour of the original object without investigating the original object itself. Scale models are used in many fields including engineering, filmmaking, military, sales and hobby model building. To be considered a true scale model, all important aspects must be accurately modelled: not just the scale of the object, but also the material properties. An example could be an aerospace company wanting to test a new wing design. They could construct a scaled-down model and test it in a wind tunnel under simulated conditions.

One famous model is that of a space shuttle by Dr Maxime Faget from NASA. He needed a model to demonstrate to his colleagues that a space shuttle would be able to glide back to Earth without power — a concept unknown until then.

KEY CONCEPTS

This topic covers the following key concepts from the VCE Mathematics Study Design:
- similar shapes including the conditions for similarity
- similar objects and the application of linear scale factor $k > 0$ to scale lengths, surface areas and volumes with practical applications.

Source: VCE Mathematics Study Design (2023–2027) extracts © VCAA; reproduced by permission.

13.2 Similar objects

13.2.1 Conditions for similarity

Objects are called **similar** when they are exactly the same shape but have different sizes. Objects that are exactly the same size and shape are called **congruent**.

Similarity is an important mathematical concept that is often used for planning purposes in areas such as engineering, architecture and design. Scaled-down versions of much larger objects allow designs to be trialled and tested before their construction.

Two-dimensional objects are similar when their internal angles are the same and their side lengths are **proportional**. This means that the ratios of corresponding side lengths are always equal for similar objects. We use the symbols ~ or ||| to indicate that objects are similar.

tlvd-5054

WORKED EXAMPLE 1 Proving two objects are similar

Show that these two objects are similar.

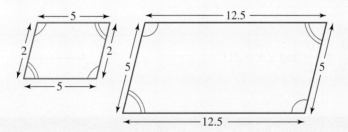

THINK	WRITE
1. Confirm that the internal angles for the objects are the same.	The diagrams indicate that all angles in both objects match.
2. Calculate the ratio of the corresponding side lengths and simplify.	Ratios of corresponding sides: $\dfrac{12.5}{5} = 2.5$ and $\dfrac{5}{2} = 2.5$
3. Write the answer in a sentence.	The two objects are similar as their angles are the same and the ratios of the corresponding side lengths are equal.

13.2.2 Similar triangles

The conditions for similarity apply to all objects, but not all of them need to be known in order to demonstrate similarity in triangles. If pairs of triangles have any of the following conditions in common, they are similar.

1. Angle–angle–angle (AAA)

> **If two different-sized triangles have all three angles identified as being equal, they are similar.**

$$\angle A = \angle D, \ \angle B = \angle E, \ \angle C = \angle F$$

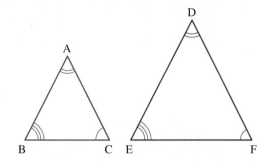

2. Side–side–side (SSS)

> **If two different-sized triangles have all three sides identified as being in proportion, they are similar.**

$$\frac{DE}{AB} = \frac{DF}{AC} = \frac{EF}{BC}$$

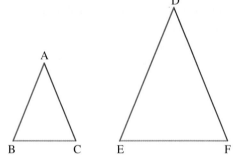

3. Side–angle–side (SAS)

> **If two different-sized triangles have two pairs of sides identified as being in proportion and their included angles are equal, they are similar.**

$$\frac{DE}{AB} = \frac{DF}{AC} \text{ and } \angle A = \angle D$$

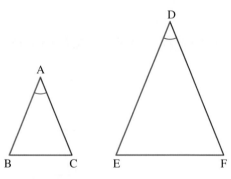

WORKED EXAMPLE 2 Proving triangles are similar

Show that these two triangles are similar.

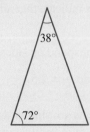
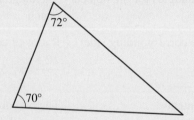

THINK	WRITE
1. Identify all possible angles and side lengths.	The angles in the blue triangle are: 38°, 72° and $180 - (38 + 72) = 70°$. The angles in the red triangle are: 70°, 72° and $180 - (70 + 72) = 38°$.
2. Use one of AAA, SSS or SAS to check for similarity.	The two triangles have all three angles identified as being equal.
3. Write the answer in a sentence.	The two triangles are similar as they satisfy the condition AAA.

WORKED EXAMPLE 3 Determining an unknown length in similar triangles

Calculate the value of d required to make the pairs of objects similar.

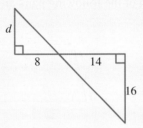

THINK

WRITE

1. Identify the two separate triangles and ensure they are orientated the same way.

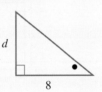

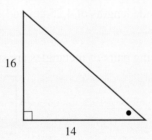

2. The ratio of corresponding side lengths in similar objects must be equal.
Write down a ratio statement comparing corresponding side lengths.

$\dfrac{d}{16} = \dfrac{8}{14}$ or equivalent

3. Solve this equation for d. Use a calculator to assist with the arithmetic if required.

$$\dfrac{d}{16} \times 16 = \dfrac{8}{14} \times 16$$
$$d = \dfrac{64}{7}$$

5. Write the answer.
Note: Because d is a non-terminating decimal, the fractional answer must be given.

For the triangles to be similar, the value of d must be $\dfrac{64}{7}$.

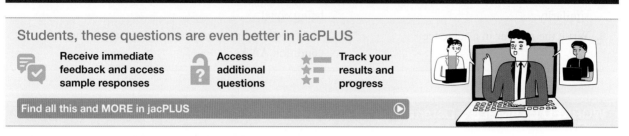
13.2 Exercise

1. **WE1** Show that the two objects in each of the following pairs are similar.

a.

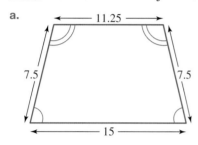
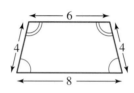

b.

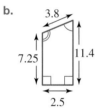
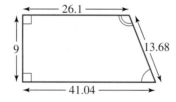

2. Show that a rectangle with side lengths of 4.25 cm and 18.35 cm will be similar to one with side lengths of 106.43 cm and 24.65 cm.

3. Identify which of the following pairs of rectangles are similar.

a.

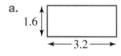
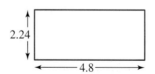

b.

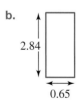
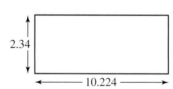

c.

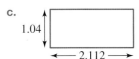

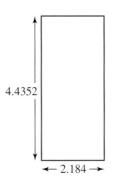

d.

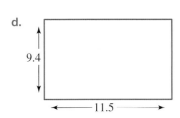

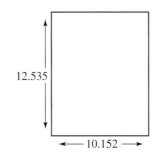

4. Identify which of the following pairs of polygons are similar.

a.

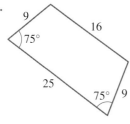

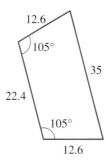

b.

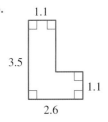

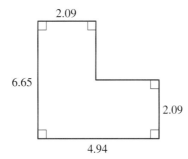

c.

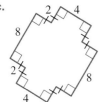

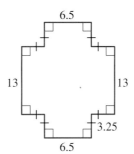

d.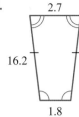

2.7

16.2

1.8

3.24

19.44

2.16

5. **WE2** Show that the two triangles in each of the following pairs are similar.

a.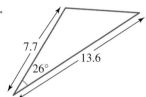

7.7

26°

13.6

78.2

44.275

26°

b.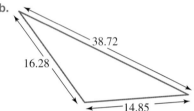

38.72

16.28

14.85

7.4

6.75

17.6

6. Identify which of the following pairs of triangles are similar.

a.

12

14

18

21.6

23.8

32.4

b.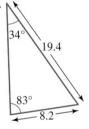

34°

19.4

83°

8.2

22.96

63°

34°

54.32

7. In each of the following groups, state which two triangles are similar.

a.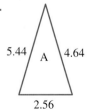

5.44

A

4.64

2.56

8.84

B

6.96

4.16

3.4

C

2.9

1.6

b.

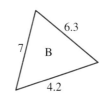

c.

d.

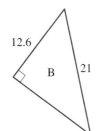

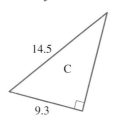

8. Explain why each of the following pairs of objects must be similar.

a.

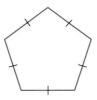

b.

c.

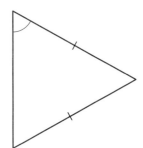

9. Calculate the ratios of the corresponding sides for the following pairs of objects.

a.

13.85

3.4

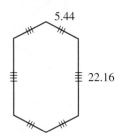
5.44

22.16

b.

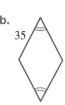
35

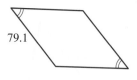
79.1

c.

12   13

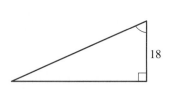
18

10. Evaluate the ratios of the corresponding side lengths in the following pairs of similar objects.

a.

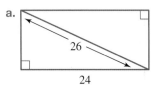
26

24

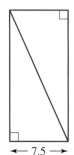
← 7.5 →

b.

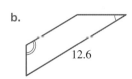
12.6

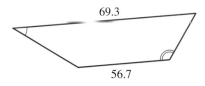
69.3

56.7

11. Evaluate the unknown side lengths in the following pairs of similar objects.

a. 1.2

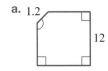
12

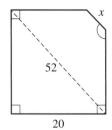
x

52

20

b.

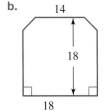
14

18

18

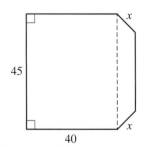
x

45

40

x

12. Verify that the following are similar.

    a. A square of side length 8.2 cm and a square of side length 50.84 cm

    b. An equilateral triangle of side length 12.6 cm and an equilateral triangle of side length 14.34 cm

13. **WE3** Calculate the value of x required to make the pairs of objects similar in each of the following diagrams.

    a.   **b.**

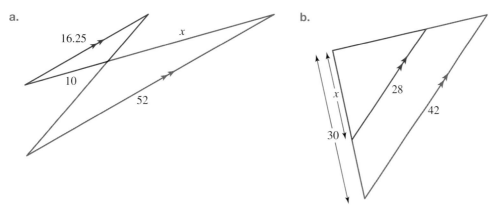

14. Calculate the value of x for the following similar shapes. Give your answer correct to 1 decimal place.

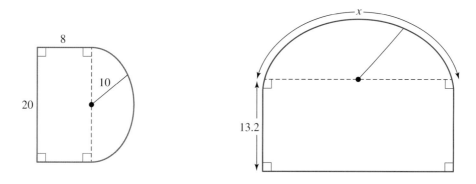

15. Calculate the values of x and y in the diagram.

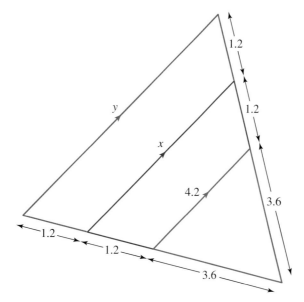

16. For the polygon shown, draw and label a similar polygon where:

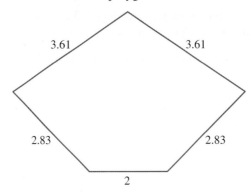

a. the corresponding sides are $\dfrac{4}{3}$ the size of those shown

b. the corresponding sides are $\dfrac{4}{5}$ the size of those shown.

13.2 Exam questions

▶ **Question 1 (1 mark)**

Source: VCE 2019, Further Mathematics Exam 1, Section B, Module 3, Q4; © VCAA.

MC Triangle M, shown below, has side lengths of 3 cm, 4 cm and 5 cm.

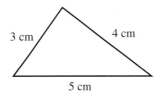

Four other triangles have the following side lengths:
- Triangle N has side lengths of 3 cm, 6 cm and 8 cm.
- Triangle O has side lengths of 4 cm, 8 cm and 12 cm.
- Triangle P has side lengths of 6 cm, 8 cm and 10 cm.
- Triangle Q has side lengths of 9 cm, 12 cm and 15 cm.

The triangles that are similar to triangle M are
- **A.** triangle N and triangle O.
- **B.** triangle N, triangle O and triangle P.
- **C.** triangle O and triangle P.
- **D.** triangle O and triangle Q.
- **E.** triangle P and triangle Q.

▶ **Question 2 (1 mark)**

Source: VCE 2016, Further Mathematics Exam 1, Section B, Module 3, Q2; © VCAA.

MC Triangle ABC is similar to triangle DEF.

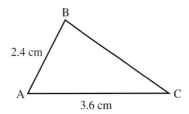 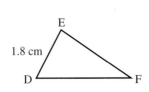

The length of DF, in centimetres, is
- **A.** 0.9
- **B.** 1.2
- **C.** 1.8
- **D.** 2.7
- **E.** 3.6

Source: VCE 2013, Further Mathematics, Exam 1, Section B, Module 2, Q6; © VCAA.

MC In triangle *MNR*, point *P* lies on side *MR* and point *Q* lies on side *NR*.

The lines *PQ* and *MN* are parallel.

The length of *RQ* is 4 cm, the length of *QN* is 6 cm and the length of *PQ* is 5 cm.

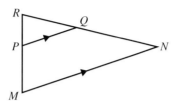

The length of *MN*, in cm, is equal to
   A. 7.5   **B.** 8.3   **C.** 12.0   **D.** 12.5   **E.** 15.0

More exam questions are available online.

13.3 Linear scale factors

LEARNING INTENTION

At the end of this subtopic you should be able to:
- calculate the linear scale factor for a pair of triangles.

13.3.1 Linear scale factors

Consider the pair of similar triangles shown in the diagram.

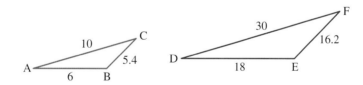

The ratios of the corresponding side lengths are:

$$DE : AB = 18 : 6$$
$$= 3 : 1$$

$$EF : BC = 16.2 : 5.4$$
$$= 3 : 1$$

$$DF : AC = 30 : 10$$
$$= 3 : 1$$

Note: In this topic we will put the image first when calculating ratios of corresponding lengths.

The side lengths of triangle DEF are all three times the lengths of triangle ABC. In this case, we would say that the **linear scale factor** is 3. The linear scale factor for similar objects can be evaluated using the ratio of the corresponding side lengths.

$\triangle ABC \sim \triangle DEF$ Linear scale factor: $\dfrac{DE}{AB} = \dfrac{EF}{BC} = \dfrac{DF}{AC} = k$

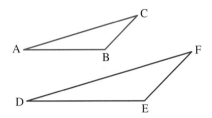

Linear scale factor

$$\text{Linear scale factor}\,(k) = \dfrac{\text{length of image}}{\text{length of object}}$$

A linear scale factor greater than 1 indicates enlargement, and a linear scale factor less than 1 indicates reduction.

tlvd-5055

WORKED EXAMPLE 4 Calculating a linear scale factor

Calculate the linear scale factor for the pair of similar triangles shown.

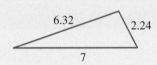

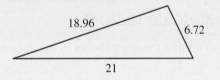

THINK

1. Calculate the ratio of the corresponding side lengths and simplify.

2. Write the answer.

WRITE

$k = \dfrac{21}{7} = \dfrac{18.96}{6.32} = \dfrac{6.72}{2.24} = 3$

The linear scale factor is 3.

WORKED EXAMPLE 5 Determining an unknown length in similar triangles

For the pair of similar triangles shown, calculate the unknown side lengths.

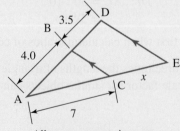

All measurements in metres

THINK	WRITE
1. Identify the two separate triangles and include measurements.	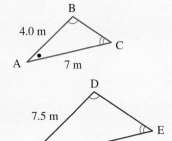
2. Determine the linear scale factor, k, by calculating the ratio of the corresponding side lengths.	$k = \dfrac{7.5}{4} = \dfrac{15}{8}$ (or 1.875)
3. Multiplying the side length of 7 metres in the blue coloured triangle by the linear scale factor will give the length of the corresponding side in the pink triangle, that is, $(7+x)$. Write down this mathematical statement.	$7 \times k = 7 + x$
4. Substitute the value of k into this equation and solve for x.	$\begin{aligned} 7 \times 1.875 &= 7 + x \\ 13.125 &= 7 + x \\ x &= 13.125 - 7 \\ x &= 6.125 \end{aligned}$
5. Write the answer	The length of the unknown side is 6.125 metres.

Note: The above worked example could have been solved using ratios by solving $\dfrac{7+x}{7} = \dfrac{7.5}{4}$, or equivalent.

13.3 Exercise

Students, these questions are even better in jacPLUS

Receive immediate feedback and access sample responses

Access additional questions

Track your results and progress

Find all this and MORE in jacPLUS

1. **WE4** Calculate the linear scale factors for the pairs of similar triangles shown.

a.
4.4   6.3
10

9.68   13.86
22

b.
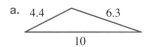
6.8   7.21
10.11

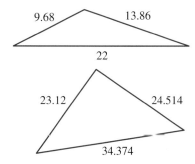
23.12   24.514
34.374

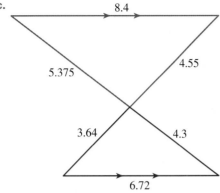

c.

d.

2. Calculate the linear scale factors for the pairs of similar objects shown.

a.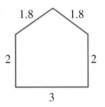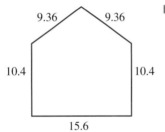

b.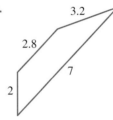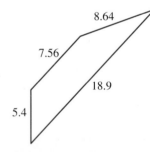

c.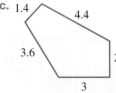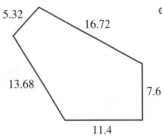

d.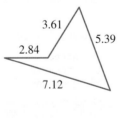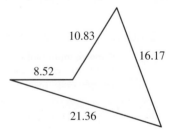

3. Calculate the linear scale factors for the following ratios of corresponding side lengths.

   a. $3:2$ **b.** $12:5$ **c.** $3:4$ **d.** $85:68$

4. Calculate the missing values for the following.

   a. $\dfrac{3}{\Box} = \dfrac{\Box}{12} = 6$ **b.** $\dfrac{5}{\Box} = \dfrac{44}{11} = \Box$

   c. $\dfrac{\Box}{7} = \dfrac{81}{9} = \Box$ **d.** $\dfrac{2}{\Box} = \dfrac{\Box}{2} = 0.625$

5. Calculate the unknown side lengths in the pairs of similar shapes shown.

a.

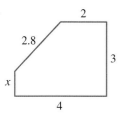

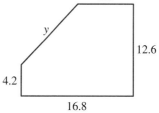

b.

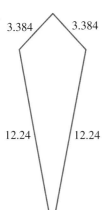

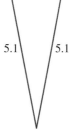

c.

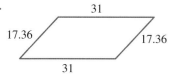

d.

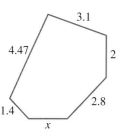

6. WE5 Calculate the unknown side lengths in the diagrams shown.

a.

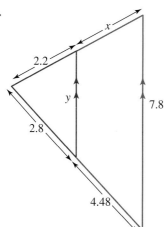

b.

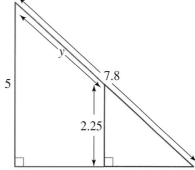

7. Calculate the length of BC in the diagram shown.

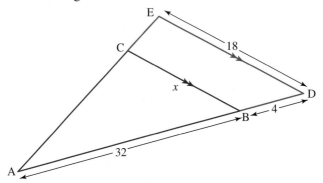

8. Calculate the value of x in the diagram shown.

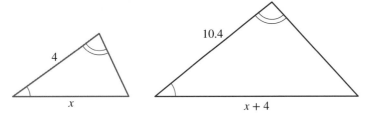

9. The side of a house casts a shadow that is 8.4 m long on horizontal ground.

   a. At the same time, an 800-mm vertical garden stake has a shadow that is 1.4 m long. Calculate the height of the house.

   b. When the house has a shadow that is 10 m long, calculate the length of the garden stake's shadow.

10. Calculate the value of x in the diagram.

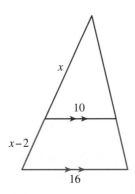

11. A section of a bridge is shown in the diagram. Calculate how high point B is above the roadway of the bridge.

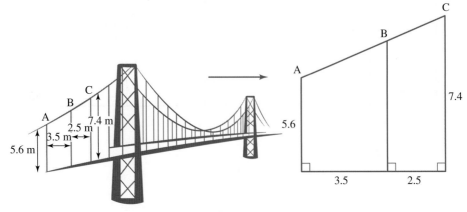

12. To calculate the distance across a ravine, a surveyor took a direct line of sight from point B to a fixed point A on the other side and then measured out a perpendicular distance of 18 m. From that point the surveyor measured out a smaller similar triangle as shown in the diagram.
Calculate the distance across the ravine along the line AB.

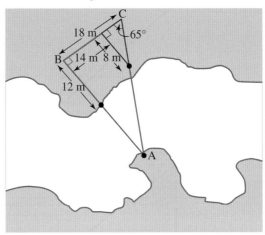

13. Over a horizontal distance of 6.5 m, an escalator rises 12.75 m. If you travel on the escalator for a horizontal distance of 4.25 m, calculate the vertical distance you have risen.

14. In a game of billiards, a ball travels in a straight line from a point one-third of the distance from the bottom of the right side and rebounds from a point three-eighths of the distance along the bottom side. The angles between the bottom side and the ball's path before and after it rebounds are equal.

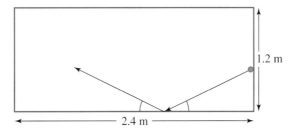

a. Calculate the perpendicular distance, correct to 2 decimal places, from the bottom side after the ball has travelled a distance of 0.8 m parallel with the bottom side after rebounding.
b. If the ball has been struck with sufficient force, determine at what point on an edge of the table it will next touch. Give your answer correct to 2 decimal places.

13.3 Exam questions

Question 1 (1 mark)

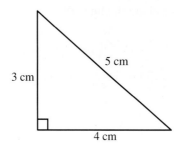

MC When enlarged by a factor of 2, the shape above becomes

A.

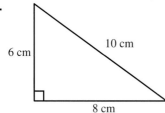

B.

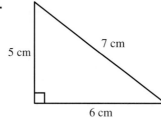

C.

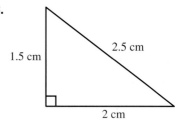

D.

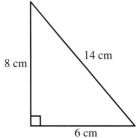

E.

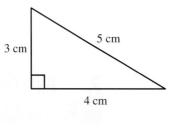

Question 2 (1 mark)

MC The missing lengths in this diagram are

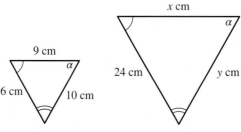

A. $x = 18$ cm, $y = 20$ cm   **B.** $x = 40$ cm, $y = 36$ cm   **C.** $x = 36$ cm, $y = 40$ cm
D. $x = 18$ cm, $y = 30$ cm   **E.** $x = 30$ cm, $y = 18$ cm

Question 3 (1 mark)

MC When considering similar triangles, select the *incorrect* statement.
- **A.** Whether enlarged or reduced, similar triangles have identical angles.
- **B.** When enlarged by a scale factor of 3, all side lengths will be multiplied by 3.
- **C.** When enlarged by a factor of 2, the size of each angle in the new triangle will be double the size of the corresponding angle in the original triangle.
- **D.** When reduced by a scale factor of 2, all side lengths will be multiplied by $\dfrac{1}{2}$.
- **E.** When reduced by a scale factor of 2, all side lengths will be divided by 2.

More exam questions are available online.

13.4 Area and volume scale factors

LEARNING INTENTION

At the end of this subtopic you should be able to:
- calculate the area scale factor for a pair of similar objects
- calculate the volume scale factor for a pair of similar objects.

13.4.1 Area scale factor

Consider three squares with side lengths of 1, 2 and 3 cm. Their areas are 1 cm², 4 cm² and 9 cm² respectively.

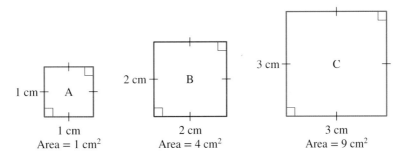

The linear scale factor between square A and square B is 2, and the linear scale factor between square A and square C is 3. When we look at the ratio of the areas of the squares, we get 4 : 1 for squares A and B, and 9 : 1 for squares A and C. In both cases, the **area scale factor** is equal to the linear scale factor raised to the power of two.

	Linear scale factor	Area scale factor
B : A	2	4
C : A	3	9

Comparing squares B and C, the ratio of the side lengths is 2 : 3, resulting in a linear scale factor of $\frac{3}{2}$ or 1.5.

From the ratio of the areas we get 4 : 9, which once again indicates an area scale factor $\frac{9}{4} = 2.25$, that is, the linear scale factor to the power of two.

Area scale factor

In general, if the linear scale factor for two similar objects is k, the area scale factor will be k^2.

---

tlvd-5056

WORKED EXAMPLE 6 Calculating the area scale factor for a pair of similar triangles

Calculate the area scale factor for the pair of similar triangles shown in the diagram.

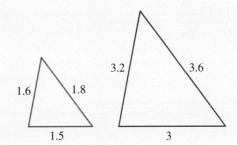

1. Calculate the linear scale factor, that is, the ratio of the corresponding side lengths.

$$k = \frac{3}{1.5} = \frac{3.6}{1.8} = \frac{3.2}{1.6} = 2$$

2. Square the linear scale factor to obtain the area scale factor.

$$k^2 = 2^2 = 4$$

3. Write the answer in a sentence.

The area scale factor is 4.

13.4.2 Volume scale factor

Three-dimensional objects of the same shape are similar when the ratios of their corresponding dimensions are equal. When we compare the volumes of three similar cubes, we can see that if the linear scale factor is k, the **volume scale factor** will be k^3.

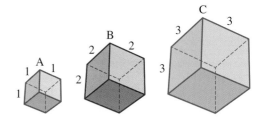

	cube B : cube A	scale factor
Linear	2 : 1	2
Area	4 : 1	$2^2 = 4$
Volume	8 : 1	$2^3 = 8$

Volume scale factor

If the linear scale factor for two similar objects is k, the volume scale factor will be k^3.

tlvd-5057

WORKED EXAMPLE 7 Calculating the volume scale factor for a pair of spheres

Calculate the volume scale factor for the pair of spheres shown in the diagram.

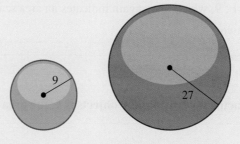

1. Calculate the linear scale factor, that is, the ratio of the corresponding side lengths.

$$k = \frac{27}{9} = 3$$

2. Cube the linear scale factor to obtain the volume scale factor.

$$k^3 = 3^3 = 27$$

3. Write the answer.

The volume scale factor is 27.

13.4 Exercise

Students, these questions are even better in jacPLUS

Receive immediate feedback and access sample responses

Access additional questions

Track your results and progress

Find all this and MORE in jacPLUS

1. **WE6** Calculate the area scale factor for each pair of similar triangles shown.

a.

b.

c.

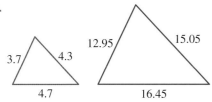

2. Calculate the area scale factor for each pair of similar objects shown.

a.

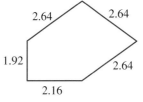

b.

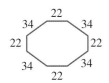

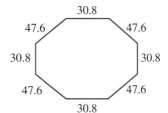

3. a. Calculate the areas of the two similar triangles shown in the diagram.

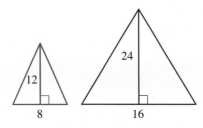

b. State how many times larger in area the bigger triangle is.
c. Calculate the linear scale factor.
d. Calculate the area scale factor.

4. A hexagon is made up of six equilateral triangles of side length 2 cm. If a similar hexagon has an area of $24\sqrt{3}$ cm^2, calculate the linear scale factor.

5. A rectangular swimming pool is shown on the plans for a building development with a length of 6 cm and a width of 2.5 cm. If the scale on the plans is $1:250$:

a. calculate the area scale factor
b. calculate the surface area of the swimming pool.

6. The area of the triangle ADE in the diagram is 100 cm^2, and the ratio of $DE:BC$ is 2:1.

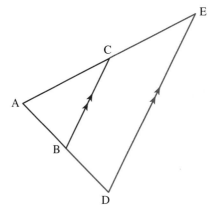

Calculate the area of triangle ABC.

7. The floor of a square room has an area of 12 m^2. Calculate the area that the room takes up in a diagram with a scale of $1:250$.

8. **WE7** Calculate the volume scale factor for the pair of spheres shown.

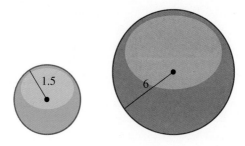

9. Calculate the volume scale factor for each pair of similar objects shown.

a.

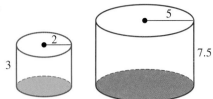

b.

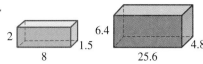

c.

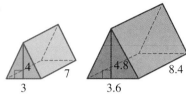

10. An architect makes a small scale model of a house out of balsa wood with the dimensions shown in the diagram.

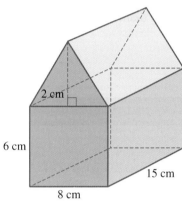

a. If the actual length of the building is 26.25 m, calculate the scale of the model.
b. Calculate the ratio of the volume of the building to the volume of the model.

11. Two similar cylinders have volumes of 400 cm³ and 50 cm³ respectively.

a. Calculate the linear scale factor.
b. If the length of the larger cylinder is 8 cm, calculate the length of the smaller one.

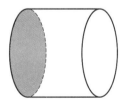

12. If a cube has a volume of 25 cm³ and is then enlarged by a linear scale factor of 2.5, calculate the new volume.

13. Calculate the linear scale factor between two similar drink bottles if one has a volume of 600 mL and the other has a volume of 1.25 L.

14. If an area of 712 m² is represented on a scale drawing by an area of 44.5 cm², calculate the actual length that a distance of 5.3 cm on the drawing represents.

15. A model car is an exact replica of the real thing reduced by a factor of 12.

    a. If the actual surface area of the car that is spray painted is 4.32 m², calculate the equivalent painted area on the model car.
    b. If the actual storage capacity of the car is 1.78 m³, calculate the equivalent volume for the model car.

16. A company sells canned fish in two sizes of similar cylindrical cans. For each size, the height is four-fifths of the diameter.

    a. Write an expression for calculating the volume of a can of fish in terms of its diameter.
    b. If the dimensions of the larger cans are 1.5 times those of the smaller cans, write an expression for calculating the volume of the larger cans of fish in terms of the diameter of the smaller cans.

13.4 Exam questions

▶ **Question 1 (1 mark)**

Source: VCE 2021, Further Mathematics Exam 1, Section B, Module 3, Q3; © VCAA.

MC A photograph was enlarged by an area scale factor of 9.

The length of the original photograph was 12 cm.

The original photograph and the enlarged photograph are similar in shape.

The length of the enlarged photograph, in centimetres, is

   A. 4
   B. 9
   C. 27
   D. 36
   E. 108

▶ **Question 2 (1 mark)**
Source: VCE 2016, Further Mathematics Exam 1, Section B, Module 3, Q8; © VCAA.

MC A string of seven flags consisting of equilateral triangles in two sizes is hanging at the end of a racetrack, as shown in the diagram.

The edge length of each black flag is twice the edge length of each white flag.

For this string of seven flags, the total area of the black flags would be
- **A.** two times the total area of the white flags.
- **B.** four times the total area of the white flags.
- **C.** $\frac{4}{3}$ times the total area of the white flags.
- **D.** $\frac{16}{3}$ times the total area of the white flags.
- **E.** $\frac{16}{9}$ times the total area of the white flags.

▶ **Question 3 (1 mark)**
Source: VCE 2013, Further Mathematics Exam 1, Section B, Module 2, Q4; © VCAA.

MC A cafe sells two sizes of cupcakes with a similar shape.

The large cupcake is 6 cm wide at the base and the small cupcake is 4 cm wide at the base.

6 cm   4 cm

The price of a cupcake is proportional to its volume.

If the large cupcake costs $5.40, then the small cupcake will cost
- **A.** $1.60
- **B.** $2.32
- **C.** $2.40
- **D.** $3.40
- **E.** $3.60

More exam questions are available online.

13.5 Review

13.5 Exercise

Multiple choice

1. **MC** The triangle that is similar to ΔABC is:

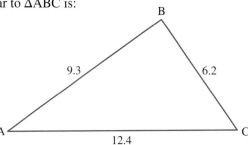

A.

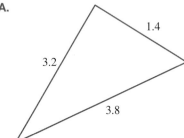

B.

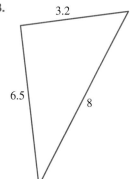

C.

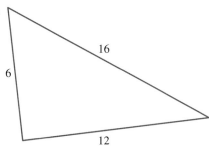

D.

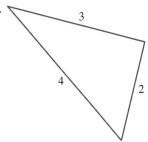

E.

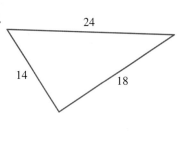

2. **MC** If a map has a scale factor of 1 : 50 000, an actual distance of 11 km would have a length on the map of:

   A. 21 cm   **B.** 22 cm   **C.** 21.5 cm   **D.** 11 cm   **E.** 5.5 cm

3. **MC** If the areas of two similar objects are 12 cm^2 and 192 cm^2 respectively, the linear scale factor will be:

   A. 3   **B.** 4   **C.** 16   **D.** 2   **E.** 9

4. **MC** If the volumes of two similar solids are 16 cm^3 and 128 cm^3 respectively, the area scale factor will be:

    A. 2   **B.** 3   **C.** 4   **D.** 8   **E.** 9

5. **MC** A tree casts a shadow that is 5.6 m long. At the same time a 5 m light pole casts a shadow that is 3.5 m long. The height of the tree is:

    A. 4.4 m   **B.** 7.5 m   **C.** 8.0 m   **D.** 6.0 m   **E.** 5.6 m

6. **MC** The value of x in the diagram is:

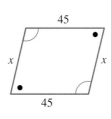

    A. 38   **B.** 41   **C.** 43   **D.** 36   **E.** 40

7. **MC** If the scale factor of the volumes of two similar cuboids is 64 and the volume of the larger one is 1728 cm^3, the surface area of the smaller cuboid is:

    A. 64 cm^2 **B.** 27 cm^2 **C.** 54 cm^2 **D.** 9 cm^2 **E.** 16 cm^2

8. **MC** An enlargement diagram of a very small object is drawn at a scale of 18 : 1. If the diagram has an area of 162 cm^2, the equivalent area of the actual object is closest to:

    A. 0.5 cm^2 **B.** 0.4 cm^2 **C.** 9 cm^2 **D.** 0.6 cm^2 **E.** 18 cm^2

9. **MC** The plans of a house show the side of a building as 12.5 cm long. If the actual building is 15 m long, the scale of the plan is:

    A. 1 : 250   **B.** 1 : 150   **C.** 1 : 220   **D.** 1 : 120   **E.** 1 : 175

10. **MC** The plans for a building show a concrete slab covering an area of 12.5 cm × 8.4 cm to a depth of 0.25 cm. If the plans are drawn to a scale of 1 : 225, the actual volume of concrete is closest to:

    A. 26.25 m^3 **B.** 262.5 m^3 **C.** 5906.25 m^3 **D.** 299 m^3 **E.** 29.9 m^3

Short answer

11. Calculate the value of x in the diagram of two similar objects shown.

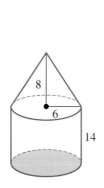
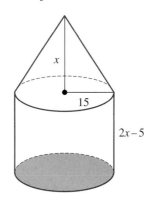

12. The volume of a solid is 1600 cm^3. If the ratio of the corresponding dimensions between this solid and a similar solid is 4 : 5, calculate the volume of the similar solid.

13. Calculate the values of the pronumerals in the following diagrams.

a.

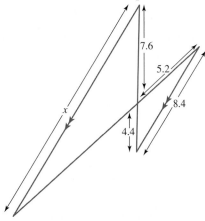

b.

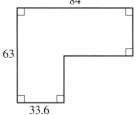

c.

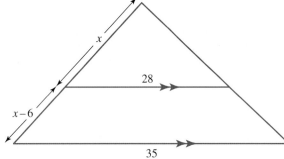

14. A triangle has side lengths of 32 mm, 45 mm and 58 mm.

    a. Calculate the side lengths of a larger similar triangle using a corresponding side ratio of 2 : 3.

    b. Calculate the side lengths of the larger triangle in a drawing with a scale of 5 : 2.

15. A farmer divides paddock ABCD into two separate paddocks along the line FE as shown in the diagram. All distances are shown in metres.

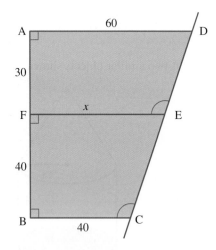

    a. Express the ratio of the corresponding sides of the original paddock, ABCD, with paddock BCEF in its simplest form.

    b. Calculate the length of fencing required to separate the two paddocks along the line FE. Give your answer correct to the nearest centimetre.

16. The top third of an inverted right cone is removed as shown in the diagram.

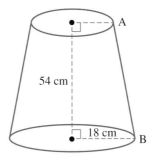

54 cm

18 cm

A

B

  a. Calculate the height of the cone that has been removed.
  b. Calculate the distance along the edge of the remaining part of the cone from A to B.

Extended response

17. On a map that is drawn to a scale of 1 : 225 000, the distance between two points is 88 cm.

  a. Calculate the actual distance between the two points.

  A boat sets out to travel from one point to the other, but the navigator makes an error. After travelling 100 km, the crew realise they are directly south of a point that they should have reached after travelling 90 km in a direct line to their destination.

  b. If they continue in their current direction, calculate how much further they have to travel to be directly south of the intended destination.
  c. Calculate how far away from the intended destination the boat will be when it reaches the point on their course that is directly to the south.
  d. When the boat is at the point directly to the south of their intended destination, determine how far away they will be on the map.

18. A rectangular box with dimensions 94 cm × 31 cm leans against a wall as shown in the diagram. A larger box leans against the first box.

  a. If the ratio of the corresponding side lengths between the two boxes is 4 : 7, calculate the dimensions of the larger box.
  b. If the smaller box touches the floor at a point that is 52 cm from the base of the wall, calculate how far up the wall it reaches.
  c. Calculate how far up the wall the larger box reaches.

19. A swimmer is observed in the water from the top of a vertical cliff that is 5 m tall. A line of sight is taken from a point 2 m back from the edge of the cliff.

  a. If the swimmer, the edge of the cliff and a point 1.8 m above the cliff are in line, calculate how far the swimmer is from the base of the cliff.
  b. If the swimmer moves a further 2.45 m away from the cliff, determine from how far above the cliff the new line of sight should be taken.

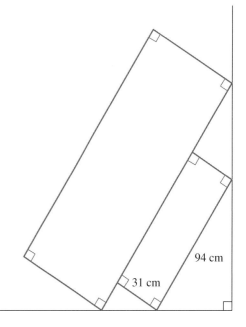

94 cm

31 cm

20. A caterer sells takeaway coffee in three different-sized cups whose dimensions are in proportion.

  a. If the small cup has a capacity of 200 mL and the medium cup has a capacity of 300 mL, calculate the linear scale factor correct to 2 decimal places.
  b. If the ratio of corresponding dimensions between the small and large cups is 4 : 5, calculate the capacity of the large cup.
  c. If the caterer charges $3.00 for a small cup, $3.50 for a medium cup and $4.00 for a large cup, determine which is the better value for the customer.

13.5 Exam questions

Question 1 (1 mark)

Source: VCAA 2020, Further Mathematics Exam 1, Section B, Module 3, Q9; © VCAA.

MC Shot-put is an athletics field event in which competitors throw a heavy spherical ball (a shot) as far as they can.

The size of the shot for men and the shot for women is different.

The diameter of the shot for men is 1.25 times larger than the diameter of the shot for women.

The ratio of the total surface area of the women's shot to the total surface area of the men's shot is

  A. 1 : 4   B. 1 : 25   C. 4 : 5   D. 5 : 4   E. 16 : 25

Question 2 (4 marks)

Source: VCE 2020, Further Mathematics Exam 2, Section B, Module 3, Q2; © VCAA.

Khaleda has designed a logo for her business.

The logo contains two identical equilateral triangles.

The side length of each triangle is 4.8 cm, as shown in the diagram below.

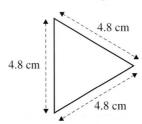

  a. Write a calculation to show that the area of one of the triangles, rounded to the nearest centimetre, is $10 \, \text{cm}^2$.   **(1 mark)**
  b. In the logo, the two triangles overlap, as shown below. Part of the logo is shaded and part of the logo is not shaded.

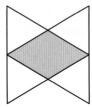

    What is the area of the entire logo?
    Round your answer to the nearest square centimetre.   **(1 mark)**

c. What is the ratio of the area of the shaded region to the area of the non-shaded region of the logo? **(1 mark)**

d. The logo is enlarged and printed on boxes for shipping.
The enlarged logo and the original logo are similar in shape.
The area of the enlarged logo is four times the area of the original logo.
What is the height, in centimetres, of the enlarged logo? **(1 mark)**

Question 3 (3 marks)

Source: VCE 2017, Further Mathematics Exam 2, Section B, Module 3, Q1; © VCAA.

Miki is planning a gap year in Japan.

She will store some of her belongings in a small storage box while she is away.

This small storage box is in the shape of a rectangular prism.

The diagram below shows that the dimensions of the small storage box are $40 \text{ cm} \times 19 \text{ cm} \times 32 \text{ cm}$.

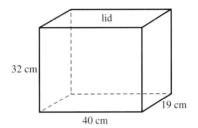

The lid of the small storage box is labelled on the diagram above.

a. i. What is the surface area of the lid, in square centimetres? **(1 mark)**
ii. What is the total outside surface area of this storage box, including the lid and base, in square centimetres? **(1 mark)**

b. Miki has a large storage box that is also a rectangular prism.
The large storage box and the small storage box are similar in shape.
The volume of the large storage box is eight times the volume of the small storage box.
The length of the small storage box is 40 cm.
What is the length of the large storage box, in centimetres? **(1 mark)**

Question 4 (5 marks)

Source: VCE 2014, Further Mathematics Exam 2, Module 2 Q3; © VCAA.

The chicken coop contains a circular water dish.

Water flows into the dish from a water container.

The water container is in the shape of a cylinder with a hemispherical top.

The water container and the dish are shown in the diagrams below.

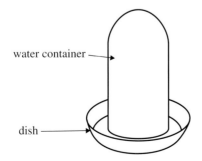

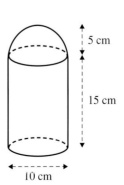

The cylindrical part of the water container has a diameter of 10 cm and a height of 15 cm.

The hemisphere has a radius of 5 cm.

   a. What is the surface area of the hemispherical top of the water container?
      Write your answer, correct to the nearest square centimetre.   **(1 mark)**

   b. What is the maximum volume of water that the water container can hold?
      Write your answer, correct to the nearest cubic centimetre.   **(2 marks)**

   c. The eating space of the chicken coop also has a feed container.
      The feed container is similar in shape to the water container.
      The volume of the water container is three-quarters of the volume of the feed container.
      The surface area of the water container is 628 cm^2.
      What is the surface area of the feed container?
      Write your answer, correct to the nearest square centimetre.   **(2 marks)**

 Question 5 (2 marks)
Source: VCE 2013, Further Mathematics Exam 2, Module 2, Q4; © VCAA.

Competitors in the intermediate division of the discus use a smaller discus than the one used in the senior division, but of a similar shape. The total surface area of each discus is given below.

intermediate discus   senior discus

total surface area 500 cm^2 total surface area 720 cm^2

By what value can the volume of the intermediate discus be multiplied to give the volume of the senior discus?

More exam questions are available online.

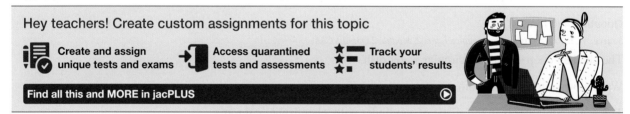

Answers

Topic 13 Similarity and scale

13.2 Similar objects

13.2 Exercise

1. a. $\dfrac{15}{8} = \dfrac{11.25}{6} = \dfrac{7.5}{4} = 1.875$, and all angles are equal.

   b. $\dfrac{41.04}{11.4} = \dfrac{26.1}{7.25} = \dfrac{9}{2.5} = \dfrac{13.68}{3.8} = 3.6$, and all angles are equal.

2. $\dfrac{106.43}{18.35} = \dfrac{24.65}{4.25} = 5.8$, and all angles are equal.

3. a. $\dfrac{4.8}{3.2} = 1.5$, $\dfrac{2.24}{1.6} = 1.4$; not similar

   b. $\dfrac{10.224}{2.84} = 3.6$, $\dfrac{2.34}{0.65} = 3.6$; similar

   c. $\dfrac{4.4352}{2.112} = 2.1$, $\dfrac{2.184}{1.04} = 2.1$; similar

   d. $\dfrac{12.535}{11.5} = 1.09$, $\dfrac{10.152}{9.4} = 1.08$; not similar

4. a. $\dfrac{35}{25} = 1.4$, $\dfrac{22.4}{16} = 1.4$, $\dfrac{12.6}{9} = 1.4$ and all angles are equal; similar

   b. $\dfrac{4.94}{2.6} = 1.9$, $\dfrac{6.65}{3.5} = 1.9$, $\dfrac{2.09}{1.1} = 1.9$ and all angles are equal; similar

   c. $\dfrac{13}{8} = 1.625$, $\dfrac{6.5}{4} = 1.625$, $\dfrac{3.25}{2} = 1.625$ and all angles are equal; similar

   d. $\dfrac{3.24}{2.7} = 1.2$, $\dfrac{19.44}{16.2} = 1.2$, $\dfrac{2.16}{1.8} = 1.2$ and all angles are equal; similar

5. a. $\dfrac{44.275}{7.7} = \dfrac{78.2}{13.6} = 5.75$, SAS

   b. $\dfrac{38.72}{17.6} = \dfrac{16.28}{7.4} = \dfrac{14.85}{6.75} = 2.2$, SSS

6. a. $\dfrac{32.4}{18} = \dfrac{21.6}{12} = 1.8$, $\dfrac{23.8}{14} = 1.7$ not similar

   b. $\dfrac{54.32}{19.4} = \dfrac{22.96}{8.2} = 2.8$ and all angles are equal; similar

7. a. A and C   b. B and C   c. A and C   d. A and B

8. a. All angles are equal and side lengths are in proportion.

   b. All measurements (radius and circumference) are in proportion.

   c. All angles are equal and side lengths are in proportion.

9. a. 1.6 : 1   b. 2.26 : 1   c. 3.6 : 1

10. a. 4 : 3   b. 5.5 : 1

11. a. 4.8   b. 7.07

12. a. $\dfrac{50.84}{8.2} = 6.2$, all side lengths are in proportion and all angles equal.

   b. $\dfrac{14.34}{12.6} = 1.138$, all side lengths are in proportion and all angles are equal.

13. a. 32   b. 20

14. 51.8

15. $x = 5.6$, $y = 7$

16. a.
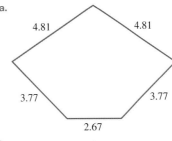

   b.
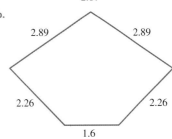

13.2 Exam questions

Note: Mark allocations are available with the fully worked solutions online.

1. E

2. D

3. D

13.3 Linear scale factors

13.3 Exercise

1. a. 2.2   b. 3.4   c. 1.25   d. 1.6

2. a. 5.2   b. 2.7   c. 3.8   d. 3

3. a. 1.5   b. 2.4   c. 0.75   d. 1.25

4. a. $\dfrac{3}{\boxed{0.5}} = \dfrac{\boxed{72}}{12} = 6$

   b. $\dfrac{5}{\boxed{1.25}} = \dfrac{44}{11} = \boxed{4}$

   c. $\dfrac{\boxed{63}}{7} = \dfrac{81}{9} = \boxed{9}$

   d. $\dfrac{2}{\boxed{3.2}} = \dfrac{\boxed{1.25}}{2} = 0.625$

5. a. $x = 1$, $y = 11.76$ b. 1.41
   c. 2.8   d. $x = 2$, $y = 1.6$, $z = 3.576$

6. a. $x = 3.52$, $y = 3$ b. $y = 4.29$

7. 16

8. 2.5

9. a. 4.8 m   b. 1.67 m

10. 5

11. 6.65 m

12. 24 m

13. 8.34 m

14. a. 0.36 m

  b. 0.67 m from the bottom of the left side

13.3 Exam questions

Note: Mark allocations are available with the fully worked solutions online.

1. A

2. C

3. C

13.4 Area and volume scale factors

13.4 Exercise

1. a. $\dfrac{49}{16} = 3.0625$ **b.** $\dfrac{3136}{625} = 5.0176$

  c. $\dfrac{49}{4} = 12.25$

2. a. $\dfrac{64}{25} = 2.56$ **b.** $\dfrac{49}{25} = 1.96$

3. a. 48 and 192 square units

  b. 4

  c. 2

  d. 4

4. 2

5. a. 62 500   **b.** 93.75 m^2

6. 25 cm^2

7. 1.92 cm^2

8. 64

9. a. 15.625   **b.** 32.768   **c.** 1.728

10. a. 1 : 175   **b.** 5 359 375 : 1

11. a. 2   **b.** 4 cm

12. 390.625 cm^3

13. 1.28

14. 21.2 m

15. a. 300 cm^2 **b.** 1030 cm^3

16. a. $V = \dfrac{\pi D^3}{5}$ **b.** $V_2 = \dfrac{27\pi D^3}{40}$

13.4 Exam questions

Note: Mark allocations are available with the fully worked solutions online.

1. D

2. D

3. A

13.5 Review

13.5 Exercise

Multiple choice

  1. D

  2. B

  3. B

  4. C

  5. C

6. E

7. C

8. A

9. D

10. D

Short answer

11. 20

12. 3125 cm^3

13. a. 14.51

  b. $x = 20,\ y = 33.6$

  c. 8

14. a. 48 mm, 67.5 mm, 87 mm

  b. 120 mm, 168.75 mm, 217.5 mm

15. a. 7 : 4   **b.** 5143 cm

16. a. 27 cm   **b.** 55.32 cm

Extended response

17. a. 198 km **b.** 120 km **c.** 95.90 km **d.** 42.62 cm

18. a. 164.5 cm $\times$ 54.25 cm

  b. 78.31 cm

  c. 137.04 cm

19. a. 5.6 m   **b.** 1.24 m

20. a. 1.14   **b.** 390.6 mL   **c.** Large cup

13.5 Exam questions

Note: Mark allocations are available with the fully worked solutions online.

1. E

2. a. Please see the worked solution.

  b. 15 cm^2

  c. 1 : 2

  d. 9.6 cm

3. a. **i.** 760 cm^2

    ii. 5296 cm^2

  b. 80 cm

4. a. 157 cm^2 **b.** 1440 cm^3 **c.** 761 cm^2

5. $\dfrac{216}{125}$ or 1.728

GLOSSARY

adjacency matrix a square matrix representing the number of edges joining each pair of vertices in a graph

algorithm a finite set of steps or instructions for performing a computation or for solving a problem

angle of depression the angle measured down from the horizontal line (through the observation point) to the line of vision

angle of elevation the angle measured up from the horizontal line (through the observation point) to the line of vision

annulus the area between two circles with the same centre. The formula is $A = \pi(R^2 - r^2)$, where R is the radius of the outer circle and r is the radius of the inner circle.

area the two-dimensional space taken up by an object

area scale factor the ratio of the corresponding areas of similar objects. It is equal to the linear scale factor raised to the power of 2.

arithmetic sequence a sequence in which the difference between any two successive terms is the same

back-to-back stem plot a stem plot used to compare two different sets of data. Back-to-back stem plots share the same stem, with one data set appearing on the left of the stem and the other data set appearing on the right.

bar chart a display of the categories of data on one axis (usually the horizontal axis) and the frequency of the data on the other axis (usually the vertical axis)

bearing the direction of a fixed point, or the path of an object, from the point of observation

boxplot a graphical representation of the five-number summary

break-even point the point at which revenue begins to exceed the cost of production

bridge an edge in a connected graph whose deletion will no longer cause the graph to be connected

capacity the amount of liquid an object can hold

cardinal point one of the four main directions or standard bearings: north (N), south (S), east (E) and west (W)

categorical data data that can be organised into groups or categories and is often an 'object', 'thing' or 'idea'. Examples include brand names, colours, general sizes and opinions.

causation the measure of how much the change in one variable is caused by the other

circuit a trail beginning and ending at the same vertex

circumference the perimeter of a curved figure

coefficient of determination the square of the product-moment correlation coefficient (r^2). This is a measure of how much a change in the response variable is influenced by a change in the explanatory variable.

column matrix a matrix that has only one column

common difference the difference between each term in an arithmetic sequence: $d = t_{n+1} - t_n$

common ratio the ratio between two consecutive terms in a geometric sequence: $r = \dfrac{t_{n+1}}{t_n}$

compound interest interest calculated on the changing value throughout the time period of a loan or investment. Interest is added to the balance before the next interest calculation is made. The amount of the loan or investment can be calculated using $A = P\left(1 + \dfrac{r}{100}\right)^n$.

cone a solid object in which one end is circular and the other end is a single vertex. Its cross-section is a series of circles that gradually get smaller as they approach the vertex.

congruent describes objects that are exactly the same size and shape

connected graph a graph with a path between each pair of vertices that makes it possible to reach every vertex of the graph by moving along the edges

constant of proportionality or **constant of variation** denoted as k, a value that is equal to the ratio of y to x for any data pair. It is the rate at which y varies with x, or the gradient.

constant of variation denoted as k, a value that is equal to the ratio of y to x for any data pair. It is the rate at which y varies with x, or the gradient.

Consumer Price Index (CPI) a measure of changes, over time, in retail prices of a constant basket of goods and services representative of consumption expenditure by resident households in Australian metropolitan areas

continuous data numerical data that can take any value that lies within an interval. Continuous data values are subject to the accuracy of the measuring device being used.

conventional (compass) bearings a bearing that is measured first in terms of north and south, then in terms of east and west

correlation a measure of the strength of the linear relationship between two variables

cosine the ratio of the side length adjacent to an internal angle of a right-angled triangle with the hypotenuse: $\cos(\theta) = \dfrac{A}{H}$

cosine rule a rule that can enable unknown side lengths and angles of triangles to be determined if either two side lengths and an included angle, or all three side lengths are known: $a^2 = b^2 + c^2 = 2bc \times \cos(A)$ or $\cos(A) = \dfrac{b^2 + c^2 - a^2}{2bc}$

credit card a method of purchasing whereby a financial institution loans an amount of money to an individual up to a pre-approved limit

cycle a path beginning and ending at the same vertex

cylinder a solid object with ends that are identical circles and a cross-section that is the same along its length

debit card a method of purchasing in which the money is debited directly from a bank account or a pre-loaded amount

degree (graph theory) the number of edges that are directly connected to a vertex

determinant In the 2×2 matrix $A = \begin{bmatrix} a & b \\ c & d \end{bmatrix}$, the determinant (det A or $|A|$) is the difference between the product of the diagonal elements: $ad - bc$. If $A = 0$, the inverse matrix will not exist.

digraph *see* directed graph

direct variation when quantities are proportional to each other; for example, doubling one doubles the other

directed graph a graph in which it is only possible to move along the edges of the graph in one direction

disconnected graph a graph in which it is not possible to reach every vertex by moving along the edges

discrete data numerical data that is counted in exact values, with the values often being whole numbers

dot plot a display of discrete numerical data or categorical data. A scaled horizontal axis is used and each data value is indicated by a dot above this scale, resulting in a set of vertical 'lines' of evenly-spaced dots.

edge a line of a graph

element an entry in a matrix

elimination the process of simplifying a mathematical expression by removing a variable; common when solving simultaneous equations

equilibrium *see* steady state

Euler circuit a circuit in which every edge is used once

Euler trail a trail in which every edge is used once

explanatory variable (EV) the variable represented on the x-axis, used in a bivariate analysis to explain or predict a difference in the response variable

extrapolation making a prediction from a line of best fit that appears outside the parameters of the original data set

face a distinct area created by the non-intersecting edges and vertices of a planar graph

first-order recurrence relation a relation that relates a term in a sequence to the previous term in the same sequence

frequency table a display that tabulates data according to the frequencies of predetermined groupings

future value the value of an asset at a time in the future, based on the original cost less depreciation

geometric sequence a pattern of numbers whose consecutive terms increase or decrease in the same ratio. Each term after the first is found by multiplying the previous one by a fixed non-zero number called the common ratio.

goods and services tax a tax that is charged on most purchases; also known as GST

gradient also known as the slope; determines the change in the y-value for each change in x-value. This measures the steepness of a line as the ratio $m = \dfrac{\text{rise}}{\text{run}}$. If (x_1, y_1) and (x_2, y_2) are two points on the line, $m = \dfrac{y_2 - y_1}{x_2 - x_1}$.

gradient-intercept form the form of a linear equation expressed as $y = mx + c$, where m is the gradient and c is the y-intercept

graph a visual representation of the connections and/or relationships that exist between things. In graph theory, a graph is a series of points and lines that are used to represent the connections that exist in various settings.

greedy algorithm an algorithm that seems to make the best choice at each step

GST goods and services tax; a tax that is charged on most purchases

Hamiltonian cycle a cycle that reaches all vertices of a network

Hamiltonian path a path that reaches all vertices of a network

Heron's formula a method of calculating the area of the triangle if you are given all three side lengths: $A = \sqrt{s(s-a)(s-b)(s-c)}$, where a, b and c are the side lengths and s is half the perimeter

histogram a display of continuous numerical data similar to a bar chart, in which the width of each column represents a range of data values and the height of the column represents that range's frequency

horizontal asymptote Asymptotes are lines a graph approaches but never reaches. A horizontal asymptote shows the long-term behaviour as, for example, $x \to \infty$.

hypotenuse the longest side of a right-angled triangle. The hypotenuse is opposite the right angle.

identity matrix a square matrix in which all of the elements on the diagonal line from the top left to bottom right are 1 and all of the other elements are 0

inflation a general increase in prices over time that effectively decreases the purchasing power of a currency. This is an application of compound interest.

initial-state matrix a column matrix, denoted by S_0, used to represent the starting value of a system

interpolation making a prediction from a line of best fit that appears within the parameters of the original data set

interquartile range the difference between the upper and lower quartiles of a data set

interval data numerical data found on a scale with no true zero

inverse matrix A matrix that when multiplied with another, results in the identity matrix. In other words, if $AA^{-1} = A^{-1}A = I$, then A^{-1} is called the inverse matrix of A or the multiplicative inverse of A.

inverse variation when increasing one variable decreases another

isomorphic graph a graph that has the same number of vertices and edges, with corresponding vertices having identical degrees and connections

Kruskal's algorithm a set of logical steps that can be used to identify the minimum spanning tree for a weighted connected graph

least-squares regression line an accurate linear trend line that is calculated by minimising the vertical distances between the data points and the line of best fit. When taking the squares of these vertical distances, this line would represent the smallest possible sum of all of these squares.

line of best fit a straight line that best represents the general trend of the data in a scatterplot

line of good fit *see* line of best fit

linear equation an equation in which the highest power of any variable is 1

linear model a model that uses a linear equation to represent a real situation

linear relation a relation between up to two variables of degree 1 that produces a straight line

linear scale factor the ratio of the corresponding side lengths of similar objects

linearise to create a more linear (straight line) form

literal equation an equation that contains pronumerals rather than numerals as terms or coefficients

logarithmic scale a scale that uses the exponent (power) of a base number. In base 10 numbers, each unit increase or decrease represents a tenfold increase or decrease in the quantity being measured.

logarithm the power to which a fixed number (the base) must be raised to produce a given value. The relationship between exponential and logarithmic functions is $y = A^x \Leftrightarrow \log_A (y) = x$.

loop an edge in a graph that joins a vertex to itself

lower fence the lower boundary beyond which a data value is considered to be an outlier: $Q_1 - 1.5 \times \text{IQR}$

lower quartile the median of the lower half of an ordered data set

matrix (plural *matrices*) a rectangular array of rows and columns that is used to store and display information

mean commonly referred to as the average; a measure of the centre of a set of data. The mean is calculated by dividing the sum of the data values by the number of data values.

median the middle value of the ordered data set if there are an odd number of values, or halfway between the two middle values if there are an even number of values

minimum spanning tree a spanning tree with the lowest total weighting

modality the description of a graph or data set in terms of its modes. A graph or data set may have one mode, two modes (called bimodal) or many modes (called multimodal).

mode the category or data value(s) with the highest frequency. It is the most frequently occurring value in a data set.

multiple edge occurs when two vertices are connected by more than one edge

negatively skewed describes distributions with higher frequencies on the right-hand side of the graph

net a 2-dimensional plan of the surfaces that make up a 3-dimensional object

nominal data categorical data that has no natural order or ranking

numerical data data that can be counted or measured

order the number of rows, m, and columns, n, of a matrix expressed as $m \times n$

ordinal data categorical data that can be placed into a natural order or ranking

outlier a data point that has a very different value compared to the rest of the data

parallel boxplot a display where two or more boxplots share the same scale to enable comparisons between data sets

path a walk in which no vertices are repeated, except possibly the start and finish

Pearson's product-moment correlation coefficient a measure of the strength of a linear trend that is designated with the letter r and associated with a numerical value between -1 and $+1$. Values close to $+1$ or -1 indicate strong linear trends; values close to zero indicate weak or no linear trends.

perimeter the distance around an object

personal loan a loan made by a lending institution to an individual. A personal loan will usually have a fixed interest rate attached to it, with the interest paid by the customer calculated on a reduced balance.

planar graph a graph that can be redrawn to represent information so that it has no intersecting edges

polygon a 2-dimensional shape consisting of at least three straight sides

positively skewed describes distributions with higher frequencies on the left side of the graph

Prim's algorithm a set of logical steps that can be used to identify the minimum spanning tree for a weighted connected graph

principal an amount that is borrowed or invested

prism a solid object that has identical ends that are joined by flat surfaces and a cross-section that is the same along its length

pronumeral a letter or symbol representing a number (usually a variable) in a mathematical expression or equation

proportional when two quantities have the same ratio; therefore, they always have the same size in relation to each other

pyramid a solid object whose base is a polygon and whose sides are triangles that meet at a single point

Pythagoras' theorem The square of the hypotenuse of a right-angled triangle is equal to the sum of the squares of the other two sides.

quartiles These divide a set of data into quarters. The lower quartile Q_1 is the median of the lower half of an ordered data set. The upper quartile Q_3 is the median of the upper half of an ordered data set. The middle quartile Q_2 is the median of the whole data set.

range a measure of the spread of a numerical data set determined by calculating the difference between the smallest and largest values

ratio data numerical data found on a scale with a true zero

recurrence relation a relation that relates a term in a sequence to a combination of previous terms in the same sequence

reducing balance depreciation a method of depreciation, sometimes called 'diminishing value' depreciation, in which the value of an asset is reduced by a fixed percentage of its previous book value. This is an application of compound interest. The depreciated future value can be calculated using $FV = P\left(1 - \dfrac{r}{100}\right)^n$.

regression line a line that describes the behaviour of the points in a scatterplot

response variable (RV) the variable represented on the y-axis that is explained or predicted by the explanatory variable in a bivariate analysis

row matrix a matrix that has only one row

scalar quantity can be any real number (as opposed to a matrix or matrix elements), such as negative or positive numbers, fractions or decimals

scatterplot a visual display of bivariate data

scientific notation When a number is written as $a \times 10^b$, the product of a number a where $1 \le a < 10$ and a power of 10, it is expressed in scientific notation.

sector fraction of a circle. The area of a sector can be calculated using $A = \dfrac{\theta}{360}\pi r^2$.

sequence a related set of objects or events that follow each other in a particular order

significant figures the number of digits that would occur in a if the number was expressed in scientific notation as $a \times 10^b$ or as $a \times 10^{-b}$

similar describes objects that are exactly the same shape but have different sizes

simple graph a graph in which pairs of vertices are connected by one edge at most, so there are no loops or multiple edges

simple interest interest calculation based on the original amount borrowed or invested; also known as, 'flat rate' as it is a constant amount: $I = \dfrac{PrT}{100}$

simultaneous equation an equation belongs to a system of equations in which the solutions for the values of the unknowns must satisfy each equation

sine the ratio of the side length opposite an internal angle of a right-angled triangle with the hypotenuse: $\sin(\theta) = \dfrac{O}{H}$

sine rule The ratios of a side length with the sine of the angle opposite it are equal throughout a triangle: $\dfrac{a}{\sin(A)} = \dfrac{b}{\sin(B)} = \dfrac{c}{\sin(C)}$. This can be used to find the unknown side length or angle in non-right-angled triangles.

SOH–CAH–TOA a mnemonic to assist in remembering the trigonometric ratios: Sine is Opposite over Hypotenuse; Cosine is Adjacent over Hypotenuse; Tangent is Opposite over Adjacent

spanning tree a tree that is a sub-graph which includes all of the vertices of the original graph

spending power also known as 'purchasing power'; the number of goods or services that can be purchased with a unit of currency

sphere a solid object that has a curved surface such that every point on the surface is the same distance (the radius of the sphere) from a central point

square matrix a matrix that has the same number of rows and columns

standard deviation the most common measure of the spread of data around the mean; found by taking the square root of the variance

standard form the form of a linear equation expressed as $Ax + By = C$

stem plot an arrangement used for numerical data in which data points are grouped according to their numerical place values (the 'stem') and then displayed horizontally as single digits (the 'leaf')

steady state or **equilibrium** the state that a series of transitions reaches when there is no longer any change

step graph a graph formed by two or more linear graphs that have zero gradients

substitute replace with an equivalent value or expression

surface area the combined total of the areas of each individual surface that forms a solid object

tables of values tables of x- and y-values that are true for a line, used for solving problems and plotting graphs

tangent the ratio of the side length opposite an internal angle of a right-angled triangle with the side length adjacent to it: $\tan(\theta) = \dfrac{O}{A}$

time payment also known as 'hire purchase'; a method of purchasing in which the customer receives an item by paying a small amount as a deposit followed by weekly or monthly instalments

trail a walk in which no edges are repeated

transition matrix a square matrix that describes the way that transitions are made between two states

transition table a table that describes the way that transitions are made between two states

transpose to rearrange an expression or formula

tree a simple connected graph with no circuits

trigonometric ratio a ratio of the lengths of two sides of a right-angled triangle

true bearing a bearing that is measured in a clockwise direction from the north–south line. A true bearing is written as a three-digit expression of the angle.

undirected graph a graph in which it is possible to move along the edges of the graph in any direction

unit cost the cost of a single item

unit cost depreciation a method of depreciating an asset according to its use — the more it is used, the faster it will depreciate

upper fence the upper boundary beyond which a data value is considered to be an outlier: $Q_3 + 1.5 \times IQR$

upper quartile the median of the upper half of an ordered data set

variable a quantity that can take on a range of values depending on its relationship to other values; typically represented by pronumerals

vertex (plural *vertices*) a point on a network that may be connected to other nodes via edges

volume the amount of space that is taken up by any solid or 3-dimensional object

volume scale factor the ratio of the corresponding volumes of similar objects. This is equal to the linear scale factor raised to the power of 3.

walk any route taken through a network, including those that repeat edges and vertices

weighted graphs a graph with values attached to the edges

write-off value the value at which an asset is removed from the books of a company as it is considered effectively worthless; also called the 'scrap value'

***x*-intercept** the point where the graph of an equation crosses the *x*-axis. This occurs when $y = 0$.

***y*-intercept** the point where the graph of an equation crosses the *y*-axis. This occurs when $x = 0$.

zero matrix a square matrix that consists entirely of '0' elements

INDEX

W

walk   516
weighted graphs   526–527
write-off value   363

X

x-intercept   267

Y

y-intercept   267

Z

zero matrix   199
zeros   625–626